ADAPTIVE MECHANISMS IN THE ECOLOGY OF VISION

Adaptive Mechanisms in the Ecology of Vision

Edited by:

S. N. Archer

International Marine Centre, Oristano, Sardinia, Italy

M.B.A. Djamgoz

Department of Biology, Imperial College, London, U.K.

E.R. Loew

Department of Physiology, Cornell University, Ithaca, NY, USA

J.C. Partridge

School of Biological Sciences, University of Bristol, Bristol, U.K.

and

S. Vallerga

International Marine Centre, Oristano, Sardinia, Italy

KLUWER ACADEMIC PUBLISHERS

DORDRECHT / BOSTON / LONDON

Library of Congress Cataloging-in-Publication data is available.

ISBN 0-7923-5319-6

Published by Kluwer Academic Publishers,
P.O. Box 17, 3300 AA Dordrecht, The Netherlands.

Sold and distributed in North, Central and South America
by Kluwer Academic Publishers,
101 Philip Drive, Norwell, MA 02061, U.S.A.

In all other countries, sold and distributed
by Kluwer Academic Publishers,
P.O. Box 322, 3300 AH Dordrecht, The Netherlands

Printed on acid-free paper

Cover pictures:

Top left: Eye from the deep sea dragon fish, *Malacosteus niger*
Top right: A mantis shrimp eye, *Lysiosquilla maculata*
Bottom left: A wrasse eye, *Thalassoma lutescens*
Bottom right: A goatfish eye, *Parupeneus cyclostomus*

Picture credit: Justin Marshall.

Printed and bound in Great Britain by MPG Books, Bodmin, Cornwall.

Contents

Foreword

John Lythgoe was one of the pioneers of the 'Ecology of Vision', a subject that he ably delineated in his classic and inspirational book published some 20 years ago [1]. At heart, the original book aimed generally to identify inter-relationships between vision, animal behaviour and the environment. John Lythgoe excelled at identifying the interesting 'questions' in the ecology of an animal that fitted the 'answers' presented by an analysis of the visual system.

Over the last twenty years, however, since Lythgoe's landmark publication, much progress has been made and the field has broadened considerably. In particular, our understanding of the 'adaptive mechanisms' underlying the ecology of vision has reached considerable depths, extending to the molecular dimension, partly as a result of development and application of new techniques. This complements the advances made in parallel in clinically oriented vision research [2]. The current book endeavours to review the progress made in the ecology of vision field by bringing together many of the major researchers presently active in the expanded subject area. The contents deal with theoretical and physical considerations of light and photoreception, present examples of visual system structure and function, and delve into aspects of visual behaviour and communication. Throughout the book, we have tried to emphasise one of the major themes to emerge within the ecology of vision: the high degree of adaptability that visual mechanisms are capable of undergoing in response to diverse, and dynamic, environments and behaviours.

The idea for this book arose initially from a conference on the 'Ecology of Vision' that was held at the International Marine Centre, Sardinia, in 1994. As with the conference, the book is dedicated to the memory of John Lythgoe who died in 1992. John was an inspiration to us all, both professionally and as a friend and is much missed. We hope that this book will continue to inspire fellow scientists and students.

We would like to thank the staff at Chapman & Hall (especially Jon Evans and Rachel Young), and later Kluwer (John Martin), and all the contributing authors for cooperating so patiently with us during the preparation of this book. We also thank Heather McDonald for all her help throughout the editorial process.

Simon N. Archer, Mustafa B.A. Djamgoz, Ellis R. Loew, Julian C. Partridge and Silvana Vallerga.

References

1. Lythgoe, J.N. (1979) *The Ecology of Vision*. Oxford University Press, Oxford.
2. Djamgoz, M.B.A., Archer, S.N. and Vallerga, S., eds. (1995) *Neurobiology and Clinical Aspects of the Outer Retina*. Chapman and Hall, London.

John Lythgoe (1937 – 1992)

Preface:
An Essay on the Ecology of Vision in honour of John Lythgoe

JOHN WOODS
Imperial College, London University, SW7 2AZ

This essay addresses two closely-linked aspects of the ecology of vision. The first is a personal memoir of John Lythgoe, one of the founders of the scientific discipine of the *Ecology of Vision* and inspiration for the research programme of the International Marine Centre at Oristano. The second picks up one of the themes of John Lythgoe's research, the visual environment and shows how mathematical simulation is contributing to the problem of predicting what fish see under different conditions of water quality.

John Lythgoe (1937 – 1992)

John came from a distinguished line of scientists. He was the fourth generation in his family to graduate from that most famous of Cambridge University's colleges, Trinity, where Newton invented dynamics. In John's day Trinity still boasted more Nobel Prizes than France. His grandfather was the pioneer of modern ecology, Sir Arthur Tansley, who had gone on to become Professor of Botany in Oxford University and Biological Secretary of the Royal Society. His father had created the Medical Research Council's Vision Research Unit before dying young.

After his student years in Cambridge John moved to America as a postdoc ending up at La Jolla, Southern California, where he learnt to dive and refocused his research goal from botany to fish behaviour and vision. Returning to England in 1963 John joined the Medical Research Council as a member of the MRC Vision Research Unit then directed by Dr. Dartnall. The Unit was based at the University of London Institute of Ophthalmology in Judd Street, near the British Museum, but later moved to the campus of the new University of Sussex, between Brighton and Lewes. When Herbert Dartnall retired the Medical Research Council closed the Vision Unit and John was offered an MRC-funded research position, equivalent to a research readership, in the Zoology Department of the University of Bristol. That personal position provided his research base for the rest of his career.

In 1966 John married Gillian Yeates, who was then employed at the Natural History Museum. They met in Malta when Gillian was a member of my diving

S.N. Archer et al. (eds.), Adaptive Mechanisms in the Ecology of Vision, ix–xix

team. They produced two daughters Charlotte and Katrina. Katrina has followed in the family tradition of ecological research, working with me on plankton demography while an undergraduate at Balliol College, Oxford University, and subsequently on population dynamics of parasites in Edinburgh University. The family lived first in London, then near Chailey in Sussex and later in Bristol. Their Sussex home, set in a bluebell wood, came closest to John's dream of a naturalist's retreat, like nearby Down House where Charles Darwin had lived while experimenting on animals and developing the theory of natural selection. Visiting John and Gillian with their screeching peacocks and other animals was a special treat for a young physicist with little previous experience of biology. This fascination with all aspects of nature continued throughout their happy married life, culminating in the purchase of a property in Costa Rica, their own rain forest.

Mixing home and research, business with pleasure became a way of life for the young Lythgoes. As they became increasingly involved in diving and underwater photography, they and their friends founded the British Society of Underwater Photographers. BSOUP, commonly pronounced Pea Soup, a fair description of visibility in British waters which forced the members to become experts in close-up photography. Photographs by BSOUP members illustrated John and Gillian's best-selling book 'Fishes of the Sea', which was published around the world in various languages. Increasing demand for reproduction of BSOUP photographs in journals and books led John and Gillian to create a photographic agency, which flourished and over the years broadened to include all aspects on nature. By 1995, when Gillian sold it, Planet Earth Pictures had become one of the world's leading natural history photographic agencies, notable for the high quality of both the images and the biological knowledge behind the documentation of every photograph. Another of their secrets of commercial success was the special relationship John and Gillian developed with the photographers who trusted them to sell their treasured pictures.

Diving

After learning to dive in California, John returned to Cambridge where he made contact with Dr. C.C. (Bill) Hemmings who had founded the University of Cambridge Underwater Exploration Research Group with Dr. N.C. Flemming. John and Bill discovered that they shared a passion for understanding fish vision, colouration and behaviour. They established a close research collaboration which lasted for over a decade. When he moved to the MRC Vision Research Unit John joined the London Branch of the BSAC. I remember meeting him for the first time in 1962 at the Oasis Centre in Holborn, a swimming pool used by the British Sub Aqua Club. I was then President of the Imperial College Underwater Club. I had just returned from leading a diving expedition to Malta, where my team had discovered a sunken fleet of Roman merchant ships that had been carrying supplies to a Roman Army attacking Carthage in 212 BC. The central Mediterranean island of Malta was soon to become the base for a growing association of scientific diving groups from Cambridge, London and Oxford Universities.

As John's circle of university diving friends in Cambridge and London grew, he embarked on a series of field studies in partnership with first one and then another. In 1963 I joined him for a short trip to Marseilles, at the Station Marine d'Endoume which I had visited the year before with Nic Flemming. John wanted to collect the eyes of the tiny black fish *Chromis chromis* which the French call Demoiselle. His research aim was to find out whether the retinas of these black fish contained a visual pigment that would help them see their bright blue young, which they studiously herded. The technical problem was to capture live chromis and put them into black plastic bags on the sea floor, so that their retinas would not be bleached by the sun as they were brought to the surface. John decided to catch the tiny fish with a speargun, which required special permission from the Marseille chief of police; the guns had been banned after the prolific fish life of the Cote d'Azur had been destroyed by spear-fishermen using aqualungs.

Malta

In 1965 John started a research programme that involved spending each summer in Malta, the island located between Italy and North Africa, whose calm, clear waters provided the ideal underwater laboratory for vision experiments. He was awarded a grant from the Royal Society which paid for accommodation, equipment and a compressor to fill aqualung bottles. I arranged support from the Navy's diving centre. During the next few years, Malta became the summer location for many research groups from British universities. We formed the Underwater Association which developed into the vehicle for presenting our scientific results at an annual symposium at the Zoological Society and publishing the proceedings. John was a founder member of the Association and later became its President. One of the landmark events was the publication by Oxford University Press of a monograph *Underwater Science* edited by John and myself. The book contained chapters by leading members of the Association each showing how diving was being used in his scientific discipline. Thanks to these activities, the subject grew rapidly in Britain. Its leading members, including John, resisted government plans to impose onerous commercial safety rules on diving scientists. The result was the Underwater Association's 'Code of Practice for Scientific Diving' that was adopted by UNESCO and is now used around the world.

One of the ambitions of John and his colleagues in the Underwater Association was to create in Malta a permanent field station for scientific diving where one of the research topics would be ecology of vision. This proposal was reviewed favourably by a government commission but sadly it never received funding and research in Malta continued on an ad hoc basis until the early 1970s. However, John did not abandon his dream, and eventually it came true in Oristano,

Sardinia, where he helped Silvana Vallerga to establish the International Marine Centre. He encouraged a close working relationship between his working group in the University of Bristol and the staff in Oristano, and focused his own marine fieldwork there. At the IMC John inspired a multi-disciplinary research programme in what was coming to be known as the ecology of vision. His monograph on the subject was published in 1979, and ten years later he was writing a much larger definitive treatise on the ecology of vision. Meanwhile he helped to build the new scientific discipline through a steady flow of original scientific papers and by inspiring his students and research colleagues.

The continuing quest

John Lythgoe was a biologist deeply rooted in the traditions of natural history. His scientific curiosity extended over a wide range of topics, from fish vision to the ecology of tropical rain forests. He was a scientific entrepreneur with an urge to build novel infra-structure that would make research easier. He did so first in Malta, then in Sardinia and finally in Costa Rica. He inspired others to work with him, and to share the hard work needed to make his dreams come true. He was remarkably successful in his wide-ranging ventures, but he did not live to reap the public recognition that would surely have come to him for his diverse contributions to science. He was a modest man who did not seek rewards, beyond the friendship, help and love of those with whom he worked. John was the model of a modern major generalist in an era when biology had become increasingly specialized. His approach was right for ecology which his grandfather had helped create, the inter-disciplinary subject par excellence. A broad approach was essential for the young sub-discipline he championed, ecology of vision.

John Lythgoe had the good fortune to hit the right time to pursue his goal of combining the specialist disciplines, each at the leading edge of their subject, into a coherent attack on the ecology of fish vision. At the end of the 20th century new tools coming available were making it possible to combine the exciting results gained from each line of attack into predictions about the ecology of vision. The prospect of quantitative predictions suitable for comparison with observations was promising to carry ecology forward from an exploratory into an experimental science. John subscribed to Arthur Eddington's (1932) famous view:

> Observation and theory get on best when they are mixed together, both helping one another in the pursuit of truth. It is a good rule not to put overmuch confidence in a theory until it has been confirmed by observation..... it is also a good rule not to put overmuch confidence in the observational results that are put forward until they are confirmed by theory.

One of those new tools that will help ecology of vision make this transition is mathematical simulation, which has already had a major impact in other disciplines that involve great complexity, like the weather or financial markets. In the second half of this essay I shall show how the techniques of mathematical simulation of complex systems can be applied to the scientific problems of the ecology of vision. This is offered as a tribute to my friend John Lythgoe who did so much to advance the subject. In fact, John's last major project was the development of a model that would predict the optimum pair of visual pigments that a dichromatic fish should possess in a given spectral environment in order to maximise contrast between an object and the background spacelight (Lythgoe & Partridge, 1989; Lythgoe & Partridge, 1991; Lythgoe *et al.*, 1994). The model makes some interesting and accurate predictions and is a perfect example of the Lythgoe approach to the Ecology of Vision.

Simulating the visual environment

The challenge

Before we can answer the question 'What does a fish see?' it is necessary to ask 'What is its ambient optical environment?' and 'How does that environment change with location, season and time of day?'. It is now possible to simulate the optical environment inside the sea sufficiently accurately to answer the latter question. It is possible to predict the spectrum of radiation directed towards the fish and the spectrum of light scattered from a target, perhaps another fish, in its field of view. In principle, that can be repeated for a school of fish, each looking at the other fishes around it in a virtual ecosystem. The modeller can introduce algorithms based on laboratory experiments for the information that each fish extracts from its visual environment, and for the way it changes its behaviour as the result. In that way he will soon be able to simulate the schooling of fish, and predict how a fish interacts with its prey, competitors and predators. Having created such a virtual ecosystem, it will be possible to test conjectures about each link in the chain from the changes in a fish's visual environment to its behaviour in response to those changes, by comparing patterns of behaviour with observations.

That may sound like an impossible dream, but it will soon be reality. The complexity of the system I have just described is no greater than that of many systems currently being routinely simulated in fields as diverse as climate, traffic engineering, the insurance business and stockmarkets. If you are interested in the subject you might like to read Casti (1997). The growth of super-computers has made it possible to simulate systems with over one million agents. In the simulation of traffic flow, the agents are vehicles and people. In ecology of vision, the agent is a fish. These agents roam around their simulated environment. For traffic in towns it is a precise map of the roads, buildings, bridges, parking places, traffic lights and so on. As Einstein said ‘my environment is everything but me’. For a pelagic fish it includes not only the water, but also other fish and the plankton that affect the optical properties of the water. An agent’s environment includes all the other agents.

Simulating the visual environment

For several years I have been developing a capability to simulate the upper ocean ecosystem by mathematical models in which the agents are plankton. The aim of this research is to improve our understanding of the way that plankton populations interact with their environment under the controlling influence of the sun and the weather. Using this approach it has proved possible to predict the properties of the upper ocean ecosystem, given the changing weather, and the oceanic conditions some years earlier at a location which may be thousands of kilometres upstream. Inspired by the conference that created the idea for this book, I decided to see whether it would be possible to predict the visual environment from these simulations of the ecosystem. Thanks to the hard work of my research student, Liu Cheng Chien, the work has been successful and it is now possible to predict the ambient visual environment encountered by a pelagic fish at any depth, looking in any chosen direction. So far as I know, this has not been done before. The work will be published formally elsewhere; here I shall briefly describe the method and present some results to illustrate the new capability. The next step will be to use the simulated optical field to illuminate a target, and so to compute the visibility of that target, when observed by a fish with specified visual pigments.

Method

Simulation of the visual environment involves three steps: (1) predicting plankton distribution, suspended sediment and dissolved organic matter, etc, and therefore the intrinsic optical properties of the seawater, (2) computing the spectrum of irradiance in the chosen direction, and (3) filtering that spectrum of irradiance according to the observer’s visual pigments.

The optical properties of the seawater depend on the concentration of plankton which absorb and scatter the light, and the properties of particle-free water. (At least they do in the open ocean: it gets more complicated in shallow coastal waters, where it is necessary to take account of suspended sediment and dissolved organic matter.) We can use a one-dimensional model to simulate the variation of plankton concentration with depth and time. There can be more than one billion plankton in a vertical water column one kilometre deep by one square metre, but it is possible to generate an accurate profile of plankton concentration with 100,000 agents or particles. Each particle behaves like a single plankter while carrying information about a sub-population whose number varies demographically according to the birth and death rates computed as the particle follows its own unique trajectory through the environment in response to motion with the water (advection and turbulent mixing) and motion through the water (sinking or swimming). This individual-based approach is known to generate a realistic description of population dynamics. In my method, the influence of the plankton on their environment is computed in layers one metre thick using the ensemble mean concentration of plankters in each layer. The same statistical procedure is used to compute the consequences of predation. We give the name Lagrangian Ensemble method to this combination of individual trajectories plus statistical feedback and predation (Woods and Barkmann 1994). Integrating the model in this way creates a virtual ecosystem comprising the vertical distributions of environmental variables, including the concentration of each plankton species. The integration proceeds in half time steps under the influence of (1) an initial condition derived from Levitus's (1994) ocean climatology and (2) an upper boundary condition which describes the surface fluxes of solar radiation, heat, water and momentum at each time step, derived from Isemer & Hasse's (1986) atmospheric climatology. Figure 1 shows the profile of phytoplankton concentration off the Azores on the following dates in the simulated year: 17 January when the population has its annual minimum and on 13 May at its maximum during the spring bloom.

Having simulated the vertical profile of phytoplankton (expressed as chlorophyll concentration), the next step is to compute the influence it has on solar radiation entering the water column. This is done by a Monte Carlo method (see Mobley 1994). It involves computing the trajectory of a billion photons as they pass through the atmosphere and then the ocean from an initial direction determined by the solar elevation. Each photon's trajectory depends on scattering in the atmosphere, refraction at the sea surface and scattering inside the ocean. It may also be absorbed by the water or by a plankter. The details of the trajectory of an individual photon are computed by applying randomly a probability of it being scattered through a given angle or absorbed metre by metre through the water column. This is repeated for the billion photons, to build up the flux of light in all directions at every depth in the water column, for one wavelength of light. The computation is then repeated for 31 wavebands covering the solar spectrum. The result is a simulation of the spectral distribution of solar energy in any direction at any depth in the water column (Figure 2). We have verified this technique against observations, using Mobley's (1994) synthesis.

The third stage of the simulation takes account of the observer's visual pigments. The spectrum of incident light along the direction in which the observer is looking is filtered to simulate the colour of the background illumination. To illustrate the results we have used the CEI numerical colour classification to show the colour of the light that a human observer would see on the 17th January

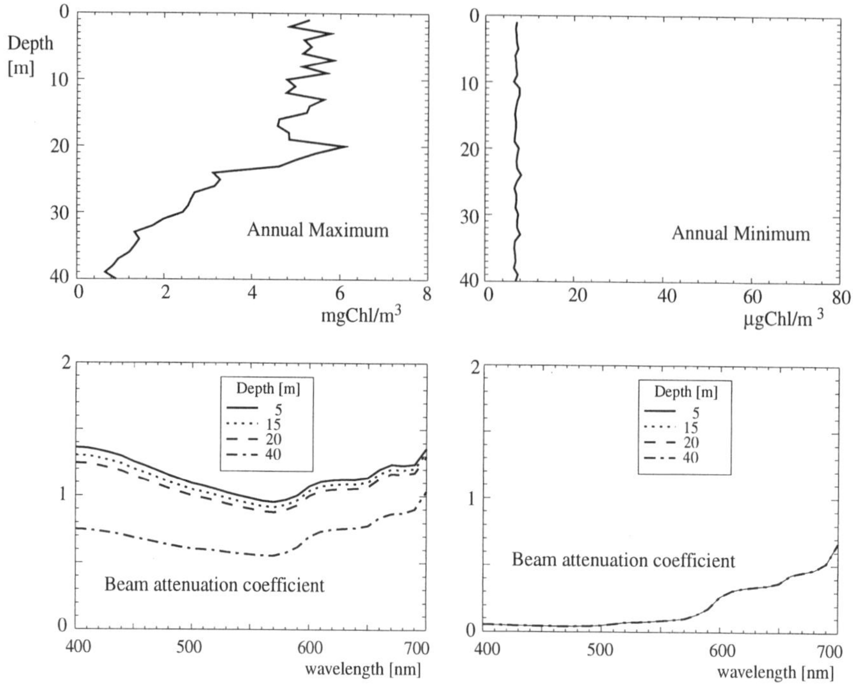

Figure 1 Variation with depth of the chlorophyll concentration and beam attenuation off the Azores (27°N 41°W) on 17 January and 13 May predicted by the LE'97 model integrated by the Lagrangian Ensemble method. [Reproduced from C.C. Liu 1997, unpublished report]

(plankton minimum) and the 13th of May (plankton maximum) when looking directly up or down while swimming at a number of depths (Table 1).

Independent verification of this technique comes from simulation of ocean colour as seen from a satellite. In this case the observer is in space and the visual pigments are replaced by the filters in the Coastal Zone Colour Scanner (CZCS), a three colour band instrument on an American satellite. There exists an empirical calibration of the CZCS data against ship measurements of plankton chlorophyll concentration in the surface mixed layer of the ocean, but there has been no simulation of ocean colour from an individual-based primitive equation model of the upper ocean ecosystem. We have now done that by analysing our virtual ecosystem to measure the energy in each CZCS waveband carried by the few photons in each of our 31 spectral bands that enter the water and are scattered by plankton until they emerge from the sea in the direction of the satellite. Figure 3 shows the comparison of the seasonal variation of plankton concentration derived from ocean colour in the two ways: (1) from the CZCS, and (2) from our virtual ecosystem with 100,000 plankton particles exposed to five million photons per waveband. The comparison is encouraging. The combination

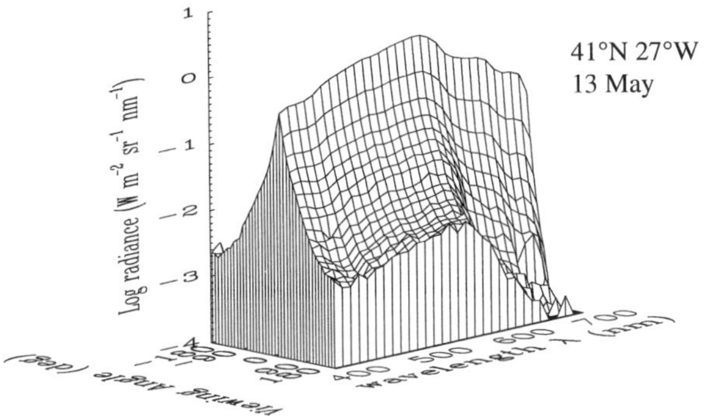

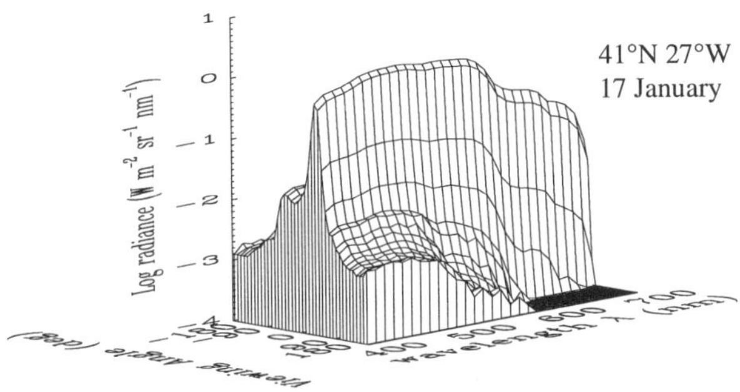

Figure 2 Spectral irradiance as a function of viewing direction in the solar plane at a depth of 5 m. on the 13 May and 17 January computed by the Monte Carlo method for the plankton profiles in fig.1 [Reproduced from C.C. Liu 1997, unpublished report]

Table 1: Perceived colour (CEI%) looking up and down off the Azores on 13 May & [17 January]

Depth	0 m	1 m	5 m	10 m	15 m	20 m
Looking up						
Red	70	68	48	23	19	18
	[52]	[47]	[28]	[19]	[17]	[16]
Green	30	32	47	23	9	2
	[40]	[42]	[41]	[28]	[21]	[12]
Blue	0	0	5	54	72	80
	[8]	[11]	[31]	[53]	[62]	[72]
Looking down						
Red	26	25	20	18	17	17
	[17]	[17]	[17]	[17]	[17]	[17]
Green	28	25	10	2	1	1
	[2]	[2]	[1]	[1]	[1]	[1]
Blue	46	50	70	80	82	82
	[81]	[81]	[82]	[82]	[82]	[82]

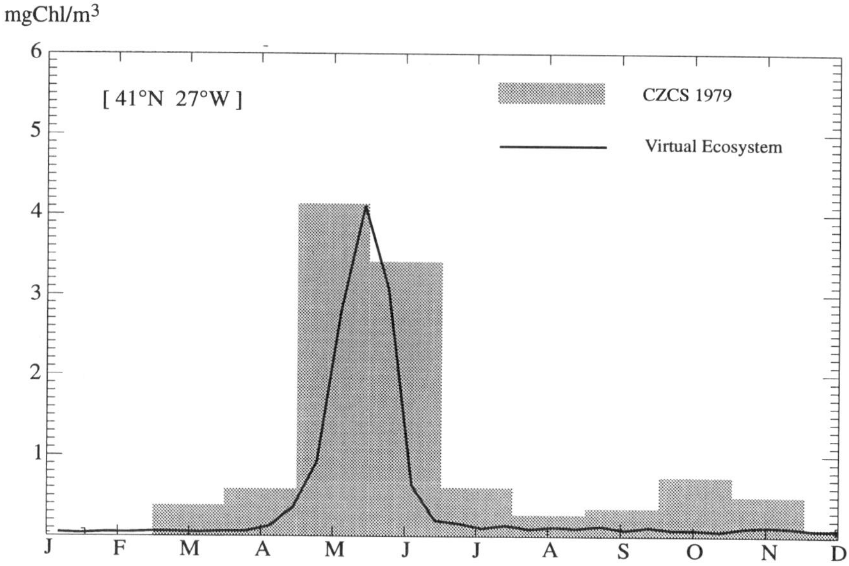

Figure 3 Seasonal variation of ocean colour off the Azores measured in 1979 by the CZCS and predicted by the Monte Carlo method applied to the seasonally-varying chlorophyll profile from the Lagrangian Ensemble method. The ocean colour index has been converted to chlorophyll concentration using the empirical calibration of Gordon & Morel 1983); see also fig.1. [Reproduced from C.C. Liu 1997, unpublished report]

of the Lagrangian Ensemble method of simulating the upper ocean ecosystem, and the Monte Carlo method of computing the visual environment in that virtual ecosystem produces quite realistic results.

Conclusion

I believe that these new techniques will provide an effective way to link advances in the biology of fish vision to observations by divers of the behaviour of fish in the sea. That is what John Lythgoe was striving to achieve in his research, and what the International Marine Centre is continuing in its research programme.

References

Casti, J. (1997) *Would be Worlds: How simulation is changing the frontiers of science.* New York, John Wiley. 242 pp.

Eddington, A. (1935) *New Pathways in Science.* London, MacMillan.

Gordon, H. R. and Morel, A. Y. (1983). *Remote assessment of ocean color for interpretation of satellite imagery: a review.* New York, Springer-Verlag.

Isemer, H. J. and Hasse, L. (1986) *The Bunker Climate Atlas of the North Atlantic Ocean.* Berlin, Springer-Verlag. 218pp.

Levitus, S. and Boyer, T. P. (1994) *World Ocean Atlas 1994*. Washington, D.C., Government Printing Office.
Liu, C. C. (1997) *Simulation of ocean colour.* Imperial College, Huxley School, Ocean Resources group. (Unpublished Report)
Lythgoe, J. N. (1979) *The Ecology of Vision*. Oxford, Clarendon Press, 244pp.
Lythgoe, J. N. and Lythgoe, G. (1971) *Fishes of the Sea.* London, Blandford Press, 320pp.
Lythgoe, J. N. and Partridge, J. C. (1989) Visual pigments and the acquisition of visual information. *J. Exp. Biol.* 146, 1–20.
Lythgoe, J. N. and Partridge, J. C. (1991) The modelling of optimal visual pigments of dichromatic teleosts in green coastal waters. *Vision Research*, 31, 361–371.
Lythgoe, J. N., Muntz, W. R. A. and Partridge, J. C. *et al.* (1994) The ecology of the visual pigments of snappers (Lutjanidae) on the Great Barrier Reef. *J. Comp. Physiol.* A, 174, 461–467.
Mobley, C. T. (1994) *Light and Water: Radiative Transfer in Natural Waters*. San Diego, Academic Press. 592 pp.
Woods, J. D. and Barkmann, W. (1994) Simulating the upper ocean plankton ecosystem by the Lagrangian Ensemble method. *Phil. Trans. Roy. Soc. London B.,* 343, 27–31
Woods, J. D. and Lythgoe, J. N. (Eds) (1977) *Underwater Science,* Oxford, Oxford University Press, 325pp.

Section I

General Principles

1a
Light and photoreception: Physical principles

E. LOEW[1]
[1]Department of Physiology, Cornell University, Ithaca, 14853 New York, USA

1.1 Introduction

Life on earth evolved within an all-pervasive electromagnetic field (EMF) that represented not only a potential source of energy, but also a source of information about the local environment and, in fact, the universe in general. The first autotrophs developed molecular mechanisms for tapping into the energy of the EMF and evolved means for detecting the most fundamental properties of the field – energy, intensity and direction – and using these to orient, regulate exposure and tell time. Living systems responding to these field properties are said to be *photosensitive*. *Vision*, while including the basic photosensitivity functions, is concerned with extracting information about the interactions of the EMF with the surrounding media and targets within that media. Single cells and, in some cases, even subcellular organelles may be photosensitive. However, vision is a multicellular capability and required the evolution of appropriate cellular arrays and analytical channels. In this chapter we will deal first with the equations defining the EMF and how the field interacts with matter. We will then look at basic photoreceptive principles and, lastly, at the kinds of adaptations.

1.2 The physics of light

1.2.1 The nature of electromagnetic radiation

One of the most fundamental concepts in physics is the dual nature of electromagnetic radiation. It can be described as a particulate entity, the quantum, and a self-propagating electromagnetic wave. These are not separate descriptions – both are necessary for an understanding of electromagnetic radiation. When dealing with most biological systems, the quantum is called a photon and all of its energy can be assigned to a single point in space. This energy is related to its frequency by

$$E = h\nu \qquad [1]$$

S.N. Archer et al. (eds.), Adaptive Mechanisms in the Ecology of Vision, 3–23

where h is Planck's constant (6.626×10^{-34}) and ν is the frequency. As for any wave, the wavelength (λ) is equal to the speed of the wave divided by its frequency. Thus,

$$E = h\nu = \frac{hc}{\lambda} \qquad [2]$$

where c equals the speed of quantum propagation in the medium. For a vacuum, $c = 2.998 \times 10^{8}\ m\ s^{-1}$. The unit of energy for the quanta is the Joule (J). From the above formula, a photon having a wavelength of 500 nanometres (nm) would have an energy of 3.973×10^{-19} J in a vacuum. As is seen, the energy of a quantum is inversely proportional to its wavelength. In some situations it is more convenient to use units of ergs (1 Joule = 10^{7} ergs), calories (1 calorie = 4.184 Joules) or electron volts (1 eV = 1.6022×10^{-19} Joules or $E(eV) \approx 1240/\lambda$). The quantum is the fundamental unit of the electromagnetic force. As such, energy transfer between the electromagnetic field and matter can only occur in integral steps defined by the photon energy.

As a wave-like entity, the quantum can also be thought of as a localized region within time-varying electric and magnetic fields. According to Maxwell's equations, a time-varying electric field generates an orthogonal time-varying magnetic field and vice versa. These propagate through free space in a direction orthogonal to the planes of oscillation. From the point of view of the quantum, at some times its total energy is carried by the electric field, while at others it is carried by the magnetic field. As one field collapses, the other proportionately increases. The ratio of the two field energies, which, when combined yield the total quantum energy, is the sum of two vectors – the electric (e) vector and the magnetic (m) vector. The plane of vibration of the e-vector within a given reference frame defines its state of polarization.

The dual nature of light is exemplified by the relationship between a particle-like property (its momentum *p*) and a wave property λ as given by the Einstein relationship

$$\lambda = \frac{h}{p} \qquad [3]$$

Thus, one can think of electromagnetic radiation as a collection of photons of momentum $h\nu/c$ guided by a wave and requiring no media for propagation (Abrahamson & Japar, 1972).

1.2.2 Geometric optics

The science of optics studies the propagation of electromagnetic energy through space and interactions of the EMF with the media and targets. It can be divided into geometric optics, physical optics and quantum optics. For visual scientists, an understanding of geometric optics is obligatory as it most conveniently describes the operation of lens systems and is useful in understanding radiative transfer and absorption. Geometric optics is concerned with a theoretical entity called the *light ray*. While lacking a precise definition, for this discussion a light ray will be considered as the path taken by a photon. Groups of light rays are called *beams*. It should not be forgotten that a photon is also a wave and other definitions of light rays are possible within this context. The optical characteristics of the media are completely specified by the index of refraction n that is defined as

$$n = \frac{\mathrm{c}}{\mathrm{v}} \qquad [4]$$

where c is the speed of light in a vacuum and v is the speed of light in the medium. In a homogeneous medium, n is isotropic for photons travelling in all directions and with any polarization state. In a heterogeneous medium, n varies with position.

Light rays have the following characteristics (after Goodman, 1995):

1. They follow straight paths in homogeneous media and curved paths in heterogeneous media.
2. They are fully reversible. That is, the path is the same in either direction.
3. They carry energy. The total power passing through a given area of media is approximately equal to the number of rays passing through the area or the ray density.
4. They have a unique direction at each position where n is continuous given by the *direction vector* r whose Cartesian components are direction cosines such that

$$r = (\alpha, \beta, \gamma) \qquad [5]$$

where

$$|r|^2 = \alpha^2 + \beta^2 + \gamma^2 = 1 \qquad [6]$$

A cornerstone of geometric optics is *Fermat's Principle* which states that *a ray of light follows the path of 'least time'* (Kohen, *et al.* 1995). Thus, in going from A to B in a homogeneous medium a ray will always be a straight line. From this can be derived the laws of reflexion and refraction, as well as equations describing ray path and geometric wavefronts in all kinds of media.

1.2.2(a) Reflexion

This is the change in direction of a light ray striking a totally opaque boundary and follows the well-known relationship

$$i = r \qquad [7]$$

where i is the angle of incidence of a ray hitting a target and r is the angle of reflexion, both measured relative to the normal drawn at the intersection point of the surface and the ray. A specular surface is one having irregularities much less than the wavelength of the impinging light. In this case parallel rays in the incident beam remain parallel in the reflected beam. Diffuse reflexion is seen when surface irregularities act so that different rays in a beam 'see' different surface normals. Each ray behaves according to the above principle, however parallel incident rays are not necessarily parallel in the reflected beam.

1.2.2(b) Refraction

This is seen when a ray moves across the boundary separating media that differ in refractive index. The change in direction of the ray, again relative to the normal passing through the boundary, is given by Snell's Law

$$n_1 \sin \mathrm{i} = n_2 \sin \mathrm{t} \quad [8]$$

where n_1 and n_2 are the respective indices of refraction and i and t are the angles of the incident and transmitted ray from the boundary normal. In almost all real situations, a light beam incident on a boundary is both reflected and refracted as seen in Figure 1.

We can arrange Snell's Law to give

$$\frac{\sin \mathrm{i}}{\sin \mathrm{t}} = \frac{n_2}{n_1} = \mathrm{K} \quad [9]$$

where K is the relative refractive index between the media. This equation is valid for every value of i as long as the wave is travelling from the medium of lesser to greater n (K > 1). However, in the reverse situation K will be less than one and it will be seen that if i exceeds a certain value, sin t will become greater than 1. Since there is no sin value greater than one, there can be no refracted ray and thus

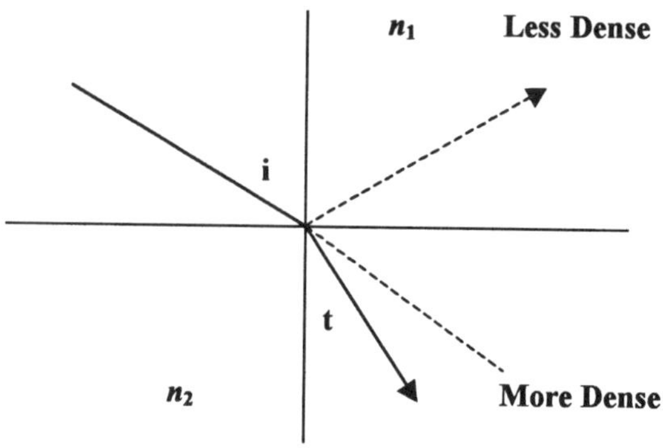

Figure 1 Schematic depiction of Snell's Law

the incident ray will be totally reflected. The value of i at which this occurs is called the *critical angle*. In cases where the less dense medium is air ($n \approx 1$), the critical angle can be found using the formula sin i = K. For example, the refractive index of water is approximately 1.33. Thus, the critical angle is 48.75°.

In travelling from one media to another, not all of the light is refracted at the interface, some is reflected. For the case of *normal incidence*, the reflexion coefficient I_r (i.e. the fraction of the incident beam reflected) is given by

$$I_r = \frac{(n_2 - n_1)^2}{(n_2 + n_1)^2} \qquad [10]$$

where n_2 is the refractive index of the dense medium and n_1 is that of the rarer medium. For glass and air I_r works out to be approximately 0.046 which means that at normal incidence approximately 4% of the light is reflected. This relationship is polarization independent. However, as the angle of incidence is increased from zero the reflected component takes on a polarization angle dependence such that at some angle the reflected component is totally plane polarized. This angle is called the Brewster angle.

1.2.2(c) Dispersion

The refractive index is wavelength dependent such that in going from a less dense to a more dense medium, short wavelengths are refracted more than long wavelengths. This phenomenon is called dispersion and must be considered whenever broad-spectral beams traverse refractive boundaries. The change in refractive index with wavelength is not linear, with $dn/d\lambda$ increasing with decreasing wavelength. Thus, the dispersion is not linear and the distance between diverging rays is greatest for the shortest wavelengths.

1.2.3 Wave (physical) optics

While geometric optics can be well described within the definition of light as a particle taking the shortest distance between two points while tracing out a ray, physical optics deal with laws best understood within the context of light as a wave. This separation is really an historic one as both are correct representations of light. While reflexion and refraction deal with the interactions between light and the physical environment, *interference* and *diffraction* are phenomena resulting from the interactions between electromagnetic waves. In a biological context, interference is responsible for such things as the structural colors of butterfly wings and iridescent corneas of fishes.

1.2.3(a) Interference

As a wave, light propagates through space as a sine wave having amplitude A and a phase ϕ relative to an arbitrary frame of reference. For two interfering waves having the same wavelength, the resultant energy distribution will be related to the sum of the amplitudes and their phase relationship (i.e. constructive interference). If the waves have zero phase difference the total amplitude A_t will be the sum of the individual wave amplitudes (for small amplitudes) (Wood, 1936) with the general equation

$$A_t = \sum_i A_i \qquad [11]$$

If the waves are 180° out of phase the amplitudes AT THAT POINT IN SPACE cancel (i.e. destructive interference) and there is darkness. Since each wave carries energy that cannot be destroyed, darkness at one point must be compensated for with brightness at some other point. Thus, interference is a redistribution of energy in space. The relationship between the phase difference *e* and the intensity due to interference is given by

$$I = A_1^2 + A_2^2 + 2A_1A_2 \cos e \qquad [12]$$

All optical systems producing interference can be characterized by the way in which phase differences are introduced and interactions occur. Fermat realized a number of ways to produce interference using both reflective and refractive systems. The only requirement was that two beams of light must emanate from two similar sources and converge in space such that they can interfere (see description in Wood, 1936). For example, we can imagine the light from two point sources s_1 and s_2 separated by distance x allowed to interfere on a surface AD parallel to the line joining the two sources and placed some distance (P) away from them (Figure 2). The light arriving at point C is in phase having travelled an equal distance from the sources and will have maximum intensity at all wavelengths. If we now move along line AD to a point (B) where distance r_1 is exactly ½ λ less than r_2 we will observe destructive interference and a dark band. The distance from C of bright (constructive) and dark (destructive) bands is related to the wavelength and given by

$$d = \left(P/x\right) n \left(\lambda/2\right) \qquad [13]$$

Even values of n are bright bands while odd values are black. Since the spacing is wavelength dependent, the bright bands will be colored and there will not be any real black bands.

Interference can occur between waves reflected at the interfaces between media of varying refractive index with the intensity of the interfering reflected beam dependent on the distance between the reflecting boundaries. Maximum intensity is seen when the 'optical thickness' *x* (the distance between refractive

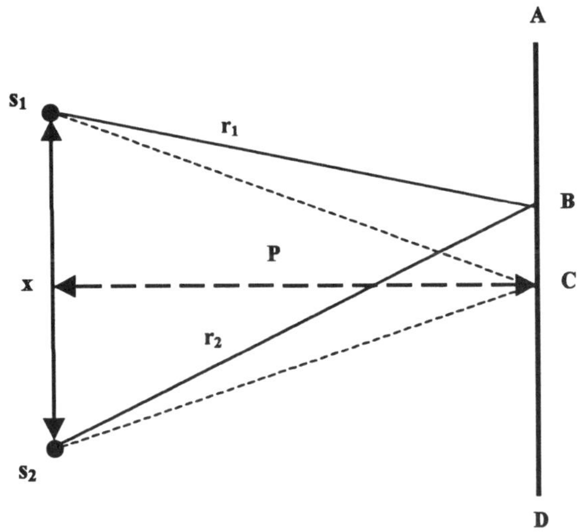

Figure 2 Fresnel interference scheme

boundaries times the refractive index) produces a phase change of $\lambda/2$ between the beam reflected at the first interface and that reflected from the second (Figure 3). The intensity of the reflected component is wavelength and angle-of-incidence dependent, and is quite small for a single layer. However, by stacking a number of thin layers of appropriate thickness and refractive index, the reflected component at a given wavelength can be made quite high. Note that the light remaining in the transmitted component will change in spectral characteristics due to wavelength-specific reflexion at the interfaces. Through the use of multiple stacks and appropriate coatings it is possible to reduce or enhance reflexion over particular spectral regions and create wavelength-selective 'interference' transmissive filters. The use of interference phenomena by organisms to reduce reflexion and produce color filters is discussed thoroughly in Land (1981).

1.2.3(b) Diffraction

This is the bending of light around the edges of objects thus illuminating the region of the shadow with a system of maxima and minima. Thomas Young recognized these fringes as an interference phenomenon, but it was Fresnel who gave the first mathematical treatment of the subject using the wave theory of Huygens as a starting point (Wood, 1936). An example of diffraction are the rings of decreasing intensity seen surrounding a pinhole made in a black film when viewed through a microscope lens (Figure 4). For a circular aperture 83.9% of the energy is in the central disk (the Airy disk). The separation between dark bands is as seen in the figure and is related to the wavelength and the numerical

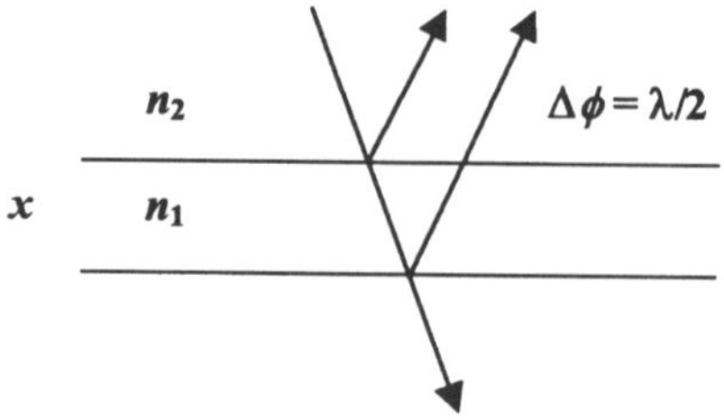

Figure 3 Interference by reflexion from two layers (after Land, 1981)

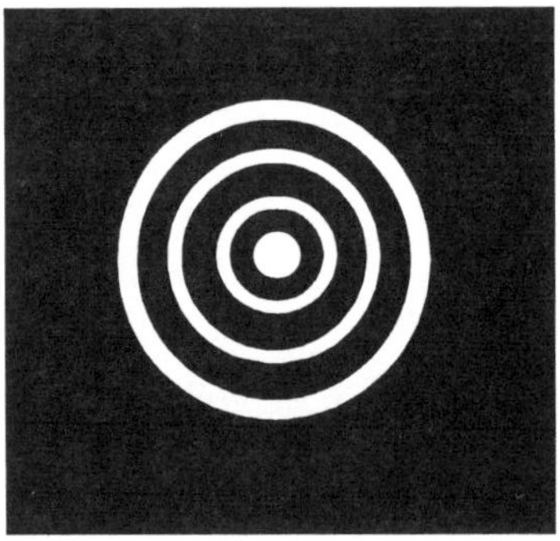

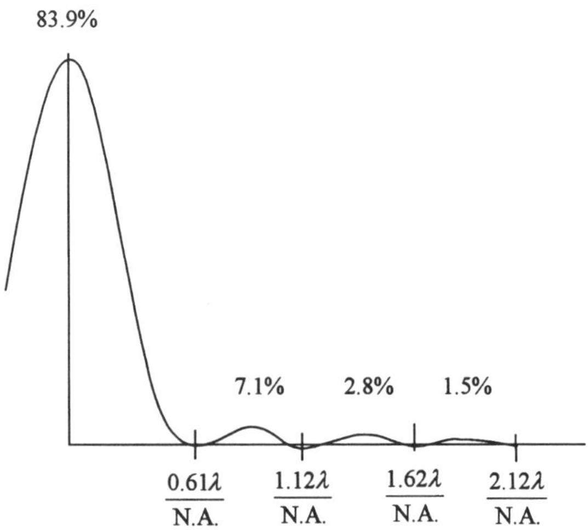

Figure 4 The image at the top represents the diffraction pattern for a point source showing the central, Airy disk and surrounding rings. The energy distribution (percentage) and separation of the rings is shown at the bottom

aperture (N.A.) of the imaging lens where N.A. = $n\sin u$ (the refractive index times the sine of the half angle of light in that medium). For the same focal length, the N.A. of lenses are proportional to their diameter as is the light-gathering power. As seen, the width of the central disk is proportional to the wavelength and inversely proportional to the numerical aperture. As two point sources are brought together in space, their imaged diffraction patterns begin to overlap. At some point the overlap will be such that it becomes impossible to resolve the two points. The Rayleigh criteria states that two point sources can be resolved up to the point where the first dark ring of one pattern coincides with the center of the other pattern. In other words, the minimum separation is $0.61\lambda/\mathrm{N.A.}$ In object space this becomes $1.22\lambda/D$ where D is the effective aperture of the imaging lens. In a visual context this means that the optical limit of resolution is related to the size of the eye (i.e. its entrance aperture) and its focal length.

1.2.3(c) Scatter

Scatter is here defined as the redirection of an electromagnetic (EM) wave by interaction with particulate matter. Before proceeding it should be stated that reflexion and diffusion can equally be thought of as scattering phenomena. The distinction made here concerns the particulate nature of the scatterer.

An EM wave consists of an oscillating electric field that can induce similar oscillations in appropriate ‘antenna’ systems. If we think of a scattering particle as a collection of electric dipoles, an impinging EM wave sets these dipoles into harmonic motion. These oscillating dipoles then act as radiators emitting their own EM waves. The characteristics and directionality of these re-radiated waves depends on the geometry and composition (e.g. refractive index) of the particles and the nature of the irradiant.

It is to be expected that the total power scattered by a particle (W_{sca}) should be proportional to the intensity (I_i) of the incident irradiant. An equation is formed with the addition of a constant. Using the symbology of Bohren (1995)

$$W_{sca} = C_{sca} I_i \qquad [14]$$

C_{sca} has the dimensions of area and is known as the *scattering cross-section*. This equation is identical in form to the power of the incident irradiant absorbed by the particle

$$W_{abs} = C_{abs} I_i \qquad [15]$$

where C_{abs} is the *absorption cross-section*. The total amount of power of the incident beam affected by the interaction with the particle is the sum of the two cross-sections and is called the *extinction cross-section* C_{ext}. Of import here is the amount of power lost from a beam in traversing a volume containing N identical scattering particles. This could be a volume of atmosphere or water, or the ocular media. This can be calculated from values for the incident intensity I_i and the transmitted intensity I_t using the formula.

$$I_t = I_i e^{-NC_{ext}h} \tag{16}$$

where N is the number of particles and h is the path-length. For a medium containing particles of different size, shape, composition or orientation, the NC_{ext} term in the equation must be replaced by the number and extinction cross-section for each particle class.

The wavelength dependence of scattering is related to the character of the incident beam (e.g. polarization state, wavelength) and the geometry of the scattering particle. Lord Rayleigh observed that the scatter from a beam of light passing through a medium containing small particles relative to the wavelength of the illuminant was blue and plane polarized at 90° to the direction of the incident beam (Wood, 1936). This type of scatter is often called molecular scatter. The intensity of scattered light I_{sca} at an angle β relative to the direction of the incident beam is given by the Rayleigh equation

$$I_{sca} = I_i^2 \frac{(D'-D)^2}{D^2} (1 + \cos^2\beta) \frac{N\pi V^2}{\lambda^4 r^2} \tag{17}$$

where D′ and D are the optical densities, related to the refractive index, of the particles and media, N is the number of particles, V is the particle volume and r is the distance between the particle and the measurement point. Note in particular that the intensity varies inversely as λ^4.

Scatter by larger particulates and aerosols, often called Mie scattering, is quite complex and beyond the scope of this treatment. For particles on the order of the wavelength of the irradiant the scatter was fairly symmetrical. However, with increasing size the scatter increased in the forward direction. The scattered light is usually of the same wavelength as the irradiant.

1.2.3(d) Polarization

As previously mentioned, the e-vector of a light wave vibrates in a direction transverse to the direction of propagation. If we look head-on at the propagating wave the e-vector describes a curve on a surface normal to the direction of propagation. If this curve is a straight line, the wave is said to be plane polarized. The angle of this line relative to a reference axis is the polarization angle Φ. A beam of light is said to be non-polarized if Φ for all rays in the beam are totally symmetric around the direction of propagation – that is, the e-vector is vibrating in all possible angles in the transverse plane. Thus, the time-average energy of the beam is equal for all Φ. This condition is met for thermal radiation emitted from diffuse surfaces. The direct solar beam above the atmosphere is non-polarized. However, most light beams do not show e-vector symmetry and exhibit some degree of polarization.

Polarization is produced by a difference in the phase and/or magnitude of the orthogonal components of the e-vector which is, you will remember, the electric field strength. If there is a phase difference between the orthogonal components, the end of the e-vector will be seen to follow an elliptical path end-on (i.e. a spiral path in the direction of propagation). The handedness of the spiral will depend on the lead/lag relationship between the vectors. Elliptical polarization is characterized by its ellipticity β and the angle of the major axis, where tan β = minor axis/major axis. In this context, plane polarization is achieved when there is no phase difference, or the phase difference is an integral multiple of π. The ellipse becomes a circle when the phase difference is an odd multiple of $+/-\pi/2$.

It is rare to find any natural source that is 100% polarized. The parameter P is defined as the intensity of polarized light divided by the total intensity of the beam and should include all polarization states. In order to fully characterize a light beam it is necessary to know its total intensity I, its ellipticity β, its degree of polarization P and its polarization angle Φ. A more convenient way of characterizing the polarization state of a beam was developed in 1852 by Sir George Stokes and are called the Stoke's parameters. They all have the dimensions of intensity (i.e. wave amplitudes) and are additive such that the final state of a beam produced by adding together other beams of known polarization is obtained by simply adding the individual Stoke's parameters. A full treatment of Stoke's parameters can be found in any optics textbook (see list at end of this chapter).

1.3 Measurement systems and units

1.3.1 Radiometry

The considerations above are valid over the entire electromagnetic spectrum – from very long radio waves to ultrashort cosmic rays. However, only a limited bandwidth of the available spectrum is utilized for photoreception and vision. This spans the region from the far-ultraviolet to the near infra-red. The science of Radiometry is usually concerned with quantitative measurements of the electromagnetic field over this range although the term may often be used to include measurements over any spectral range. While not strictly correct, we will refer to the range of photon energies used by photoreceptive and visual systems of any species as *Light*. For an in-depth discussion of radiometric terms and measurement schemes see Mobley (1994) or Kohen *et al.* (1995). For a more vision-centered discussion see Loew & McFarland (1990).

The common radiometric terms are:

1. Radiant flux (Φ or power P) is the rate of energy flow and has the units of Watts (1 W = 1 J s^{-1}) or the more visually relevant quanta per second. Wattage is usually measured as the temperature rise in a black body. This has the advantage of summing all photon energies over a broad bandwidth. Obtaining

total power from quantal measurements requires that the spectral distribution of the quantal flux be taken into account. Thus, the same energy per unit time can be delivered using fewer short wavelength photons than long wavelength photons.

2. Radiant exitance (*M*) or radiant flux density is the radiant flux emitted per unit area from a source. Its units are Watts m^{-2} or quanta s^{-1} m^{-2}.
3. Radiant intensity (*I*) is the radiant flux per unit solid angle whose vertex is at the source. The units are Watts per steradian (sr) or quanta s^{-1} sr^{-1}. Note that there are 4π steradians in a sphere.
4. Irradiance (*E*) is the radiant flux density incident on a unit area of surface and has units of Watts m^{-2} or quanta s^{-1} m^{-2}.
5. Radiance (*L*) is the radiant flux per unit solid angle and has units of Watts sr^{-1} m^{-2} or quanta s^{-1} sr^{-1} m^{-2}.

The most commonly reported metrics are irradiance and radiance. Both can be measured using the same basic instrument as long as appropriate detector headstage input optics are used. Of more use to visual scientists are the irradiance and radiance measured at some spectral intervals across a given bandwidth. The curves obtained are the spectral irradiance and radiance. The total value is obtained by integrating the spectral curve over the measured bandwidth.

1.3.2 Photometry

Radiometry is 'vision neutral' and simply reports the quantal flux of the electromagnetic field over a given bandwidth using a particular detector geometry. The spectral range may have nothing to do with visual sensation. However, it is often essential to know the characteristics of the field in relation to its ability to excite visual sensation. Photometry is the study of the response of the human visual system to the radiometric values. To get from radiometric to photometric values requires knowledge of the human luminosity functions. These relate the relative ability of radiant energy of equal photon flux at given wavelengths to evoke the sensation of brightness and are seen in Figure 5. The photopic luminosity curve, $\bar{y}(\lambda)$ is the one employed when dealing with bright light or color conversions. When dealing with dim light the scotopic curve is employed. Note that the luminosity curves are relative. Absolute values must be derived from some kind of agreed upon radiance standard and brightness assignation by a 'standard observer'. The standard defining the basic unit of brightness, the *lumen* (lm), is the surface of a platinum body at its melting point of 2042°K. By definition this surface has a luminance of 6.10^5 lumens m^{-2} sr^{-1}. This value is tied historically to the 'standard candle' definition of the lumen. The spectral radiance for melting platinum can be calculated using Planck's law for a blackbody radiator. Conversions from spectral radiance to luminance are made using the following formula (Mobley, 1994)

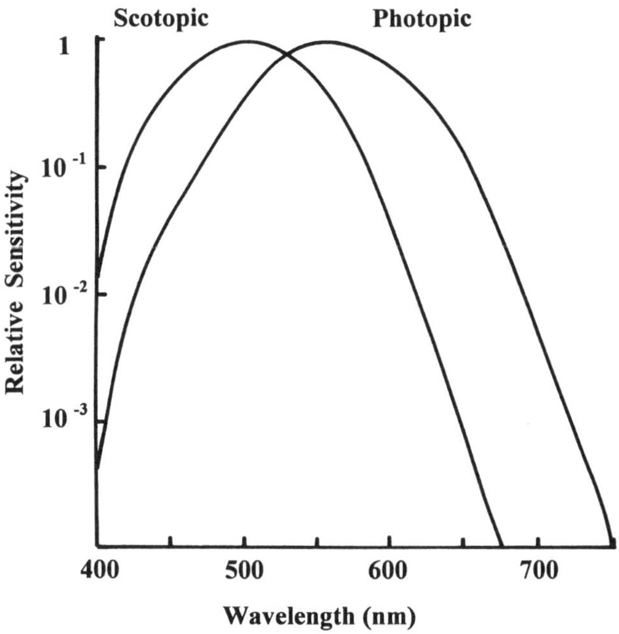

Figure 5 Scotopic and photopic luminosity functions

$$\mathrm{lm\ m^{-2}\ sr^{-1}} = K_m \int_0^\infty L(\lambda)\bar{y}(\lambda)d(\lambda) \qquad [18]$$

which, when solved for the platinum standard using the photopic luminosity function, gives

$$K_m = 683\ \mathrm{lm\ W^{-1}} \qquad [19]$$

Having determined K_m, it is possible to determine the subjective brightness of a given radiance distribution. If $L(\vec{x};\hat{\xi};\lambda)$ is the spectral radiance over the entire spectrum $0 \leq \lambda \leq \infty$, then the corresponding luminance is denoted by $L_v(\vec{x};\hat{\xi})$ and is defined as

$$L_v(\vec{x};\hat{\xi}) \equiv K_m \int_0^\infty L(\vec{x};\hat{\xi};\lambda)\bar{y}(\lambda)d\lambda \qquad [20]$$

The subscript *v* denotes a photometric quantity. Table 1 compares radiometric and photometric quantities while Table 2 gives typical luminance values for various sources.

It is possible to convert any radiometric term into its comparable photometric one using a generalization of the above formula. A general operator is derived using K_m and the radiometric term from which the photometric term is derived.

Table 1: Radiometric and photometric quantities and units

Radiometric Quantities		
Radiant power	Watt	Φ
Radiant intensity	Watts/meter^2	I
Radiance	Watts/steradian/meter^2	L
Irradiance	Watts/meter^2	E
Photometric Quantities		
Luminous power	Lumen	Φ_V
Luminous intensity	Lumens/steradian (=candles)	I_v
Luminance	Lumens/meter^2/steradian (=candles/meter^2)	L_v
Illuminance	Lumens/meter^2	E_v

Table 2: Some typical luminances (taken from Mobley, 1994)

Source	Luminance
Solar disk at earth's surface	1×10^9
60 W frosted light bulb	1×10^5
Full moon's disk	6×10^3
Clear blue sky (anti-sun)	3×10^3
Twilight sky	3
Clear sky, moonlit night	3×10^{-2}
Overcast sky, moonless night	3×10^{-5}

1.3.3 Colorimetry

This is the science that quantifies radiant stimuli on the basis of their ability to excite the perception of color in humans. In much the same way that photopic and scotopic spectral sensitivities operating on radiometric terms define photometric brightness, spectral functions based on the human trichromatic color vision system are used to derive colorimetric values. The functions adopted by the CIE (*Commission Internationale de L'Eclairage*) are the tristimulus functions $\bar{x}$, $\bar{y}$, and $\bar{z}$ as seen in Figure 6. By convention, the green tristimulus function ($\bar{y}$) is identical to the photopic luminosity curve. Where $F(\lambda)$ is a radiometric function defined over the spectrum, we can obtain photometric 'brightnesses' for the three human color channels

$$\mathrm{X} \equiv K_{\mathrm{m}} \int_0^\infty F(\lambda)\bar{x}(\lambda)d\lambda \qquad [21]$$

$$\mathrm{Y} \equiv K_{\mathrm{m}} \int_0^\infty F(\lambda)\bar{y}(\lambda)d\lambda \qquad [22]$$

$$\mathrm{Z} \equiv K_{\mathrm{m}} \int_0^\infty F(\lambda)\bar{z}(\lambda)d\lambda \qquad [22]$$

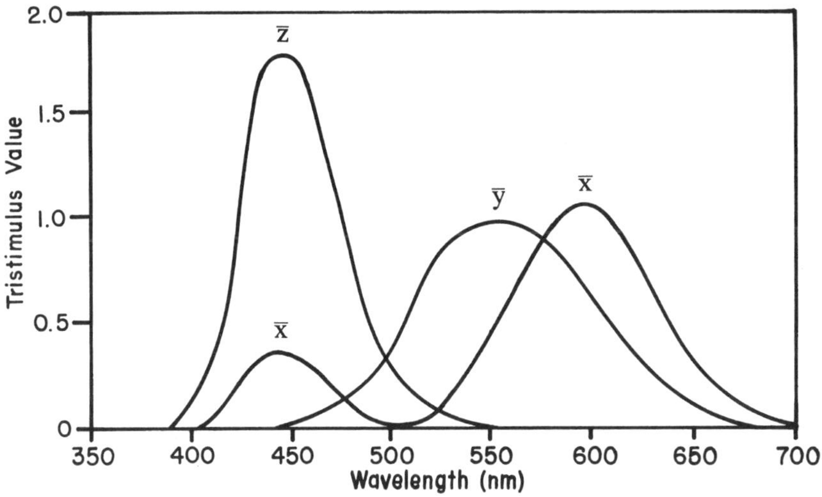

Figure 6 Human tristimulus values

The color sensation (C_v) produced by $F(\lambda)$ is related to these three photometric brightnesses such that

$$Cv \equiv X\bar{x}(\lambda) + Y\bar{y}(\lambda) + Z\bar{z}(\lambda) \qquad [23]$$

In this system, the color is the superposition of weighted amounts of photometric quantities where X, Y and Z are the color components. In addition, the vector (X, Y, Z) can be used to locate a particular color in a color space. Such a color space has been adopted by the CIE and allows the specification of any color in terms of *x* and *y* values (Figure 7).

1.3.4 From humans to animals

Photometry and colorimetry are by definition homocentric. However, it would be very useful if one could construct photometric and colorimetric counterparts for other animals. A photometry would require scotopic and photopic spectral sensitivities for the animal in question. These would be applied to the radiometric factors, as was done for humans, to yield species-specific photometric transforms. The trick here is to somehow come up with meaningful scotopic and photopic spectral sensitivity curves. The scotopic curve can be constructed directly for animals having rods as it will match the visual pigment absorption spectrum (after taking into account any ocular filtering). Unfortunately, the photopic curve is not so easily derived. The best data would be a behaviorally derived spectral sensitivity function. Such curves are available for several species (see Jacobs, 1981; Neumeyer, 1991) and have been used to define color spaces.

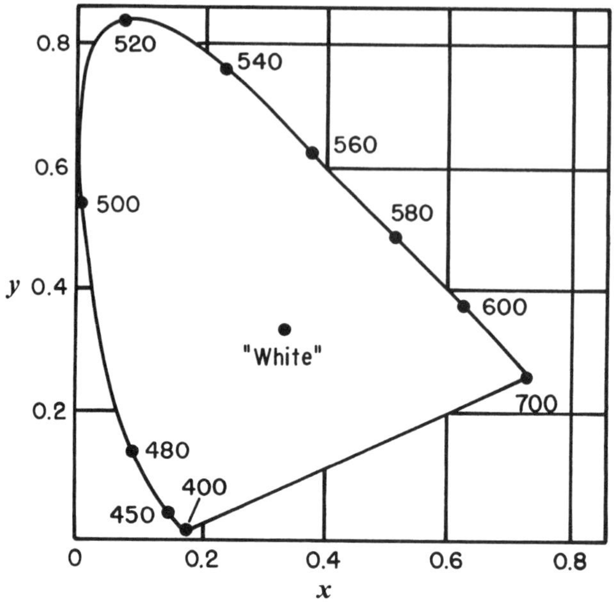

Figure 7 The 1931 CIE chromaticity diagram

However, the shape of such curves can vary depending on the characteristics of the behavioral test. Another approach uses the absorption spectra of the cone (or color vision-mediating) visual pigments to construct the color space (See Chapter 17 of Vorobyev & Menzel). Since these spectra set the boundary conditions for spectral sensitivity, this is not an unreasonable method. In the absence of data concerning scaling of the photoreceptor outputs and the exact nature of the color-encoding algorithms, results from these models are highly speculative. Yet they have proved useful in calculations concerning the 'fitness' of particular animal color spaces for discriminating among colored targets falling within the calculated space.

1.4 Absorption and emission of light

Biological systems can act as EM radiation sinks and sources – that is, they can both absorb and emit EM radiation. These actions must obey quantum physics, which means that radiation is absorbed or emitted as discrete photons of a given energy. Going hand in hand with this is the fact that matter itself is also quantized and can be described by wave-like parameters as first suggested by de Broglie (see Born & Wolf, 1980).

In determining the energy state of an atom at a given time, it is necessary to know the permissible states and the probability that the atom is in a particular allowed state. The Schrödinger equation provides solutions with inputs such as the internal energy of the atom and probability functions (wave functions) locating the positions of the electrons. For a given electron, the wave function ψ can be thought of as the probability that the electron exists at a certain distance from the nucleus. For an electron occupying a stationary energy state this function has the form of a standing wave, while it is a traveling wave for free electrons or those in a time-varying state. In the convention of electron orbits around the nucleus this translates to Bohr's rule that only orbits with a circumference equal to an integral number of electron wavelengths are allowable. Solutions of the Schrödinger equation are quantized in that only certain energy states are allowed. The absorption and emission of light by an atom or molecule (referred to as excitation and relaxation, respectively) involves a jump from one quantized energy state to another. The energy difference between states is the energy of the absorbed or emitted photon.

The allowable states for electrons are characterized by a number of parameters. The electrostatic force between the positive nucleus and an electron, as well as electron-electron repulsion lead to a classification of the major energy states as was made by Niels Bohr. The *total quantum number n* is an integer and corresponds to the classical electron shells (K, L, M...with K being closest to the nucleus). Electrons in the K shell have $n = 1$, while $n = 2$ for the L shell and so on, with each increase in n corresponding to a shell further from the nucleus. In addition to electrostatic interactions, an electron may also interact magnetically with the nucleus and other electrons. This capacity is called the *orbital magnetic moment* and is specified by another integer l – the higher the integer, the greater the magnetic moment. l is restricted to whole number values no greater than $n - 1$. Thus for the K shell $n = 1$ and $l = 0$. States are often identified according to the value of l such that $l = 0$ is called s, $l = 1$ is called p, $l = 2$ is called d, etc. This is often put together with the n state such that 1s electrons have $n = 1$ and $l = 0$ while 2p electrons have $n = 2$ and $l = 1$. An electron can also be characterized by the orientation of its wave function relative to the atomic reference frame. This is quantified by the *magnetic quantum number* m_l that can have integral values of $-l < 0 < +l$. The last quantum number reflects the direction of the magnetic moment of an electron due to its spin, which can be parallel or anti-parallel relative to other magnetic moments. The *spin magnetic quantum number,* m_s, can be either +½ or –½.

In assigning quantum states to atomic electrons it must be remembered that no two electrons in a given atom may occupy the same quantum state (this is an application of the famous Pauli exclusion principle). For example, helium has two electrons in the K shell with quantum states of $n = 1$, $l = 0$, $m_l = 0$ and $m_s = +/- ½$. No other states are allowed without violating the Pauli principle, so the maximum number of electrons in the K shell is two.

In the ground, or lowest energy state of an atom all lower shells are filled before electrons populate higher shells and usually all electrons are paired.

Considered in shell pairs, electrons with opposite spins cancel so that the total spin for an unexcited atom having an even number of electrons is zero. For an odd number the total spin is ½. When an atom absorbs a photon of proper energy to boost one electron of a parallel/anti-parallel pair to a higher energy level (i.e greater *n*), the spin of this excited electron is no longer necessarily constrained by the exclusion principle and may flip. If it does not, the *total spin quantum number* S remains the same. However, if it does flip S will equal 1. S = 0, S = ½ and S = 1 are called singlet, doublet and triplet states, respectively. This is graphically summarized in Figure 8 for helium and lithium.

If photons of different wavelengths are fired at an atom, only those having energies corresponding to possible electronic transitions will be absorbed. Plotting absorption against wavelength produces a line spectrum. However, not all-possible transitions have the same probability so the lines do not necessarily have the same amplitude for a constant quantal irradiating flux. The time needed for absorption is on the order of one photon period (period is the reciprocal of frequency). For a 500 nm photon this is approximately 1.6×10^{-15} seconds. The downward transition resulting in the emission of a photon can be stimulated by another photon of the same frequency, or may occur spontaneously. For stimulated de-excitation the process 'looks' like the absorption of one photon and the emission of two. As with absorption, this stimulated process is intensity dependent while spontaneous de-excitation is not photon dependent and can occur in the dark. The number of atoms in the excited state at any given time represents an equilibrium between the rate of photon absorption and the rates of the two de-excitation processes. Of particular importance to photochemists and photobiologists is the triplet state as this is particularly reactive and long lived.

The situation for molecules is much more complex since one must take into account the vibrational and rotational modes of nuclei and the interactions

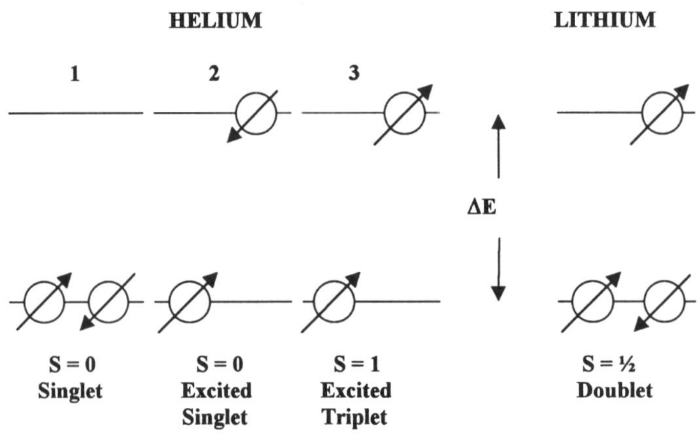

Figure 8 Various electronic states for helium and lithium (after Clayton, 1977)

between electrons and nuclei other than that of the 'parent'. One can still define ground states and excited states, but these are multiplied by the various interactions into a number of possible excitation states. In defining the energy state of a molecule it is convenient to use molecular orbital theory where the individual wave functions of electrons are combined to produce various orbital patterns and energy configurations. In some cases, individual electrons can no longer be associated with a particular atom and are best described in molecular terms. An example used by Clayton (1977) involves the interaction between the 2p electrons in separate atoms. The wave function for 2p electrons is a figure-of-eight where the two lobes represent +/– ψ. The wave functions can be combined in a number of ways. In one such case the wave function is expanded to encompass both atoms with the new ψ having the same figure-of-eight shape. In this case the electrons are shared equally by the two atoms and a covalent bond is formed. It is also possible for the interaction to produce a state where the two atoms are in an 'antibonding' configuration. Here there is a plane between the atoms where ψ equals zero due to repulsive forces. These are called the π and $\pi*$ orbitals. An electron in the π orbital has an energy less than the original 2p state while the $\pi*$ orbital has a greater energy than 2p and is an excited state. $\pi*$ can be singlet or triplet depending on the spin states of the electrons.

The alternating single and double bonds of molecules like vitamin A and retinal lead to an extended π orbital system with electron delocalization. The more extensive the system, the closer and more numerous are the $\pi - \pi*$ orbitals. Since orbital distance is the same as photon energy, closer spacing means that photons of longer wavelength (i.e. less energy) can be absorbed. The multiplicity of states leads to a broad, essentially continuous absorption spectrum.

1.4.1 Absorption measurement

Photosensitive molecules like visual pigments are normally characterized by their absorption spectrum. The process involves illuminating a volume of solution of known path-length with light and measuring the light emerging from the volume. The difference is the amount of light absorbed. This is done at different wavelengths with the resultant absorption values producing an absorption spectrum. Using the terminology of Dartnall (1957), $I_{absorbed} = I_{incident} - I_{transmitted}$ and the percentage of light absorbed is $I_{abs}/I_{inc} \times 100$. A problem with this type of measurement is that while it certainly characterizes the solution as a whole, the shape of the spectrum is dependent on concentration so that only for very dilute solutions can the absorption spectrum for a single molecule be approached. This problem can be overcome by expressing the results in a different form. For a homogeneous solution the probability of a photon being absorbed at any plane normal to the path is the same – that is the fraction of light absorbed d*I*/*I* is the same. Thus, the total amount of light absorbed is proportional to the concentration of absorbers C and the number of planes defining the path, i.e. the path length x. To form an equality requires a proportionality constant known as the absorption coefficient α. Integrating over pathlength gives

$$I_{trans} = I_{inc} e^{-\alpha Cx} \quad [24]$$

or

$$\ln \frac{I_{inc}}{I_{trans}} = \alpha Cx \equiv \log \frac{I_{inc}}{I_{trans}} = \frac{\alpha Cx}{2.303} \quad [25]$$

The term log I_{inc}/I_{trans} is the optical density and when measured over a wavelength range yields an *absorbance* as opposed to an *absorption* spectrum. The two main advantages of density spectra are that densities are additive, and when the spectrum is normalized to its highest value, the shape is concentration and path-length independent.

Conclusions

The main goal of this chapter has been to cover radiation, light, optics and related topics to a level consistent with application to the following chapters. As stated in the *Introduction*, life had to evolve in an all-pervasive electromagnetic field and develop visual systems operating according to the laws governing radiation in our local universe. As will be seen in the succeeding chapters, in extracting information from the EM field, visual systems are usually operating right at the theoretical limits dictated by these laws. How remarkable this is can be appreciated by comparing our technological solutions to those of living systems. As our abilities to describe and model visual systems improve, the feats of evolution will become even more obvious and awe inspiring.

Bibliography

Born, M. and Wolf, E. (1980) *Principles of Optics*. Pergamon, New York, 808 p.

Coulson, K.L. (1988) *Polarization and Intensity of Light in the Atmosphere*. Deepak Publishing, Hampton, Virginia (USA), 596 p.

Guenther, R.D. (1990) *Modern Optics*. Wiley & Sons, New York, 696 p.

Hurvich, L.M. (1981) *Color Vision*. Sinauer Assoc., Sunderland, Massachusetts (USA) 328 p.

Jenkins, F.A. and White, H.E. (1976) *Fundamentals of Optics*. McGraw-Hill, New York, 746 p.

Lythgoe, J.N. (1979) *The Ecology of Vision*. Oxford University Press, Oxford, 244 p.

Newton, I. (1952) *Opticks*. Dover Publications, New York, 406 p.

Overheim, R.D. and Wagner, D.L. (1982) *Light and Color*. Wiley & Sons, New York, 269 p.

Pedrotti, F.L. and Pedrotti, L.S. (1993) *Introduction to Optics*. Prentice-Hall, Englewood Cliffs, New Jersey (USA), 602 p.

Stimson, A. (1974) *Photometry and Radiometry for Engineers*. Wiley & Sons, New York, 446 p.

Wasserman, G.S. (1978) *Color Vision: An Historical Introduction*. Wiley & Sons, New York, 224 p.

Walsh, J.W.T. (1965) *Photometry*. Dover Publications, New York, 544 p.

Wyszecki, G. and Stiles, W.S. (1982) *Color Science*. Wiley & Sons, New York, 950 p.

References

Abrahamson, E. W. and Japar, S. M. (1972) Principles of the interactions of light and matter. In *The Handbook of Sensory Physiology* VII/1 (H.J.A. Dartnall, ed.), Springer-Verlag, Berlin, pp 1–32.

Bohren, C. F. (1995) Scattering by particles. In *Handbook of Physics* Vol. 1 (M. Bass, ed.-in-chief), McGraw-Hill, New York, pp 6.1–6.21.

Clayton, R. K. (1977) Light and Living Matter Vol. 1, Krieger, New York, 148 p.

Dartnall, H. J. A. (1957) The Visual Pigments, Methuen, London, 216 p.

Goodman, D. S. (1995) General principles of geometric optics. In *Handbook of Physics* Vol. 1 (M. Bass, ed.-in-chief), McGraw-Hill, New York, pp 1.3 –1.109.

Jacobs, G. H. (1981) *Comparative Color Vision*. Academic Press, New York, 209 p.

Kohen, E., Santus, R. and Hirschberg, J.G. (1995) *Photobiology*. Academic Press, New York, 506 p.

Land, M. F. (1981) Optics and vision in invertebrates. *In The Handbook of Sensory Physiology* VII/6B (H. Autrum, ed.), Springer-Verlag, Berlin, pp 471–594.

Loew, E. R. and W. N. McFarland (1990). The underwater visual environment. In *The Visual System of Fish* (R. Douglas & M. Djamgoz, eds.) Chapman & Hall, London pp 1–44.

Mobley, C. D. (1994) Light and Water. Academic Press, New York, 592 p.

Neumeyer, C. (1991) Evolution of colour vision. In *Evolution of the Eye and Visual System* (J. R. Cronly-Dillon & R. L. Gregory, eds.) CRC Press, Boca Raton, Florida, USA, pp 284 – 305.

Wood, R.W. (1936) *Physical Optics*. Macmillan, New York, 846 p.

1b
Light and photoreception: Visual pigments and photoreception

S.N. ARCHER
International Marine Centre, 09072 Torregrande, Oristano, Italy

1.6 Introduction

It is perhaps inevitable that organisms on our planet have evolved the ability to perceive light since light provides the very energy that has supported life. For humans it would be surprising to find a world bathed in light where organisms were not light-sensitive. In fact, on earth many animals that cannot 'see' well inhabit environments where light does not penetrate very much and in some of these dark places organisms even make their own bioluminescent light (see chapter by Widder). Plants and simple animals, which may only need to orientate themselves with respect to the light source, possess very basic photon detecting systems but, as life has evolved into more complex forms, the ability of organisms to sample light energy has become more sophisticated.

The majority of higher animals with more developed behaviour have evolved photon detecting systems which are capable of image formation. This has allowed for the visual representation of the external environment but in turn has necessitated the development of major sensory systems with dedicated processing power to interpret the sensory input. The visual sensory systems generally become more sophisticated as the animal performs more complicated visual tasks, which can include predator, prey and mate detection. To perform these tasks, the visual system becomes adapted to function well in the surrounding environmental light conditions and according to the behavioural needs of the animal. Different types of environment have different light conditions (e.g. deep blue oceans or red-stained rivers) and these conditions can vary with the time of day or year. In addition, the environmental light source may not be constant in all viewing directions and this can be further complicated if the animal moves. The behaviour of the animal can also change during development and this obviously changes the visual tasks. Across species, the visual system has adapted to a wide range of environmental light conditions and behaviours. Within species, there is also the possibility for the visual system to demonstrate a certain amount of plasticity in adaptation to varying conditions and behaviour, but this is of course restrained by physiology. As John Lythgoe wrote in 1979, '...the visual sense must work within performance limits set by the physical nature of light and by their own physiological performance'. It is the consideration of these

S.N. Archer et al. (eds.), Adaptive Mechanisms in the Ecology of Vision, 25–41

limits and the attempt to understand them in terms of behaviour that is now represented by the concept of the ecology of vision. In this chapter we will introduce some of the fundamental elements of this concept upon which rests much of the visual adaptation described in detail in the rest of the book.

1.7 Photoreception and Phototransduction

Photoreception is not always associated with vision or image formation. Extraretinal photoreceptors can occur in the skin, cornea and various regions of the brain including the pineal (see chapters by Foster and by Shand and Foster). These diverse receptors are involved in regulating skin pigmentation, screening pigment migration and circadian rhythms. In one deep-sea shrimp a collection of photoreceptors on the carapace appear to act as heat sensors to avoid close exposure to thermal vents (see chapter by Marshall *et al.*). As far as we know, however, all of these photoreceptors contain light-sensitive visual pigments which are similar in composition to those of the rod and cone photoreceptors involved in vision and, in describing photoreception and phototransduction, we will focus upon what is known for the visual pigments in retinal photoreceptor cells.

1.7.1 Photoreception

The photoreceptors of the retina (rods and cones in vertebrates, see chapter by Locket, and rhabdoms in invertebrates, see chapters by Land and by Marshall *et al.*) are packed together in an array of efficient photon detectors. Photoreceptors in vertebrates are composed of an inner- and an outer-segment, nucleus and synapse. The inner segment contains the bulk of the cellular 'machinery', including many mitochondria, and the outer segment is composed of a dense stack of membrane folds (cones) or discs (rods). The outer-segment membrane is packed with the light-sensitive visual pigment molecules which are rapidly turned over by production in the inner-segment. Photons of light pass through this membrane where they are captured and transduced by the visual pigments.

As soon as the light signal is transduced, the resulting electrical signal from the photoreceptors enters a neural network where it is processed and analysed to recreate an image of the external environment. All the visual information available for this process must pass though the photoreceptor cells and in particular the group of light-sensitive molecules known as the visual pigments. These molecules (of which there are only four types in humans) act as the visual information interface between the environment and the organism and can convert information on light intensity and spectral composition. Neural processing can then extract further information such as frequency of stimulus, contrast, movement, light polarization etc. and much of this basic processing occurs in the retina itself.

The visual pigments are the light-sensitive molecules in photoreceptors. They are membrane-associated receptors which are composed of a protein component (opsin) coupled with a light sensitive chromophore. The chromophore in most animals is either an aldehyde of vitamin A1 or A2, and is called retinal or 3-dehydroretinal respectively, although some invertebrates have been shown to also utilize 3- and 4-hydroxyretinal (Kito *et al.*, 1992). Visual pigments which are composed of retinal are called rhodopsins and those with 3-dehydroretinal are called porphyropsins. The opsin protein is a member of the super-family of G-protein linked receptors (see Watson and Arkinstall, 1994) which all appear to share the characteristics of being integral membrane proteins that contain seven hydrophobic, transmembrane (TM) domains. In the visual pigments, the receptor ligand is replaced by the chromophore which is bound in the inactive state. The opsin protein threads through the outer-segment disc membrane seven times with the N-terminal starting in the intradiscal space and the C-terminal region ending in the cytoplasm. The transmembrane bundles form a pocket surrounding the chromophore (see Figure 9) which is linked via a Schiff's base to a lysine in the seventh α-helix. Comparison of G-protein coupled receptor sequences (Baldwin, 1993) and analysis of crystal structure (Schertler *et al.*, 1993) has predicted that the transmembrane bundles are arranged in a specific formation (Figure 9A) with respect to each other. The bundles are packed closer together on the cytoplasmic side and are more widely spaced on the extracellular side to form the binding pocket opening. The chromophore lies roughly horizontal in the pocket (parallel to the plane of the membrane) with the terminal carbon ring extending out towards the fourth and fifth helix. It is suggested that the shape of the binding pocket accommodates well the bent shape of the inactivated 11-cis chromophore and crystal structure analysis has demonstrated that the better characterised bacteriorhodopsin has a different pocket conformation in the ground state because it has all-*trans* retinal bound (Baldwin, 1993). It is this close interaction between the chromophore and the opsin protein in the binding pocket that forms the basis for visual pigment spectral sensitivity tuning and the mechanisms for this will be described briefly here (see also Chapter by Bowmaker and Hunt for more detail).

Much of the visual pigment structure is highly conserved between rods and cones and across species. This not only reflects the general conservation of G-protein linked receptors, but also the need to create the correct binding domain for the chromophore and provide the structural features required to transduce the light signal. Many of the conserved features of opsin are shown schematically in Figure 10. In this figure, opsin is shown as a chain of amino acid circles which begins at the N-terminal on the intradiscal (or extracellular) side of the membrane, passes through the membrane seven times (rectangles 1-7) and ends in the cytoplasm with the C-terminal. The precise organization of the amino acids with respect to the membrane must remain speculative, but the general arrangement (and numbering) depicted here is based on the current working model for bovine rhodopsin taking account of available data (see Hargrave and McDowell, 1992).

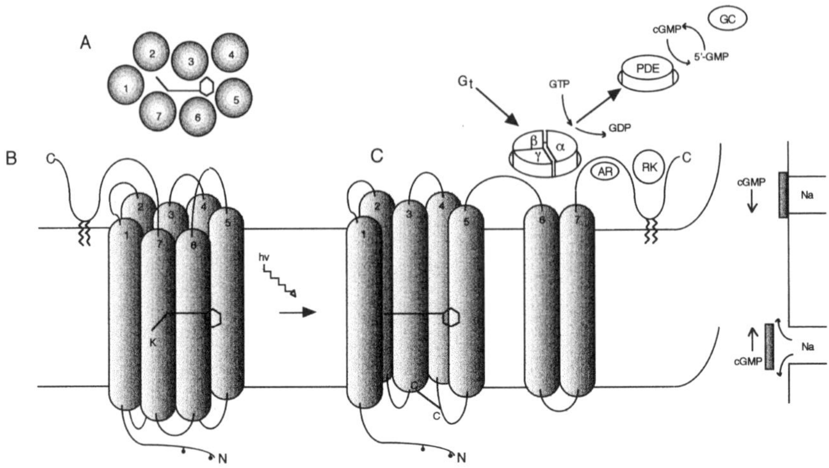

Figure 9 A schematic representation of visual pigment in the photoreceptor membrane. A. Looking from above on the cytoplasmic side, opsin has seven transmembrane domains which form a pocket surrounding the chromophore. B/C. The transmembrane domains are connected by three intradiscal loops and three cytoplasmic loops. The C-terminals of many rhodopsins become palmitoylated (zigzag line) forming a fourth cytoplasmic loop. Intradiscal loop 2 and the base of transmembrane 3 are held together in configuration by a disulphide bridge between two conserved cysteines. The chromophore lies roughly perpendicular to the membrane bound at a lysine in transmembrane 7. There are two highly conserved glycosylation sites in the N-terminal region (indicated by line and dot). On absorbing light, the chromophore isomerises from the 11-*cis* to the all-*trans* form which initiates the transduction cascade. The G-protein, transducin (Gt), binds to opsin and GTP replaces GDP on the alpha subunit which activates and dissociates. The activated alpha subunit then activates cGMP-phosphodiesterase (PDE) which hydrolyses cGMP to 5′-GMP and falling cGMP levels close sodium channels (Na) in the cell membrane causing hyperpolarization of the cell. Inactivation is brought about by rhodopsin kinase (RK) mediated phosphorylation of the opsin and binding of arrestin (AR). The positions of RK and AR in the figure are not meant to imply their actual binding positions but RK is known to phosphorylse sites on the C-terminal (see text for details). The transducin and PDE subunits reassociate and guanylate cyclase (GC) restores cGMP levels opening Na channels (see text for details)

The figure shows amino acids that are absolutely conserved in vertebrate rods and cones and has been generated by comparing the sequence of 38 rod opsins and 20 cone opsins. Amino acids conserved in rods are shown in open circles and those conserved in both rods and cones are shown in shaded circles. It is immediately apparent that the opsins are very highly conserved across the whole protein, but there are specific areas of interest for photoreception and phototransduction. Interestingly, of the 44 rod opsin mutation sites known to be associated with autosomal dominant retinitus pigmentosa (adRP) in humans, 28 sites (64%) are fully conserved in the rods and are indicated in figure 10 by an asterisk next to the amino acid (for a full list of RP sites, see Swissprot database entry accession no. P08100). The fact that such a high percentage of these sites are conserved across all rod opsins indicates how important this conservation is

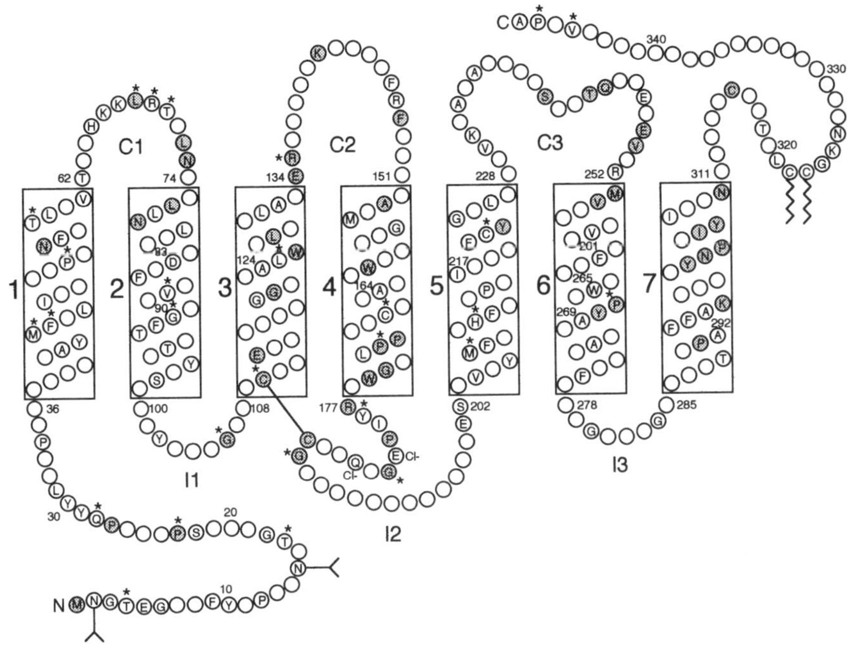

Figure 10 A schematic map of ospin showing conserved amino acids among rods and cones and conserved sites of interest. The map is based upon the bovine rhodopsin model (Hargrave and McDowell, 1992) and has been generated by comparing the sequence alignments of the vast majority of currently known vertebrate rod and cone opsins. The transmembrane helices are numbered 1–7, connected by the intradiscal (I1– I3) and cytoplasmic loops (C1– C3). Amino acid sites conserved in all rods are indicated with their one-letter code in open circles. Those conserved in all rods and cones are in shaded circles. Conserved sites which are also mutation sites linked with human autosomal dominant retinitis pigmentosa are indicated with an asterisk. Two highly conserved glycosylation sites occur in the N-terminal region at sites 2 and 15. Two conserved cysteines at 110 and 187 form a disulphide bridge. Two sites associated with a chloride ion sensitive region in I2 are labelled (Cl–). The lysine retinal binding site is numbered at position 296 and its counterion is at 113. Amino acids within the transmembranes which are implicated in wavelength tuning are numbered (83, 90, 124, 164, 217, 261, 265, 269 & 292). Two highly conserved cysteines occur at 322 and 323 which can become membrane associated through palmitoylation. The map was generated from comparison of the following opsin sequences (all available on the GenBank, Swiss Protein or National Biomedical Research Foundation Protein Sequence databases, unless indicated): *Pomatoschistus minutus* (sand goby rod opsin), *Poecilia reticulata* (guppy rod opsin), *Gambusia affinis holbrooki* (mosquitofish rod opsin), *Carassius auratus* (goldfish rod, blue, green1, green2 & red opsins), *Cyprinus carpio* (carp rod opsin), *Anguilla anguilla* (european eel, freshwater and deep sea rod opsins), *Conger conger* (conger eel rod opsin, Archer and Hirano, 1996), the rod opsins from four species of deep-sea fishes (Hope *et al.*, 1997), the rod opsins from 11 species of Lake Baikal fishes (Hunt *et al.*, 1996), the rod opsins from 3 holocentrid fish species (*Myripristis berndti, Neoniphon sammara* and *Sargocentron diadema*), *Lampetra japonica* (japanese lamprey rod opsin), *Bos taurus* (bovine rod & blue opsins), *Mus musculus* (mouse rod opsin), *Rattus norvegicus* (rat rod opsin), *Cricetulus griseus* (chinese hamster rod opsin), *Homo sapiens* (human rod, blue, green & red opsins), *Oryctolagus cuniculus* (rabit rod opsin), *Ovis orientalis aries* (sheep rod opsin), *Canis lupus familiaris* (dog rod opsin), *Gallus gallus* (chicken rod, violet, blue, green & red opsins), *Rana pipiens* (northern leopard frog rod opsin), *Xenopus laevis* (african clawed frog rod & violet opsins), *Gecko gecko* (gecko blue & green opsins), *Astyanus mexicanus* (mexican tetra blue), *Astyanus fasciatus* (cave fish green1, green2 & red opsins), *Anolis carolinensis* (american chameleon red opsin)

to maintain correct wild-type function and many of these sites are strictly associated with known opsin characteristics.

In the N-terminal region there are two highly conserved glycosylation sites at residues 2 and 15 that are probably involved in cellular trafficking of the opsin (Watson and Arkinstall, 1994) and substitutions of the threonine residues at sites 4 and 17 remove the motifs, which gives rise to adRP. The intradiscal loops of the opsin (I1-I3) appear to be important for confirming correct tertiary structure in the protein (Doi *et al.*, 1990). Two conserved cysteines (C110 & C187) form a disulphide bridge which anchors I2 to the base of TM3. This bridge is absolutely necessary to form a functional visual pigment (Karnik and Khorana, 1990) and the mutation C110Y is linked to adRP. Glycine is associated with turns and kinks in peptide chains and there are five highly conserved glycine residues in the intradiscal loops. Three of the glycines are associated with adRP and two of these are located in a well conserved region of I2 where the disulphide bridge interacts with the loop. All these factors point towards the important role that the intradiscal domain plays in visual pigment structure. Mutation studies in this domain have revealed that substitutions or deletions in many regions, and in particular I2, lead to defective opsin protein (Doi *et al.*, 1990). The authors of that study conclude that the intradiscal domain folds and holds the opsin in a specific configuration which is essential for correct function. Recent studies with other 7TM receptors have shown that mutant human thrombin receptors can be constitutively activated by chimeric substitutions in I2, and certain known 'activating antibodies' have been shown to be directed towards I2 in adreno-receptors and bradykinin receptors (for review, see Schwartz and Rosenkilde, 1996). In addition, opsin I2 also contains two residues (E181 & Q184) which form part of a chloride ion sensitive region. These sites distinguish rod and short-wave sensitive cone pigments from long-wave sensitive cone pigments which have H181 and K184 and are chloride-sensitive (see chapter by Bowmaker and Hunt for more detail). So it appears that the intradiscal domain (especially I2) is very important for tertiary structure and that tertiary structure has a big influence on receptor function.

The seven TM regions are composed of mainly hydrophobic amino acids arranged in α-helices. The most important role of the TM amino acids in opsin appears to be interaction with the chromophore. The chromophore is bound by a Shiff's base at the conserved site K296 and it has been shown that this binding site has a conserved, stabilizing counterion (E113) in TM3 (Nathans, 1990). Opsin on its own absorbs in the UV but when combined with retinal this shifts to around 500 nm. The chloride ion sensitive region in I2, mentioned above, can further shift the absorption up to 570 nm. Within these spectral bands, further fine tuning occurs by changing the energy difference between the ground and excited states of the molecule. This is achieved by altering the protein-chromophore interaction by substituting charged for non-charged or polar for non-polar amino acids in the opsin. Because these residues interact with the chromophore they are located in positions in the α-helix which face into the binding pocket. So far, a number of sites have been implicated in spectral tuning from sequence comparisons and mutagenesis studies (including sites 83, 90, 124, 164, 217, 261, 265, 269 & 292) and these are discussed in more detail in the chapter by Bowmaker and Hunt. Obviously, these sites are not conserved across visual pigments with different absorbances because they change to achieve the tuning, but it is worth noting that their positions in TM2-7 form a roughly horizontal line around the middle of the helices, consistent with the theoretical positioning of the chromophore. Only one of these potential tuning sites corresponds with a conserved rod opsin adRP site and this occurs at site 90 where all wild-type rod opsins have a glycine but most short wave cone pigments have a serine. Recently it has also been shown that mutation at a second potential tuning site (W265F) can lead to heterogenous opsin distribution in the membrane (Decaluwe and Degrip, 1996). At present, however, it is not clear if these mutations reflect the involvement of the sites in tuning or some other, unknown function.

Of course, another important requirement of the TMs is that they can be arranged in a formation which produces the chromophore binding pocket. It is not simply enough that the TMs surround the chromophore in a 3+4 arrangement, they must also be able to bend relative to each other so that the pocket has the correct 3-dimensional form. Proline and glycine residues can fulfil the role of providing the necessary kinks in the helices and there are 13 conserved proline and glycine residues in TM1-7, 7 of which are conserved in all opsins examined and 4 of these are associated with adRP. These highly conserved proline and glycines are, therefore, strong candidates for allowing the necessary bends in the helices.

On the cytoplasmic side there are several conserved regions. The first cytoplasmic loop (C1) is very highly conserved and has three adjacent adRP sites although the exact function of this loop is not known. The three conserved adRP sites form part of a four amino acid deletion linked to adRP and more recently it has been shown that substitution of one of these amino acids (R69H) leads to the formation of a truncated protein, indicating that this position is important for correct translation (Decaluwe and Degrip, 1996). In the C-terminal region there is a section of conserved amino acids in the rod opsins around two adjacent cysteines which become palmitoylated forming a fourth cytoplasmic loop. Competitive binding assays between transducin and anitbodies for various opsin domains has implicated the fourth loop along with the C2 and C3 loops in transducin binding (Nathans, 1992). Furthermore, mutation studies have shown that there is a conserved charged pair (E134/R135) in C2 which is required for transducin activation and that substitutions at conserved sites S240 and T243 (involved in light-induced phosphorylation) significantly reduce activation (Franke *et al.*, 1992). The same mutation study also demonstrated that deletions to loops C2 and C3 affected transducin activation and highlighted the carboxyl-terminal region of C3 as being particularly important. Among opsins this region is highly conserved and the same region in adrenoreceptors has also been found to be important in G-protein interaction (Watson and Arkinstall, 1994). It should also be noted that the C-terminal region of opsin is rich in serine and threonine residues that are potential sites of phosphorylation events during inactivation of the visual pigment. However, because the exact position of these residues varies among different opsins, the sites do not present themselves as being conserved in this type of analysis.

1.7.2 Phototransduction

In darkness, vertebrate photoreceptors have an inflowing dark current through open, cGMP-gated cation channels. On receiving photons of light, the isomerization of the photoreceptor chromophore leads to the closure of these channels and the hyperpolarization of the cell which decreases transmitter release at the photoreceptor synapse and propagates the neural signal. Between isomerization and closure of the ion channels there is a cascade of secondary messengers which are involved in the transduction of the light energy into electrical energy. This chain of enzymes communicates the signal and amplifies it at the same time, ensuring that very few photons can elicit a significant response. Once the light signal has terminated, additional mechanisms turn off the response and reset the components of the cascade. The major mechanisms of the phototransduction cascade are shown in Figure 9C. As a result of chromophore isomerization, a conformational change occurs in the opsin protein which initiates this cascade. In Figure 9C, TM6 and TM7 are depicted 'swung' away from the other TMs to make the presentation clearer and not necessarily to imply that this is the change in conformation that occurs. Here we will review the major mechanisms involved but the reader is also referred to the recent review by Kawamara (1995) for more details.

Upon absorption of light, the retinal chromophore isomerizes from the 11-*cis* to the all-*trans* form. During this transition rhodopsin passes through two intermediate forms called metarhodopsin I and II (MI and MII) before releasing the all-*trans* retinal. It is the photoactivated, MII form of rhodopsin that is able to activate the G-protein transducin (Gt). There is some evidence that deprotonation of the Schiff's base is an integral part of the photoactivation of rhodopsin and that this is coupled with protonation of several histidine residues (H65, H152 and H 211) (see Nathans, 1992). Mutagenesis has demonstrated that one histidine in particular (H211) is important in driving the photoactivation transition towards MII. Of these three histidine sites, two are conserved among rod opsins (see Figure 10) including H211 which is also an adRP site. However, an unprotonated Schiff's base mutant has been shown to activate rhodopsin in response to light and this implies other mechanisms are also involved (Fahmy and Sakmar, 1993). It has been suggested that a salt bridge exists between the chromophore binding site (K296) and its counterion (E113). This bridge would hold rhodopsin in the inactive state but isomerization of the chromophore would break the bridge leading to a conformational change and activation (Cohen et al., 1993). Mutation studies by Cohen *et al.* (1993) also showed that a negative charge at E134 is important for correct activation and they hypothesise that this charge helps to stabilize the inactive state of opsin. E134 is at the border of TM3 and C2 (see Figure 10) and is conserved in all opsins and most G protein-linked receptors.

The activation of rhodopsin and the resulting conformational change allows Gt to bind. As mentioned in section 3.1, the opsin cytoplasmic loops C2, C4 and especially the carboxyl-half of C3 have been shown to be important for Gt binding. Recent studies have shown that mutation substitutions at sites S240, T242, T243, K248, V250, T251 and R252 in loop C3 all reduce the ability of rhodopsin to activate Gt (Franke *et al.*, 1992; Shi *et al.*, 1995). Three of these (S240, T243 and V250) are conserved in all opsins and one (R252) in all rods (see Figure 10). Mutation at K248 has also been shown to affect membrane translocation (Decaluwe and Degrip, 1996). Deletions in the central region of C2 and the carboxyl-half of C3 abolish the ability to activate Gt (Franke *et al.*, 1992) and both of these regions contain well conserved amino acids (see Figure 10). Gt is a protein composed of three subunits; Gtα (40kDa), Gtβ (35kDa) andGtγ (5kDa). The activation of rhodopsin and the binding of Gt causes the exchange of GDP for GTP which binds to Gtα. This exchange activates Gt causing the Gtα subunit to disassociate. One activated rhodopsin molecule can activate thousands of Gt molecules and thereby amplify the photoreception signal. The disassociated Gtα activates cGMP phosphodiesterase (PDE) which then hydrolyses cGMP to 5′-GMP. The subsequent fall in cGMP levels leads to the closure of cGMP-gated channels. There does not appear to be any amplification between Gt and PDE but one PDE molecule can hydrolyse hundreds of cGMP molecules, further amplifying the signal.

When the light stimulus stops, rhodopsin and the phototransduction cascade are disactivated. Rhodopsin becomes inactivated by phosphorylation by rhodopsin kinase. There are multiple opsin sites for phosphorylation which are located in the cytoplasmic loops and the C-terminal. It is assumed that these sites become exposed to rhodopsin kinase when rhodopsin is activated. Cytoplasmic loop C3 again appears to play a major role in this respect since cleavage of the loop reduces phosphorylation. There are a series of three alanine residues in C3 (A233, A234 and A235) where substitutions have been shown to affect levels of phosphorylation (as well as Gt activation) and two of these (A233 and A235) are conserved in all rods (see Figure 10) (Shi *et al.*, 1995). As mentioned previously, the C-terminal is a major site for phosphorylation although the position of exact sites are not well conserved. In bovine rhodopsin, major phosphorylation sites in the C-terminal are S339, S343 and T336 (McDowell *et al.*, 1993). The importance of the C-terminal for phosphorylation has recently been demonstrated elegantly in transgenic mice with rod cells expressing a rhodopsin truncated at S334 (S334ter) (see Lem and Makino, 1996). This mutant demonstrated a prolonged light response, presumably due to the missing C-terminal phosphorylation sites. Loops C1 and C2 are also involved since substitutions at several sites in these loops (T62, V63, Q64, R147, F148 and G149) also affect phosphorylation (Shi *et al.*, 1995). Rhodopsin kinase becomes autophosphorylated during phosphorylation of rhodopsin which presumably disactivates the enzyme. Once rhodopsin is phosphorylated another protein, called arrestin, binds to rhodopsin and stops further binding of Gt which becomes inactivated by its own GTPase activity. Once phosphorylation and arrestin binding have occurred, the inactivated MII rhodopsin finally disassociates into opsin and all-*trans* retinal. Opsin then becomes dephosphorylated by a phosphatase and arrestin is removed allowing the opsin to recombine with 11-*cis* retinal and regenerate rhodopsin. Finally, the cGMP levels are restored by cGMP synthesis by guanylate cyclase (GC). GC is either soluble or membrane bound and there is some debate as to which forms are present in photoreceptor outer-segments (see Kawamura, 1995).

1.8 Visual pigment function and the ecology of vision

The nature of the biochemical machinery of the visual pigment necessary for photoreception and phototransduction inevitably introduces certain performance restrictions with respect to visual pigment sensitivity functions. These restrictions are the essence of the physiological limits that John Lythgoe pointed out in 1979 and almost certainly have significant consequences for visual ecology. The ecology of vision is concerned with understanding the adaptation of an animal's visual system to performing visual tasks within a given light environment. Of course, some of this adaptation will be achieved at other levels and involve evolutionary change of optical structure (e.g. chapters by Land; Sivak; Martin) or neural processing (e.g. chapters by Djamgoz and Wagner; Frishman), for example. Light arriving at the visual pigment can be spectrally modified by other optical structures in the eye (see chapter by Douglas and Marshall) and the signals from the photoreceptors can be combined and modified by the neural network to alter the sensitivity function. This means that it is not enough, and sometimes perhaps even flawed, to simply measure the spectral sensitivity of a visual pigment and try to relate this to available light. However, historically a central theme in studying the ecology of vision has been the identification of correlations between visual pigment spectral sensitivity and the wavelengths of light available (in the environment) for vision. This approach has been generally successful and has identified some overall trends (see chapter by Partridge and Cummings). The identification of these trends is important because it identifies functional frameworks but, perhaps more importantly, it also points the way to studying the potentially more interesting trend-breakers. Normally, where there is a trend-breaker, interesting behaviour and adaptation will also be found. Here we will briefly look at some of the physiological restrictions that may be placed upon spectral sensitivity due to visual pigment biochemistry.

1.8.1 Visual pigment spectral tuning and visual tasks

Visual pigments in vertebrates have maximum spectral sensitivities that range from around 350 to 635 nm. Within this range, the exact location of the maximum sensitivity can be tuned by alterations at specific sites in the opsin sequence mentioned in section 1.7.1 and described in detail in the chapter by Bowmaker and Hunt. So far a maximum of around 20 sites have been indicated as potential tuning sites in opsin and it appears that effects at different combinations of these sites can produce a variety of spectral sensitivity shifts. In the European eel, for example, there are substitutions at three potential sites (83, 124 & 292) which could be responsible for tuning differences between green- and blue-sensitive rod pigments (Archer *et al.*, 1995). Mutations at two of these sites (83 & 292) in bovine rhodopsin have been shown to cause 8 –10 nm blue-shifts (Nathans, 1990; Nakayama and Khorana, 1991) and the third site has been correlated with a blue shift of between 7–10 nm in Lake Baikal fish (Hunt *et al.*, 1996). These sites combined are thought to be responsible for the shift between the 501 nm green-sensitive rod pigment and the 482 nm blue-sensitive rod pigment (see also chapter by Partridge and Cummings). The conger eel has a single, blue-sensitive rod pigment with a spectral sensitivity maximum at 487 nm. This is 5 nm longer than the blue-sensitive rod pigment in the European eel, but it has been found that only two of the three tuning site substitutions identified in the European eel are conserved in the conger eel and lack of the third substitution may be responsible for the red shift (Archer and Hirano, 1996).

From this example, it can be seen that one simple source of restriction upon the possible spectral sensitivity of a visual pigment may be derived from a limited number of potential tuning sites in the opsin molecule. Long before molecular biology had begun to provide opsin sequence and uncover tuning sites, it was recognised that particular configurations between opsin and the chromophore could produce a discrete family of visual pigments with spectral sensitivities grouped at specific cluster points (Dartnall and Lythgoe, 1965; Lythgoe, 1979). The ‘cluster hypothesis’ has subsequently been re-examined in several studies where the visual pigments from significant numbers of species have been compared and the results confirm the original findings and predict a cluster separation distance of 5 – 8 nm (Partridge *et al.*, 1989; Jacobs, 1993; Hunt *et al.*, 1996).

Changing the opsin in the photoreceptor visual pigment can ‘jump’ the spectral sensitivity to a new ‘cluster’ position and changes in opsin gene expression could be used as a means to achieve adaptation to changing conditions or behaviour. Changes in opsin gene expression and opsin turnover can occur rapidly (Wood and Partridge, 1993; Archer *et al.*, 1995) and in rainbow trout there is the suggestion that UV opsin gene expression may be reversible since UV sensitivity may be regained in adults migrating back to shallow water (Browman and Hawryshyn, 1994). But there is no evidence that opsin gene expression can be turned on and off reversibly and repeatedly to provide continuous adaptation to the changing light conditions that may be experienced seasonally by freshwater fish, for example. In these conditions it is possible that a less biochemically expensive solution could be to change the ratio of rhodopsin and porphyropsin usage providing a rapid plastic response. Changes in opsin gene expression may be reserved for more developmentally well-defined periods of adaptation.

Within the opsin 'jump' points, spectral sensitivity could also be fine-tuned with the use of optical filters. Different combinations of coloured photoreceptor filters (oil droplets) and visual pigments could increase hue discrimination in birds (see chapter by Douglas and Marshall). The use of filters to 'fine-tune' spectral sensitivity also creates the possibility for an animal to shift its visual communication to more 'private' or less 'crowded' wavelengths. Thus, species-specific signals could be developed in the form of body colouration or bioluminescence which could co-evolve with specific combinations of visual pigments and filters. This type of mechanism will also put evolutionary pressure on predators to develop 'code-cracking' devices and the yellow lenses of some deep-sea fish may serve as a device to enhance sensitivity to bioluminescence used in down-ward pointing photophore camouflaging (see chapter by Douglas and Marshall).

1.8.2 Noise as a limiting factor

The implications for the ecology of vision of a definite set of visual pigment spectral sensitivities are obvious. Many authors have commented on the theoretical reasons for why single and multiple visual pigments in animals should have their observed spectral locations (Loew and Lythgoe, 1978; Lythgoe, 1979; Partridge, 1990; Loew, 1995; Bowmaker, 1995). The emerging theories have been called the Sensitivity and Contrast Hypotheses and predict that in certain situations visual pigments will best detect contrast if their spectral sensitivity matches the background spacelight while in others an 'offset' visual pigment will be better. Generally, matched pigments are better for detecting dark or distant objects in low light conditions, whereas bright objects in shallow water are best detected with offset pigments. From this it should be expected that deep-living fish with single visual pigments will be matched to the narrow-band, predominantly blue background spacelight and indeed several large studies on the visual pigments of deep-sea fishes have shown this to be the case (Partridge *et al.*, 1992; Douglas *et al.*, 1995; Douglas and Partridge, 1997). The same studies have also identified species with offset visual pigments which may be tuned into bioluminescent light or whose overlapping sensitivity functions could be providing colour vision capabilities. Fish living in shallower water tend to have at least two overlapping visual pigments, and hence the ability for colour vision, and generally these are often matched to and offset from the water spacelight.

There is little doubt that in many situations the theory of the sensitivity hypothesis operates in practice and the single rods of numerous species of deep-sea fishes are a good example of this ecological adaptation (Douglas and Partridge, 1997). However, the deep sea is a light limiting environment and it is perhaps not surprising to find that in low-light conditions many species have chosen to maximise sensitivity to the narrow band of light available for vision. In environments with a broader spectrum of light, or those that transmit more longer-wavelength light, the relationship between rod spectral sensitivity and the maximum spectral energy of the environmental light is not so obvious. In general, the maximum scotopic spectral sensitivity in fish docs increase as water becomes more transparent to longer wavelengths, but as water becomes more red the rod spectral sensitivity fails to match (Loew and Lythgoe, 1978; Lythgoe, 1979; Loew, 1995). Of course, there may be many subtle behavioural and ecological reasons why this may occur in certain circumstances. As day changes to night, many vertebrates switch from the photopic to the scotopic system and employ a rhodopsin with a maximum spectral sensitivity around 500 nm. But this is theoretically too short compared with moonlight which actually has relatively more long-wavelength light than daylight. However, Munz and McFarland (1977) have suggested that if twilight is considered then the scotopic sensitivity match is good because the twilight sky is relatively richer in blue light. They further argue that in many fish species, the twilight period is often a critical period during which diurnal and nocturnal species are changing behaviour patterns and positions and when prey detection and predator avoidance tasks become highly significant (Munz and McFarland, 1977; see also chapter by McFarland *et al.*). Whereas behavioural pressures such as these may account for some of the observed mismatch, another theory, called the 'noise hypothesis', has been put forward which attempts to provide more fundamental biochemical reasons for the consistency of the finding.

The noise hypothesis is based on the fact that visual pigments produce a 'dark' noise signal due to thermally induced isomerizations. These random events will set a lower limit for photon sensitivity below which it will be impossible to distinguish the 'light' signal from the 'dark' noise. Experimentally, it has been confirmed that performance limits of rods are set by thermal isomerizations and that visual sensitivity is greater (due to fewer thermal isomerization events) in cold-blooded animals (Aho *et al.*, 1988). It has been shown that the frequency of thermally-induced noise events decreases with a shift to shorter wavelengths (Firsov and Govardovskii, 1990) and that porphyropsins are more 'noisy' than rhodopsins (Donner *et al.*, 1990). Thus, it is predicted that visual pigments absorbing at shorter wavelengths are less prone to thermal noise and that this could form the basis for increasing sensitivity in low light levels or at higher temperatures. The idea that certain configurations of the visual pigment should be more thermally stable has been strengthened at the molecular level by the finding that the adRP mutation G90D causes a reduction in dim-light sensitivity due to increased basal 'dark' noise while leaving bright light function normal (in figure 10, G90 is conserved in all rods) (Sieving *et al.*, 1995). It is presumed that the mutation contributes to partial deprotonation of the retinal Schiff's base and increases noise due to increased thermal isomerizations and this agrees with theoretical arguments proposed by Barlow *et al.* (1993).

If the noise hypothesis is correct then we may expect to see examples where spectral sensitivity can be correlated with body temperature, or where spectral sensitivity has been reduced to provide better signal to noise ratios in low-light environments. There are examples of this type where the noise hypothesis may be in operation. Many species of freshwater fishes can posses mixtures of rhodopsins and porphyropsins in the retina. The coupling of the A2 based chromophore to the opsin will shift the maximum absorption of the visual pigment to longer wavelengths and this may confer an adaptive advantage in turbid water stained green/red with plant products. Some species, such as rudd and roach, have been shown to change their retinal rhodopsin/porphyropsin ratios seasonally with a higher proportion of rhodopsin being present in the summer (for reviews see Beatty, 1984; Bowmaker, 1995). Freshwater colour can change with the season but tends to transmit more long-wavelength light in the summer and autumn when staining by plant chlorophylls and tanins increases. An alternative explanation for why these fishes shift the ratio towards the shorter-wavelength absorbing chromophore in the summer may be provided by the noise hypothesis. As the water temperature increases in the summer it is possible that visual pigments composed of A2 absorbing at longer wavelengths become too susceptible to thermal isomerization and image quality is compromised. In these conditions it may be better for the fish to alter the A1/A2 ratio to make use of the rhodopsin chromophore which is perhaps more thermally stable. Another example of the noise hypothesis may exist in the Lake Baikal cottoid fish that have been studied by Bowmaker *et al.* (1994). This study found that the spectral sensitivity of the rods of the Baikal cottoids shifts towards shorter wavelengths with increasing depth, as is observed for many other deep-living fishes. At first sight it may appear that this is an adaptation to the spacelight becoming more blue with depth, but in fact much of the surface water of Lake Baikal is yellow and the water remains blue-green to significant depths, implying that the rod shift to shorter wavelengths goes too far. Consequently, the authors suggest that the shift in spectral sensitivity with depth is not simply a product of the sensitivity hypothesis, but may also be due to the need to use short-wave absorbing visual pigments with more thermal stability in environments where light levels are low and thermal noise will be more significant. This raises the possibility that the rods of deep-sea fishes are sensitive to blue light not only because it is the predominant waveband, but also because a short-wave absorbing pigment reduces thermal noise in the photon-limiting environment. However, this type of argument requires further consideration since several deep-sea fish species have been found that have multiple rod visual pigments including both rhodopsins and porphyropsins and with maximum spectral sensitivity extending up to 590 nm (Partridge and Douglas, 1995; see also chapter by Partridge and Cummings).

From section 1.7 it is evident that photoreception and phototransduction are processes that involve many biochemical steps before a photon is finally converted into an electrical signal. Any series of linked biochemical processes will have inherent noise associated with it and not all noise associated with the visual process will be generated by thermal isomerizations of the chromophore. Noise may be linked to other parts of the phototransduction mechanism, such as the spontaneous activation of PDE (Riecke and Baylor, 1996), or the random opening and closing of light-sensitive ion channels in the membrane (Lamb, 1987). However, it must also be remembered that not all noise will be biochemical in nature. Light arriving at the eye may contain components that are 'noisy' with respect to the light that the organism is actually interested in 'seeing'. For example, we can think of many situations where we shade the strong, downwelling sunlight from our eyes so that we can see something in the horizontal direction more clearly. Many shallow living fish faced with exactly the same problem have developed shading structures which include; a flap called the operculum that extends down from the dorsal region of the pupil, yellow pigmentation concentrated in the dorsal cornea, and corneal structures for producing iridescence which can reflect different spectral components of the downwelling light (see chapters by: Douglas and Marshall; Shand and Foster). The wavelength-dependent scatter of light can also be a source of noise that can degrade a visual image. The scattering of light by small particles affects short-wavelength light more and is a significant problem underwater where many small, suspended particles can contribute to the effect. The effect of scattering will be to reduce the range at which an object can be seen and, in bright light conditions, the scattered light will cause dazzle. The simple way to reduce the noise associated with these effects, and to improve the image quality, is to insert a short-wave filter in the optical system and this appears to be the solution that many shallow-living fish have adopted (for more detail, see chapter by Douglas and Marshall).

1.9 Conclusions

In the second part of this chapter we have looked in some detail at the structure of the visual pigment and its relationship with photoreception and phototransduction. Many structural features are now known to be specifically involved in different functional aspects and this is being confirmed with mutation studies. The importance of these features is also highlighted by the location of known human adRP sites and highly conserved amino acid sites among species. The advance of the molecular knowledge over the last ten years has greatly helped in arriving at this level of understanding and will, no doubt, continue to do so. This level of functional understanding is also helping us to understand more clearly some of the relationships that have been seen between vision and ecology. The molecular dissection of the visual pigment allows us to determine the mechanisms of spectral sensitivity adaptation. At the same time it makes us aware of the physiological limitations placed upon the visual system that, as John Lythgoe pointed out in 1979, are important considerations in the ecology of vision.

Acknowledgements

I would like to thank Julian Partridge and Molly Cummings for helpful comments on the manuscript. I also thank Joe Hirano for help with the graphics. Simon Archer was supported by grants from the EC (STRIDE programme) and the Sardinian Regional Government.

References

Aho, A. C., Donner, K., Hyden, C., Larsen, L. O. & Reuter, T. (1988) Low retinal noise in animals with low body temperature allows high visual sensitivity. *Nature (London),* 334, 348 – 350.

Archer, S.N., Hope, A. and Partridge, J.C. (1995) The molecular basis for the blue-green sensitivity shift in the rod visual pigments of the European eel. *Proceedings of the Royal Society of London* B, 262, 289 – 295.

Archer, S. N. and Hirano, J. (1996). Absorbance spectra and molecular structure of the blue-sensitive rod visual pigment in the conger eel (*Conger conger*). *Proceedings of the Royal Society of London B*, 263, 761–767.

Baldwin, J.M. (1993) The probable arrangement of the helices in G protein-coupled receptors. *EMBO Journal,* 12, 1693 –1703.

Barlow, R. B., Birge, R. R., Kaplan, E. and Tallent, J. R. (1993) On the molecular origin of photoreceptor noise. *Nature (London)*, 366, 64 – 66.

Beatty, D. D. (1984) Visual pigments and the labile scotopic visual system of fish. *Vision Research*, 24, 1563 –1573.

Bowmaker, J. K. (1995) The visual pigments of fish, in *Progress in Retinal and Eye Research* Vol.15, No. 1, Elsevier.

Bowmaker, J. K., Govardovskii, V. I., Zueva, L. V., Hunt, D. M. and Sideleva, V. (1994) Visual pigments and the photic environment: the cottoid fish of Lake Baikal. *Vision Research*, 34, 591– 605.

Browman, H. I. and Hawryshyn, C. W. (1994) The developmental trajectory of ultraviolet photosensitivity in rainbow trout is altered by thyroxine. *Vision Research*, 11, 1397–1406.

Cohen, G. B., Yang, T., Robinson, P. R. and Oprian, D. D. (1993) Constitutive activation of opsin: influence of charge at position 134 and size at position 296. *Biochemistry*, 32, 6111– 6115.

Dartnall, H. J. A. and Lythgoe, J. N. (1965) The clustering of fish visual pigments around discrete spectral positions, and its bearing on chemical structure. Ciba Foundation Symposium on; Physiology and Experimental Psychology of Colour Vision (Wolstenholme, G. E. W. and Knight, J. eds), 3-21. J. & A. Churchill Ltd, London, U.K.

Decaluwe, G. L. J. and Degrip, W. J. (1996) Point mutations in bovine opsin can be classified in four groups with respect to their effect on the biosynthetic pathway of opsin. *Biochemical Journal*, 320, 807– 815.

Doi, T., Molday, R. S. and Khorana, H. G. (1990) Role of the intradiscal domain in rhodopsin assembly and function. *Proceedings of the National Academy of Science USA*, 87, 4991– 4995.

Donner, K., Firsov, M. L. and Govardovskii, V. I. (1990) The frequency of isomerization-like 'dark' events in rhodopsin and porphyropsin rods of the bull-frog retina. *J. Physiol. (London),* 428, 673 – 692.

Douglas, R. H. and Partridge, J. C. (1997) On the visual pigments of deep-sea fish. *Journal of Fish Biology,* 50, 68 – 85.

Douglas, R. H., Partridge, J. C. and Hope, A. J. (1995) Visual and lenticular pigments in the eyes of demersal deep-sea fishes. *Journal of Comparative Physiology A*, 177, 111–122.

Fahmy, K. and Sakmar, T. P. (1993) Light-dependent transducin activation by an ultraviolet-absorbing rhodopsin mutant. *Biochemistry,* 32, 7229 –7236.

Firsov, M. L. and Govardovskii, V. I. (1990) Dark noise of visual pigments with different absorption maxima. *Sensory Systems*, 4, 25 – 34.

Franke, R.R., Sakmar, T.P., Graham, R.M. and Khorana, H.G. (1992) Structure and function in rhodopsin. Studies of the interaction between the rhodopsin cytoplasmic domain and transducin. *J. Biol. Chem.*, 267, 14767–14774.

Hargrave, P. A. and McDowell, J. H. (1992) Rhodopsin and phototransduction. *Int. Rev. Cyt.*, 137 B, 49 – 97.

Hope, A. J., Partridge, J. C., Dulai, K. S. and Hunt, D. M. (1997) Mechanisms of wavelength tuning in the rod opsins of deep-sea fishes. *Proceedings of the Royal Society of London B*, 264, 155 –163.

Hunt, D.M., Fitzgibbon, J., Slobodyanyuk, S. and Bowmaker, J.K. (1996) Spectral tuning and molecular evolution of rod visual pigments in the species flock of cottoid fish in Lake Baikal. *Vision Research*, 36, 1217–1224.

Jacobs, G. H. (1993) The distribution and nature of colour vision among the mammals. *Biological Reviews*, 68, 413 – 471.

Karnik, S.S. and Khorana, H.G. (1990) Assembly of functional rhodopsin requires a disulfide bond between cysteine residues 110 and 187. *J. Biol. Chem.*, 265, 17520 –17524.

Kawamura, S. (1995) Phototransduction, excitation and adaptation, in *Neurobiology and clinical aspects of the outer retina* (eds. M. B. A. Djamgoz, S. N. Archer and S. Vallerga), Chapman & Hall, London, pp 105 –131.

Kito, Y., Partridge, J. C., Seidou, M., Narita, K., Hamanaka, T., Michiomae, M., Sekiya, N. and Yoshihara, K. (1992) The absorbance spectrum and photosensitivity of a new synthetic 'visual pigment' based on 4-hydroxyretinal. *Vision Research*, 32, 3 –10.

Lamb, T. D. (1987) Sources of noise in photoreceptor transduction. *Journal of the Optical Society of America*, 4, 2295 – 2300.

Lem, J. and Makino, C. L. (1996) Phototransduction in transgenic mice. *Current Opinion in Neurobiology*, 6, 453 – 458.

Loew, E. R. (1995) Determinants of visual pigment spectral location and photoreceptor cell spectral sensitivity, in *Neurobiology and clinical aspects of the outer retina* (eds. M. B. A. Djamgoz, S. N. Archer and S. Vallerga), Chapman & Hall, London, pp 57–77.

Loew, E. R., McFarland, W. N., Mills, E. L. and Hunter, D. (1993) A chromatic action spectrum for plantonic predation by juvenile yellow perch, *Perca flavescens. Canadian Journal of Zoology*, 71, 384 – 386.

Loew, E. R. and Lythgoe, J. N. (1978) The ecology of cone pigments in teleost fishes. *Vision Research*, 18, 715 –722.

Lythgoe, J.N. (1979) *The Ecology of Vision.* London: Oxford University Press.

McDowell, J. H., Nawrocki, J. P. and Hargrave, P. A. (1993) Phosphorylation sites in bovine rhodopsin. *Biochemistry*, 32, 4968 – 4974.

Munz, F. W. and McFarland, W. N. (1977) Evolutionary adaptations of fishes to the photic environment, in *Handbook of Sensory Physiology* VII/5 (ed. F. Crescitelli), Springer-Verlag, Berlin, pp193 – 274.

Nakayama, T.A. and Khorana, H.G. (1991) Mapping of the amino acids in membrane-embedded helices that interact with the retinal chromophore in bovine rhodopsin. *J. Biol. Chem.*, 266, 4269 – 4275.

Nathans, J. (1990) Determinants of visual pigment absorbance: identification of the retinylidene Schiff's base counterion in bovine rhodopsin. *Biochemistry,* 29, 9746 – 9752.

Nathans, J, (1992) Rhodopsin: structure, function and genetics. *Biochemistry*, 31, 4923 – 4931.

Partridge, J. C. (1990) The colour sensitivity and vision of fishes, in *Light and Life in the Sea* (eds. P. J. Herring, A. K. Cambell, M. Whitfield and L. Maddock), Cambridge University Press.

Partridge, J. C. and Douglas, R. H. (1995) Far-red sensitivity of dragon fish. *Nature (London)*, 375, 21– 22.

Partridge, J. C., Shand, J., Archer, S. N., Lythgoe, J. N. and van Groningen-Luyben, W. A. H. M. (1989) Interspecific variation in the visual pigments of deep-sea fishes. *Journal of Comparative Physiology A*, 164, 513 – 529.

Partridge, J.C., Archer, S.N. and Van Oostrum, J. (1992) Single and multiple visual pigments in deep-sea fishes. *J. mar. biol. Ass. UK*, 72, 113 –130.

Rieke, F. and Baylor., D. A. (1996) Molecular origin of continuous dark noise in rod photoreceptors. *Biophys. Journal,* 71, 2553 – 257.

Schertler, G. X., Villa, C. and Henderson, R. (1993) Projection structure of rhodopsin. *Nature (London)*, 362, 770 –772.

Schwartz, T. W. and Rosenkilde, M. M. (1996) Is there a 'lock' for all agonist 'keys' in 7TM receptors? *Trends in Pharmacological Science,* 17, 213 – 216.

Shand, J. (1993) Changes in the spectral absorption of cone visual pigments during the settlement of the goatfish, *Upeneus tragula*: the loss of red-sensitivity as a benthic existence begins. *Journal of Comparative Physiology A*, 173, 115 –121.

Shand, J., Partridge, J. C., Archer, S. N., Potts, G. W. and Lythgoe, J. N. (1988) Spectral absorbance changes in the violet-blue sensitive cones of juvenile pollack, *Pollachius pollachius. Journal of Comparative Physiology A*, 163, 699 –703.

Shi, W., Osawa, S., Dickerson, C. D. and Weiss, E. R. (1995) Rhodopsin mutants discriminate sites important for the activation of rhodopsin kinase and Gt. *J. Biol. Chem.*, 270, 2112 – 2119.

Sieving, P. A., Richards, J. E., Naarendorp, F., Bingham, E. L., Scott, K. and Alpern, M. (1995) Dark-light: model for nightblindness from the human rhodopsin Gly-90-Asp mutation. *Proceedings of the National Academy of Science USA*, 92, 880 – 884.

Watson, S. and Arkinstall, S. (1994) *The G-protein linked receptor factsbook*. London: Academic Press.

Wood, P. and Partridge, J.C. (1993) Opsin substitution induced in retinal rods of the eel (Anguilla anguilla (L.)): a model for G-protein-linked receptors. *Proceedings of the Royal Society London B*, 254, 227– 232.

Section II

Biophysical Adaptations

Introduction

J.C. PARTRIDGE

The Ecology of Vision rests firmly on two foundations: the physiology of animal eyes and the physical nature of light. The interaction between these is the theme of much of this book, and particularly that of Section II.

Nowhere is this more evident than in Chapter 2, in which Mike Land reviews the diversity of invertebrate eyes with an emphasis on the way in which optics have constrained their evolution. Such is the importance of optics that understanding the underlying physics enables the biologist to interpret much of the life of insects and crustaceans from the gross anatomy and micro-structure of their compound eyes. All compound eyes have relatively low resolution and untenably large eyes are required if resolution is to approach that of vertebrates. Many arthropods have compromised by having high resolution restricted to discrete parts of their visual fields and served by ommatidia with markedly enlarged facets. These acute zones are associated with specific visual tasks which are reflected in the behavioural ecology of different species. Thus some insects are found with ommatidial facets with sizes that vary over the surface of the eye in direct correlation with the velocity flow-field experienced in normal forward movement, especially in flight. Similarly, other species have adaptations for high spatial resolution that are highly specific to the need to capture mates or prey, particularly by aerial chase. Among these are the acute zones of the eyes of mantids which co-operate to form a binocular zone of high resolution for accurately striking at prey. Similarly the high acuity zones of male hover-flies and drone bees seem to be exclusively devoted to a role in nuptial chases. Acute zones are taken to an extreme in some deep-sea crustaceans, such as *Phronima*, in which the lateral eyes are divided into two parts: one with extremely high resolution and directed upwards, presumably involved in the detection of silhouettes against downwelling sunlight, and a low resolution lateral lobe serving the rest of the visual field. Bands of high resolution around the ocular equator are found in animals associated with unusual visual environments dominated by horizontal features, such as crabs living on sandflats and water striders hunting on the air-water interface of ponds.

Optical constraints also determine that the two basic forms of compound eyes, apposition and superposition, are markedly different in terms of brightness of retinal image, with the latter performing 2 – 3 orders of magnitude better. Perhaps not surprisingly superposition eyes are found in nocturnal insects and deep-sea crustaceans, both of which live in photon-limited environments. More surprising is the diversity of superposition eyes; natural selection having produced radically

S.N. Archer et al. (eds.), Adaptive Mechanisms in the Ecology of Vision, 45–49

different solutions to the common problem of how to evolve a superposition eye from an ancestral apposition type, with the retinal image of different types of superposition eyes being formed variously by ommatidia containing telescopes, reflection optics, or even combinations of lenses and mirrors.

In Chapter 3, Jake Sivak tackles the optics of the vertebrate eye and demonstrates that, despite the lack of superficial diversity compared with invertebrates, the camera eye design still has scope for variation that reflects both species' ecology of vision and the interaction of biology and physics. The vertebrate eye evolved underwater and the ancestral eye must have shared a feature common to modern fish, namely that the cornea is ineffective at focusing light and all focusing is thus done by the lens. For maximum image brightness the lens must be spherical and of high refractive index (RI) which approaches 1.65 in fish. To minimise spherical aberrations, the RI must follow a gradient from nucleus to edge but, because of the hardness of such a protein-packed lens, dynamic accommodation is possible only by movement of the lens. This turns out to be surprisingly complex. For instance, some fishes accommodate independently in the naso-temporal and medio-lateral directions, allowing accurate focusing on different parts of the retina in the course of complex visual tasks.

In terrestrial vertebrates the cornea is in contact with air and in most (amphibia are an exception) it has become the main optical element, with the lens relegated to fine-tuning the focus. Being of relatively low RI (e.g. 1.41 in humans) the lens is flexible and accommodation usually occurs by changes in lens shape. Nevertheless, this is a generality and focusing is effected by movement of the lens in most snakes, and changes in shape of both cornea and lens in birds. The latter group shows a particularly wide diversity in the relative importance of the two mechanisms, with different muscles being present in different species and these anatomical differences result in diverse degrees of deformation of corneal and lenticular surfaces in the course of accommodation. Thus, for example, pigeons and kestrels focus by changing the shape of their corneas, whilst the hooded merganser (an aquatic species) accommodates predominantly by lenticular deformation.

In the final part of Chapter 3, the problem of how a vertebrate eye attains the correct shape in the course of development is considered. The discussion is centred on the way in which the rate of eye growth is affected by the focus, or lack of it, of the retinal image, but the implication is clear: if retinal image quality directly determines eye development, what more complex features of the natural visual scene might influence ocular development?

The meaning of biophysical adaptation takes a new slant in Chapter 4 in which Ron Douglas and Justin Marshall exhaustively review the manifold ways in which the spectrum of light involved in vision is modified by optical filtering within the eye. Although radically different in structure the spectral sensitivities of the eyes of invertebrates and vertebrates are both dependent on: (i) the way in which their photoreceptor visual pigments absorb light of different frequencies, but also (ii) the way in which this light is spectrally modified by intraocular filters before arrival at the visual pigment. Filters are found within virtually all parts of the vertebrate eye: the cornea, lens, and ocular media being the most notable before the retina, but retinal blood vessels and macular pigments are also important pre-receptor filters. Within the photoreceptors themselves further filtering is done by oil droplets and the ellipsosomes of cone inner segments, whilst coloured tapeta behind the retina further modify spectral sensitivity.

Ocular filters have various putative functions including protection of the retina from photo-oxidative damage, but clearly they also have diverse roles in vision. These include: the enhancement of colour vision by fine-tuning photoreceptor spectral sensitivities; the improvement of retinal image quality by the removal of scattered light within the eye; increasing spatial resolution by maintenance of the optical isolation of retinal photoreceptors; and the dynamic control of light flux to the photoreceptors by movement of screening pigments. The significance of spectral filters in the ecology of animal vision is emphasised by various correlations with environment and behaviour, the best known being the coloured oil droplets of bird cone photoreceptors; different species having droplets of different colours in differing proportions. Other correlations are also recognised. For example the retinal tapeta of fish vary markedly in spectral reflexion and there is a general trend for these to match the spectral irradiance of the underwater environment. Thus deep-sea sharks and lantern fishes have shortwave reflecting tapeta, and freshwater fish have longwave reflectors. Exceptions to such trends serve to emphasise the role of tapeta in sensitivity enhancement; for instance the deep-sea fish *Malacosteus niger* has a bright red tapetum, which will considerably enhance its sensitivity to its own far-red bioluminescence.

It is hardly surprising, given the high diversity of eye design, that invertebrates have a wide range of types of ocular filters. Many are similar to those of vertebrates, with pigmented corneas, reflecting tapeta and intracellular coloured filters having direct analogues in the vertebrate eye. However, some types of filter are unique to invertebrates and exploit either the unusual properties of the arthropod cuticle or ommatidia of the compound eye. For example, the corneal surface of some nocturnal moths is covered in a forest of cuticular 'nipples' which, by virtue of their refractive index and precise morphology, reduce the reflection of the cornea, thereby increasing corneal transmission and, as a result, visual sensitivity. The arthropod cuticle is particularly easily modified to produce interference filters, made from multi-layer stacks of materials having different refractive indices, and this is used as the basis for coloured corneal filters in many dipteran flies, and in the retinal tapeta of butterflies. Because of their small size many rhadoms act as waveguides but a considerable proportion of the trapped light travels the length of the photoreceptor outside, rather than inside, the rhabdom. Pigment in the surrounding retinula cells is thus ideally placed to attenuate the light and, by movement of the pigment, to regulate the light flux and act as a longitudinal pupil mechanism that contributes to the adaptation of the photoreceptors to different levels of light. If coloured, such lateral pigments also modify the spectrum of the light absorbed, and so modify the spectral sensitivity of the photoreceptor. This mechanism is well developed in fireflies, in which lateral filters have been shown to tune precisely the spectral sensitivity of the retinal photoreceptors to conspecific bioluminescence.

In Chapter 5, the biophysical design of vertebrate retinal photoreceptors, particularly the size and shape of retinal rods and cones, is considered by Adam Lockett. This delineation of photoreceptor types has proved surprisingly robust since it was first proposed, but there is, nevertheless, considerable variation in cellular morphology. Thus cones occur as similar (twin) and dissimilar double cones, as triple cones, and even as quadruple cones in which the inner segments are in close apposition and the cells appear to be functionally unified. Unusual cone morphology reaches an extreme in anchovies in which two types of cone are present: one wedge shaped, the other short and bifid. These cones have disk lamellae arranged at right angles and thus are well adapted for the determination of the plane of polarised light.

In many vertebrates retinal rods and cones undergo circadian retinomotor movements by cellular expansion and contraction. These movements occur in conjunction with movement of retinal pigments and serve to position preferentially the different cell types for optimal function: thus during the day cone cells are laterally screened with melanin pigment for best optical isolation and rods are protected in pigment, whereas at night the cones are kept away from the rods which are clear of their pigment screen for maximum sensitivity. Retinomotor movements in teleost fishes also affect the mosaic-like arrangement of photoreceptor cells in the retina. Typically these mosaics consist of repeating elements of squares of double and single cones and the re-arrangement of the mosaic in the course of retinomotor movements or during ontogeny (e.g. at metamorphosis) will theoretically affect visual resolution, although this remains to be demonstrated.

To some extent visual resolution is also set by the size of photoreceptors and this varies considerably in vertebrates. For instance, the diameters of lungfish rods can be up to 14 μm (subtending 0.1° in visual space), whereas the cones of many birds are only 1.5 μm in diameter (subtending 0.04°) and thus close to the theoretical limit imposed by the diffraction of light. Diffraction will tend to increase cross-talk between small, closely packed photoreceptors but this is not a problem for some deep-sea fish in which retinal rods are arranged into discrete groups. In *Scopelarchus guntheri*, for instance, these bundles consist of 23 cells surrounded by a silvery tapetal cup formed from pigment epithelial cells and thus act as a low resolution but highly sensitive macroreceptor.

Deep-sea fish photoreceptors are also notable, in some species, for extreme length and, for example, the rods of *Diretmus argenteus* can be up to 900 μm long, conferring greater and broadened spectral sensitivity. Such cells are unusual and most rods are much smaller, typically being 20 – 50 μm long even in deep-sea species, but some deep-sea species attain high sensitivity by arranging such cells into multibanks. Animals with such tiered retinae also have the potential for colour vision, albeit based on rods, due to the spectral screening of overlying (vitread) layers altering the spectrum of light (and hence effective spectral sensitivity) of underlying more sclerad cells.

The final two chapters of Section II review the range of extraretinal photoreceptors that are found in non-mammalian vertebrates and how invertebrate biological clocks are regulated by light. The function of these receptors is to sense the environment and regulate physiology and behaviour and, although not serving a visual function, they share many features of eyes. Notably they are also subject to many of the same biophysical constraints as eyes and thus it is not surprising that similar ecological adaptations are present.

In vertebrates the best known extraretinal photoreceptors are probably the pineal and parietal eye, structures of variable morphology which are located on the roof of the brain, usually under a semi-transparent window in the skull. In most cases several types of photoreceptor-like cells are apparent but these are particularly well developed in the parietal eye of lizards. In these organs the morphological evidence for two types of photoreceptor is supported by electrophysiological measurements that reveal both chromatic and achromatic responses, suggesting that the different cells also have different spectral sensitivities. Arguably this arrangement aids the detection of the changes in spectral irradiance that occur at twilight and provide a potentially stable signal for the entrainment of circadian rhythms.

Although the ecology of extraretinal photoreceptors associated with the brain is poorly understood, light detection by dermal and corneal receptors generally has clear adaptive advantages. For example, illumination of the skin of *Xenopus* causes changes in melanophore contraction and concomitant lightening of the dermis. Similarly, direct illumination of the iridophores of the neon tetra (*Paracheirodon innesi*) causes rapid changes in skin colour by changes in the spacing of plates of guanine within the cells, thus altering the interference of light from the intracellular multilayer stack. In this the presumed ecological adaptation is that the fish are rendered highly visible in daytime as an aid to schooling, but become inconspicuous when vulnerable to predation at twilight.

In summary, Section II emphasises the way in which both visual and non-visual photoreceptors provide the essential sensory link between animals and their physical surroundings.

2
Compound eye structure: Matching eye to environment

M.F. LAND
Sussex Centre for Neuroscience, School of Biological Sciences, University of Sussex, Brighton BN1 9QG, UK

2.1 Introduction

In what ways is an animal's lifestyle reflected in the design of its eyes? A superficial look at the eyes of any vertebrate group – fish, birds or mammals – reveals very little, except perhaps some relative differences in overall eye size. It is not until the retina is examined in detail that ecologically related differences start to appear, particularly in the distribution of ganglion cells (Collin, this volume). With compound eyes the situation is quite different, and it is often possible to learn much about the life of an insect or crustacean from the gross anatomy of the eyes, as well as their microscopic structure. This chapter explores why this is so, and offers a guide to the interpretation of the variation one finds.

The compound eye design, in its various forms, has been extremely successful in evolution. It is the dominant type of eye in most arthropods (Nilsson, 1989) and has also evolved independently on at least one occasion in both the annelids and the bivalve molluscs (Land, 1981a, Nilsson, 1994). Notwithstanding this success, all compound eyes suffer from a fundamental design flaw not shared by the simple (or 'camera-type') eyes of vertebrates and cephalopod molluscs. The small size of the individual optical elements means that they suffer from diffraction – a loss of resolution resulting from the interference of the focussed light waves. The severity of this problem varies inversely with lens diameter, so that the 25 μm ommatidium in the eye of a bee, for example, can only resolve lines spaced about 1° apart in the visual field, whereas a human eye with a 2.5 mm pupil can resolve detail 100 times finer than this. To improve resolution by increasing the sizes of the ommatidial lenses would seem an obvious solution, but it can't be done, because it would mean an increase in ommatidial numbers as well in order to exploit the improved resolution. This combined effect of increasing both size and number of elements means that compound eyes must grow to an absurd size to attain resolution comparable with the simple eyes of vertebrates. Mallock (1894) estimated that for resolution similar to that of the human eye (ca 1 minute of arc), a compound eye would need a diameter of many metres.

S.N. Archer et al. (eds.), Adaptive Mechanisms in the Ecology of Vision, 51–71

The practical outcome of this limitation is that compound eyes are cramped for space. This is particularly true when resolution better than 1° is needed, as for example when capturing prey on the wing, or finding a mate at a distance. In an eye of supportable size, an increase in resolution in one direction must be paid for by decreases in others. In other words, arthropods are under great pressure to 'tune' the distribution of resolution across the surface of the eye to match the ecological needs of the animal, and this tuning often shows up as distortions in the pattern of ommatidia across the eye surface. 'Acute zones' of various kinds are common features of compound eyes (Horridge, 1978, Land, 1989a). These regions often have enlarged facets, and image the parts of the world around the animal that it has prioritised. Thus for a drone bee the acute zone is in the fronto-dorsal region where the female is imaged during the nuptual chase, whereas in the water-strider *Gerris* which hunts on water surfaces, there is an acute strip a few degrees high around the eye's horizon. There are also many examples of double eyes, where the need for an acute region has led to the splitting of the eye into separate specialized regions. The 'turbanate' eyes of male mayflies and the double eyes of male bibionid flies are insect examples, and in the ocean similar double eyes are seen in hyperiid amphipods and euphausiids. From the distribution of these various kinds of acute regions across the compound eyes of both sexes, one can learn a good deal about an animal's lifestyle and the role played by vision.

Resolution, in the sense of fine spatial sampling, is not the only factor contributing to good visual performance. At low light levels the capacity of an eye to secure sufficient photons is just as important. This is because small photon numbers make for a statistically noisy signal, which in turn limits the eye's ability to detect contrast. This is a real problem for eyes at all intensity levels lower than bright daylight (Snyder, 1979, Land, 1981a). Thus for use in twilight, or the dim light of the ocean, it is important that eyes have a high sensitivity, which means a structure adapted to give a high image brightness, as well as large receptors to capture high photon samples. Whilst all types of eye can be tuned to some extent to improve sensitivity, there is a real difference between the two main types of compound eye in this respect. The superposition design, in which the optical elements combine to form a single image, (Figure 1b) allows for a wider pupil and larger receptors than eyes of the apposition type, where each ommatidial element is optically separated from its neighbours (Figure 1a). Superposition eyes are thus better suited to dim environments. Here there seems to be a rather nice match between physics and ecology.

In the rest of the chapter I will first provide a minimal amount of theory to firm up some of the ideas outlined in this introduction, then go on to discuss in more detail the different types of acute zones, and the match between the distribution of resolution and lifestyles of different animals.

2.2 Theoretical background

2.2.1 Resolution and eye size

A good measure of the resolution of an eye is the spatial frequency (in cycles per radian or degree) of the finest grating it can just resolve. As it takes two receptors, or in the case of an apposition compound eye, two ommatidia, to resolve one dark/light cycle of a grating, this *sampling frequency*, is given by:

$$\nu_S = 1/(2\Delta\phi) \qquad [1]$$

where $\Delta\phi$ is the inter-receptor or inter-ommatidial angle. Diffraction imposes another kind of limit on the eye's resolution in the form of the '*cut-off frequency*'. This is the highest spatial frequency (ν_{CO}) that the eye's optics will pass – finer gratings than this have no contrast in the image. This is given by:

$$\nu_{CO} = D/\lambda \qquad [2]$$

where D is the lens diameter, and λ the wavelength of light (ca 0.5 μm). If we simply multiply equations 1 and 2 together we obtain:

$$\nu_S \bullet \nu_{CO} = 1/(2\Delta\phi) \bullet D/\lambda \qquad [3]$$

In the fovea of the human eye the sampling and cut-off frequencies are known to be almost exactly matched (about 60 $c.deg^{-1}$), and if we assume that this match applies generally we can substitute ν_S^2 for the left hand side of the above equation. Also, since in an apposition compound eye the inter-ommatidial angle $\Delta\phi$ is equal to the subtense of a single facet at the centre of the eye (i.e. D/R, see Figure 1) we can rewrite the above equation

$$\nu_S^2 = R/(2\lambda) \quad \text{or} \quad R = 2\lambda\nu_S^2 \qquad [3]$$

the D's having cancelled. This is the result given in the Introduction: in a compound eye the size (radius) is proportional to the *square* of the resolution (ν_S). Putting numbers in equation 3 gives the following results. When $\Delta\phi = 1°$, ν_S is 28.7 cycles per radian (Eq. 1; 1 radian = 57.3°) and R is 0.82 mm, which is a typical size for an apposition eye. However, if $\Delta\phi = 0.5'$, the human foveal value, ν_S is 3438 cycles per radian, and R is 11.8 metres! Kirschfeld (1976) pointed out that in the human eye resolution falls dramatically away from the fovea, and he came up with a more reasonable compound eye equivalent with a diameter of about 1 m, which, however, would still look pretty peculiar on a fly. In a camera-type eye with only one lens, diffraction need never be limiting, and it is receptor size that sets a lower limit to eye size. Waveguide effects limit receptor diameters and separations to about 2 μm (see Snyder, 1979), and as the inter-receptor angle $\Delta\phi$ is the receptor separation divided by the focal length (f), Eq. 1 gives $\nu_S = f/4$ cycles per radian, when f is in μm. This gives a focal length of 115 μm for $\Delta\phi = 1°$, and 13.8 mm for $\Delta\phi = 0.5'$. The latter figure is close to the actual focal length of the human eye (16.7 mm), and nearly a thousand times smaller than the 'worst case' compound eye (we assume here that f and R are comparable dimensions in terms of eye size).

Equation 3 also shows how difficult high acuity regions are to accommodate in compound eyes. To double the resolution in a part of the eye requires a local increase in the eye's radius of 4 times. This must somehow be squeezed into the overall spherical form of the eye by flattening the surface (as in praying mantids) or distorting the alignment of the ommatidial axes (as in hoverflies) or, as a last resort putting the acute zone in a physically separate structure (as in the double eyes of male bibionid flies, Figure 4b). Stavenga (1979) has discussed the way these adaptations are fitted into eyes.

2.2.2 *Sensitivity and eye type*

Since the work of Exner (1891), compound eyes have been divided into two basic types, apposition and superposition, with many variations within each group (Nilsson, 1989). In apposition eyes (Figure 1a) each receptor cluster has its own optical system and operates as a single spatial sampling station, although within the 8 or so receptors making up each ommatidium there are often

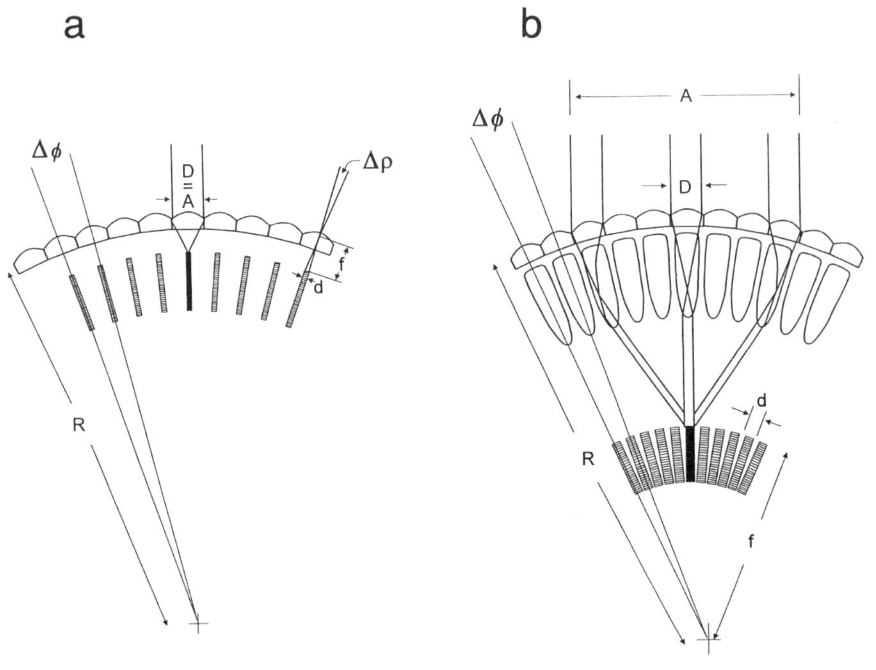

Figure 1 Optics of (**a**) apposition and (**b**) superposition eyes. Symbols: $\Delta\phi$, inter-ommatidial angle; $\Delta\rho$, rhabdom acceptance angle; A, aperture diameter; D, facet diameter; R, radius of curvature; d, rhabdom diameter; f, focal length of optical system. Further explanation in text

differences in spectral and polarization sensitivity. In superposition eyes (Figure 1b) the receptors form a single deep-lying retina, rather as in a single-lens eye except that the retina is convex rather than concave, and the optical elements jointly form a single erect image. To do this these elements must behave as telescopes (Exner, 1891), as mirrors (Vogt, 1980) or lens-mirror combinations (Nilsson, 1988). Apposition eyes have evolved into superposition eyes a number of times, and possibly vice-versa, and these transitions are discussed by Nilsson (1989). Both types of eye are capable of resolution at the diffraction limit (Land, 1984), and size-for-size the potential resolution is the same. They do, however, differ markedly in their light-capturing power, and hence their performance in dim light. The sensitivity of an eye (*S*), which can be defined as the number of photons caught by a receptor viewing a scene of unit brightness (strictly, Radiance), is related to the anatomical features of an eye by:

$$S = 0.62\,(A/f)^2 d^2 p \qquad [4]$$

where 0.62 is a correction for the area of a circular pupil, A is the aperture diameter (equal to D in apposition but not superposition eyes), f is the focal length of the optical system, d is the diameter of a receptor, and p is the proportion of photons entering a receptor that are actually absorbed and registered, which we will assume for convenience to be unity. A derivation of this formula can be found in Land (1981a).

Figure 1 shows how these definitions apply to apposition and superposition eyes. If we now consider an eye of each type, of identical size (radius 2 mm) and the same resolution ($\Delta\phi = 1°$), we can use Eq. 4 to compare sensitivities (Table 1). Typical values for A/f are 0.5 in apposition eyes, and >1 in superposition eyes. The other dimensions follow from the eye geometry. The resulting sensitivities are 0.23 and 189, a ratio in favour of the superposition eye of 1:822. Other things being equal, a superposition eye can provide the same numbers of photons per receptor as an apposition eye, but in illumination conditions almost 3 orders of magnitude dimmer (Table 1). Indeed, the difference may be greater than this, as superposition eyes often employ a reflecting tapetum behind or around the receptors, potentially doubling sensitivity. Although there are a few examples of apposition eyes adapted to dim light conditions (e.g. the ostracod *Macrocypridina*, Land and Nilsson, 1990), it is not surprising that most nocturnal and deep-water arthropods have superposition eyes. A variation of the apposition eye type is the so-called 'neural superposition' eye, found in dipteran flies and possibly some hemipterans. Here 6 – 8 receptors view the same point in space, providing a modest sensitivity gain (Kirschfeld, 1979, Nilsson & Ro, 1994). This, however, is nothing like the gain provided by optical superposition. The main advantage of apposition eyes, in daylight conditions, seems to be the improved contrast that results from reduced scattering, brought about by the screening pigment surrounding each ommatidium.

2.3 Types of resolution distribution in compound eyes

In an earlier review (Land, 1989a) I suggested that there were 3 main patterns of acuity distribution that one could identify in compound eyes. These were: a) patterns related to the velocity flow-field encountered in forward locomotion, especially flight; b) acute zones associated with predation or sex, these zones sometimes developing into separate components of a double eye; and c) horizontal strips of high resolution in animals living in environments such as water surfaces and sand-flats where almost all important activity takes place around the horizon. Figure 2 illustrates these situations. More information is now available both supporting this categorization, and providing interesting exceptions to it, and it seems appropriate to look again at these ideas.

Table 1 Sensitivity of apposition and superposition eyes of the same size $\Delta\phi = 1°$, $R = 2$ mm. A, f, d in μm. S from Eq. 4.

Type	A	f	d	p	S
Apposition	34.9	69.8	1.22	1	0.23
Superposition	1000	1000	17.45	1	188.8

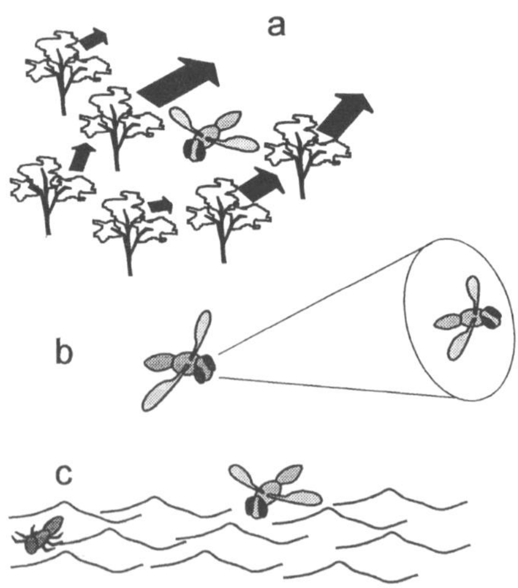

Figure 2 Three reasons why an animal's behavioural ecology may result in modifications of eye structure. (**a**) Forward flight, producing differences in velocity across the retina, indicated by the broad arrows. (**b**) The detection and pursuit of small moving targets requiring acute forward or upward vision. (**c**) Environments in which objects of interest – in this case a drowning insect – are restricted to a narrow horizontal band

2.3.1 *The forward flight pattern*

When an animal is moving through the world, the objects in it appear to move backwards across the eye, in a pattern that has become known as a velocity flow field (Figure 2a). Objects to the sides move faster than those in front, and there is a point in the direction of the animal's travel (the 'pole') where there is no image motion. Objects further away move more slowly than near objects. The structure of the flow-field is summed up in the following expression:

$$d\alpha/dt = V \sin\alpha \,/\, s \qquad [5]$$

where $d\alpha/dt$ is the angular speed of the image on the retina, V the animal's actual velocity, α the angle between the particular object and the animal's heading direction, and s the object's distance. Clearly, near objects to the side are likely to move so fast across the retina as to cause blurring, and if this is the case it would be economical to employ fewer receptors there, as high resolution is not usable. A butterfly or bee spends much time flying past foliage, and reasonable values for V might be 2 $m.s^{-1}$, and s 0.5 m. If α is 90°, then Equation 5 gives the speed across the retina as 4 radians or 229° per second. A typical response time of a light-adapted insect photoreceptor is 10 ms, which means that in one response time the image will have moved 229 × 0.01°, or put another way the length of the blur on the retina is about 2.3°. It follows that there is little point in having lateral-pointing receptors closer together than 2 – 3°, however good the resolution at the front of the eye. This seems to be borne out in practice. In the butterfly *Heteronympha merope*, for example the horizontal inter-ommatidial angle decreases from 1.4° in front to 2.6° at the side (Figure 3).

Bees, butterflies and acridid grasshoppers are flying insects, and their eyes all show decreasing horizontal inter-ommatidial angles from front to rear, consistent with the preceding discussion. Non-flying insects, for example many tettigonid grasshoppers, have more or less spherical eyes, without this gradient. In all the flying groups there is another, separate gradient. The vertical inter-ommatidial angles also vary (Figures 3 and 8a); they are smallest around the eye's equator, and increase towards both dorsal and ventral poles. This results in a band around the equator with enhanced vertical acuity. This pattern was first noticed by Baumgärtner (1928) in bees and in various other insects by Horridge (1978) and Land (1989a). The most likely reason for this vertical gradient is that the region around the eye's equator contains the highest density of information important to the animal, especially if it is an insect that feeds on flowers. The need for higher acuity is obviously greatest in that part of the field. Vertebrates that live in open landscapes (rabbits, cheetahs) often have elongated regions of high ganglion cell density that correspond to the horizon, and the relationship of these 'visual streaks' to the structure of the surroundings is sometimes known as the 'terrain hypothesis' (Hughes, 1977). More extreme versions of the vertical gradient are found in animals from really flat environments such as beaches, but these are discussed later (Figure 8b).

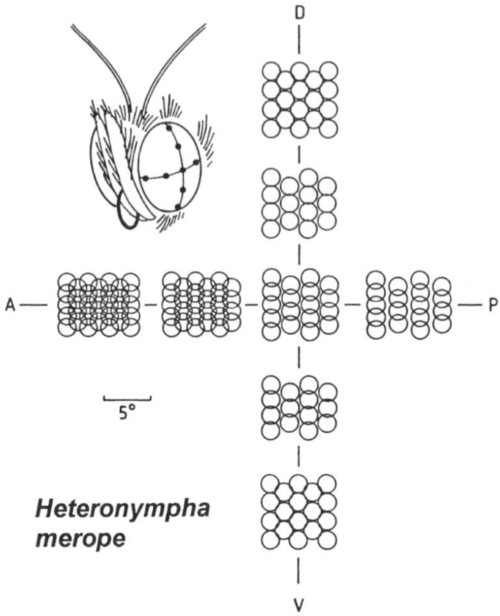

Figure 3 The pattern of ommatidial receptive fields in different regions of the eye of the Australian butterfly *Heteronympha merope* (from Land, 1989a). The circles represent the dark-adapted acceptance angles ($\Delta\rho = 1.9°$) measured optically (Nilsson et al., 1988), which are nearly constant across the eye. There is also little variation in facet diameter (21–26 μm) indicating a similar diffraction limit everywhere. There are, however, major variations in both $\Delta\phi_h$ and $\Delta\phi_V$ (see Figure 5b) and hence in the pattern of the lattice

The combined effects of these two gradients on the overall density of ommatidial axes is shown for a locust in Figure 8a. A similar pattern has been documented in worker honey bees (Seidl, 1980, Land, 1989a) where the horizontal variation is smaller than in butterflies, and the vertical variation is rather greater, as in locusts. Interestingly the highest value for horizontal inter-ommatidial angle is not in the forward direction, as it is in butterflies, but about 40° from the midline. Possibly this is related to the amount of sideways manoeuvring bees undertake compared to butterflies. Female blowflies (*Calliphora*) also show a pattern similar to that in butterflies (Land & Eckert, 1985), although in the males, as in drone honey bees, this pattern is distorted to give a more pronounced acute zone concerned with mate capture (see Figure 6).

In describing acute zones it is helpful to indicate how densely the ommatidial array samples different regions of the surroundings (Figure 5a). The measure adopted in Figures 6 and 8 is the number of ommatidial axes per square degree. This is easily calculated from the partial inter-ommatidial angles $\Delta\phi_h$ & $\Delta\phi_v$ as defined in Figure 5b (see Stavenga 1979). The axis density is then $1/(2\Delta\phi_h\Delta\phi_v)$, or $1/(\sqrt{3}\Delta\phi^2/2)$ if the array is symmetrical.

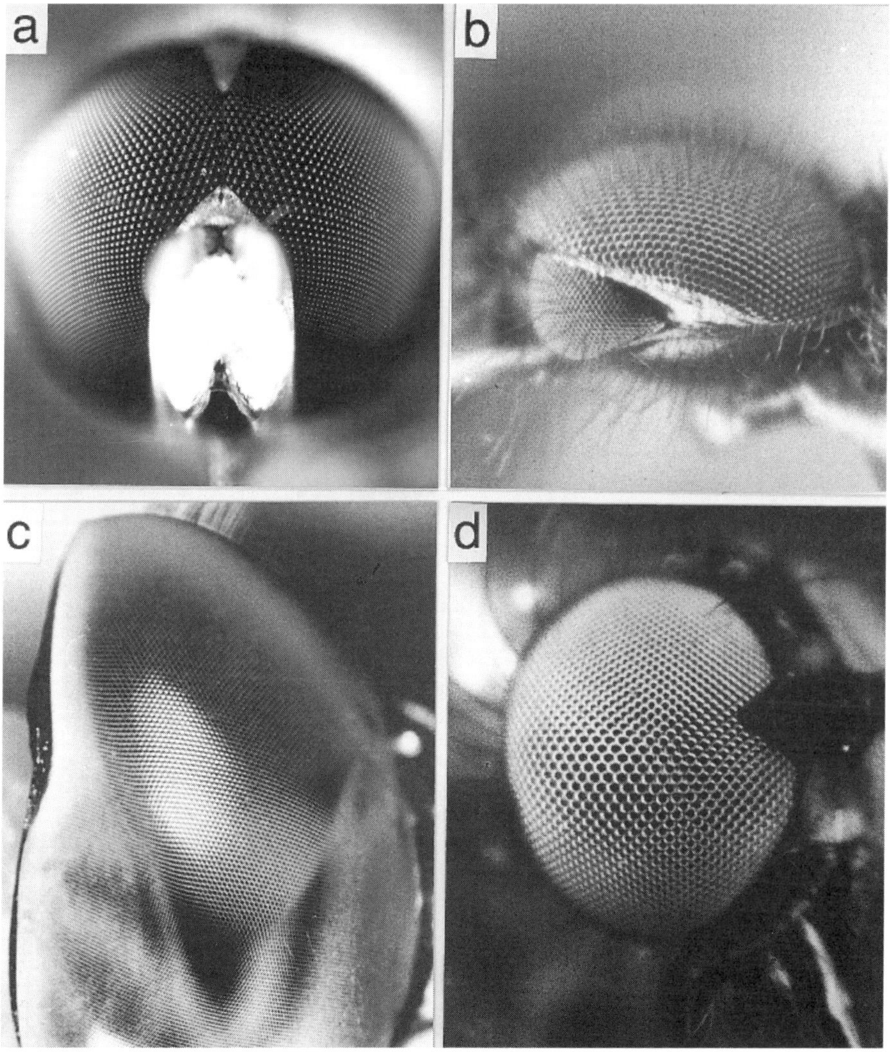

Figure 4 Asymmetries in the eyes of various insects, linked to their life-styles. (**a**) Male *Syritta pipiens*, with a dorso-frontal acute zone of larger facets (Collett & Land, 1975). (**b**) Male *Dilophus* sp., a Bibionid fly with a clearly divided eye. The female lacks the upper eye. (**c**) Upper part of right eye of *Aeschna multicolor*, a dragonfly with a dorsal-pointing wedge of enlarged facets (Sherk, 1978, see also Figure 6c). (**d**) An empid, *Hilara* sp., which hunts close to water surfaces (see Figure 2c) and has a horizontal band of enlarged facets and high resolution. Photograph courtesy of J. Zeil. See Zeil *et al.* (1989)

2.3.2 Acute zones concerned with prey capture and mating

Many insects and crustaceans have a forward or upward-pointing region of high acuity, related either to the capture of other insect prey, or to the pursuit in flight

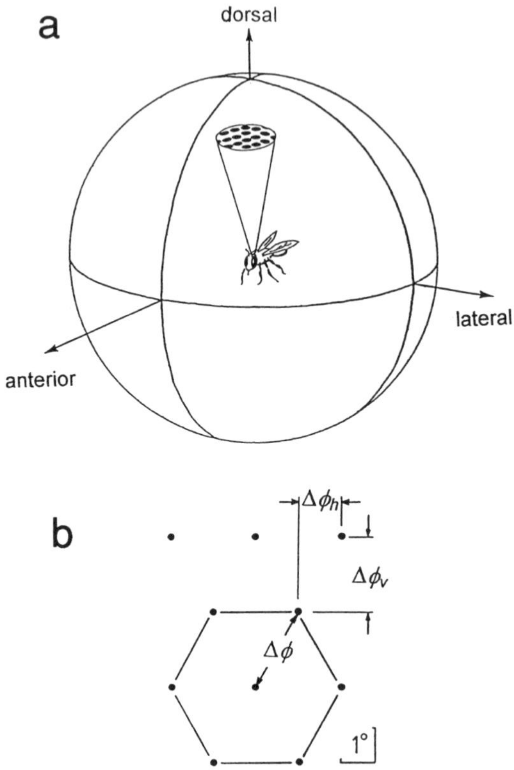

Figure 5 (**a**) Explanation of axis density. This is the number of ommatidial axes (or receptive fields) intersecting a unit solid angle on the surface of a sphere around the animals. The field depicted here represents about 100 square degrees, whereas the contour lines on Figures 6 and 8 show numbers of axes per 1 square degree (i.e. 1° × 1°). (**b**) The convention used here to designate inter-ommatidial angles (Stavenga, 1979). $\Delta\phi_h$ and $\Delta\phi_v$ are the partial inter-ommatidial angles measured between neighbours in adjacent horizontal and vertical rows. The conventional inter-ommatidial angle $\Delta\phi$ is equal to $\sqrt{(\Delta\phi_h^2 + \Delta\phi_v^2)}$

of females by males. Where both sexes have the specialization (mantids, dragonflies, robber-flies, hyperiid amphipods, euphausiids) predation is the reason, but more commonly it is only the male that has the acute zone (simuliid midges, hoverflies, mayflies, drone bees) indicating a role in sexual pursuit. The acute zones vary considerably. In male houseflies and blowflies (Figure 6a) they may be little more than a local increase in the acuity of the 'forward flight' acute zone common to both sexes (Land & Eckert, 1985). However, in other insects and many crustaceans the acute zone may be in a separate eye, as is the case with the dorsal eyes of male bibionid flies (Figure 4b, see Zeil, 1983a), or the upper eyes of krill (Figure 7d, see Land *et al.*, 1979). In these more extreme double eyes, the upward-pointing part is often specialized for detecting other small animals against the sky, or the residual downwelling daylight in the sea.

Frontal and dorso-frontal acute zones

Praying mantids are ambush predators. In both sexes there are large, binocularly overlapping acute zones which are used to centre potential prey before it is struck with the spiked forelegs (Rossel, 1980). Mantids provide the only known example in insects where prey distance is determined by binocular triangulation (Rossel, 1983). In a comprehensive study of *Tenodera australasiae*, Rossel (1979) found that $\Delta\phi$ varied from 0.6° in the acute zone centre, to 2.5° laterally, and that in all regions of the eye $\Delta\rho$ was virtually identical to $\Delta\phi$. Facet diameters decreased from 50 μm in the acute zone to 35 μm peripherally, but this is less of a decrease than would be expected from diffraction considerations alone.

Amongst crustaceans there are few known examples of frontal acute zones concerned with predation, but perhaps the best documented is in the carnivorous cladoceran *Polyphemus*, which uses its single fused compound eye to locate and track swimming prey (Young & Taylor, 1988). The *Polyphemus* eye has 130 ommatidia, and includes a distinct acute zone of 22 ommatidia, where inter-ommatidial angles are as small as 2°, which is remarkable in a 0.2 mm eye. The eye also incorporates 5 different varieties of optical system, and 4 anatomical types of rhabdom structure (Odselius & Nilsson, 1983).

In many male brachyceran flies an acute zone is associated with sexual pursuit, and is typically situated 20 to 30° above the flight direction (Figures 4a & 6a). In *Calliphora erythrocephala* it is characterised by a low value for $\Delta\phi$ of 1.07° compared with 1.28° in the female. The facet size is also larger, as expected from diffraction considerations: 37 μm compared with 29 μm in the female (Land and Eckert, 1985). In houseflies (*Musca domestica*) and probably in other flies there are also important anatomical differences at the receptor level. One of the central pair of receptors (R7), which in females and in other parts of the male eye sends its axon directly to the medulla, behaves more like one of the outer receptors (R1-6) in the male acute zone, sending its receptor to the lamina (Hardie *et al*, 1981). It is thought that this helps to improve sensitivity slightly, and hence the detectability of flying females. Central to the lamina, a number of 'male specific' interneurons have been discovered, which are undoubtedly involved in the organization of pursuit behaviour (Strausfeld 1991).

In the small hoverfly *Syritta pipiens* the sex difference is more striking (Collett & Land, 1975). In the male's acute zone $\Delta\phi$ is about 0.6°, nearly three times smaller than elsewhere in the eye, or anywhere in the female eye (Figure 4a). A male can use this acuity advantage to track a female closely, whilst remaining beyond her visual detection range. In other dipterans there is a very wide variety of sex-specific acute zones, beautifully described by Dietrich (1909). Drone bees have a similar antero-dorsal acute zone, where the density of ommatidial axes is three to four times greater than anywhere in the female eye (Figure 6b; see Seidl, 1980, and Land, 1989a). They use this region when they chase the queen, or a dummy queen on a string (van Praagh *et al.*, 1980). Drones will pursue a target subtending 0.32°, much smaller than the ommatidial acceptance angle of 1.2° (Vallet & Coles, 1991), implying that the trigger for pursuit is a brief decrease in the intensity received by single rhabdoms of about 6%. Tsetse flies (*Glossina*) seem to need a much larger signal to evoke chasing, about 1.6°, which is similar to the foveal acceptance angle (Brady, 1991).

An increase in the detectability of small objects can be achieved either by reducing the rhabdom acceptance angle ($\Delta\rho$, Figure 1) so that a small target causes a large change in the signal on the rhabdom that images it, or by increasing the numbers of photons available to the rhabdoms, thereby reducing the noise against which the signal must be detected. Either course of action requires a larger facet diameter D. In most of the examples discussed it seems that the increased facet diameter in the acute zone is 'spent' on reducing $\Delta\rho$, but in one well-documented case that is not so. The male blowfly *Chrysomyia megalocephala* has a 'bright zone' rather than an acute zone, where $\Delta\rho$ is similar to the rest of the eye, but the photon catch per rhabdomere is enhanced by an increase in both facet and rhabdomere diameter (van Hateren *et a*l. 1989). This increase, compared with elsewhere in the eye, is about ten-fold. It is not known why this fly has taken this particular route, but one would guess that it mates in dim conditions.

There is an interesting exception to the rule that it is always the males that have the acute zone. In pipunculid flies of the genus *Chalarus* the females have greatly enlarged ommatidia in the fronto-dorsal region (Jervis, 1992). These flies parasitize leafhoppers (Homoptera: Auchenorrhyncha) and the females have to locate these on the undersides of leaves in order to lay their eggs. The males have no equivalent need for keen eyesight. A somewhat similar situation occurs in certain phorid flies, which follow ants or termites by flying above them, prior to laying eggs on them (Disney and Schroth, 1989). In the most extreme case (*Apocephalus laceyi*) the ventral facets are 35 μm across, and the dorsal ones 26 μm. This is a rare example of a downward pointing acute zone.

Dorsal acute zones and double eyes

Most of the animals just discussed have to detect their prey or mates against a background of foliage. This is a far from easy task, as the target usually only differs from the background by virtue of its motion, not because of static qualities such as brightness or texture, and colour is rarely a usable cue. However, many insects and crustaceans have simplified the problem by using the sky as a background, against which any non-luminous object becomes a dark spot. Thus one finds both upward-pointing acute zones, as well as double eyes with one component directed skyward – or in the ocean towards the surface.

Dragonflies hunt other insects on the wing, and have acute zones with a variety of configurations (Sherk, 1978). Many in fact have two acute zones, one forward pointing, and presumably concerned with forward flight as discussed above, and another directed dorsally and used to detect prey (Figures 4c & 6c). Primitive zygopterans have only a weak forward pointing zone, whereas in the faster-flying corduliids this is more pronounced, and there is a second zone directed vertically. In perching libellulids the frontal zone is minimal, but there is an acute zone about 40° wide in the fronto-dorsal region. The migratory, fast-flying aeschnids have the largest eyes and most impressive acute zones. Sherk (1978) counted 28,672 ommatidia in one eye of Anax *junius,* which has the smallest inter-ommatidial angles of any insect (0.24° in the dorsal acute zone), and facets of corresponding size (62 μm). The dorsal acute zone takes the form

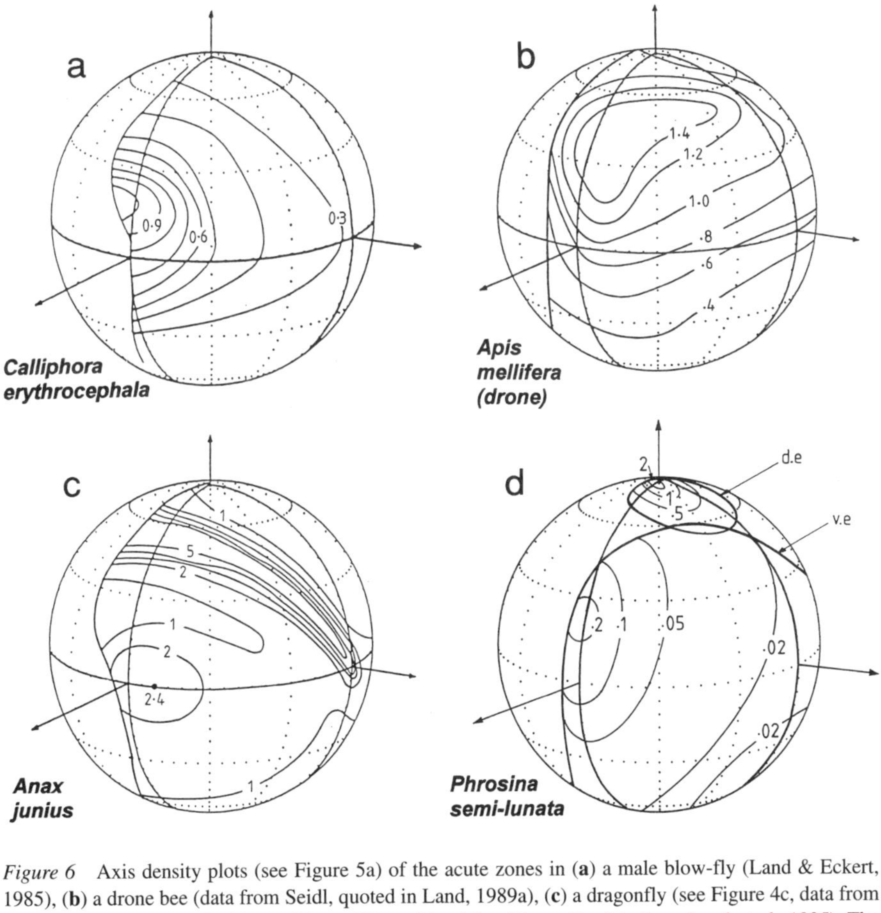

Figure 6 Axis density plots (see Figure 5a) of the acute zones in (**a**) a male blow-fly (Land & Eckert, 1985), (**b**) a drone bee (data from Seidl, quoted in Land, 1989a), (**c**) a dragonfly (see Figure 4c, data from Sherk, 1978) and (**d**) a double-eyed hyperiid amphipod (see Figure 7c, data from Land *et al.*, 1995). The significance of the high-density regions is discussed in the text

of a narrow band of high resolution extending across the upper eye along a great circle, 50 – 60° up from the forward direction (Figure 6c). The axis density (5 per square degree) is twice that in the forward acute zone, and five times higher than in male *Calliphora.* The dorsal acute zone is easily visible as a wedge of enlarged facets (Figure 4c), although the frontal zone is less obvious. Presumably the great high-acuity stripe in *Anax* is used to trawl through the air, picking out insects against the sky much as the scan line on a radar set picks up aircraft. Labhart and Nilsson (1995) studied the divided eyes of *Sympetrum,* and found that the yellow pigment in the dorsal region increases the conversion rate of meta-rhodopsin to rhodopsin, thus maximizing sensitivity and hence the detectability of prey.

Simuliid flies have divided eyes, and use the upper part to detect potential mates against the sky. They can do this at a distance of 0.5 m, when a female subtends an angle of only 0.2°. As in drone bees, this is a small fraction of an acceptance angle (Kirschfeld, 1979). The facets in the dorsal part are much larger (25 – 40 μm compared with 10 – 15 μm). The dorsal rhabdomeres are also very long, reaching almost to the roof of the mouth. Like simuliids, the eyes of male bibionid flies are divided (Figure 4b, see Zeil, 1983a) with larger facets and smaller inter-ommatidial angles in the 'dorsal eye (1.6° compared with 3.7°, in *Bibio marci*). The upper eyes are used exclusively for the detection of females, and Zeil (1983b) found that whilst movement of stripes around the lower eye evokes a strong optomotor turning response – the almost universal visual behaviour used by insects to prevent involuntary rotation – the dorsal eye is quite unresponsive to this kind of stimulus. Other insect eyes with separate upward-directed components include the 'turbanate' eyes of male mayflies (Horridge & McLean, 1978), and the eyes of the neuropteran owl fly *Ascalaphus* (Schneider *et al.*, 1978)

Amongst the crustaceans, mid-water representatives of three groups have specialized in double eyes: the hyperiid amphipods with apposition eyes, and the euphausiids and mysids with superposition eyes (Figure 7). The hyperiid eyes present an extraordinary range of eye anatomy, from surface-living forms with single eyes *(Lestrigonus* and *Thamneus* spp), to mid-water species with double eyes of various kinds (*Parapronoe*, *Platyscelus*, *Phrosina* and *Phronima* spp.), and finally to the deep-living *Cystisoma,* a large and very transparent animal which only has the upward-pointing component of the eye (Land, 1981b, 1989b). The logic of this trend seems to be that the deeper an animal lives, the more important it becomes to devote as much photon-catching power as possible to the residual downwelling light, because it is against this dim background that potential food can be sighted from the silhouette it casts. It is interesting in this context that a great many midwater animals disguise their silhouettes with ventral photophores, whose luminance may in some cases be adjusted to match the downwelling light. In all the double-eyed forms the upper part has larger facets and smaller inter-ommatidial angles than the lower part. In *Phronima sedentaria* this angle ($\Delta\phi$) is only 0.25° in the dorsal eye, compared with 10° in the ventral. Interestingly, the acceptance angles of ommatidia ($\Delta\rho$, Figure 1) in the dorsal eye are about 9 times the $\Delta\phi$ values which in an ordinary terrestrial eye would imply huge over-sampling of the image. However, where the problem is to detect single dot-like objects and not to resolve texture, it can be shown that this mismatch does not matter, because the apparent contrast loss can be recovered by neural pooling later on (Land, 1981b). *Phronima* is also unique in that the light focussed by the lenses of the dorsal eye is conveyed the 5 mm distance to the ventrally-situated retina by light-guides 18 μm wide (Figure 7b). It has yet to be demonstrated unequivocally that the dorsal eyes of hyperiids are used for silhouette detection as suggested here, but it is known that when *Phrosina* tracks a self-luminous object (such as a luminescing animal) it does so with its *ventral* eye (Figure 6d & 7c, see Land *et al.*, 1995).

In superposition eyes major departures from spherical symmetry are rare because the geometry of the eye is constrained by the shared optics (Figure 1b). However, in euphausiids (krill) bilobed eyes are common (Figure 7d). The two components are usually optically separate structures, so that each can be regarded as a separate eye with its own symmetry. The focal length of the dorsal eye is usually longer than that of the ventral eye, and this results in smaller inter-ommatidial angles: in *Stylocheiron maximum* $\Delta\phi$ is 1.2° in the dorsal eye compared with 2.6° in the ventral. In some species (e.g. *Nematoscelis atlantica*, and *Nematobrachion boopis*) the upper eye has an interesting non-concentric structure, which has the apparent function of providing a local correction for the spherical aberration inherent in the refracting superposition design (Land *et al.*, 1979). In *N. boopis* the lower eye is almost absent – a parallel with the amphipod *Cystisoma.* It has been shown in the double-eyed *Nematoscelis atlantica* that the dorsal eye is always kept pointing upwards, towards the daylight, while the animals themselves swim at various angles to the vertical (Land, 1980). Thus it seems very likely that the role of the dorsal eyes is to look for dark silhouettes, as suggested for the hyperiid amphipods.

In the genus *Stylocheiron* some species show a reduction in the numbers of facets in the dorsal eye, and this seems to be related to the depths at which the animals swim. In deeper-living species there are typically several hundred facets altogether. In *S.elongatum*, which lives at 180 – 420 m in daytime, each row contains 13 – 16 facets, but in *S.affine* (40 – 140 m) this reduces to 4 – 8, and in *S.suhmi* (0 – 50 m) there are only 3. This appears to be a crude but effective way of ensuring that the different eyes provide similar retinal illumination levels. The ultimate reduction of the superposition eye design, to only one facet, is seen in a tropical shallow-water mysid *Dioptromysis paucispinosa* (Figure 7e & f). Nilsson & Modlin (1994) describe this shrimp as 'carrying binoculars', which is an almost exact analogy. The eyes are double: the main part is a conventional super-position eye, but the accessory region is quite different. It has only one large facet with a single giant crystalline cone, beneath which is a retina of 120 rhabdoms, compared with 800 – 900 in the rest of the eye. This accessory eye is thus a

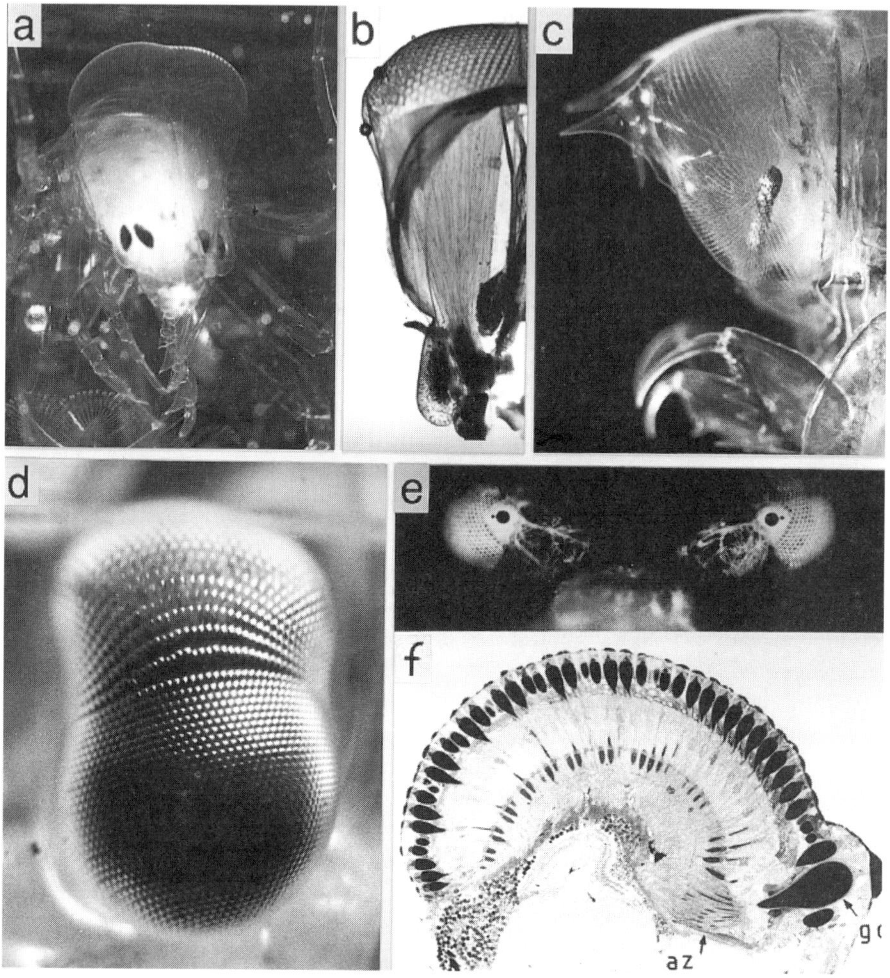

Figure 7 Double eyes in Crustacea. (**a**) Head of *Phronima sedentaria*, a hyperiid amphipod with a double apposition eye. The inner pair of black dots are the retinae of the upper eyes, which cover the dorsal surface of the head, and the outer pair are the retinae of the tear-drop shaped lower eyes. (**b**) Left eyes of *Phronima* from the rear, showing the 5 mm long light guides joining the upper eye surface to the ventrally located retina. (**c**) *Phrosina semi-lunata*, a double-eyed hyperiid with a less extreme division than in *Phronima*. A plot of the visual fields of dorsal and ventral eyes is given in Figure 6d. (**d**) Double superposition eyes of the euphausiid *Nematoscleis megalops* (Land *et al.*, 1979). (**e**) The head of the mysid *Dioptromysis paucispinosa*, viewed from behind showing the single giant facet in each eye. (**f**) Horizontal section of the eye of *Dioptromysis*, showing the conventional superposition eye, with the acute zone (az) of narrow rhabdoms on the right, with the single giant facet and crystalline cone (gc). Details in Nilsson & Modlin (1994). e) and f) courtesy of D.-E. Nilsson

unique example of a single lens superposition eye – effectively a simple eye but with erecting rather than inverting optics. The accessory eye is an acute zone with a minimum separation of rhabdom axes ($\Delta\phi$) of 0.64° – impressive in an eye so small. The 44 μm diameter of the giant facet means that the diffraction limit is much lower than in the rest of the eye, which has 16 μm facets. The most peculiar feature of these already strange eyes is that the giant facet and its acute zone normally point backwards! Nilsson & Modlin (1994) found that occasionally the eyes are rotated, directing the acute zones forwards, where they are presumably used for a higher resolution scrutiny of potential food or mates. Unlike other double-eyed arthropods, there are no good reasons for thinking that these special eyes normally point upwards.

2.3.3 Horizontal acute zones

As we have seen, many flying insects have a zone of increased vertical acuity around the horizon, no doubt reflecting the visual importance of this region in their lives. The visual field of the locust in Figure 8a shows this clearly. Interestingly, non-flying desert ants of the genus *Cataglyphis* show a very similar pattern (Zollikofer *et al.*, 1995), presumably because in the relatively flat desert terrain there is little of interest outside the 10 or 20 degrees above and below the horizon. There are environments where this region is even more important. Sand and mud flats are good examples, and many of the crabs that inhabit them have a narrow band of high vertical acuity around the equators of the eyes (Zeil *et al,* 1986). In the ghost crab, *Ocypode ceratophthalmus*, this band is about 30° high, with vertical inter-ommatidial angles as low as 0.5°; by contrast the horizontal inter-ommatidial angles are four times larger (Horridge, 1978, Zeil *et al.*, 1986). There are interesting differences in this respect between crabs of the flat beach and those of the rocky upper shore. The former tend to have tall eyes close together, with a very pronounced equatorial band, whereas the latter have rounder eyes far apart, with a weakly developed band. Zeil *et al.* (1986, 1989) suggest that the tall-eyed crabs measure distance by the angle down from the horizon to the feet or base of object they are looking at – a strategy which will only work on a flat surface – whereas the upper shore crabs with their wide spread eyes use conventional triangulation. Another feature of the horizon is that objects that penetrate above it are necessarily larger than the crab itself, and are thus likely to be predators. Land & Layne (1995) showed that fiddler crabs (*Uca pugilator*) react defensively to moving objects above the horizon, but not to objects of similar angular size or speed below it.

Water surfaces provide a similarly constrained field of view, and water-striders (*Gerris*), which hunt prey stranded in the surface film, have a narrow acute band imaging this region (Dahmen, 1991), as shown in Figure 8b. This has a height of only about 10°, centred on the horizon, and within this $\Delta\phi_v$ in the frontal region is only 0.55°, which is close to the diffraction limit, and impressive in an eye with only 920 ommatidia. Frontally, $\Delta\phi_h$ is about 1.9°, and both $\Delta\phi_h$ and $\Delta\phi_v$ increase towards the rear, as in other eyes. The backswimmer, *Notonecta,* is in some ways more remarkable. Living just below the surface film, it looks up at the water surface, and can view potential prey in two ways. With the ventral part of the eye – *Notonecta* hangs upside down – it can look through the water surface to view the top of the prey in air, or it can look below the surface to see the bottom of the same prey through the water. The two views are separated by about 30° in the sagittal plane. In an impressive demonstration of the way natural selection and physics sometimes agree, Schwind (1980) found two acute bands in *Notonecta,* imaging these two corresponding views. $\Delta\phi$ (= $2\Delta\phi_v$) was between 1 and 2° in both acute bands, with a low acuity zone where $\Delta\phi$ decreased to 4° between them.

Insects that fly over water surfaces have a similarly narrow equatorial field of interest. Empid flies hunt close to the surfaces of ponds, again looking for stranded insects, and they have a horizontal acute zone that can be recognised by a linear region of enlarged facets around the eye (Figure 4d). In *Rhamphomyia tephraea,* Zeil *et al.* (1989) found inter-ommatidial angles of 0.5° in this 15° high region, rising to 2° above and below it.

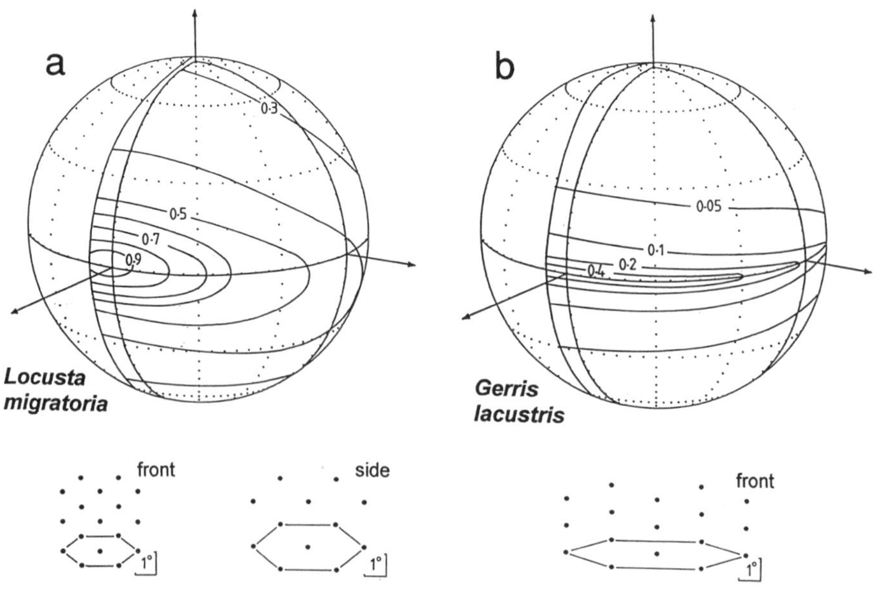

Figure 8 Axis density plots of horizontal acute zones in (**a**) a locust (new pseudopupil measurements) and (**b**) a water-strider (data from Dahmen, 1991). Details in the text. The inserts below show how the lattice of aommatidial axes is distorted in the equatorial region to give enhanced vertical resolution

2.4 Conclusions

Because of the diffraction limit, compound eyes have to be carefully tailored to the behavioural ecology of their bearers, or they will become unsupportably large. This has resulted in several recognisable patterns of resolution which show up in the overall distribution of density of ommatidial axes, as well as the ratios of vertical to horizontal inter-ommatidial angle. The three most striking patterns are:

A) a reduction of horizontal resolution from front to rear, related to the blurring effect of forward motion, especially flight.
B) acute zones pointing in a frontal, fronto-dorsal, or dorsal direction, associated with the need to chase and capture mates or prey. Although small in angular extent, these zones can be physically large and are often housed in a separate eye.
C) bands of high vertical resolution around the eye's equator, imaging the horizon. These are particularly pronounced in arthropods from flat environments such as water surfaces and sand flats.

Although both apposition and superposition eyes can be adapted to high or low light conditions, a superposition eye provides 2 – 3 orders of magnitude more light to the receptors than an apposition eye of the same size. In dim conditions (nocturnal or deep sea animals) superposition eyes predominate.

In this chapter I have only considered the distribution of resolution and sensitivity in compound eyes. Light has other important qualities, in particular its wavelength distribution and polarization structure. These too have ecologically interesting consequences, to which eyes are tuned in various ways. These adaptations are considered in Chapters 4, 9 and 17.

References

Baumgärtner, H. (1928) Der Formensinn und die Sehschärfe der Bienen. *Z. Vergl Physiol.* 7, 56 –143.

Brady, J. (1991) Flying mate detection and chasing by tsetse flies (*Glossina*). *Physiol. Entomol.* 16, 153 –161.

Collett, T. S., Land, M.F. (1975) Visual control of flight behaviour in the hoverfly *Syritta pipiens* L. *J. Comp. Physiol.* 99, 1– 66.

Dahmen, H. (1991) Eye specialisation in waterstriders: an adaptation to life in a flat world. *J. Comp. Physiol. A.*, 169, 623 – 632.

Dietrich, W. (1909) Die Facettenaugen der Dipteren. *Z. Wiss. Zool.*, 92, 465 – 539.

Disney, R. H. L., Schroth, M. (1989) Observations on *Megaselia persecutrix* Schmitz (Diptera: Phoridae) and the significance of ommatidial size- differentiation. *Entomologist's Monthly Magazine*, 125, 196 –174.

Exner, S. (1891) Die Facettenaugen der Dipteren von Krebsen und Insecten. Leipzig: Deuticke.

Hardie, R. C., Franceschini, N., Ribi, W., Kirschfeld, K. (1981) Distribution and properties of sex-specific photoreceptors in the fly *Musca domestica. J. Comp. Physiol. A.*, 145, 139 –152.

Horridge, G. A. (1978) The separation of visual axes in apposition compound eyes. *Phil. Trans. Roy. Soc. Lond. B.*, 285,1– 59.

Horridge, G. A., McLean, M. (1978) The dorsal eye of *Atalophlebia* (Ephemeroptera). *Proc. Roy. Soc. Lond. B.*, 200,137–150.

Hughes, A. (1977) The topography of vision in mammals of contrasting lifestyle: comparative optics and organization. In: Crescitelli, F. (ed.) Handbook of Sensory Physiology vol. VII/5, pp. 613–756. Berlin: Springer.

Jervis, M. A. (1992) A taxonomic revision of the pipunculid fly genus *Chalarus* Walker, with particular reference to the European fauna. *Zoo. J. Linnean Soc.,* 105, 243–252.

Kirschfeld K (1976) The resolution of lens and compound eyes. In: Zettler F, Weiler R (eds) Neural Principles in Vision pp 354-370. Berlin: Springer.

Kirschfeld K (1979) The visual system of the fly: physiological optics and functional anatomy as related to behaviour. In: Schmitt, F. O., Worden, F. G. (eds) The Neurosciences 4th Study Program. pp 297–310. Cambridge Mass: MIT Press.

Labhart, T. and Nilsson, D.-E. (1995) The dorsal eye of the dragonfly *Sympetrum:* specializations for prey detection against the sky. *J. Comp. Physiol. A.* 176, 437–453.

Land, M. F. (1980) Eye movements and the mechanism of vertical steering in euphausiid crustcea. *J., Comp., Physiol.,* 137, 255-265.

Land, M. F. (1981a) Optics and vision in invertebrates. In: Autrum, H. (ed.) Handbook of Sensory Physiology. vol. VII/6B, pp 472 – 592. Berlin: Springer.

Land, M. F. (1981b) Optics of the eyes of *Phronima* and other deep-sea amphipods. *J. Comp. Physiol.,* 145, 209 – 226.

Land, M. F. (1984) The resolving power of diurnal superposition eyes measured with an ophthalmoscope. *J. Comp. Physiol.,* 154, 515 – 533.

Land, M. F. (1989a) Variations in the structure and design of compound eyes. In: Stavenga, D. G., Hardie, R. (eds) Facets of Vision pp. 90-111. Berlin: Springer.

Land, M. F. (1989b) The eyes of hyperiid amphipods: relations of optical structure to depth. *J. Comp. Physiol. A.,* 164, 751–762.

Land, M. F. and Eckert, H. (1985) Maps of the acute zones of fly eyes. *J. Comp. Physiol. A.,* 156, 525 – 538.

Land, M. and Layne, J. (1995) The visual control of behaviour in fiddler crabs. I. Resolution, thresholds and the role of the horizon. *J. Comp. Physiol. A.*, 177, 81– 90.

Land, M. F. and Nilsson, D.-E. (1990) Observations on the compound eyes of the deep-sea ostracod *Macrocypridina castanea. J. Exp. Biol.,* 148, 221– 233.

Land, M. F., Burton, F. A. and Meyer-Rochow, V. B. (1979) The optical geometry of euphausiid eyes. *J. Comp Physiol* 130: 49–62.

Land, M. F., Marshall, N. B. and Diebel, C. (1995) Tracking of blue lights by hyperiid amphipods. *J. Mar. Biol. Assn. UK,* 75, 71–81

Mallock, A. (1894) Insect sight and the defining power of compound eyes. *Proc. R. Soc. London B.,* 55, 85–90

Nilsson, D.-E. (1988) A new type of imaging optics in compound eyes. *Nature (Lond.),* 332, 76 –78.

Nilsson, D.-E. (1989) Optics and evolution of the compound eye. In: Facets of Vision (eds Stavenga, D. G. and Hardie, R. C.) pp. 30 –73. Berlin: Springer-Verlag.

Nilsson, D.-E. (1994) Eyes as optical alarm signals in fan worms and ark clams. *Phil. Trans. R. Soc. Lond. B.* 346, 195–212.

Nilsson, D.-E. and Modlin, R. F. (1994) A mysid shrimp carrying a pair of binoculars. *J. Exp. Biol.,* 189, 213–236.

Nilsson, D.-E. and Ro, A.-I. (1994) Did neural pooling for night vision lead to the evolution of neural superposition eyes? *J. Comp. Physiol. A.* 175, 289 – 302.

Odselius, R. and Nilsson, D.-E. (1983) Regionally different ommatidial structure in the compound eye of the water flea *Polyphemus* (Cladocera, Crustacea). *Proc. R. Soc. Lond. B.,* 217, 177–189.

Rossel, S. (1979) Regional differences in photoreceptor performance in the eye of the praying mantis. *J. Comp. Physiol.,* 131, 95 –112.

Rossel, S. (1980) Foveal fixation and tracking in the praying mantis. *J. Comp. Physiol.* 139, 307– 331.

Rossel, S. (1983) Binocular stereopsis in an insect. *Nature (Lond.),* 302, 821– 822.

Schneider, L., Gogala, M., Draslar, K., Langer, H. and Schlecht, P. (1978) Feinstruktur und Schirmpigment-Eigenschaften der Ommatidien des Doppelauges von *Ascalaphus* (Insecta, Neuroptera). *Cytobiologie,* 16, 274 – 307.

Schwind, R. (1980) Geometrical optics of the *Notonecta* eye: adaptations to optical environment and way of life. *J. Comp. Physiol.,* 140, 59–68.

Seidl, R. (1980) Die Sehfelder und Ommatidien-Divergenzwinkel der drei Kasten der Honigbiene (*Apis mellifica*). *Verh. Dtsch. Zool. Ges.* 1980, 367. Stuttgart: Gustav Fischer.

Sherk, T. E. (1978) Development of the compound eyes of dragonflies. III. Adult compound eyes. *J. Exp. Zool.* 203, 61–80.

Snyder, A. W. (1979) Physics of vision in compound eyes. In: Autrum H (ed.) Handbook of Sensory Physiology VII/6A. Springer, Berlin-Heidelberg-New York, pp. 225–313.

Stavenga, D. G. (1979) Pseudopupils of compound eyes. In: Autrum, H. (ed.) Handbook of Sensory Physiology VII/6A. Springer, Berlin-Heidelberg-New York, pp 357–439.

Strausfeld, N. J. (1991) Structural organization of male-specific visual neurons in calliphorid optic lobe. *J. Comp. Physiol. A.,* 169, 379–393.

Vallet, A. M. and Coles, J. A. (1991) A method of estimating the minimum visual stimulus that evokes a behavioural response in the drone, *Apis mellifera* male. *Vision Res.,* 31, 1453–1455.

van Hateren, J. H., Hardie, R. C., Rudolph, A., Laughlin, S. B., Stavenga, D. G. (1989) The bright zone, a specialised dorsal eye region in the male blowfly *Chrysomyia megalocephala. J. Comp. Physiol. A.,* 164, 297–308.

van Praagh, J. P., Ribi, W., Wehrhahn, C. and Wittmann, D. (1980) Drone bees fixate the queen with the dorsal frontal part of their compound eyes. *J. Comp. Physiol.,* 136, 263–266.

Vogt, K. (1980) Die Spiegeloptik des Flusskrebsauges. The optical system of the crayfish eye. *J. Comp. Physiol.,* 135, 1–19.

Young, S. and Taylor, V. A. (1988) Visually guided chases in *'Polyphemus pediculus. J. Exp. Biol.,* 137, 387–398.

Zeil, J. (1983a) Sexual dimorphism in the visual system of flies: the compound eyes and neural superposition in Bibionidae (Diptera). *J. Comp. Physiol.,* 150, 379–393.

Zeil, J. (1983b) Sexual dimorphism in the visual system of flies: the free flight behaviour of male Bibionidae (Diptera). *J. Comp. Physiol.,* 150, 395–412.

Zeil, J., Nalbach, G. and Nalbach, H.-O. (1986) Eyes, eye stalks and the visual world of semi-terrestrial crabs. *J. Comp. Physiol. A.,* 159, 801–811.

Zeil, J., Nalbach, G. and Nalbach, H.-O. (1989) Spatial vision in a flat world: optical and neural adaptations in arthropods. In: Singh, R. N., Strausfeld, N. J. (eds) Neurobiology of Sensory Systems, pp. 123–137. New York: Plenum.

Zollikofer, C. P. E., Wehner, R. and Fukushi, T. (1995) Optical scaling in conspecific *Cataglyphis* ants. *J. Exp. Biol.,* 198, 1637–1646.

3
Vertebrate optical structure

J.G. SIVAK, M.E. ANDISON and M.T. PARDUE
School of Optometry, University of Waterloo, Waterloo, Ontario, Canada N2L 3G1

3.1 Introduction

The vertebrate eye is a sophisticated optical instrument with optical characteristics that are primarily determined by whether it is used in air or water (or both) and under diurnal or nocturnal conditions (Walls, 1942). The evolutionary change from an aquatic to a terrestrial environment is the prime factor in the development of the optics of the vertebrate eye. The cornea of the eye is a transparent structure consisting of two curved and almost parallel surfaces so that its refractive function is dependent on the existence of media of unequal refractive index internally (aqueous humor) and externally (air) (Sivak, 1978). The aquatic habitat of ancestral vertebrates limited the original function of the cornea to that of a transparent window. Existing aquatic species, be they fishes or mammals (cetaceans, sirenians), possess an eye characterized by the presence of a spherical, or near spherical, crystalline lens of rigid consistency and high refractive index. The lens, as the only refractive structure of these eyes, is responsible for accommodative changes in refractive power. The various accommodative mechanisms proposed for aquatic species always involve lens movement (Sivak, 1980).

The lens is a completely cellular structure located inside the eye within an acellular capsule. Cell multiplication is restricted to the periphery so that older lens material is concentrated toward the centre as the number of lens cells increases throughout life (Mann, 1969). In order to focus light on to the retina, the lens must have a refractive index that is substantially higher than its surrounding media; hence the extremely high protein concentration characteristic of all lens tissue. In the case of the spherical fish lens, an equivalent refractive index of 1.65 is usually quoted (e.g. Walls, 1942). By contrast, the adult human lens, for example, is elliptical in shape, soft in consistency and has a relatively low refractive index (1.41) (Bennett and Francis, 1962). Accommodation is usually a function of the lens in terrestrial eyes, although a role for the cornea has been demonstrated in some species. When accommodation exists, it normally involves changes in lens shape brought about by a variety of intraocular mechanisms (Sivak, 1980).

S.N. Archer et al. (eds.), Adaptive Mechanisms in the Ecology of Vision, 73–94

Good resolving power is important to diurnal vision whereas sensitivity to low light levels is the primary concern in nocturnal conditions. Adaptations that enhance visual acuity, whether optical or anatomical, generally reduce sensitivity and vice versa. A nocturnal eye may therefore be characterized as having poor resolving power, a large nearly spherical lens with substantial amounts of spherical aberration, a cornea with relatively large radius of curvature and a virtually absent accommodative mechanism (Sivak, 1980).

The optical properties of the eye are determined by the size and shape of the globe, the curvature of the cornea and the focal properties and position of the lens. In spite of over 100 years of intense study, the specific causes of refractive errors are still unknown. While it is frequently assumed that ocular development is controlled both by genetic and environmental factors, their relative importance is not clear, particularly with respect to the effect(s) of early visual experience.

Research using animal models of refractive development carried out during the past decade have shown that early visual experience is a very important component to the search for the causes of refractive errors, at least in higher vertebrates. This research has been carried out with a variety of mammals, including monkeys, cats, tree shrews and others (see Irving *et al.*, 1995). However, the majority of work has involved the use of young chicks because; 1) the chick eye is large and easy to measure, even right after hatching, 2) ocular development is very rapid, with most eye growth occurring within two weeks and, 3) chickens are precocial birds and this means that the manipulation of early visual experience can begin as soon as they hatch.

Earlier research involved either suturing the eyelids of one eye together or applying a translucent goggle over one eye (Pickett-Seltner *et al.*, 1987). Invariably, this type of manipulation, referred to as form deprivation, results in the development of myopia. In fact, it was concluded that the process of depriving the young eye of clear vision always produces myopia.

More recent research has shown that the effects of early manipulation of visual experience can be more complex. It is possible to produce variable amounts of myopia, hyperopia and astigmatism by using lenses to defocus the retinal image of a young chick eye (Schaeffel *et al.*, 1988; Irving *et al.*, 1992; 1995). Removal of the inducing lens after one or two weeks results in a rapid return to emmetropia. By making the young eye artificially myopic or hyperopic with lenses, the eye grows faster or slower to compensate. In only one week, or less, the eye becomes myopic or hyperopic by an amount equal to the power of the inducing lens, up to limits of approximately ± 20 diopters. This work has also shown that the choroid can play a short term role by rapidly decreasing the axial length of the eye when eye growth is slowed in response to a convex inducing lens or when the eye is recovering from induced myopia (Irving *et al.*, 1992; 1995).

Concurrent with the optical studies described above, studies have concentrated on efforts to determine the retinal chemistry associated with induced refractive errors (Guo *et al.*, 1995). Dopamine and its associated compounds, are common retinal transmitter agents. Furthermore, dopamine may be associated with transverse (horizontal) activity by amacrine cells of the retina in response to retinal image focus. In the study of Guo *et al.* (1995), retinal dopamine and its metabolite 3,4-dihydroxyphenylacetic acid (DOPAC) were reduced (relative to controls) in myopic eyes and elevated in hyperopic ones.

Reviews concerning the optical characteristics of the vertebrate eye have appeared in several conference proceedings and book chapters during the past two decades. These include, among others, a survey of vertebrate accommodative mechanisms (Sivak, 1980), reviews describing aspects of optical performance of the mammalian (Hughes, 1977), avian (Glasser and Howland, 1995) and fish (Sivak, 1990) eyes and general overviews of vertebrate optics (Charman, 1991). The bulk of this work deals with topics such as refractive state, range of accommodation, ocular aberrations, and the relationship between ocular optics and visual acuity. Less attention has been directed toward the morphology and embryology of the tissues associated with optical function. While it is true that the eye is an optical instrument, it is also a living organ subject to fundamental biological laws and limitations related to structure.

The chapter which follows will focus on four topics that concern vertebrate optical structure. The first will describe the anatomy and embryology of the cornea and crystalline lens of the vertebrate eye in order to provide a basis for understanding the relative importance of the structural factors that limit the optical quality of these two refractive elements. A second section will deal specifically with recent research concerning the cellular architecture of the mammalian lens, the formation of sutures, and the effect of suture morphology on lens optical function. For comparative purposes, an example of structural/ embryological limitations affecting an invertebrate lens, that of cephalopods, will be described as well. A third topic will describe recent research on the functional morphology of accommodative mechanisms of aquatic and aerial vertebrates, with teleost accommodation as the aquatic case and avian accommodation as the aerial example. Finally, the importance of environmental factors in ocular development will be discussed, with some emphasis on comparison between higher and lower vertebrates.

The work described below is part of an effort to understand 1) the effect of environmental factors on the overall refractive plan of the vertebrate eye, 2) the optical limitations imposed by the biological solutions adopted by vertebrates and 3) how vertebrates cope with the visual demands of their life histories.

3.2 Ocular embryology and its relevance to optical function

3.2.1 Vertebrate corneal anatomy and development

The cornea is a relatively thin avascular structure consisting of two curved and nearly parallel surfaces. As noted earlier, corneal refractive power is substantial in air but virtually eliminated when the eye is under water. Authors commonly attribute this effect to the similarity in refractive indices between the cornea and water (Walls, 1942; Graham, 1972). In fact, although some variation in refractive index is found among vertebrate species, corneal refractive index is always appreciably greater than that of water. However, the refractive index of aqueous humor *is* very close to that of water. Since the radii of curvature of the two corneal surfaces are similar, all or most of the refractive effect of the external surface is negated by the internal surface when the eye is in water. The typical aquatic cornea has a relatively flat shape, presumably an adaptation related to the need for streamlining body shape in a dense medium such as water (Jamieson, 1971). In addition, the radius of curvature of the terrestrial cornea is relatively short, often shorter than that of the globe as a whole, as in the human case (8 vs. 12 mm). The relatively short radius of curvature of the terrestrial cornea may be related to the need for a significantly arched configuration in order to maintain a smooth refractive surface in air.

The cornea and the crystalline lens are the two optical elements responsible for focusing an image on the retina. Both face the same basic problem of maintaining adequate conditions for living tissue while being part of a sophisticated optical instrument. Thus, both structures have evolved metabolic processes which are adequate for nonvascular conditions. Anatomically, both structures exhibit adaptations designed to minimize light scatter, but despite these and other obvious similarities, fundamental differences in embryonic development, location, and function result in differences with far-reaching consequences.

The cornea can be described as layers of collagen sandwiched between epithelial tissue (Mann, 1969; Hogan, *et al.*, 1971). The external layer, the epithelium, consists of stratified squamous epithelial cells approximately five to seven cell layers thick. A mitotic basal layer of columnar cells rests on a basement membrane, whereas the superficial squamous layer is subject to a process of erosion, or sloughing. In humans the full regenerative cycle takes about seven days. The fact that corneal epithelial cells adhere tightly to one another by means of desmosomes and other junctional devices minimizes intercellular space and potential light scatter. Unlike the epidermis, the external and internal epithelial surfaces are relatively smooth and parallel to one another. The innermost layer of the cornea, the endothelium, consists of a single layer of epithelial cells. In the adult eye, the endothelial cells do not exhibit a measurable regenerative cycle (Hogan *et al.*, 1971). The rest of the cornea – Bowman's membrane, the stroma, and Descemet's membrane – is made up of collagen fibres. The few existing cells, mainly fibroblasts, are located far apart. In the stroma, the collagen fibres are uniformly thick and are arranged parallel to the corneal surface. The characteristic regularity of interfibre spacing and the dimensions of the fibres and spaces is commonly believed to produce destructive light interference, a prerequisite for corneal transparency (Hogan *et al.*, 1971). The corneal stroma has a very slow rate of turnover, if any (Rodrigues *et al.*, 1982).

Embryologically the cornea originates from a single layer of surface ectoderm cells which reforms over the developing eye when the surface attachment of the lens vesicle has disappeared (Hogan *et al.*, 1971). These cells are joined by an internal layer of mesodermal material, which develops in conjunction with the condensation of mesodermal tissue around the optic cup. These cells, Descemet's mesothelium, ultimately produce the corneal endothelium and Descemet's membrane. The surface layer then becomes two cells thick. A second mesodermal formation develops between the cells of the surface ectoderm and the mesothelium. The final multilayered structure of the surface ectoderm, the epithelium, develops at the same time.

In addition to its refractive function, the external location of the cornea is important for another reason. The epithelium is the only corneal component that undergoes a significant regenerative cycle. Basal columnar cells reach the surface as squamous cells in approximately 7 days and are sloughed off to the external environment in a manner identical to the regenerative process of the epidermis. Older cells are therefore not retained.

3.3.2 Vertebrate lens anatomy and development

Like the cornea, lens morphology is also noted for regularity of structure (Hogan *et al.*, 1971). However, the lens is a completely cellular structure. A single layer of epithelial cells covers its anterior surface, and extremely elonated cells (fibres) fill the interior, extending to its posterior surface. The lens fibres are arranged in concentric shells of increasing diameter.

The lens fibres contain relatively few organelles, and those in the central core lack nuclei. This adaptation is probably related to the need for transparency. The fibres are intricately interconnected by means of numerous ball-and-socket articulations, which produce a highly ordered geometric pattern (Worgul, 1982). The tight articulation between fibres reduces interfibre space and minimizes light scatter (Hogan *et al.*, 1971).

The early stages of lens development coincide with optic cup development (Mann, 1969; Worgul, 1982); whereas the cup develops as an outpunching of the neural ectoderm, forming the early forebrain, the lens forms as an invagination of the surface ectoderm. The ectoderm invaginates from the surface to form a hollow spherical vesicle in the optic cup. The posterior cells of the vesicle elongate to fill the vesicle and form the primary lens fibres. The epithelial cells at the lens equator continue to undergo cell division before birth and throughout life (Hogan *et al.*, 1971). The new equatorial cells elongate toward the anterior and posterior lens poles; the anterior extension travels underneath the original epithelial surface. These cells are the secondary lens fibres. Their size, shape, and interconnections determine the ultimate size and shape of the lens.

That the lens is located within the eye, and that it is composed of cells of surface origin that continue to multiply throughout life, results in a unique set of problems. First, because the lens is surrounded by a liquid or nearly liquid medium its refractive function depends on the existence of a refractive index significantly higher than water. The equivalent refractive index of the human lens is given as 1.41, considerably greater than that of water or corneal tissue (1.333 –1.337) (Bennett and Francis, 1962). The cells of the lens have the highest concentration of protein of any body cells, presumably as an adaptation to the need for a high refractive index.

Second, the internal location of the lens, couple with the peripheral development and growth of new fibres around the equator, means that old lens tissue must be retained. In fact, the pattern of lens growth directs older tissue toward the center. The result is a lens with a refractive index gradient, which is maximal at the center and minimal at the periphery. This gradient has the important optical consequence of helping to minimize lens spherical aberration (Kreuzer and Sivak, 1984).

In most diurnal vertebrates the lens also has the important function of providing the eye with a variable focal mechanism or accommodation. In the aerial case, accommodative change in lens shape usually takes place in response to neural directives given to the ciliary muscle-zonule apparatus. The shape of the lens may therefore change hundreds of times each day. As noted earlier, aquatic species that accommodate do so by moving the lens.

The embryologic descriptions provided here make it possible to understand why the crystalline lens is the most fragile and sensitive of the two optical components of the eye, despite the fact that the external location of the cornea makes it very susceptible to mechanical damage. The high protein content of the lens creates potential molecular instability, which may lead to loss of transparency while the loss of accommodation with age is a common consequence related to lens development and aging in humans.

3.3 Lens development and optical limitations

3.3.1 Vertebrates

As already noted, continued mitotic activity at the lens equator is responsible for the formation of new lens fibres around the posterior outer lens and under the anterior epithelial cell layer. The anterior and posterior lens fibre terminations meet at sutures of varying complexity depending on age and species. The point of central interest is that the lens sutures are regions of disorder in an otherwise ordered lens. Paradoxically, the sutures are located along the optical axis of the lens where they can most detract from the optical quality of the lens (and eye).

The optical quality (focal length variability) of lenses has been analyzed with a computer-assisted scanning laser. The technique incorporates a scanning laser (helium-neon) beam initially positioned at the optical center of the lens. The beam passes through the lens in predetermined steps and records the focal length at each step. The relationship between rabbit and bovine lens sutures and lens optical quality was examined by correlative scanning electron microscopy and laser scan analysis (Kuszak *et al.*, 1991; Sivak *et al.*, 1994a). Areas of lens scanned that produced the least amount of focal variability were characterized by uniform fibre cells arranged in parallel, radial cell columns. In contrast, optical analysis with the incident beam directed along a lens suture showed a significant increase in focal length variability.

The sutures of the rabbit lens are relatively simple, consisting of an anterior vertical line and a posterior horizontal one. Bovine (and primate) lenses are more complex. In both cows and young primates, the sutures are in the form of a Y, erect on the anterior surface and inverted on the back. In primates and cows this arrangement becomes more complex (Figure 1) with age, as the suture terminals develop branches. The adult human lens, for example, consists of a nine branched suture arrangement (star suture).

A typical bovine lens of 40 mm focal length shows an increase in back vertex distance from less than 2 mm between sutures to about 20 mm along sutures (Figure 2). In addition, bovine lenses of increasing age show greater levels of focal variability along sutures and this variation can be shown to be directly related to an increasingly complex sutural architecture (Figure 2). Similar suture effects were noted for rabbit lenses and both optical and anatomical measures showed a difference between albino and pigmented rabbits. However, in the case of monkey lenses (*Macaque nemestrina*) the difference between lens optical variability along and between sutures was found to be reduced (Kuszak *et al.*, 1994). Anatomical studies showed that primate lenses may be characterized by the absence of continuous suture planes perpendicular to the optical (or visual) axis. Rather, distinct generations of progressively more complex suture branching results in an offset arrangement that minimizes the distance incident light rays have to travel through a disordered region of the lens.

It is of interest to note that the secondary lens fibres of teleost and avian lenses are tapered sufficiently so that their ends converge at the anterior and posterior poles, forming 'umbilical' sutures (Kuszak, 1995). The absence of linear sutures, as in mammals, may contribute to lens optical quality, although this point remains to be demonstrated experimentally.

3.3.2 Cephalopods

Lenses of both cephalopod and vertebrate eyes display anatomical adaptations designed to minimize light scatter (Hogan *et al.*, 1971; Arnold, 1967). In both instances, growth and development continues well beyond the embryogenic phase and results in the formation of a lens with a gradient refractive index which minimizes lens aberrations, especially spherical aberration (Sivak, 1990; 1991).

In spite of these similarities, significant differences between the lenses of the two unitary eyes have been noted. This is particularly obvious with respect to

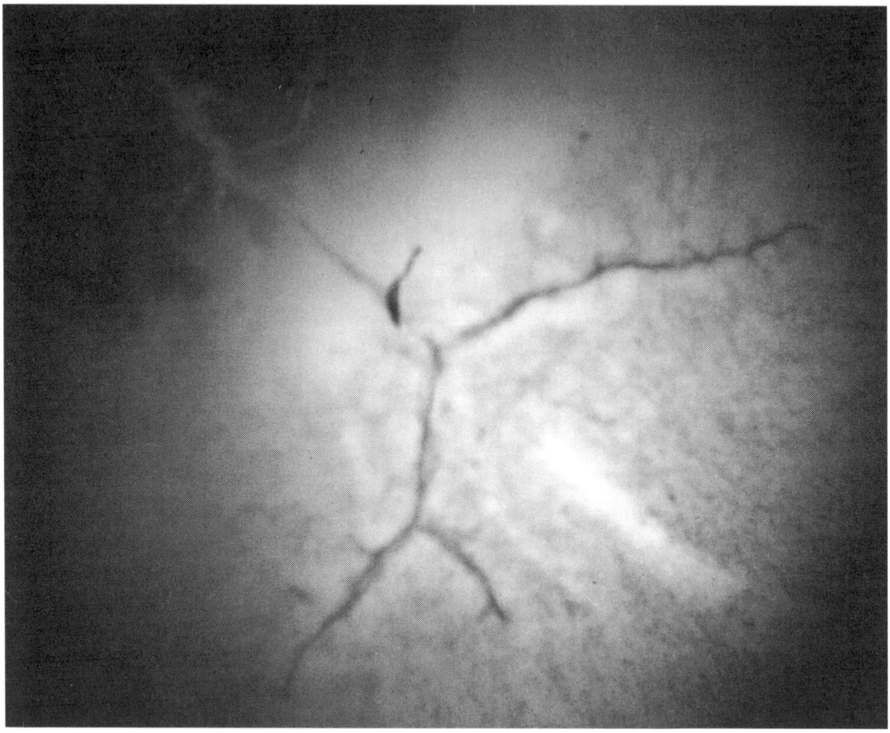

Figure 1 Photograph showing suture branching on the anterior surface of an aged (5 – 8 yrs) bovine lens (magnification approx. 53 X)

development. The cephalopod lens develops as two roughly hemispheric halves from two separate ectodermal sources (West *et al.*, 1995). The anterior and posterior lens primordia are derived from cellular processes that extend from cells of the lentigenic body. The few optical studies that have been carried out indicate that a refractive index gradient, reminiscent of the teleost lens, must exist (Sivak, 1991; Sivak *et al.*, 1994b).

The fact that the cephalopod lens develops as two roughly hemispheric halves from separate sources is a factor which may limit its optical quality. Measurements of spherical aberration indicate that the aberration is minimized. In general however, cephalopod lenses are not as well corrected as those of visual teleosts (Kreuzer and Sivak, 1984). Focal length measurement for lenses from a sight feeder such as the northern rock bass (*Ambloplites rupestris rupestris*) show less variation (spherical aberration) between lenses and within a single lens. For example, lenses from two year old rock bass with average diameters of 4.27 mm showed a variation in focal length, expressed as one standard deviation, of 0.13 mm. However, it is worth noting that the teleost iris is usually immobile.

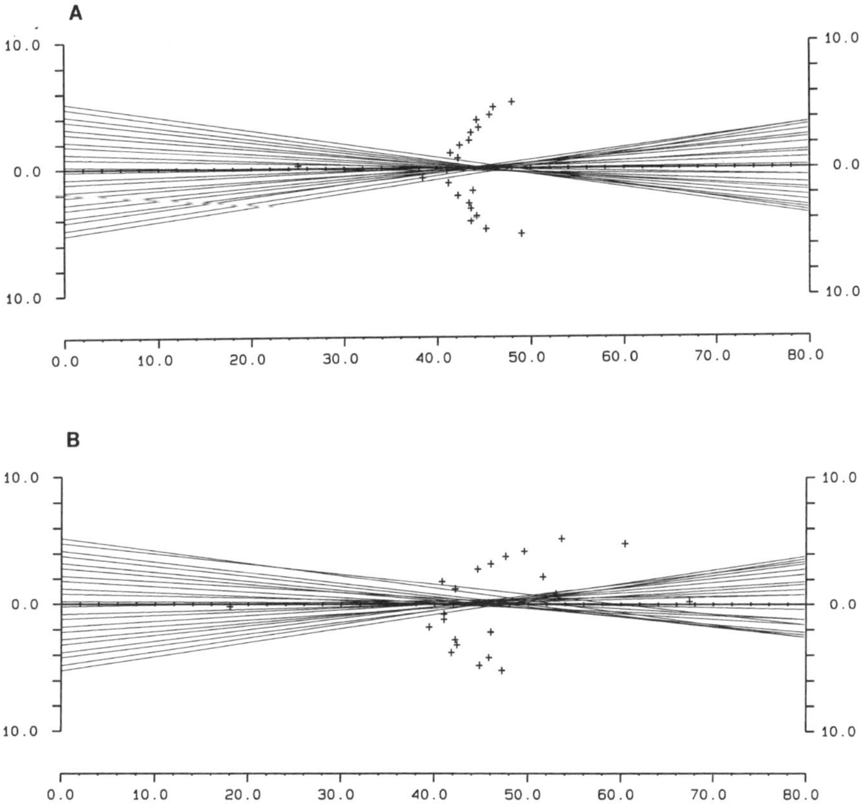

Figure 2 Focal length profile for a typical old bovine lens in culture medium. The horizontal axis (abscissa) indicates equivalent focal length (mm). The vertical axis (ordinate) refers to laser beam position (mm) from the optical center (0 – 0) of the lens. The plus (+) signs indicate the focal length for each incident beam position away from the optical center. The array of 22 lines coming to a sharp focus (**A**) represents the directions of 22 refracted laser-beam positions produced by scanning the lens along an axis that did not coincide with one of the lens suture branches or planes. The array of 22 lines that do not come to a sharp focus (increased variability of focal length; **B**) represents the directions of 22 refracted laser-beam positions produced by scanning the lens along an axis that either coincided with or passed close to one of the lens sutures

In fact, pupil diameter often exceeds lens diameter (Sivak, 1990). Thus the entire lens participates in retinal image formation. By contrast, the cephalopod iris is mobile (Muntz, 1977) and while the mechanism controlling pupil size and shape is uncertain, it is possible that lenticular spherical aberration does not seriously affect resolution ability because of pupil constriction.

The teleost lens is usually spherical in shape whereas, the shape and relative proportions of the two halves of cephalopod lenses vary considerably (Sivak *et al.*, 1994b). Typically, the posterior half of the lens accounts for 55 to 65% of axial diameter. However, this proportion varies, with animal and species, from 50 to 65%. Similarly, lens shape varies considerably. For example, in *Loligo opalescens*, the equatorial diameter of the lens is from 2.3 to 10.5% greater than its axial diameter (Sivak, 1991). It is possible that this variability implies a developmental variability caused by the fact that the lens develops as two components from different germinal sources, in comparison to the single source vertebrate lens.

The two components of the cephalopod lens meet at a septum composed of radially arranged elongated fibres (West *et al.*, 1995). The septum may have a separate origin from the rest of the lens and its precise role in cephalopod lens development is not known. However, its orientation perpendicular to the optical axis of the lens suggests that it may limit lens optical performance. Perhaps a correlative study involving lens optical quality and septum architecture will reveal a relationship similar to the one described earlier between suture structure and optics of certain vertebrate lenses.

3.4 Accommodation – aerial and aquatic

3.4.1 Functional anatomy and physiology of teleost accommodation

In teleosts, the lens is not deformed during accommodation but is moved within the eye by a smooth muscle, the retractor lentis. This muscle originates in the anterior end of the falciform process and inserts via a transparent ligament into the lens capsule (Walls, 1942). Beer (1894) observed that electrical stimulation caused contraction of the retractor lentis, resulting in retraction of the lens toward the retina. Similar conclusions were reached by Somiya and Tamura (1973) and Somiya (1987). Meader (1936) summarized the reflex pathways of accommodation in the teleost genus *Holocentrus*, using stained serial sections, and reported that the retractor lentis muscle is innervated by postganglionic parasympathetic nerve fibres. These results were supported by Sivak (1973), who used pilocarpine (a parasympathomimetic) and atropine (a parasympatholytic) to induce contraction and relaxation, respectively, of the accommodative system in a number of teleost species (Figure 3). This author also observed that the magnitude and direction of the accommodative lens movement was strongly correlated with the feeding behavior of the species studied (Sivak, 1973) Somiya (1987) confirmed that the retractor lentis is controlled by parasympathetic nerve fibres of the oculomotor nerve via the radix brevis in bass and mackerel. He noted that electrical stimulation of the short ciliary nerve, which contains both parasympathetic and sympathetic fibres, caused the retractor lentis muscle to contract, whereas stimulation of the sympathetic ganglion did not.

Histological study demonstrates that the retractor lentis is composed of smooth muscle tissue (Meader, 1936). Using stained serial sections, Munk

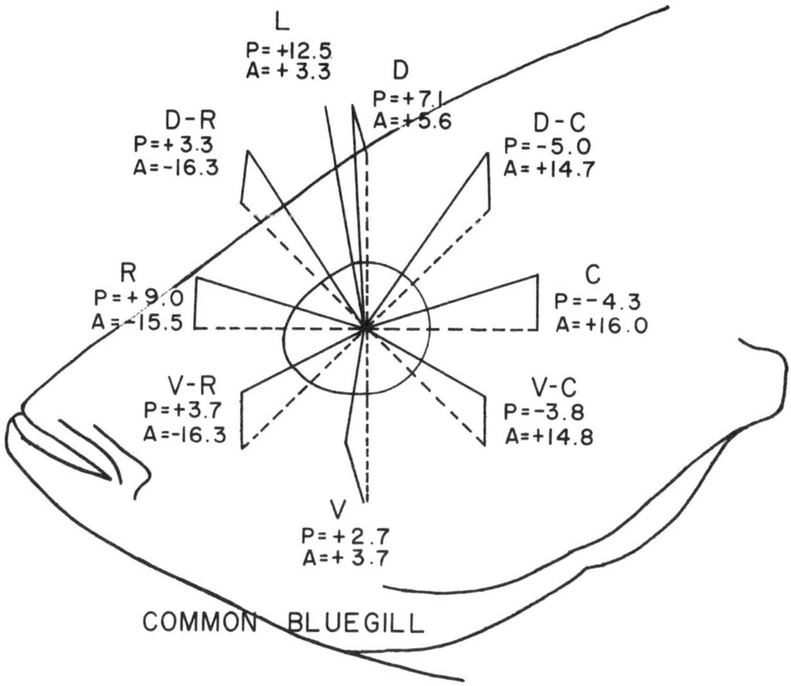

Figure 3. Mean refractive errors under the influence of maximal dose levels of atropine and pilocarpine and the mean accommodation ranges of the common bluegill (*Lepomis macrochirus macrochirus)* in several different directions of regard. Dotted lines represent various directions in the pupil plane. Solid lines indicate that the actual measurements were made in the same direction but 20° ± 2° away from the pupil plane (except for the lateral direction in which the measurement was made perpendicular to the pupil plane). Values indicated with a + or – are the mean refractive errors (myopia or hyperopia) at maximal doses of atropine and pilocarpine (designated as A or P respectively). Abbreviations for directions of regard are as follows: R = rostral, D–R = dorsal–rostral, V–R = ventral–rostal, L = lateral, D = dorsal, V = ventral, D–C = dorsal–caudal, V–C = ventral–caudal, C = caudal

(1971) observed two separate intraocular muscles, the lateral and medial lens muscles. He suggested that these are responsible for moving the lens during accommodation in certain visual teleosts, including several from the families Blenniidae and Clinidae. Munk (1971) correlated the orientation of the medial and lateral lens muscles with the lenticular movement during accommodation; although he did not report observing this movement himself. Recent work involved the opportunity to observe the naturally occurring accommodative response of a visual teleost, the oscar (*Astronotus ocellatus*) by using a television videotape monitoring system. Both an axial and a nasotemporal component in the accommodative response of this species were noted. How the morphology of the retractor lentis muscle supports this observation was examined as well (Andison and Sivak, 1994).

In *A. ocellatus*, the retractor lentis muscle is a single, pigmented, triangular muscle originating from along the ventral iris border just anterior to the *pars retinae terminalis* and inserting via a transparent ligament into the ventral surface of the lens. In a 10 cm specimen, the lens measured 5.2 mm in diameter and the base and height of the muscle measured 2.8 and 3.8 mm, respectively. Interestingly, there was an additional iris insertion from the midpoint of the muscle to just below the inferior edge of the pupil that did not appear to resemble the pigmented coverlet described by Somiya (1987), but perhaps serves a similar stabilizing function during muscle contraction. Innervation of the retractor lentis muscle is via the short ciliary nerve which pierces the sclera with the optic nerve and does not branch as it traverses the eyeball in the choroid. It branches at the point where it pierces the retina to enter the vitreous chamber; the main branch terminates on the posterior surface of the refractor lentis, where it fans out over the broad surface of the muscle. Stimulation of the short ciliary nerve results in contraction of the retractor lentis muscle and lenticular movement, but no pupillary constriction.

Light micrographs reveal two distinct smooth muscle fibre orientations within the retractor lentis. Both fibre orientations are contained within the same muscle sheath and are not visibly distinct from each other in the freshly dissected muscle. In scanning electron microscopy, the muscle ligament appears to fan out at its insertion point in an anteroposterior direction, with long filaments extending in these two directions along and into the lens capsule (Figure 4).

Dynamic study of the naturally occurring accommodative response of *A. ocellatus* was carried out using a CCD camera mounted above a specially-designed tank to record lens movements. These were simultaneously displayed on a video monitor. Goldfish were used as a visually-meaningful food target for the oscars. Goldfish were introduced into the tank at a fixed distance (48 cm) from the oscar and moved to within 20, 12 and 4 cm from the oscar via a series of clear trap doors. The video information was analyzed frame-by-frame using a custom computer program. The results show that the lens is capable of movement in nasotemporal and mediolateral directions during accommodation. These components have been observed to occur independently and/or together. Significant differences between the accommodated and unaccommodated states were found for target distances of 4, 12 and 20 cm for the nasotemporal component. The nasotemporal movements measured more than twice the mediolateral movements, on average. The greatest difference was found at a target distance of 4 cm. For example, for an oscar 14 cm in length, a target 4 cm away produced an average nasotemporal movement of 0.33 ± 0.04 mm and an average mediolateral movement of 0.12 ± 0.02 mm. These changes correspond to the dioptric needs imposed by the target distances and indicate that this highly visual species is capable of maintaining image quality over a range of target distances.

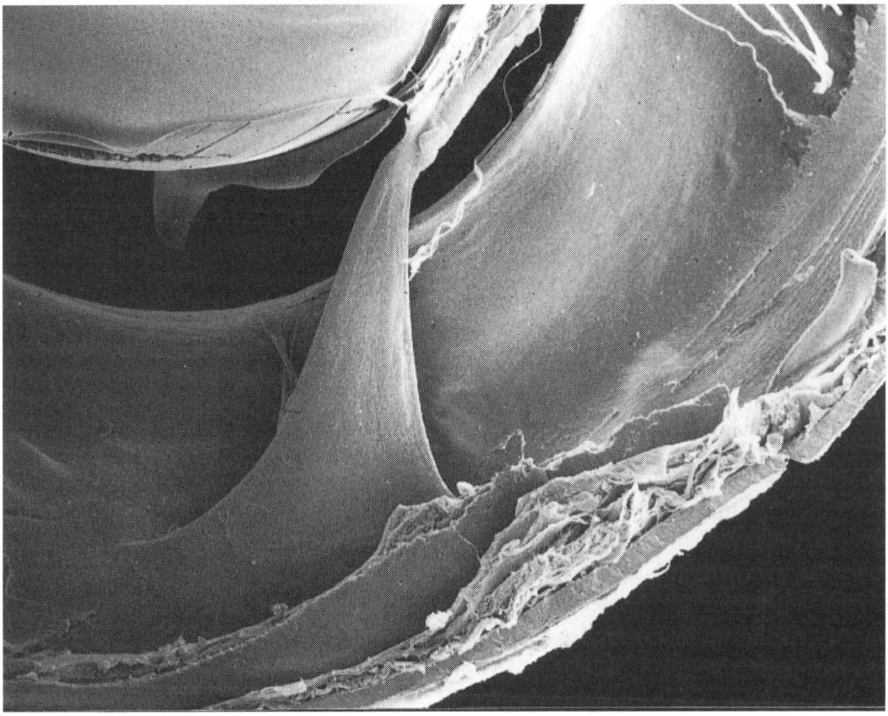

Figure 4 Scanning electron micrograph showing retractor lentis muscle originating from the ventral-temporal retina and inserting into the lens capsule. Muscle length is about 3.8 mm

3.4.2 Functional anatomy of the avian ciliary muscle

In the aerial case, vertebrates accommodate by directly or indirectly applying force to change the curvature of the lens, by moving the lens (snakes), or by changing the curvature of the cornea (Sivak, 1980). The mechanism of accommodation has been difficult to define since the apparatus involved is located within the eye and is not easily visible to an observer. In avian species the ciliary muscle is not continuous with the epithelial layers of the ciliary body and ciliary processes as in humans (Tripathi, 1974). Rather, a ciliary cleft separates the ciliary body into an outer leaf composed of ciliary muscle, trabecular meshwork, and aqueous sinus and an inner leaf formed by the fibrous baseplate of the ciliary body (Tripathi, 1974). The ciliary epithelial layers and ciliary processes are positioned on the internal surface of this baseplate. The avian ciliary muscle is striated, although it is still innervated by the autonomic nervous system (Walls, 1942). The lens is surrounded by a ring of columnar epithelial cells called the annular pad (Walls, 1942) and the ciliary processes make direct contact with the annular pad. The avian eye contains a ring of overlapping bones called the scleral ossicles which are positioned over the ciliary region and which are responsible for the formation of the cornea-scleral sulcus (Walls, 1942).

Birds can accommodate by changing the shape of the lens and cornea. Lenticular accommodation can be described as the process by which the ciliary muscles contract, forcing the ciliary processes to squeeze the lens, decreasing its radius of curvature and thus increasing its power (Meyer, 1977; West *et al.*, 1991). However, muscle fibres in the iris periphery have also been shown to play a role in lenticular accommodation (Schaeffel and Howland, 1987; Glasser *et al.*, 1995).

The anatomy of the ciliary muscle varies among avian species (Lord, 1956). However, three divisions of the ciliary muscle are usually described: Crampton's muscle, Müller's muscle and Brücke's muscle (see Glasser and Howland, 1995 for a historical review). Crampton's muscle was first described in eagles and ostriches as a muscle arising from the internal surface of the sclera under the sclera ossicles and inserting into the internal surface of the cornea. Brücke examined owls and emus and verified the existence of Crampton's muscle and described a second group of fibres posterior to Crampton's muscle that course backward in the eye, originating under the scleral ossicles and inserting into the choroid. He referred to this muscle as the *tensor choroidea*, theorizing that it tenses the choroid in order to pull it anteriorly toward the lens. In addition to Brücke's *tensor choroidea*, Müller described a second posterior fibre group in the falcon eye that runs from the choroid to insert indirectly into the inner lamellae of the cornea.

Since the initial description of these muscle groups in the 1800's, confusion has arisen as to the divisions of the avian ciliary muscle, in part because of differences among species. Some authors have generalized in suggesting that the ciliary muscle in all avian species consists of two divisions: Crampton's and Brücke's muscle (Sivak, 1980). Other authors have described the avian ciliary muscle as having three parts: Crampton's, Brücke's and Müller's muscles (Gundlach *et al.*, 1945). More confusion arises as to the exact location of Müller's muscle. It has been referred to both as the posterior portion of Crampton's muscle (Walls, 1942) and as the anterior portion of Brücke's muscle (Meyer, 1977). It is unclear whether Müller's muscle is a separate muscle group or simply a subdivision of Crampton's or Brücke's muscles.

Lord (1956) reported on six species of birds with varying numbers of muscle divisions. These included a single group of ciliary muscle fibres in the meadow-lark to four groupings in the sparrow-hawk and red-tailed hawk; consisting of Crampton's, Müller's, Brücke's and a newly described temporal muscle. The diving ducks are reported to have few or no anterior fibres (Walls, 1942). Slonaker (1918) described the ciliary muscle of the English sparrow as one unit without divisions. The chicken ciliary muscle has been described as having an anterior and posterior muscle group (West *et al.*, 1991), or having two regional groups with five distinct arrangements or muscle fibres (Murphy *et al.*, 1995).

In recent work, the ciliary muscle of the chicken (*Gallus gallus domesticus*), the pigeon (*Columbia livia*), the kestrel (*Falco sparverius*), and the hooded merganser (*Mergus cucullatus*) (four species with differing accommodative needs) was examined histologically in both the relaxed and contracted states (Pardue and Sivak, 1995). The ciliary muscle of all four species can be divided into three main muscle fibre groups based on insertion and origin: anterior, posterior and internal. The anterior muscle fibre group originates at the sclera under the scleral ossicles and inserts into the inner lamellae of the cornea. During accommodation these fibres pull the cornea posteriorly, changing the curvature of the cornea. The posterior muscle fibre group originates on the sclera and inserts posteriorly onto the baseplate of the ciliary body. The posterior fibre group acts on the baseplate of the ciliary body, pulling it forward to change the curvature of the lens during accommodation. The internal muscle fibre group extends from the baseplate of the ciliary body to the inner lamallae of the cornea and thus has a role in both corneal and lenticular accommodation. Species differences do exist as seen in subgroups determined by change in orientation of the fibres between the relaxed and contracted states and by the relative proportions of fibres within the muscle fibre groups (Figure 5, Table 1). In general, most of the ciliary muscle fibres of the chicken, pigeon, and kestrel are in the anterior muscle fibre group, suggesting an emphasis of corneal accommodation. The ciliary muscle of the hooded merganser is largely made up of fibres in the internal and posterior muscle fibre groups, indicating that lenticular accommodation is the predominant form of accommodation (Sivak *et al.*, 1985).

3.5 Visual environment and ocular development

As noted in the introduction, visual experience can influence the refractive development of the eyes of higher vertebrates. A number of experiments have demonstrated that it is possible to induce ametropias in the eyes of young animals by distorting early visual experience. Much of the work has involved the use of chicks (*Gallus gallus domesticus*) because of their precocial nature and because they grow and develop rapidly. The earlier studies generally involved depriving

Table 1: The number of muscle fibres (and relative percentages) in the ciliary muscle of four avian species, for three fibre groups

Muscle fibre groups	Chicken	Pigeon	Kestrel	Hooded merganser
Anterior	35.53 ± 0.28 (66%)	27.47 ± 0.33 (55%)	24.74 ± 0.32 (62%)	7.03 ± 0.24 (28%)
Posterior	6.75 ± 0.15 (13%)	10.75 ± 0.23 (22%)	4.85 ± 0.14 (12%)	11.04 ± 0.52 (29%)
Internal	11.45 ± 0.25 (21%)	11.33 ± 0.21 (23%)	10.53 ± 0.25 (26%)	20.61 ± 0.63 (53%)

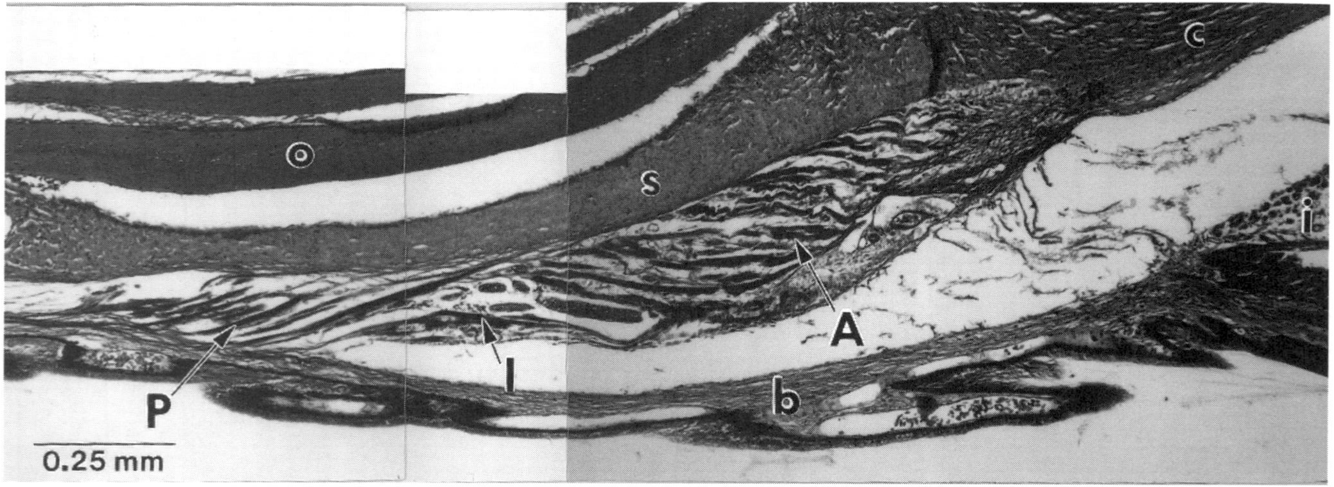

Figure 5 A light micrograph showing the ciliary muscle of the pigeon in the contracted state. The ciliary muscle of the pigeon, chicken, kestrel, and hooded merganser can be divided into three main muscle fibre groups based on fibre origin and insertion. The anterior muscle fibre group (A) originates at the sclera (s) under the scleral ossicles (o) and inserts into the inner lamellae of the cornea (c). The posterior muscle fibre group (P) originated on the sclera and inserts onto the baseplate of the ciliary body (b). The internal muscle fibre group (I) extends from the baseplate of the ciliary body of the inner lamellae of the cornea. The root of the iris (i) is also visible

the retina of a clear image by either suturing the eyelids together or applying some kind of translucent or light scattering device over the eye during development (e.g. Pickett-Seltner *et al.*, 1987). This kind of treatment leads to varying levels of myopia, often referred to as form deprivation or experimental myopia, through the axial elongation of the eye.

In 1988, Schaeffel *et al.* showed that it is possible to induce both myopia and hyperopia in the eyes of young chicks by defocussing the retinal image with hood-mounted convex and concave spectacle lenses. The range of errors induced was extended through the use of either special hydrogel (soft) contact lenses mounted directly on the cornea (Irving *et al.*, 1992), or by using rigid contact lenses inserted into a translucent goggle base and mounted over the eye (Irving *et al.*, 1992, Irving *et al.*, 1995). Use of hydrogel lenses was discontinued because of pressure induced corneal flattening, a confounding problem which is also associated with experiments in which form deprivation is induced by lid suture (eg. McBrien and Norton, 1992).

Experiments using the lens/goggle combination, with powers ranging from +30 to −20 diopters, have demonstrated that the eyes of newly hatched chicks will compensate accurately (within 4 to 7 days) to defocus of between −10 and +15 diopters (Figure 6) by developing a refractive error equal in sign and amount to the inducing lens (Irving *et al.* 1992). Beyond these limits there is a levelling off followed by a decrease in induced error. The refractive states induced are a result of enhanced or diminished ocular growth which results in relative increases and decreases in axial eye length. Corneal flattening is found associated with higher levels of hyperopia. Astigmatic refractive states can be produced with cylindrical inducing lenses (convex and concave). The astigmatic response displays a meridional sensitivity and is due to induced corneal toricity possibly coupled with some lenticular astigmatism (Irving *et al.*, 1992; Irving *et al.*, 1995).

Two major effects of the lenses used in defocus experiments are blur and magnification/minification. Blur circle diameters are a likely source of information as to the magnitude but not the sign of the defocus. However, the magnification (and minification) produced by the lenses can be a possible cue. In fact, small magnification differences between the two eyes are discernable to humans and retinal image size difference as small as 1% can produce troublesome aniseikonic effects. The magnification and minification of the lenses used in the Irving *et al.* studies mentioned earlier are considerably greater than 1%. Intraocular differences in magnification may be cues used in controlling ocular growth, although such sensitivity may require a well developed binocular system. However, when newly hatched chicks were treated with special goggles containing afocal iseikonic lenses designed to produce 10% retinal image magnification without changing the refractive state, no effect on ocular growth was produced (Sivak *et al.*, 1995).

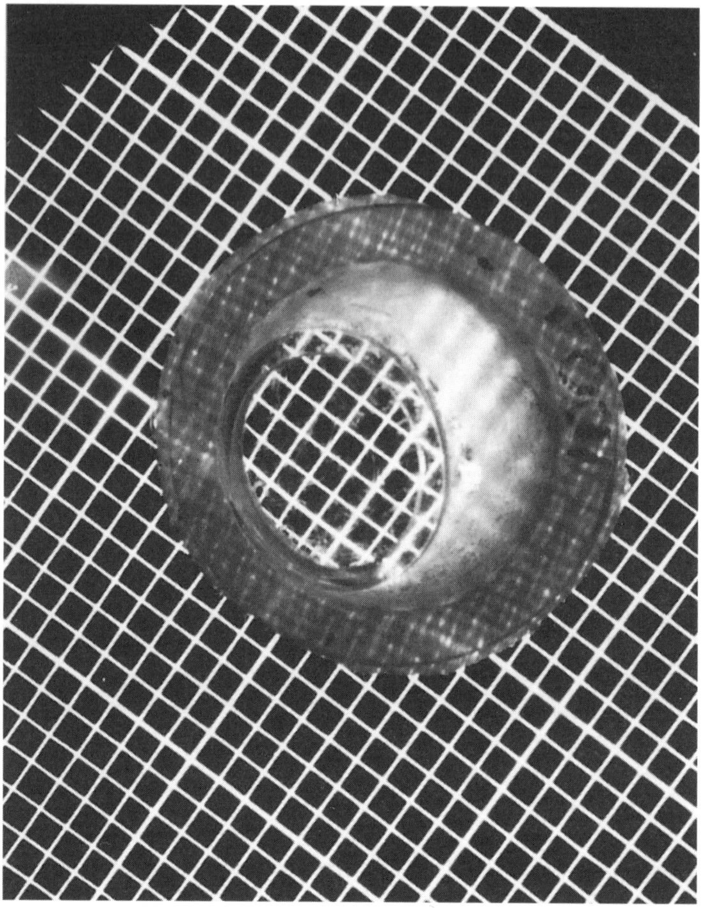

Figure 6 Photo of lens/goggle combination (used to study the refractive development of the chick eye) showing a millimetre grid through a concave defocussing lens

While these examples clearly show a relationship between visual environment and visual development in higher vertebrates, there is much less information available about the question of visual experience and development in lower vertebrates and/or during embryogenesis. These questions are particularly interesting because growth rate does not level off with maturity in lower species to the extent that it does in birds and mammals and because ocular development during embryonic periods normally takes place in the absence of photic stimulation.

The clearnose skate (*Raja eglanteria*) develops in an almost opaque eggcase, lays its eggs in pairs and can be removed from the eggcase without any apparent detrimental effects to normal development. Thus the clearnose skate provides an opportunity to easily change the normal visual environment during embryonic development. In a recent study (Pardue *et al.*, 1995), one sibling from each of eight pairs of skates was removed from its eggcase during embryonic development, while the other sibling developed inside the eggcase. The refractive development of the eyes at hatching was examined to determine if ambient light exposure during embryonic development could influence the refractive states of hatchlings. Measurements included refractive states, ocular dimensions and lens focal properties. The differences in measurements between the two groups were not significant, which indicates that environmental light does not influence the refractive development of the embryonic skate eye. It is possible that the plastic period of refractive development only begins after hatching. In fact, when chicks in the latter stages of embryonic development are exposed to strobe light stimulation, after a window is cut in the eggshell, refractive development is not altered (Andison, unpublished).

Another study (Kröger and Fernald, 1994) examined the effect of the photic environment on eye growth in a fish, the cichlid *Haplochromis burtoni*. In this case, visual stimulation was controlled from hatching for a period of eight months. Eye growth was accelerated in fish raised in dim light or in total darkness, although no significant differences in ocular dimensions were noted in the case of fish raised in a variety of monochromatic or white light regimens. Unfortunately, the effect of these changes on refractive state or on lens focal properties was not measured. Nevertheless, the results indicate that ocular growth in fish is a photopically regulated phenomenon.

3.6 Conclusions

The study of physiological optics, is, as the name suggests, a hybrid area of research in that it involves elements of both biology and physics. Since individuals who are equally adept at both sciences are rare, there is a tendency in the literature for an imbalance in one direction or another. For example, reports on the optical properties of the crystalline lens may neglect some of the biological limitations mentioned above while studies which deal with the biochemistry and protein make-up of the lens often ignore its principal function – refraction.

The four segments of this chapter are designed to draw attention to four structural features of the eye which may not have been given sufficient attention in the past. Comparisons and discussion include reference to aerial and aquatic situations as well as to higher and lower groups. The first section describes the anatomy and embryological development of the lens and cornea, emphasizing how the location of each affects its optical function. Thus, the cornea assumes a primary refractive role when the eye is in air but is relatively unimportant, in this sense, when the eye is in water. In addition, the continued sequence of lens development, coupled with an internal ocular location, has resulted in a gradient index lens, along with a susceptibility to damage.

The second section examines the question of the extent to which lens optical function is limited by development. Attention is focussed on the junctional zones (or sutures) of the lens and the extent to which their axial location interferes with lens optical properties. While the primary interest concerns the sutures of the vertebrate lens, an invertebrate example, that of the dual origin cephalopod lens, is examined as well.

The third topic concerns the functional anatomy of the accommodative apparatus and how a better understanding of this area can help explain the process of accommodation. In the case of the fish eye, the morphology of the retractor lentis muscle, as well as the suspensory lens material, can help explain the direction of lens movement during accommodation. In the avian eye, study of the effect of contraction and relaxation on the different groupings of ciliary muscle fibres can provide a more accurate description of that muscle's effect on lens shape.

The final area deals with the question of how susceptible the refractive development of the young vertebrate eye is to interference with the normal retinal stimulus pattern. It is clear that the post-embryonic development of eyes of higher (and lower) vertebrates are, to at least some degree, visually guided. The limited evidence available suggests that the embryonic eye may not be sensitive to photic stimulation, although this point requires much more study. Similarly, the relative effect(s) of the visual environment on anterior vs. posterior segment ocular growth, and on lens development, needs clarification.

Acknowledgements

The authors' research reported in this review was supported by the Natural Sciences and Engineering Research Council of Canada and by Ciba Vision Canada. The photograph used for Figure 6 was taken by Dr. M.G. Callender.

References

Andison, M. E. and Sivak, J. G. (1994) The functional morphology of the retractor lentis muscle of a teleost fish, *Astronotus ocellatus*. *Canadian Journal of Zoology*, 72, 1880 –1886.

Arnold, J. M. (1967) Fine structure of the development of the cephalopod lens. *Ultrastructure Research*, 17, 527– 543.

Bennett, A. G. and Francis, J. L. (1962) The eye as an optical system, in *The Eye Vol. 4, Visual Optics and the Optical Space Sense*, (ed H. Davson), Academic Press, New York, pp. 103 – 32.

Beer, T. (1894) Die Accommodation des Fischauges. *Pflugers Archiv für die Gesamte Physiologie des Menschen und der Tiere,* 58, 523 – 650.

Charman, W. N. (1991) The vertebrate dioptric apparatus, in *Vision and Visual Dysfunction, Vol. 2, Evolution of the Eye and Visual System* (ed. J. R. Cronly-Dillon), CRC Press, Boca Raton, pp. 82 –117.

Glasser, A. and Howland, H. C. (1995) The history of studies of avian accommodation. *Quarterly Review of Biology,* (in press).

Glasser, A., Murphy C. J., Troilo D., and Howland H. C. (1995) The mechanism of lenticular accommodation in chicks. *Vision Research*, 35, 1525 –1540.

Graham, J. B. (1972) Aerial vision in amphibious fishes. *Fauna*, 3, 14 – 23.

Gundlach, R. N., Chard, R. D., and Skahen, J. R. (1945) The mechanism of accommodation in pigeons. *Journal of Comparative Psychology*, 38, 27– 42.

Guo, S. S., Sivak, J. G., Callender, M. G. *et al.* (1995) Retinal dopamine and lens induced refractive errors in chicks. *Current Eye Research*, 14, 385 – 89.

Hogan, M. J., Alvarado, J. A. and Weddell, J. E. (1971) *Histology of the Human Eye*, W.B. Saunders, Philadelphia.

Hughes, A. (1977) The topography of vision in mammals of contrasting life style: comparative optics and retinal organization, in *Handbook of Sensory Physiology, vol II/5, The Visual System in Vertebrates*, (ed F. Crescitelli), Springer-Verlag, Berlin, pp. 613 –756.

Irving, E. L., Sivak, J. G. and Callender, M. G. (1992) Refractive plasticity of the developing chick eye. *Ophthalmic and Physiological Optics*, 12, 448 – 56.

Irving, E. L., Callender, M. G. and Sivak, J. G. (1995) Inducing ametropias in hatchling chicks by defocus-aperture effects and cylindrical lenses. *Vision Research*, 9, 1165 –1174.

Jamieson, G. S. (1971) The functional significance of corneal distortion in marine mammals. *Canadian Journal of Zoology*, 49, 421– 423.

Kreuzer, R. O. and Sivak, J. G. (1984) Spherical aberration of the fish lens: interspecies variation and age. *Journal of Comparative Physiology A*, 154, 415 – 422.

Kröger, R. H. and Fernald, R. D. (1994) Regulation of eye growth in the African cichlid fish *Haplochromis burtoni. Vision Research*, 34, 1807–1814.

Kuszak, J. R. (1995) The development of lens sutures, in *Progress in Retinal and Eye Research, Vol. 14, No. 2*, (eds N. Osborne and G. Chader), Elsevier, Oxford, pp. 567– 591.

Kuszak, J. R., Sivak, J. G and Weerheim, J. A. (1991) Lens optical quality is a direct function of lens sutural architecture. *Investigative Ophthalmology & Visual Science*, 32, 2119 – 29.

Kuszak, J. R., Peterson, K. L., Sivak, J. G. *et al.* (1994) The interrelationship of lens anatomy and optical quality. II. primate lenses. *Experimental Eye Research*, 59, 521– 35.

Lord, R. D. Jr. (1956) A comparative study of the eyes of some falconiform and passeriform birds. *American Midland Naturalist*, 56, 325 – 44.

Mann, I (1969) *The Development of the Human Eye*. Grune and Stratton, New York.

McBrien, N. A. and Norton, T. T. (1992) The development of experimental myopia and ocular component dimensions in monocularly lid-sutured tree shrews. *Vision Research*, 32, 843 – 52.

Meader, R. G. (1936) The innervation of the muscle of accommodation in the eye of the teleost, *Holocentrus. Journal of Morphology*, 59, 163 –172.

Meyer, D. B. (1977) The avian eye and its adaptations, in *Handbook of Sensory Physiology, Vol.VII/5, The Visual System in Vertebrates* (ed. F. Crescitelli), Springer-Verlag, Berlin, pp. 549 – 611.

Munk, O. (1971) On the occurrence of two lens muscles within each eye of some teleosts. *Videnskabelige Meddeleser fra Dansk Naturhistorisk Forening*, 134, 7–19.

Muntz, W. R. A. (1977) Pupillary response of cephalopods. *Symposia of the Zoological Society of London*, No. 38, pp. 277– 285.

Murphy, C. J., Glasser, A. and Howland, H. C. (1995) The anatomy of the ciliary region of the chicken eye. *Investigative Ophthalmology & Visual Science*, 36, 889 – 96.

Pardue, M. T. and Sivak, J. G. (1995) The functional anatomy of the ciliary muscle in four avian species. *Brain, Behavior and Evolution* (in press).

Pardue, M. T., Luer, C. A., Callender, M. G. *et al.* (1995) The absence of a photopic influence on the refractive development of the embryonic eye of the clearnose skate (*Raja eglanteria*). *Vision Research*, 35, 1675 –1678.

Pickett-Seltner, R. L., Weerheim, J., Sivak, J. G. *et al.* (1987) Experimentally induced myopia does not affect post-hatching development of the chick lens. *Vision Research*, 27, 1779 – 82.

Rodrigues, M. M., Warring III, G. O., Hackett, J. *et al.* (1982) Cornea, in *Ocular Anatomy, Embryology and Teratology* (ed F. A. Jakobiec), Harper and Row, Philadelphia, pp. 182.

Schaeffel, F. and Howland, H. C. (1987) Corneal accommodation in chick and pigeon. *Journal of Comparative Physiology A*, 160, 375 – 84.

Schaeffel, F., Glasser, A. and Howland, H. C. (1988) Accommodation, refractive error and eye growth in chickens. *Vision Research*, 28, 639 – 57.

Sivak, J. G. (1973) Interrelation of feeding behavior and accommodative lens movement in some species of North American freshwater fishes. *Journal of the Fisheries Research Board of Canada*, 30, 1141– 46.

Sivak, J. G. (1978) Vertebrate strategies for vision in air and water, in *Sensory Ecology*, (ed. M. A. Ali), Plenum, New York, pp. 503 – 20.

Sivak, J. G. (1980) Accommodation in vertebrates: a contemporary survey, in *Current Topics in Eye Research*, Vol. 3, (eds H. Davson and J. Zadunaisky), Academic Press, New York, pp. 281– 330.

Sivak, J. G. (1990) Optical variability of the fish lens, in *The Visual System of Fish*, (eds R. H. Douglas

and M. B. A. Djamgoz), Chapman and Hall, London, pp. 63 – 80.
Sivak, J. G. (1991) Sharpe and focal properties of the cephalopod lens. *Canadian Journal of Zoology*, 69, 2501– 06.
Sivak, J. G., Hildebrand, T. and Lebert, C. (1985) Magnitude and rate of accommodation in diving and non-diving birds. *Vision Research*, 25, 925 – 33.
Sivak, J. G., Herbert, K. L., Peterson, K. L. *et al.* (1994a) The interrelationship of lens anatomy and optical quality. 1. non-primate lenses. *Experimental Eye Research*, 59, 505 – 20.
Sivak, J. G., West, J. A. and Campbell, M. C. (1994b) Growth and optical development of the ocular lens of the squid (*Sepioteuthis lessoniana*). *Vision Research*, 34, 2177– 87.
Sivak, J. G., Curry, T. A., Callender, M. G. *et al.* (1995) Is magnification a cue in defocus induced ametropia in chicks? *Optical Society of America, Technical Digest Series, Vol. 1*, 84 – 87.
Slonaker, J. R. (1918) A physiological study of the anatomy of the eye and its accessory parts in the English sparrow (*Passer domesticus*). *Journal of Morphology*, 31, 351– 459.
Somiya, H. (1987) Dynamic mechanism of visual accommodation in teleosts: structure of the lens muscle and its nerve control. *Proceedings of the Royal Society, London, B Biological Sciences*, 230, 77– 91.
Somiya, H. and Tamura, T. (1973) Studies on the visual accommodation in fishes. *Japanese Journal of Ichthylogy*, 20, 193 – 206.
Tripathi, R. C. (1974) Comparative physiology and anatomy of the outflow pathway, in *The Eye, Vol. 5, Comparative Physiology*, (eds H. Davson and L. T. Graham Jr.), Academic Press, New York, pp. 163 – 356.
Walls, G. L. (1942) *The Vertebrate Eye and its Adaptive Radiation*. Cranbook Institute of Science, Bloomfield Hills, Michigan.
West, J. A., Sivak, J. G. and Doughty, M. J. (1991) Functional morphology of lenticular accommodation in the young chicken *(Gallus domesticus). Canadian Journal of Zoology,* 69, 2183 – 93.
West, J. A, Sivak, J. G. and Doughty, M. J. (1995) Microscopical evolution of the crystalline lens of the squid (*Loligo opalescens*) during embryonic development. *Experimental Eye Research,* 60, 19 – 35.
Worgul, B. V. (1982) The lens, in *Ocular Anatomy, Embryology and Tertology*, (ed F. A. Jakobiec), Harper and Row, Philadelphia, pp. 355 – 89.

4
A review of vertebrate and invertebrate ocular filters

R.H. DOUGLAS* and N.J. MARSHALL†
*Department of Optometry & Visual Science, City University, 311–321 Goswel Road, London, EC1V 7DD, U.K.
†Vision Touch and Hearing Research Centre, The University of Queensland, Brisbane, Queensland, 4072, Australia.

4.1 Introduction

The spectral information available to an animal's visual system depends both on the wavelengths reaching its outer segments and the visual pigments contained within them. The former is governed not only by the chromatic stimuli present in the environment, but also by the degree to which these are modified through intraocular filters, before being absorbed by the visual pigments. Although the ocular media of the majority of animals are transparent to light above about 310 nm, their primary function being either refractive or nutritive, in some, pigments are present that filter the spectral content of the light reaching the retina. Light impinging on the visual pigments may have its spectrum further modified by both filters within the retina itself and by reflective structures behind the retina.

This review sets out to describe the spectral characteristics of such ocular filters in both vertebrates and invertebrates, and outlines their location, the biochemical nature of the filtering pigments, and their function. In the following sections, filters are described by sequentially following the path of light through the eye's optical components from the outside world into the photoreceptors and, where present, reflection back by a tapetum. Most vertebrate filters, whatever their origin and placement, remove short wavelengths. Their function is therefore dealt with in a separate section. The eyes of invertebrates are more varied in their design and consequently possess a greater variety of functionally distinct filters. Their suggested functions are detailed within each section.

4.2 Vertebrate ocular filters

4.2.1 General spectral characteristics

Even in the absence of a specific short-wave absorbing filter, no ocular structure will transmit significant amounts of radiation below about 310 nm due to absorption by its nucleic acids and various structural protein components, such as aromatic amino acids. In such unpigmented structures the decline in

S.N. Archer et al. (eds.), Adaptive Mechanisms in the Ecology of Vision, 95–162

transmission in the UV is smooth and the 'cut-off' wavelength (often defined as the wavelength of 50% transmission or the wavelength of half maximum optical density; Lipetz 1984a; Douglas & Thorpe 1994) is generally between 310 – 350 nm and depends primarily on the structure's thickness. The presence of specific short-wave absorbing pigments will result in further absorption in the UV/blue part of the spectrum. The transmission spectrum of a structure at short wavelengths is, therefore, the product of both 'background' absorption by its proteins and nucleic acids, and light loss caused by any specific filtering pigments. At longer wavelengths, transmission is generally above 90%, especially if living tissue free from any possible *post-mortem* artefacts is examined (e.g. van Best *et al.* 1988). In the infra-red, the spectral characteristics of most ocular media resemble those of pure water (e.g. Brainard *et al.,* 1994; van den Berg and Spekreijse 1997).

4.2.2 Cornea

The corneas of most animals contain no significant amounts of short-wave absorbing pigments and consequently transmit most light down to around 310 nm (e.g. *Mammals* – Hemmingsen & Douglas 1970, Chou & Cullen 1984, van den Berg & Tan, 1994; Ambach *et al.* 1994; *Birds* – Hemmingsen and Douglas 1970, Emmerton *et al.,* 1980, Govardovskii and Zueva 1977, Maier 1994; *Reptiles* – Thorpe 1991, Ellingson *et al.,* 1995; *Amphibia* – Thorpe 1991). Figure 1 (curve i) shows the smooth and rapid decline in transmission of a monkey cornea at short wavelengths, which is characteristic of all such unpigmented corneas.

Pigmented corneas are almost entirely restricted to fish, among whom they are quite common, the only exceptions being a small amount of pigmentation, which may have little or no effect on vision, found in the corneas of the frog, *Rana pipiens*, and the ground squirrel, *Citellus beecheyi* (Figure 1, curves ii & iv). The degree of pigmentation in fish differs markedly between species. While some have corneas which, like most other vertebrates, contain no short-wave absorbing pigments (e.g. McCandless *et al.,* 1969, Douglas 1989, Douglas and McGuigan 1989), and others have heavily pigmented, visibly yellow, corneas absorbing radiation above 400 nm (Heinermann 1984; Kondrashev *et al.,* 1986; for reviews, Figure 1, curve v), a significant number have a small, possibly inconsequential, amount of pigmentation, betrayed by irregularities in their absorption spectra in the near UV (e.g. Douglas and McGuigan 1989; Figure 1, curve iii).

Teleost corneal pigments are often restricted to, or are densest in, a dorsal 'eye-shade' region above the pupil (e.g. Muntz 1973, 1976a, 1982; Douglas 1989; Plate 1h*). Since the distribution of light underwater is very directional (Lythgoe 1975, 1976), such 'eyeshades' will selectively reduce the amount of

*Colour plates will be found between pages 124 and 125.

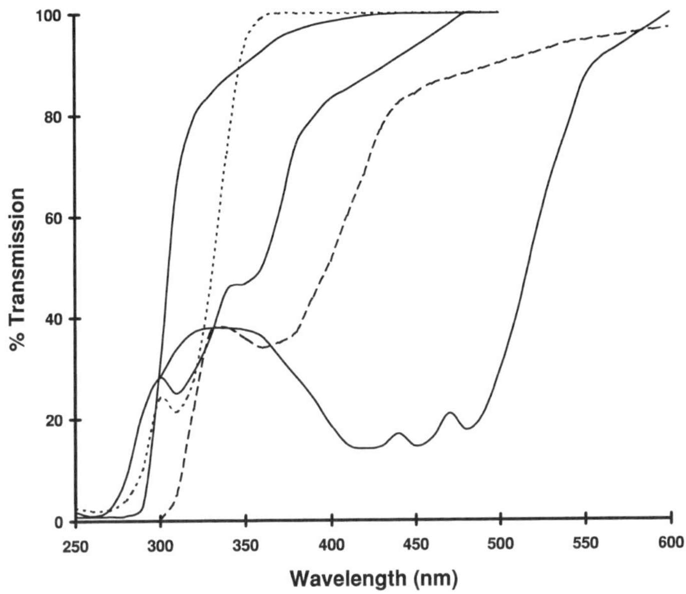

Figure 1 Transmission spectra of intact vertebrate corneas. (from left to right) (i) Primate, *Macaca fascicularis* (Thorpe 1991), with no detectable amounts of pigmentation, (ii) Frog, *Rana pipiens* containing small amounts of an unidentified pigment (Thorpe 1991, dotted line), (iii) Fish, *Herotilapia multispinosa*, displaying 2 absorbance maxima at 325 and 360 nm characteristic of the mycosporine-like amino acids palythine and palythene (Thorpe 1991), (iv) California ground squirrel, *Citellus beecheyi* (Hemmingsen & Douglas 1970, dashed line) (v) Dorsal cornea of the fish *Ctenopoma oxyrhynchus*, containing a carotenoid pigment (Thorpe 1991)

bright downwelling light, while leaving the less intense light, entering the eye along the optic axis, relatively unaffected.

The spectra of heavily pigmented corneas are usually characterised by 3 absorbance maxima around 425, 450 and 480 nm (Figure 1, curve v), suggesting these pigments are carotenoids. However, in some species, both intact corneal absorbance spectra (Thorpe 1991) and methanol extracts (Dunlap *et al.,* 1989) indicate the presence of mycosporine-like amino acids (MAAs), identical to those acting as short-wave absorbing filters in the fish lens (see 4.2.3) (Figure 1, curve iii). In several, both carotenoid pigments and MAAs are observed in the same cornea; the MAAs being uniformly distributed throughout and the carotenoid forming the dorsal eye-shade (Thorpe 1991), thus effectively screening all wavelengths below 500 nm; the carotenoid-like pigments absorbing light between 400 – 500 nm, and the MAAs absorbing below 400 nm.

Usually, corneal pigments, if present in sufficient quantities, confer a yellow colouration on the cornea. The occlusable corneas (see below) of a small number of species, such as *Hexagrammos octogrammus*, however, have both a yellow carotenoid pigment, and a second red pigment, probably astaxanthin, contained within a separate group of chromatophores (Orlov and Gamburtseva 1976a&b, Gamburtseva *et al.,* 1979). The orange corneal dorsal eye shade of the goldfish also contains a pigment absorbing maximally around 460 – 480 nm (Figure 14), which is likely to be astaxanthin (Douglas 1989). Astaxanthin is also one of the major pigments in the goldfish skin (Hata and Hata 1971). Similarly in puffer fish, the pigments extracted from the cornea and the skin are indistinguishable (Appleby and Muntz 1979). Thus, there seems to be a close relationship between dermal and corneal pigmentation (Walls and Judd 1933b).

Although having a cornea absorbing significant amounts of light below 500 nm potentially has several advantages in photopic conditions (section 4.2.7), it results in an inevitable, and undesirable, loss of sensitivity in low light levels. Consequently, the levels of pigmentation observed in most fish corneas (usually an optical density of less than 1.0 at wavelengths 400 – 500 nm) probably represent a compromise. However, some fish have overcome this problem and can tolerate much higher levels of corneal pigmentation, by developing 'occlusable' corneas, in which pigment distribution depends on ambient light levels (Orlov and Gamburtseva 1976a&b, Appleby and Muntz 1979, Muntz 1982, Gamburtseva *et al.,* 1979, Shand 1988). In dim light pigment is aggregated in chromatophore cell bodies arranged around the periphery of the cornea. On exposure to higher light levels, the pigment migrates into the central cornea through chromatophore processes, conferring a yellow/orange colouration to the whole cornea. These pigment migrations are produced, as in skin melanophores, using microtubules (Orlov and Gamburtseva 1976a) and are regulated locally within the eye (Appleby and Muntz 1979, Kondrashev and Khodtsev 1984).

Another method of selectively reducing the amount of radiation entering the eye, rather than absorbing it, is to reflect it. Consequently, the corneas of many fish, like those of several invertebrates (section 4.3.1a), contain a variety of reflective structures causing corneal iridescence (Lythgoe 1971, 1975, 1976; Locket 1972; Best and Nicol 1980; Collin and Collin 1996; Figure 2). This is the result of constructive interference of light reflected from multilayer stacks of different refractive index (see section 4.3.1a for explanation) arranged so as to preferentially reflect oblique rays coming from above the animal. As a result, like the dorsal pigmentation described above, they will only reduce the bright downwelling light and do not greatly compromise on-axis vision (Lythgoe 1976). Furthermore, again as in some pigmented corneas and presumably for similar reasons, the iridescence changes with ambient illumination, either completely disappearing, reducing in degree or shifting towards shorter wavelengths during dark adaptation (Shand and Lythgoe 1987; Shand 1988; Lythgoe and Shand 1989; Figure 2). Such short term changes in iridescence are likely to be the result of varying the degree of corneal hydration through the use of epithelial ionic pumps (Ashcroft and Lythgoe 1991).

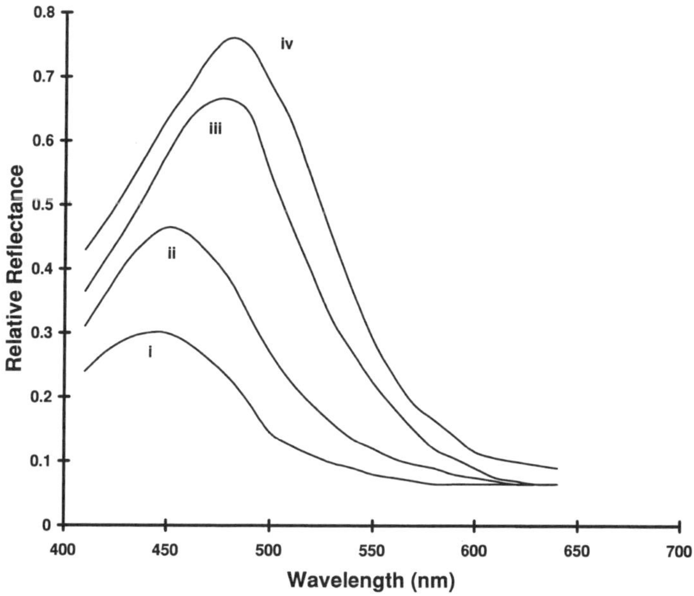

Figure 2 Spectral reflectance from the cornea of the sand goby (*Pomatoschistus minutus*) after (i) 1, (ii) 5, (iii) 20 and (iv) 30 min light exposure following 2 hrs in darkness. (Lythgoe & Shand 1989)

4.2.3 Lens

The lens is the most common vertebrate ocular filter and, with the exception of some pigmented fish corneas, usually determines the spectral absorption of the whole eye. Although, many species have lenses containing no detectable amounts of pigment and hence transmit short wavelengths well (e.g. *Mammals* – Gorgels & van Norren 1992, Brainard *et al.,* 1994; *Birds* – Emmerton *et al.,* 1980; Maier 1994; *Fish* – McCandless *et al.,* 1969; Douglas and McGuigan 1989; Thorpe *et al.,* 1993; *Amphibia and Reptiles* – Thorpe 1991; Figure 3), many animals have lenses with varying degrees of pigmentation. Pigmented lenses are common in fish (Heinermann 1984; Douglas and McGuigan 1989; Thorpe *et al.,* 1993 for reviews; Plate 1e), as well as primates (Boettner and Wolter 1962; Dartnall *et al.,* 1965; Cooper and Robson 1969b, Thorpe 1991) and Sciurid mammals (Walls and Judd 1933b; Cooper and Robson 1969a; Jacobs and Yolton 1972; Yolton *et al.,* 1974; Chou and Cullen 1984; Zigman *et al.,* 1985b), but also occur in some other mammals (Thorpe 1991), amphibia (Merker 1934, Denton 1955; Kennedy and Milkman 1956), and reptiles (Walls and Judd 1933b; Ellingson *et al.,* 1995) (Figure 3). Highly pigmented lenses have never been described in birds, although the ocular media of some species of duck do absorb strongly at wavelengths below 370 nm, suggesting the presence of some lenticular pigmentation (Jane & Bowmaker, 1988).

The identity of the short-wave absorbing lens pigments is known for only a few species. In some animals, these are a number of freely diffusible, water soluble, closely related tryptophan derivatives with absorbance maxima in the UV at 365 – 375 nm and 260 – 265 nm (Cooper & Robson 1969a&b; Thorpe 1991; Figure 4a). In young primates, the major lens pigment is 3-hydroxy-kynurenine-O-β-glucoside, which is found in the human lens together with lower levels of kynurenine and 3-hydroxy-kynurenine (van Heyningen 1971a&b, 1973a; Bando *et al.* 1981; Thorpe 1991; Wood & Truscott 1993). Human lenses, especially those from older individuals, contain, in addition to their tryptophan derived pigments, a number of other chromophores (Zigman 1985; Sen *et al.,* 1992), including low concentrations of the same two carotenoids that make up the primate macular pigment (section 4.2.5b); lutein and zeaxanthin (Yeum *et al.,* 1995). Tryptophan-derived kynurenine pigments are also found in the grey squirrel (van Heyningen 1971b, 1973a; Zigman *et al.,* 1985b; Zigman and Phaxia 1988; Nie *et al.,* 1990; Figure 5) and some species of fish (Thorpe *et al.,* 1992, 1993; Truscott *et al.,* 1992; Figure 4a).

Not all lens pigments are tryptophan-derived, however. Although their absorption spectra often resemble those of the tryptophan-related pigments (Zigman *et al.,* 1985a; Zigman 1991), most fish pigments are in fact a series of dietary-derived MAAs (Dunlap *et al.,* 1989; Thorpe *et al.* 1993; Figure 4b). Spectrally distinct short-wave absorbing pigments, that are probably neither MAAs nor derived from tryptophan, have also been extracted from a variety of other

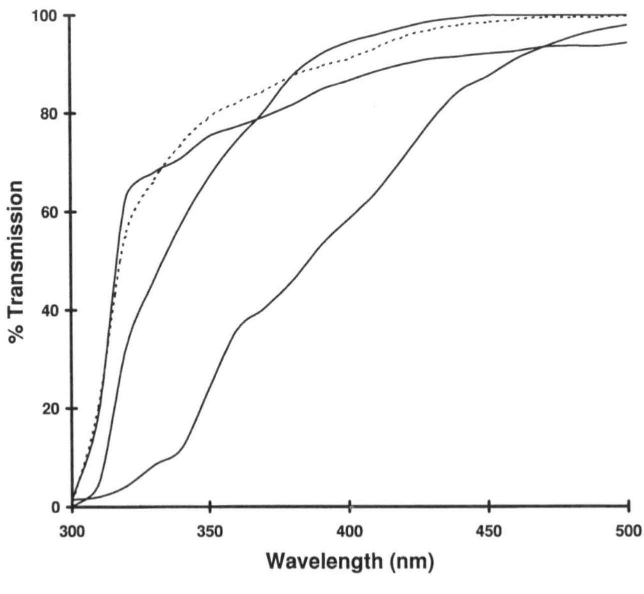

a

Figure 3 (Continued on next page)

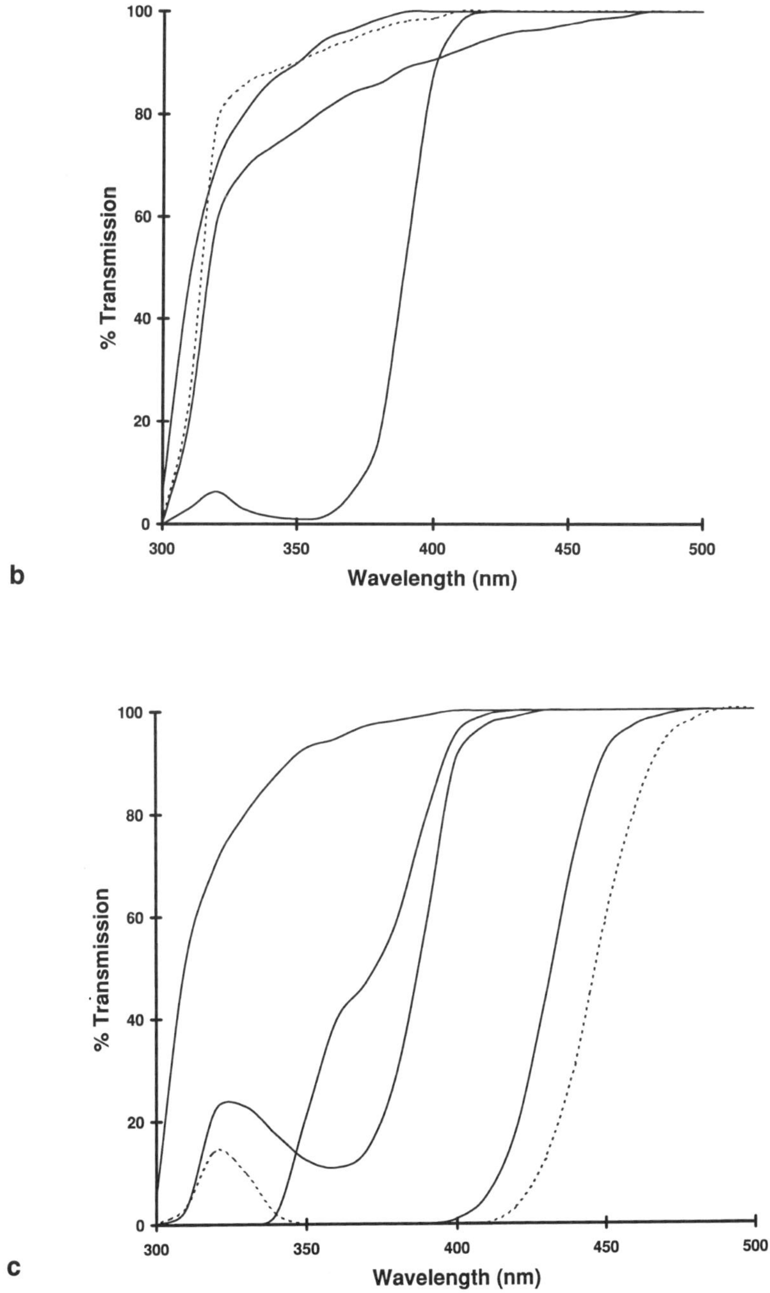

Figure 3 (Continued on next page)

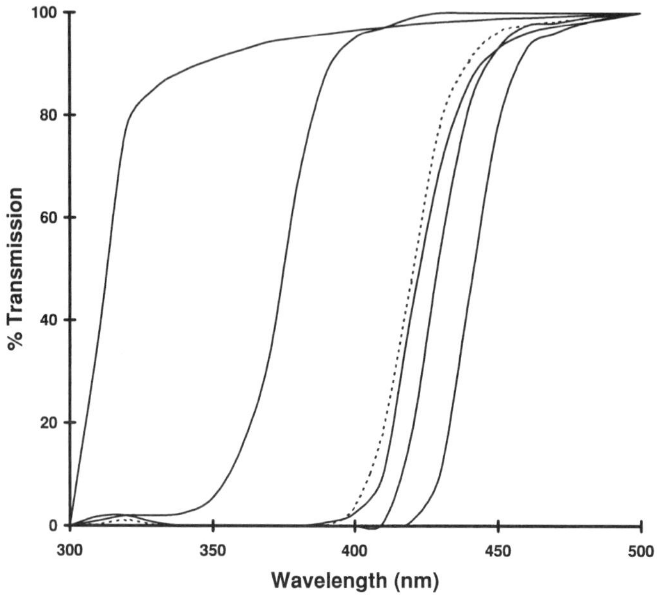

d

Figure 3 Spectral transmission of vertebrate lenses. (**a**) Birds and reptiles (left to right); i – Robin (*Erithacus rubecula*, 2.6 mm, Douglas unpubl), ii – Water dragon (*Physignathus leseueri*, dotted line, 2.52 mm), iii – average of 10 chicken lenses (*Gallus gallus*, av 4.79 mm), iv – Rat snake (*Elaphe guttata*, 2.89 mm) (**b**) Amphibia (left to right); i – Salamander (*Salamandra salamandra*, dotted line, 2.01 mm), ii – newt (*Cynops pyrrhogaster*, 0.88 mm), iii – Axolotl (*Ambystoma mexicanum*, 1.63 mm) , iv - Frog (*Rana pipiens*, 3.22 mm) (**c**) Fish (left to right); i – *Leuciscus idus* (1.80 mm), ii – *Platytaenoidus degeni* (1.61 mm), iii – *Merlangius merlangus* (3.98 mm), iv – *Herotilapia multispinosa* (2.08 mm), v – *Aequidens maronii* (1.58 mm) (**d**) Mammals (left to right); i – Golden hamster (*Mesocricetus auratus*, Brainard *et al.* 1994), ii – Guinea pig (*Cavia porcellus*, 3.70 mm), iii – Macaque (*Macaca fascicularis*, dotted line, 2.79 mm), iv – Marmoset (*Callithrix jacchus*, 2.97), v – Capuchin (Cebus appelus, 3.92 mm), vi – Grey squirrel (*Sciurus carolinensis*, 3.60 mm). All data are from Thorpe 1991, unless otherwise indicated. The numbers in brackets represent lens diameters along the optic axis

vertebrates. Aqueous extracts of the frog, *Rana pipiens*, lens reveal a pigment unlike that of primates or fish, with a λ_{max} at 347 nm (Kennedy and Milkman 1956; Thorpe 1991, Figure 5). Guinea pig lenses, on the other hand, contain small amounts of a pigment absorbing maximally around 325 nm (Thorpe 1991; Figure 5). As yet unidentified pigments have also been observed in ground squirrel and chipmunk lenses (Nie *et al.,* 1990). The large variety of chemically distinct short-wave absorbing pigments within the vertebrate lens is a clear testament to their importance and indicates that this trait has evolved separately on many occasions.

The vertebrate lens grows continually throughout life. Therefore the path-length the light has to traverse will increase, leading to an inevitable reduction in short-wave transmission in older animals. This accounts for the age related decrease in lens transmission reported in several fish (Douglas 1989; Thorpe and

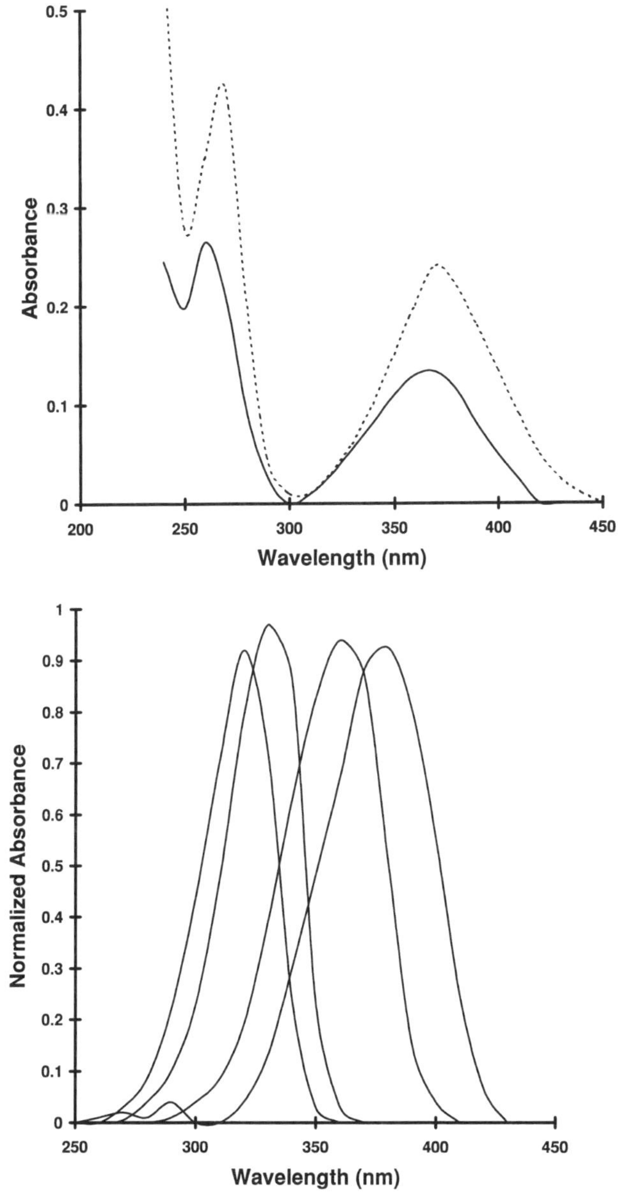

a

b

Figure 4 Absorbance spectra of HPLC purified lens pigments. (**a**) Monkey (*Macaca fascicularis*, solid line, Thorpe 1991) and Gourami (*Trichogaster trichopterus*, dotted line, Truscott *et al.* 1992). Both pigments had an HPLC retention time and an absorbance spectrum identical to that of authentic 3-hydroxykynurenine. Absolute absorbance differences between the curves are arbitrary. (**b**) Four mycosporine-related pigments extracted from the lenses of various species of fish (left to right); i – Palythine (λ_{max} 320 nm) from *Oreochromis niloticus*, ii – Asterina-330 (λ_{max} 330 nm) from *Chaetodon* sp., iii – Palythene (λ_{max} 360 nm) from *Clupea harengus*, iv – an unidentified mycosporine-like compound (λ_{max} 385 nm) from *Exocoetus obtusirostris*. (Thorpe *et al.* 1993)

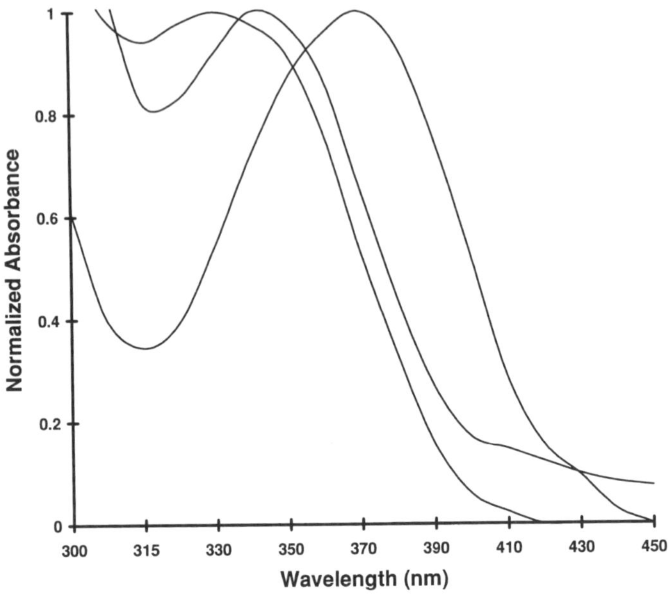

Figure 5 Absorbance spectra of crude water soluble pigments extracted from vertebrate lenses. (left to right); i - Guinea pig (*Cavia porcellus*), ii - Frog (*Rana pipiens*), iii - Grey squirrel (*Sciurus carolinensis*). Extracts have not been purified making a detailed analysis of absorbance below 300 nm impossible. (Thorpe 1991)

Douglas 1993; Figure 6a triangles) and may explain similar phenomena noted in cats (Dodt and Walther 1958) and cows (Merker 1934), and will occur in all species with unpigmented lenses. In pigmented lenses, however, a number of things can happen during ageing depending on the rate of pigment production, resulting in a variety of ageing patterns, all of which can be found in various species of fish (Thorpe and Douglas 1993 for review). In some species, for instance, there is a steady increase in pigment production with age, resulting in decreased short-wave transmission in older lenses (Figure 6b). In others, an initial rapid rise in pigment production is followed by a levelling off (Figure 6a circles) or even a decrease in the amount of UV removed (Figure 6c). Furthermore, some species are born with highly pigmented lenses, that then become more transparent to short wavelengths as they age (Figure 6d). These surprising increases in UV transmission with age in several species (Figures 6c&d) indicate that pigment accumulation has ceased in older animals, and that the inevitable increases in lens volume effectively dilute the freely diffusible lens pigments (Thorpe and Douglas 1993). Why fish should display such a wide variety of ageing patterns is unclear, but may relate to factors such as changes in diet and habitat (Thorpe and Douglas 1993) and possibly exposure to UV radiation (see section 4.2.7d).

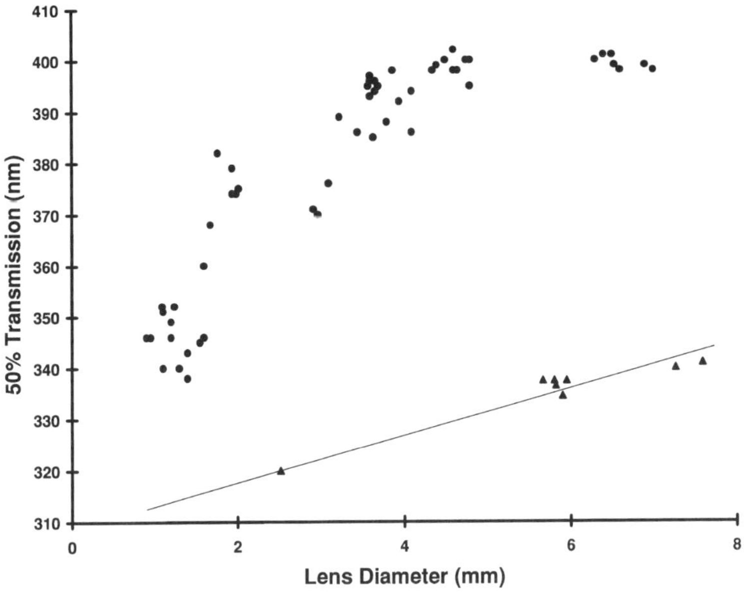
410
400
390
380
370
360
350
340
330
320
310
50% Transmission (nm)
0
2
4
6
8
Lens Diameter (mm)

a

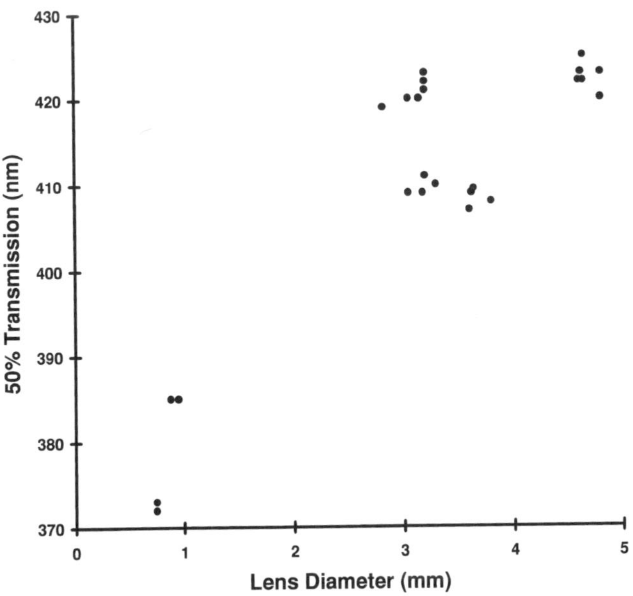
430
420
410
400
390
380
370
50% Transmission (nm)
0
1
2
3
4
5
Lens Diameter (mm)

b

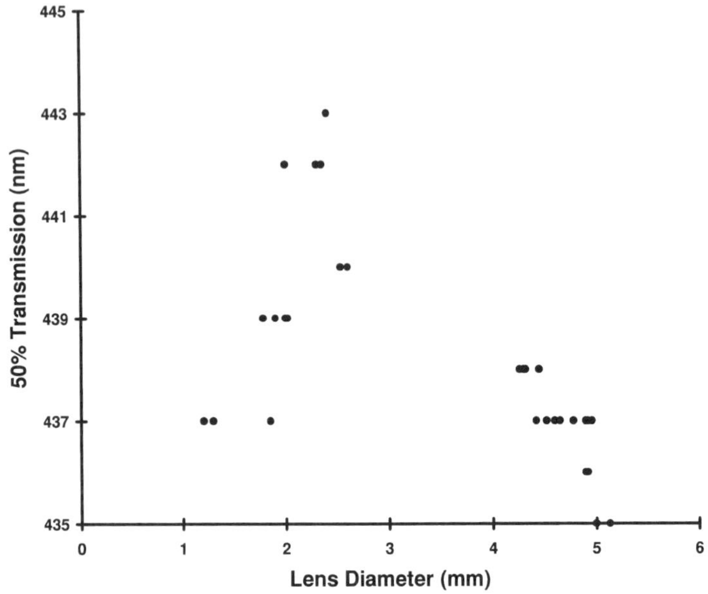

c

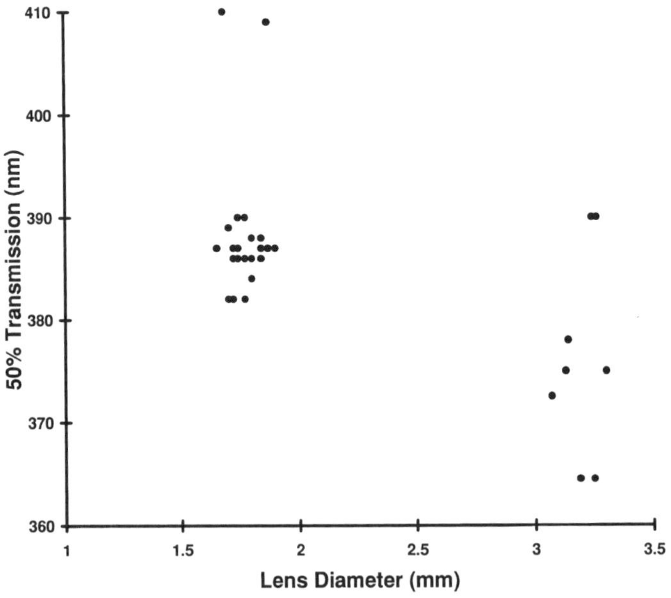

d

Figure 6 Wavelength of 50% transmission as a function of lens diameter for various species of fish. The wavelength of 50% transmission indicates the amount of UV removed by the lens, while the lens diameter is a reliable indicator of fish age (Douglas 1987). (**a**) i – Unpigmented lenses of the Hake, *Merluccius merluccius* (triangles, the line, which fits the data well, is the expected relationship if lens transmission were determined solely by an increase in lens diameter without the accumulation of a specific short-wave absorbing pigment), ii – pigmented lenses of Tilapia, *Oreochromis niloticus*. (**b**) Pigmented lenses of Herring, *Clupea harengus*. (**c**) Pigmented lenses of Flying fish, *Exocoetus obtusirostris*. (**d**) Pigmented lenses of Kissing gourami, *Helostoma temmincki*. (Thorpe & Douglas 1993)

Not surprisingly, the most studied lenticular ageing is the reduced short-wave transmission of older human lenses (e.g. Said and Weale 1959; Boettner and Wolter 1962; Grover and Zigman 1972; Weale 1988). Such increased 'yellowing' with age has a marked effect on short-wave colour perception (Ruddock 1972; Werner 1982), and Monet's increased use of blue colours as he aged is, for instance, blamed on lenticular ageing. The mechanisms responsible for this process are complex, beyond the scope of this review and still not completely resolved, but it is clear that the human lens proteins undergo considerable modifications with age, often characterised by changes in fluorescence, which, at least in part, result from exposure to short-wave radiation (e.g. Lerman 1980, 1987; Zigman 1983, 1985; Sen *et al.,* 1992), although it is unlikely that all the pigmentary changes in the ageing human lens are related to prolonged exposure to light (Yu *et al.,* 1988). Perhaps surprisingly, the concentrations of all 3 tryptophan-derived short-wave filtering pigments so far described in the human lens, appear to decrease with age (Cooper and Robson 1969b; van Heyningen 1973b; Bando *et al.,* 1981; Wood and Truscott 1993, 1994). Consequently, the diminished age-related short-wave transmission of the human lens is probably attributable to the accumulation of one or more additional pigments, probably bound to lens proteins within the lens nucleus (e.g. Cooper and Robson 1969b; Zigman 1985; Yu *et al.,* 1989; Weale 1995).

Although, lens pigments are usually associated with animals living in relatively high light levels, a striking exception to this general trend are the large variety of pigments observed in several species of fish inhabiting the deep ocean, an environment with comparatively low levels of illumination (Plate 1e). Of the over 150 deep-sea fish examined to date 27 contain detectable amounts of pigmentation in their lens (Douglas and Thorpe, 1992; Douglas *et al.* 1995 for reviews). These pigments, as in the lenses and corneas of shallow water species, are either carotenoids, kynurenine derivatives, MAAs, or a variety of as yet unidentified substances (Douglas and Thorpe 1992 for review; Figure 7 and section 4.2.7e).

4.2.4 Humours

Dalton incorrectly attributed his own colour blindness to a blue tinted vitreous (Hunt *et al.,* 1995) and in general, there is little evidence for specific light-absorbing filters in either the aqueous or vitreous humour (e.g. *Fish* – McCandless *et al.,* 1969; *Mammals* – Merker 1934; Ambach *et al.,* 1994; *Birds* – Emmerton *et al.,* 1980, Govardovskii and Zueva 1977; Figure 8). The only exceptions are qualitative reports of a yellow vitreous in 2 species of fish (Lythgoe 1971; van Heyningen and Linklater 1976). Experience suggests, however, that these may be *post mortem* artefacts.

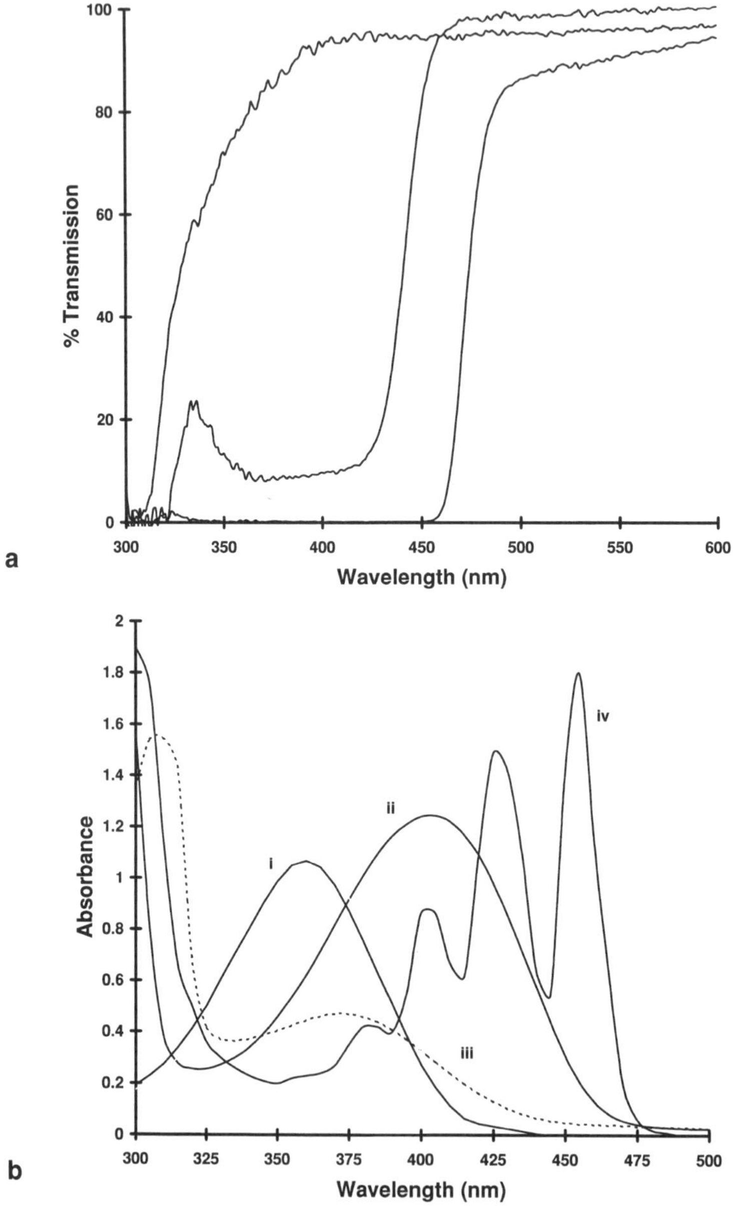

Figure 7 Spectral characteristics of some deep-sea fish lenses (**a**) Transmission spectra through the intact lenses of (left to right); i – *Diretmus argenteus* (diameter 4.0 mm), ii – *Evermanella balbo* (3.6 mm), iii – *Scopelarchus analis* (4.8 mm) (**b**) Absorbance spectra of pigments extracted from the lenses of; i – *Stylephorus chordatus* (λ_{max} 365nm, identified as kynurenine, Thorpe *et al.* 1992), ii – *Scopelarchus analis* (λ_{max} 403.5 mm, unidentified), iii – *Evermanella balbo* (λ_{max} 373 nm, unidentified, dotted line), iv – Intact lens of *Argyropelecus sladeni* displaying a profile typical of a carotenoid (Douglas & Thorpe 1992). All data are from Douglas (unpublished observations), except where indicated

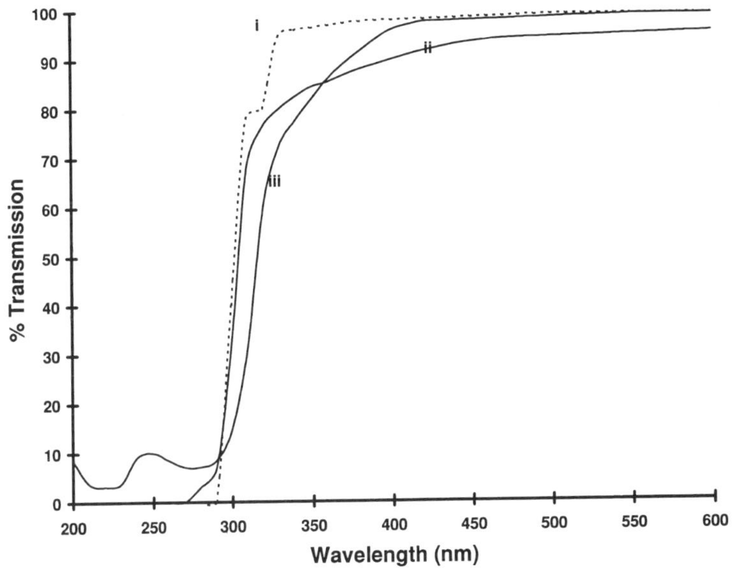

Figure 8 Spectral transmission of vertebrate humours. i – Ground squirrel aqueous humour (Chou & Cullen 1984, dotted line), ii – Human aqueous humour (Boettner & Wolter 1962), iii – Goldfish vitreous humour (Thorpe 1991)

4.2.5 Retina

(a) Blood Vessels The vascular supply to the retina differs markedly between species, ranging from an extensive capillary network throughout most of the retina to the complete absence of an intraretinal circulation (François and Neetens 1974). In animals with extensive retinal capillaries, such as many mammals, the blood, which has a pronounced absorbance maximum at around 420 nm, will inevitably act to some degree as a short-wave absorbing filter as it overlies the retina (Muntz 1972, Sivak and Roth 1978). Such filtering is likely to be uniform, as the vessels will be out of focus (Walls and Judd 1933a), and, due to the small amount of blood in the capillaries, its effect will be slight in comparison to the animal's overall dynamic range. Significant UV absorption by larger vessels has, however, been shown to protect the underlying retina from light induced damage (Collier *et al.,* 1989).

(b) Macular Pigment The central retina of diurnal primates, including man, contains a yellow region centred around the fovea, known as the *macula lutea* (Nussbaum *et al.,* 1981). The yellow colouration is due to a diffuse photostable short-wave absorbing pigment, located mainly within the Henle fibres and to a lesser extent in the inner plexiform layer (Snodderly *et al.,* 1984a,b).

Pigmentation is densest in the central fovea and declines towards the periphery (Snodderly *et al.,* 1984a,b, 1991; Weiter *et al.,* 1988; Abadi and Cox 1992).

Although there is some debate as to the exact form of the macular pigment absorption spectrum (Pease *et al.,* 1987; Stark 1987; Bone *et al.,* 1992), determined both spectroscopically and by psychophysical means, it is typical of a carotenoid (Figure 9, dotted line) and is thought to be a mixture of the xanthophylls, lutein & zeaxanthin (Bone *et al.,* 1985, 1988, 1992, 1993; Handelman *et al.,* 1988, 1991, 1992). As for other carotenoid-based pigments, such as retinal oil droplets (4.2.5c), it is possible to manipulate the level of macular pigmentation by dietary means (Malinow *et al.,* 1980). However, the frequently cited individual variation in the macular pigment density, is explained by both genetic and dietary factors (Hammond *et al.,* 1995). Absolute optical densities reported for humans range from 0 to 1.22, with average densities from 0.13 to 0.77 (Pease *et al.,* 1987 for review, also Werner *et al.,* 1987; Abadi and Cox 1992; Hammond and Fuld 1992; Wild and Hudson 1995).

A somewhat surprising feature of the macular pigment is that it is dichroic, possibly due to its close association with tubulin (Bernstein *et al.*, 1997), thereby causing the entoptic phenomena of Haidinger's brushes and Maxwell's spot (Nussbaum *et al.,* 1981; Bone and Landrum 1984; Snodderly *et al.,* 1984b), and possibly making it a more effective filter (Handelman *et al.,* 1991). The macular

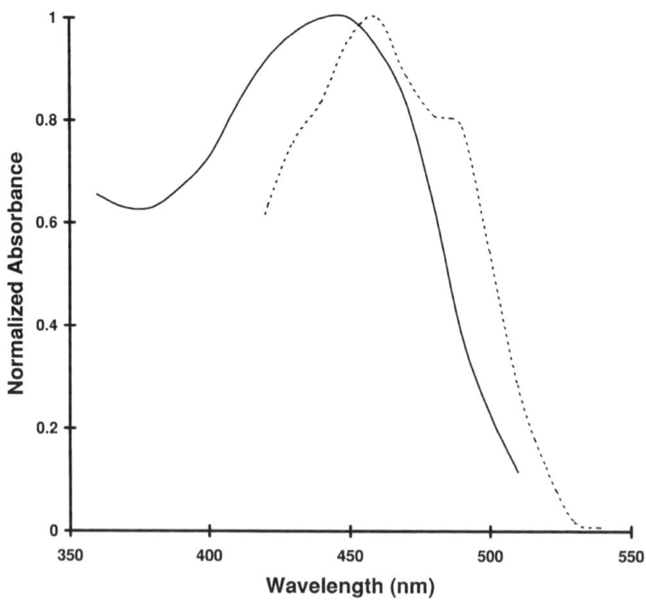

Figure 9 Absorbance spectra of photostable retinal pigments. i – yellow pigment extracted from the fish *Crenicichla lenticulata* (Muntz 1973, solid line), ii – human macular pigment (Vos 1972, in Bone *et al.,* 1992, dotted line)

pigment has two probable functions; the improvement of visual acuity by reducing the amount of chromatic aberration and atmospherically scattered light (section 4.2.7b), and the protection of the retina by both removing short wavelengths, which are those most likely to cause retinal damage, and by acting as a scavenger for cytotoxic free radicals (section 4.2.7a)

Although true macular pigment is restricted to primates, the retinae of some shallow (Muntz 1973, 1976a, 1982) as well as deep-water (Denton & Locket 1989) fish appear visibly yellow, due to the presence, possibly in the outer segments, of high concentrations of photostable short-wave absorbing pigments (Figure 9, solid line). As is the case for oil droplets (section 4.2.5c), the advantage of having the photostable pigment in the outer segment, is that it can act locally on individual receptors. Thus, in the deep-sea fish *Diretmus argenteus*, in a manner reminiscent of corneal eyeshade pigmentation (section 4.2.2), it is only the rods in the ventral retina, which receive the relatively brighter downwelling sunlight, that contain the photostable yellow pigment (Denton & Locket 1989).

The outer segments of another species of deep-sea fish, *Malacosteus niger*, also contain a photostable pigment (Bowmaker *et al.,* 1988; Partridge *et al.,* 1989). However, this absorbs maximally at 670 nm, which is at significantly longer wavelengths than the visual pigments of this species. Its function therefore, is not as a selective filter and it probably serves as a photosensitiser, enabling these animals to see their own far-red bioluminescence (Bowmaker *et al.,* 1988; Douglas *et al.*, 1998; see also section 4.2.6b).

(c) Oil droplets The ellipsoids of cone inner segments of many birds (e.g. Goldsmith *et al.,* 1984; Partridge 1989; Plate 1i), and reptiles (e.g. Liebman and Granda 1975; Ohtsuka 1985; Lipetz 1984b), as well as some amphibia (Hailman 1976), monotreme mammals (Ahnelt *et al.,* 1995), and 'primitive' (Chondrostei and Dipnoi) fish (Govardovskii *et al.,* 1992; Robinson 1994), contain clear or coloured oil droplets (Muntz 1972; Young and Pettigrew 1991; Bowmaker 1991b for reviews). Oil droplets are absent in placental mammals and teleost fish. Since this structure is situated immediately in front of the light absorbing outer segment, most light absorbed by the visual pigments (> 90% – Wortel and Nuboer 1986; 60% – Schneeweis and Green 1995) will pass through it, inevitably altering the spectral sensitivity of the photoreceptor. Oil droplets usually occur as a single large structure filling the whole outer part of the ellipsoid, but more rarely can be composed of several smaller droplets (Pedler and Boyle 1969). The accessory members of the double cones of some birds and reptiles, as well as the inner segment of some lungfish cones, do not contain an oil droplet as such, yet the ellipsoid region nevertheless contains a similar short-wave absorbing pigment (Bowmaker and Knowles 1977: Jane and Bowmaker 1988; Maier and Bowmaker 1993; Kolb and Jones 1982; Figure 10b). Oil droplets generally act as edge filters transmitting long wavelengths and absorbing short-wave radiation, with a sharp cut-off at some intermediate wavelength that determines the colour of the oil droplet (Figure 10a).

The flat top of most published oil droplet absorbance spectra (Figure 10a) results either from the limited sensitivity of the measurement system or from light leaking around the oil droplets. In turtles, for example, peak optical densities of intact droplets may be as high as 90 (Liebman and Granda 1975). However, thin layers or dilute extracts of oil droplets have absorption spectra typical of carotenoids (Liebman and Granda 1975; Lipetz 1984b; Goldsmith *et al.,* 1984; Figure 10b), as do the relatively lightly pigmented oil droplets and the droplet-free ellipsoids of some species (Bowmaker and Knowles 1977; Jane and Bowmaker 1988; Maier and Bowmaker 1993; Figure 10b). Thus, oil droplets are composed almost entirely of neutral lipids (Johnston and Hudson 1976) and dietary derived carotenoids such as zeaxanthin, astaxanthin and galloxanthin (Wallman 1979; Goldsmith *et al.,* 1984; Schiedt *et al.,* 1991; Bowmaker *et al.,* 1993).

Early workers attributed a variety of functions to oil droplets (Walls and Judd 1933a&b, Walls 1963, Muntz 1972, Wolbarsht 1976 for reviews), most notably perhaps that they were a means by which animals, assumed to have only a single visual pigment, could produce a system for colour vision (e.g. King-Smith 1969). However, many factors have since made this suggestion untenable, most obviously the discovery of more than one cone visual pigment in most animals, including those with coloured oil droplets, and the maintenance of colour vision following dietary oil droplet pigment depletion (Wallman 1979). Consequently, numerous other roles have been assigned to oil droplets. They might, for instance, perform some of the functions attributed to other intraocular short-wave filters (sections 4.2.7a&b), such as reducing chromatic aberration, increasing contrast and protecting the retina. Rather like the filtering pigments found in some fish outer segments (section 4.2.5b), they have an advantage over more global filters, such as the lens, as they act on individual photoreceptors. In general, the longer the λ_{max} of a cone's visual pigment, the more short-wave radiation is removed by the oil droplet associated with it (Ohtsuka 1985; Jane and Bowmaker 1988; Bowmaker 1991a&b; Bowmaker *et al.,* 1993; Maier and Bowmaker 1993). Thus, long-wave sensitive cones often have red coloured oil droplets, thereby exposing their outer segments only to 'optically advantageous' long wavelengths. UV-sensitive cones, on the other hand, have completely unpigmented oil droplets. Consequently, within a single retina, oil droplets are tuned to remove as much 'harmful' short-wave radiation from each cone type as possible, yet still allow an overall broad retinal spectral sensitivity.

Oil droplets, especially those that are transparent to UV and which therefore cannot perform a filtering role, could also act as microlenses to enhance outer segment photon capture (Walls and Judd 1933a; Young and Martin 1984). Alternatively it has been suggested that they contribute to polarisation sensitivity (Young and Martin 1984), and function as chemical storage organs (Hailman 1976). The most attractive function of coloured oil droplets, however, is that they enhance colour vision either by increasing the available number of spectrally

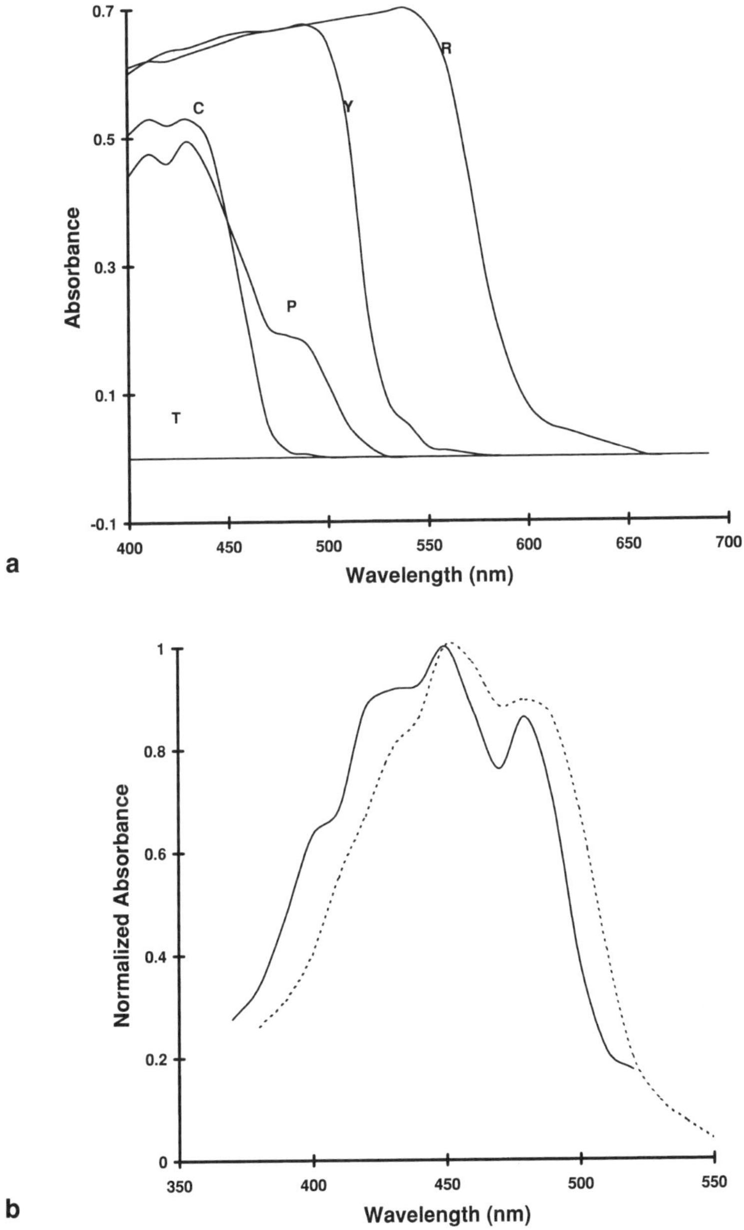

Figure 10 Spectral characteristics of oil droplets (**a**) Absorbance spectra of Duck (*Anas platyrhynchos*) retinal oil droplets; T – translucent, C – clear, P – pale, Y – yellow, R – red (Jane & Bowmaker 1988). (**b**) Absorbance spectra of; i – 'thinned' yellow oil droplets from the retina of the Turtle (*Emydoidea blandingii*) (Lipetz 1984b, dotted line), ii – the pigment from the oil droplet free accessory cone ellipsoid of the Peking robin (*Leiothrix lutea*) (Maier & Bowmaker 1993, solid line)

distinct photoreceptor types or by sharpening/tuning the response of existing spectral channels. Classically, the pigeon retina, for example, was described as containing 3 different cone visual pigments and 5 oil droplet types, which combined to form 6 spectrally distinct cone types (Bowmaker 1977, 1980). Similarly, the three visual pigments of the turtle that have been described by microspectrophotometry could combine with a variety of oil droplets to give at least six spectrally distinct cones (Schneeweis and Green 1995 for review). Recently, however, the picture, as far as the pigeon is concerned at least, has become somewhat more complex, since pigeons most likely have 5 cone visual pigments (λ_{max}s ca366 nm, 409 nm, 453 nm, 507 nm and 568 nm; Bowmaker *pers com*; Vos *et al.*, 1994). Furthermore, it is probable that not all the possible 7 cone/oil droplet combinations are used in colour vision. It is likely that only single cones, which contain their own exclusive oil droplet types, are involved in chromatic vision, with the double cones possibly coding for luminosity (Maier and Bowmaker 1993). Thus, the most likely function for oil droplets in this situation is not to increase the number of spectral channels available but to narrow existing spectral sensitivities of visual pigments. Such spectral 'tuning' has been shown both theoretically (Bowmaker and Knowles 1977; Bowmaker 1977, 1980; Govardovskii 1983; Jane and Bowmaker 1988) and practically (Martin and Muntz 1979; Neumeyer and Jäger 1985; Maier and Bowmaker 1993) and is accompanied by a shift in wavelength of peak absorption towards longer wavelengths, and a reduction in its overall sensitivity and may serve to enhance hue discrimination (Barlow 1982; Govardovskii 1983; Bowmaker 1991a). A modulatory role for oil droplets in colour vision is supported by the observation that colour vision is altered in birds whose oil droplets have been depleted of carotenoids (Wallman 1979), while visual pigments are unaffected by similar treatment (Bowmaker *et al.,* 1993).

Oil droplets in some (Muntz 1972, Lythgoe 1979), but not all (Jane and Bowmaker 1988) retinae, are unevenly distributed. The pigeon and a number of other herbivorous birds have a posterior dorsal retinal area (the region for looking forward and downward) dominated by single cones associated with red oil droplets (Pedler and Boyle 1969; Bowmaker 1977). The resulting regional variation in spectral sensitivity potentially allows fine spectral discrimination of green foliage (Bowmaker 1979; Martin and Muntz 1978, 1979; Remy and Emmerton 1989; Lythgoe 1979).

Light coloured sea-birds have a general abundance of red and orange oil drops in their retinae. It is argued that this enables them to cut through atmospheric haze and detect other light coloured birds, which may be forming feeding clusters, a long way off (Lythgoe 1979). For birds feeding from the surface red oil drops may also help to filter the scattered light from the surface, enabling them to see further into the water to detect prey (Muntz 1972). Other diving birds and birds which feed underwater possess relatively fewer long-wave filtering oil drops which may reduce sensitivity to the short wavelengths needed for prey detection beneath the surface (Lythgoe 1979 and see section 4.2.7). However, such ecological explanations for inter-specific variation in oil droplet content may not always be applicable (Begin & Handford 1987).

(d) Ellipsosomes Several teleost fish possess structures which superficially resemble the oil droplets of other animals (e.g. Berger 1966; MacNichol *et al.,* 1978; Nag 1995). Like oil droplets, these ellipsosomes are positioned within the outer segment of some cone types immediately before the outer segment. Although these structures resemble classical oil droplets in certain ways, they are in fact quite distinct organelles since they have, for instance, different staining properties (Kunz and Regan 1973, Nag and Bhattacharjee 1995), develop later (Kunz and Wise 1973, Nag and Bhattacharjee 1995), and possess certain components such as cristae-like structures (Kunz and Wise 1973; Borwein and Hollenberg 1973; Anctil and Ali 1976), which oil droplets do not. Ellipsosomes almost certainly have a mitochondrial origin and contain a pigment similar in its spectral absorbance characteristics to reduced cytochrome c (λ_{max} 415 nm) (MacNichol *et al.,* 1978; Avery and Bowmaker 1982; Figure 11). Since this pigment can have a peak optical density of 0.5 or more, it appears visibly pink in fresh preparations and will prevent much of the incident short-wave (400 – 430 nm) radiation impinging on the outer segments of cones possessing ellipsosomes. Thus, like oil droplets, ellipsosomes will alter the spectral sensitivity functions of the different cone types and possibly enhance wavelength discrimination (Avery and Bowmaker 1982; Bowmaker 1990). Although, true ellipsosomes, containing high densities of cytochrome c are restricted to the cones of teleost fish, primate photoreceptor inner segments have also been found to contain smaller amounts of similar photostable pigments (Snodderly *et al.,* 1984a; Bowmaker *et al.,* 1991).

(e) Visual pigments The visual pigments themselves and their photoproducts could act as selective chromatic filters, as light has to pass through them in order to get to the more scleral parts of the outer segments (Goldstein and Williams 1966, Muntz 1972; Alpern *et al.,* 1987). Although, photoproducts are probably too short-lived to have a significant influence in vertebrate receptors, the longer lasting invertebrate photoproducts may have a greater role to play (section 4.3.3). Self screening by the visual pigment, however, may be of some importance in vertebrates, especially if the optical density of the pigment is high and the outer segments are long. It may, for instance, play a significant role in some species of deep-sea fish which are known to have several layers of rod outer segments. Such animals, could theoretically have a form of colour vision based on just a single photopigment, as the spectral sensitivity of each layer of rods will be significantly different because of modification of the incident light by more vitread layers (Denton and Locket 1989).

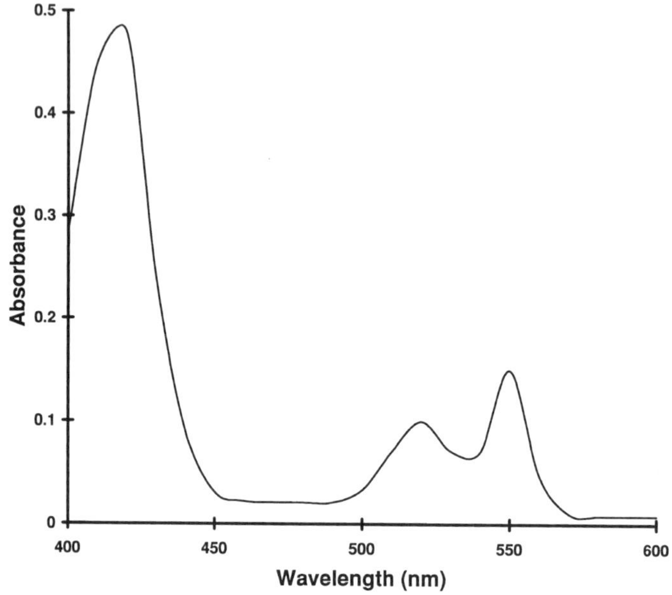

Figure 11 Mean absorbance spectrum of six *Anableps anableps* ellipsosomes. (Avery & Bowmaker 1982)

4.2.6 Postreceptoral filters

Much of the light that reaches the level of the retinal photoreceptors is not absorbed by the photopigments but passes beyond the rods and cones and strikes the retinal pigment epithelium (RPE). In most animals the RPE contains the pigment melanin, which absorbs most of this stray light. However, some species living in low light levels, have highly reflective tapeta, either within the RPE or choroid, that direct this light back towards the photoreceptors. Even in animals with no tapetum, a small amount of light will be reflected from the RPE onto the photoreceptors. As neither melanin nor tapetal reflecting pigments are spectrally neutral, they will act as wavelength selective filters.

(a) Melanin The main function of the melanin within the RPE is usually thought to be for the absorbtion of stray light, thereby, along with similar pigments in the choroid, ciliary body and iris, contributing to the eye's black box effect, thus enhancing image quality. However, melanin almost certainly also protects the retina from light damage, either by simply decreasing the amount of scattered light or, in a way analogous to the macular pigment, acting as a lipid anti-oxidant by scavenging free radicals and singlet oxygen (section 4.2.7a), although there is some suggestion that melanin might also actually contribute to retinal damage by being a cause of oxidative free radical stress (Sarna 1992 for review).

In 'lower' vertebrates, such as fish and amphibia, the melanin within the RPE, along with the photoreceptors, undertakes diurnal migrations, aggregating near the choroid in dark adapted conditions and expanding to fill most of the rod and cone layer, thereby completely covering the rod outer segments, during light adaptation (Wagner *et al.,* 1992 for review). These 'retinomotor' movements serve both to position the rods and cones optimally for respectively scotopic and photopic vision, and to protect the rods from excessive bleaching in daytime light levels (Douglas 1982).

The absorbance spectrum of RPE melanin has been measured for a number of vertebrates, most commonly humans, and generally absorbance increases with decreasing wavelengths (Geeraets and Berry 1968; Hunold and Malessa 1974; Boulton *et al.,* 1990; Figure 12a) contributing to greater reflection at longer wavelengths (Figure 12b). Although the predominant pigment in the RPE of most species is melanin, in some fish it is replaced by a red pigment (Ito *et al.,* 1975; Best and Nicol 1984; Figure 12a) leading to enhanced long-wave reflectance (Figure 12b).

(b) Tapetum lucidum Tapeta are widespread among nocturnal animals and those inhabiting dimly lit environments (Walls 1963; Nicol 1981). The most commonly held view is that tapeta serve to reflect light back through the retina, giving the photoreceptors a second chance of absorbing the light, thereby enhancing overall retinal sensitivity, often at the expense of image quality. However, in many cases the improvement in sensitivity is surprisingly small (Muntz 1990). Furthermore, Denton and Nicol (1964) showed that in deep-sea fish, the photoreceptors of species with tapeta had approximately half the visual pigment density of non-tapetal animals, resulting in a similar quantum capture in all species. The advantage of a tapetum in these animals is therefore not a simple increase in the number of photons absorbed, but might arise from a reduction in the amount of thermal noise resulting from lower visual pigment densities.

There are a variety of structural types of tapeta both within the RPE and the choroid (Walls 1963; Muntz 1972; Nicol 1981 for reviews) containing a number of different reflecting materials, most commonly guanine. Retinal tapeta are composed of either small densely packed reflective particles or stacks of thin reflective plates within the RPE. Choroidal tapeta are of two types; the tapetum cellulosum composed of tiers of reflecting cells, and the tapetum fibrosum, which is made up of densely packed connective tissue fibres, such as collagen. Tapeta are often restricted to the dorsal half of the eye, which receives the lowest intensity radiation. Some form of reflecting layer has been described in most vertebrate groups (Walls 1963), but specialised tapeta are particularly well developed in most classes of fish (Denton and Nicol 1964; Nicol and Arnott 1973; Locket 1977; Best and Nicol 1980; Somiya 1980; Nicol 1989 for reviews), many mammals (e.g. Dartnall *et al.,* 1965; Braekevelt 1993; Wen *et al.,* 1985) and some reptiles (Laurens and Detwiler 1921) and birds (Nicol and Arnott 1974). Since

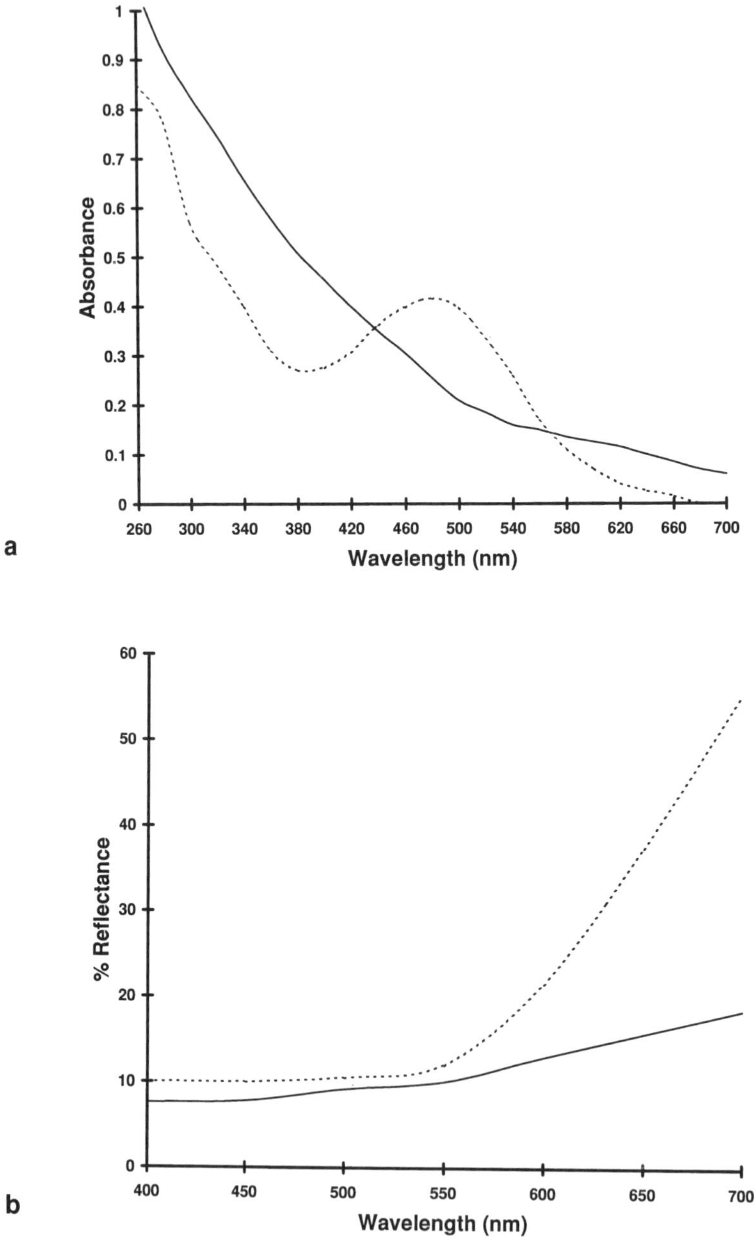

Figure 12 Spectral characteristics of retinal pigment epithelium (**a**) Absorbance spectra of; i – dissolved RPE melanin from humans aged 30 – 49 years (Boulton *et al.* 1990, solid line), ii – red pigment extracted from the RPE of the cunner (Best & Nicol 1984, dotted line), (**b**) Spectral reflectance of; i – human RPE and choroid (Geeraets & Berry 1968, solid line), ii - RPE of the chain pickerel (Best & Nicol 1984, dotted line)

tapeta are usually only beneficial in low light levels and may cause image degradation in brighter light, some fish have developed occlusable tapeta in which melanin migrates to cover the reflective material during light adaptation (Denton and Nicol 1964; Heath and Hindman 1988).

Surprisingly, given the wealth of structural and biochemical data, information on spectral reflectance characteristics of vertebrate tapeta are relatively sparse. However, it is clear, given the large range in colours of observable eyeshine, that spectra will differ between species (Figure 13 and see Nicol 1981 for review). Furthermore, significant differences between the eyeshine of members of the same species (Weale 1953) and even regional differences within a single retina (Wang *et al.,* 1980), are not uncommon. Spectral reflection curves, which can range from a wide spectral bandwidth, usually indicated by a white/silvery tapetum (Figure 13, curve i), to more narrowly defined spectra, giving rise to specific colours such as blue, green or red, have been obtained from; cat (Weale 1953), bush baby (Dartnall *et al.,* 1965), and various species of fish (Denton and Nicol 1964; Arnott *et al.,* 1974; Wang and Nicol 1974; Nicol *et al.,* 1975; Wang *et al.,* 1980, 1981; Figure 13).

It seems probable that the wavelengths reflected by a selective tapetum are those that are in some way most relevant to the animal. For example, in cartilaginous fishes, deep-water species have shorter wavelength reflecting tapeta than shallow water animals, which matches both the predominant wavelengths in the deep-sea and the shorter-wave absorbing visual pigments of the fish living there (Denton and Nicol 1964). Similarly, tapeta of fish living in freshwater, in which long wavelengths usually predominate, often reflect red light most efficiently (e.g. Wang and Nicol 1974). However, such simple relationships are far from universal (Nicol 1981), and in general, correlating an animal's environment with its visual system is fraught with difficulties, primarily because the environmental factors most relevant to an animal are usually not known in enough detail.

The deep-sea is optically a relatively simple environment, with a narrow spectral band of downwelling illumination centred around 470 – 490 nm and bioluminescent emissions which are often well characterised (section 4.2.7e). Furthermore, many deep-sea fish have comparatively simple visual systems using just a single visual pigment (e.g. Douglas *et al.,* 1995). Consequently, the spectral characteristics of deep-sea fish tapeta provide good examples of adaptations to specific functions. *Malacosteus niger*, for example, is unusual in having far-red bioluminescent emissions peaking around 706 nm, which are invisible to most animals in the deep-sea (Widder *et al.,* 1984; Denton *et al.,* 1985). To facilitate the detection of its own red bioluminescence, it has both long-wave-shifted visual pigments in comparison to most other deep-sea fish (Bowmaker *et al.,* 1988; Partridge *et al.,* 1989) which may be linked to a long-wave photosensitising pigment (section 4.2.5b), and a bright red, astaxanthin-based, long-wave reflecting tapetum (Denton and Herring 1971; Locket 1977; Somiya 1982; Bowmaker *et al.,* 1988; Figure 13 curve iii). A 'normal' deep-sea fish, such as *Diaphus rafinesquei*, on the other hand, has a tapetum, which appears blue, reflecting maximally around 460 nm (Figure 13 curve ii), that both matches the predominant wavelength of the remaining downwelling illumination, and the emission maxima of most bioluminescent stimuli (section 4.2.7e) and is similar to the absorption spectrum of this animal's single visual pigment (λ_{max} 488 nm, Partridge *et al.,* 1988).

4.2.7 *Function of short-wave absorbing filters*

As many vertebrate filters act primarily as short-wave absorbing filters, they probably have similar functions (Walls & Judd 1933a; Muntz 1972; Wolbarsht 1976; Kirschfeld 1982; for reviews). Unfortunately, many of the proposed functions remain speculative as they are difficult to prove experimentally. Furthermore, the various theories should not be seen as mutually exclusive, and many short-wave absorbing filters probably perform more than one function.

Most proposed functions for short-wave absorbing filters relate to the fact that, with the obvious exception of deep-sea species, animals with such filters tend to be diurnal, inhabiting relatively high light level environments. Furthermore, if a pigment is unevenly distributed, such as is the case in many fish corneas (section 4.2.2), it is the dorsal area receiving most illumination that is most heavily pigmented. Thus, if animals are exposed to high light levels, there seems to be some benefit derived from decreasing the intensity of short-wave radiation

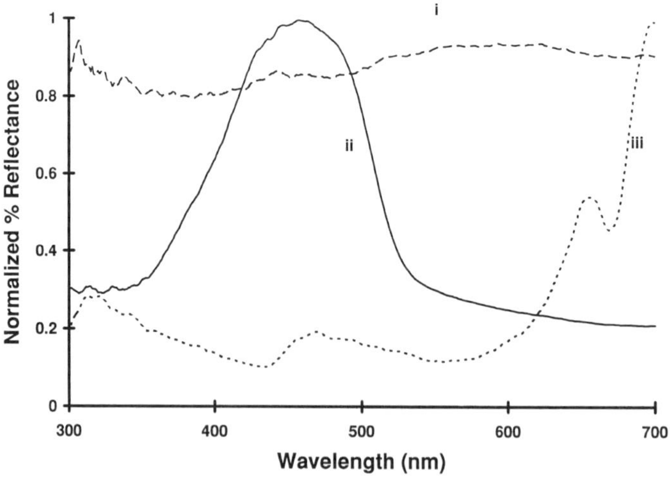

Figure 13 Examples of tapetal reflectance curves from some deep-sea teleosts. i – *Rinoctes sp.*, ii – *Diaphus rafinesquei* and iii – *Malacosteus niger*. All data from Marshall (unpublished observations). Data were obtained using intact eyes measured through aphakic gaps so include any corneal but not lens absorbance. Measurements were made with 'Sub-Spec' an Oriel Instruments/Andor Technology reflectance meter, a Xenon arc lamp and a 'Spectralon' white standard

reaching the retina. Nocturnal animals, or animals living in low light levels, on the other hand, generally do not have short-wave absorbing filters in their eyes, presumably because they would derive no benefit from doing so and would be severely disadvantaged by the inevitable reduction in sensitivity.

It should also be stressed that although short-wave filters are common in diurnal animals, many such species do not possess them, potentially enabling them to perceive UV wavelengths (Douglas and McGuigan 1989; Jacobs 1992). In general, the ocular media set the short wavelength limit to vision. For example, the human 'blue' cones could absorb significant amounts of radiation in the UV (Dartnall *et al.,* 1983). However, they are prevented from doing so by the absorption of these wavelengths by the lens. If the lens is removed, UV becomes perceptible to humans (Stark 1987; Griswold and Stark 1992). Surprisingly however, the possession of intraocular short-wave absorbing filters and UV vision are not always mutually exclusive since a species of diurnal gecko has both a yellow lens and a UV-sensitive retina (Ellingson *et al.,* 1995). In this case, the relatively low optical density of the pigment (1.3) probably represents a compromise, being low enough to allow sufficient UV to reach the retina to enable short-wave sensitivity, yet high enough to bestow on the retina the benefits of diminished levels of short-wave radiation.

(a) Protection Light is potentially damaging to the eye. The degree of actinic photochemical retinal damage in vertebrates, which is probably mediated through photoxidative processes (Ham *et al.,* 1984; Gerster 1991), is related to the wavelength of the incident illumination. In mammals, the incidence of retinal damage rises dramatically at wavelengths below 500 nm and can be particularly severe in the UV (Ham *et al.,* 1976, 1979, 1982; Organisciak and Winkler 1994 for review). UV has also been shown to have similar deleterious effects on the non-primate retina (Zigman and Bagley 1971; van Norren and Schellekens 1990; van Norren 1991). Any short-wave absorbing pigment, regardless of where it is located in the eye, will protect the retina from these potentially damaging wavelengths.

Both light and oxygen can be toxic to cells. Together their toxicity is enhanced (Ham *et al.,* 1984). Thus, the retina, more so than any other tissue, is very susceptible to such photoxidative damage, as light is focused on highly oxygenated cells. This light, apart from being absorbed by the visual pigment, will interact with various photosensitisers such as porphyrins, which are abundant within the retina, resulting in the production of free radicals and excited molecules, such as singlet oxygen, which can cause extensive cellular damage (Krinsky 1979; Kirschfeld 1982; Ham *et al.,* 1984; Handleman and Dratz 1986; Gerster 1991; Schalch 1992).

Carotenoids, such as β-carotene, are known to be highly effective quenchers of various stages of the photoxidative reaction and can prevent harmful reactions such as lipid peroxidation (Krinsky 1979; Burton and Ingold 1984). It is therefore likely that retinal carotenoids, such as the macular pigment, apart from simply absorbing the most damaging wavelengths, also protect the retina against photoxidative damage by acting as free radical and singlet oxygen scavengers (Kirschfeld 1982; Ham *et al.,* 1984; Handleman and Dratz 1986; Gerster 1991; Schalch 1992). Apart from the macular pigment, other carotenoids, such as those found in the oil droplets of many vertebrates (section 4.2.5c), and in some invertebrate retinae (section 4.3.3b), may have a similar protective role. Furthermore, it is possible, that the carotenoids in the corneas of some species (section 4.2.2) and in the human lens (Yeum *et al.,* 1995) could also help to protect these structures from photoxidative damage.

Although most of the proposed functions for short-wave absorbing filters are difficult to prove experimentally, there is some evidence supporting a protective role. Such a function for the primate macular pigment, for instance, is indicated by the observation that the loss of short-wave sensitivity normally associated with human ageing is higher in the relatively unpigmented periphery than in the pigmented foveola (Haegstrom-Portnoy 1988) and that the macular area is less prone to phototoxic damage than the periphery (Ham *et al.,* 1978; Jaffe and Wood 1988; Weiter *et al.,* 1988). Dietary manipulation of carotenoid levels also supports a protective role for the macular pigment (Snodderly 1995). When monkeys are deprived of carotenoids, their retinae show degenerative changes possibly related to the lack of macular pigmentation (Malinow *et al.,* 1980), while the experimental administration of carotenoids protects the mammalian retina to some extent from photochemical damage (Ham *et al.,* 1984; Tso 1989). A protective role for lens pigments is indicated by the observation that retinae of squirrel eyes which have had their highly pigmented lens removed, suffer considerable damage following exposure to UV radiation, while the retinae of intact companion eyes remain unaffected (Collier *et al.,* 1989).

(b) Enhancement of image quality Short wavelengths are not only potentially harmful to the retina, but they also degrade the retinal image in comparison to longer wavelengths. Their removal should therefore result in increased image quality.

The refractive index of an optical structure, which determines its focal length, will vary with the wavelength of incident light. Different wavelengths will therefore be focused at different levels within the retina. As short wavelengths are most prone to such chromatic aberration, their removal will increase image quality (Walls and Judd 1933a; Reading and Weale 1974; Muntz 1976a; Sivak and Bobier 1978).

Short wavelengths are also more prone to some forms of scatter than longer wavelengths. Scattering caused by small particles (Rayleigh scattering) is greatly increased at short wavelengths, while scattering from large particles is wavelength independent (Muntz 1976a). Light can be scattered by both substances within the eye and in the environment. While environmental wavelength-dependent Rayleigh scatter is widespread, the majority of intraocular scatter, at least in humans, probably arises from wavelength-independent scatter (Wooten and Geri 1987; van den Berg and Ijspeert 1995).

Short-wave absorbing filters will reduce the amount of Rayleigh scatter. This can have several beneficial effects. Firstly, the overall image quality will be enhanced. Secondly, in bright light, scatter will result in 'glare' and 'dazzle', which will be reduced by short-wave filters. Yellow intraocular filters can also increase the contrast and hence the distance at which objects become visible, by reducing the 'blue haze', caused by short-wave environmental scatter, intervening between an object and the observer (Walls and Judd 1933a; Walls 1963; Muntz 1972, 1976a).

Scattered light is a special problem for animals living underwater, as it can be caused both by the water molecules themselves and by small particles, such as plankton, suspended in the water column. This scattered light, will reduce the range at which objects will be visible underwater (Lythgoe 1975, 1976; Muntz 1976a) and might account for the widespread occurrence of short-wave filters in fish.

As well as enhancing the contrast of objects by removing the scattered 'veiling brightness', short-wave absorbing filters can also decrease the intensity of blue backgrounds, such as the sky and oceanic waters. This increases the contrast of light objects seen against such backgrounds and therefore improves their visibility (Walls and Judd 1933a; Walls 1963; Muntz 1972). A special case of such contrast enhancement is provided by deep-sea animals (section 4.2.7e).

(c) Functions unrelated to spectral filtering Just because a structure in the eye is not spectrally neutral in its absorption one cannot *a priori* assume that its primary purpose is to act as a wavelength selective filter. Vertebrate oil droplets, for example, might act as microlenses or serve as nutritive stores (section 4.2.5c), while carotenoids, such as the macular pigment, are scavengers of free radicals and singlet oxygen (section 4.4.1b), functions quite unrelated to their absorbance spectra. Furthermore, the corneal iridescence of some fish (section 4.2.2; Lythgoe 1976) and invertebrates (4.3.1a), might be primarily a device for camouflaging the eye and it has been suggested that the macular pigment could be a secondary oxygen carrying mechanism for the avascular fovea (Nussbaum *et al.,* 1981). Similarly, although retinal blood vessels will inevitably filter out some light (4.2.5a), this is obviously not their primary function. Cytochromes are also primarily involved in respiration, although they will filter light to some extent (4.2.5d). However, there is likely to be great selective pressure against any structure that significantly impairs visual performance and it is therefore likely that the wavelength-selective effect of these filters is either small or their filtering effect is beneficial to the system even if it is not their primary function.

(d) Functionless result of light exposure It is conceivable that some ocular filters serve no direct function and are simply the inevitable by-product of some

other process. For example, since ocular filters are usually found in animals living in high light levels, it is possible that they are the direct result of light exposure. The age-related 'yellowing' of the mammalian lens, for instance, is probably a consequence, at least in part, of prolonged exposure to short-wave radiation resulting in the accumulation of short-wave absorbing waste products which cannot be disposed of (section 4.2.3). It is certainly difficult to imagine a function for such decreased short-wave transmission with age. Similarly, a direct effect of light on goldfish corneal transmission is likely, since fish raised in high light levels have significantly more skin colouration and greater levels of corneal pigmentation than animals raised in lower light levels (Figure 14). Interestingly, Muntz (1973) also noted that the corneas of fish kept in captivity were much paler than those of fish caught in the wild.

Although light can have a direct effect on ocular media transmission, it is unlikely that short-wave absorbing pigments in general are a direct response to high light levels. For instance, it cannot account for the fact that some fish are born with highly pigmented lenses (section 4.2.3) and that primate macular pigment density changes little with age (Bone *et al.*, 1988). It is also inconsistent

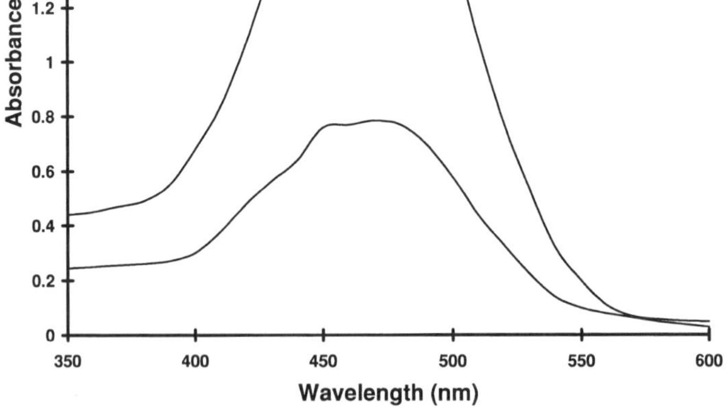

Figure 14 Relative spectral absorbance of acetone extracts of the corneal dorsal eye-shade pigment from *Carassius auratus* raised in different light levels; i – high light (1660 lux for 22 hrs out of every 24 hrs for 11 months, upper curve) and ii – low light (0.94 lux for 2 hrs out of every 24 hrs for 11 months, lower curve) levels. Traces are corrected for differences in extract volume and number of corneas in each sample. Although the degree of dorsal carotenoid eye shade pigmentation varied with rearing conditions, the MAA pigment (λ_{max} 325nm) present throughout the cornea was unaffected (Thorpe 1991)

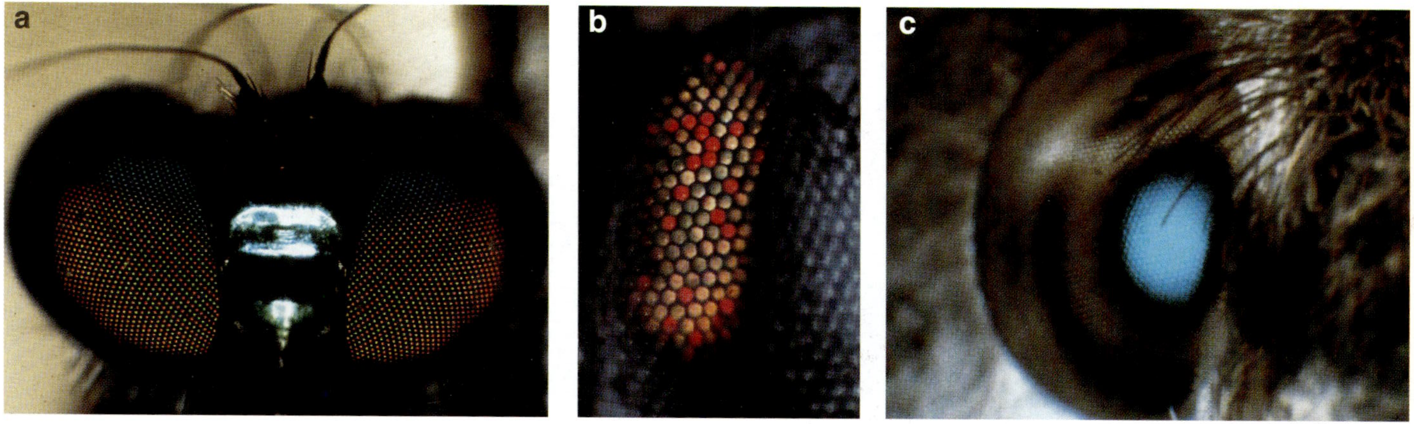

Plate 1 (**a**) Alternate rows of green and yellow facets in the eye of the dolichopodid fly *Condylostylus mundus*. These colours are produced by 1/4λ interference and may influence visual mechanisms in the retina beneath in a number of ways (Photograph – Gary Bernard). (**b**) Intensely coloured light re-emerging from the facets of the eye of the Carolina Satyr butterfly. This is a result of reflection from individual 1/4λ tracheal tapeta at the base of each photoreceptor (Photograph – Gary Bernard and see Figure 23). (**c**) Frontal aspects of the right eye of the diurnal hawkmoth *Macroglossum stellatarum* showing broad-band, sky blue tapetal eye glow (Photograph – Justin Marshall & Claudia Gunther). (**d**) Hemisected head of the dragonfly *Sympetrum* showing yellow pigment in the 'dorsal eye' for photoreconversion of metarhodopsin (Photograph Tom Labhart and see Labhart & Nilsson 1995). (**e**) The yellow lenses of the tube eyed deep-sea fish *Scopelarchus analis*, a possible adaptation for breaking ventral bioluminescent camouflage (Photograph – Justin Marshall). (**f**) Intrarhabdomal filters from the mid-band retina of stomatopod crustaceans, top two in transverse section and bottom two longitudinal section (Photographs – Justin Marshall). (**g**) The head of the tabanid fly *Tabanus lineola*. Alternate bands of colour are the result of 1/4λ interference filters situated in the cornea (Photograph – Gary Bernard). (**h**) Pigmented dorsal corneas in the eyes of two goldfish (Photograph –Ron Douglas). (**i**) Oil drops overlying cone outer segments in the retina of the starling *Sturnus vulgaris* (Photograph – Nathan Hart)

d
e
f
g
h
i

with the decreased levels of pigmentation observed in some older fish lenses (Figure 6). We have also tried raising several species of fish in different light levels, and while we have been able to influence both corneal pigmentation (see above) and retinal structure (Thorpe 1991), we have been unable to affect the level of lens pigmentation.

(e) Deep-sea animals; a special case The spectral region of highest intensity downwelling light in the deep sea is a narrow band between 470 and 480 nm (Jerlov 1976). Not only is the spectral bandwidth of light available for vision severely restricted, but its intensity also decreases rapidly with depth, until at around 1000 m insufficient sunlight penetrates to allow vision in even the most sensitive fish (Denton 1990). The yellow filters in the lenses (section 4.2.3) and retina (section 4.2.5b) of some deep-sea fish and cephalopods (section 4.3.1d) are therefore surprising, since they decrease both the spectral bandwidth and the intensity of the already restricted illumination even further, removing up to 80% of all downwelling light (Douglas and Thorpe 1992). They must therefore confer a significant adaptive advantage. Given the low level of illumination in the deep-sea and the high degree of convergence in the pure rod retinae of most deep-sea fish, it is unlikely that these filters serve to either protect the retina from excessive short-wave radiation or to enhance the quality of the image.

Many deep-sea animals produce their own light, and below about 1000 m this bioluminescence is the only visual stimulus available. In the mesopelagic zone (200 –1000 m), where both bioluminescence and downwelling sunlight are present, there is a potential conflict in tuning visual systems to the perception of these two sources. Although the wavelength of peak emission of most bioluminescent photophores lies between 450 – 500 nm, matching the wavelengths that most readily penetrate the water column, there are several exceptions to this and peak emissions at longer wavelengths are not uncommon (Herring 1983; Widder *et al.*, 1983, 1984). Furthermore, the emission spectra of the bioluminescence often contain more long-wave radiation than the surrounding spacelight. Short-wave absorbing filters will therefore decrease the intensity of downwelling sunlight more than that of the relatively long-wave rich bioluminescence, thereby enhancing the contrast of the bioluminescence and making it more visible (Somiya 1976; Muntz 1976b; Douglas and Thorpe 1992). The absence of significant amounts of intraocular pigmentation in the lenses of fish living deeper than 1000 m, where insufficient sunlight penetrates to provide a visible background (Denton 1990), is consistent with such a function (Douglas *et al.*, 1995).

Bioluminescence in the deep-ocean has a variety of potential uses, probably acting as; an intra and inter-specific signal, a means of startling predators, an attractant to prey, and as a means of simply illuminating their darkened world. Of all the proposed functions, however, perhaps the most interesting is the counter-illumination camouflage employed by many animals. A dark animal observed from below by a potential predator will inevitably cast a silhouette against the residual sunlight. Consequently, many animals have photophores on their ventral surface matching the intensity of the downwelling illumination, thereby obliterating this shadow. Yellow lenses would effectively break this counterillumination camouflage by removing much of the background light thereby enhancing the visibility of the bioluminescence and making the animal an easy and well illuminated target (Muntz 1976b).

4.3 Invertebrate ocular filters

Eye design in the invertebrates is more diverse than the standard pattern found in vertebrates and this gives rise to a larger variety of filter types. Simple eyes, similar in plan to the vertebrate eye, are found in cephalopods, spiders, some crustaceans and a variety of worms. Many crustaceans and insects, possess compound eyes made up of many units called ommatidia. These have two basic designs: a) The apposition compound eye where each ommatidium acts as an isolated optical unit and the cornea is assigned the job of focusing light on the photoreceptors beneath, b) The superposition eye where the dioptric parts of neighbouring ommatidia are shared in order to form a single image on the retina, either by reflection or refraction of light rays (Land 1981; Nilsson 1990).

At an even simpler level, filters may be found in eyes of unicells and the worm-like animals, often forming eye-cups to give the photoreceptors directionality. For many unicells, especially the phytoflagellates, the basic function of 'vision' is to enable phototaxes away from light to avoid damage or towards it for photosynthesis. Not surprisingly therefore the most complex unicell 'eyes' are found in the phytoflagellates. At their most complex these structures include cornea, lens, humour, retina, pigment cup and tapetum and may contain filters constructed from carotenoids, $1/4\lambda$ interference reflectors and other substances similar to those found in the higher organisms (Couillard 1984). Ocular filters have therefore probably been present since the moment vision evolved.

4.3.1 Cornea and lens

(a) Corneal interference filters In the hard bodied invertebrates, a cornea made of modified chitin is the first structure light passes through on its way to the photoreceptors. Though, like many vertebrates, often clear down to just above 300 nm (Figure 15), corneas of some dipteran insects, for instance, are spectacularly coloured (Plate 1g; Bernard and Miller 1968). Ommatidial facets may be coloured individually or more commonly grouped in clusters or stripes which may be red, orange, yellow, green or blue. Examples are known from 23 of the 60 families of diptera, most notably in the tabanids which, as with the example in Plate 1, generally have alternating stripes of coloured facets (Bernard and Miller 1968).

In common with the iridescent corneas found in some vertebrates (section 4.2.2), these intense colours are the result of thin film interference. Several alternating layers of high and low refractive index cuticle, whose thickness is close to the wavelength of light, are found just below the corneal surface. At every interface a certain fraction of the incident light is reflected back and undergoes constructive or destructive interference with light from other layers. The colour of the reflected light depends on the *optical* thickness of the layers (thickness × refractive index), the wavelength in question and the angle of viewing or illumination. For any wavelength (λ), the most efficient constructive interference, and therefore rejection of the colour, occurs when the thickness of the layers is $1/4\lambda$ (Land 1972). Only a fraction of light of the chosen wavelength is reflected at each layer, however, and a total of around 10 layers are needed for almost 100% reflection. This ideal situation of $1/4\lambda$ layer thickness and 6 –12 layers is often found in nature (Land 1972; Bernard and Miller 1968) and such coloured regions are commonly called $1/4\lambda$ reflectors.

Dipteran flies achieve a variety of corneal colours by adjusting the layer thickness to reject different wavelengths of light. Light not reflected by corneal interference filters is transmitted to the retina beneath and it has been postulated that one or more such wavelength selective filters could enhance colour contrast, photoreconvert metarhodopsins (see section 4.3.2) or even directly mediate colour vision (Lunau & Knüttel 1995). This function has yet to be positively demonstrated and critically depends on the photoreceptors beneath and their neural interconnections. Diptera possess ‘open rhabdoms’ with a central pair of photoreceptors surrounded by 6 others arranged in a trapezium. These six rhabdomeres (R1-6) are used for spatial vision (Hardie 1986) and in *Musca, Drosophila* and a variety of other species contain a single visual pigment with maximum sensitivity typically around 490 nm. The central two photoreceptors (R7 and R8) are usually more complex, containing different visual pigments that constitute the fly’s colour vision system (Hardie 1986; section 4.3.3b). Any one of these photoreceptor types in flies with coloured corneas would have their spectral sensitivity altered by overlying filters, the different colours possibly increasing the number of spectral channels or tuning existing ones. As the photoreceptor spectral sensitivities of any dipteran with a coloured cornea have not been determined, this possibility has not been fully investigated. Alternatively, as the colours in a variety of compound eyes are often arranged in stripes or other patterns, they may simply be for camouflage or display (Stavenga 1979; Lunau and Knüttel 1995; Figure 21b and Plate 1g).

One example is known in which the underlying retinal structure is different depending on the colour of the overlying cornea, strongly suggesting a real visual function for the colour filters. The ‘long-legged’, Dolichopodid flies possess rows of facets alternately coloured red and yellow/green (Bernard and Miller 1970; Trujillo-Cenóz and Bernard 1972; Plate 1a). In ommatidia with red facets the R7 cells of the central pair of photoreceptors have microvilli oriented

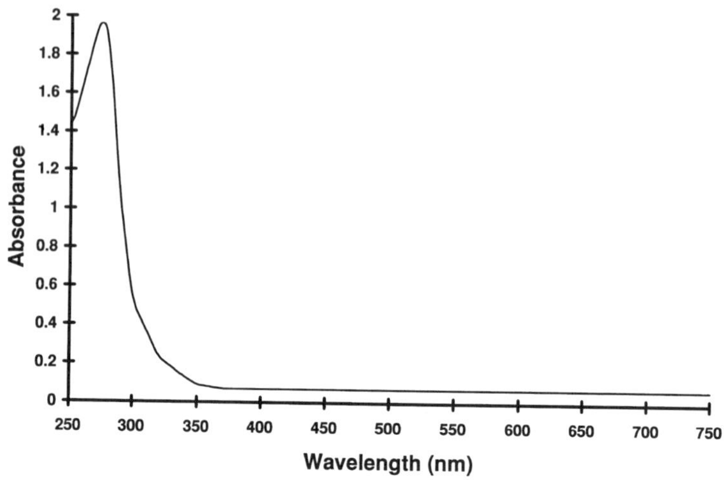

Figure 15 Spectral absorbance of the cornea of a moth showing good transmission to 300 nm (after Miller 1979)

vertically and those beneath yellow/green facets are oriented horizontally (Trujillo-Cenóz and Bernard 1972). The exact spectral sensitivities of these cells are not known, but two plausible suggestions for the function of this filter/photoreceptor combination have been made.

Trujillo-Cenóz and Bernard (1972) suggest these are the first elements of a colour enhancement system and that red facets with vertical microvilli would be good for prey detection against shiny, horizontally polarized surfaces (and see Land 1993a). Photoreceptors with vertical microvilli are less sensitive to horizontally polarized surfaces such as ponds or shiny leaves, due to the intrinsic dichroism of the microvilli, such that maximum photopigment absorption occurs when light is polarised parallel to the microvilli. Alternatively, Land (1993b) suggests, that the eyes of *Poecilobothrus nobilitatus* are specialized Dolichopodid detectors and aid their elaborate sexual and territorial displays. *P. nobilitatus* bodies are coloured with shiny metallic interference colours which also polarize the light reflected from their surface. As the colour seen on the body depends on the angle of view (as it does with all interference colours), different colours will also be polarized to various degrees, depending on the reflection angle. A set of photoreceptors which combine a colour difference signal with a polarization difference signal may be ideal for spotting bodies reflecting different colours and E-vector intensities.

(b) Corneal nipple arrays The corneal surface of the compound eyes of many nocturnal or crepuscular moths and some diurnal butterflies is studded with thick forests of small nipples (Bernard 1967; Figure 16). The dimensions of these

structures are around 1/2λ of light in the middle of the spectrum and they render the eyes non-reflective. Man-made lenses and eyes with smooth corneas, exhibit specular reflection of light due to the sudden interphase between air and the higher refractive index glass or chitin. One way round this is to graduate the change in refractive index from outside to inside and this is exactly what the conically-shaped corneal nipples do. The small size of corneal nipples relative to the wavelength of light is critical, as this means they do not interfere with the image-forming action of the lens (Bernard *et al.,* 1965). A gradual refractive index change is hard to achieve in photographic and optical lenses, so lens coating is used as a compromise. A 1/4λ layer of material, whose refractive index is between that of air and glass, is used to create reflections at the air/coating, coating/glass interfaces, which eliminate each other by constructive interference and cause improved transmission.

The function of the nipple arrays seems to be twofold: sensitivity increase and camouflage (Bernard 1967). The presence of nipple arrays mainly on the corneas of nocturnal insects such as moths (Lepidoptera) and netwings (Neuroptera) suggests they may help in dim light vision. Light, which would be reflected by a smooth cornea, is absorbed by these eyes. Although this is only around 4 – 5% of the total light incident on the eye (Bernard 1967; Muntz 1972), for an insect working at the threshold of vision this is a real advantage.

The shiny eyes of diurnal insects such as dragonflies and bees are conspicuous and may become relatively more so in dim light. Many lepidoptera hide themselves on coming to rest by folding their wings to reveal a camouflaged pattern. A large shiny eye would ruin this camouflage and therefore lusterless eyes will no doubt help when hiding from possible predators. Camouflage is probably the main function of corneal nipples where present in diurnal lepidoptera, as any relative sensitivity increase for these species would be negligible.

a 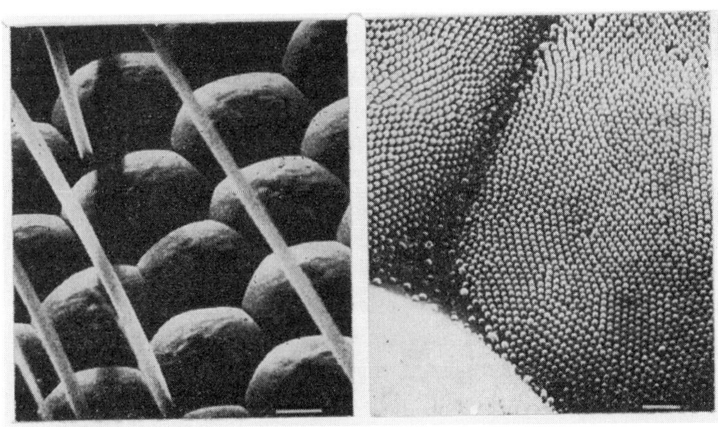 b

Figure 16 Scanning electron micrographs (SEM) of the eye of a night moth. (**a**) The hexagonal facet array (scale 10 μm). (**b**) Detail of two facets showing corneal nipples (After W.H. Miller *et al.* 1966)

(c) Scattering filters for polarisation vision This category of corneal filters is found associated with areas of compound eyes specialized for the detection of polarized light. These are the 'dorsal rim' areas found in many hymenopteran and a variety of other insects (Rossel and Wehner 1984; Labhart *et al.,* 1992). Ommatidia from this region bear microvilli specifically arrayed for efficient absorption of polarised light, an inherent property of microvillar receptors (Snyder *et al.,* 1973; section 4.3.1a). Along the dorsal rim, microvillar direction varies in a fan-like pattern and this can be matched directly to the pattern of polarised light in the sky at various times of the day (Rossel and Wehner 1984). In this way insects such as ants and bees, for instance, navigate over long distances from home to food sources and back.

The corneal facets of the dorsal rim in the cockchafer, for example, are cloudy or grey in appearance (Labhart *et al.,* 1992). This is due to the presence of small air bubbles or inhomogeneities in the cornea, which scatter light, making the facets seem opalescent. As a result the acceptance angle of the underlying ommatidia is increased in a manner similar to the action of cosine receivers found on some light meters. The function of this, and other adaptations which increase acceptance angle in the dorsal rim ommatidia, seems to be to extend the field of view of the receptors over much of the available sky, rather than increase sensitivity (Labhart *et al.,* 1992). In this way more of the available polarised sky pattern is analysed, and localised disturbances in the pattern, such as clouds or tree branches, are integrated out of the final signal. Fine resolution in this eye area is unimportant.

(d) Lens filters of cephalopod simple eyes Short-wave absorbing pigments are common in various parts of the vertebrate eye (section 4.2). Since the simple eyes of cephalopods (squid, cuttlefish and octopus) are remarkably convergent with the vertebrate design (Packard 1972), the presence of varying degrees of pigmentation in the lenses of cephalopod eyes resulting, as in vertebrates, in widely differing UV transmission between species (Figure 17a), is perhaps no surprise.

It has been suggested (Denton and Warren 1968) that, as for fish (4.2.3), deeper living squid usually have less pigmentation in their lenses than those living nearer the surface. However, in a manner also reminiscent of some deep-sea fish (section 4.2.3), and presumably for similar reasons (section 4.2.7e), the most pigmented lens belongs to *Histiotheuthis meleagroteuthis*, a mesopelagic animal living at depths of 500 –1000 m (Denton and Warren 1968; Muntz 1976b). This species is unusual in having eyes of differing size; a large eye, with a more heavily pigmented lens, that may be orientated in an upward pointing direction and could therefore be involved in breaking bioluminescent camouflage (section 4.2.7e), and a smaller, less heavily pigmented, downward facing eye.

The cephalopod lens splits easily into three different regions; an anterior and a posterior cortex and a central nucleus. In *Histiotheuthis meleagroteuthis*, pigmentation is densest in the posterior portion of the lens (Muntz 1976b), while in

Sepia officinalis and *Loligo forbesi* pigmentation is most intense anteriorly (Thorpe 1991).

Thin sections of *Histiotheuthis meleagroteuthis* lens indicate that the heavily pigmented posterior portion of the large lens contains a pigment absorbing maximally at just over 400 nm, while the more anterior areas of the large eye and all of the less pigmented small eye, contain a pigment with λ_{max} at 385 nm (Muntz 1976b) (Figure 17b). Extracts reveal that most areas of the *Sepia officinalis* lens, on the other hand, contain a pigment with λ_{max} around 360nm (Figure 17b), although the anterior cortex may contain one or more additional pigments. HPLC purified extracts of all parts of the *Loligo forbesi* lens show a more complex absorption profile with a main peak at 320 nm (Figure 17b). The identity of all these pigments is unknown. There is some evidence that the pigment in *Histiotheuthis meleagrotheuthis* accumulates as a function of age (Denton and Warren 1968; Muntz 1976b). Similar increases in lens pigmentation have been observed in *Sepia officinalis* (Thorpe 1991).

4.3.2 Filters between the crystalline cones

In many apposition compound eyes, such as those in crabs or bees, pigment is found between the crystalline cones, housed in the primary (distal) pigment cells. Its main function is to help optically isolate ommatidia and thus maintain image quality. Its colour and broad absorbance has a similar function to melanin in the vertebrate eye (section 4.2.6a), however invertebrates generally use the dark brown ommachrome pigments for this job (Figure 18; Hallberg and Elofsson 1989; Stavenga 1979). This is what gives the eyes of bees, crabs and other invertebrates their dark brown or black appearance. Pigment between the cornea and retina in superposition eyes may directly adjust image brightness over 2 – 3 log units by varying the extent to which they curtain off individual ommatidia (Nilsson 1990; Warrant and McIntyre 1996).

Primary pigment in a number of insects and crustaceans may be coloured and arranged in spots or stripes over the eye. A possible function for these patterns is camouflage or display (Stavenga 1979). Examples of this can be found in various grasshoppers, flowerflies, some praying mantids, porcellainid crabs and mantis shrimps (Figure 21b). It is not known if these colour patterns have any real visual function as is supposed for corneal interference colours (section 3.3.2). Some incomplete evidence does exist for the clear and brown striped eye of the grasshopper *Phlaeoba* in which lateral filtering by this pigment may differentiate photoreceptors into two types (Kong *et al.,* 1980; section 4.3.3).

In a few instances, brown ommochrome pigments between the crystalline cones have been replaced by a more selective filter which leaks light beyond 550 nm, giving the eye a yellow or orange hue. This apparently poorly designed filter is actually part of an elegant strategy to increase sensitivity in these ommatidia. Pigments like these are found only in specialised dorsal regions of

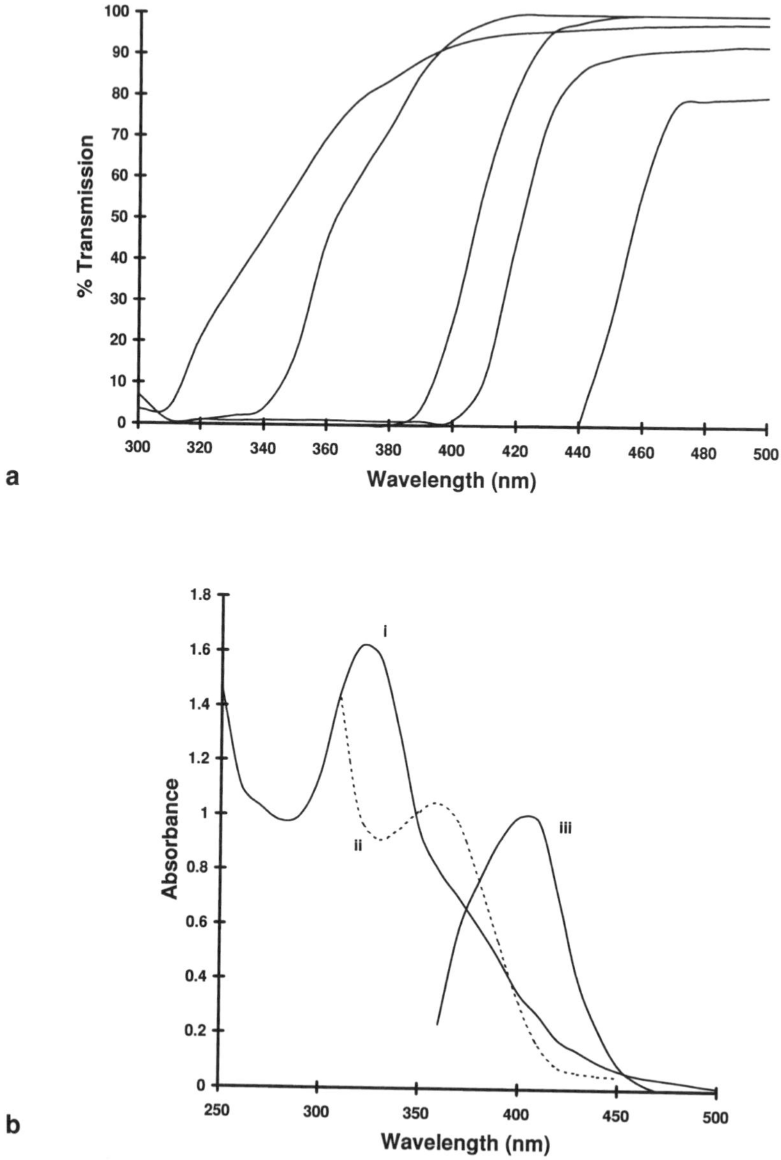

Figure 17 Spectral characteristics of intact cephalopod lenses. (**a**) Whole lens transmission spectra. (left to right); i – *Pyroteuthis* sp. (lens diameter along optical axis 5.26 mm; Douglas unpubl), ii – *Alloteuthis subulata* (3.38 mm; Thorpe 1991), iii – *Sepia officinalis* (9.05 mm; Thorpe 1991), iv – *Todarodes sagittatus* (13.6 mm; Douglas unpubl), v – Large eye of *Histioteuthis meleagroteuthis* (ca 10 mm; Muntz 1976b). (**b**) Absorbance spectra of lens pigments; i – HPLC purified lens pigment of *Loligo forbesi* (Thopre 1991), ii – unpurified extract of posterior cortex of *Sepia officinalis* lenses (Thorpe 1991), iii – thin slice of the posterior cortex of a *Histioteuthis meleagroteuthis* lens (Muntz 1976b). Differences between curves in absolute absorbance are arbitrary

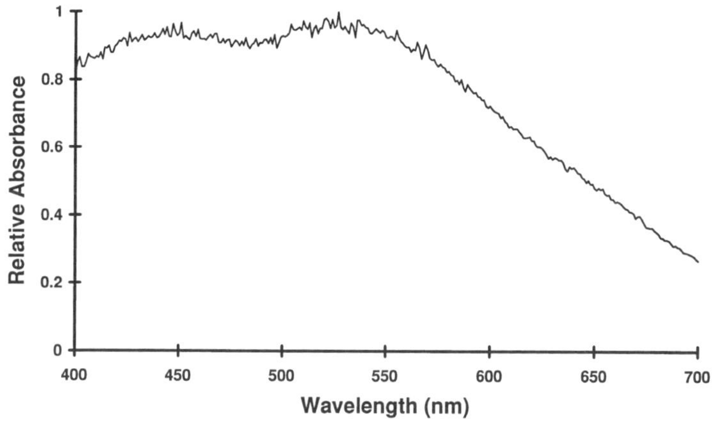

Figure 18 Broad-band absorbance of ommachrome, the dark screening pigment found in many invertebrate eyes (Cronin & Marshall 1989b)

compound eyes which may exist only in the male of the species. The dorsal regions consist of enlarged ommatidia, sometimes entirely separate to the rest of the eye, and are used for the difficult task of detecting small flying objects such as prey or mates against the sky (Figure 21c). Prominent examples are found in mayflies (Ephemeroptera – Horridge 1976; Horridge and McLean 1978), dragonflies (Odonata – Labhart and Nilsson 1995; Figure 19; Plate 1d), owlflies (Neuroptera – Schneider and Langer 1975), simuliids (Diptera – Kirschfeld & Wenk 1976), bibionids (Diptera – Zeil 1983) and drone honey bees where the pigment is red but performs the same function (Hymenoptera – Bertrand *et al.*, 1979; for reviews see Wehner 1981; Land 1989).

For good image detection against the predominantly blue sky, the photoreceptors of dorsal eye regions contain visual pigments with blue or UV sensitivity. All invertebrates have bistable visual pigments: when a molecule of visual pigment (rhodopsin) absorbs a photon, it is converted to a thermostable metarhodopsin which can in turn be reconverted back to rhodopsin by the absorption of another photon. Visual pigments with a short wavelength maximum sensitivity (λ_{max}) usually have metarhodopsins whose λ_{max} is shifted to longer wavelengths (Labhart and Nilsson 1995; Figure 19). For instance in the drone honey bee, the λ_{max} of the dorsal eyes rhodopsin is 446 nm and it's metarhodopsin peak absorption is at 505 nm (Menzel *et al.*, 1991). For the owlfly the equivalent rhodopsin/metarhodopsin λ_{max} values are 345 nm and 475 nm (Hamdorf *et al.*, 1973). Metarhodopsin is 'deactivated' in the sense that it has no affinity for molecules in the phototransduction cascade which convert light to nervous energy.

For light to be seen, as much rhodopsin as possible is needed in the photoreceptors. The yellow, orange or red screening pigments are used in order to 'leak' long wavelengths throughout the dioptric system, and indeed the rest of the eye including the retina (section 4.3.3a describes similar pigments in the retinular cells). This light is absorbed by the long λ_{max} sensitive metarhodopsin, reconverting it to rhodopsin, and is at the same time not visible to the short λ_{max} rhodopsin (Figure 19). The eyes' optical resolution is not degraded, as the yellow/red pigment is an effective screen for the short wavelength sensitive rhodopsin and simultaneously the rhodopsin/metarhodopsin equilibrium is driven almost 100% towards the rhodopsin side under natural conditions. For the dragonfly *Sympetrum*, this increases the sensitivity of the eye by about 0.4 Log units, a critical advantage for an eye straining to see small dark objects against the sky (Labhart and Nilsson 1995).

4.3.3 Retina

Invertebrate photoreceptors (rhabdoms) are typically constructed from relatively long 'rods' of microvilli, contributed to by several retinular cells. These may be fused together in one unit (crustaceans, cephalopods and insects such as bees, butterflies and dragonflies) or exist as separate entities (dipteran insects and a few crustaceans such as the isopod *Ligia*). Once light enters the rhabdom it is kept from leaving the photoreceptor by its inherent light guiding properties (Snyder 1975; Snyder and Horridge 1972) or by retinular cell pigment, usually ommochrome, drawn close to the rhabdom. As well as preventing light entering neighbouring photoreceptors, retinular cell pigment 'bleeds' light from the rhabdom and regulates light flux within it. This longitudinal pupil mechanism contributes to the adaptation of photoreceptors which may be required to respond to intensity ranges of as much as 5 log units over the course of the day (Land 1981).

By replacing the broad-band ommachrome sheath with coloured pigments, light travelling down photoreceptors can be spectrally modified. The pigment acts as a lateral filter, absorbing some wavelengths preferentially and reflecting others back into the rhabdom, thus modifying the spectral sensitivity of the photoreceptor itself. This is a relatively inefficient filter, as only a small proportion of the light within the rhabdom is filtered. In rhabdoms less than 5 μm in diameter, however, waveguide effects mean a substantial fraction of light travels outside the central photoreceptor (Snyder *et al.*, 1973; Snyder 1975). This places the light within the lateral screen itself, making lateral filtering more pronounced, especially if the rhabdom is long, and effectively like a serial filter.

To maximise sensitivity, many invertebrate photoreceptors are long. This however introduces the potential problem of self screening, where the rhodopsin and metarhodopsin of distal regions, filters light reaching the more proximal zone. In a single long photoreceptor, this invariably means the overall sensitivity

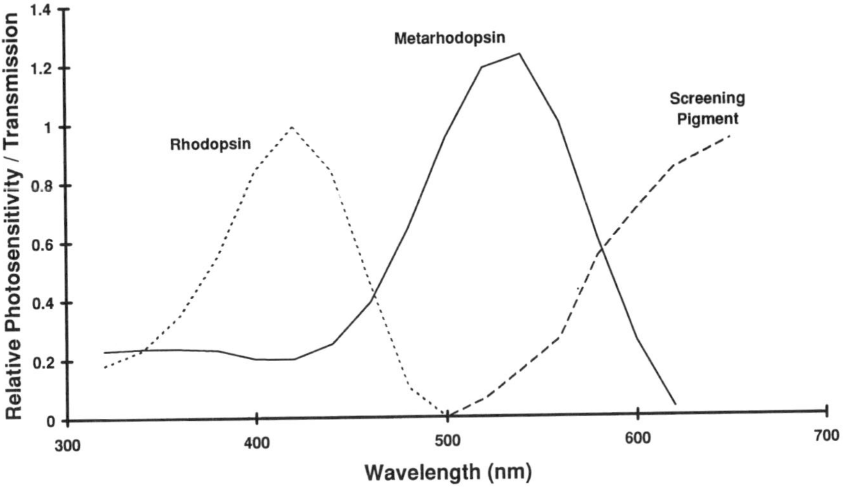

Figure 19 Comparison of rhodopsin and metarhodopsin sensitivities and the transmission of the 'leaky' screening pigment in the Dragonfly *Sympetrum.* The light let through by this yellow pigment (see Plate 1) reconverts metarhodopsin to rhodopsin, while still acting as an effective screening pigment for the rhodopsin, and increases photoreceptor sensitivity (After Labhart & Nilsson 1995)

broadens considerably (and may be modulated by the different peak absorption of metarhodopsin – see section 4.3.2), a desirable feature perhaps for a system seeking general sensitivity but undesirable if spectral sensitivities must be preserved (Snyder *et al.,* 1973).

Much of the remainder of this section examines serial and lateral filters that tune the spectral sensitivity of invertebrate photoreceptors. Firstly however, it may be recalled from section 4.3.2 that a variety of compound eyes, such as the dorsal eye of male owlflies, are specifically adapted for the reconversion of metarhodopsin to rhodopsin in order to increase sensitivity. This adaptation requires yellow, orange or red distal pigments to leak long wavelength light throughout the eye. The same pigments are present in retinular cells, extending the extent of leakage throughout the whole eye including the retina, accelerating photoreconversion (Stavenga 1996).

(a) Lateral screens The fly (Vogt *et al.,* 1982), pierid butterflies (Ribi 1979), the digger wasp *Sphex cognatus* (Ribi 1978), fireflies (Lall *et al.,* 1982, 1987), stomatopods (Marshall *et al.,* 1991b) and a variety of crabs (Leggett 1979; Stowe 1980; Lall and Cronin 1987) have all been shown to possess coloured pigment in close apposition to their rhabdoms and may use these to alter the spectral sensitivity of their photoreceptors. In crabs it has been implied that different coloured lateral screens could work in conjunction with a single visual pigment

to construct colour vision (Leggett 1979; Hyatt 1975; Stowe 1980). This has yet to be convincingly demonstrated and it seems more likely that the job of lateral filters is to tune already existing spectral sensitivities derived from different visual pigment types.

Fireflies present a particularly interesting example, where the colour of lateral screening pigments varies from yellow through pink, magenta and red in different species. Pigment colour is correlated with the species peak activity time around dusk (Lall *et al.,* 1982, 1987). The filter colours sharpen the spectral absorption characteristics of an existing green photoreceptor (λ_{max} 550 nm), maximising the individual species' sensitivity to the bioluminescent emissions of their conspecifics. The absorption profile of the yellow pigment, which is present in the night active *Photinus* sp, is strongly indicative of a carotenoid pigment, while the chemical identity of the red pigments remains unknown (Lall *et al.,* 1987).

The retinular cell pigment in the mid-band photoreceptors of the mantis shrimps (stomatopods) are often coloured differently and provide a good example of how such spectral sharpening works (Marshall *et al.,* 1991b). This highly specialized compound eye region contains more than ten visual pigments, and in concert with a variety of filter types, constructs many sharply tuned spectral sensitivities for colour vision (Marshall 1988; Cronin and Marshall 1989a & b; Marshall 1994; for further explanation see section 4.3.3b). Coloured lateral screens are made from retinular pigment or oil droplets which may be yellow or red in different mid-band rows. Stomatopod photoreceptors in the mid-band are long to increase sensitivity (Marshall and Land 1994). A penalty paid by long photoreceptors is broadening of the photoreceptors spectral sensitivity through self screening (Snyder *et al.,* 1973), an undesirable quality in a chromatic system with many colour channels to fit within the spectrum (Bowmaker 1983). One way to counteract this is to couple optically the photoreceptor to a structure, such as a lateral filter sheath, which absorbs wavelengths offset from the visual pigment λ_{max} and which therefore 'extracts' the broadening wavelengths from light in the photoreceptor. There may also be a shift in the λ_{max} position of the photoreceptor, as occurs in the firefly, depending on the spectral absorption of the filter and the visual pigment it combines with (Marshall *et al.* 1991b). Section 4.3.3b details why such precise spectral sensitivity positioning is critical for stomatopods.

The longitudinally fused rhabdoms in bees or butterflies (Menzel 1981) result in a similar spectral sharpening of their photoreceptors by using visual pigments in place of inert coloured absorbing pigments. Their photoreceptors abut each other along their entire length and are therefore optically coupled. They contain visual pigments of different spectral sensitivity (the bee has a UV (340 nm), blue (440 nm) and green (540 nm) receptor for example) which behave as lateral colour filters for each other. The broad spectral sensitivities expected if these long rhabdoms were separate become three sharply tuned responses with far less spectral overlap and redundancy (Figure 20a). Both here, and where coloured screening sheaths are used, photoreceptors maintain high absolute sensitivity while sharp individual spectral identity is retained for good colour contrast (Snyder *et al.,* 1973).

(b) Photoreceptor tiering, intrarhabdomal and other serial filters The λ_{max} of photoreceptors can be both sharpened and shifted by serial filters, again largely in order to tune for increased colour contrast rather than create colour vision systems. The penalty paid for this arrangement is a loss of sensitivity in proximal areas due to the light absorbed by the filter above. For animals operating in plenty of light, however, this may not be a problem (Snyder *et al.,* 1973; Marshall *et al.,* 1991b; Cronin and Marshall 1989b). The filter may consist of only the overlying visual pigment in tiered photoreceptors, photostable pigments or both. Where photostable coloured pigments are used, they may be discrete filter units, similar to the coloured oil drops in vertebrate eyes (section 4.2.5c), or dispersed throughout the photoreceptor membrane.

The central photoreceptors of the open rhabdoms of fly eyes (R7 and R8) are tiered and in some instances contain a photostable pigment. This pair of photoreceptors are not a homogenous population throughout the eye but come in 4 different types assigned a variety of functions such as colour vision (two types), sensitivity to polarized light and mate detection (Hardie 1986). R7 overlies R8 so light entering R8 is filtered by whatever is found in R7.

In one of the four R7R8 variants, called R7yR8y due to the yellow colour of light they transmit, R7y contains a remarkably complex cocktail of pigments consisting of two rhodopsins (λ_{max}s at 430 nm and 340 nm) and a blue absorbing photostable pigment dispersed throughout the photoreceptor microvilli (Kirschfeld *et al.,* 1978, 1988). The measured sensitivity of R7y at 355nm is the result of an interaction between all three pigments. Direct absorption of light by the blue sensitive 430 nm visual pigment is largely blocked by the filtering action of the photostable C_{40} – carotenoid pigment (zeaxanthin and/or lutein) the same pigments found in the macula lutea of vertebrate eyes (section 4.2.5b). The high UV sensitivity of the cell (Figure 20b) therefore comes from the 340 nm visual pigment which transfers its absorbed energy to the 430 nm pigment. It is argued that this antennal mechanism increases the sensitivity of the photoreceptor by broadening its spectral sensitivity slightly (Kirschfeld *et al.,* 1988). The carotenoid pigment may also protect the R7y cell from damaging UV radiation (Zhu and Kirschfeld 1984), an important function suggested for similar pigments in vertebrate and cephalopod eyes (section 4.2.7a).

The spectral sensitivity of the R8y photoreceptor is the result of filtering by both visual pigments and to a large extent the carotenoid in R7y, which changes the wide R8 rhodopsin peak at 520 nm to a narrow one at 530 nm (Figure 20b). This, and R7pR8p cells (p stands for pale – the transmission characteristics of these rhabdoms), which have relatively simple sensitivities based only on visual pigments with λ_{max} values at 335 nm and 460 nm respectively, provide the basis of colour vision in the fly (Troje 1993). R8p is filtered by the R7p visual pigment, but only on its short wavelength limb, resulting in little change.

Stomatopods from the gonodactyloid and lysiosquilloid superfamilies also possess tiered photoreceptors in a specialised zone of 'mid-band' ommatidia (Marshall 1988; Figure 21a). Many crustaceans have a two tiered retina with a single short-wave sensitive retinular cell (R8 – λ_{max} around 400 nm) overlying a rhabdom contributed to by 7 green sensitive cells (R1-7 – λ_{max} around 500 nm and Chapter 9 of this book). Generally the R8 cell possesses a rather short rhabdom, so its filtering effect on the underlying cells is negligible (see Gaten *et al.,* 1992 and below for exceptions to this). The mid-band may contain 6 rows of enlarged ommatidia and in 4 of these rows (numbers 1– 4 counting from the top down; Figure 21a) the rhabdom is divided into three tiers. This is achieved by dividing the R1-7 cells into two tiers of 3 or 4 cells underneath the R8 cell (Figure 22a). Rows 2 & 3 of the mid-band contain remarkable intrarhabdomal colour filters of purple, blue, yellow, orange or red blocks of membrane bound vesicles (Figure 22; Plate 1f). There may be up to four of these in each species situated between the photoreceptor tiers and almost all are at least partially constructed from carotenoid pigments (Marshall 1988; Cronin *et al.,* 1994a; Marshall *et al.,* 1991b).

With one or more colour filter types, theoretically a 'one (visual) pigment screened' colour vision system could be constructed. In practise however, as with the analogous oil droplet structures in vertebrates (section 4.2.5c), and elsewhere in the invertebrates, this does not happen. Intrarhabdomal filters help tune and sharpen photoreceptors, containing different visual pigments, to certain spectral loci (Cronin and Marshall 1989b). Astonishingly, stomatopod species with 6 row mid-bands possess at least 12 different visual pigments, 8 of these in the various R1-7 rhabdomeric portions of rows 1– 4. Although all different in λ_{max}, the sensitivity functions of even the theoretical rhodopsins, not accounting for broadening through self screening, overlap heavily (Figure 22d&e). By an elegant system of sequential serial filtering, utilising the long-pass colour filters of rows 2 and 3 and the filtering properties of the rhabdom tiers themselves, a series of very sharply tuned photoreceptors is constructed (Cronin and Marshall 1989a). This gives the potential for extremely fine spectral discrimination over the whole spectrum, not just a part of it (Troje 1993) and is good for maintaining colour constancy with varying illumination.

The sensitivity loss in this sequential filtering mechanism is enormous. In row 3, for instance, less than 1% of incident light reaches PR1-7 cells in some species and in light limited circumstances, such a photoreceptor would not work. Stomatopods with six-row mid-bands are found in murky, coastal waters and deep waters (10 –100 m, where the spectrum of light is attenuated particularly at short and long wavelengths – Jerlov 1976) and this heavy filtering could present a problem. However species like *Odontodactylus scyllarus* or *Lysiosquilla*

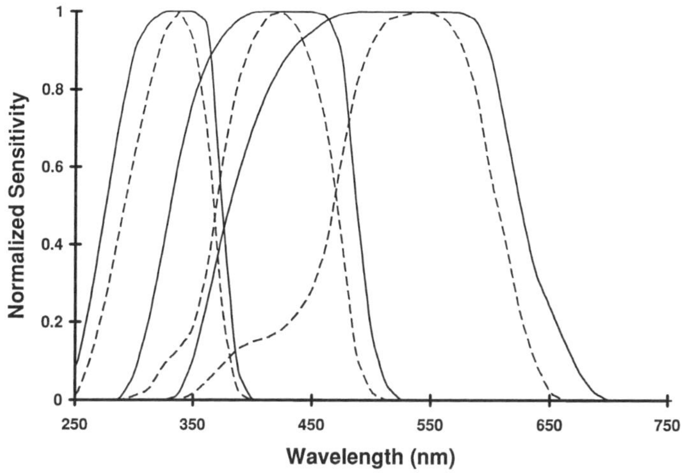

a

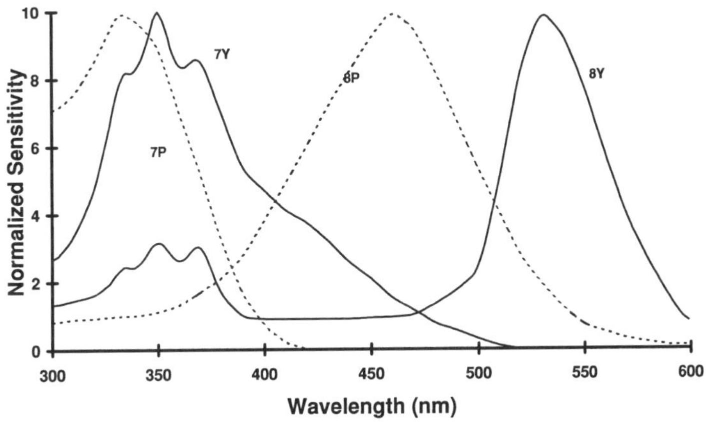

b

Figure 20 Serial and lateral filtering in the eyes of insects. (**a**) Lateral filtering in three optically coupled photoreceptors in the eye of the bee keeps the spectral sensitivities of photoreceptors narrow (dotted curves) which if not coupled have broad sensitivities (solid curves) and poor spectral resolution due to self screening (After Snyder *et al.* 1973). (**b**) Serial filtering in the central two-tiered rhabdoms (7 and 8) of the fly eye. The 'yellow' (y) and 'pale' (p) photoreceptor populations are found scattered throughout the eye. The relatively narrow spectral sensitivities in the underlying R8 cells are the result of the filtering effect of the R7 cells (After Hardie 1986)

sulcata, which live in these spectrally limited photic environments, restrict their spectral coverage to match the ambient illumination (Cronin *et al.*, 1994b&c). This is achieved by using visual pigments with λ_{maxs} closer to the wavelength of light best transmitted in sea water (475 nm), tuning filter transmission to allow through more light, or even 'removal' of filters.

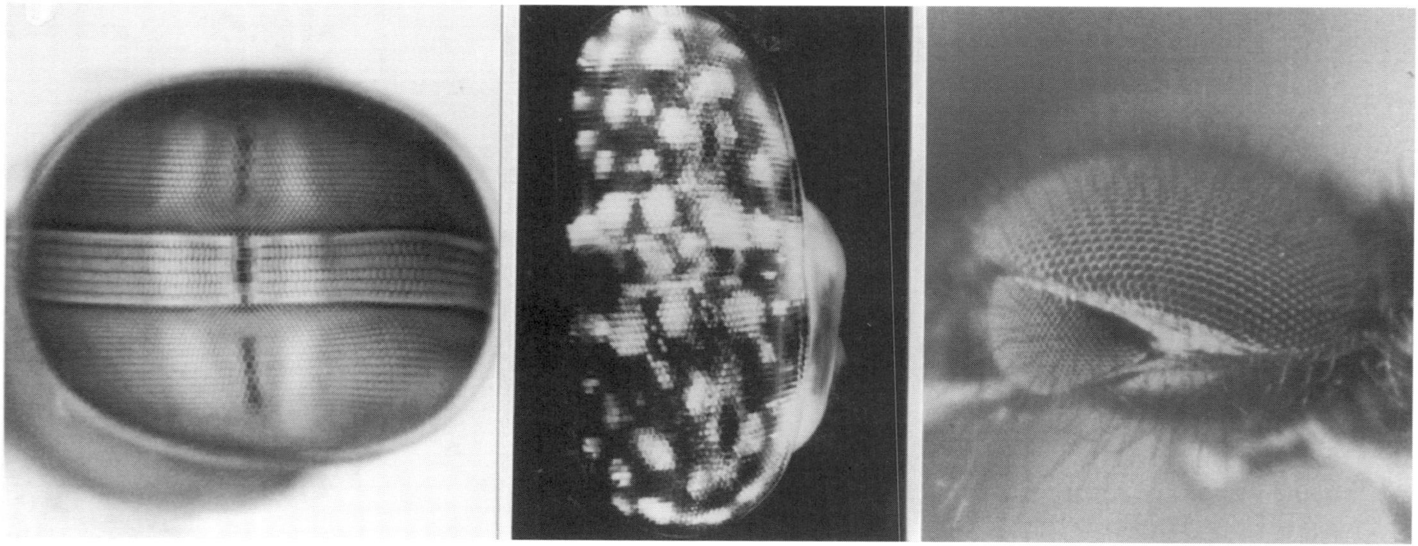

Figure 21 Compound eyes showing various adaptations involving filters. (**a**) The eye of the stomatopod *Odontodactylus scyllarus* showing the enlarged mid-band ommatidia. It is this modified eye region which contains the colour filters and tiered photoreceptors of the stomatopod multispectral colour vision system (Figure 22 and Plate 1). (**b**) The mottled appearance of the eye of the stomatopod *Lysiosquilla maculata*. Although these distally placed pigments may have a function in filtering light reaching the photoreceptors below, it is more likely they camouflage the eyes, which are the only part of the animal protruding through a hole in their sandy burrow as it waits to ambush prey (see also Stavenga 1979). (**c**) The head of a male Bibionid fly in lateral aspect. Each eye is split into a small ventrally directed region and a massively expanded dorsal region used for hunting females (Zeil 1983). These dorsal 'eyes' contain leaky filter screens for metarhodopsin reconversion to gain sensitivity (Photograph J. Zeil and M.F. Land see also Plate 1)

a

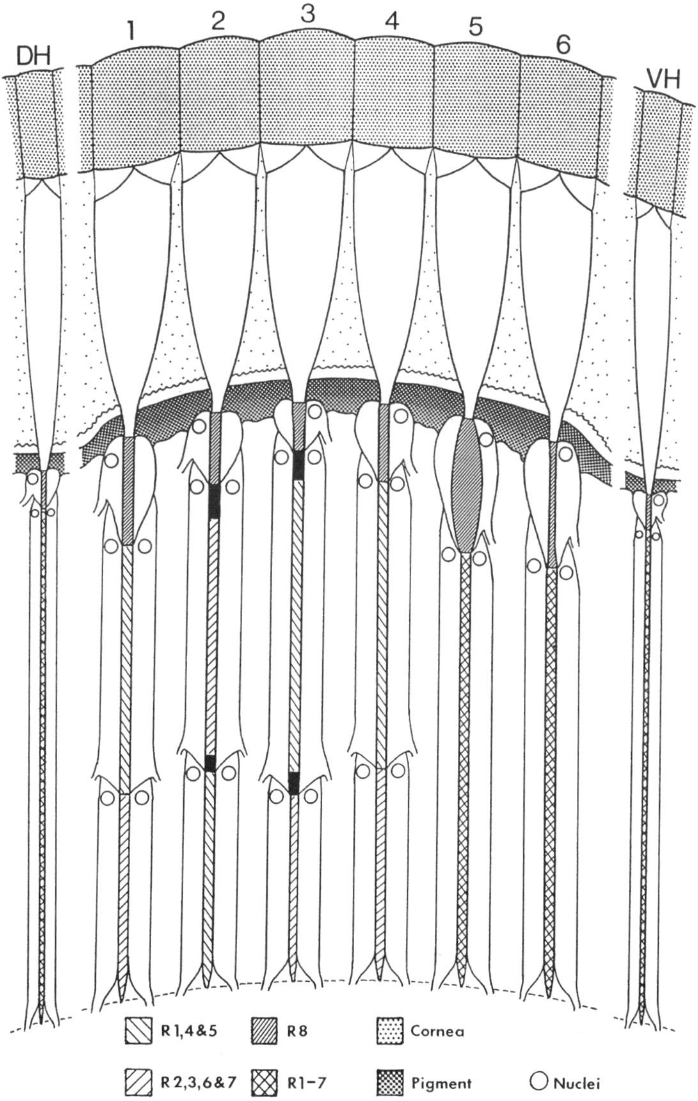

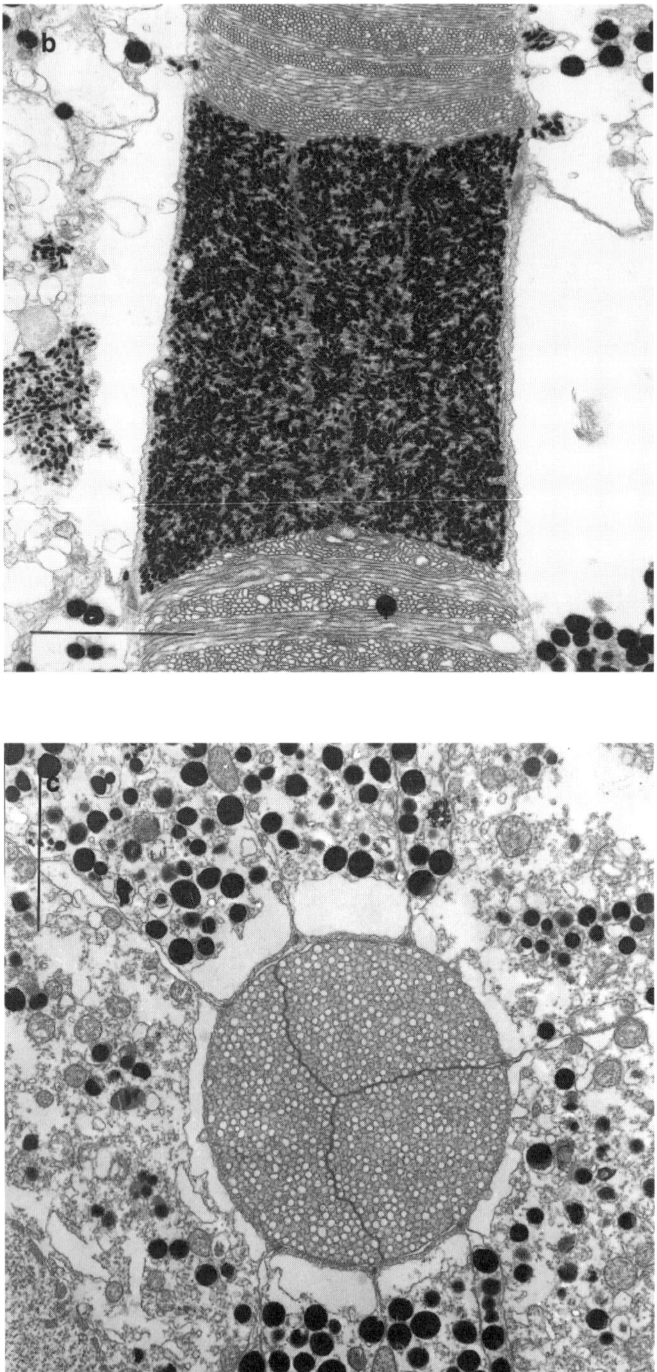
b
c

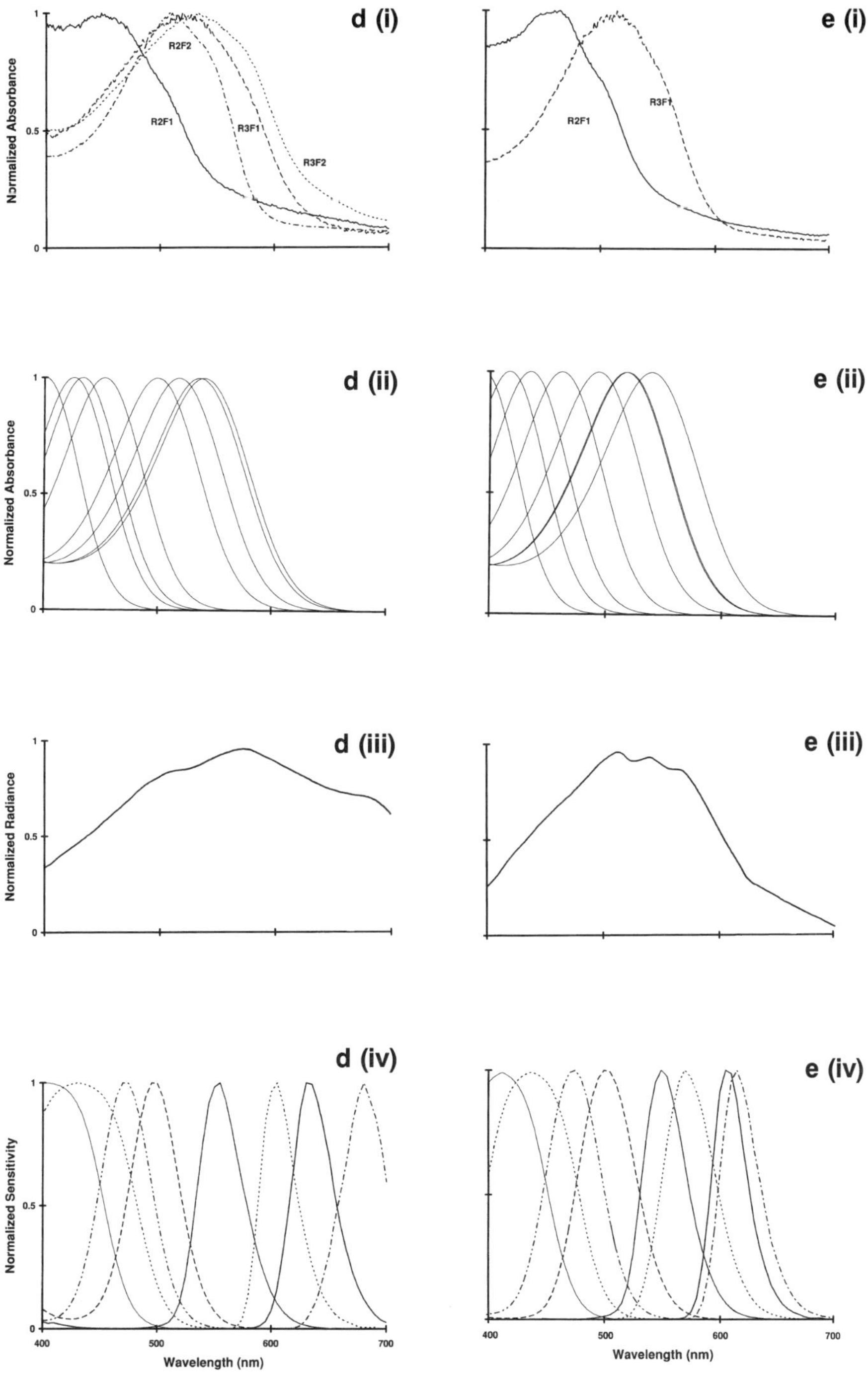
d (i)
R2F2
R2F1
R3F1
R3F2
e (i)
R3F1
R2F1
d (ii)
e (ii)
d (iii)
e (iii)
d (iv)
e (iv)
Normalized Absorbance
Normalized Radiance
Normalized Sensitivity
Wavelength (nm)
1
0.5
0
400
500
600
700

Figure 22 (previous pages) Serial filtering in the eyes of stomatopod crustaceans. (**a**) Diagrammatic representation of the specialised ommatidia of the mid-band. Rows 1–4 compose the colour vision system. They are triple tiered and between the tiers in rows 2 and 3 intrarhabdomal carotenoid filter blocks may be present (Plate 1f). (R1,4&5, R8, R1–7, R2,3,6&7 – retinular cells; after Marshall *et al.* 1991a). (**b** and **c**) TEMs of blue and yellow coloured filters in longitudinal and transverse section respectively. The small vesicles which may be round or oval shaped contain carotenoid pigment (Marshall *et al.*, 1991b, Cronin *et al.*, 1994a). (**d**) *Pseudosquilla ciliata.* Four panels show, from the top down: normalised absorbances of the four filters present in rows 2 and 3 of the mid-band (each acts as a cut-off filter passing long wavelengths Cronin *et al.*, 1994a), nomograms (G.D. Bernard unpublished) of the 8 visual pigments found in the proximal (non R8) regions of rows 1–4 (Cronin *et al.*, 1994c), normalised radiance conditions in typical *P. ciliata* habitat at 1 m (Cronin *et al.*, 1994c) and calculated sensitivities of the 8 retinal regions based on sequential serial filtering of visual pigments and intrarhabdomal filters (for details of calculations see Cronin & Marshall 1989b). This stomatopod dwells near the surface and therefore has the full spectrum of light available to it. Apparently as a result it 'spreads' its sensitivities across the whole spectrum – bottom panel. (**e**) *Lysiosquilla sulcata.* Four panels equivalent to (**d**). This stomatopod species lives in slightly deeper, murkier water as can be seen in the decreased radiance where it lives at long and short wavelengths. To compensate for this and to boost sensitivity the two proximally placed filters are 'lost' and the visual pigments are clustered more around the middle of the spectrum where light is available. Resulting sensitivities, bottom panel, cover this centralised spectral region and not the red or UV end where light is limited (and see Cronin *et al.*, 1994c)

Tiered retinae nearly always contain photoreceptors with different spectral sensitivities (see section 4.2.5e for an exception to this in vertebrates), the short wavelengths sampled by the distal tiers and progressively longer wavelengths by more proximal layers. Other examples of this are found in jumping spiders (Salticidae), where a 4 layered retina exists with spectral sensitivities at 360 nm, 480 – 500 nm and 580 nm (Yamashita and Tateda 1976) and a variety of insects, other than the diptera discussed above. The layering of the salticid retina, with short wavelength receptors placed closer to the lens, may be arranged to counteract chromatic aberration in which short wavelength rays are brought to a focus in front of longer wavelengths. Although constructed from fused rhabdoms, such as those found in bees, in some lepidopterans (the moths *Dielephila* and *Spodoptera* – Schlecht *et al.*, 1978; Langer *et al.*, 1979: and butterflies – Bandai *et al.*, 1992; Arikawa and Uchiyama 1996), dragonflies (Laughlin and McGuinness 1978; Yang and Osorio 1996), mayflies (Horridge and McLean 1978; Horridge 1976) and the lacewing *Chrysopa* (Horridge 1976), the rhabdoms contribute differing amounts of microvilli at different levels. This effectively makes the photoreceptors tiered and capable of serial sharpening of spectral sensitivity using the visual pigment of distally placed rhabdom portions.

A tiered rhabdom design and multiple visual pigments are found in some members of the enoploteuthid squid family such as *Watasenia scintillans* (Matsui *et al.*, 1988). *Watasenia* has five photoreceptor types in its eyes and three visual pigments with λ_{max} values at 470, 484 and 500 nm (Michinomae *et al.*, 1994). The ventral, upward looking, retinal region is tiered and also possesses yellow and red pigment granules. The sensitivity of the proximal region has been modelled and is sharpened and shifted from 500 nm to 550 nm by the filtering action of the overlying retina (Michinomae *et al.,* 1994). This is a particularly remarkable finding as all other cephalopods are probably monochromats, possessing only a single visual pigment housed in a monomorphic population of photoreceptors. Stranger still, *Watasenia* and its relatives are deep-sea squid, living below 250 m in the daytime, where the spectral range of light is restricted largely to short wavelengths only. *Watasenia* does however come to surface waters for communal spawning in spring and emits two colours of intense biolu-minesce from organs on its arms and body (Tsuji 1985). For this reason it is known as the firefly squid. These bioluminescent displays are particularly notable during surface water spawning, clearly an important event for the survival of these animals. Perhaps this vital group behaviour and whatever spectral discrimination it requires, is enough to explain the unusual retinal complexity of this cephalopod.

(c) Tiered polarizing filters For many arthropods, polarization vision may be as important as colour vision, and there are a variety of adaptations in their eyes which can be classified as polarising filters. The corneal light scattering filters found in the dorsal rim of some dipteran eyes is one example and is discussed in section 4.3.1c. There are also a variety of retinal adaptations which enhance polarisation vision. Photoreceptors constructed from microvilli are intrinsically dichroic, such that they are maximally sensitive to E-vectors of light parallel to the microvilli they carry. Cells with regular arrays of microvilli therefore have high polarisation sensitivity (PS). Where tiered photoreceptors exist the PS of the upper rhabdom may increase the PS of the lower photoreceptor by acting as a polarising filter (Snyder 1975). This filtering effect is exploited in the dorsal rim area of the compound eyes of many insects. Recall that this region of ommatidia are specifically adapted for navigation using the natural pattern of polarised light in the sky (section 4.3.1c; Rossel and Wehner 1984).

Crustaceans often possess rather long photoreceptors which might tend to lose PS through self screening (Snyder 1975). They have therefore adopted an interdigitating arrangement of narrow microvillar layers which exploits the E-vector filtering properties of microvilli to overcome this problem (also note that lateral filtering in optically coupled fused rhabdoms may have the same effect - Snyder *et al.,* 1973; Snyder 1975; Waterman 1981). Alternate layers carry microvilli that are perpendicular to each other, so light passing from one layer to another on its journey down the rhabdom is alternately switched from having vertical to horizontal E-vectors extracted, and therefore remains multidirectional to the bottom of the photoreceptor. While the photoreceptors extract the E-vector information needed for polarisation vision (Bernard and Wehner 1977), the PS reducing self screening, found in long rhabdoms of unidirectional microvilli, is minimised (Snyder 1975).

4.3.4 Postreceptoral filters

A variety of reflective structures exist at the base of invertebrate photoreceptors which increase photon capture by redirecting light back up the rhabdom. In common with the vertebrate tapeta (section 4.2.6b), the primary aim here is to increase sensitivity in the eye without increasing receptor length. It is not surprising therefore, that such adaptations are mostly found in nocturnal or crepuscular invertebrates and in those animals living in the depths of the sea. Beautifully coloured, spectrally narrow, tapeta are also found in the ommatidia of certain diurnal butterflies (Miller and Bernard 1968). Tapeta can be functionally divided into two different types: light coloured scattering pigments and those using thin film interference.

(a) Pigmentary tapeta The eyes of a variety of crustaceans, in particular those from the deep-sea like the *Penaeid* shrimps, exhibit spectacular eyeshine when illuminated (Zyznar and Nicol 1971; Shelton *et al.,* 1992; Gaten *et al.,* 1992). Superposition optics are used in such low-light environments (Land 1981) and the light seen is the 'luminous pseudopupil' (Exner 1891; Stavenga 1979) emerging from the facets involved in forming the superposition image. Light is reflected back through the rhabdom by white or light coloured pigments, generally purines (guanine is one, often used in vertebrate and spider tapeta – section 4.2.6b and below) and/or pteridines, situated round the base of the rhabdom. The reflection, although less efficient at short wavelengths (Miller 1979), is effective over much of the spectrum and can almost double light capture in superposition eyes, clearly an asset for night vision or vision in deep-ocean environments.

Curiously, the cells that form the tapetum in mid-water decapods are not always evenly distributed over the eye. There may be holes in the tapetum or gradations in reflecting efficiency around the eye (Shelton *et al.,* 1992). Typically, the part of the eye which looks into the lightless depths, where ambient brightness is 2 log units less than downwelling light (Denton 1970, 1990), has a thicker tapetum and brighter eyeshine than the regions looking up. The tapetum is in fact matched to the oceanic irradiance distribution with eye regions requiring greater sensitivity served by a thicker tapetum. The gradient is formed relative to the shrimps' natural body position in life. Oplophorid species, such as *Systelaspis debilis*, which swim with their body near horizontal most of the time, have dorso-ventral gradients of bright to dimly reflecting tapeta. Conversely, the sergestid *Sergestes henseni* swims with the body oriented vertically and its bright to dim tapetal gradient is rostro-caudal (Shelton *et al.,* 1992).

(b) Thin film interference tapeta The physics behind thin film interference was explained earlier (section 4.3.1a) and this mechanism is used in a number of invertebrates to construct reflective tapeta. Simple eyed invertebrates known to possess tapeta of this design include lycosid (wolf) spiders (Land 1981, 1985),

some copepod crustaceans (Fahrenbach 1964; Vaissière 1961), rotifers (Eakin 1972), ostracod crustaceans (Land 1981), scallops (Land 1966), and another bivalve mollusc, the cockle *Cardium edule* (Barber and Wright 1969). In two of these examples the tapetum is actually used instead of a lens to image light on the retina. This allows resolution of spatial detail in the scallop *Pecten*, while in the deep-sea ostracod *Gigantocypris*, it is for increasing sensitivity (to around 6000 × that of the human eye! – M.F. Land, pers. comm), an important requirement of life below 600 m in the ocean (Land 1981).

The lycosid spiders also require sensitive vision, as they hunt moving prey at twilight, pouncing on them in a single well executed jump. In common with many vertebrates, spider tapeta are made from thin-layered stacks of guanine crystals. Also like the vertebrates, those spider eyes containing tapeta (each spider has 8 eyes of two different designs – Land 1985) often arrange their retinae in an inverted manner with the receptors pointing inwards, towards the tapetum. This makes the absorption of reflected light more efficient and, as vertebrate eyes evolved in nocturnal animals which probably possessed tapeta, suggests one reason for the curious inverted arrangement of vertebrate eyes left today (Miller 1979).

The lepidoptera construct tapetal reflectors of a quite unique design by modifying part of their tracheal (air breathing) system. This allows them to use air spaces as the low refractive index layers, interspersed with chitinous cytoplasmic plates, giving the highest attainable refractive index change for light passing between layers. This results in the maximum possible percentage reflection of light at each interface. The tapetum in nocturnal moths is a relatively disorganised blanket of tracheal tissue reflecting green or bluish tinted light, relatively broad-band in its spectral characteristics, but possibly matched to the photoreceptors λ_{max}, for maximum photon reflux (Miller 1979). This may be achieved by combining several narrow-band reflector stacks, or by simply constructing multilayers with varying thicknesses and separations.

Surprisingly for day-time active animals, butterflies and some diurnal moths also possess tracheal tapeta consisting of exquisitely regular chitin plates (Figure 23). The result is a distinctively monochromatic or saturated looking colour reflection (Plate 1b). Each rhabdom has its own private tapetum and these may reject different eyeshine colours depending on the actual $1/4\lambda$ thickness of the layers and may also change with position round the eye. Although there is much variation and a number of patterns, dorsal eye facets usually emit blue and those in the ventral eye region, red (Miller and Bernard 1968; Bernard and Miller 1970). Occasionally there is a complete gradation with ommatidia in between these two extremes emitting green, yellow and orange (Miller 1979).

The presumed function of the narrow-band tracheal tapeta, like lateral or serial filtering, is tuning of photoreceptor sensitivity. The tapetal method has the disadvantage that all wavelengths of light enter the rhabdom at the top and are partially absorbed on their way down the photoreceptor, before being colour tinted and returned up the rhabdom by the tapetum. Nevertheless, light of the

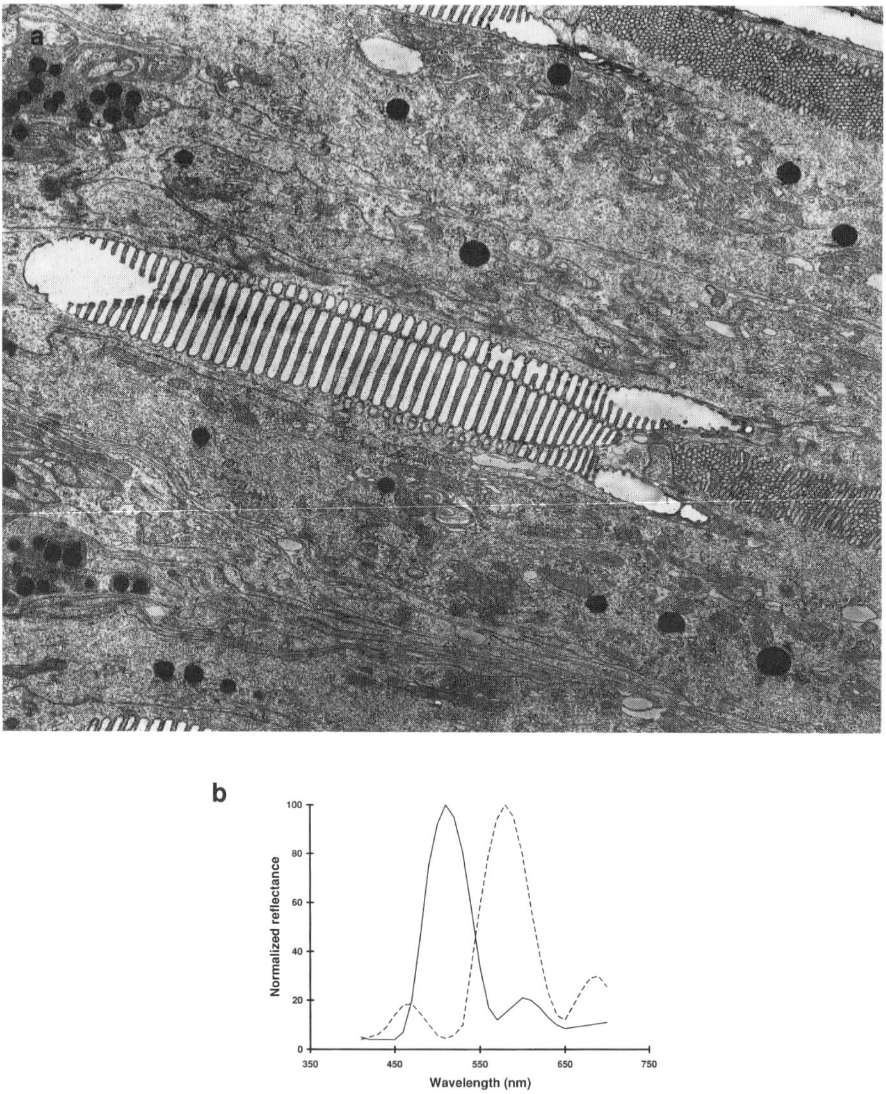

Figure 23 (**a**) Transmission electon micrograph (TEM) of the individual tracheal tapeta of the skipper butterfly. The very regular layering of material of different refractive indices (in this case, chitin and air) and spacing of layers at 1/4λ intervals mean that narrow band colour is reflected back up the photoreceptor. Such colours can be seen emerging from the eye in Plate 1b. (air space thickness between tracheal plates is 0.123 μm) (**b**) Normalised reflectance of a different 1/4λ reflector, in this case the alternating green and yellow corneal facets in the eyes of Dolichopodid flies (Plate 1a). The light not reflected here is transmitted to the photoreceptors below and may help in spectral sensitivity tuning (After Miller 1979)

wavelength reflected by the tracheal interference filter passes twice through the rhabdom, and this may be enough to render ommatidia with different eye glow colours spectrally distinct. Whether this directly affects colour vision or simply enhances colour contrast remains an enigma.

Interestingly red oil drop filters predominate in the ventrally directed dorsal region of bird retinae (Muntz 1972; section 4.2.5c). One explanation for this is that such filtering increases object contrast in a background of green foliage (Lythgoe 1979), the sort of visual signal experienced by flying or perched birds. The same hypothesis could be extended for the red filtered facets in the ventral eye of butterflies and enlarged to include the blue tint of the dorsal facets which possibly increases their efficiency at viewing things in a predominantly blue sky. The tracheal filter distribution is therefore matched to the butterflies' visual world.

One further complication to this story is that not all red eye glow is due to the tapetum alone. Some facets in the eyes of pierid butterflies look red when illuminated but this is the combined result of broad-band light reflected from a tracheal tapetum (which changes the spacing of its plates along its length to reject more than one wavelength – Miller 1979) and specialised laterally placed red retinular cell pigment. The red pigment granules are drawn in so close towards the rhabdom that some actually enter the microvilli, greatly influencing the colour emerging from the facet (Ribi 1978, 1979). The reasons for using a combination filter rather than relying on one method are obscure.

4.4 Conclusions

The importance of intraocular filters is indicated by their widespread occurrence. Some form of filter is, for instance, found in almost every anatomical region of both vertebrate and invertebrate eyes. Furthermore, many structures, such as the lenses of a variety of vertebrates, contain a plethora of spectrally, and hence functionally, similar pigments which are biochemically quite distinct, suggesting such filters have evolved independently on a number of occasions. Finally, even the most phylogentically ancient eyes contain filters, suggesting they have been present since vision first evolved.

This widespread occurrence of intraocular filters, and the fact that they deprive animals of potentially useful spectral information, implies that they must have one or more important functions. Not surprisingly when one considers the greater diversity of eye design in the invertebrates compared to vertebrates, there are some filters whose functions are unique to invertebrates. The functions proposed for intraocular filters in both groups include:

1. P*rotection from light damage and the enhancement of image quality*: These are assumed to be the most widespread functions of both corneal and lens filters in vertebrates since most such filters remove short wavelengths, which

are those most prone to optical degradation and most likely to cause structural damage. Similar functions are probably performed by the vertebrate macular pigment, invertebrate corneal nipple arrays and a variety of other coloured retinal inclusions in both vertebrates and invertebrates. The seemingly greater involvement of vertebrate filters in protection against the damaging effects of short wavelengths compared to invertebrates is perhaps due to the generally superior photon capture of most vertebrate eyes, the result of the larger apertures and generally larger sizes of simple eyes.

2. *Enhancement of colour vision/spectral sensitivity modification*: Individual photoreceptors, or groups of photoreceptors, in invertebrates and vertebrates may have their spectral sensitivities tuned by coloured pigments, in or next to the photoreceptors, or by interference filters positioned before or after the retina. In both groups such colour filters exist primarily to narrow existing spectral sensitivities rather than increase the number of spectral channels. Sharpened spectral sensitivities provide a basis for enhanced colour vision (Govardovskii 1983; Cronin and Marshall 1989a; Marshall *et al.,* 1991b). It should be noted that several of these mechanisms, especially in the insects, are still only hypothetical.
3. *Maintenance of photoreceptor isolation to preserve spatial resolution*; is performed by the melanin and ommachrome within the vertebrate and invertebrate retina and by sheaths of pigment round crystalline cones in invertebrate eyes.
4. *Dynamic control of light flux*; using corneal pigment migration and RPE retinomotor movements in vertebrates, and by radial movement of retinular cell pigment and longitudinal pigment migration in invertebrates.
5. *Sensitivity increase/photoreceptor noise reduction*: This is achieved through the use of broad and narrow band reflecting tapeta in both vertebrates and invertebrates. In addition to this, invertebrates have developed a series of measures that also serve to increase sensitivity; photo-reconversion of metarhodopsin to rhodopsin, corneal nipple arrays, and sensitising pigments (such a pigment may also exist in one fish retina). The number of filter mechanisms displayed only by invertebrates to increase sensitivity may be explained by their generally poor light capturing abilities noted above. It should not be forgotten however, that the most sensitive eye belongs to an invertebrate (section 4.3.4), the result of exploiting image formation by mirror optics, a mechanism not found in the vertebrates (Land 1981).
6. *Filters associated with polarisation vision*; such as corneal dispersion and coloured corneal filters are only found in invertebrates. Oil drops in bird eyes may be involved in this visual capability.
7. *Increasing the visibility of bioluminescence*; is probably the main function of yellow filters in the lens, and sometimes the retina, of many deep-sea fish and cephalopods.

References

Abadi, R. V. and Cox, M. J. (1992) The distribution of macular pigment in human albinos. *Investigative Ophthalmology and Visual Science,* 33(3), 494 – 497.

Ahnelt P. K., Hokoc, J. N. and Rohlich, P. (1995) Photoreceptors in a primitive mammal, the South American Opossum, *Didelphis marsupialis aurita* – characterization with anti-opsin immunolabeling. *Visual Neuroscience,* 12(5), 793 – 804.

Alpern, M., Fulton, A. B. and Baker, B. N. (1987) 'Self-screening' of rhodopsin in rod outer segments. *Vision Research,* 27(9), 1459 –1470.

Ambach, W., Blumthaler, M., Schöpf, T., Ambach, E., Katzgraber, F., Daxecker, F. and Daxer, A. (1994) Spectral transmission of the optical media of the human eye with respect to keratitis and cataract formation. *Documenta Ophthalmologica,* 88, 165 –173.

Anctil, M. and Ali, M. A. (1976) Cone droplets of mitochondrial origin in the retina of *Fundulus heteroclitus* (Pisces: Cyprinodontidae). *Zoomorphologie,* 84, 103 –111.

Appleby, S. J. and Muntz, W. R. A. (1979) Occlusable yellow corneas in Tetradontidae. *Journal of Experimental Biology*, 83, 249 – 259.

Arikawa, K. and Uchiyama, H. (1996) Red receptors dominate the proximal tier of the retina in the butterfly *Papilio xuthus. Journal of Comparative Physiology, A* 178, 55 – 61.

Arnott, H. J., Best, A. C. G., Ito, S. and Nicol, J. A. C. (1974) Studies on the eyes of catfishes with special reference to the tapetum lucidum. *Proceedings of the Royal Society of London B,* 186, 13 – 36.

Ashcroft, N. R. and Lythgoe, J. N. (1991) Ionic basis for colour changes in the iridescent cornea of the sand goby (*Pomatoschistus minutus*). *Pigment Cell Research,* 4, 209 – 215.

Avery, J. A. and Bowmaker, J. K. (1982) Visual pigments in the four-eyed fish, *Anableps anableps. Nature,* 298, 62 – 64.

Bandai, K., Arikawa, K., and Eguchi, E. (1992) Localisation of spectral receptors in the omatidium of butterfly compound eye determined by polarisation sensitivity *Journal of Comparative Physiology A,* 171, 289 – 297.

Bando, M., Nakajima, A. and Satoh, K. (1981) Spectrophotometric estimation of 3-OH L-Kynurenine O-β-Glucoside in the human lens. *Journal of Biochemistry,* 89, 103 –109.

Barber, V. C. and Wright, D. E. (1969) The fine structure of the eye and optic tentacle of the mollusc *Cardium edule. Journal of Ultrastructure Research,* 26, 515 – 528.

Barlow, H. B. (1982) What causes trichromacy? A theoretical analysis using combfiltered spectra. *Vision Research,* 22, 635 – 643.

Begin, M. T. and Handford, P. (1987) Comparative study of retinal oil droplets in grebes and coots. *Canadian Journal of Zoology*, 65, 2105 – 2110.

Berger, E. R. (1966) On the mitochondrial origin of oil droplets in the retinal double cone inner segments. *Journal of Ultrastructure Research*, 14, 143 –157.

Bernard, G. D. (1967) Structural and functional adaptation in a visual system. *Endeavour*. 26, 79 – 84.

Bernard, G. D. and Miller, W. H. (1968) Interference filters in the corneas of diptera. *Investigative Opthalmology,* 7(4), 416 – 434.

Bernard, G. D. and Miller, W. H. (1970) What does antenna engineering have to do with insect eyes? *IEEE Student Journal,* Jan-Feb.

Bernard, G. D. and Wehner, R. (1977). Functional similarities between polarisation vision and colour vision. *Vision Research,* 17:1019 –1028.

Bernard, G. D., Miller, W. H. and Moller, A. R. (1965) The insect corneal nipple array. *Acta Physiologica Scandinavia,* 63 (243), 1–77.

Bernstein, P. S., Balashov, N. A., Tsong, E. D. and Rando, R. R. (1997) Retinal tubulin binds macular carotenoids. *Investigative Ophthalmology & Visual Science,* 38(1), 167–175

Bertrand, D., Fuortes, G. and Muri, R. (1979) Pigment transformation and electrical responses in retinula cells of drone, *Apis mellifera. Journal of Physiology,* 296, 431– 441.

Best, A. C. G. and Nicol, J. A. C. (1980) Eyeshine in fihses. A review of ocular reflectors. *Canadian Journal of Zoology,* 58(6), 945 – 956.

Best, A. C. G. and Nicol, J. A. C. (1984) Red pigment epithelium of fish eyes. *Journal of the Marine Biological Association of the United Kingdom,* 64, 909 – 917.

Boettner, E. A. and Wolter, J. R. (1962) Transmission of the ocular media. *Investigative Ophthalmology,* 1(6), 776 –783.

Bone, R. A. and Lamdrum, J. T. (1984) Macular pigment in Henle fiber membranes – a model for Haidinger brushes. *Vision Research,* 24(2), 103 –108.

Bone, R. A., Landrum, J. T. and Tarsis, S. L. (1985) Preliminary identification of the human macular pigment. *Vision Research,* 25(11), 1531–1535.

Bone, R. A., Landrum, J. T. and Cains, A. (1992) Optical density spectra of the macular pigment *in vivo* and *in vitro. Vision Research,* 32(1), 105 –110.

Bone, R. A., Landrum, J. T., Fernandez, L. and Tarsis, S. L. (1988) Analysis of the macular pigment by HPLC: retinal distribution and age study. *Investigative Ophthalmology and Visual Science,* 29(6), 843 – 849.

Bone, R. A., Landrum, J. T., Hime, G. W., Cains, A. and Zamor, J. (1993) Stereochemistry of the human macular carotenoids. *Investigative Ophthalmology and Visual Science,* 34(6), 2033 – 2040.

Borwein, B. and Hollenberg, M. J. (1973) The photoreceptors of the 'four-eyed' fish *Anableps anableps* L. *Journal of Morphology*, 140, 405 – 441.

Boulton, M., Docchio, F., Dayhew-Barker, P., Ramponi, R. and Cubeddu, R. (1990) Age-related changes in the morphology, absorption and fluorescence of melanosomes and lipofuscin granules of the retinal pigment epithelium. *Vision Research,* 30(9), 1291–1303.

Bowmaker, J. K. (1977) The visual pigments, oil droplets and spectral sensitivity of the pigeon. *Vision Research,* 17, 1129 –1138.

Bowmaker, J. K. (1979) Visual pigments and oil droplets in the pigeon retina, as measured by microspectrophotometry, and their relationship to spectral sensitivity, in *Neural Mechanisms of Behaviour in the Pigeon* (ed A. M. Granda and J. H. Maxwell), Plenum Press, New York, pp. 327– 351.

Bowmaker, J. K. (1980) Colour vision in birds and the role of oil droplets. *Trends in Neuroscience,* 3, 196 –199.

Bowmaker, J. K. (1983) Trichromatic colour vision: why only three receptor channels? *Trends in Neuroscience,* 6, 41– 43.

Bowmaker, J. K. (1990) Visual pigments of fishes. In *The Visual System of fish* (Douglas, R. H. & Djamgoz, M. B. A.), Chapman & Hall, London. pp. 81–107.

Bowmaker, J. K. (1991a) The evolution of vertebrate visual pigments and photoreceptors, in *Vision and Visual Dysfunction volume 2: Evolution of the Eye and Visual System*, (eds J. R. Cronly-Dillon and R. L. Gregory), CRC Press, Boca Raton, pp. 63 – 81.

Bowmaker, J. K. (1991b) Visual pigments, oil droplets and photoreceptors, in *Vision and Visual Dysfunction volume 6: The Perception of Colour* (ed P. Gouras), CRC Press, Boca Raton, pp. 108 –127.

Bowmaker, J. K. and Knowles, A. (1977) The visual pigments and oil droplets of the chicken retina. *Vision Research,* 17, 755 –764.

Bowmaker, J. K., Dartnall, H. J. A. and Herring, P. J. (1988) Longwave-sensitive visual pigments in some deep-sea fishes: segregation of 'paired' rhodopsins and porphyropsins. *Journal of Comparative Physiology A,* 163, 685 – 698.

Bowmaker, J. K., Astell, S., Hunt, D. M. and Mollon, J. D. (1991) Photosensitive and photostable pigments in the retinae of old world monkeys. *J. Exp. Biol.,* 156, 1–19.

Bowmaker, J. K., Kovach, J. K., Whitmore, A. V. and Loew, E. R. (1993) Visual pigments and oil droplets in genetically manipulated and carotenoid deprived quial: a microspectrophotometric study. *Vision Research,* 33(5/6), 571– 578.

Braekevelt, C. R. (1993) Fine structure of the tapetum Lucidum of the Paca (*Cuniculus paca*). *Acta Anatomica,* 146(4), 244 – 250.

Brainard, G. C., Barker, F. M., Hoffman, R. J., Stetson, M. H., Hanifin, J. P., Podolin, P. L. and Rollag, M. D. (1994) Ultraviolet regulation of neuroendocrine and circadian physiology in rodents. *Vision Research,* 34(11), 1521–1533.

Burton, G. W. and Ingold, K. U. (1984) β-carotene: an unusual type of lipid antioxidant. *Science,* 224, 569 – 573.

Chou, B. R. and Cullen, A. P. (1984) Spectral transmittance of the ocular media of the thirteen-lined ground squirrel (*Spermophilus tridecemlineatus*). *Canadian Journal of Zoology,* 62, 825 – 830.

Collier, R. J., Waldron, W. R. and Zigman, S. (1989) Temporal sequence of changes to the Grey squirrel retina after near-UV exposure. *Investigative Ophthalmology and Visual Science,* 30(4), 631– 637.

Collin, H. B. and Collin, S. P. (1996) The fine structure of the cornea of the Salamanderfish, *Lepidogalaxias salamandroides* (Lepidogalaxiidae, Teleostei). *Cornea,* 15(4), 414 – 426.

Cooper, G. F. and Robson, J. G. (1969a) The yellow colour of the lens of the grey squirrel (*Sciurus carolinensis leucotis*). *Journal of Physiology,* 203, 403 – 410.

Cooper, G. F. and Robson, J. G. (1969b) The yellow colour of the lens of man and other primates. *Journal of Physiology,* 203, 411– 417.

Couillard, P. (1984) Photoreception in Protozoa, an overview. In: *Photoreception and vision in invertebrates.* (ed. Ali, B). Plenum, New York.

Cronin, T. W. and Marshall, N. J. (1989a) A retina with at least ten spectral types of photoreceptors in a mantis shrimp. *Nature*, 339, 137–140.
Cronin, T. W. and Marshall, N. J. (1989b) Multiple spectral classes of photoreceptors in the retinas of gonodactyloid stomatopod crustaceans. *Journal of Comparative Physiology A,* 166, 261– 275.
Cronin, T. W., Marshall, N. J. and Caldwell, R. L. (1994a) The intrarhabdomal filters in the retinas of mantis shrimps. *Vision Research,* 34, 279 – 291.
Cronin, T. W., Marshall, N. J. and Caldwell, R. L. (1994b) The retinas of mantis shrimps from low light environments (Crustacea; Stomatopoda; Gonodactylidae). *Journal of Comparative Physiology A,* 174, 607– 619.
Cronin T. W., Marshall N. J., Caldwell R. L. and Shashar N (1994c) Specialisation of retinal function in the compound eyes of mantis shrimps. *Vision Research,* 34, 2639 – 2656.
Dartnall, H. J. A., Arden, G. B., Ikeda, H., Luck, C. P., Rosenberg, M. E., Pedler, C. M. H. and Tansley, K. (1965) Anatomical, electrophysiological and pigmentary aspects of vision in the bush baby: an interpretative study. *Vision Research,* 5, 399 – 424.
Dartnall, H. J. A., Bowmaker, J. K. and Mollon, J. D. (1983) Human visual pigments: microspectrophotometric results from the eyes of seven persons. *Proceedings of the Royal Society B* 220, 115 –130.
Denton, E. J. (1955) Absorption du cristallin de *Rana esculenta et d' Anguilla vulgaris. Bulletin du Muséum, 2e série, t. XXVII,* 5, 418 – 425.
Denton, E. J. (1970) On the organization of reflecting surfaces in some marine animals. *Philosophical Transactions of the Royal Society of London B,*. 258, 285 – 313.
Denton, E. J. (1990). Light and vision at depths greater than 200m. In: *Light and life in the sea* (ed Herring, P. J. *et al*) Cambridge University Press, Cambridge, pp. 127–148.
Denton, E. J. and Nicol, J. A. C. (1964) The choroidal tapeta of some cartilaginous fishes (Chondrichthyes). *Journal of the Marine Biological Association of the United Kingdom,* 44, 219 – 258.
Denton, E. J. and Warren, F. J. (1968) Eyes of the Histioteuthidae. *Nature* 219, 400 – 401.
Denton, E. J. and Herring, P. (1971) Report to the council. *Journal of the Marine Biological Association of the United Kingdom,* 51, 1035.
Denton, E. J. & Locket, N. A. (1989) Possible wavelength discrimination by multibank retinae in deep-sea fishes. *Journal of the Marine Biological Association of the United Kingdom,* 69, 409 – 435.
Denton, E. J., Herring, P. J., Widder, E. A., Latz, M. F., and Case, J. F. (1985) The role of filters in the photophores of oceanic animals and their relation to vision in the oceanic environment. *Proceedings of the Royal Society of London B,* 225, 63 – 97.
Dodt, E. and Walther, J. B. (1958) Netzhautsensitivität, Linsenabsorption und physikalische Lichstreuung: der skotopische Dominator der Katze in sichtbaren und ultravioletten Spektralbereich. *Pflügers Archive,* 266, 167–174.
Douglas, R. H. (1982) The function of the photomechanical movements in the retina of the rainbow trout (*Salmo gairdneri*). *Journal of Experimental Biology,* 96, 389 – 403.
Douglas, R. H. (1987) Ocular lens diameter as an indicator of age in brown trout, *Salmo trutta. Journal of Fish Biology,* 31, 835 – 836.
Douglas, R. H. (1989) The spectral transmission of the lens and cornea of the brown trout (*Salmo trutta*) and Goldfish (*Carassius auratus*): effect of age and implications for ultraviolet sensitivity. *Vision Research,* 29(7), 861– 869.
Douglas, R. H. and McGuigan, C. M. (1989) The spectral transmission of freshwater teleost ocular media: an interspecific comparison and a guide to potential ultraviolet sensitivity. *Vision Research,* 29(7), 871– 879.
Douglas, R. H. and Thorpe, A. (1992) Short wave absorbing pigments in the ocular lenses of deep-sea teleosts. *Journal of the Marine Biological Association of the United Kingdom,* 72, 93 –112.
Douglas, R. H. and Thorpe, A. (1994) Absorbance vs transmission: species and age – a response. *Vision Research,* 34(19), 2503 – 2504.
Douglas, R. H., Partridge, J. C. and Hope, A. J. (1995) Visual and lenticular pigments in the eyes of demersal deep-sea fishes. *Journal of Comparative Physiology A,* 177, 111–122.
Douglas, R. H., Partridge, J. C., Dulai, K., Hunt, D., Mullineaux, C. W., Tauber, A. and Hynninen, P. H. (1998) Dragon fish see using chlorophyll. *Nature*, 393, 423–424.
Dunlap, W. C., Williams, D. Mc.B., Chalker, B. E. and Banaszak, A. T. (1989) Biochemical photoadaptation in vision: UV-absorbing pigments in fish eye tissues. *Comparative Biochemistry and Physiology,* 93B, 601– 607.
Eakin, R. M. (1972) Structure of invertebrate photoreceptors. In: *Handbook of sensory physiology*. VII/1. (ed. Dartnall, H. J. A.), Springer, Berlin, Heidelberg, New York. pp. 625 – 684.

Ellingson, J. M., Fleishman, L. J. and Loew, E. R. (1995) Visual pigments and spectral sensitivity of the diurnal gecko *Gonatodes albogularis*. *Journal of Comparative Physiology A,* 177, 559 – 567.

Emmerton, J., Schwemer, J., Muth, I. and Schlecht, P. (1980) Spectral transmission of the ocular media of the pigeon (*Columba livia*). *Investigative Ophthalmology and Visual Science,* 19(11), 1382 –1387.

Exner, S. (1891) Die physiologie der facettirten augen von krebsen und insecten. Deuticke. Liepsig, Wein.

Fahrenbach, W. H. (1964) Fine structure of a nauplius eye. *Zeitschrift für Zellfoschung,* 62, 182 –197.

François, J. and Neetens, A. (1974) Comparative anatomy of the vascular supply of the eye in vertebrates, in *The Eye, volume 5* (eds H. Davson and L. T. Graham, Jr.), Academic Press, New York and London. pp. 1–70.

Gamburtseva, A. G., Gnyubkina, V. P., Kondrashev, S. L. and Orlov, O. Yu. (1979) Chromatophores and colouration of cornea of fishes of the sea of Japan. *Soviet Journal of Marine Biology,* 5 (6), 495 – 503.

Gaten, E., Shelton, P. M. J. and Herring, P. J. (1992) Regional morphological variations in the compound eyes of certain mesopelagic shrimps in relation to their habitat. *Journal of the Marine Biological Association of the UK,* 72, 61–75.

Geeraets, W. J. and Berry, E. R. (1968) Ocular spectral characteristics as related to hazards from lasers and other light sources. *American Journal of Ophthalmology,* 66(1), 15 – 20.

Gerster, H. (1991) Review: antioxidant protection of the ageing macula. *Age and Ageing* 20, 60 – 69.

Goldsmith, T. H., Collins, J. S. and Licht, S. (1984) The cone oil droplets of avain retinas. *Vision Research,* 24(11), 1661–1671.

Goldstein, E. B. and Williams, T. P. (1966) Calculated effects of 'screening pigments'. *Vision Research* 6, 39 – 50.

Gorgels, T. G. M. F. and van Norren, D. (1992) Spectral transmittance of the rat lens. *Vision Research,* 32(8), 1509 –1512.

Govardovskii, V. I. (1983) On the role of oil drops in colour vision. *Vision Research* 23(12), 1739 –1740.

Govardovskii, V. I. and Zueva, L. V. (1977) Visual pigments of chicken and pigeon. *Vision Research,* 17, 537– 543.

Govardovskii, V. I., Rohlich, P., Szel, A. and Zueva, L. V. (1992) Immunocytochemical reactivity of rod and cone visual pigments in the sturgeon retina. *Visual Neuroscience,* 8(6), 531– 537.

Griswold, M. S. and Stark, W. S. (1992) Scotopic spectral sensitivity of phakic and aphakic observers extending into the near ultraviolet. *Vision Research,* 32(9), 1739 –1743.

Grover, D. and Zigman, S. (1972) Colouration of human lenses by near ultraviolet photo-oxidized tryptophan. *Experimental Eye Research,* 13, 70 –76.

Haegerstrom-Portnoy, G. (1988) Short-wavelength-sensitive-cone sensitivity loss with aging: a protective role for macular pigment? *Journal of the Optical Society of America A,* 5 (12), 2140 – 2144.

Hailman, J. P. (1976) Oil droplets in the eyes of adult anuran amphibians: a comparative study. *Journal of Morphology,* 148, 453 – 468.

Hallberg, E. and Elofsson, R. (1989) Construction of the pigment shield of the crustacean compound eye: a review. *Journal of Crustacean Biology*, 9(3), 359 – 372.

Ham, W. T., Mueller, H. A. and Sliney, D. H. (1976) Retinal sensitivity to damage from short wavelength light. *Nature,* 260, 153 –155.

Ham, W. T., Mueller, H. A., Ruffolo, J. J. and Clarke, A. M. (1979) Sensitivity of the retina to radiation damage as a function of wavelength. *Photochemistry and Photobiology,* 29, 735 –743.

Ham, W. T., Mueller, H. A., Ruffolo, J. J., Millen, J. E., Cleary, S. F., Guerry, R. K. and Guerry, D. (1984) Basic mechanisms underlying the production of photochemical lesions in the mammalian retina. *Current Eye Research,* 3(1), 165 –174.

Ham, W. T., Ruffolo, J. J., Mueller, H. A., Clarke, A. M. and Moon, M. E. (1978) Histologic analysis of photochemical lesions produced in rhesus retina by short-wavelength light. *Investigative Ophthalmology and Visual Science,* 17 (10), 1029 –1035.

Hamdorf, K., Gogala, M. and Schwemer, J. (1973) Photoregeneration and sensitivity control of photoreceptors of invertebrates. In: (ed. Langer, H.) *Biochemistry and physiology of visual pigments.* Springer, Berlin, Heidelberg, New York, pp. 156 –166.

Hammond, B. R. and Fuld, K. (1992) Interocular differences in macular pigment density. *Investigative Ophthalmology and Visual Science,* 33(2), 350 – 355.

Hammond, B. R., Fuld, K. and Curran-Celentano, J. (1995) Macular pigment density in monozygotic twins. *Investigative Ophthalmology and Visual Science,* 36(12), 2531– 2541.

Handelman, G. J. and Dratz, E. A. (1986) The role of antioxidants in the retina and retinal pigment epithelium and the nature of prooxidant-induced damage. *Advances in Free Radical Biology and Medicine,* 2, 1– 89.

Handelman, G. J., Dratz, E. A., Reay, C. C. and van Kuijk, F. J. G. M. (1988) Carotenoids in the human macula and whole retina. *Investigative Ophthalmology and Visual Science,* 29(6), 850 – 855.

Handelman, G. J., Snodderly, D. M., Adler, A. J., Russett, M. D. and Dratz, E. A. (1992) Measurement of carotenoids in human and monkey retinas. *Methods in Enzymology,* 213, 220 – 230.

Handelman, G. J., Snodderly, D. M., Krinsky, N. I., Russett, M. D. and Adler, A. J. (1991) Biological control of primate macular pigment; biochemical and densitometric studies. *Investigative Ophthalmology and Visual Science,* 32(2), 257– 267.

Hardie, R. C. (1986) The photoreceptor array of the dipteran retina. *Trends in Neuroscience,* 9(9), 419 – 423.

Hata, M. and Hata, M. (1971) Carotenoid pigments in goldfish (*Carassius auratus*) I. Composition and distribution of carotenoids. *International Journal of Biochemostry,* 2, 11–19.

Heath, A. R. and Hindman, H. M. (1988) The role of cyclic AMP in the control of elasmobranch ocular tapetum lucidum pigment granule migration. *Vision Research,* 28(12), 1277–1285.

Heinermann, P. H. (1984) Yellow intraocular filters in fishes. *Experimental Biology,* 43, 127–147.

Hemmingsen, E. A. and Douglas, E. L. (1970) Ultraviolet radiation thresholds for corneal injury in antarctic and temperate-zone animals. *Comparative Biochemistry and Physiology,* 32, 593 – 600.

Herring, P. J. (1983) The spectral characteristics of luminous marine organisms. *Proceedings of the Royal Society of London B,* 220, 183 – 217.

Horridge, G. A. (1976) The ommatidium of the dorsal eye of *Cloen* as a specialization for photoreisomerisation. *Proceedings of the Royal Society of London B,* 193, 17– 29.

Horridge, G. A. and McLean, M. (1978) The dorsal eye of the mayfly *Atalophlebia* (Ephemeroptera). *Proceedings of the Royal Society of London B.*, 200, 137–150.

Hunold, W. and Malessa, P. (1974) Spectrophotometric determination of the melanin pigmentation of the human ocular fundus *in vivo. Ophthalmic Research,* 6, 355 – 362.

Hunt, D. M., Dulai, K. S., Bowmaker, J. K. and Mollon, J. D. (1995) The chemistry of John Dalton's colour blindness. *Science,* 267, 984 – 988.

Hyatt, G. W. (1975) Physiological and behavioural evidence for colour discrimination by fiddler crabs (Brachiura, Ocypodidae, genus *Uca*). In: *Physiological ecology of estuarine organisms.* (ed. Vernberg, F. J.). pp. 333 – 365. University of South Carolina Press. Columbia SC.

Ito, S., Thurston, E. L. and Nicol, J. A. C. (1975) Melanoid tapeta lucida in teleost fishes. *Proceedings of the Royal Society of London B,* 191, 369 – 385.

Jacobs, G. H. (1992) Ultraviolet vision in vertebrates. *American Zoologist,* 32, 544 – 554.

Jacobs, G. H. and Yolton, R. L. (1972) Some characteristics of the eye and the electroretinogram of the prairie dog. *Experimental Neurology,* 37, 538 – 549.

Jaffe, G. J. and Wood, I. S. (1988) Retinal phototoxicity from the operating microscope: a protective effect by the fovea. *Archives of Ophthalmology,* 106, 445 – 446.

Jane, S. D. and Bowmaker, J. K. (1988) Tetrachromatic colour vision in the duck (*Anas platyrhynchos* L.): microspectrophotometry of visual pigments and oil droplets. *Journal of Comparative Physiology A,* 162, 225 – 235.

Jerlov N. G. (1976) *Marine Optics.* Elsevier Scientific: Amsterdam, Oxford, New York, pp. 231.

Johnston, D. and Hudson, R. A. (1976) Isolation and composition of varotenoid containing oil droplets from cone photoreceptors. *Biochimistry and Biophysics Acta,* 424, 235 – 245.

Kennedy, D. and Milkman, R. D. (1956) Selective light absorption by the lenses of lower vertebrates, and its influence on spectral sensitivity. *Biological Bulletin,* 111, 375 – 386.

King-Smith, P. E. (1969) Absorption spectra and function of the coloured oil drops in the pigeon retina. *Vision Research,* 9, 1391–1399.

Kirschfeld, K. (1982) Carotenoid pigments: their possible role in protecting against photooxidation in eyes and photoreceptor cells. *Proceedings of the Royal Society of London B,* 216, 71– 85.

Kirschfeld, K and Wenk, P. (1976) The dorsal compound eye of simuliid flies: An eye specialization for the detection of small, rapidly moving objects. *Zeitschrift für Naturforschung,* 31c, 764 –765.

Kirschfeld, K., Feiler, R. and Franceschini, N. (1978) A photostable pigment within the rhabdomere of fly photoreceptors No. 7. *Journal of Comparative Physiology A,* 125, 275 – 284.

Kirschfeld, K., Hardie, R. C., Lenz, G. and Vogt, K (1988). *Journal of Comparative Physiology A,* 162, 421– 433.

Kolb, H. and Jones, J. (1982) Light and electron microscopy of the photoreceptors in the retina of the red-eared slider *Pseudemys scripta elegans. Journal of Comparative Neurology,* 209, 331– 338.

Kondrashev, S. L. and Khodtsev, A. S. (1984) Light-dependent and humoral control of pigment transport in corneal chromatophores in marine fishes. *Zoological Journal of Physiology,* 88, 317– 325.

Kondrashev, S. L., Gamburtseva, A. G., Gnjubkina, V. P., Orlov, O. J. and My, P. T. (1986) Colouration of corneas in fish: A list of species. *Vision Research,* 26, 287– 290.

Kong, K-L., Fung. Y. M. and Wasserman, G. S. (1980) Filter-mediated colour vision with one visual pigment. *Science,* 207, 783 –786.

Krinsky, N. I. (1979) Carotenoid protection against oxidation. *Pure and Applied Chemistry* 51, 649 – 660.

Kunz, Y. W. and Regan, C. (1973) Histochemical investigations into the lipid nature of the oil-droplet in the retinal twin-cone of *Lebistes reticulatus* (Peters). *Revue Suisse de Zoologie,* 80, 699 –703.

Kunz, Y. W. and Wise, C. (1973) Ultrastructure of the 'oil-droplet' in the retinal twin-cone of *Lebistes reticulatus* (Peters). *Revue Suisse de Zoologie,* 80, 694 – 698.

Labhart, T. and Nilsson, D.-E. (1995) The dorsal eye of the dragonfly *Sympetrum*: specializations for prey detection against the blue sky. *Journal of Comparative Physiology A,* 176, 437– 453.

Labhart, T, Meyer, E. P. and Shenker, L. (1992) Specialized ommatidia for polarization vision in the compound eye of cockchafers, *Melolontha melolontha* (Coleoptera, Scarabaeidae). *Cell and Tissue Research,* 268, 419 – 429.

Lall A. B. and Cronin T. W. (1987) Spectral sensitivity of the compound eyes in the purple land crab *Gecarcinus lateralis* (Freminville). *Biological Bulletin,* 173, 398 – 406

Lall, A. B., Lord, E. T. and Trouth, C. O. (1982) Vision in the firefly *Photuris lucicrescens* (Coleoptera: Lampyridae): spectral sensitivity and selective adaptation in the compound eye. *Journal of Comparative Physiology A,* 147, 195 – 200.

Lall, A. B., Strother, G. K., Cronin, T. W. and Seliger, H. H. (1987) Modification of spectral sensitivities by screening pigments in the compound eyes of twilight-active fireflies (Coleoptera: Lampyridae). *Journal of Comparative Physiology A,* 162(1), 23 – 34.

Land, M. F. (1966) A multilayer interference reflector in the eye of the scallop, *Pecten maximus. Journal of Experimental Biology*, 45, 433 – 447

Land, M. F. (1972) The physics and biology of animal reflectors. In: *Progress in Biophysics and Molecular Biology*. 24: pp. 75 –106. (ed. Butler, J. A. V. & Noble, D.), Pergamon Press. Oxford and New York.

Land, M. F. (1981) Optics and vision in invertebrates. In: *Handbook of sensory physiology. VII/6B*. (ed. Autrum, H.), pp. 471– 594. Springer-Verlag. Berlin, Heidelberg, New York.

Land, M. F. (1985) The morphology and optics of spider eyes. In: *Neurobiology of arachnids*. (ed. Barth F. G.) Springer-Verlag, Berlin, Heidelberg. pp. 53 –78.

Land, M. F. (1989) Variations in the structure and design of compound eyes. In: *Facets of Vision*. (eds Stavenga, D. G. & Hardie, R. C.), Springer-Verlag, Berlin, Heidelberg. pp 90 –111.

Land, M. F. (1993a) Old twist in a new tale. *Nature,* 363, 581– 582.

Land, M. F. (1993b) The visual control of courtship behaviour in the fly Poecilobothrus nobilitatus. *Journal of Comparative Physiology A,* 173, 595 – 603.

Langer, H., Hamann, B. and Meinecke, C. C. (1979) Tetrachromatic visual system in the moth *Spodoptera exempta* (Insecta: Noctuidae). *Journal of Comparative Physiology A,* 129, 235 – 239.

Laughlin, S. and McGuinness, S. (1978) The structures of dorsal and ventral regions of a dragonfly retina. *Cell and Tissue Research,* 188, 427– 447.

Laurens, H. and Detwiler, S. R. (1921) The structure of the retina of *Alligator mississippiensis* and its photomechanical changes. *Journal of Experimental Zoology,* 32, 207– 234.

Leggett. L. M. W. (1979) A retinal substrate for colour discrimination in crabs. *Journal of Comparative Physiology A,* 133, 159 –166.

Lerman, S. (1980) *Radiation energy and the Eye*. MacMillan, New York.

Lerman, S. (1987) Chemical and physical properties of normal and aging lens: Spectroscopic (UV, fluorescence, phosphorescence, and NMR) analyses. *American Journal of Optometry and Physiological Optics,* 64, 11– 22.

Liebman, P. A. and Granda, A. M. (1975) Super dense carotenoid spectra resolved in single cone oil droplets. *Nature,* 253, 370 – 372.

Lipetz, L. E. (1984a) A new method for determining peak absorbance of dense pigment samples and its application to the cone oil droplets of *Emydoidea blandingii. Vision Research,* 24(6), 597– 604.

Lipetz, L. E. (1984b) Pigment types, densities and concentrations in cone oil droplets of *Emydoidea blandingii. Vision Research,* 24(6), 605 – 612.

Locket, N. A. (1972) The reflecting structure in the iridescent cornea of the serranid teleost *Nemanthias carberryi. Proceedings of the Royal Society B,* 182, 249 – 254.

Locket, N. A. (1977) Adaptations to the deep-sea environment. In *Handbook of Sensory Physiology* 7(5) (ed F. Crescitelli). Springer, Belin Heidelburg New York. pp. 67– 192.

Lunau, K. and Knüttel, H. (1995) Vision through coloured eyes. *Naturwissenschaften* 82: 432 – 434.

Lythgoe, J. N. (1971) Iridescent corneas in fishes. *Nature,* 233, 205 – 207.
Lythgoe, J. N. (1975) The structure and function of iridescent corneas in teleost fishes. *Proceedings of the Royal Society B,* 188, 437– 457.
Lythgoe, J. N. (1976) The ecology, function and phylogeny of iridescent multilayers in fish corneas. In *Light as an ecological factor: II* (eds G. C. Evans, R. Bainbridge & O. Rackham) Blackwell scientific Publications, Oxford pp. 211– 247.
Lythgoe, J. N. (1979) The ecology of vision. Clarendon press. Oxford.
Lythgoe, J. N. and Shand, J. (1989) The structural basis for iridescent colour changes in dermal and corneal iridophores in fish. *Journal of Experimental Biology,* 141, 313 – 325.
MacNichol, E. F., Kunz, Y. W., Levine, J. S., Harosi, F. I. and Collins, B. A. (1978) Ellipsosomes: Organelles containing a cytochrome-like pigment in the retinal cones of certain fishes. *Science,* 200, 549 – 552.
Maier, E. J. (1994) Ultraviolet vision in a passeriform bird: from receptor spectral sensitivity to overall spectral sensitivity in *Leiothrix lutea. Vision Research,* 34(11), 1415 –1418.
Maier, E. J. and Bowmaker, J. K. (1993) Colour vision in the passeriform bird, *Leiothrix lutea*: correlation of visual pigment absorbance and oil droplet transmission with spectral sensitivity. *Journal of Comparative Physiology A,* 172, 295 – 301.
Malinow, M. R., Feeney-Burns, L., Peterson, L. H., Klein, M. L. and Neuringer, M. (1980) Diet-related macular anomalies in monkeys. *Investigative Ophthalmology and Visual Science,* 19(8), 857– 863.
Marshall, N. J. (1988) A unique colour and polarization vision system in mantis shrimps. *Nature.* 333, 557– 560.
Marshall, N. J. (1994) The 'six-eyed' stomatopod. *Endeavour,* 18(1), 17– 26.
Marshall, N. J. and Land, M. F. (1994) Some optical features of the eyes of stomatopods I & II. *Journal of Comparative Physiology A,* 173, 565 – 594
Marshall N. J., Land M. F., King C. A. and Cronin T. W. (1991a) The compound eyes of mantis shrimps (Crustacea, Hoplocarida, Stomatopoda). I. Compound eye structure: the detection of polarised light. *Philosophical Transactions of the Royal Society of London B,* 334, 33 – 56
Marshall N. J., Land M. F., King C. A. and Cronin T. W. (1991b) The compound eyes of mantis shrimps (Crustacea, Hoplocarida, Stomatopoda). II Colour pigments in the eyes of stomatopod crustaceans: polychromatic vision by serial and lateral filtering. *Philosophical Transactions of the Royal Society of London B,* 334:57– 84.
Martin, G. R. and Muntz, R.A. (1978) Spectral sensitivity of the red and yellow oil droplet fields of the pigeon (*Columba livia*). *Nature,* 274, 620 – 621.
Martin, G. R. and Muntz, W. R. A. (1979) Retinal oil droplets and vision in the pigeon (*Columba livia*), in *Neural Mechanisms of Behaviour in the Pigeon* (ed A. M. Granda and J. H. Maxwell), Plenum Press, New York, pp. 307– 325.
Matusi, S., Seidou, M., Horiuchi, S., Uchiyama, I. and Kito, Y. (1988) Adaptation of a deep-sea cephalopod to the photic environment. *Journal of General Physiology,* 92, 55 – 66.
McCandless, R. L., Hoffert, J. R. and Fromm, P. O. (1969) Light transmission by corneas, aqueous humor and crystalline lenses of fishes. *Vision Research,* 9, 223 – 232.
Menzel, R. (1981) Spectral sensitivity and colour vision in invertebrates. In: *Handbook of sensory physiology, vol VII/6A* (ed Autrum, H.) pp 503-580. Springer, Berlin Heidelberg New York.
Menzel, R., Wunderer, H. and Stavenga, D. G. (1991) Functional morphology of the divided compound eye honeybee drone (*Apis mellifera*). *Tissue Cell,* 23, 525 – 535.
Merker, E. (1934) Die Sichtbarkeit ultravioletten Lichtes. *Biological Reviews* 9(1), 49 –78.
Michinomae, M., Masuda, H., Seidou, M. and Kito, Y. (1994) Structural basis for wavelength discrimination in the banked retina of the firefly squid *Watasenia scintillans. Journal of Experimental Biology,* 193, 1–12.
Miller, W. H. (1979) Ocular optical filtering, in *Handbook of Sensory Physiology VII/6A* (ed H. Autrum), Springer, Berlin, pp 69 –143.
Miller, W. H. and Bernard, G. D. (1968) Butterfly glow. *Journal of ultrastructure Research,* 24, 286 – 294.
Miller, W. H., Møller, A. R. and Bernard, C. G. (1966) The corneal nipple array, in *The functional organisation of the compound eye* (ed C. G. Bernard) Pergamon, Oxford, pp. 21– 33.
Muntz, W. R. A. (1972) Inert absorbing and reflecting pigments, in *Handbook of Sensory Physiology VII/1* (ed H. J. A. Dartnall), Springer, Berlin, pp 529 – 565.
Muntz, W. R. A. (1973) Yellow filters and the absorption of light by the visual pigments of some Amazonian fishes. *Vision Research,* 13, 2235 – 2254.
Muntz, W. R. A. (1976a) The visual consequences of yellow filtering pigments. In *Light as an ecological factor: II* (eds G. C. Evans, R. Bainbridge & O.Rackham) Blackwell Scientific Publications, Oxford pp. 271– 287.

Muntz, W. R. A. (1976b) On yellow lenses in mesopelagic animals. *Journal of the Marine Biological Association of the United Kingdom,* 56, 963 – 976.
Muntz, W. R. A. (1982) Visual adaptations to different light environments in Amazonian fishes. *Revue Canadienne de Biologogie Experimentale,* 41 (1), 35 – 46.
Muntz, W. R. A. (1990) Stimulus, environment and vision in fishes, In *The Visual System of Fish* (eds R. H. Douglas and M. B. A. Djamgoz), Chapman and Hall, London, pp. 491– 511.
Nag, T. C. (1995) Ultrastructure of ellipsosomes in the retina of *Garra lamta. Journal of Electron Microscopy,* 44, 405 – 407.
Nag, T. C. and Bhattacharjee, J. (1995) Retinal ellipsosomes: morphology, development, identification, and comparison with oil droplets. *Cell and Tissue Research,* 279, 633 – 637.
Neumeyer, C. and Jäger, J. (1985) Spectral sensitivity of the freshwater turtle *Pseudemys scripta elegans*: evidence for the filter-effect of coloured oil droplets. *Vision Research,* 25(6), 833 – 838.
Nicol, J. A. C. (1981) Tapeta lucida of vertebrates. In *Vertebrate Photoreceptor Optics* (eds J. M. Enoch and F. L. Tobey), Springer-Verlag, Berlin pp. 401– 431.
Nicol, J. A. C. (1989) *The Eyes of Fishes.* Oxford University Press, Oxford. pp. 308.
Nicol, J. A. C. and Arnott, H. J. (1973) Tapeta lucida in bony fishes (Acinopterygii): a survey. *Canadian Journal of Zoology,* 51, 69 – 81.
Nicol, J. A. C. and Arnott, H. J. (1974) Tapeta lucida in the eyes of goatsuckers (Caprimulgidae). *Proceedings of the Royal Society of London B.,* 187, 349 – 352.
Nicol, J. A. C., Zyznar, E. S., Thurston, E. L. and Wang, R. T. (1975) The tapetum lucidum in the eyes of cusk-eels (Ophidiidae). *Canadian Journal of Zoology,* 53, 1063 –1079.
Nie, S., Castillo, C. G., Bergbauer, K. L., Kuck, J. F. R., Nabiev, I. R. and Yu, N.-T. (1990) Surface-enhanced Raman spectra of eye lens pigments. *Applied Spectroscopy,* 44(4), 571– 575.
Nilsson, D.-E. (1990) From cornea to retinal image in invertebrate eyes. *Trends in Neuroscience,* 13(2), 55 – 64.
Nussbaum J. J., Pruett, R. C. and Delori, F. C. (1981) Historic perspectives -- Macular yellow pigment – the first 200 years. *Retina,* 1, 296 – 310
Ohtsuka, T. (1985) Relation of spectral types to oil droplets in cones of turtle retina. S*cience*, 229, 874 – 877.
Organisciak, D. T. and Winkler, B. S. (1994) Retinal light damage: practical and theoretical considerations, in *Progress in Retinal and Eye Research, volume 13(1)* (eds. N. N. Osborne and G. J. Chader), Pergamon Press, Oxford, pp. 1– 29.
Orlov, O. Yu. and Gamburtseva A. G. (1976a) Changeable colouration of cornea in the fish *Hexagrammos octogrammus. Nature,* 263: 405 – 427.
Orlov, O. Yu. and Gamburtseva, A. G. (1976b) Dynamics of the colour of the cornea of the Atka fish. *Biofizika,* 21 (2), 362 – 365.
Packard, A. (1972) Cephalopods and fish: the limits of convergence. *Biol. Rev.*, 47:241– 307.
Partridge, J. C. (1989) The visual ecology of avian cone oil droplets. *Journal of Comparative Physiology A,* 165, 415 – 426.
Partridge, J. C., Archer, S. N. and Lythgoe J. N. (1988) Visual pigments in the individual rods of deep-sea fishes. *Journal of Comparative Physiology A,* 162, 543 – 550.
Partridge, J. C., Shand, J., Archer, S. N., Lythgoe, J. N. and van Groningen-Luyben, W. A. H. M. (1989) Interspecific variation in the visual pigments of deep-sea fishes. *Journal of Comparative Physiology A,* 164, 513 – 529.
Pease, P. L., Adams, A. J. and Nuccio, E. (1987) Optical density of human macular pigment. *Vision Research,* 27(5), 705 –710.
Pedler, C. M. H. and Boyle, M. (1969) Multiple oil droplets in the photoreceptors of the pigeon. *Vision Research,* 9, 525 – 528.
Reading, V. M. and Weale, R. A. (1974) Macular pigment and chromatic aberration. *Journal of the Optical Society of America,* 64 (2), 231– 234.
Remy, M and Emmerton, J. (1989) Behavioural spectral sensitivities of different retinal areas in pigeons. *Behavioural Neuroscience,* 103 (1), 170 –177.
Ribi, W. A. (1978) A unique hymenopteran compound eye. The retina fine structure of the digger wasp *Sphex cognatus* Smith (Hymenoptera, Sphecidae). *Zoologisches Jahrbuch der Anatomie,* 100, 299 – 342.
Ribi, W. A. (1979) Coloured screening pigments cause red eye glow in Pierid butterflies. *Journal of Comparative Physiology A,* 132, 1– 9.
Robinson, S. R. (1994) Early vertebrate colour vision. *Nature,* 367, 121.

Rossel, S. and Wehner, R. (1984) How bees analyse the polarisation patterns in the sky. Experiments and model. *Journal of Comparative Physiology A,* 154, 607– 615.

Ruddock, K. H. (1972) Light transmission through the ocular media and macular pigment and its significance for psychophysical investigation, in *Handbook of Sensory Physiology VII/4* (ed ?), Springer, Berlin, pp 455 – 469.

Said, F. S. and Weale, R. A. (1959) The variation with age of the spectral transmissivity of the living human crystalline lens. *Gerontologia,* 3, 213 – 231.

Sarna, T. (1992) Properties and function of the ocular melanin – a photobiophysical view. *Journal of Photochemistry and Photobiology. B: Biology,* 12, 215 – 258

Schalch, W. (1992) Carotenoids in the retina – a review of their possible role in preventing or limiting damage caused by light and oxygen, in *Free Radicals and Aging* (eds I. Emerit and B. Chance), Birkhäuser Verlag, Basel, pp. 280 – 298.

Schiedt, K., Bischof, S. and Glinz, E. (1991) Recent progress on carotenoid metabolism in animals. *Pure and Applied Chemistry,* 63(1), 89 –100.

Schlecht, P., Hamdorf, and Langer, H. (1978) The arrangement of colour receptors in a fused rhabdom of an insect. A microspectrophotometric study on the moth *Dielephila. Journal of Comparative Physiology A,* 123, 239 – 243.

Schneeweis, D. M. and Green, D. G. (1995) Spectral properties of turtle cones. *Visual Neuroscience*, 12, 333 – 344.

Schneider, L. and Langer, H. (1975) Electron microscopic investigations on the structure of the photoreceptor cells in the compound eye of *Ascalaphus macaronius* (Insecta: Neuroptera). In: *Photoreceptor optics* (eds. Snyder, A. W. & Menzel, R) Springer, New York, Heidelberg, Berlin, pp. 410 – 412.

Sen, A. C., Ueno, N. and Chakrabarti, B. (1992) Studies on human lens: I. Origin and development of fluorescent pigments. *Photochemistry and Photobiology,* 55(5), 753 –764.

Shand, J. (1988) Corneal iridescence in fishes: light-induced colour changes in relation to structure. *Journal of Fish Biology*, 32, 625 – 632.

Shand, J. and Lythgoe, J. N. (1987) Light-induced changes in corneal iridescence in fish. *Vision Research* 27(2), 303 – 305.

Shelton, P. M. J., Gaten, E. and Herring, P. J. (1992) Adaptations of tapeta in the eyes of mesopelagic decapod shrimps to match the oceanic irradiance distribution. *Journal of the Marine Biological Association of the UK,* 72, 77– 88

Sivak, J. G. and Bobier, W. R. (1978) Chromatic aberration of the fish eye and its effect on refractive state. *Vision Research,* 18(4), 453 – 455.

Sivak, J. G. and Roth, P. I. (1978) Possible role of fundus circulation as an intraocular colour filter in certain fishes. *Reviews in Canadian Biology*, 37(2), 85 – 90.

Snodderly, D. M. (1995) Evidence for protection against age-related macular degeneration by carotenoids and antioxidant vitamins. *American Journal of Clinical Nutrition,* 62(suppl), 1448S –1461S.

Snodderly, D. M., Auran, J. D. and Delori, F. C. (1984b) The macular pigment. I. Spatial distribution in primate retinas. *Investigative Ophthalmology and Visual Science,* 25(6), 674 – 685.

Snodderly, D. M., Handelman, G. J. and Adler, A. J. (1991) Distribution of individual macular pigment carotenoids in central retina of macaque and squirrel monkeys. *Investigative Ophthalmology and Visual Science,* 32(2), 268 – 279.

Snodderly, D. M., Brown, P. K., Delori, F. C. and Auran, J. D. (1984a) The macular pigment. I. Absorbance spectra, localisation, and discrimination from other yellow pigments in primate retinas. *Investigative Ophthalmology and Visual Science,* 25(6), 660 – 673.

Snyder, A. W. (1975) Optical properties of invertebrate photoreceptors.In: *The Compound Eye and Vision of Insects*. (ed. Horridge, G. A.). pp. 179 – 235. Clarendon. Oxford.

Snyder, A. W. and Horridge, G. A. (1972) The optical function of changes in the medium surrounding the cockroach rhabdom. *Journal of Comparative Physiology A,* 81, 1– 8.

Snyder, A. W., Menzel, R. and Laughlin, S. B. (1973) Structure and function of the fuzed rhabdom. *Journal of Comparative Physiology A,* 87, 99 –135.

Somiya, H. (1976) Functional significance of the yellow lens in the eyes of *Argyropelecus affini. Marine Biology,* 34, 93 – 99.

Somiya, H. (1980) Fishes with eye shine: functional morphology of guanine type tapetum lucidum. *Marine Ecology Progress Series,* 2, 9 – 26.

Somiya, H. (1982) 'Yellow lens' eyes of a stomatoid deep-sea fish, *Malacosteus niger. Proceedings of the Royal Society B,* 215, 481– 489.

Stark, W. S. (1987) Photopic sensitivities to ultraviolet and visible wavelengths and the effects of the macular pigments in human aphakic observers. *Current Eye Research,* 6(4), 631– 638.

Stavenga, D. G. (1979) Pseudopupils in compound eyes. In: *Handbook of Sensory Physiology. vol. VII 6A.* (ed. Autrum, H.). pp. 5 – 24. Springer. Berlin, Heidelberg, New York.

Stavenga, D. (1996) Insect retinal pigments: spectral characteristics and physiological functions. (in press).

Stowe, S. (1980) Spectral sensitivity and retinal pigment movement in the crab *Leptograpsus variegatus* (Fabricus). *Journal of Experimental Biology,* 87, 73 – 98.

Thorpe, A. (1991) Spectral trtansmission and short-wave absorbing pigments in the lenses of fish and other animals. Ph.D. thesis, City University, London.

Thorpe, A. and Douglas, R.H. (1993) Spectral transmission and short-wave absorbing pigments in the fish lens. II. Effects of age. *Vision Research,* 33, 301– 307.

Thorpe, A., Truscott, R. J. W. and Douglas, R. H. (1992) Kynurenine identified as the short-wave absorbing lens pigment in the deep-sea fish, *Stylephorus chordatus. Experimental Eye Research,* 55, 53 – 57.

Thorpe, A., Douglas, R. H., and Truscott, R. J. W. (1993) Spectral transmission and short-wave absorbing pigments in the fish lens – I. Phylogenetic distribution and identity. *Vision Research,* 33, 289 – 300.

Troje N (1993) Spectral categories in the learning behaviour of blowflies. *Zeitschrift der Naturforschung,* 48c, 96 –104.

Trujillo-Cenoz, O. and Bernard, G. D. (1972) Some aspects of the retinal organisation of *Sympyncus lineatus* (Diptera, Dolichopodidae). *Journal of Ultrastructure Research,* 38, 149 –160.

Truscott, R. J. W., Carver, J. A., Thorpe, A. and Douglas, R. H. (1992) The identification of 3-hydroxykynurenine as the lens pigment in the gourami, *Trichogaster trichopterus. Experimental Eye Research,* 54, 1015 –1017.

Tso, M. O. M. (1989) Experiments on visual cells by nature and man: in search of treatment for photoreceptor degeneration. *Investigative Ophthalmology and Visual Science,* 30(12), 2430 – 2460.

Tsuji, F. I. (1985) ATP-dependent bioluminescence in the firefly squid, *Watasenia scintillans. Proceedings of the National Academy of Science U.S.A.,* 82, 4629 – 4632.

van Best, J. A., Bollemeijer, J. G. and Sterk, C. C. (1988) Corneal transmission in whole human eyes. *Experimental Eye Research,* 46(5), 765 –768.

van den Berg, T. J. T. P. and Ijspeert, J. K. (1995) Light scattering in donor lenses. *Vision Research,* 35(1), 169 –177.

van den Berg, T. J. T. P. and Spekreijse, H. (1997) Near infrared light absorption in the human eye media. *Vision Research,* 37(2), 249 – 253.

van den Berg, T. J. T. P. and Tan, K. E. W. (1994) Light transmittance of the human cornea from 320 to 700 nm for different ages. *Vision Research,* 34(11), 1453 –1456.

van Heyningen, R. (1971a) Fluorescent Glucoside in the human lens. *Nature,* 230, 393 – 394.

van Heyningen, R. (1971b) Fluorescent derivatives of 3-Hydroxy-L-kynurenine in the lens of man, the baboon and the Grey Squirrel. *Proceedings of the Biochemical Society,* 123, 30P.

van Heyningen, R. (1973a) Assay of fluorescent glucosides in the human lens. *Experimental Eye Research,* 15, 121–126.

van Heyningen, R. (1973b) The glucoside of 3-hydroxykynurenine and other fluorescent compounds in the human lens. In *The human lens in relation to cataract. Ciba Foundation symposium* 19. Elsevier, Amsterdam. pp. 151–171.

van Heyningen, R. and Linklater, J. (1976) Serine and threonine ethanolamine phosphate diesters, and some other unusual compounds in the lens of the cod fish (*Gadus morhua*) and haddock (*Gadus aeglefinus*). *Experimental Eye Research,* 23, 29 – 34

van Norren, D. and Schellekens, P. (1990) Blue light hazard in rat. *Vision Research,* 30(10), 1517–1520.

van Norren, D. (1991) Photochemical damage to the eye. *News in Physiological Sciences,* 6, 232 – 234.

Vassière, R. (1961) Morphologie et histologie comparées des yeux des crustacés copépodes. *Archives Zoologie Experimentale Génerale,* 100, 1–125.

Vogt, K., Kirschfeld, K. and Stavenga, D. G. (1982) Spectral effects of the pupil in fly photoreceptors. *Journal of Comparative Physiology A,* 146, 145 –152.

Vos, J. J. (1972) *Literature review of human macular absorption in the visible and its consequences for the cone receptor primaries.* Institute for Perception, RVO-TNO, IZF 1972 –17, Soesterberg, The Netherlands. (from Bone *et al.* 1992).

Vos, J. J., Coemans, M. A. J. M. and Nuboer, J. F. W. (1994) The photopic sensitivity of the yellow field of the pigeon's retina to ultraviolet light. *Vision Research,* 34(11), 1419 –1425.

Wagner H.-J., Kirsch, M. and Douglas, R. H. (1992) Light dependent and endogenous circadian control of adaptation in teleost retinae. In '*Rhythms in fishes*' (ed M. A. Ali). Plenum Press: New York. pp. 255 – 291.

Wallman, J. (1979) Role of the retinal oil droplets in the colour vision of Japanese quail, in *Neural Mechanisms of Behaviour in the Pigeon* (ed A. M. Granda and J. H. Maxwell), Plenum Press, New York, pp. 327– 351.

Walls, G. L. and Judd, H. D. (1933a) The intraocular colour filters of vertebrates. *British Journal of Ophthalmology,* 17, 641– 675.

Walls, G. L. and Judd, H. D. (1933b) The intraocular colour filters of vertebrates. *British Journal of Ophthalmology,* 17, 705 –725.

Walls, G. L. (1963) *The Vertebrate Eye and its Adaptive Radiation.* Hafner Publishing Co., New York.

Wang, R. T. and Nicol, J. A. C. (1974) The tapetum lucidum of gars (Lepisosteidae) and its role as a reflector. *Canadian Journal of Zoology,* 52,1523 –1530.

Wang, R. T., Nicol, J. A. C. and Arnott H. J. (1981) Diffuse reflectance of retinal tapeta lucida, with special reference to drums (Sciaenidae). *Canadian Journal of Zoology,* 59, 271– 284.

Wang, R. T., Nicol, J. A. C., Thurston, E. L. and McCants, M. (1980) Studies on the eyes of big eyes (Teleostei Priacanthidae) with special reference to the tapetum lucidum. *Proceedings of the Royal Society of London B,* 210, 499 – 512.

Warrant, E. J. and McIntyre, P. D. (1996) The visual ecology of pupillary action in superposition eyes. *Journal of Comparative Physiology A,* 178, 75 – 90.

Waterman, T. H. (1981) Polarization sensitivity. In: Handbook of sensory physiology. VII/6C (ed. Autrum, H.) pp. 281– 469. Springer Verlag. Berlin, Heidelberg, New York.

Weale, R. A. (1953) The spectral reflectivity of the cats tapetum measured *in situ. Journal of Physiology,* 119, 30 – 42.

Weale, R. A. (1988) Age and the transmittance of the hman crystalline lens. *Journal of Physiology,* 395, 577– 587.

Weale, R. A. (1995) Is the crystalline lens a calender or a photometer? *Points de Vue,* 33, 8 –13.

Wehner, R. (1981) Spatial vision in arthropods. In: *Handbook of sensory physiology. VII/6C* (ed. Autrum, H.). Springer, Berlin, Heidelberg, New York. pp. 287– 645.

Weiter, J. J., Delori, F. and Dorey, C. K. (1988) Central sparing in annular macular degeneration. *American Journal of Ophthalmology,* 106, 286 – 292.

Wen, G. Y., Sturman, J. A. and Shek, J. W. (1985) A comparative study of the tapetum, retina, and skull of the ferret, dog and cat. *Laboratory Animal Science,* 35(3), 200 – 210.

Werner, J. S. (1982) Development of scotopic sensitivity and the absorption spectrum of the human ocular media. *Journal of the Optical Society of America,* 72(2), 247– 258.

Werner, J. S., Donnelly, S. K. and Kliegl, R. (1987) Aging and human macular pigment density – appended with translations from the work of Max Schultze and Ewald Hering. *Vision Research,* 27(2), 257– 268.

Widder, E. A., Latz, M. I. and Case, J. F. (1983) Marine Bioluminescence spectra measured with an optical multichannel detection system. *Biological Bulletin,* 165, 791– 810.

Widder, E. A., Latz, M. I., Herring, P. J., and Case, J. F. (1984) Far red bioluminescence from two deep-sea fishes. *Science,* 225, 512 – 514.

Wild, J. M. and Hudson, C. (1995) The attenuation of blue-on-yellow perimetry by the macular pigment. *Ophthalmology,* 102(6), 911– 917.

Wolbarsht, M. L. (1976) The function of intraocular colour filters. *Federation Proceedings,* 35(1), 44 – 50.

Wood, A. M. and Truscott, R. J. W. (1993) UV filters in human lenses: Tryptophan catabolism. *Experimental Eye Research,* 56, 317– 325.

Wood, A. M. and Truscott, R. J. W. (1994) Ultraviolet filter compounds in human lenses: 3-hydroxy-kynurenine glucoside formation. *Vision Research,* 34(11), 1369 –1374.

Wooten, B. R. and Geri, G. A. (1987) Psychophysical determination of intraocular light scatter as a function of wavelength. *Vision Research,* 8, 1291–1298.

Wortel, J. F. and Nuboer, J. F. W. (1986) The spectral sensitivity of blue sensitive pigeon cones: evidence for complete screening of the visual pigment by oil droplets. *Vision Research,* 26(6), 885 – 886.

Yamashita, S. and Tateda, H. (1976) Spectral sensitivities of jumping spider eyes. *Journal of Comparative Physiology A,* 105, 29 – 41.

Yang, En-C. and Osorio, D. (1996) Spectral responses and chromatic processing in the dragonfly lamina. *Journal of Comparative Physiology A,* 178, 543 – 550.

Yeum, K.-Y., Taylor, A., Tang, G. and Russell, R. M. (1995) Measurement of carotenoids, retinoids and topopherols in human lens. *Investigative Ophthalmology and Visual Science,* 36(13), 2756 – 2761.

Yolton, R. L., Yolton, D. P., Renz, J. and Jacobs, G. H. (1974) Preretinal absorbance in Sciurid eyes. *Journal of Mammalogy,* 55(1), 14 – 20.

Young, H. M. and Pettigrew, J. D. (1991) Cone photoreceptors lacking oil droplets in the retina of the echidna, *Tachyglossus aculeatus* (Monotremata). *Visual Neuroscience,* 6, 409 – 420.

Young, S. R. and Martin, G. R. (1984) Optics of retinal oil droplets: a model of light collection and polarization detection in the avian retina. *Vision Research,* 24(2), 129 –137.

Yu, N.-T., Barron, B. C. and Kuck, J. F. R. (1989) Distribution of two metabolically related fluorophors in human lens measured by laser microprobe. *Experimental Eye Research,* 49, 189 –194.

Yu, N.-T., Cai, M.-Z., Ho, D. J.-Y. and Kuck, J. F. R. (1988) Automated laser-scanning-microbeam fluorescence/Raman image analysis of human lens with multichannel detection: Evidence for metabolic production of a green fluorophor. *Proceedings of the National Academy of Sciences USA,* 85, 103 –106.

Zeil, J. (1983) Sexual dimorphism in the visual system of flies: the compound eyes and neural superposition in Bibionidae (Diptera). *Journal of Comparative Physiology A,* 150, 379 – 393.

Zhu, H. and Kirschfeld, K. (1984) Protection against photodestruction in fly photoreceptors by carotenoid pigments. *Journal of Comparative Physiology A,* 154, 153 –156.

Zigman, S. (1983) The role of sunlight in human cataract formation. *Survey of Ophthalmology,* 27(5), 317– 326.

Zigman, S. (1985) Photobiology of the lens. In The ocular lens: Structure, function and pathology (Ed Maisel H.). Dekker Inc., New York. pp 301– 347.

Zigman, S. (1991) Comparative biochemistry and biophysics of elasmobranch lenses. *The Journal of Experimental Zoology, Supplement* 5, 29 – 40.

Zigman, S. and Bagley, S. J. (1971) Near ultraviolet light effects on dogfish retinal rods. *Experimental Eye Research,* 12, 155 –157.

Zigman, S. and Phaxia, T. (1988) The nature and properties of squirrel lens yellow pigment. *Experimental Eye Research*, 47, 819 – 824.

Zigman, S., Paxhia, T. and Waldron, W. (1985a) Properties and functions of near-UV absorbing pigments in marine animal lenses. *Biological Bulletin,* 169, 564.

Zigman, S., Paxhia, T. and Waldron, W. (1985b) Biochemical features of the grey squirrel lens. *Investigative Ophthalmology and Visual Science,* 26, 1075 –1082.

Zyznar, E. S. and Nicol, J. A. C. (1971) Ocular reflecting pigments of some malacostraca *Journal of Experimental Marine Biology and Ecology,* 6, 235 – 248.

5
Vertebrate photoreceptors

N.A. LOCKET
Department of Anatomical Sciences, University of Adelaide, South Australia 5005

5.1 Introduction

Practically all vertebrates have functional eyes, and in all these the site of the photochemical reactions that start the process of vision are within specialised neurons in the retina. These cells are the photoreceptors or visual cells, which include the two broad classes, rods and cones. The visual pigments, molecules of which undergo reversible structural changes in response to light, at least in rods, are located in the membranes of bilayer discs, stacks of which make up most of the portion of the cell known as the outer segment. The spectral properties of the pigments have been correlated with environmental parameters, notably by Lythgoe (1979), and the connectivities and arrangements within the retina and visual pathways have been studied in depth by more authors than can be referred to here. Retinal photoreceptors, while all possessing common features, vary widely in size and arrangement between different retinas, and show various features that are developed to a greater or lesser extent. Are all these differences concerned with phylogeny, or are some responses to environmental factors? Some adaptations certainly correlate well with environmental conditions, such as the devices for optimising sensitivity in deep-sea fishes. Other adaptations have developed in a single group only, though in response to a problem common to many groups, e.g. cone mosaics in teleosts. This chapter will describe the features of visual cells and their sometimes specialised arrangements within retinas, and attempt to relate the structural features to the habitat of the animals in which they are found. Though some conclusions can be drawn with confidence, it will be apparent that neat answers are often lacking.

Rods and cones, which are highly specialised and polarised neurons, share a clearly discernable common structure, and their arrangement within the eye is also very uniform among the vertebrates. This uniformity largely reflects their common embryological derivation, now briefly reviewed.

At an early stage of development the neural tube forms two lateral evaginations, the primary optic vesicles. The centre of these grow slowly while the rims grow faster, outgrowing the centre and leading to the formation of a two-layered optic cup. Figure 1. The outer layer of this cup begins to form melanin pigment early on and is destined to become the pigment epithelium (PE) of the retina and its forward continuation over the ciliary body and back of the iris. The inner layer

S.N. Archer et al. (eds.), Adaptive Mechanisms in the Ecology of Vision, 163–196

of the optic cup gives rise to all the cell families of the neuroretina. The space between the outer and inner layers of the optic cup is continuous at first with the ventricular cavity of the developing brain, and the cells abutting onto this space correspond to ependyma, which lines the ventricles and with which the PE cells and visual cells, though they become structurally and functionally different, share affinities. The original ventricular space persists as the intercellular space between the vitread projections of the PE cells and the sclerad portions of the visual cells, sclerad to the outer limiting membrane (OLM). Due to the formation of the optic cup by invagination, the outer segments of the visual cells eventually face sclerally, and the light forming the image must traverse the rest of the neuroretina before reaching the light sensitive part. The outer limiting 'membrane' is formed by the junctional complexes between visual cells and the sclerad extremities of radial (Müller's) fibres, it delimits the ventricular space between the opposing cells of the pigment epithelium and the neuroretina where the cells are in close contact. Not all the neuroretina is occupied by retinal neurons; the gaps between them are occupied by expansions of radial fibres.

5.2 The visual cell

A general description of an unspecialised vertebrate visual cell is now given, and variations on this general model follow. Before proceeding, attention is drawn to the reviews of Crescitelli (1972), Rodieck (1973) and Borwein (1981), which

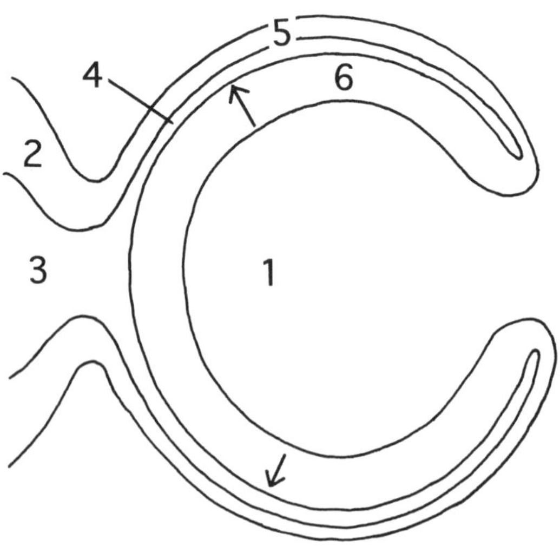

Figure 1 Optic cup, 1, forming from evagination of forebrain, 2. The cavity of the forebrain ventricle, 3, is continuous with the ventricular space between the two layers of the cup, 4. Outer layer of the cup, 5, will become pigment epithelium and its forward continuations. Inner layer of cup, 6, will give rise to the neuroretina. Visual cells are derived from ependyma, so face sclerad into the ventricular space, arrows

deal with many of the features now described in more detail, with many references.

The sclerad portion of the receptor, the outer segment, is specialised to contain visual pigment, and it is here that the initial events in phototransduction occur. The pigment molecules are located in the membranes of closely stacked bilaminar discs, arranged (in rods) like piles of coins, and so in a plane normal to the incoming light. The thickness and spacing of the disc membranes (c16 nm) is much less than the wavelength of light (c500 nm), so that the whole outer segment is optically homogeneous; it is also optically dense. The photolabile visual pigment is constantly being broken down and reformed in photopic conditions, a process in which the PE cells play a major role in most species. The outer segments, particularly those of rods, also undergo physical turnover, being eroded at the scleral end and formed at the vitread. The close relations of the outer segments, structural and functional, to the PE, and outer segment turnover are considered further below.

Immediately vitread to the outer segment is the inner segment, extending as far as the outer limiting membrane. The sclerad part of the inner segment is called the ellipsoid, and always contains an aggregation of mitochondria. In normal retinas the outer segment is always found in close proximity to the mitochondria of the ellipsoid, a modified cilium with the 9 + 0 pattern of tangential microtubule doublets characteristic of sensory cilia joining the two. In some cases the cilium appears to be the sole connection between the inner and outer segments, though exceptions, discussed further below, are known. The close contact between the ellipsoid and outer segment results in these behaving optically as one. Between the ellipsoid-outer segment complex and OLM there is usually a portion known as the myoid, still within the ventricular space, and so called because of its capacity for elongation and contraction with light and dark adaptation in certain animals, particularly teleost fishes. In the extremes of these states the myoid may appear virtually absent, with the ellipsoid immediately adjacent to the OLM, or much elongated. Alteration in length implies a corresponding alteration in diameter, and in some cases the elongated myoid becomes a filament only c 0.5 μm across. Within the myoid, i.e. vitread to the mitochondria, the synthetic apparatus of the cell, the Golgi complex and varying amounts of rough endoplasmic reticulum, are located, both concerned with the renewal of outer segment material.

Vitread to the OLM the cell continues as a conducting fibre, of varying thickness, and containing, somewhere along its length, the nucleus. The visual cell nuclei, of varying size and shape are located in the outer nuclear layer. This layer is usually several nuclei thick; the layering is a packing phenomenon, due to the oval nuclei being wider than the cylindrical inner-outer segment complexes. In retinas with more than one population of receptors, the nuclei of rods are typically smaller than those of cones, and located more vitreally. Cone nuclei are often paler staining, e.g. with toluidine blue, with a tigroid pattern of chromatin, and located immediately adjacent to the OLM, or in some cases sclerad to it.

The extension of the conducting fibre vitread to the nucleus passes to the outer plexiform layer, where the photoreceptor terminates as a synaptic ending. In some instances the synapse is immediately adjacent to the nucleus, without intervening constriction; such sessile synapses occur in e.g. alligators (Kalberer and Pedler 1963). The synaptic terminal varies considerably in size and complexity, but does show some uniform features right across the chordates. The terminal is simple and called a spherule in the case of rods, or complex and called a pedicle in cones. The conducting fibre ends in a swelling, invaginated at the vitread end by processes from second order cells. The terminal contains elongated densities called synaptic ribbons, typical of chordate photoreceptors, which are aligned with the sites of contact of the second order cells and are usually surrounded by a cluster of synaptic vesicles. The ribbons are not static structures; they have been shown to disassemble in the dark, when the exocytosis of vesicles is occurring, reforming in the light, when synaptic transmission is low (Schnapf and Baylor 1987, Wagner 1973, Schmitz *et al.* 1989). Schmitz and Drenckhahn (1993) studied this disassembly by enhancing it with Li^+, finding the ribbons disintegrated into spheres before dissolving into aggregates of fine granules. Polarised against the pedicle membrane at the site of triad contacts, where an arciform density bulges the synaptic cleft, the ribbons are platelike structures to which synaptic vesicles appear attached by thin threads. The ribbons are believed to control synaptic vesicles, guiding them to the synaptic cleft (Dowling 1987). These vesicles, also scattered rather evenly through the cytoplasm of the terminal, contain and release glutamate (Sarantis and Mobbs (1992) for references).

An undifferentiated receptor thus includes a pigment-containing outer segment, closely associated with the mitochondria of the inner segment. A nucleus is present at some location along the conducting fibre, and there is a synaptic terminal; probably all these features are necessary for a receptor to function. We may now examine variations, consider how they may affect function and so be related to environmental conditions and lifestyles.

5.3 Rods and cones

It has been known since the time of Schultze (1866) that many animals have two main populations of receptors, called from the shape of their outer segments rods and cones. From their predominance in the retinas of nocturnal animals Schultze proposed that rods subserve scotopic vision, while cones, dominant in some diurnal animals, function in photopic conditions. This duplicity theory has stood the test of time, though his further assertion that only cones subserve colour vision, based on the loss of colour vision in dim light, has not. The differences have been refined over the years, using various characteristics of parts of the receptor; some of these differences appear fundamental, but others are variable. A major distinguishing feature is the presence of closed versus open discs. The bimembrane discs in rod outer segments, except for a few at the base, lie separate from the surrounding membrane, so that each is a closed sac. Cone discs, however, open for at least part of their circumference to the extracellular space. Exceptions to these relationships occur at the base of rod outer segments, where new discs are being formed. Here some discs in continuity with the plasma membrane may be seen, though this is not invariably so. Figure 2.

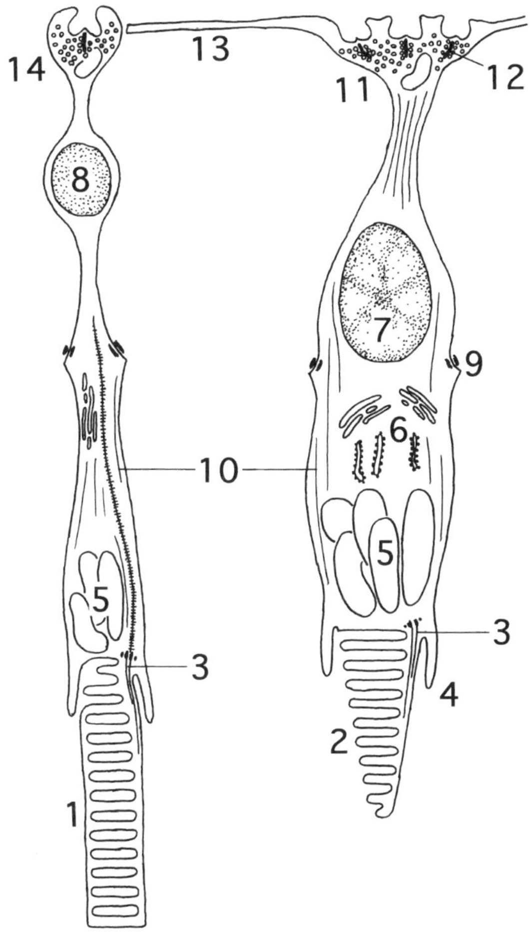

Figure 2 Diagrams of features of rod and cone. Rod outer segment, 1, has closed discs except at base, where new ones are formed. Cone outer segment discs 2, open to the ventricular space. Cilium, 3. is shown as sole communication between inner and outer segments; this is not always so. Rod cilium shows striated rootlet. Calycal processes, 4, surround base of outer segment. Mitochondria form ellipsoid, 5. Golgi apparatus and endoplasmic reticulum occupy myoid, 6. Cone nucleus, 7, is larger than that of rod, 8, and nearer OLM, 9. Microtubules and filaments are present in both, 10. Cone pedicle, 11, is complex, with many synaptic ribbons, 12, and basal filaments, 13. Rod spherule, 14, is simple

5.3.1 Rods

Typically cylindrical, hence the name, the rod outer segment is composed of a stack of free bimembrane discs of equal size, the whole enclosed within the plasma membrane. The discs may have incised or scalloped edges, in which case these are aligned, so that the intact outer segment appears grooved or fluted. The plasma membrane in such cases generally bridges the incisures, so that the outer surface is smoothly curved over the flutings beneath. The incisures may be single, when they are related to the microtubules derived from the connecting cilium, or multiple, as in frogs, where the outer segment appears like of a fluted column. The disc membranes house the visual pigment, a complex of a protein, opsin, and a chromophore, in the case of rhodopsin 11-cis retinal. Most of the discs are regular and arranged transversely, but some at the base of the outer segment may be irregular, appearing too wide for the outer segment and folded alongside the discs further sclerad. Some basal discs may also appear irregular or vesiculated, and may be closer spaced than those more sclerally.

A constant finding at the base of the outer segment is the modified cilium, with the 9 + 0 pattern of microtubule doublets that characterises nonmotile sensory cilia. This cilium takes origin from a basal body, with the 9 triplet pattern of the centriole, located at one side of the inner segment. The tubules of the cilium extend along the side of the outer segment, beneath the plasma membrane, but their symmetrical pattern is soon lost. Isolated microtubules may be recognisable for several microns, but no trace remains at the sclerad tip of the rod. Sometimes, but by no means in all rods, there is a second centriole, located in the inner segment and at right angles to that which gives rise to the cilium. Associated with the centriole(s) may be a sheaf of striated ciliary rootlets which extend for a variable distance vitreally, though they may be absent. Those in guinea pig receptors extend as far as the synaptic pedicle (Spira and Millman 1979). In some rods a series of calycal processes project from the shoulders of the inner segment in a sclerad direction, forming a pallisade parallel to the outer segment. In such cases, some appear to have constant relations with the cilium.

Closely associated with and immediately vitread to the outer segment is invariably an aggregate of mitochondria forming the ellipsoid. These mitochondria are often arranged parallel to the long axis of the rod, or radiating, aligned with the basal body at the base of the cilium. Frequently the outer and inner segments appear joined only by the cilium, though this separation is not invariable; sometimes complete continuity between the cytoplasm of the inner and outer segments is seen. Richardson (1969) demonstrated that rods and cones in four mammalian species had distinct cytoplasmic connections apart from the cilium, though there was always a space between the cilium and the cytoplasmic bridge, so that the cilium was cylindrical, with its own plasma membrane surrounding the axoneme. In tangential sections of retina, so transverse through the visual cells, he invariably found such cytoplasmic bridges in the ground squirrel, and suggested that this is a more likely route for exchange of metabolites between the inner and outer segments than had been assumed from the cilium contact only model. Townes-Anderson (1995) has explored experimental conditions which result in such fusion, finding that it is almost entirely restricted to rods, and that retinal detachment markedly increases the proportion of cells showing fusion. The rods of most vertebrates show little specialisation within the inner segment; oil droplets, found in a wide variety of cones, are uncommon, though they do occur in lungfish.

Outer segment turnover

Early experiments using radioautography showed that rod outer segments undergo renewal, outer segment discs being formed at the vitread end, and shed in packets from the sclerad end, and that this process shows circadian rhythmicity (Young 1978 for references). Further evidence for cyclical disc formation is given by alternating bands of bi-refringence in *Rana pipiens* outer segments, which did not appear in that part of the outer segment formed while an animal was kept in darkness (Kaplan 1981). The shed discs are phagocytosed by the PE cells, in which they appear as phagosomes before their complete destruction.

Besharse (1986) discussed the formation and incorporation of outer segment discs, of which the rate is rapid. He (1982) had measured the rate of disc formation as one in 7.5 minutes, and later (1986) calculated 6×10^4 opsin molecules per minute per rod in *Xenopus*. Besharse (1986) further showed that opsin, which accounts for approximately 90% of the disc protein, is formed in association with rough endoplasmic reticulum in the myoid. From here it migrates to the Golgi apparatus, where it becomes associated with the membranes of vesicles, in which form it is transported to the distal plasma membrane of the inner segment. Opsin, being a transmembrane protein, will not be transported through the cytoplasm of the cilium. It might be transported along the cilium membrane, though the area available compared with the known rate of incorporation, casts doubt on this idea. Though Richardson (1969) had described fusion of inner and outer segments, Besharse (1986) did not think this a likely mechanism since such fusion is so seldom seen. The latter described disc formation as an evagination from the ciliary membrane, not an invagination of the outer segment plasma membrane, accounting for the eventual inclusion of the discs by a membrane fusion process extending from the cilium in both directions round the outer segment, meeting and fusing opposite the cilium. This pattern of disc formation would result in the original outer surface of the plasma membrane becoming the intradisc surface of the closed discs.

Besharse and Wetzel (1995) further investigated the incorporation of opsin into *Xenopus* outer segments, finding it present in outer segment disc and plasma membrane, and inner segment distal membrane and vesicles, but not the ciliary membrane, suggesting that earlier ideas of transport entirely via the cilium may be flawed. They did observe labelled projections of inner segment apical membrane and basal outer segment discs in the space between inner and outer segments, postulating that transport may be directly from apical inner segment membrane to new outer segment discs. In well fixed material these are often in very close proximity, without the gap seen in those authors' cultured material.

The process of disc formation cannot be divorced from the removal of discs at the tip of the outer segment, a process both light related and under circadian influences. The length of rod outer segments at any given time will be determined by the relation between rate of formation and rate of removal. The experiments discussed by Young and Bok (1969) showed that the pulse of radiolabel incorporated into rod outer segment eventually became detached and incorporated into the PE cells. Work since then, reviewed by Besharse (1986) has shown that components derived from both rod and PE cells are involved in the process. Packets of discs become detached or partially so, the zone of separation being marked initially by a band of vesiculation. The tip then separates, either spontaneously or due to phagocytic action by the PE cell, in which a zone of cross-linked actin develops. The separated tip is finally surrounded by pseudopodial processes of PE cells, when the fragment becomes a membrane-bound phagosome, its origin from discs still recognisable initially. The ingested portion then moves to the base of the PE cell, where it fuses with a membrane-bound lysosome, undergoing degradation to an amorphous residual body before being eliminated from the cell altogether. Herman and Steinberg (1982) showed that colchicine supressed phagosome movement through the PE cells in the opossum, suggesting that microtubules were involved.

Experiments by LaVail (1976) showed that rod discs shedding was induced in rats by exposure to light after a period in the dark. This observation triggered a volume of work, including the major contributions of Besharse and colleagues, on the cyclical nature of rod disc shedding. An episode of nearly synchronous disc shedding occurs in the early morning, even in conditions of darkness. The ongoing rhythm in constant conditions is evidence for a circadian element in disc shedding, the zeitgeber in this case being the onset of light. Though light triggers the shedding episode, constant light suppresses the shedding cycle; a dark process is required before this can take place. Besharse (1986) has observed shedding in cultured eye cups of *Xenopus*, showing that the mechanism resides within the eye, and is not dependent on systemic factors. This was already suspected since shedding can be induced in one eye of an intact frog by light to one eye only.

In early experiments with radio labelled outer segments, it was thought that cones did not shed discs, since the label did not form discrete bands. Besharse (1986) states that shedding and reassembly do in fact occur in *Xenopus* cones, some 30 – 60 min after light onset.

5.3.2 *Cones*

The retinas of animals of diurnal habit, including most birds, many teleosts, reptiles and mammals, contain a large proportion of cones. These photoreceptors were originally named from their conical outer segments, though some receptors with the properties of cones have cylindrical outer segments. Cone outer segments differ from those of rods in the relations of their discs. Those of rods, except for the basal few, lie free within the plasma membrane, but cone discs retain their connection with that membrane, so that each opens to the ventricular space. Many cones share features in which they differ from rods, some morphological, but also the possession of different visual pigments, discussed elsewhere in this volume. Cone outer segments are often smaller than those of rods in the same retina, but the inner segments are larger. A cilium is present at the base of the outer segment as in rods, and similar, but often more numerous calycal processes. The cones of some fishes have a feature variously called the accessory outer segment or lateral sac. This is a projection of plasma membrane from the sclerad end of the ellipsoid, running parallel with the outer segment for a variable distance, and containing the extensions of the ciliary microtubules. The membrane is constricted between the lateral sac and outer segment, so that the two are only connected by a strip of cytoplasm, which may be interrupted, giving a series of bridges (Borwein 1981). The presence of this strip distinguishes the sac from the more usual continuation of the ciliary tubules along the outer segment, beneath the plasma membrane which is not raised from the outer segment surface. The lateral sac when present forms a cytoplasmic connection between inner and outer segments, of greater diameter than the ciliary shaft. It may reach to and make contact with the pigment epithelium. Though its function remains obscure, it could be a pathway for metabolites independent of, and wider than, the cilium.

The cone ellipsoid resembles that of rods, but the mitochondrial aggregation is wider than the outer segment, and considerably wider than that of rods in the same retina, containing many more mitochondria. In cones of certain taxa, notably birds, there is a refractile droplet in the sclerad part of the ellipsoid, just proximal to the outer segment, which is composed either of lipid or altered mitochondria (See below for discussion). Vitread to the ellipsoid, but still sclerad to the outer limiting membrane, may be a body called the paraboloid, which is composed of abundant smooth endoplasmic reticulum enmeshing granules of glycogen. Such paraboloids may be highly refractile, as shown by microdissection. The glycogen of the frog paraboloid did not appear to be altered by starvation or light or dark exposure (Majima 1925), suggesting that it is not primarily a store of nutrient. Glycogen is also present in primate myoids, though it does not form a discrete paraboloid; this glycogen is depleted by interruption of the choroidal circulation (Mizuno and Takei 1975), so may be accessible for metabolism. The base of the cone myoid, just sclerad to the OLM, may be surrounded by a series of fins, which extend for variable distances along the inner segment, becoming lower and fading out in some cases, or becoming continuous with the calycal processes in others. The fins of adjacent cones may intermesh with each other near the OLM, and microvilli from radial fibres may also be closely associated with them at this level. The function of these fins is not clear, but they certainly present a large surface area of plasma membrane to the surrounding ventricular fluid, perhaps concerned with diffusion of metabolites between the cone and its surroundings.

The nuclei of cones are usually situated close to the OLM, but within the neuroretina, though occasionally they may be on the sclerad side of the OLM. Cone nuclei in duplex retinae are commonly larger and less densely stained than those of the rods. Cone connecting fibres are usually thicker than their rod counterparts, and the synaptic pedicles show marked differences. Cone pedicles are complex, spreading into a flattened disc indented by horizontal cell and bipolar processes making contact. Whereas rod spherules typically have from one to five contacts, cones may have twelve or more. Besides the indenting contacts, bipolar processes, often from midget bipolars, make surface contacts with the pedicle. The complexity of cone pedicles is further shown by the basal filaments which project in the tangential plane, to make contacts with neighbouring receptors.

5.3.3 Multiple cones

In the retinas of teleosts and reptiles, multiple cones of two main patterns are found. In double cones, typical of reptiles, the two constituent cones are dissimilar. One, the chief cone, has an oil droplet between its ellipsoid and the outer segment, a small paraboloid and thin myoid. This is located against the very large paraboloid of the accessory cone, which has no oil droplet, and of which the ellipsoid is located more vitread than that of the chief cone. The retinas of teleost fishes also contain multiple receptors, including double cones in which the two members are unalike, though in teleosts neither have oil droplets. Paired cones in teleosts usually appear identical, so are known as twin cones; the two members may contain different visual pigments. The outer segments of twin cones, though in close proximity, are not in contact. The ellipsoids do lie in close contact with a subsurface cistern located beneath the adjacent plasma membranes. Such cisterns also occur in other retinal neurons, where they approach other cells, neural or glial, closely. Borwein (1981; also for further references) suggests these cisterns originate from rough endoplasmic reticulum, pointing out that they may be associated with ribosomes. The close contact may extend right to the synaptic pedicles; in the unequal double cones of a turtle one is wrapped partly round the other (Goede and Kolb 1994). In other instances the fibres may be separated from each other by radial fibre material, and the synaptic pedicles discrete. Twin cones are usually arranged in geometrical mosaics, treated further below. Scattered in retinae containing double or twin cones may be aggregates of three or even four cones. In cyprinid triple cones the three members are arranged in a line, but in gadids and most other taxa which have them the three are in a triangle. Quadruple cones are unusual, and when they do occur are usually in a square, as in *Cottus* and the related *Agonus*. In the minnow *Phoxinus* and certain flatfish there is a central cone with three others arranged around it. (Engström 1963 for discussion). Figure 3.

Anchovies: a special case A most unusual association of cones occurs in anchovies (Fineran and Nicol 1976, 1977, 1978). The anchovy retina appears vertically striated when viewed from externally, and light and transmission microscopy show that this is due to alternating rows of rods and cone complexes. Both rods and cones participate in retinomotor movements, in which the PE is also closely involved, its movements linked to that of the cones. All the cones and the PE cells move in synchrony, but the rods, of which the structure is orthodox, move independently. Unlike the rods, the anchovy cones are highly specialised. They occur in two types, one long, with a wedge shaped outer segment, and one short, with a bifid outer segment, the two types being arranged in intimate contact to form a succession of cone units. The outer segments of the long cones are oriented with the flat faces and tip of the wedge horizontal, and at a more sclerad level than the bifid cones. Vitread to the long cone outer segment the cell is indented on upper and lower sides by arms of outer segment material belonging to the bifid cones above and below it in the row. The tip of each bifid

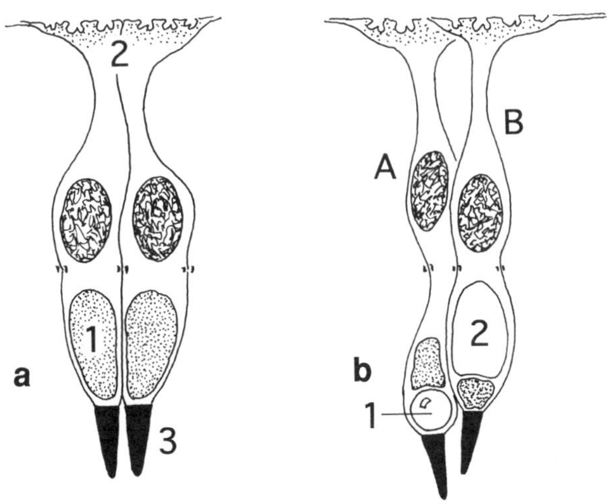

Figure 3 a. Twin cone, as found in teleosts. The two members are in close contact from ellipsoid, 1, to pidicle, 2. The outer segments, 3, though they look identical may contain different visual pigments. b. Double cone, as in reptiles, amphibians and birds. The chief cone, A, has an oil droplet, 1, but no paraboloid. The accessory cone, B, has no droplet, but a large paraboloid, 2. Conducting fibres and pedicles are separate, but inner segments are in close contact

cone, consisting of outer segment material, is partially split by the wedge-shaped tip of a PE cell, and also abuts the long cone outer segment. The arrangement is best visualised from a diagram, (Figure 4). The spacing of cones within a row appears very regular, and alternates between rows, so that tangential sections show an approximately hexagonal pattern.

The cone outer segments are unusual not only in shape; the lamellae, corresponding to the discs of conventional receptors, are not oriented transversely to the long axis of the cone, but almost longitudinally. This is so in both long and short cones, but the long cone lamellae are parallel to the ventral flat face of the wedge, so lie horizontally with respect to the long axis of the fish, while the bifid cone lamellae are at right angles to these, in the plane of the vertical rows of cones. The lamellae of both lobes of bifid cone outer segments are in continuity vitreally. The long cone lamellae are open to the extracellular space along the nasal aspect of the outer segment, while a lateral sac projects from the inner segment and lies beside the temporal side of the outer segment. Calycal processes are also present on the temporal and nasal faces of the long cones. The bifid cone lamellae are open at the faces opposed to the PE cell wedge interposed between the lobes.

The inner segments of both cone types contain aggregates of mitochondria, and where the inner segments abut those of adjacent cones, subsurface cisterns of endoplasmic reticulum, similar to those described in typical (e.g. Fineran and Nicol 1974) double cones occur. The alternation of long and bifid cones means that each vertical row is composed of cones of the two types with no significant extracellular space between them at the ellipsoid level.

The anchovy rods contain a typical connecting cilium between inner and outer segments, but the cones have a connecting structure which, though probably

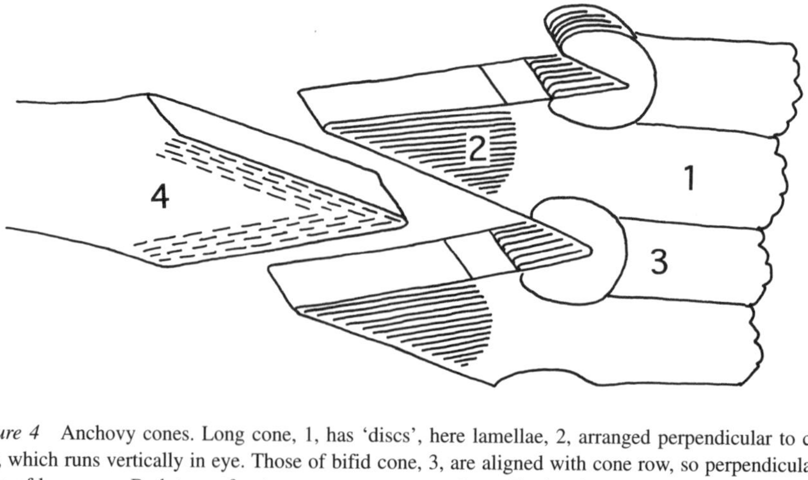

Figure 4 Anchovy cones. Long cone, 1, has 'discs', here lamellae, 2, arranged perpendicular to cone row, which runs vertically in eye. Those of bifid cone, 3, are aligned with cone row, so perpendicular to those of long cone. Both type of outer segment are exposed to reflection from crystals in interposed pigment epithelium cell, here withdrawn for clarity, 4. After Fineran and Nicol (1978)

derived from a cilium, has lost the microtubule doublet pattern. Single filaments pass through the connector, associated at the vitread end with densities, but the latter do not show the triple tubule pattern of a typical basal body.

Reflecting tapetal plates within the PE cells move with the cones in retinomotor movements, retaining their orientation, but in the dark adapted position the cones will be largely blanked off from incoming light. In this situation the plates probably reflect light back into the rods, now vitread to the cones. In this condition the illumination of the rods is further enhanced by the presence of diffusely reflecting crystallites in the bodies of the PE cells, sclerad to the oriented plates. This mass of diffuse reflecting material, which bridges the gaps between the rows of cones in the dark adapted condition, probably gives rise to the prominent eyeshine described by the authors.

Fineran and Nicol (1978) discuss the function of the cone units and their reflectors, indicating that such an array of visual pigment-containing lamellae orientated at right angles and in relation to a reflecting surface of platelets arranged parallel to the long cone lamellae is well adapted to analyse the angle of plane polarised light. Lythgoe (1972) pointed out that underwater illumination is partially plane polarised due to Rayleigh scattering, and noted that a diffusely reflecting object depolarises the light it reflects. Most vertebrates do not have the ability to detect polarised light, though some fishes and many invertebrates do, the latter with rhabdomal microvilli rather than the lamellae characteristic of vertebrates containing the oriented visual pigment molecules. Waterman (1981) gives an extensive review of polarisation sensitivity, discussing Fineran and Nicol's findings in anchovies. The presence of a mechanism which could detect the plane of polarisation does not necessarily mean that capability is present, but it might be advantageous if it is, since viewing a scattering object by its depolarised light against a polarised background could increase contrast. (See also under cone mosaics, below, and Cameron and Easter 1995).

5.4 Oil droplets and ellipsosomes

The retinas of birds, some reptiles, some primitive fishes, monotremes and marsupials contain refractile and often densely coloured spheres in the ellipsoid, adjacent to the base of the outer segment. These bodies, referred to in early literature, e.g. Walls (1942), as oil droplets, have been shown to contain carotenoid pigments in some cases. Bowmaker *et al.* (1993) showed that Japanese quail kept on a carotenoid deficient diet lost the oil droplet pigmentation, as measured by microspectrophotometry, the colours returning when the birds were returned to a normal diet.

Though the droplets appear similar in fresh preparations or by older techniques of light microscopy, they are not identical in all cases. Kunz and Regan (1973) examined the droplets in the guppy retina, showing that they do not stain with lipid stains, but do, like the adjacent mitochondria, with Janus green. MacNichol *et al.* (1978) showed that the droplets in guppies and some related fresh-water teleosts have an absorption spectrum typical of mitochondrial cytochrome c, and coined the name ellipsosomes for these mitochondria-derived droplets. The pigment, present in great density, will block a narrow band of wavelengths from 400 – 430 nm from reaching the outer segments, and, interacting with the cone pigments present, enhance contrast in the blue-violet region of the spectrum.

The droplets in bird retinas have different characteristics. Muntz (1972) lists the colours found in certain species, noting that there are red, orange, light orange, yellow, green and colourless in various taxa. He states that colourless droplets occur in chondrosteans, dipnoans, amphibians, reptiles, birds, monotremes and marsupials. Yellow in amphibians, reptiles and birds, but red in birds and turtles only. This list may be extended by droplets of unknown colour in a few cones of *Latimeria* (Locket 1973) and red in *Protopterus* cones.

It was long thought that the coloured droplets would act as cutoff filters, enhancing contrast. Muntz (1972) examined those in the cones of certain seabirds, noting that those which catch their prey under water, e.g. razorbill, shearwater and shag, have few red and orange droplets, while gulls and terns, which feed at or near the surface, have many such droplets. Muntz suggested that the red and orange droplets would help terns and gulls to see through the water surface, though Lythgoe (1979) points out that their presence would increase the apparent brightness of reflections from the surface, increasing glare. Lythgoe proposed that the droplets enhanced vision at long distances in hazy conditions, advantageous to birds that are attracted to others which have found a patch of feed. He also pointed out that red droplets would markedly reduce the sensitivity of a retina at even shallow depths in sea water, which filters out the longer wavelengths preferentially. The cone pigments and oil droplets act together to determine the spectral characteristics of the cone. Such pairs were reported by Maier and Bowmaker (1993) in *Leiothrix lutea*, a passerine bird, which has four pairs of combinations, including one in the UV. They were able to show that the physical data, modified by neural opponent mechanisms, would mediate the observed behavioural spectral sensitivity function.

5.5 Retinomotor movements

Walls (1942) summarised the early findings on the movements which rods, cones and pigment granules make in certain lower vertebrates, notably fishes and amphibians. These movements occur in response to light and dark, and, like disc shedding, also show circadian changes. The well marked movements in the retina of the catfish, *Ictalurus nebulosus*, demonstrate the events: in the light adapted retina the cone myoids are short, so that the cones are close against the OLM. The rod myoids are elongated, and their ellipsoid-outer segment complexes located sclerad to the cones, amongst the processes of the PE cells. In these cells the pigment granules are in the tips of the processes, thus vitread to the rods but surrounding the cone outer segments. These are thus shielded from each other by the pigment, maximising resolution in photopic conditions, while the rods are shielded, almost or entirely, from incoming light. In the dark, the rod myoids shorten, bringing their outer segments close to the OLM, where the cones were in the light. The cone myoids elongate, so that the cone inner-outer segments are now sclerad to those of the rods, and the pigment granules withdraw sclerally to the base of the PE cells. In this condition the rods are fully exposed to any incoming light, but are not shielded from each other. This situation is well adapted to dim illumination, as sensitivity not resolution may be most important in scotopic conditions. In teleosts with twin cones, both move together, but in the case of double cones, only the principal cone moves, the accessory one remaining the same length whatever the conditions.

The mechanisms underlying these movements have been investigated by Burnside and colleagues (Burnside *et al.* 1993, and for references). Teleost cone myoids contain a central bundle of microtubules and peripheral actin filaments; the microtubules are involved in myoid retraction, and the actin in elongation. Burnside *et al.* (1993) have shown, using isolated inner-outer segment preparations, that both microtubules and filaments can be generated from the ellipsoid in vitro. Forskolin, an adenylate cyclase activator, produced marked myoid elongation, while light-induced cone contraction is mimicked by dopamine, via D2 type receptors, antagonists to which block circadian cone contraction, suggesting a role for dopamine and D2 receptors in cone contraction. These experiments show that the transduction mechanism for cytoskeletal movements are present in the inner-outer segment part of the receptor cell.

5.6 Cone mosaics

Many teleost fish have cones arranged in well marked mosaics; in some the cones are arranged in rows, in others patterns based on squares. The patterns include double or twin cones, of which the pairs may be arranged in an alternating pattern, and single cones, of one or more kinds. The morphological varieties of cones and their patterns were shown and discussed for many taxa of fishes by Engström (1963), who stressed the importance of using tangential as well as radial sections when examining them. Ali and Anctil (1976) give illustrations of retinas from 167 families of fish, showing not only cone mosiac patterns but the extreme variability of visual cells among this vertebrate group. Engström described and illustrated two basic types of mosaic, one consisting of alternating rows of double cones with rows of single cones. In this type, found in cyprinid fishes, the orientation of the double cones alternates within the row, and long and short single cones alternate within their rows. The second type, which may be derived from the first by altering spacings and angles of the various constituent cones, is the more common, being found in many families of fish. Engström illustrates four different arrangements which could be regarded as basic to the pattern, with doubles orientated in different ways to the singles interspersed among them. He states that of the two types of single cones, the long, which occur in the centre of the squares of doubles, are commonest, while short or additional singles are rare or absent in many taxa (Figure 5). Not present in all fishes, the better developed mosaics are found in those which rely on vision, e.g. diurnal fish which take fast moving prey. In benthal, nocturnal fish it is usual to find the mosaics disintegrated.

The mosaic may change with retinomotor movements, as shown by Kunz (1980, 1983) in guppies, and in such cases it would be important to fix eyes for study under appropriate conditions. Light and dark adaptation should be considered, and also time of day, since retinomotor movements are subject to circadian influences. In such a case, the cones would be in their mosaic positions when they are in the light adapted (day) position, close to the OLM. When their myoids elongate, the pattern may well be obscured as the rods take up their dark position. Engström notes that the single and double cones of mackerel migrate differently, and so do the two elements of the unequal doubles in the cyprinids, as in amphibians, reptiles and birds. Lythgoe (1979) further discussed such mosaics, in some of which the cone pigments have been identified, and gave diagrams of the arrangements of cones by maximal absorbance in several fish species, relating the types to the spectral properties of the waters in which they live. He showed oceanic species with a preponderance of green and blue cones but no red; shallow marine fish with mainly greens and yellow, but with one species (*Spinachia*) having red/green pairs, and fresh water species with red cones well represented, in *Perca* almost to the exclusion of others. These fish have red fins, more intense in the breeding season, of which the spectral reflectance is negligible below c500 nm.

As Wagner (1990) points out, the patterns may change with the growth and development of the fish. Evans and Fernald (1993) describe a single population of receptors, which they characterise as cone-like, forming a hexagonal array in the epipelagic larvae of winter flounders. With metamorphosis and descent to the bottom the uniformity vanishes; cones are in greatest density in the dorsal retina, while single and double cones are now present, in a square pattern with four doubles surrounding one single. Rods also appear at metamorphosis, but are most numerous in the ventral retina. This unequal distribution may subserve light sensitivity and perhaps movement detection in the water above the fish, while the cones are concerned with more detailed vision on the bottom, where prey is likely to be found. Beaudet, Browman and Hawryshyn (1993) describe a UV-sensitive cone mechanism in young trout, but not larger fish, correlated with accessory corner cones, which disappear from the retina with growth.

The mosaic pattern of cones may well be implicated in the sensitivity to polarised light shown by some fishes. Cameron and Easter (1995) examined the orientation of the elliptical cross-section of the double cones of the green sunfish,

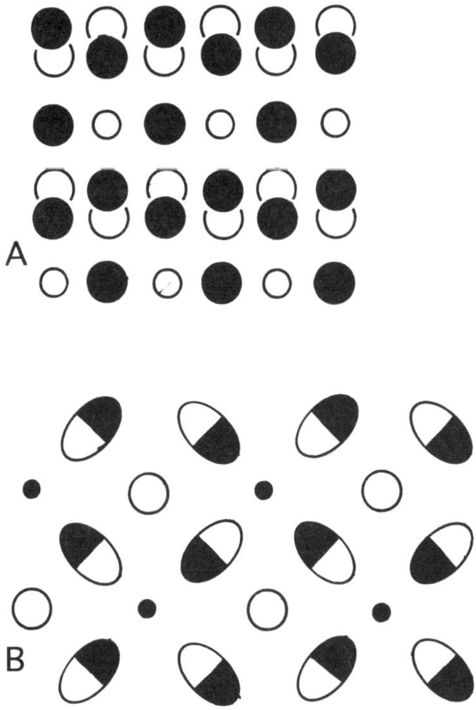

Figure 5 Examples of cone mosaics. A. alternating rows of long and short single cones and double cones, as found in cyprinids. B. Square mosaic, with long and short single, and twin cones, as found in labrids. Different cone types may contain different visual pigments. After Engström (1963)

Lepomis cyanellus, with respect to the psychophysical details of the polarisation sensitivity, finding that the ellipses are arranged in a rhombic mosaic, with the axes at +/– 45° to the nearest retinal edge. Cone densities are highest in the temporal retina, presumably subtending prey in front of the fish, but the mosaic pattern is the same throughout the retina.

5.7 Receptor size and spacing

Amongst the vertebrates the large variation in cell size is reflected in the size of photoreceptors. Those of urodele amphibians, e.g. *Necturus*, and lungfish are particularly large; the rods of *Neoceratodus* are 14 μm in diameter, and subtend c0.1°. This may be compared with the foveal cones of birds of prey, c1.5 μm, subtending c0.04°. The limits of receptor size are set largely by considerations of resolution. Since the best resolution of a retina is ultimately determined by the receptor diameter and spacing, large outer segments will limit resolution. The lower limit of resolution is not limited by receptor diameter, but by the wavelength of light. A point source is not imaged as a point image, but as the Airy disc, of which the diameter is determined by the wavelength and the f number of the system (Land 1981 for discussion). With receptors of diameter less than c4 times the wavelength, geometrical optics cease to apply, and waveguiding phenomena take over. Due to these constraints, the useful lower limit of diameter is about 2 μm, and few outer segments are found below this size.

The ellipsoid-outer segment complex, located within the ventricular space, is separated from its neighbours by ventricular fluid, or in some cases, processes of PE cells. Dunn (1969) showed outer segments about 5 –10 μm diameter in gecko retina, separated from each other by PE processes containing abundant melanosomes. In such a case each outer segment, or pair in the case of double cones, is optically isolated from the next. The separation of outer segments then determines the resolution attainable by that retina. In the gecko retinas the presence of PE processes means that the outer segments occupy only a fraction of the available area, c67% in the case of *Gecko gecko* (Dunn, loc. cit.). Some deep-sea fishes show rods of uniform diameter so closely packed that they form a regular hexagonal array in tangential sections. Such packing results in as much photoreceptive surface in a given area as possible with cylindrical outer segments.

Fine resolution depends on the optics and retinal grain of the eye, but the need for such quality depends to a large extent on the surroundings in which the animal lives and the use to which it puts its eyes. A bird of prey, flying in clear air, can take advantage of the full resolution of which its eyes are capable, and the magnifying effect of the deep fovea of such birds (Snyder and Miller 1978) is a further adaptation to this end. Deep-sea fishes may have little use for fine resolution; often their problem is catching sufficient light to see at all, and adaptations to this are common, and sometimes complex (Locket 1977). Deep ocean water, below the depth to which sunlight penetrates, is not always totally dark; many bioluminescent displays are bright, and since they are often evoked by physical disturbance, may occur close to a fish's eye, when the light flux to the retina will be considerable. Ocean water is clear, but coastal and inland waters may be turbid, to the extent that light may not penetrate more than a few centimeters below the surface. Fishes living deep in such turbid river waters, where there is no bioluminescence, may inhabit darker habitats than deep-sea fishes, and in some the eyes are degenerate, a situation discussed further below. Not only among such fishes is vision a relatively unimportant sense; microchiropteran bats often have small eyes with poorly developed retinas, relying on their highly developed sonar to catch flying insects at night.

5.7.1 *Outer segment length*

Rod outer segments in many species approximate to 20 – 50 μm in length, but much longer outer segments occur in some. In an imperfect quantum absorber, which visual pigments are, it clearly makes sense to have a light path through the visual pigment which gives as good a chance of quantum capture as possible. In photopic conditions there is no need for great length, as there will be ample light to activate the photoreceptor under its operating conditions, but in scotopic conditions, and particularly those close to absolute threshold, a greater length of outer segment might increase the chance of quantum capture. In rod outer segments, subject to turnover, their length will be determined by the balance between formation and removal. Schremser and Williams (1995) have shown that the outer segments of rats become longer if the animals are transferred from a bright to a dark light regime. The change, which is complex, occurs over several weeks.

Long outer segments are not uncommon in deep-sea fishes, even up to 900 μm long in *Diretmus argenteus* (Munk (1966) and personal observation). In some fishes the outer segments, though long, are all of similar length, e.g. *Platytroctes* in which they are c150 μm long. Some fishes however show marked and in some cases abrupt differences in the length of outer segments in different parts of the same retina. This is well shown in *Scopelarchus sagax*, a close relative of *S. guntheri*, and like that species having grouped and non-grouped regions in the main retina. *S. sagax* however has a rostral region in which the rods are c230 μm long, compared to 40 μm caudal to that. The change from one to the other is abrupt, and the resulting step in retinal thickness filled in by PE cells (Locket 1971).

The visual pigments of *Scopelarchus analis* have been investigated using microspectrophotometry by Partridge *et al.* (1992). They measured from the centre of the main and accessory retinae (see below; tubular eyes), but unfortunately did not distinguish between grouped and non-grouped main retina. A most interesting finding was that some main retina outer segments contained two pigments, VP445 in the vitread part, with VP507 sclerad to this. Other outer segments contained only VP445. The accessory retina also contained two-pigment outer segments, the pigments in this case being VP443 and VP503. Though the transition from one pigment to the other was abrupt, the distance along the outer segment was not constant, ranging from 30 –180 μm from the inner segment in 200 μm long outer segments. The accessory retina, all of which is grouped (see below), contained a third pigment, at 479 nm intermediate between the other two, and always continued in very short outer segments, (or fragments of them). The authors suggest that the immature adult fish (76 mm SL) on which they worked may have been in the process of changing over from a longer wavelength pigment to a shorter, but unusually by substitution of opsin rather than chromophore.

5.8 Multibank retinas

Many fishes and some other animals have receptors in their retina arranged in more than one layer, apart from the alternating positions in retinomotor movements, discussed above.

Miller and Snyder (1977) discussed two cases in which the rods and cones, though present as two layers, do not undergo migration, calling this arrangement the tiered retina. They showed that in the pig, with cones vitread, at the level of their ellipsoids these occupy 70% of the light capture area. The sclerad rods occupy 90% of the light capture area, thus both populations, though present in the same retina, occupy much more than half the area each. The cat-eyed snake has its receptors the other way round; rods vitread and occupying 80% of the area while the sclerad cones still occupy almost all the available area. The authors showed that the light which had passed through the vitread bank, of rods or cones as may be, was reradiated into the second bank, with little attenuation in photopic conditions, since the visual pigment was in the bleached state.

In certain deep-sea fish retinas, apparently uniform populations of receptors are present in depth, though the individual outer segments are of normal length. In some, e.g. *Mora*, there is little order to the assemblage, but ellipsoid-outer segment complexes, each about 40 μm long, are connected to their cell bodies by conducting fibres of varying length. In other cases, e.g. *Chauliodus*, the receptors are present in well defined banks (Figure 6). Munk (1966) described and figured such multiple banks in a number of species; Brauer (1908), though he examined fishes which possess multiple banks, had not recognised them. Figures for number of banks of rods appear in the literature, but looking at other examples of the same species produced different figures; it seemed possible that new banks are added as the fish grows. To test this idea Locket (1980) examined a series of *Chauliodus* of increasing size, and confirmed that the number rose from one in a 29 mm fish to 6 in a 250 mm specimen, with examples of all numbers in between. Fröhlich, Negishi and Wagner (1995) have shown that such new banks are produced with high temporal and spatial order from a population of precursor cells just vitread to the outer limiting membrane.

What then are these multiple banks doing? Several ideas have been canvassed. In dim light it may well be advantageous to have a long light path through visual pigment, since quanta are few, and quantum capture not certain, the pigment not being a perfect absorber. It may also be necessary to have the inner segment fairly close to the site of quantum capture if transmission of the effects of capture are to be relayed to the more vitread parts of the receptor. Arranging the relatively short outer segments, each about 30 μm in *Chauliodus*, in multiple banks would ensure a long total light path without the inner segments being too far from the sites of quantum capture.

Consider also the light absorbance of the pigment in the vitread rods. These outer segments will subtract the light they absorb from the spectrum to which

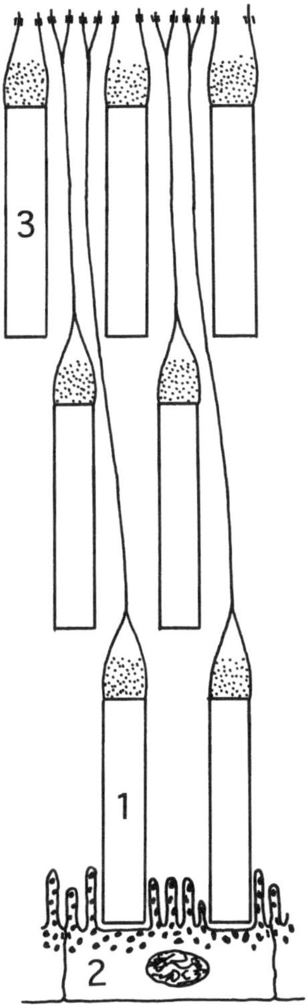

Figure 6 Multibank retina, here with three banks. Sclerad bank OS, 1, are only ones to make contact with PE, 2; middle and vitread banks do not. Vitread bank, 3, is close to OLM, and myoids are short; Middle and sclerad banks have long myoids. See text for implications

they are exposed, and the more sclerad banks of outer segments will thus be exposed to a different spectrum. Denton and Locket (1989) discuss the possibility that this arrangement could lead to wavelength discrimination using a single pigment.

Outer segment formation and erosion, with approximate maintenance of length, was discussed above. Suppose a receptor loses the capability for renewal by this process. If the visual pigment were to become degraded in some way such that the chemical regeneration of the pigment *in situ* did not continue, then the receptor would become a less efficient light absorber. Production of a population of new receptor cells would insert a bank of fully efficient absorbers once more. Counts of the ellipsoid-outer segment complexes, nuclei in the outer nuclear layer, presumed to be those of visual cells only, and synaptic pedicles show that there are significantly fewer synapses than complexes or nuclei. This suggests that the old receptors lose their synaptic connections, while the newly formed receptors must establish new connections. There is indirect evidence to support this idea. Shapley and Gordon (1980) recorded ERGs from the isolated retina of conger eels, which have multibank retinas. They found that when light came from the vitread side, as it would in life, the density spectrum and ERG spectral response matched well. Illuminating from the sclerad side gave a distorted spectral response curve, with the normal peak at 492 nm replaced by a trough. They suggest that only the vitread bank is responding, and in the inverted situation the sclerad rods are shielding the more vitread from the light.

The multibank arrangement itself poses problems for renewal and phagocytosis of outer segments. In the case of a three bank retina, (Figure 7), the vitread rod tips are far separated from the PE, and phagocytosis appears unlikely. That bank is however in the normal position with regard to the rest of the retina, and the inner segment is close to the outer limiting membrane. The sclerad rod tips are in contact with the PE, so well positioned for phagocytosis. Their connection to the cell body is via a slender conducting filament, which could provide a functional link, enabling activation at the synapse. The middle bank is, like the sclerad, connected only by the filament to the cell body, and also, like the vitread bank, remote from the PE.

Howella, a bathypelagic teleost, presents a special case of a multibank retina, and raises several points of interest. The retina at first sight appears to contain a population of rods with long, transversely banded, outer segments. Electron microscopy shows that the bands are composed of mitochondria, mostly only one mitochondrion thick, interspersed between regions with the typical structure of rod outer segments. Best and Nicol (1978) noted the orderly stacking of the rods and that there were only 10 –16 ganglion cells for 24 files of rods, giving a high summation. From their appearance it might be that each stack represents a single outer segment, interspersed with mitochondria, or that each is a true stack of ellipsoid-outer segment complexes. The latter idea is confirmed by the presence of connecting cilia at each inner-outer segment junction, and of connecting fibres between the stacks. These, not obvious in radial sections, show well in tangential sections, and more are present between the vitread end of the stacks than the sclerad, as would be expected. At the vitread end of the stacks short lengths of outer segment material can be seen, each in association with mitochondria, and presumably representing the sites of new outer segment formation (Locket 1992).

Questions of function arise in connection with this argument; the precise alignment of the files of outer segments will cause the whole stack to behave as one optically, and if each member were connected synaptically to the same bipolar, there would be opportunity for summation without loss of resolution. Counts of outer segments, nuclei and synaptic pedicles in radial sections suggest that there are fewer synapses than there are outer segments. Probably the outer segments displaced sclerad by the new ones being formed lose their synapses, presumably ceasing to be functional receptors. The newly formed receptors will then form pedicles and establish functional connections with bipolars. These deductions are made from anatomical findings only, since *Howella* and other deep-sea fishes are seldom alive when caught, and do not survive long if they are.

Jin *et al.* (1994) showed that isolated bleached cones can recover sensitivity when exposed to 11-cis retinal, but that rods under these conditions do not. If the *Howella* rods behave similarly, being unable to recover sensitivity even if supplied with the necessary visual pigment precursor, this might explain the apparent need to produce new rods rather than recycle the old. It is also interesting that Partridge *et al.* (1989) found two pigments in *Howella sherborni* by MSP, though they were unable to state whereabouts in the outer segment stack they were measuring. In view of the finding of two pigments within a single outer segment in *Scopelarchus* it would be well worth looking for such differences in other species, particularly those with multiple bank retinas.

5.9 Tubular eyes

The distribution of light in the sea is highly asymmetrical, downwelling light being much stronger than upwelling or light from laterally (Denton 1990). Many shallow water or epipelagic fishes have reflective surfaces containing crystals set at an angle to the surface of the fish, which cause the reflected light to match that from beyond the fish, making the silvery fish hard to see against the background. Though such reflectors produce an efficient camouflage when viewed from laterally or above, the amount of light available for reflection is not sufficient to match the much brighter downwelling light when seen from below. Thus many organisms can be seen by silhouetting them against the downwelling light. Not surprisingly, adaptations to take advantage of this are found.

A number of mesopelagic fishes have more or less tubular shaped eyes, in which the lens forms a focussed image on a part of the retina, termed main retina, which is located in the ventral part of the eyecup, so looking upwards. (Figure 7). The anterior, medial and posterior walls of the eyecup are lined with accessory retina, which is too close to the lens to receive a focussed image. The structures of these two regions of retina differ considerably, the main retina being well ordered, with dense populations of nuclei and usually a close packed and uniform population of rods. Such a retina appears well adapted to receiving a good image, and corresponds to the retina in a normal fish eye. The accessory retina is much less developed, with few nuclei and less complex synaptic layers. The rods often have shorter outer segments and are more loosely packed than those in the main retina. Such eyes are found in *Argyropelecus*, where the retinae are as described, but other fishes have different arrangements of rods. *Stylephorus* has tubular eyes, directed forwards so that an object in front of the fish is imaged on both main retinas simultaneously. These retinas, unlike the simple arrangement in *Argyropelecus*, contain multiple banks of rods, the number of banks probably increasing as the fish grows. Other fishes, notably scopelarchids and evermannellids, have specialised regions of retina in their tubular eyes, discussed below.

5.10 Grouped receptors

Certain fishes have been known for some time to have their receptors arranged in bundles, not the usual uniform array. Franz (1921) found that mormyrids, African fresh water fishes now known to use electroreception, have a few cones, but more numerous rods aggregated into bundles by the PE cells. These cells contain pigment in their tips, but the bulk of the cell is occupied by what he believed to be guanine crystals. Ophthalmoscopy showed a bluish reflection, and Franz proposed that the crystals formed a retinal tapetum. He compared his findings with Brauer's (1908) figure of *Evermannella* retina; this fish is discussed further below. The retina of *Hiodon*, a fresh water fish from north America, was examined by Moore (1944). He noted that the pattern of nuclei suggested a duplex retina, but saw only rods arranged in definite bundles,

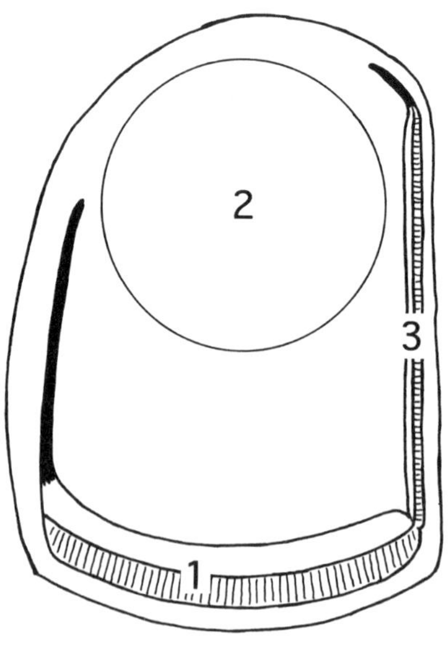

Figure 7 Coronal section through tubular eye, as found in many mesopelagic fishes. Main retina, 1, is at focus of lens, 2, and is well developed. Accessory retina, 3, is not in focus, and is less developed

embedded in the PE in a regular array. The rods undergo retinomotor movements, shielded by day by pigment surrounding the outer segment bundle. At night the pigment migrates, but to the tip of the PE cell, where it surrounds the vitread part of the group, segregating light for each bundle. Moore compared the arrangement to that of *Elops* and *Megalops*, described and compared with mormyrids by McEwan (1938).

Though most deep-sea fishes have an even array of uniform rods in their retinas, there are some in which the receptors are grouped. This grouping has functional implications, considered with each type.

5.10.1 Scopelarchids (Figure 8)

The eyes of *Scopelarchus guntheri* were examined by Locket (1971), who found two areas in the main retina with different rod arrangements. The rostral part has a uniform population, with outer segments about 30 μm long and a low PE containing pigment granules but no reflective material. The caudal part is very different; here the rods are present in groups of about 23, each group surrounded by a pit formed from PE cells, recalling the situation in *Hiodon*, outlined above. These cells contain layers of oriented crystals, with pigment granules confined to the vitread part of the cell. The thickness and separation of the crystals are such that they will function as a reflector, confining light to the group which it enters, while reflecting the light within that group. These receptors, grouped optically by the reflecting pit, also show intercellular junctions at the level of the inner segment, suggesting that they are linked electrically as well. Such a group will function as a 'macro-receptor', behaving as a single entity with regard to light reception and thus sensitivity. Each rod within the group has its own synaptic pedicle, so there is an opportunity for convergence of the synapses of each group onto a single bipolar. From the material available it has not been possible to confirm this suggestion, but in the case of *Notolepis*, another fish with groups of rods, the synapses are certainly grouped (Locket 1977).

The finding of Partridge *et al.* (1992) on *Scopelarchus analis* main and accessory retinas, in both of which two pigments occur within single outer segments, was mentioned above. They did not state whether the main retina outer segments they measured were from the grouped or non-grouped part, but the accessory retina is grouped throughout. The longer wavelength pigment, VP507, was always in the sclerad part of outer segments with VP445 in the vitread part, or was sometimes the only pigment.

Benthalbella infans, another scopelarchid fish, also has rod groups in main and accessory retina, but here the arrangements are rather more complex (A. Murphy, 1994, unpublished). The main retina groups have squared off outer segment tips, and adjacent to each group is a plate of crystals arranged parallel to each other and normal to the axis of the group. This plate behaves as a specular mirror, as shown by examination of the fresh retina, in which dark blue reflective spots are

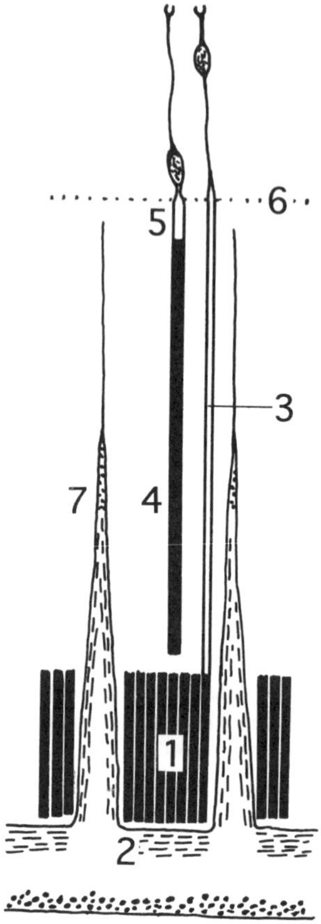

Figure 8 Rod arrangement in *Benthalbella*. Sclerad outer segments, 1, are tightly grouped and abut multilayer reflector, 2. Their ellipsoids, 3, are long, and separate vitread outer segments, 4, from each other. Vitread ellipsoids, 5, are short, and just sclerad to OLM, 6. They are not grouped at this level. PE processes, 7, separate groups sclerally

visible when viewed from the sclerad surface. Besides these mirrors, each group is set in a reflective pit as in *Scopelarchus*. These arrangements mean that light will not only be confined to the group by the walls of the pit, but any which traverses the outer segments without being absorbed will be reflected back the way it came, minimising degradation of the image by scatter.

The *Benthalbella* retina shows a major difference from that of *Scopelarchus*. The latter has only a single population of rods within each group, but in *Benthalbella* there is a second. The vitread rods have short inner segments, just sclerad to the outer limiting membrane, and long outer segments which are mingled with the long inner segments of the sclerad group Figure 8. At the level of their ellipsoid-outer segment junctions the vitread rods are not arranged in groups, though the outer segments extend far into the retina to fall within a group more sclerally. This retina then has two populations of rods, one arranged in groups and the other not. The retina can also be considered multibank, as the two populations are certainly in such a relationship to each other. No details of synaptic connectivity could be made out, so we do not know how the information from the two types of rod may be processed, but certain possibilities arise. An image falling on the retina might be resolved by the vitread rods before it is divided up amongst the groups. It would be interesting if there were more than one visual pigment present, one in the sclerad and one in the vitread rods. Partridge *et al.* (1989) found clear evidence of only one pigment, VP451, in *Benthalbella*, but did note one poor quality MSP record of a rod with VP506. These compare with VP445 and VP507 in *Scopelarchus*, and the 451 nm pigment accords well with the blue reflection from the crystal stacks. Wavelength discrimination is theoretically possible in *Benthalbella* using a single pigment, since the light reaching the sclerad rods will have passed through, and been partially absorbed by, the vitread population. Thus the spectrum presented to the groups will be different to that arriving at the vitread rods. (Denton and Locket 1989 for further discussion). In both *Scopelarchus* and *Benthalbella* the accessory retina is also grouped. This retina, not being at the focal distance of the lens, will in any case not receive a sharp image; groups to maximise sensitivity at the expense of resolution are well adapted in such a case.

5.10.2 Scopelosaurus

Munk (1977) described the retina of *Scopelosaurus lepidus* and found separate regions of visual cells. Cones, including many twins, were confined to the temporal region, subserving vision in front of the fish, and also forming the receptors in the deep fovea. The rest of the retina contained grouped rods, inserted into pits in the PE cells, much as in *Scopelarchus*. *Scopelosaurus* groups, however, contain only two to six outer segments, each c2 μm diameter. The groups are thus 8 –10 μm diameter, separated from each other by c4 μm of PE cell cytoplasm. This contains crystal layers surrounding the group, and pigment granules mostly in the tips of the PE cell processes, further isolating the groups optically. The arrangement of the PE cells with respect to visual cells is continued into the cone region; here each pair of cone outer segments has its own sheath of crystals and pigment granules, the outer segment pair measuring c1.5 × 2.2 μm diameter, and being separated from its neighbour by c1 μm of sheath. The spacing of the pair centres is c4 μm, and the focal length of the lens derived from measurements in Munk's paper and Matthiessen's ratio 670 μm. From this, the angular resolution of the foveal retina is 0.34°, compared with 0.74° for the rod groups, calculated in the same way.

5.10.3 Evermannellids Figure 9

The evermannellid fishes *Evermannella*, *Coccorella* and *Odontostomops* show groups of a different kind (Locket 1976). Here, as in *Scopelarchus*, is an area of single rods, c2 μm diameter and arranged rather loosely, not in the close packed array often seen in deep-sea retinae. Radial sections elsewhere in the retina show an apparently irregular array of 'untidy' looking rods, of even length and not arranged in tapetal pits like those of the scopelarchids. Tangential sections show very narrow rods, only about 0.6 μm diameter, but arranged in small groups, of 2 – 6 rods, in very close contact. The groups are separated by fluid, like the thicker single rods. This tight grouping of thin outer segments will have optical implications; the individual rods are of comparable diameter to the wavelength of the light they encounter (c470 nm), and at these dimensions light does not behave according to geometrical optics, but instead exhibits waveguiding properties (for discussion see Land, 1981). Under these circumstances some of the energy travels outside the outer segment, where it might excite adjacent receptors if they were close enough. The small groups, each about 2 – 3 μm diameter, will behave as a single entity as far as waveguiding is concerned. We thus have an optically single receptor, the group, with as many synaptic outputs as there are individual rods in that group. If excitation of the retina is enhanced by convergence, the sensitivity may well be increased without significantly decreasing resolution by such an arrangement.

5.11 Foveas

When we wish to examine something in our visual environment carefully we look at it, i.e. direct the visual axis to the object of interest. Indeed we usually direct both visual axes to the same point, with extraordinary precision. The retina

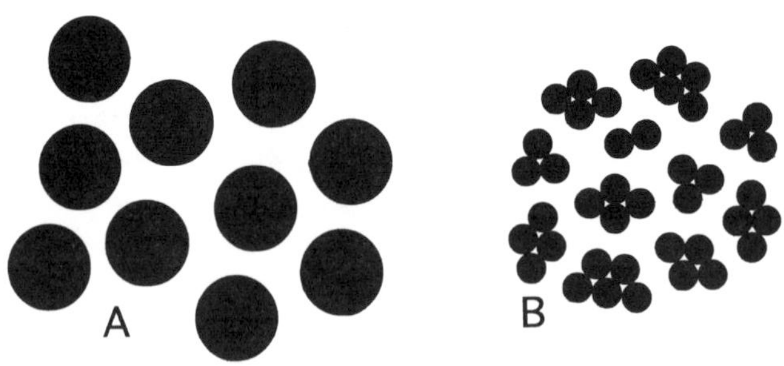

Figure 9 Outer segment arrangement in Evermannellids. A. Single outer segments, c2 μm diameter, are not in close contact, so will be optically separate. B. Small groups of thin outer segments, each c0.6 μm, form groups, each of similar size to single rod. Each group will act as one optical entity, but with several synaptic outputs

at the posterior pole of our eyes is specialised to form a fovea, of shallow or concaviclivate (Walls 1942) form. The human shallow fovea, a feature of primate eyes, is approximately 570 μm diameter, and in the centre is a foveola 350 μm across. The fovea is largely devoid of rods, and the foveola entirely so. The cones are slender, with outer segments 45 μm long by 2 μm diameter. These foveolar cones are aligned perpendicular to the outer limiting membrane, so along the visual axis; they are separated by c0.3 μm, and form a close mosaic, presenting the finest grain to the incoming image. This image has passed through less retinal tissue than elsewhere in the retina, since most of the retinal components are displaced from the foveal centre. The fibres of the foveal cones turn abruptly as they pass through the OLM, to run obliquely away from the centre. These oblique fibres, which contain the nucleus somewhere along their course, together form Henle's fibre layer. The nuclei tend to be aligned in rows, which also slope away from the centre. As they approach the outer plexiform layer, the fibres turn vitread once more so that the whole fibre is S-shaped. This displacement results in the absence of synaptic pedicles from the centre 200 μm. The inner nuclear, inner plexiform and ganglion cell layers are also displaced, with corresponding thickness of these layers in the foveal shoulders. The tissue between the Henle fibres and the inner limiting membrane, here thinned, is almost entirely of radial fibre origin. Blood vessels are largely absent from the centre of fovea, though there is a rich capillary plexus surrounding it, in the macula (Hogan *et al.* 1971). Each foveolar cone makes contact with 2 – 3 bipolar cells, so resolution may be determined solely by the dimensions of the receptors. As discussed above, diffraction limits the extent to which an image can be sharpened by reducing outer segment diameter. In the primate fovea, with outer segments about 2.0 μm diameter, this limit is approached.

The visual acuity of certain raptorial birds was shown behaviourally by Fox, Lehmkuhle and Westendorf (1976) to be up to 2.6 times better than that of humans, though their cone outer segments are of comparable diameter. Such birds also possess a fovea, but of a different kind to the shallow primate sort. Birds of prey, swallows, kingfishers and various lizards including chameleons, possess a deep contoured fovea, called by Walls (1942) convexiclivate. Deep foveas are also reported in blennies, seahorses and pipefishes amongst shallow water fishes and in a number of deep sea fishes, mostly searsid and alepocephalid. These foveas all share similar contours, and are surprisingly similar in absolute size, though from a wide variety of taxa and size of individual. Several theories on the function of such foveas have been advanced (Locket 1992 for review). They occur mainly in a position in the retina that would subtend the same point from both eyes, and this is usually where prey would have to be for the animal to make an effective strike. Such binocular positioning suggests that the foveas are concerned with rangefinding, and watching a feeding pipefish gives a strong impression that the position of the snout with respect to the prey is critical to a successful attack. Pipefish feed on free-swimming plankton, e.g. mysids, by a brief but powerful suck through a small mouth situated at the tip of the snout. Clearly for the suck to catch the prey, the mouth must be in the right place. The size of prey acceptable to a pipefish varies, as does that for a falcon or swallow, and in many cases the prey will be seen against a blank or distant background. Parallax will not give reliable distance clues in such a case, nor may there be other reference points in the field.

Walls recognised that the steep curvature of the foveal surface and the differences in optical density between retina and vitreous would produce refractive effects, and postulated that the image would be magnified. Pumphrey (1948) thought that the image would be distorted, reducing any advantage from magnification, and Snyder and Miller (1978) showed how the concave curve at the bottom of the foveal pit would act as the concave component of a telephoto system with the lens of the eye, producing a magnified but clear retinal image and so explaining the demonstrated acuity of the falcon. The shape and optical function of the deep fovea depend on the contours of the vitreo-retinal boundary and the optical densities of the two media. But what of the receptors underlying such a fovea? Those of birds of prey are narrow cones, diffraction limited. The American chameleon has been thought on morphological grounds to have a pure cone retina, but doubt was cast on this by McDevitt *et al.* (1993) who demonstrated a rod-specific protein in a sparse population of receptors in the periphery, but in high density in the central fovea. Diurnal fishes, e.g. pipefishes, also have narrow cones, but in those deep-sea fishes with deep foveas the receptors are usually all rods. An exception is *Scopelosaurus*, shown by Munk (1975, 1977) to have twin cones in its deep fovea, located in the far temporal periphery. The searsids, e.g. *Platytroctes*, mostly have foveas containing narrow but long single rods. The narrowness may give improved acuity, and the increased length could compensate for the possible loss of sensitivity in a narrow outer segment. The apogonid *Howella*, with the extreme multibank retina described above, has narrow but longer files of shorter rods at the fovea, with c45 rods per file there compared with c12 in the periphery. The increased population of rods in this fovea causes it to project sclerally, forming a fovea externa. Such projections can be prominent; that of *Bajacalifornia*, an alepocephalid, is long enough to distort the sclera, so is visible from outside the eye. (Locket 1985).

5.12 Degenerate retinas

It is self evident that retinas are related to light reception, but some animals are adapted to conditions of little or no light. Such environments are found in the ocean depths, in turbid waters and particularly in caves. All these habitats show examples of degeneration or loss of retinal elements.

Most pelagic fishes have well developed eyes, but in the bathypelagic zone, below the depth to which sunlight penetrates, live examples with reduced eyes. Many deep-sea angler fishes have very small eyes, in some cases buried beneath the skin, and frequently with lenses which do not conform to Mattheissen's ratio. The retinas of such fishes cannot receive a focussed image, and the inner retina is correspondingly poor in cells, with thin or disorganised synaptic layers. The receptors, all rods, are not degenerate, though they do appear simple. The outer segments are well formed and long; those of a *Melanocetus* 57 mm long, of which the eye was only 1.5 mm diameter, located beneath a clear patch in the otherwise deeply pigmented skin, are 55 μm long. The PE is prominent, though it does not send processes between the outer segments. The latter are in contact with the epithelial cells, and what appear to be phagosomes are present in the PE cells, suggesting that rod turnover does occur in this fish. The inner segment is small, and the synaptic pedicle simple. The simplicity of the inner retina is reflected in the number of optic nerve fibres, only c1080.

The ocean bottom, though well below the photopause, is not totally devoid of light, since many bottom living organisms are bioluminescent. Even here some fishes are blind, or with greatly reduced eyes containing degenerate or disorganised retinas. *Ipnops murrayi*, obtained by the Challenger and Valdivia expeditions, was shown from preserved material by Munk (1959) to have lost the sclera, cornea and lens, the retina being spread out as a plate on the top of the head. This retina contains only rods, of which outer segments are certainly recognisable. Attached to the base of these is a dark staining mass, which may represent the inner segment, nucleus or both. The rest of the retina is extremely tenuous and cells hard to make out. Such a retina could function as an undifferentiated light receptor at best. Other bathybenthal fishes, e.g. *Bathypterois*, have small eyes, their retinas containing few cells, and with the outer segments much reduced.

At first glance one might expect that rivers in low latitudes would be well illuminated, but turbidity due to suspended solids, and dissolved organic matter, can render their water virtually opaque. Higgins (1932) found that 90% of the light falling on the Mississippi was absorbed in the first 25 mm depth. Not surprisingly some fishes living on the bottom of such rivers have degenerate eyes like those from the ocean floor.

The darkest places are in the depths of caves, to which no light ever penetrates, and which are devoid of bioluminescence. In such situations animals are found, vertebrate and invertebrate, which have lost their eyes altogether.

5.13 Conclusions

The generalised vertebrate visual cell has an outer segment, consisting mainly of bilaminar membrane discs containing the visual pigment. The outer segment is directly adjacent to the ellipsoid, always containing mitochondria and in some cases a droplet, often coloured. A myoid, sometimes contractile, joins the inner-outer segment complex to the inner retina, vitread to the outer limiting membrane. From here a conducting fibre extends to the synaptic ending, with the nucleus along its course.

Rod outer segment discs are closed from the ventricular space by the cell membrane, but cone discs are open. Rod synaptic spherules are simple, cone pedicles complex. Rod outer segments undergo clear cyclic and light-dependent erosion and renewal; cones show less obvious turnover. Multiple cones occur in some groups, in geometrical mosaics in teleosts. Bizarre cones in anchovies may be associated with polarisation sensitivity.

Rods and cones undergo reciprocal movements in some groups in response to light/dark and cyclic influences. Some animals have separate but immobile banks of rods and cones, optimising the area occupied by each class, and some deep-sea fish have multiple banks of rods, the number increasing with size. Some teleosts have receptors arranged in optically isolated groups, increasing sensitivity at the expense of resolution. Particularly narrow receptors, in some cases probably diffraction limited, occur in regions specialised for acuity, including shallow and deep foveas. Some animals from dark environments, e.g. deep sea, turbid water, caves, have wholly or largely lost their retinal photoreceptors.

References

Ali, M. A. and Anctil, M. (1976) *Retinas of fishes*. Berlin, Heidelberg, New York. Springer-Verlag.

Beaudet, L., Browman, H. I. and Hawryshyn, C. W. (1993) Optic nerve response and retinal structure in rainbow trout of different sizes. *Vision Res.* 33, 1739 –1746.

Besharse, J. C. (1982) The daily light-dark cycle and rhythmic metabolism in the photoreceptor-pigment epithelial complex. In: *Progress in retinal research*. Vol. 1. Osborne, N and Chader, G. eds. Oxford, Pergamon Press.

Besharse, J. C. (1986) Photosensitive membrane turnover: Differentiated membrane domains and cell-cell interactions. In *The retina: A model for cell biology studies*: 1. Adler, R. and Farber, D. eds. pp. 297– 352 Orlando, Academic Press.

Besharse, J. C. and Wetzel, M. G. (1995) Immunocytochemical localization of opsin in rod photoreceptors during periods of rapid disc assembly. *J. Neurocytol.* 24, 371– 388.

Best, A. C. G. and Nicol, J. A. C. (1978) Notes on the retina and tapetum lucidum of *Howella* (Teleostei: Cheilodipteridae). *J. Mar. Biol.* Ass. *U.K.* 58, 735 –738.

Borwein B, (1981) The retinal receptor: A description. In: *Vertebrate photoreceptor optics*. Enoch, J. M. and Tobey, F. L. Jr., eds. Berlin, Heidelberg, New York. Springer-Verlag.

Bowmaker, J. K, Kovach, J. K., Whitmore, A. V. and Loew, E. R. (1993) Visual pigments and oil droplets in genetically manipulated and carotenoid deprived quail.: A microspectrophotometric study. *Vision Res.* 33, 571– 578.

Brauer, A. (1908) Die Tiefseefische. II. Anatomischer Teil. B. Augen. Wissenschaftliche Ergebnisse der Deutschen Tiefsee-Expedition auf dem Dampfer ''Valdivia' 1898 –1899. Vol. 15. Jena. Fischer Verlag.

Burnside, B, Wang E., Pagh-Roehl, K. and Rey, H. (1993) Retinomotor movements in isolated teleost retinal cone inner-outer segment preparations (CIS–COS): Effects of light, dark and dopamine. *Exp. Eye Res.* 57, 709 –722.

Cameron, D. A. and Easter, S. S. Jr. (1995) Cone photoreceptor regeneration in adult fish retina: Phenotypic determination and mosaic pattern formation. *J. Neurosci.* 15, 2255 – 2271.

Crescitelli, F. (1972) The visual cells and visual pigments of the vertebrate eye. In. *Handbook of Sensory Physiology. Vol. VII/1.* Dartnall, H. J. ed. pp. 245 – 363. Berlin, Heidelberg, New York. Springer-Verlag.

Denton, E. J. (1990) Light and vision at depths greater than 200 metres. In: *Light and life in the sea.* Herring, P. J., Campbell, A. K., Whitfield, M. and Maddock, L. eds. pp. 127–148. Cambridge, Cambridge University Press.

Denton, E. J. and Locket, N. A. (1989) Possible wavelength discrimination by multibank retinae in deep-sea fishes. *J. Mar. Biol. Ass. U.K.* 69, 409 – 435.

Dowling, J. E. (1987) Retinal synapses. In: Dowling, J. E. (ed.) *The retina: an approachable part of the*

brain. pp. 46 – 62. Cambridge, Mass. Cambridge University Press.
Dunn, R. F. (1969) The dimensions of rod outer segments related to light absorption in the gecko retina. *Vision Res.* 9, 603 – 609.
Engström, K. (1963) Cone types and cone arrangements in teleost retinae. *Acta Zool., Stockh.* 44, 110 –129.
Evans, B. I. and Fernald, R. D. (1993) Retinal transformation at metamorphosis in the winter flounder (Pseudopleuronectes americanus). *Vis. Neurosci.* 10, 1055 –1064.
Fineran, B. A. and Nicol, J. A. C. (1974) Studies on the eyes of New Zealand parrot-fishes (Labridae). *Proc. R. Soc. Lond. B.* 186, 217– 247.
Fineran, B. A. and Nicol, J. A. C. (1976) Novel cones in the retina of the anchovy (*Anchoa*). *J. Ultrastruct. Res.* 54, 296 – 303.
Fineran, B. A. and Nicol, J. A. C. (1977) Studies on the eyes of anchovies *Anchoa mitchilli* and *A. hepsetus* (Engraulidae) with particular reference to the pigment epithelium. *Phil. Trans. R. Soc. Lond. B.* 276, 321– 350.
Fineran, B. A. and Nicol, J. A. C. (1978) Studies on the photoreceptors of *Anchoa mitchilli* and *A. hepsetus* (Engraulidae) with particular reference to cones. *Phil. Trans. R. Soc. Lond. B* 283, 25 – 60.
Fox, R, Lehmkuhle, S. W. and Westendorf, D. H. (1976) Falcon visual acuity. *Science (NY).* 192, 263 – 265.
Franz, V. (1921) Zur mikroskopischen Anatomie der Mormyriden. *Zool. Jber.* 42, 91–148.
Fröhlich, E., Negishi, K. and Wagner, H-J. (1995) Patterns of rod proliferation in deep-sea fish retinae. *Vision Res.* 35, 1799 –1811.
Goede, P and Kolb, H. (1994) Identification of the synaptic pedicles belonging to the different spectral types of photoreceptor in the turtle retina. *Vision Res.* 34, 2801– 2811.
Herman, K. G. and Steinberg, R. H. (1982) Phagosome movement and the diurnal pattern of phagocytosis in the tapetal retinal pigment epithelium of the opossum. *Invest. Ophthal. Vis.* Sci. 23, 277– 290.
Hogan, M. J, Alvarado, J. A. and Weddell, J. E. (1971) *Histology of the human eye.* Philadelphia, London, Toronto. W. B. Saunders.
Jin, J., Jones, G. J. and Cornwall, M. C. (1994) Movement of retinal along cone and rod photoreceptors. *Vis. Neurosci.* 11, 389 – 399.
Higgins, E. (1932) Progress in biological inquiries. Appendix III. *Rep. U.S. Comm. Fish.* 1931. 441– 569.
Kalberer, M. and Pedler, C. (1963) The visual cells of the alligator: an electron microscopic study. *Vision Res.* 3, 323 – 329.
Kaplan, M. W. (1981) Light cycle-dependent axial variations in frog outer segment structure. *Invest. Ophthalmol.* Vis. Sci. 21, 395 – 402.
Kunz, Y. W. (1980) Cone mosaics in a teleost retina: changes during light and dark adaptation. *Experientia* 36, 1371–1374.
Kunz, Y. W. (1983) Diurnal changes of cone mosaic in a teleost retina. *Experientia* 39, 1049 –1050.
Kunz, Y. W. and Regan, C. (1973) Histochemical investigations into the lipid nature of the oil-droplet in the retinal twin cones of *Lebistes reticulatus* (Peters). *Rev. suisse Zool.* 80, 699 –703.
Land, M. F. (1981) Optics and vision in invertebrates. In: *Handbook of Sensory Physiology, VII/6B.* Ed. Autrum, H. pp. 471– 593. Berlin, Heidelberg, New York. Springer-Verlag.
LaVail, M. M. (1976) Rod outer segment disc shedding in relation to cyclic lighting. *Exp. Eye Res.* 23, 277– 280.
Locket, N. A. (1971) Retinal anatomy in some scopelarchid deep-sea fishes. *Proc. R. Soc. Lond. B.* 178, 161–184.
Locket, N. A. (1973) Retinal structure in *Latimeria chalumnae*. *Phil. Trans. R. Soc. Lond. B.* 266, 493 – 521.
Locket, N. A. (1976) Receptor grouping in fish retinae. *J. Anat.* 122, 713.
Locket, N. A. (1977) Adaptation to the deep-sea environment. *In Handbook of Sensory Physiology, VII/5.* Ed. Crescitelli, F. pp. 67–192. Berlin, Heidelberg. Springer-Verlag.
Locket, N. A. (1980) Variation of architecture with size in the multiple-bank retina of a deep-sea teleost, *Chauliodus sloani*. *Proc. R. Soc. Lond. B.* 208, 223 – 242.
Locket, N. A. (1985) The multiple bank rod fovea of *Bajacalifornia drakei,* an alepocephalid deep-sea teleost. *Proc. R. Soc. Lond. B.* 224, 7– 22.
Locket, N. A. (1992) Problems of deep foveas. *Aust. N. Z. J. Ophthal.* 20, 281– 295.
Lythgoe, J. N. (1972) The adaptation of visual pigment to the photic environment. In *Handbook of Sensory Physiology, VII/1.* Dartnell, H. J. A. ed. pp. 566 – 603. Berlin: Springer-Verlag.
Lythgoe, J. N. (1979) *The ecology of vision.* Oxford. Clarendon Press.
MacNichol, E. F., Kunz, Y. W., Levine, J. S., Harosi, F. I. and Collins, B. A. (1978) Ellipsosomes: organelles containing a cytochrome-like pigment in the retinal cones of certain fishes. *Science* 200, 549 – 551.
Maier, E. J. and Bowmaker, J. K. (1993) Colour vision in the passeriform bird, *Leiothrix lutea*: Correlation

of visual pigment absorbance and oil droplet transmission with spectral sensitivity. *J. Comp. Physiol A.* 172, 295 – 301.

Majima, K. (1925) Studien über die Struktur der Zehzellen und der Pigmentepithelzellen der Froschnetzhaut. *Albrecht von Graefe's Arch. Ophthal.* 115, 286 – 304.

McDevitt, D. S, Brahma, S. K., Jeanny, J. C. and Hicks, D. (1993) Presence and foveal enrichment of rod opsin in the 'all cone' retina of the American chameleon. *Anat. Rec.* 237, 299 – 307.

McEwan, M. R. (1938) A comparison of the retina of the mormyrids with that of various other teleosts. *Acta Zool.*, 19, 427– 465.

Miller, W. H. and Snyder, A. W. (1977) The tiered vertebrate retina. *Vision Res.* 17, 239 – 256.

Mizuno, K and Takei, Y. (1975) Retinal glycogen. In *Structure of the Eye*. Yamada, E. and Takei, Y. eds. pp. 341– 350. Tokyo. Tokyo University.

Moore, G. A. (1944) The retinae of two north American teleosts, with special reference to their tapeta lucida. *J. comp. Neurol.*, 80, 369 – 379.

Munk, O. (1959) The eyes of *Ipnops murrayi* Gunther 1887. *Galathea Rep.* 3, 79 – 87.

Munk, O. (1966) Ocular anatomy of some deep-sea teleosts. *Dana Rep.* 70, 1– 62.

Munk, O. (1975) On the eyes of two notosudid teleosts, *Scopelosaurus hoedti* and *Ahliesaurus berryi. Vidensk. Meddr. dansk naturh. Foren.* 138, 87–125.

Munk, O. (1977) The visual cells and retinal tapetum of the foveate deep-sea fish *Scopelosaurus lepidus* (Teleostei). *Zoomorph.* 87, 21– 49.

Muntz, W. R. A. (1972) Inert absorbing and reflecting pigments. In *Handbook of Sensory Physiology, VII/1*. Dartnall, H. J. A. ed. Berlin. Springer-Verlag.

Partridge, J. C, Shand, J., Archer, S. N., Lythgoe, J. N. and van Groningen-Luyben W. A. H. M. (1989) Interspecific variation in the visual pigments of deep-sea fishes. *J. comp. Physiol. A* 164, 513 – 529.

Partridge, J. C., Archer, S. N. and van Oostrum, J. (1992) Single and multiple visual pigments in deep-sea fishes. *J. Mar. Biol. Ass. U.K.* 72, 113 –130.

Pumphrey, R. J. (1948) The theory of the fovea. *J. Exp. Biol.* 25, 299 – 312.

Richardson, T. M. (1969) Cytoplasmic and ciliary connections between the inner and outer segments of mammalian visual receptors. *Vision Res.* 9, 727–731.

Rodieck, R. W. (1973) *The Vertebrate Retina*. San Francisco. Freeman, W. H.

Sarantis, M and Mobbs, P. (1992) The spatial relationship between Müller cell processes and the photoreceptor output synapse. *Brain Res.* 584, 299 – 304.

Schnapf, J. L. and Baylor, D. A. (1897) How photoreceptor cells respond to light. *Sci. Amer.* 256, 40 – 47.

Schremser, J. and Williams, T. P. (1995) Rod outer segment (ROS) renewal as a mechanism for adaptation to a new intensity environment. I. Rhodopsin levels and ROS length. *Exp. Eye Res.* 61, 17– 24.

Schmitz, F and Drenckhahn, D. (1993) Intermediate stages in the disassembly of synaptic ribbons in cone photoreceptors of the crucian carp, *Carassius auratus. Cell Tiss. Res.* 272, 487– 490.

Schmitz, F., Kirsch, M. and Wagner, H-J. (1989) Calcium modulated synaptic ribbon dynamics: a pharmacological and electron spectroscopic study. *Eur. J. Cell Biol.* 49, 207– 212.

Schultze, M. (1866) Anatomie und Physiologie der Netzhaut. *Arch. Mikr. Anat. Entw. Mech.* 2, 175 – 286.

Shapley, R. and Gordon, J. (1980) The visual sensitivity of the retina of the conger eel. *Proc. R. Soc. Lond. B.* 209, 317–330.

Snyder, A. W. and Miller, W. H. (1978) Telephoto lens system of falconiform eyes. *Nature, Lond.* 275, 127–129.

Spira, A. W. and Millman, G. E. (1979) The structure and distribution of the cross-striated fibril and associated membranes in guinea pig photoreceptors. *Amer. J. Anat.* 155, 319 – 338.

Townes-Anderson, E. (1995) Intersegmental fusion in vertebrate rod photoreceptors. *Invest. Ophthal. Vis. Sci.* 36, 1918 –1933.

Wagner, H-J. (1973) Darkness-induced reduction of the number of synaptic ribbons in fish retina. *Nature New Biol.* 246, 53 – 55.

Wagner, H-J. (1990) Retinal structure of fishes. In *The Visual System of Fish*. Douglas, R. and Djamgoz, M. eds. pp. 109 –157. London. Chapman and Hall.

Walls, G. L. (1942) *The vertebrate eye and its adaptive radiation*. Michigan. Cranbrook Institute of Science.

Waterman, T. H. (1981) Polarization sensitivity. In *Handbook of Sensory Physiology, Vol, VII/6B*. Autrum, H. pp. 281– 469. Berlin, Heidelberg, New York. Springer-Verlag.

Young, R. W. (1978) The daily rhythm of shedding and degradation of rod and cone outer segment membranes in the chick retina. *Invest. Ophthalmol. Vis. Sci.* 17, 105 –116.

Young, R. W. and Bok, D. (1969) Participation of the retinal pigment epithelium in the rod outer segment renewal process. *J. Cell Biol.* 42, 392 – 403.

6
The extraretinal photoreceptors of non-mammalian vertebrates

J. SHAND[1] and R.G. FOSTER*[2]
[1]Department of Zoology, University of Western Australia, Nedlands WA 6907, Australia
[2]Imperial College of Science Technology and Medicine, Department of Biology, Sir Alexander Fleming Building, Imperial College Road, London SW7 2A2, UK
* To whom correspondence should be addressed.

6.1 Introduction

When John Lythgoe published The Ecology of Vision in 1979 he did not include any discussion of extraretinal or extraocular photopigments. However, during the 1980's information about these photopigments emerged, and John's research interests also began to encompass this field. In 1984 John considered the adaptive aspects of extraretinal photoreceptors and noted that 'As yet there does not seem to be any reason to think that the spectral absorbance of extraretinal photopigments are located in the spectrum to maximise the capture of photons...' (Lythgoe 1984). In this chapter we will review this statement, in the light of recent findings, and discuss the sensory ecology of these remarkable photoreceptor systems. Our discussion of extraretinal photoreceptors has been divided between two chapters, this and the one following. The first part of this chapter will consider the diversity and role of the extraretinal photoreceptors in the non-mammalian vertebrates, and the second part of the chapter will consider the sensory tasks of extraretinal photoreceptors and the extent to which the photopigments of these systems are 'fine-tuned' to their sensory role. In Chapter 7 we concentrate on the role of extraretinal and retinal photoreceptors in the regulation of biological clocks.

6.2 The diversity of extraretinal photoreceptors and the responses they mediate

In those invertebrates that have some form of image-forming eye (Cnidaria, Mollusca, Annelida, Onychophora, Arthropoda) (Land 1991) extraretinal receptors also exist. These extraocular photoreceptors are anatomically diverse and regulate many different aspect of physiology and behaviour. For example, in crustaceans, insects and molluscs the photoreceptors that regulate circadian rhythms are located within the brain (Page 1982; Helfrich 1986); dermal photoreceptors mediate the control of dermal coloration in many species (for

S.N. Archer et al. (eds.), Adaptive Mechanisms in the Ecology of Vision, 197–222

review see Steven 1963); caudal photoreceptors in crayfish are used to determine the emergence from burrows (Fernandez-de-Miguel and Arechiga 1992); corneal receptors control the migration of screening pigment in the eyes of moths (Land 1987); photoreceptors located in the abdominal ganglion of opistobranch molluscs provide wavelength-dependent responses in the regulation of locomotor patterns (Nishi and Gotow 1992).

Vertebrates, like the invertebrates, also have photoreceptors located in a number of different regions of the body. Those arising from the diencephalic forebrain are classified as: 1) Intracranial pineal (*epiphysis cerebri*) and parapineal organs; 2) Extracranial 'third eyes', variously called frontal organs (frogs) or parietal eyes (lizards); 3) Deep brain photoreceptors; and 4) Lateral eyes. In addition to the diencephalic photoreceptors, many vertebrates also possess dermal, corneal, and iris photoreceptors. Below we consider the diversity of these extraocular and extraretinal photoreceptors.

6.2.1 The pineal organ

The pineal organ (or *epiphysis cerebri*) is present in some form in most vertebrates, but varies greatly in structure (Figures 1 and 2). There is a continuous spectrum of types from a combined photoreceptor and neuroendocrine organ in the lower vertebrates to an exclusively neuroendocrine organ in mammals whose activity is controlled by indirect photic input from the eyes. The pineal arises embryologically as an evagination from the roof of the diencephalon and in non-mammalian vertebrates is always situated near the surface of the brain. The pineal remains attached to the brain by some form of stalk that terminates in a blood plexus within the third ventricle (for review see Vollrath 1981). There is often a translucent area or 'window' in the skull of fish which allows approximately 10% of incident light to reach the pineal organ (Meissl and Brandstatter 1992). This window is particularly well developed in lampreys, where the pineal vesicle sits close to the skull and lies below a 'spot' which lacks pigment. Because of this pronounced pineal window, the pineal organ in lampreys is often mistakenly thought to be an extracranial 'third eye'. In amphibians, reptiles and birds this window is either absent or less transparent. However, large amounts of filtered light still enter the brain, with 0.1– 0.3% of incident light penetrating the pigmented skin and skull and reaching the pineal (Dodt and Meissl 1982). In animals that lack a pineal window the spectral quality of the light reaching the pineal is dominated by the absorption characteristics of haemoglobin. As a result, the light environment of most pineals is relatively enriched in wavelengths beyond 600 nm, although there is a small 'window' in transmission between 460 and 550 nm (wavelength of maximum transmission around 490 nm) (Hartwig and van Veen 1979; Dodt and Meissl 1982; Foster and Follett 1985).

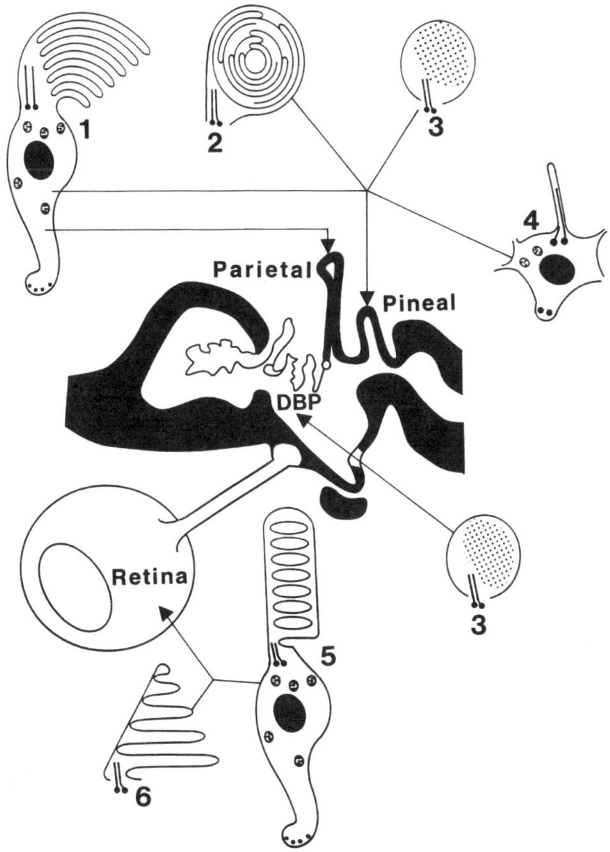

Figure 1 Schematic drawing of the vertebrate fore-brain showing the diversity of photoreceptor types within the retina, pineal complex (pineal and parietal organs) and basal brain (DBP). Photoreceptors can be broadly classified as follows: 1 Cell with an outer segment resembling a retinal cone; 2 Cell with an outer segment with 'disorganised' membrane whorls; 3 Cell with an outer segment containing few or no membrane structures; 4 Cell with no outer segment; 5 Retinal rod cell; 6 Retinal cone cell. These photoreceptors show the following taxonomic distribution: In the parietal organ of lacertilians (lizards) or frontal organ of anurans (frogs), the photoreceptors most closely resemble 1; The pineal organ is present in some form in almost all vertebrates. In the fish and amphibia = cell type 1 dominates; reptiles = cell types 1 & 2; birds = cell types 2 & 3; mammals = cell type 4. Note that in mammals, opsin has been identified within pineal cells, but these cells are not thought to be photoreceptive as they seem to lack photopigment chromophore. When present the parapineal of fish (not shown in this diagram) contains few photoreceptors that most closely resemble cell types 1 & 2. Opsin immunoreactive cells have only recently been identified within the basal brain of vertebrates, and it is assumed that these cells are the deep brain photoreceptors of the non-mammalian vertebrates. There may be more than one cell type within the brain capable of phototransduction, CSF-contacting neurones and neurosecretory cells of the NMPO. CSF-contacting neurones most closely resemble cell type 3. The retina of all vertebrates contains either rod or cone photoreceptors or more usually some mixture of both (cell types 5 & 6). See text for further details. Reprinted from Foster and Menaker 1993, Copyright (1993), Page 77. With kind permission from Elsevier Science Ltd, The Boulevard, Langford Lane, Kidlington OX5 1GB, UK

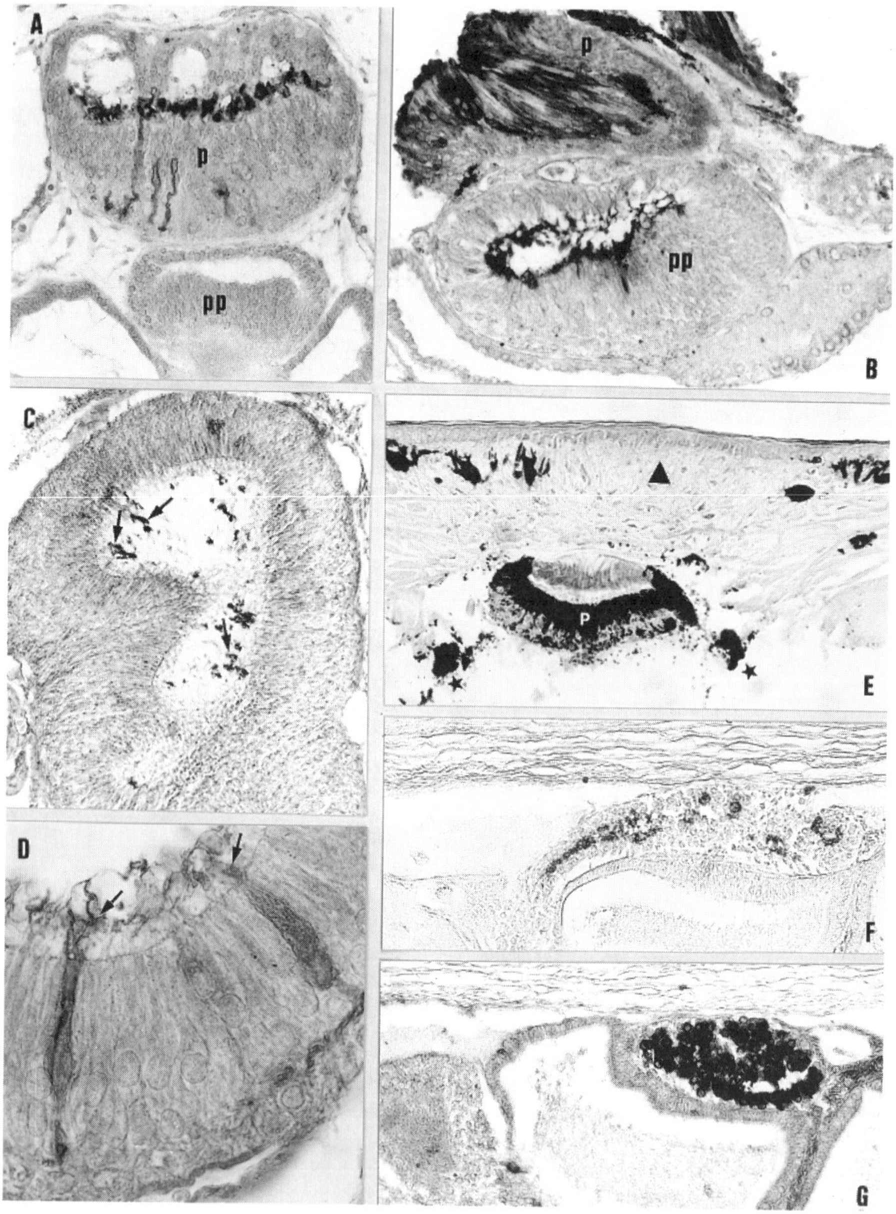

Figure 2 Immunocytochemical localisation of rod and cone opsins within the pineal complex of various vertebrates: A) Frontal section through the pineal complex of a larval sea lamprey (*Petromyzon marinus*) immunolabelled with anti-cone opsin antisera (CERN-874). (p) Pineal organ with immunolabelling of pinealocyte outer segments. Note that some photoreceptors show labelling throughout all portions of the

(continued overleaf)

Figure 2 (continued from previous page) cell. (pp) The parapineal organ has relatively few opsin labelled cells. In this section no cells were labelled with anti-cone opsin antibodies (Garcia-Fernandez and Foster 1994); B) Sagittal section through the pineal complex of an adult river lamprey (*Lampetra fluviatilis*) showing intense immunolabelling in both the (p) pineal and (pp) parapineal organs using antibodies specific for rod opsins (CERN-858) (Garcia-Fernandez, Jimenez *et al.*, 1997); C) Pineal of the lizard *Anolis carolinensis*, labelled with rod opsin specific antisera (CERN-858). This antibody only labelled pinealocyte outer segments (arrows) (Foster, Garcia-Fernandez *et al.*, 1993); D) Pineal of the lizard *Anolis carolinensis* immunolabelled with cone opsin antisera (CERN-874). In this case the outer segments (arrows), cell bodies and basal processes of the pinealocytes have been labelled (Foster, Garcia-Fernandez *et al.*, 1993); E) Frontal section through the parietal organ/eye of the lizard *Anolis carolinensis*. Photoreceptors have not been labelled in this section. Black pigmentation is seen in the connective tissue that surrounds the parietal organ (black stars) and in the pigment cells of the parietal eye (p). Note the absence of pigment in the tegmentum overlaying the parietal eye (black triangle) (Foster, Garcia-Fernandez *et al.*, 1993); F) Sagittal section of the pineal organ of the bonefish eel (*Anguilla anguilla*) showing labelling with rod opsin specific monoclonal antibodies (Rho-4D2) (Garcia-Fernandez - unpublished); G) Sagittal section of the same pineal organ of the bonefish eel shown in F). In this case pinealocytes have been intensley labelled by a polyclonal rod opsin specific antibody (CERN-858) (Garcia-Fernandez - unpublished). Figure compiled by Dr. José M. Garcia-Fernandez, Departamento de Morfologia y Biologia Celular, Universidad de Oviedo, Spain

The pineal contains several different cell types, although the dominant type in most non-mammals is photoreceptor-like in appearance. The cytology of these photoreceptor-like cells has been reviewed in detail a number of times (e.g. Vigh and Vigh-Teichmann 1988). In summary (Figure 1), fish and amphibians have an outer segment resembling a retinal cone with multiple invaginations of the photoreceptive membrane. Outer segments appear to contain both rod-like and cone-like opsin proteins (Figure 2A, 2B, 2F, 2G). Reptiles may have either cone-like photoreceptors (lizards) or cells with an outer segment with disorganised membrane whorls (snakes). Like the fish and amphibians, both rod-like and cone-like opsins have been identified within the pineal of lizards (Figure 2C, 2D). Turtles appear to occupy an intermediate evolutionary position with both types of photoreceptor-like cell (Meissl and Ueck 1980), whereas alligators seem to lack a recognisable pineal (Underwood and Groos 1982). In birds the photoreceptor-like pinealocytes have either disorganised membrane whorls (as in some snakes) or have an outer segment with few or no membrane structures. The pinealocytes of adult mammals have lost most of their photoreceptor-like morphological characteristics and appear to be exclusively secretory.

All vertebrate pineals seem to make the hormone melatonin, which in non-mammals is regulated by photoreceptors within the pineal. Melatonin production is confined to the dark portion of a light:dark cycle and provides a signal that, to a varying degree is important in the regulation of circadian and/or photoperiodic behaviours (Chapter 7). In lizards, there appears to be a complex relationship between the pineal complex, melatonin and circadian physiology in the regulation of body temperature (for reviews see Underwood 1992; Tosini and Menaker 1996). In addition to its role in circadian organisation, pineal melatonin also seems to play an important role in regulating dermal colour changes, indeed melatonin receptors were first successfully characterised by screening a *Xenopus* melanophore cDNA library (Ebisawa, Karne *et al.,* 1994). In many amphibia, melatonin administration has been shown to cause blanching of the skin. For example, melatonin administered to isolated frog (*Rana*) skin causes blanching by a process of melanin concentration in melanophores. Further experiments showed that after removal of the pineal, blanching no longer occurred when animals were placed into darkness (for review see Bagnara and Hadley 1970). Pineal melatonin appears to play a similar role in fish as in amphibians, with administration causing pigment aggregation and blanching. However, the situation in fish appears a little more complex than in amphibians with marked interspecific and age-related differences (for review see Vollrath 1981). In the other classes of vertebrates there is no clear link between pineal melatonin and dermal colour changes. The only other non-circadian function for which the pineal has been strongly implicated would be fish phototaxis (Nordtug, Berg *et al.,* 1994). For example, in *Fundulus heteroclitus* blinded and sighted animals show positive phototaxis but after pinealectomy animals show negative or no phototaxis (Pang 1965). In *Carassius auratus* negative phototaxis is abolished by pinealectomy (Fenwick 1970). Note that the pineal has been implicated in regulating a host of other biological functions but these studies are either incomplete and/or contradictory (for reviews see Vollrath 1981; Groos 1982).

In addition to the pineal, some fish have another intracranial photoreceptor organ called the parapineal which also arises as a dorsal evagination from the diencephalon (for review see Vollrath 1981). In lampreys the parapineal organ consists of a vesicle that sits beneath the pineal organ (Figure 2A, 2B). However, in comparison with the pineal, parapineal photoreceptors can be rare (Garcia-Fernandez and Foster 1994). Very little is known of the parapineal organ of other fish, and few detailed descriptions exist. A study on *Salmo gairdneri* again identified very few photoreceptors within this structure (Rudeberg 1969), and similar results were obtained in *Gasterosteus aculeatus* (Ekstrom, Borg *et al.,* 1983). It is unclear whether the parapineal produces melatonin and the role of this organ remains a mystery.

6.2.2 The extracranial 'third eye'

Fossils of primitive bony fish, early amphibians and early reptiles have a dorsal medial skull socket in which a third eye is believed to have been located (Romer 1970). Today only the Sphenodon (rhynchocephalia), some lizards (squamata) and frogs (anura) possess a dorsal skull socket and an extra-cranial third eye. In contrast to the intra-cranial pineal, parapineal and deep brain photoreceptors, 'third eye' photoreceptors are exposed to unfiltered environmental light. In anuran amphibians the pineal complex is the name given to both the extra-cranial third eye (frontal organ or *Stirnorgan*) and the intra-cranial pineal proper (*epiphysis cerebri*). Note that urodeal amphibians lack a frontal organ. In anurans the pineal complex develops as evagination of the diencephalon, and this single protrusion gives rise to both the pineal and frontal organ. During metamorphosis the frontal organ separates and moves to an anterior extracranial position, and is situated below outer translucent skin layers (Stieda's spot). In many species the frontal organ remains linked to the pineal via the pineal tract. Large numbers of well developed photoreceptor cells are seen in the frontal organ and these have cone-like outer segments. While the frontal organ is clearly a photoreceptor, and has been implicated in behavioural orientation, solid information about the role of this photoreceptor is lacking (Underwood and Groos 1982).

In lizards it remains unclear whether the pineal complex (pineal organ and parietal eye) develops from a single evagination from the diencephalon or whether these structures are formed from separate evaginations (for review see Vollrath 1981). The parietal eye of some lizards is the most 'eye-like' extra-ocular photoreceptor organ of all vertebrates, possessing a transparent cornea and a lens analogous to those found in lateral eyes (Figure 2E). These receptors synapse directly onto ganglion cells, the axons of which form a parietal eye nerve projecting to the brain (for review see Engbretson 1992). The photoreceptors themselves resemble the cones of the lateral eyes, with outer segment membranes invaginated and continuous with the extracellular space (Figure 1). Morphological analysis suggests that at least two photoreceptor classes may exist within the parietal eye of some lizards (Engbretson 1992). The parietal eye of lizards is thought to play a role in thermoregulation, reproduction and the regulation of a range of other behavioural patterns (for reviews see Quay 1979; Engbretson 1992; Underwood 1992; Tosini and Menaker 1996).

6.2.3 The 'deep brain' photoreceptors

All classes of non-mammalian vertebrates possess photoreceptors located deep within the brain (Figure 3). These have been shown to entrain biological rhythms (Chapter 7) and to be involved in the regulation of other, non-rhythmic, physiological and behavioural responses (for reviews see Foster, Garcia-Fernandez *et al.,* 1993; Foster, Grace *et al.,* 1994). The first evidence for the existence of non-pineal extraretinal photoreceptors emerged as long ago as 1911 when Karl von Frisch showed that light-induced colour changes in the skin of the minnow (*Phoxinus laevis*) occurred in the absence of the eyes and pineal complex. He localised the site of the receptor to the diencephalon by showing that lesions within the basal brain blocked the melanophore light response. On the basis of these data he proposed that the ependyma of the ventricles was photosensitive (Frisch 1911). In 1928 Scharrer showed that encephalic receptors of the minnow also mediated swimming and feeding reflexes. These data, together with later observations on amphibians and reptiles, led him to propose that there were photo-neuroendocrine cells within the hypothalamic nucleus magnocellularis preopticus (NMPO) (Scharrer 1928; Scharrer 1964). Evidence for a deep-brain

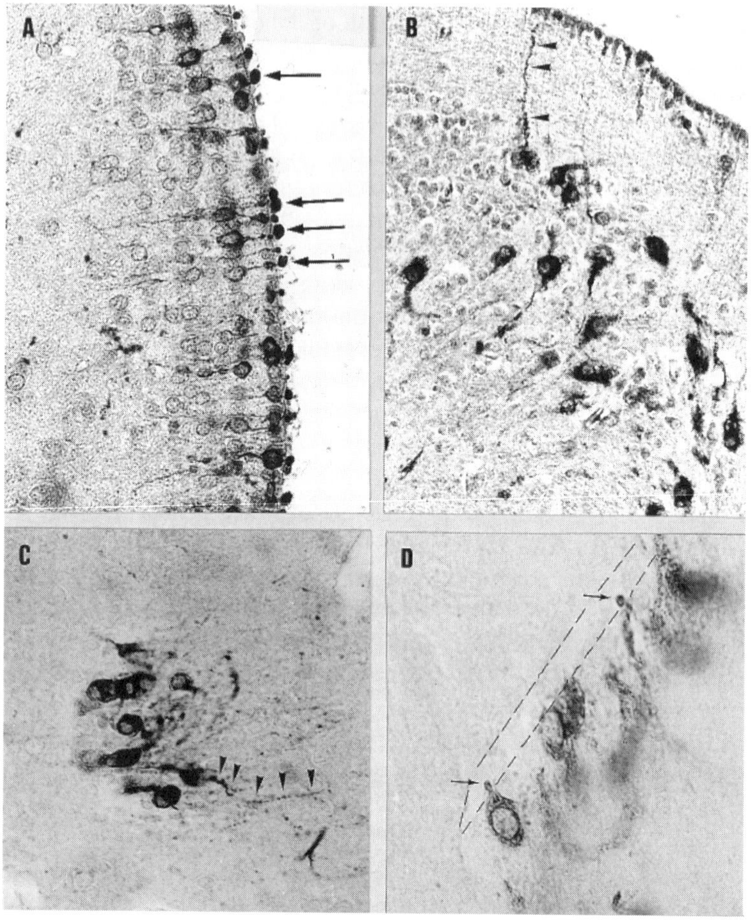

Figure 3 Localisation of rod and cone opsins within the brain of various vertebrates: A) Transverse section through the hypothalamic nucleus commissurae postopticae of an adult sea lamprey (*Petromyzon marinus*) showing large numbers of cerebrospinal fluid (CSF)-contacting neurons labelled with cone opsin antibodies (CERN-906). The strongly labelled club-like swellings of these cells protrude into the ventricle (arrows) (Garcia-Fernandez, Jimenez *et al.*, 1997); B) Transverse section through the hypothalamus (nucleus magnocellularis preopticus) of adult *Xenopus* showing cells immunolabelled with anti-cone opsin antibodies (CERN-874). Note that many of the labelled cells appear to lack a ventricular process (arrow heads) (Foster, Grace *et al.,* 1994); C) Sagital and D) frontal section of CSF-contacting neurones in the nucleus ventromedialis of the septum of the lizard *Anolis carolinensis*. CSF-contacting neurones have been labelled with anti-cone opsin antibodies (CERN-874). In D) apical processes are seen to extend from the cell body and protrude into the lateral ventricle (dashed line – surface of the lateral ventricle) (Foster, Garcia-Fernandez *et al.*, 1993). Figure complied by: Dr. José M. Garcia-Fernandez, Departamento de Morfologia y Biologia Celular, Universidad de Oviedo, Spain

receptor in lampreys and ducks followed in the 1930s. Young showed that illuminating the head of blinded and pinealectomised lampreys still triggered body movements (Young 1935a,b). Benoit demonstrated that direct illumination of the hypothalamus caused testicular growth in blinded mallards (Benoit 1935a; Benoit 1935b). This research was not followed up until the 1960s when Menaker and his colleagues began a series of experiments with sparrows that demonstrated that the regulation of circadian rhythms and the photoperiodic control of testis growth is mediated by extraretinal and extra-pineal photoreceptors (for reviews see (Menaker and Underwood 1976; Foster and Follett 1985). Although most studies have been on birds, deep-brain receptors have been demonstrated in fish, frogs and lizards (for reviews see Foster, Garcia-Fernandez *et al.,* 1993; Foster, Grace *et al.,* 1994). Note that large amounts of light penetrate deep into the brain, and as with the pineal, this light is filtered by haemoglobin so that the light environment of the deep brain and pineal photoreceptors are spectrally similar (Foster and Follett 1985).

Despite knowledge of the presence of deep-brain receptors, a definite cellular location has been less easy to establish. 'Typical' photoreceptor cells similar to retinal or pineal photoreceptors have not been identified within the brain. However, the cerebrospinal fluid (CSF)-contacting neurons of the paraventricular organ do possess some photoreceptor characteristics. For example, they have a process that resembles the inner segment of retinal and pineal photoreceptors. This consists of a single ciliated 'sensory' dendrite with a 9 + 0 arrangement of microtubules that protrudes into the ventricular lumen (Scharrer 1964; Vigh and Vigh-Teichmann 1988). In an attempt to implicate the CSF-contacting neurons as photoreceptors, antisera raised against various vertebrate rod and cone opsins have been used in immunocytochemical procedures. Unfortunately, a complicated collection of results emerged. These results can be summarised as follows: Initially CSF-contacting neurons in the avian brain were not labelled with anti-opsin antibodies, despite strong labelling of retinal and pineal photoreceptors by these antibodies in the same species (e.g. Foster, Korf *et al.,* 1987; Vigh and Vigh-Teichmann 1988). Using one particular anti-rod opsin antibody, however, Silver's group labelled CSF-contacting neurons in the septal and tuberal areas of the brain of ring dove, quail and duck (Silver, Witkovsky *et al.,* 1988). Unfortunately a conclusive opsin identification was not possible in this study because Western blots of these specific regions of the brain failed to identify opsin-like proteins. More recently antisera specific to cone opsins were shown to label CSF-contacting neurons in the lateral ventricle ependyma of the lizard *Anolis carolinensis*, and Western blots recognised a cone-like opsin in extracts from the anterior brain (Figure 3C, 3D). In addition, photopigment chromophore was also isolated from *Anolis* anterior brain extracts by using HPLC analysis (Foster, Garcia-Fernandez *et al.,* 1993). In another study CSF-contacting neurons in the hypothalamus of larval lamprey (*Petromyzon marinus*) have been successfully labelled with antibodies directed against different opsins (Garcia-Fernandez and Foster 1994) (Figure 3A). However, the same antibodies have failed to immunolabel CSF-contacting neurons in birds and certain other vertebrates (Foster, Grace *et al.,* 1994). Most recently another class of cell within the fish and amphibian brain has shown opsin-like labelling (Figure 3B). These 'other' opsin labelled cells were located within the neurosecretory cell complex of the nucleus magnocellularis preopticus (NMPO) (Foster, Grace *et al.,* 1994). This suggests that two classes of encephalic photoreceptor may exist within the vertebrates: CSF-contacting neurons in lampreys, reptiles and birds (as originally suggested by Karl von Frish, 1911) and NMPO cells in fish and amphibians (as originally suggested by Ernst Scharrer, 1928). For discussion of the novel extraretinal opsins see section 6.4.1.

6.2.4 The dermal photoreceptors

Light can regulate dermal cells either indirectly via pituitary melanocyte stimulating hormone (MSH) and pineal melatonin or directly by photon absorption within the dermal cells. Dermal photoreceptors mediate responses that fall into two broad categories: 1) The direct responses of cells to light such as the colour changes of chromatophores and iridophores; and 2) Behavioural responses which include triggering locomotor activity (for reviews see Steven 1963; Millot 1968; Wolken and Mogus 1979). The dermal chromatophores of many lower vertebrates are directly sensitive to light and, depending on species, will either aggregate or disperse pigment granules upon exposure to light (Weber 1983). The site of photoreception is not clear, and to our knowledge no sensory cilia have been reported in light-sensitive dermal chromatophores. Narrow beam illumination of the tail fin melanophores in the tadpole *Xenopus laevis* suggest that the cell membrane or cytoplasm may house the photopigments (Weber 1983). For further discussion of melanophore photosensitivity in *Xenopus* see Provencio, Jiang *et al.,* 1998, and for fish see Iga and Takabatake 1986.

In fish, the iridocytes of isolated strips of skin of at least two species (*Fundulus heteroclitus* and *Paracheirodon innesi*) are able to change their wavelengths of maximum reflection from violet to red within a few moments of illumination (Lythgoe and Shand 1982). Measures of the spectral transmission of the waters in which *P. innesi* naturally inhabit, suggest that the colour changes make this schooling fish more conspicuous during the daytime and provides camouflage at night (Lythgoe and Shand 1983). The mechanics of the colour change are thought to result from either changes in the spacing between the guanine plates of the iridocytes, with an increase in spacing causing a reddening of the reflected wavelengths (Lythgoe and Shand 1982), or alternatively, an alteration of the angle of inclination of the iridocytes to the incident light (Nagaishi, Oshima *et al.,* 1990). Opsin can be localised to the iridophore, suggesting that phototransduction appears to occur within the iridocytes themselves (Lythgoe, Shand *et al.,* 1984).

Dermal receptors are known to initiate behavioural responses in some agnathans, teleosts and amphibians (Ronan and Bodznick 1991). For example, local illumination brings about tail movement in ammocete larval lampreys (Young 1935). More recent studies have confirmed these observations and shown that electrical responses from lateral line nerves are modulated by illumination of the tail (Ronan and Bodznick 1991). Ammocete larvae live in mud and avoid light, presumably to avoid predation. Adult lamprey are also known to possess tail photoreceptors that initiate avoidance movements, although the direction of swimming requires input from the lateral eyes (Ullen, Orlovsky *et al.,* 1993). Two recent reports have also described behavioural responses to dermal photostimulation in reptiles. In the sea snake (*Aipysurus laevis*) illumination of a small dorsal region of the tail triggers movements of the tail. This is thought to facilitate efficient concealment of the elongated flexible tail in crevices during the day when the snakes are vulnerable to predation by sharks (Zimmerman and Heatwole 1990). In the lizard (*Podarcis muralis*) the selection of a basking position was found to be mediated, at least in part, by dermal photoreceptors on the dorsal surface (Tosini and Avery 1996). In neither the sea snake or lizard have these photoreceptors been further localised or described. In birds dermal photoreception appears to be functional in immature animals only. For example, illumination of the body can trigger a behaviour pattern involving head wagging in newly hatched pigeons (Harth and Heaton 1973). Once again, the detailed location of the receptors is unknown.

6.2.5 Corneal colour changes

The corneas of many diurnal teleosts possess chromatophores or iridocytes and in some cases both. The chromatophores provide a yellow filter which is thought to reduce chromatic aberration and/or enhance contrast perception by filtering out blue scattered light (for review see Muntz 1976). Iridocytes within the cornea may act by reflecting away from the eye bright downwelling light without affecting lateral light from the visual scene. Alternatively, iridocytes may camouflage the eye by obscuring the blackness of the pupil which is often used by predators to orient an initial attack (Lythgoe 1979). Corneal chromatophores are very similar in structure to dermal melanophores and may share common origins (Orlov and Gamburtzeva 1976). In a number of species, corneal chromatophores have been shown to aggregate and disperse pigment granules in a manner similar to dermal melanophores, pigments dispersing in bright light (increasing their filtering effect) and aggregating in response to darkness (Kondrashev, Gamburtzeva *et al.,* 1986). The aggregation of the yellow pigment granules in low light conditions will permit more light to enter the eye and improve sensitivity, especially in fish inhabiting clear oceanic water which transmits predominantly blue wavelengths (Appleby and Muntz 1979). This response is known to occur following optic nerve section in puffer fish (Appleby and Muntz

1979) and greenling (Kondrashev and Khodtsev 1984). However, the chromatophores in the isolated cornea of the greenling do not respond to direct illumination. This gave rise to the suggestion that chromatophores are not directly photosensitive, but may be regulated by retinal photoreceptors via some humoral signal (Kondrashev and Khodtsev 1984).

The corneal iridophores of some species have been shown to change colour in response to changes in light intensity. As with dermal iridophores, these changes always involve a shift to short wavelengths in the dark and to long wavelengths in the light (Shand and Lythgoe 1987; Shand 1988). Corneal iridophores are composed of multilayer stacks of material of alternating high and low refractive index. There are a number of phylogenetically distinct arrangements of corneal iridophores (Lythgoe 1979) and not all types are able to alter their coloration (Shand 1988) but, as with dermal iridocytes, the wavelength of maximum reflection is altered by changes in the spacing of the multilayers (Lythgoe and Shand 1989). The response to light occurs in the isolated cornea although the exact site and nature of the photoreceptor are unknown (Shand and Lythgoe 1990).

6.2.6 Iris constriction

The pupillae muscles of several vertebrates, including some non-human mammals, have been reported to be directly light sensitive. Light striking the isolated iris will cause the sphincter pupillae to contract (Zucker and Nolte 1981; Rubin and Nolte 1986; Barr and Gu 1987; Barr 1989), and an action spectrum for this response in the isolated frog iris (Barr and Alpern 1963), and eel iris (Seliger 1963) suggests the involvement of a rhodopsin-like photopigment. Anatomical studies have failed to identify photoreceptor-like structures in all species, with the exception of the eel, in which membranous whorls were noted in the sphincter pupillae (Bell 1967). In *Bufo* (frog) and *Lophius* (goosefish), light-induced calcium release was implicated in phototransduction, but levels of cGMP or cAMP were not significantly altered in the sphincter pupillae after illumination (Rubin and Nolte 1986).

6.3 Extraretinal photoreceptors and their sensory tasks

Because of the diversity of extraretinal photoreceptors generalisations are difficult. However, all seem to monitor overall levels of light in the environment (flux) and, with the exception of the extracranial third eyes and those species with a pineal window, are incapable of extracting detailed spatial or positional information from the environment. In those cases where we can define a role, extraretinal photoreceptors mediate: a) time of day cues for the regulation of the circadian system (Chapter 7); b) cues for behavioural orientation; c) cues for regulating body and/or corneal pigmentation and coloration; and d) cues for regulating pupil size.

Why the vertebrates should have evolved both image-forming visual photoreceptors and specialised extraocular and extraretinal photon flux detectors has been a topic of much discussion. Ironically, a partial explanation may have arisen from the study of mammals which lack extraretinal photoreceptors. It is often assumed that the task of irradiance detection would be 'simpler' if ocular photoreceptors were used, and that there is no selective advantage in having extraocular photoreceptors. However, recent results in retinally degenerate mice and studies in the blind mole rat (reviewed in Chapter 7) have shown that these mammals lack classical visual responses but are still capable of regulating their circadian rhythms by light with normal sensitivities. Blinding these retinally degenerate animals blocks all responses to light, demonstrating that these 'circadian photoreceptors' must be ocular. These data have led to the realisation that two functionally distinct systems (image detection and flux detection) exist for processing light information in the mammalian retina. Image detection requires a focused representation of the environment. The optical nature of the eye achieves this precision but in so doing prevents irradiance detection. For reliable irradiance measures, light must be gathered from all directions and fall uniformly on an array of photoreceptors. Because mammals lack extraretinal photoreceptors, they are faced with the sensory problem of extracting irradiance information from the environment with ocular photoreceptors. This problem seems to have been solved by evolving anatomically and organisationally distinct retinal projections to image-forming and non-image forming brain nuclei. The image-forming photoreceptor system of mammals, as in all vertebrates, constructs a representation of the environment by maintaining a topographical map between the retina and the central visual sites, such as the dorsal lateral geniculate nucleus, superior colliculus, and visual cortex. However, retinal photoreceptors also project to central structures that are not part of the image detector system (e.g. the suprachiasmatic nuclei and intergeniculate leaflet). Furthermore, this projection lacks any topographic order. The net result seems to be a form of irradiance detection analogous to that achieved by a randomized optic fibre (Cooper, Tessonneaud *et al.,* 1993) (Chapter 7).

Having argued that extraretinal photoreceptors have a selective advantage and that mammals must compensate for their loss, we must now address why the mammals have lost extraretinal photoreceptors and rely on ocular photoreceptors for both image and non-image light detection. A partial explanation may be related to the early evolutionary history of eutherian mammals and their possible passage through a 'nocturnal bottleneck'. Modern mammals seem to have been derived from nocturnal insectivorous or omnivorous animals about 100 million years ago (Young 1962). Extraretinal photoreceptors would have been adequate for monitoring changes in diurnal light conditions but may not have been sufficiently sensitive to discriminate twilight changes in primitive mammals spending the day hiding in holes or crevices. Perhaps retinal photoreceptors and central structures became modified for irradiance detection as the early mammals became nocturnal. Alternatively, and this seems more likely, extraretinal photoreceptors and retinal irradiance detectors were both present in the diurnal mammal-like reptiles and extraretinal receptors were subsequently lost when primitive mammals began to exploit the nocturnal realm (Foster and Menaker 1993).

The different sensory demands of image and non-image light detection raise the question of how the extraretinal photoreceptors may be specialised to make irradiance measures, and the extent to which retinal and extraretinal photoreceptors differ. In consideration of these questions we will consider in the last two sections of this Chapter: The photopigment characteristics of extraretinal photoreceptors (6.4), and the response characteristic of extraretinal photoreceptors (6.5).

6.4 The photopigment characteristics of extraretinal photoreceptors

In this section we will outline the diversity of extraretinal photoreceptors, and consider whether there is any correlation between their photopigment characteristics and the nature of their sensory task.

6.4.1 The diversity of extraretinal photopigments

Action spectra fitted by the Dartnall visual pigment nomogram have indicated the presence of opsin-based pigments within extraretinal photoreceptors (e.g. Deguchi 1981, Lythgoe and Shand 1984; Lythgoe and Thompson 1984; Foster, Follett *et al.,* 1985). Immunocytochemical studies have confirmed that the photopigments of many extraretinal receptors have either rod-like or cone-like opsins (e.g. Lythgoe, Shand *et al.,* 1984; Foster, Korf *et al.,* 1987; Foster, Garcia-Fernandez *et al.,* 1993), whilst HPLC analysis has identified both vitamin A_1 and A_2-derived chromophores within the non-mammalian pineal and deep-brain (e.g. Tabata, Suzuki *et al.,* 1985; Foster, Schalken *et al.,* 1989; Foster, Garcia-Fernandez *et al.,* 1993; Provencio and Foster, 1993). Most recently molecular techniques have succeeded in isolating novel extraocular opsins. These opsins are sufficiently different from the rod and cone opsins, to warrant their assignment to new opsin gene families. To date, the following novel photopigment genes have been isolated: In 1994 a cDNA from the chicken pineal was isolated that encodes a photopigment with an absorption maximum near 470 nm (Okano, Yoshizawa *et al.,* 1994). This opsin was called pinopsin (Popsin), and orthologues of Popsin have now been isolated from the pineal of several different bird (Kawamura and Yokoyama 1996), lizard (Kawamura and Yokoyama 1997), and amphibian species (Yoshikawa, Okano *et al.,* 1998). Popsin expression appears restricted to the pineal of birds, but appears to be expressed in both the pineal and retina of some reptiles, and the deep brain of some amphibia (Yoshikawa, Okano *et al.,* 1998). In 1997 two novel fish opsins were isolated. The first was VA (*vertebrate ancient*) opsin isolated from Atlantic salmon (Soni and Foster 1997).

This opsin forms a photopigment with an absorption maxima at 451 nm (with vitamin A_1) and is expressed in a sub-set of horizontal and amacrine cells of the retina (suggesting that the retina contains non-rod and non-cone photoreceptors), the pineal, and subhabenular region of the brain (Soni, Philip *et al.,* 1998). The second novel fish opsin, called parapineal opsin, was isolated from catfish. This opsin is expressed in the paprapineal and pineal organs (Blackshaw and Snyder 1997). The functional properties of parapineal opsin have yet to be determined. The most recent opsin to be discovered is melanopsin (Provencio, Jiang *et al.,* 1998). This opsin was originally isolated from the light-sensitive melanophores of *Xenopus laevis* (see 6.2.4) but is also expressed in the hypothalamus (NMPO region), iris (see 6.2.6), and the horizontal cell layer of the retina (similar to VA opsin). Again, functional studies are lacking and, at the time of writing this Chapter, we do not know if melanopsin is capable of forming a functional photopigment. The discovery of novel opsin is exciting, but the earlier immunocytochemical results should not be forgotten. For example, antibodies against rod opsin were found to label CSF-contacting neurones in the septal region of the bird brain (Silver, Witkovsky *et al.,* 1988) (see 6.2.3). The finding that these cells actually express rod opsin, and not a rod-like opsin, has been confirmed by very recent *in situ* results using rod opsin specific probes (Wada, Okano *et al.,* 1998). Thus both a novel opsin and an 'ocular' rod opsin are expressed in the avian brain. We suspect that the expression of a novel opsin as well as a rod and/or a cone opsin will be a feature of many of the extraocular photoreceptor organs.

6.4.2 Spectral discrimination

The major source of comparative information about the photopigments of extraretinal photoreceptors comes from studies of the pineal complex (pineal and third eye). As discussed by Meissl and co-workers (Meissl and Yanez 1994) the photosensitivity of the pineal complex in many lower vertebrates appears to be based on the presence of several photopigments (Table 1). Electrical recordings from second-order neurones have shown that the output can be either chromatic, where electrical activity is inhibited by short wavelengths and stimulated by longer wavelengths, or achromatic where responses always involve inhibition and are independent of the stimulus wavelength (Dodt and Meissl 1982).

The frontal organ of anurans and the parietal eye of lizards show clear chromatic responses (Table 1), and although physiological and behavioural evidence is lacking, it has been suggested that these extracranial photoreceptors may be involved in some form of twilight detection (Meissl and Yanez 1994). This suggestion is based on observations which show that during twilight there is a change in both the quantity and spectral quality of light. The relatively flat spectral composition of daylight becomes bimodal with a relative enrichment in the blue-green and a depletion in the yellow regions of the spectrum (Munz and

Table 1 Examples of photopigment wavelength of maximum absorption λmax from different extraretinal photoreceptors

Photoreceptor	Species	λ_{max} of photopigments		Ganglion cell response	Spectra determination	References
Frontal organ	Amphibia (Anura)	355	515	Chromatic	Pineal nerve recording	1
	Rana catesbeiana	500	560	Achromatic		
	Rana catesbeiana	550 – 580			MSP	2
Parietal Eye	Reptilia (Squamata)					
	Lacerta sicula campestris	450	520	Chromatic	Extracellular	3
	Xantusia vigilis	495			Intracellular	4
		440	(after long λ adaptation)			
	Uta stansburiana	495			Intracellular	4
		440	(after long λ adaptation)			
Pineal	Agnatha					
	Lampetra japonica	525			Intracellular	5 & 6
		380	525	Chromatic	Intracellular	7
	Actinopterygii					
	Esox lucius	380	620	Chromatic	Extracellular	8
		530	620	Achromatic		
	Salmo gairdneri	520			Extracellular	9
	Salmo irideus	500	533		Extracellular	10
	Salmo irideus	495	521		Intracellular	11
	Salmo irideus	$A_1$463	$A_2$561		MSP	12
	Salmo gairdneri	500			Melatonin suppression	13
	Salmo salaar	451 (VA opsin)			Expressed & reconstituted (A)	14
	Amphibia (Anura)					
	Rana catesbeiana	355	515	Chromatic	Extracellular	15
		500	560	Achromatic		
	Rana temporaria	500			MSP	16
	Xenopus laevis (larvae)	520			Extracellular	17
	Reptilia (Chelonia)					
	Pseudemys scripta elegans	606 –700		Achromatic	Extracellular	18
	Aves					
	Gallus domesticus	500			NAT suppression	19

(continued on next page)

Table 1 (continued)

Photoreceptor	Species	λ_{max} of photopigments	Ganglion cell response	Spectra determination	References
	Gallus domesticus	470 (P opsin)		Expressed & reconstituted (A_1)	20
Deep brain	Actinopterygii				
	Salmo salaar	451 (VA opsin)		Expressed & reconstituted (A_1)	14
	Aves				
	Corturnix japanica	492		LH secretion	21
Dermal iridophore	Actinopterygii				
	Paracheirdon innesi	461 545(?)		Spectral reflection	22
Dermal melanophore	Agnatha				
	Lampetra fluviatilis	530		Swimming behaviour	23
	Myxine glutinosa	520		Swimming behaviour	24
	Actinopterygii				
	Xiphophorus maculatus	420		Melanin aggregation	25
	Amphibia (Anura)				
	Xenopus laevis (larvae)	461 500		Melanin aggregation	26
	Rana temporaria	385 500		Extracellular	27
Iris	Actinopterygi				
	Anguilla	500 <400(?)		Contraction of isolated iris	28
	Amphibian (Anura)				
	Rana	500		Contraction of isolated iris	29

The λ_{max} is given in nanometers. Chromatic and achromatic responses recorded from ganglion cells are given where known. Chromatic responses refer to where short wavelength stimuli have caused an inhibitory output (hyperpolarisation) and longer wavelengths have caused an excitatory output (depolarisation). In this table the hyperpolarising wavelength is listed first. Achromatic responses refer to inhibitory responses at all wavelengths. Extracellular and Intracellular refers to the type of electrical responses recorded. *Abbreviations*: A_1 = vitamin A_1 based photopigment; A_2 = vitamin A_2 based photopigment; LH = luteinising hormone; MSP = microspectrophotometry; NAT = N-acetyl transferase (rate limiting enzyme in melatonin synthesis) P opsin = chicken pinopsin; VA opsin = salmon vertebrate ancient opsin. References cited: 1 = Dodt and Heerd 1962; 2 = Hartwig and Baumann 1974; 3 – Dodt and Scharrer 1968; 4 = Solessio and Engbretson 1993; 5 = Morita *et al.*, 1985; 6 = Tamotsu and Morita 1986; 7 = Uchida and Morita 1990; 8 = Falcon and Meissl 1981; 9 = Meissl and Ekstrom 1988; 10 = Morita 1966; 11 = Marchiafava and Kusmic 1993; 12 = Kusmic *et al.*, 1993; 13 = Max and Menaker, 1992; 14 = Soni, Philp *et al.*, 1998 15 = Dodt and Morita 1964; 16 = Hartwig and Baumann 1974; 17 = Foster and Roberts 1982; 18 = Meissl and Ueck 1980; 19 = Deguchi 1981; 120 = Okano, Yoshizawa *et al.*, 1994; 21 = Foster and Follett 1985; 22 = Lythgoe and Shand 1984; 23 = Steven 1950; 24 = Steven 1955; 25 = Wakamatsu, Kawamuro *et al.*, 1980; 26 = Lythgoe and Thompson 1984; 27 = Rayport and Wald 1978; 28 = Seliger 1963; 29 = Barr and Alpern 1963

McFarland 1977). Chromatic responses result from the inhibitory effect of short wavelengths and the excitatory effect of mid-long wavelengths. As a result, the neural output from extraretinal photoreceptors showing chromatic responses would vary as the quality of light changed during twilight. It has been suggested that this may be a mechanism for the detection of dawn and dusk and the regulation of behavioural rhythms (Meissl and Brandstatter 1992) also see Chapter 7. Such chromatic responses could originate from two populations of cells with different spectral absorption properties, or alternatively from within individual cells that contain a chromatically-dependent antagonistic mechanism. The latter type of response has been demonstrated in the parietal eye of lizards (Solessio and Engbretson 1993). A green (495 nm) depolarizing mechanism was found to be antagonised by a blue (440 nm) hyperpolarising mechanism within individual photoreceptors following adaptation to green light. In lizards, parietal eye 'twilight detection' was suggested as follows: The green-sensitive mechanism would be active at dawn causing depolarisation, followed by hyperpolarisation as the spectral composition of the light shifts to shorter wavelengths during the day, then as dusk approaches the photoreceptor would once again depolarise towards its dark resting potential.

One might expect that if chromatic responses were used for twilight detection then these responses would not be found in those extraretinal photoreceptors heavily filtered by haemoglobin and tissue. The pineal photoreceptors of some teleosts do show chromatic responses (Table 1) and in these cases we speculate that the pineal window probably allows sufficient light to enter the pineal to permit spectral discrimination and twilight detection. However, until behavioural studies are undertaken the role of these chromatic responses remains a mystery. We stress that the possession of multiple photopigments does not necessarily mean that an extraretinal photoreceptor is capable of a chromatic response or some form of colour discrimination. Some pineal organs may employ several photopigments, with different spectral responses, to obtain achromatic measures over a broad spectral range (Meissl and Brandstatter 1992; Provencio and Foster 1993).

The first evidence to suggest that the avian pineal might contain more than one photopigment came from biochemical studies on the pineal of the Japanese quail. This work suggested the presence of a rod-like photopigment and a second photopigment sensitive to long wavelengths (Foster, Schalken *et al.,* 1989). More recently a chick pineal cDNA library was screened using cDNA probes encoding chicken retinal opsins. This approach identified an opsin identical to chicken red opsin and, as discussed above, a novel opsin (P opsin, see 6.4.1) which forms a photopigment with a wavelength of maximum absorption (λ_{max}) at 470 nm (Okano, Yoshizawa *et al.,* 1994). Again we do not know how these photopigments are employed within the avian pineal but it should be noted that melatonin biochemistry is modified by light in two distinct ways. Light mediates an acute response, inhibiting melatonin synthesis, and also entrains the circadian rhythm of melatonin synthesis (Chapter 7). Perhaps these different responses are mediated via different photoreceptors (Foster, Schalken *et al.*, 1989).

6.4.3 Spectral tuning

The analysis of spectral tuning in the visual systems involves the comparison of species from different light environments to determine whether the λ_{max} is in a position to give the maximum integrated quantal catch. This approach can also be applied to the extraretinal photoreceptors which can be divided into two groups based upon whether they are directly exposed to environmental light (frontal organs, parietal eyes, pineal organs associated with a pineal window, iridophores, melaonophores, and photosensitive iris) and those screened by tissue and haemoglobin (pineal organs lacking a pineal window and deep brain photoreceptors).

Although considerable amounts of light of all wavelengths reach deep into the brain, the light environment of deep brain and most pineal photoreceptors is dominated by the spectral absorption of haemoglobin (Foster and Follett 1985). If the photopigments of these receptors were to maximise photon capture then they should have a λ_{max} situated well into the red part of the spectrum, but this is clearly not the case. An action spectrum for the deep brain photoreceptors in the quail suggests a pigment with a λ_{max} near 500 nm (Foster, Follett *et al.*, 1985), and for those pineals screened by tissue and haemoglobin, photopigment λ_{max} are around 500 nm (Table 1). As noted previously, there is a small window in haemoglobin transmission (460 – 540 nm) peaking around 490 nm. Perhaps this haemoglobin window has exerted some selection pressure on the spectral position of these intra-cranial photopigments (Foster and Follett 1985; Kusmic, Barsanti *et al.,* 1993). In the lizard *Anolis carolinensis* extraretinal photoreceptors appear to be based upon a vitamin A_2 chromophore (Provencio and Foster 1993). Unfortunately we do not know where the λ_{max} of these photopigments are situated, and whether this chromophore confers increased sensitivity to longer wavelengths.

In those extraretinal photoreceptors exposed to unfiltered light there is also no clear correlation between photopigment λ_{max} and the quality of the light environment, although it should be noted that very few comparative studies have been attempted. The one exception would be the pike, a species which inhabits long wavelength transmitting waters and possesses an exceptionally long wavelength-sensitive pineal photopigment (Table 1). We mention above the possibility that some extraretinal photoreceptors employ several photopigments, with different λ_{max}, to obtain achromatic measures of irradiance over a broad spectral range. If this is indeed the case, then perhaps we should not expect to see a correlation between the photopigment λ_{max} of extraretinal photoreceptors and their light environment, particularly in terrestrial species which inhabit a broad spectral environment. In contrast to the terrestrial environment however, the spectral composition of the aquatic environment is more restricted and we would predict that aquatic organisms would confine their achromatic responses to the spectral range available. For example, the λ_{max} of pineal photopigments would be widely spaced in aquatic organisms living close to the surface (like terrestrial species), while the λ_{max} of pineal photopigments would be closer together and moved towards the blue in those species living at greater depth in oceanic waters. In deep water the spectral environment would be so restricted that there would be no advantage in having more than one photopigment for irradiance detection. It would also be interesting to determine the dimensions and transparency of the pineal window of fish species living at increasing depths.

6.5 The response characteristics of extraretinal photoreceptors

Phototransduction mechanisms in the extraretinal photoreceptors have not been investigated to any detailed extent. Of the relatively few studies undertaken, all suggest that pineal and deep brain photoreceptors use similar phototransduction proteins to those used by retinal rods and cones (for example van Veen, Elofsson *et al.,* 1986; Foster, Korf *et al.,* 1987; Garcia-Fernandez and Foster 1994; Yoshikawa, Yashiro *et al.,* 1994; Okana and Fukada 1997; Okano, Yamazaki *et al.,* 1997; Yoshikaw and Oishi 1998.). In addition, the pinealocytes of many lower vertebrates hyperpolarise in response to light and a cyclic GMP-activated channel has been reported in chick pineal photoreceptors (Dryer and Henderson 1991). However, the extent to which the phototransduction proteins of the extraretinal and retinal photoreceptors are identical or only similar remains to be determined.

In contrast to photoreceptors of diencephalic origin (eyes and pineal) important differences have emerged between the phototransduction mechanisms of dermal and rod and cone photoreceptors. For example, cyclic AMP, not cyclic GMP, appears to be the second messenger in the phototransduction of the response of melanophores (Rollag 1993) (for further discussion see Provencio, Jiang *et al.,* 1998). In addition, the sodium channels of iridophores were found to open following light-activation, rather than close as in rods and cones (Clothier and Lythgoe 1987).

Several studies have compared the overall response characteristics of retinal and extraretinal photoreceptors. The main difference to emerge is that extraretinal photoreceptors show slower responses to light. The most detailed comparison has been between the photoreceptors of the retina and pineal of fish (Meissl and Ekstrom 1988). These studies have shown that although the basic response pattern of pineal photoreceptors is comparable to retinal photoreceptor potentials, pineal responses are much slower. The time from response onset to peak potential and the time for recovery of membrane potential after light exposure are considerably longer. On the basis of this slow time-course Meissl and co-workers have argued that these photoreceptors are highly specialised for the detection of slow gradual changes in environmental light rather than the detection of transient light stimuli (Meissl and Yanez 1994). This would be entirely consistent with the view that the pineal acts as an irradiance detector.

The slower responses of pineal photoreceptors suggest that at some level the phototransduction cascade of extraretinal photoreceptors may be altered. In this context the structure of the recently identified chicken P opsin and fish VA opsin (see 6.4.1) is very interesting. An opsin sequence defines more than the absorption characteristics of the photopigment, it also determines the nature of the interactions with the phototransduction proteins. Both P and VA opsins exhibit several structural changes that would favour their role as irradiance detectors. Foremost among these are modifications to the carboxyl terminal tail. The tail of both P and VA opsin have fewer hydroxyl containing residues than the 'image forming' visual opsins (Soni and Foster 1997). Hydroxyl residues are phosphorylated after opsin activation, which facilitates the binding of arrestin and prevents further interaction with transducin. As a result, activated P and VA opsins may trigger the phototransduction cascade for longer than their ocular counterparts. In addition, there are changes in the transducin binding domains of both opsins. For example, VA opsin contains striking differences in the third cytoplasmic loop, an area which interacts with transducin in other opsins. A comparison of VA opsin to a consensus sequence of third cytoplasmic loops reveals four non-conserved substitutions and a two amino acid deletion. Additionally, VA lacks conserved N-linked glycosylation sites which play a role in signal transduction (Soni and Foster 1997). On the basis of, admittedly limited information, it appears that the novel extraretinal opsins show specialisations which are associated with the dynamics of the phototransduction cascade rather than with spectral tuning.

6.6 Conclusions

Extraretinal photoreceptors detect environmental light to provide: a) time of day cues for the regulation of circadian rhythms; b) cues for behavioural orientation; c) cues for regulating body and/or corneal pigmentation and colouration; and d) cues for regulating pupil size. These sensory tasks are quite different from the demands imposed upon the image forming visual system which must maximise photon capture for contrast detection. These differences in sensory task have resulted in important differences between the extraretinal and retinal photoreceptors. All extraretinal photoreceptors, with the exception of some third eyes, lack optical structures for image detection and act primarily as irradiance detectors. While several of the extraretinal photoreceptor organs show some form of spectral discrimination, it seems very unlikely that this facility is used to enhance image detection. The most persuasive, but not the only explanation for this function, is that it provides a mechanism for the detection of twilight and hence time cues for the regulation of rhythmic behaviour. Conversely, multiple photopigments with different λ_{max} may also be used by extraretinal photoreceptors to provide measures of irradiance over a broad spectral range, particularly in those environments where there is unfiltered sunlight. Perhaps this forms a partial explanation why there is little evidence for a match between the photopigment λ_{max} of extraretinal photopigments and the light environment of those species so far investigated. The use of molecular approaches has led to the identification of several new photopigment gene families. Significantly, these genes are not expressed in the rods and cones of the retina, but are expressed in multiple sites within the CNS. These novel opsins have features which suggest that they may have specialisations associated with the dynamics of phototransduction rather than with spectral tuning.

Collectively the recent findings show that non-mammalian vertebrates possess multiple extraocular photoreceptors that may express novel, rod and/or cone photopigments. The future challenge will be to link these photopigments with the physiology and behaviour of the whole animal.

Acknowledgments:

John Lythgoe for introducing us to the ecology of vision. Elizabeth Downes, Malcolm von Schantz and GianLuca Tosini for discussion and editorial comments (RF). The Department of Marine Biology, James Cook University of North Queensland provided support during initial drafting of the manuscript (JS).

References

Appleby, S. J. and Muntz, W. R. A. (1979) Occlusable yellow corneas in *Tetraodontidae*. *Journal of Experimental Biology*, 83, 249 – 259.

Bagnara, J. T. and Hadley, M. E. (1970) Endocrinology of the amphibian pineal. *American Zoologist*, 10, 201– 216.

Barr, L. (1989) Photomechanical coupling in the vertebrate sphincter pupillae. *CRC Critical Review of Neurobiology*, 4, 325 – 366.

Barr, L. and Alpern, M. (1963) Photosensitivity of the frog iris. *Journal of General Physiology*, 46, 1249 –1265.

Barr, L. and Gu, F. (1987) A quantitative model of myosin phosphorylation and the photomechanical response of the isolated sphincter pupillae of the frog iris. *Biophysics Journal*, 51, 895 – 904.

Bell, A. L. (1967) Morphological and physiological investigations of the photosensitive iris of the American eel (*Anguilla rostrata*). State University of New York.

Benoit, J. (1935a) Le role des yeux dans l'action stimulante de la lumiere sur le developpement testiculaire chez le candard. *Comptes Rendus des Séances de la Société de Physique et de ses Filiales*, 118, 669 – 671.

Benoit, J. (1935b) Stimulation par la lumiere artificielle du developpement testiculaire chez des canards aveugles par section du nerf optique. *Comptes Rendus des Séances de la Société de Physique et de ses Filiales*, 120, 133 –136.

Blackshaw, S. and Snyder, S. (1997) Parapineal opsin, a novel catfish opsin localized to the parapineal organ, defines a new gene family. *The Journal of Neuroscience*, 17(21), 8083 – 8092.

Clothier, J. and Lythgoe, J. N. (1987) Light-induced colour changes by the iridophores of the neon tetra, *Paracheirodon innesi*. *Journal of Cell Science*, 88, 663 – 668.

Cooper, H. M., Tessonneaud, A., Caldani, A., Locatelli, A., Richard, S. and Viguier-Martinez, M. C. (1993) Morphology and distribution of retinal ganglion cells (RGC) projecting to the suprachiasmatic nucleus in the sheep. *Society for Neuroscience Abstracts*, 19, Abstract 701.11.

Deguchi, T. (1981) Rhodopsin-like photosensitivity of isolated chicken pineal gland. *Nature*, 290, 702–704.

Dodt, E. and Heerd, F. (1962) Mode of action of pineal nerve fibers in frogs. *Journal of neurophysiology,* 25, 405 – 429.

Dodt, E. and Meissl, H. (1982) The pineal and parietal organs of lower vertebrates. *Experientia*, 38, 996 –1000.

Dodt, E. and Morita, Y. (1964) Purkinje-verschiebung, absolute schwelle und adaptives verhalten einzelner elemente der intrakranialen anuren-epiphyse. *Vision Research*, 4, 413 – 421.

Dodt, E. and Scharrer, E. (1968) Photic responses from the parietal eye of the lizard *Lacerta sicula campestris* (De Betta). *Vision Research*, 8, 61–72.

Dryer, S. E. and Henderson, D. (1991) A cyclic GMP-activated channel in dissociated cells of the chick pineal. *Nature*, 353, 756 –758.

Ebisawa, T., Karne, S., Lerner, M. R and Reppert, S. M. (1994) Expression cloning of a high-affinity melatonin receptor from *Xenopus* dermal melanophores. *Proceedings of the National Academy of Sciences USA*, 91(13), 6133 – 6137.

Ekstrom, P., Borg, B. and van Veen, T.-H. (1983) Ontogenetic development of the pineal organ, parapieal organ, and retina of the three-spined stickleback, *Gasterosteus aculeatus* L. (Teleostei). *Cell & Tissue Research,* 233, 593 – 609.

Engbretson, A. (1992) Neurobiology of the lacertilian parietal eye system. *Ethology, Ecology and Evolution,* 4, 89 –107.

Falcon, J. and Meissl, H. (1981) The photosensory function of the pineal organ of the pike (*Esox lucius* L.). Correlation between structure and function. *Journal of Comparative Physiology A*, 144, 127–137.

Fenwick, J. C. (1970) Effects of pinealectomy and bilateral enucleation on the phototactic response and on the conditioned response to light of the goldfish *Carassius auratus* L. *Canadian Journal of Zoology*, 48, 175 –182.

Fernandez-de-Miguel, F. and Arechiga, H. (1992) Sensory inputs mediating two opposite behavioural responses to light in the crayfish *Procambarus clarkii. Journal of Experimental Biology,* 164, 153 –169.

Foster, R. G. and Follett, B. K. (1985) The involvement of a rhodopsin-like photopigment in the photoperiodic response of the Japanese quail. *Journal of Comparative Physiology A*, 157, 519 – 528.

Foster, R. G., Follett, B. K. and Lythgoe, J. N. (1985) Rhodopsin-like sensitivity of extra-retinal photoreceptors mediating the photoperiodic response in quail. *Nature*, 313(5997), 50 – 52.

Foster, R. G., Garcia-Fernandez, J. M., Provencio, I. and DeGrip, W. J. (1993) Opsin localization and chromophore retinoids identified within the basal brain of the lizard *Anolis carolinensis. Journal of Comparative Physiology A*, 172, 33 – 45.

Foster, R. G., Grace, M. S., Provencio, I., DeGrip, W. J. and Garcia-Fernandez, J. M. (1994) Identification of vertebrate deep brain photoreceptors. *Neuroscience and Biobehavioral Reviews*, 18(4), 541– 546.

Foster, R. G., Korf, H. G. and Schalken, J. J. (1987) Immunocytochemical markers revealing retinal and pineal but not hypothalamic photoreceptor systems in the Japanese quail. *Cell and Tissue Research,* 248, 161–167.

Foster, R. G. and Menaker, M. (1993) Circadian photoreception in mammals and other vertebrates. *Light and biological rhythms in man.* L. Wetterberg, Pergamon: pp. 73 – 91.

Foster, R. G. and Roberts, A. (1982) The pineal eye in *Xenopus laevis* embryos and larvae: A photoreceptor with a direct excitatory effect on behavior. *Journal of Comparative Physiology A*, 145, 413 – 419.

Foster, R. G., Schalken, J. J., Timmers, A. M. and De Grip, W. J. (1989) A comparison of some photoreceptor characteristics in the pineal and retina: I. The Japanese quail *(Coturniz cotunrix). Journal of Comparative Physiology A*, 165, 553 – 563.

Frisch, K. v. (1911) Beitrage zur Physiologie der Pigmentzellen in der Fischaut. *Pflügers Archiv für die gesamte Physiologie des Menschen und der Tiere*, 138, 319 – 387.

Garcia-Fernandez, J. M. and Foster, R. G. (1994) Immunocytochemical identification of photoreceptor proteins in hypothalamic cerebrospinal fluid-contacting neurons of the larval lamprey *(Petromyzon marinus). Cell and Tissue Research*, 275, 319 – 326.

Garcia-Fernandez, J. M., Jimenez, A. J., Gonzalez, B., Pombal, M. A. and Foster, R. G. (1997) An immunocytochemical study of encephalic photoreceptors in three species of lamprey. *Cell and Tissue Research:* 288, 267– 278.

Groos, G. (1982) The comparative physiology of extraocular photoreception. *Experientia*, 38, 989 –1128.

Harth, M. S. and Heaton, M. B. (1973) Non-visual responsiveness in newly hatched pigeons. *Science*, 180, 753 –755.

Hartwig, H.-G. and Baumann, C. H. (1974) Evidence for the photosensitive pigments in the pineal

complex of the frog. *Vision Research*, 14, 597– 598.
Hartwig, H.-G. and van Veen, T. (1979) Spectral characteristics of visible radiations penetrating into the brain and stimulating extra-retinal photoreceptors. *Journal of Comparative Physiology A*, 120, 277– 282.
Helfrich, C. (1986) Role of the optic lobes in the regulation of the locomotor activity rhythm of Drosophila melanogaster: Behavioural analysis of visual mutants. *Journal of Neurogenetics,* 3, 321– 343.
Iga, T. and Takabatake, I. (1986) Local light stimulation of melanophores of a teleost, *Zaco temmincki. Journal of Experimental Biology*, 238, 385 – 391.
Kawamura, S. and Yokoyama, S. (1996) Molecular characterization of the pigeon P-opsin gene. *Gene,* 182, 213 – 214.
Kawamura, S. and Yokoyama, S. (1997) Expression of visual and nonvisual opsins in American chameleon. *Vision Research,* 37, 1867–1871.
Kondrashev, S. L., Gamburtzeva, A. G., Gnjubkina, V. P., Orlov, O. J. and My, P. T. (1986) Colouration of corneas in fish. A list of species. *Vision Research*, 26, 287– 290.
Kondrashev, S. L. and Khodtsev, A. S. (1984) Light-dependent and humoral control of pigment transport in corneal chromatophores in marine fishes. *Zoological Journal of Physiology,* 88, 317– 325.
Kusmic, C., Barsanti, L., Passarelli, V. and Gualtieri, P. (1993) Photoreceptor morphology and visual pigment content in the pineal organ and in the retina of juvenile and adult trout, *Salmo irideus. Micron.*, 24, 279 – 286.
Land, M. F. (1987) Screening pigment migration in a sphingid moth is triggered by light near the cornea. *Journal of Comparative Physiology A*, 160, 355 – 357.
Land, M. F. (1991) Optics of the eyes of the animal kingdom. *Vision and Visual Dysfunction*. J. R. Cronly-Dillon and R. L. Gregory, CRC Press. 2, 118 –135.
Lythgoe, J. N. (1979) *The Ecology of Vision.* Oxford: Clarendon Press.
Lythgoe, J. N. (1984) Visual pigments and environmental light. *Vision Research*, 24, 1539 –1550.
Lythgoe, J. N. and Shand, J. (1982) Changes in spectral reflexions from the iridophores of the neon tetra. *Journal of Physiology*, 325, 23 – 34.
Lythgoe, J. N. and Shand, J. (1983) Diel colour changes in the neon tetra *Paracheirodon innesi. Environmental Biology of Fish,* 8, 249 – 254.
Lythgoe, J. N. and Shand, J. (1984) Action spectra for the iridophore light response in the neon tetra. *Photochemistry & Photobiology,* 40, 551– 553.
Lythgoe, J. N. and Shand, J. (1989) The structural basis for iridescent colour changes in dermal and corneal iridophores in fish. *Journal of Experimental Biology*, 141, 313 – 325.
Lythgoe, J. N., Shand, J. and Foster, R. G. (1984) Visual pigment in fish iridocytes. *Nature*, 308(5954), 83 – 84.
Lythgoe, J. N. and Thompson, M. (1984) A porphyropsin-like action spectrum from *Xenopus* melanophores. *Photochemistry & Photobiology*, 40, 411– 412.
Marchiafava, P. L. and Kusmic, C. (1993) The electrical responses of the trout pineal photoreceptors to brief and prolonged illumination. *Progress in Brain Research*, 95, 3 –13.
Max, M. and Menaker, M. (1992) Regulation of melatonin production by light, darkness, and temperature in the trout pineal. *Journal of Comparative Physiology A,* 170(4), 479 – 489.
Meissl, H. and Brandstatter, R. (1992) Photoreceptive function of the teleost pineal organ and their implications in biological rhythms. *Rhythms in fishes*. M. A. Ali. New York, Plenum, 235 – 254.
Meissl, H. and Ekstrom, P. (1988) Photoreceptor responses to light in the isolated pineal organ of the trout, *Salmo gairdneri. Neuroscience*, 24, 1071–1076.
Meissl, H. and Ueck, M. (1980) Extraocular photoreception of the pineal gland of the aquatic turtle *Pseudomys scripta elegans. Journal of Comparative Physiology A,* 140, 173 –179.
Meissl, H. and Yanez, J. (1994) Pineal photosensitivity. A comparison with retinal photoreception. *Acta Neurobiologiae Experimentalis*, 54(suppl.), 19 – 29.
Menaker, M. and Underwood, H. (1976) Extraretinal photoreception in birds. *Photochemistry and Photobiology,* 23, 299 – 306.
Millot, N. (1968) The dermal light sense, *Invertebrate receptors.* J. D. Carthy and G. E. Newells. New York, Academic Press, 1– 36.
Morita, Y. (1966) Entladungsmuster pinealer neurone der regenbogenforelle (*Salmo irideus*) bei belichtung des zwischenhirns. *Pflügers Archiv für die gesamte physiologies des Menschen und der Tiere*, 289, 155 –167.
Morita, Y., Tabata, M. and Tamotsu, S. (1985) Intracellular response and input resistance change of pineal photoreceptors and ganglion cells. *Neuroscience Research*, 2, 79 – 88.
Muntz, W. R. A. (1976) The visual consequences of yellow filtering pigments in the eyes of fishes

occupying different habitats. *Light as an Ecological Factor II*. G. C. Evans, R. Bainbridges and O. Rackham. Oxford, Blackwell Scientific, 271– 287.

Munz, F. W. and McFarland, W. N. (1977) Evolutionary adaptations of fishes to the photic environment. *Handbook of Sensory Physiology*. F. Crescitelli. Berlin: Springer-Verlag. VII/5, 193 – 274.

Nagaishi, H., Oshima, N. and Fujii, R. (1990) Light-reflecting properties of the iridophores of the neon tetra, *Parachierodon innesi. Comparative Biochemistry and Physiology A*, 35, 337– 342.

Nishi, T. and Gotow, T. (1992) A neural mechanism for processing colour information in molluscan extra-ocular photoreceptors. *Journal of Experimental Biology,* 168, 77– 91.

Nordtug, T., Berg, O. L. and Melo, T. B. (1994) Directional light transmission in the pineal window of atlantic salmon (*Salmo salar* L.) may be used for solar orientation. *The Journal of Experimental Zoology*, 269, 403 – 412.

Okana, T. and Fukada, Y. (1997) Phototransduction cascade and circadian oscillator in chicken pineal gland. *Journal of Pineal Research,* 22, 145 –151.

Okano, T., Yamazaki K., Kasahara, T. and Fukada, Y. (1997) Molecular cloning of heterotrimeric G-protein α-subunits in chicken pineal gland. *Journal of Molecular Evolution,* 44(Suppl 1), S91– S97.

Okano, T., Yoshizawa, T. and Fukada, Y. (1994) Pinopsin is a chicken pineal photoreceptive molecule. *Nature*, 372(6501), 94 – 97.

Orlov, O. Y. and Gamburtzeva, A. G. (1976) Changeable coloration of cornea in the fish *Hexagrammos octogrammus. Nature*, 263, 405 – 407.

Page, T. L. (1982) Extraretinal photoreceptors in entrainment and photoperiodism in invertebrates. *Experientia*, 38, 1007–1013.

Pang, P. K. T. (1965) Light sensitivity of the pineal gland in blinded *Fundulus heteroclitus. American Zoologist,* 5, 682.

Provencio, I. and Foster, R. G. (1993) Vitamin A_2-based photopigments within the pineal gland of a fully terrestrial vertebrate. *Neuroscience Letters*, 155, 223 – 226.

Provencio, I., Jiang, G., DeGrip, W. J., Hayes, W. P. and Rollag, M. D. (1998) Melanopsin: An opsin in melanophores, brain and eye. PNAS 95, 340 – 345.

Quay, W. B. (1979) The parietal eye-pineal complex. *Biology of the Reptilia.* C. Glans, R. G. Northcutt and P. Ulinski, Academic Press 9.

Rayport, S. and Wald, G. (1978) Frog skin photoreceptors. *Program of Abstracts of the 6th Annual Meeting of American Photobiology,* 94 – 95.

Rollag, M. D. (1993) Pertussis toxin sensitive photoaggregation of pigment in isolated *Xenopus* tail-fin melanophores. *Photochemistry & Photobiology,* 57, 862 – 866.

Romer, A. S. (1970) *The Vertebrate Body.* Philadelphia: W. B. Saunders..

Ronan, M. and Bodznick, D. (1991) Behavioural and neurophysiological demonstration of a lateralis skin photosensitivity in larval sea lampreys. *Journal of Experimental Biology*, 161, 97–117.

Rubin, L. J. and Nolte J. F. (1986) Cyclic nucleotide levels of the photosensitive irises of *Bufo Marinus* and *Lophius*. *Experimental Eye Research*, 43, 771–779.

Rudeberg, C. (1969) Structure of the parapineal organ of the adult rainbow trout, *Salmo gairdneri* Richardson. *Zeitschrift Zellforsch*, 93, 282 – 304.

Scharrer, E. (1928) Die lichtempfindlichkeit blinder Elritzen I. Untersuchungen uber das Zwischenhirn der Fische. *Zeitschrift Vergleichende Physiologie,* 7, 1– 38.

Scharrer, E. (1964) Photoneuroendocrine systems: General concepts. *Annals of the New York Academy of Science*, 117, 13 – 22.

Seliger, H. H. (1963) Direct action of light on naturally pigmented muscle fibers, I. Action spectrum for eel iris sphincter. *Journal of General Physiology*, 46, 333 – 342.

Shand, J. (1988) Corneal iridescence in fishes: light-induced colour changes in relation to structure. *Journal of Fish Biology,* 32, 625 – 632.

Shand, J. and Lythgoe, J. N. (1987) Light-induced changes in corneal iridescence in fish. *Vision Research*, 27, 303 – 305.

Shand, J. and Lythgoe, J. N. (1990) The isolated iridescent cornea of the sand goby is photoresponsive. *Photochemistry and Photobiology*, 51, 737–739.

Silver, R., Witkovsky, P., Horvath, P., Alones, V., Barnstable, C. J. and Lehman, M. N. (1988) Co-expression of opsin- and VIP-like-immunoreactivity in CSF-contacting neurons of the avian brain. *Cell and Tissue Research,* 253, 189 –198.

Solessio, E. and Engbretson, G. A. (1993) Antagonistic chromatic mechanisms in photoreceptors of the parietal eye of lizards. *Nature*, 364, 442 – 445.

Soni, B. G. and Foster, R. G. (1997) A novel and ancient vertebrate opsin. *FEBS Letters* 406, 279 – 283.

Soni, B. G., Philp, A., Knox, B. E. and Foster, R. G. (1998) Novel Retinal Photoreceptors. *Nature,* 394, 27– 28.

Steven, D. M. (1950) Some properties of the photoreceptors of the Brook lamphrey. *Journal of Experimental Zoology*, 27, 350 – 364.

Steven, D. M. (1955) Experiments in the light sense of the hag fish, *Myxine glutinosa. Journal of Experimental Biology.*

Steven, D. M. (1963) The dermal light sense. *Biological Review*, 38, 204 – 240.

Tabata, M., Suzuki, T. and Niwa, H. (1985) Chromophores in the extraretinal photoreceptor (pineal organ) of teleosts. *Brain Research,* 338, 173 –176.

Tamotsu, S. and Morita, Y. (1986) Photoreception in pineal organs of larval and adult lampreys, *Lampetra japonica. Journal of Comparative Physiology A,* 159, 1– 5.

Tosini, G. and Avery, R. (1996) Dermal receptors regulate basking behaviour in the lizard *Podarcis muralis. Physiology and Behavior*, 59, 195 –198.

Tosini, G. and Menaker, M. (1996) The pineal complex and melatonin affect the expression of the daily rhythm of behavioural thermoregulation in the green iguana. *Journal of Comparative Physiology A,* 179, 135 –142.

Uchida, K. and Morita, Y. (1990) Intracellular responses from UV-sensitive cells in the photosensory pineal organ. *Brain Research*, 534, 237– 242.

Ullen, F., Orlovsky, G. N., Delisgina, T. G. and Grillner, S. (1993) Role of dermal photoreceptors and lateral eyes initiation and orientation of locomotion in lamphrey. *Behavioral Brain Research*, 54, 107–110.

Underwood, H. (1992) Endogenous Rhythms. *Biology of the reptilia Vol. 18: Hormones, Brain, and Behaviour*. C. Gans and D. Crews. Chicago and London, The University of Chicago. Chapter 5, 229 – 297.

Underwood, H. and Groos, G. (1982) Vertebrate circadian rhythms: Retinal and extraretinal photoreception. *Experientia*, 38, 1013 –1021.

van Veen, T., Elofsson, R., Hartwig, H. G., Gery, I., Mochizuki, M., Cena, V. and Klein, D. C. (1986) Retinal S-antigen: immunocytochemical and immunochemical studies on distribution in animal photoreceptors and pineal organs. *Experimental Biology*, 45(1), 15 – 25.

Vigh, B. and Vigh-Teichmann, I. (1988) Comparative neurohistology and immunocytochemistry of the pineal complex with special reference to CSF-contacting neuronal structures. *Pineal Research Re*view, 6, 1– 65.

Vollrath, L. (1981) *The Pineal Organ*. Berlin, Heidelberg, New York, Springer-Verlag.

Wada, Y., Okano, T., Adachi, A., Ebihara, S. and Fukada, Y. (1998) Identification of rhodopsin in the pigeon deep brain. *FEBS Letters*, 424, 53 – 56.

Wakamatsu, Y., Kawamuro, S. and Yoshizawa, T. (1980) Light-induced aggregation in cultured melanophores: spectral sensitivity and inhibitory effects of theophylline and cyclic adenosine 3-5 monophosphaste. *Journal of Cell Science*, 41, 65 –74.

Weber, W. (1983) Photosensitivity of Chromatophores. *American Zoology*, 23, 495 – 506.

Wolken, J. J. and Mogus, M. A. (1979) Extra-ocular photo-sensitivity. *Photochemistry & Photobiology*, 29, 189 –196.

Yoshikawa, T. and Oishi, T. (1998) Extraretinal photoreception and cricadian systems in nonmammalian vertebrates. *Comparative Biochemistry and Physiology,* 119B(1), 65 –72.

Yoshikawa, T., Okano, T., Oishi, T. and Fukada, Y. (1998) A deep brain photoreceptive molecule in the toad hypothalamus. *FEBS Letters*, 424, 69 –72.

Yoshikawa, T., Y. Yashiro, T. Oishi, K. Kokame and Y. Fukada (1994) Immunoreactivities to rhodopsin and rod/cone transducin antisera in the retina, pineal complex and deep brain of the Bullfrog, *Rana catesbeiana. Zoological Science,* 11, 675 – 680.

Young, J. Z. (1935a) The photoreceptors of lampreys I. Light-sensitive fibers in the lateral line nerves. *Journal of Experimental Biology*, 12, 229 – 238.

Young, J. Z. (1935b) The photoreceptors of lampreys II. The functions of the pineal complex. *Journal of Experimental Biology,* 12, 254 – 270.

Young, J. Z. (1962) *The life of the vertebrates*. Oxford: The Clarendon Press.

Zimmerman, K. and Heatwole, H. (1990) Cutaneous photoreception: a new sensory mechanism for reptiles. *Copeia*, 860 – 862.

Zucker, R. M. and Nolte, J. (1981) A search for the photoreceptor in the photosensitive irises of normally pigmented and albino hamsters *(Mesocricetus auratus). Current Eye Research*, 1(1): 9 –18.

7
The regulation of vertebrate biological clocks by light

R.G. FOSTER*[1] and I. PROVENCIO[2]
[1]Reader in Zoology, Imperial College of Science Technology and Medicine, Department of Biology, Sir Alexander Fleming Building, Imperial College Road, London SW7 2AZ, UK.
[2]Department of Anatomy, Uniformed Services University of the Health Sciences, 4301 Jones Bridge Road, Bethesda MD 20814-4799, USA.
*To whom correspondence should be addressed.

7.1 Introduction

Many of the complex environmental demands, and hence selection pressures, imposed upon an organism are associated with periodic events like the day:night cycle, the seasons or the tides. An organism could simply change behaviour in response to these changing conditions but responses, by their very nature, are subject to delays. In many cases changes in the state of behaviour (e.g. from sleep to wake), require a complex realignment of physiology, and cannot be accomplished without a significant loss of time. If, however, an organism can anticipate periodic environmental events, and be ready to exploit the changed conditions, then it will be at a distinct selective advantage. In view of these considerations it is not surprising that almost all organisms, with the possible exception of certain groups of prokaryotes, have evolved some form of endogenous time-keeping system or biological clock.

Those biological clocks with an endogenous period of approximately 24 h are called circadian clocks or oscillators from the Latin 'about' (*circa*) and 'day' (*dies*). Circadian clocks are distinguished from other biological clocks on the basis of their period; those with periods close to 12.4 h are called circatidal clocks, 14.8 days – circa-spring-tide clocks, 29.5 days – circa-lunar clocks, and 1 year – circannual clocks (Aschoff 1981). This chapter will restrict the discussion to those biological rhythms that employ a circadian oscillator.

Most circadian clocks under constant conditions have periods that deviate significantly from 24 h. The freerunning period (the expressed rhythm under constant conditions) can range from 21–27 hours depending on species, although most vertebrates are normally in the 23 – 25 h range. Because biological clocks are not exactly 24 hours (and there are interesting theoretical reasons why this should be so (Pittendrigh 1981)) the circadian system must be synchronised to ensure that internal and local time coincide – the process of entrainment. Entrainment regulates not only period but also the timing or phase of the rhythm. When entrained, the circadian clock adopts a distinct phase

S.N. Archer et al. (eds.), Adaptive Mechanisms in the Ecology of Vision, 223–243

relationship with the astronomical day, with different expressed rhythms then adopting their own phase relationships with the clock. The function of the circadian system is to regulate the phases at which biological events occur, either in relation to specific features of the 24 h cycle such as dawn or dusk, or in relation to other periodic events within the organism.

For the majority of organisms the biological clock is entrained by changes in the quantity, and perhaps spectral quality, of light at dawn and dusk (photoentrainment). Other environmental cycles (e.g. temperature) can entrain circadian behaviours in some species, but the precision of the light cycle seems to have been adopted by natural selection to provide the primary entraining cue or '*Zeitgeber*' (German for 'time giver' Zeit geber) for circadian systems in general. While it is clear that photoreceptors must play a crucial role in circadian organisation, our understanding of these photosensory systems in all organisms remains surprisingly limited. In many cases the nature of the photoreceptors and their precise connections with the clock are unknown, and we have only a superficial understanding of the features of the light environment that are used for photoentrainment. Although our understanding is limited, a striking feature of all vertebrate circadian organisation is that the image-forming visual system does not seem to convey light information to the biological clock. All vertebrates, including mammals, appear to have specialised 'non-image-forming' photoreceptor systems which perform this role. The basis for this statement and discussion why this should be so, form the central theme of this chapter. In Chapter 6, we provide a general review of extraretinal photoreception in non-mammalian vertebrates. In this chapter we outline the role of extraretinal photoreceptors in circadian physiology and discuss in detail the role of retinal photoreceptors in the regulation of mammalian circadian rhythms.

7.2 Photoentrainment in non-mammalian vertebrates

Vertebrates possess a diverse complement of photoreceptors, with some reptiles having photoreceptors located within the eye, pineal gland, parietal eye, deep brain, and skin. By contrast, mammals have only ocular photoreceptors. In Chapter 6 we discussed the characterisation and localisation of extraretinal photoreceptors in general. In this section we will discuss the role of deep brain, pineal and retinal photoreceptors in the regulation of circadian physiology.

7.2.1 Deep brain photoreceptors

Although the existence of extraretinal photoreceptors in the non-mammals had been known since the turn of the century (Frisch 1911), the first report that linked these photoreceptors with photoentrainment came from the work of Menaker and colleagues in the 1960's and early 1970's. Their studies showed that house sparrows (*Passer domesticus*) would entrain to laboratory light cycles after loss of both the eyes and the pineal gland (Menaker and Underwood 1976). These results demonstrated the existence of photoreceptors somewhere in the brain, and stimulated the search for 'deep brain photoreceptors' in other vertebrates. Although the majority of subsequent studies were on birds, studies in fish, amphibians and reptiles have all demonstrated the involvement of deep brain photoreceptors in regulating rhythmic physiology (Foster, Garcia-Fernandez *et al.,* 1993). For example, a detailed study on 8 species of lizards demonstrated that blinding and removing the pineal complex did not prevent animals from entraining circadian activity rhythms to a light:dark cycle (Underwood and Menaker 1976). It should be noted that considerable amounts of light penetrate deep into the brain of vertebrates (Foster and Follett 1985), and although this light cannot provide spatial information about the environment, it can provide information about levels of illumination and hence time of day. Neural tissue is surprisingly permeable to light, indeed, it seems likely to us that the pigmented eye cup evolved as a means of providing some directional light sensitivity. Light 'leaking' into the eye from the brain would significantly increase signal to noise problems, and hence increase the difficulties of extracting spatial information from light focused onto the photoreceptors. The conventional view for the possession of a pigmented eye cup is that it evolved to prevent the intraocular reflection and scatter of light entering through the pupil.

Deep brain photoreceptors have also been shown to play an important role in the regulation of seasonal reproductive cycles in non-mammalian vertebrates. The primary factor controlling seasonal breeding in temperate-zone birds is the annual variation in daylength (photoperiod). The measurement of photoperiod requires both a photoreceptor and some form of biological clock (believed to be a circadian clock) to determine the duration of the light detected. The regulation of the reproductive cycle by photoperiod is called the photoperiodic response. In the 1930's Benoit showed that blinded ducks exposed to long spring-like photoperiods would be stimulated to breed (Benoit 1935; Benoit 1935). More recent studies, involving enucleation (removal of the eye), sectioning of the optic nerve, or shielding light from entering the brain confirmed this original finding in a range of bird species (Foster and Follett 1985). The pineal gland was originally thought to regulate reproductive responses, but several studies have shown that the pineal plays no obvious role in the avian photoperiodic response. In quail, for example, illuminating the pineal with radioluminous paint does not stimulate gonadal growth (Homma, Ohta *et al.,* 1980) and pinealectomy of either blinded or eye intact quail leaves the photoperiodic response unaffected (Simpson, Urbanski *et al.,* 1983). By contrast, local illumination of the basal brain with luminous paint or fiber optics in blinded and pinealectomised animals, causes gonadal growth at maximal rates (Yokoyama, Oksche *et al.,* 1978; Oliver, Jallageas *et al.,* 1979).

7.2.2 Pineal photoreceptors

In addition to the deep brain photoreceptors, pineal photoreceptors seem to play an important role in circadian organisation. In all cases, pineal photoreceptors are intimately associated with the regulation of pineal melatonin and in some cases the pattern of melatonin release seems to drive components of rhythmic behaviour. The first evidence for the involvement of the pineal in circadian organisation came from birds. In several passerine birds, including the house sparrow (*Passer domesticus*) (Gaston and Menaker 1968) and the Java sparrow (*Padda oryzivora*) (Ebinhara and Kawamura 1981), pinealectomy abolished circadian rhythms of locomotor activity and body temperature in animals maintained in constant dark (DD) or constant light (LL). More recent studies in the house sparrow showed that transplantation of the pineal into the anterior chamber of the eye in a pinealectomised host restored the circadian activity rhythm of animals in DD. Significantly, the phase of the restored rhythm was determined by the phase of the donor (Zimmerman and Menaker 1979). In these experiments, the distant site of transplantation, and the rapidity of rhythm restoration, suggested that the message from the pineal to the behavioural centers was hormonal. The evidence that this hormonal message is melatonin is outlined below.

The avian pineal gland (or even dispersed pineal cells) *in vitro*, shows a circadian pattern of melatonin release under constant conditions. This rhythm can be entrained to a light:dark cycle, so that melatonin synthesis is high during the night and low during the day. Light can also regulate melatonin by acutely suppressing melatonin synthesis (Takahashi, Murakami *et al.*, 1989). Melatonin has been linked to behavioural rhythms by experimental administration of exogenous melatonin. For example, infusion of melatonin for 10 hours each day, at physiological levels, into pinealectomised pigeons will restore rhythmic behaviour (Chabot and Menaker 1992). Collectively, these data show that the pineal of some birds contains all of the components needed to organise circadian behaviour: a photoreceptor, a biological clock and a rhythmic signal (melatonin) to drive behavioural rhythms. While all avian pineals seem to contain these components, the importance of the pineal in circadian organisation can vary greatly. As mentioned above, pinealectomy will abolish freerunning circadian rhythms in passerine birds (sparrows) maintained in DD or LL, but this is not the case in all birds. In the chicken and quail, pinealectomy has no obvious effects on freerunning rhythms in DD, while in starlings there are only subtle effects of pinealectomy (Takahashi and Menaker 1979). In addition, destruction (lesions) of an area of the brain called the suprachiasmatic nuclei (SCN) results in arrhythmicity in all bird species housed under conditions of DD, *including pineal intact passerine birds!* These rather confusing results have been interpreted as follows. The circadian system of birds consists of at least two distinct circadian clocks which can influence behaviour. These clocks are located in the pineal and SCN. The relative importance of these clocks in circadian organisation varies depending on species and lighting conditions. Under conditions of DD or LL, the pineal of passerine birds is the 'master' clock producing a freerunning rhythmic melatonin signal that sustains a 'weak' clock in the SCN. A signal from the SCN then drives rhythmic freerunning behaviour. In the absence of the melatonin signal from the pineal, or of an entraining light:dark cycle, the SCN clock damps out after a few cycles and behaviour becomes arrhythmic. By contrast, in chickens and quail, the SCN does not require a rhythmic melatonin signal under constant lighting conditions to sustain rhythmic behaviour. Finally, some bird species show an intermediate condition, with the pineal and SCN acting in unison to sustain normal rhythmicity under constant conditions.

Reptiles, amphibians and fish have yet to be studied in the same detail as birds, but similarities are emerging. Pinealectomy of different lizard species will have different effects: a) abolish rhythms (Underwood 1983); b) split the circadian pattern of activity into two components ('splitting') (Underwood 1977); c) change the period of the rhythm (Underwood 1983; Innocenti, Minutini *et al.*, 1993); or d) have no effect (Janik, Pickard *et al.*, 1990). In the few lizard species studied (e.g. *Sceloporus occidentalis* and *S. olivaceus*) the chronic administration of melatonin causes a lengthening of the period of the freerunning rhythm, while melatonin injections given to *S. occidentalis* will phase shift the freerunning rhythm when given at certain circadian phases (Underwood 1986). These results show that although circulating melatonin is not required for the expression of activity rhythms, exogenous melatonin can modify rhythmic behaviour. As in birds, the isolated pineal of some lizards (e.g. *Anolis carolinensis*) shows a circadian rhythm of melatonin release that can be entrained to the light:dark cycle (Menaker and Wisner 1983). But unlike the pineal of birds, melatonin synthesis in the pineal of *Anolis* cannot be acutely suppressed by light. This has suggested that the photoreceptor mechanisms which entrain circadian rhythms of melatonin synthesis may differ from those photoreceptor pathways that acutely suppress melatonin synthesis (Underwood 1986). Finally, not all lizard pineals show a circadian rhythm of melatonin synthesis in organ culture (Janik, Pickard *et al.*, 1990). This has led to the conclusion that species like the desert iguana lack a circadian clock within the pineal.

In amphibia and fish, the pineal contains photoreceptors and produces melatonin, but its role in circadian organisation remains poorly understood. The effects of pinealectomy have been reported in relatively few species. In two recent reports, the evidence for a circadian clock within the pineal is strong. Removal of the pineal in the Japanese newt (*Cynops pyrrhogaster*) and in the lamprey (*Lampetra japonica*) resulted in arrhythmicity of behavioural rhythms after several days (Morita, Tabata *et al.*, 1992; Chiba, Kikuchi *et al.*, 1993). In organ culture, the pineal of the Northern pike (*Esox lucius*; (Falcon, Guerlotte et al. 1987)) and goldfish (*Carassius auratus*; (Kezuka, Aida *et al.*, 1989)) show circadian rhythms of melatonin synthesis that can be entrained to light. By contrast, the pineal gland of salmonids (e.g. *Salmo gairdneri*) shows no rhythm of melatonin synthesis; light having only an acute suppressive effect on melatonin production *in vitro* (Max and Menaker 1992). To our knowledge, there have been no detailed studies on the effects of melatonin administration on circadian behavioural rhythms in fish and amphibians.

In considering the information presented above we can make the general statement that pineal photoreceptors in non-mammals either entrain a circadian rhythm of melatonin synthesis and/or regulate melatonin synthesis acutely (the melatonin rhythm is driven by the light:dark cycle and there is no endogenous rhythm under constant conditions). Which ever the route, melatonin production is confined to the dark portion of a light:dark cycle. However, the role of the pineal and the melatonin signal in circadian organisation varies greatly, both within and between the vertebrate classes. In some species the pineal is an essential component of the circadian system, with pinealectomy causing arrhythmicity. In other species pinealectomy has no obvious effects on circadian behaviour. Note that although the parietal eye of lizards and frontal organ of anurans (see Chapter 6) contain well differentiated photoreceptors there is only one published report that these 'third eyes' produce melatonin and contain a circadian oscillator (Menaker and Tosini 1996). We feel confident that future studies will identify both melatonin and a clock within the parietal/frontal organ of other species.

7.2.3 Ocular photoreceptors

There is now overwhelming evidence that the vertebrates, including mammals, have circadian clocks located within the eyes (Tosini and Menaker 1996). These clocks appear to regulate many different ocular rhythms, including gene transcription, pigment migration and disc shedding and renewal. These studies have been discussed recently by Cahill and Besharse (Cahill and Besharse 1995) and this excellent review will not be duplicated here. It should be noted, however, that we know little about the contribution of these ocular clocks to the overall organisation of circadian behaviour in the vertebrates. In the Japanese quail, at least, neural signals from the eye clocks are important for the normal expression of circadian behaviour (Underwood, Barrett *et al.,* 1990).

We also have little information on the role of ocular photoreceptors in photoentrainment. We know that the eyes are not required for photoentrainment in non-mammals, but the extent to which retinal photoreceptors contribute to entrainment in the non-mammalian vertebrates has been considered in only a few studies. For example, in the house sparrow the threshold for photoentrainment is lower in sighted than in blinded birds and arrhythmicity cannot be produced in bright constant light in blinded animals (Menaker & Underwood 1976). These results suggest that the eyes may add to the overall photosensitivity of the circadian system, but until both deep brain and pineal photoreceptors are lesioned, we will not know whether the eyes alone can mediate photoentrainment in non-mammals.

7.3 Photoentrainment in mammals

In mammals the circadian clock is located within the SCN and is regulated by photoreceptors within the eye. Unlike the other vertebrate classes, loss of the eyes blocks all circadian responses to light in every mammal examined (for example Nelson and Zucker 1981; Foster, Provencio *et al.,* 1991). Note that there are very few studies on photoentrainment and circadian organisation in the monotremata (platypus and echidna) and marsupials. As a result, the discussion that follows is limited to generalisations about the eutherian mammals.

The mammalian pineal was once thought to be photoreceptive but this now seems unlikely. The evidence can be summarised as follows. Mammalian pinealocytes show two ultrastructural features usually associated with photoreceptors: Synaptic ribbons (characteristic of photoreceptor synapses), and sensory-like (9 + 0) cilia with bulbous terminals. These resemble photoreceptor inner segments, and the bulbous terminals appear to be homologous with photoreceptor outer segments (Vigh-Teichmann and Vigh 1992). In addition, the mammalian pineal (like that of non-mammals) contains a range of photoreceptor proteins. These include: rhodopsin kinase, arrestin, interphotoreceptor retinoid-binding protein, cellular retinal-binding protein, and opsin (Foster, Timmers *et al.,* 1989; Araki & Taketani, 1992; Zhao *et al.*, 1997). Although these phototransduction proteins have been identified, the mammalian pineal contains no chromophore (11-*cis* or all-*trans* retinaldehyde). As a result, the opsins present are probably not part of a functional photopigment (Foster, Timmers *et al.,* 1989; Tsin, Philips *et al.,* 1989; Korf, Kramm *et al.,* 1991). These results are in marked contrast to the findings in the non-mammalian vertebrates, where 11-*cis* and all-*trans* retinoid have been identified within the pineal (for example Foster, Schalken *et al.,* 1989; Provencio and Foster 1993).

While we know that the entraining photoreceptors are ocular, the few studies undertaken all show that there are significant differences between the processing of light information for image-forming visual responses and photoentrainment. The route by which light information reaches the circadian system, the retinohypothalamic tract (RHT), is anatomically, developmentally and physiologically distinct from the visual projections (Moore & Klein 1974; Pickard, Turek *et al.,* 1982; Moore, Speh *et al.,* 1995). In addition, circadian responses to light seem to differ from visual responses; the threshold for phase shifting circadian rhythms is relatively high, and intensity-duration reciprocity holds for stimuli of very long durations (up to 45 minutes in the golden hamster) (Nelson & Takahashi 1991; Nelson and Takahashi 1991).

In the mouse retina as few as 0.001% of retinal ganglion cells (RGCs) project exclusively to the SCN (Provencio, Cooper *et al.,* 1998). This illustrates what a small proportion of the eye forms the photoentrainment pathway, and that it is obscured by the large number of photoreceptors and inner retinal neurones devoted to image formation. As a result, the role of rods, cones and other retinal neurones in entrainment has been difficult to study. In recent years, retinally degenerate models have been used as reduced preparations, correlating the loss of retinal elements with any effects on photoentrainment. The first experiments in this area used mice homozygous for *retinal degeneration* (*rd/rd*). These mice experience a massive degeneration of the rods and cones. By 60 days of age all rod cells have degenerated, and between 90 and 150 days of age even the crudest electrophysiological and behavioural responses to bright light have disappeared (Provencio, Wong *et al.,* 1994). In the mouse retina, approximately 97–98% of all photoreceptors are rods, and although all rods degenerate in the *rd/rd* retina a few cone cells remain in animals over one year of age. These cones lack outer segments and constitute only 2 – 5% of the cone cells found within the normal (+/+) retina. As a result, the *rd/rd* eye at about one year of age contains approximately 0.09% of its original number of photoreceptor cell bodies (perhaps 4,500 cells), and with increasing age, cone cell loss continues. To determine the impact of photoreceptor loss on the circadian system, the effects of a standardised light pulse on phase shifting the freerunning locomotor rhythm were determined in three genotypes of mice (*rd/rd*, *rd*/+, +/+) from the same C57BL strain. Despite the loss of visual photoreceptors in *rd/rd* mice, these animals showed circadian responses to light that were indistinguishable from mice with phenotypically normal retinas (*rd*/+, +/+). The irradiance required to produce both saturating and half-saturating responses was the same for all groups (Figure 1). It is important to stress that not only does some photosensitivity remain in mice with degenerate retinas, but the circadian photosensitivity shown by these animals is not different from the sensitivity of animals with normal retinas. In addition, unattenuated sensitivities are maintained in *rd/rd* mice greater than two years of age. These data demonstrate that the sensitivity of the circadian system to light does not parallel loss of either rod or cone photoreceptors in the *rd/rd* mouse (Provencio, Wong *et al.,* 1994).

The impact of the *rd/rd* mutation on the entrainment pathway (rather than overt circadian behaviour) has been examined by using the expression of the immediate early gene *c-fos* within the SCN. Studies from a range of laboratories have demonstrated that *c-fos* is induced within the SCN upon retinal illumination at times when the circadian system is sensitive to light. In addition, levels of *c-fos* expression parallel the strength of the photoentrainment signal. Light stimuli that result in a large phase shift also induce large levels of *c-fos* expression within the SCN (Kornhauser, Nelson *et al.,* 1990; Rusak, Robertson *et al.,* 1990). Experiments using the *rd/rd* mouse have shown that the distribution, number and density of Fos immunostained cell nuclei within the SCN were identical in *rd/rd* and +/+ mice, further demonstrating that the *rd/rd* mutation has no impact on the photoentrainment pathway (Colwell and Foster 1992). In addition, viral tract

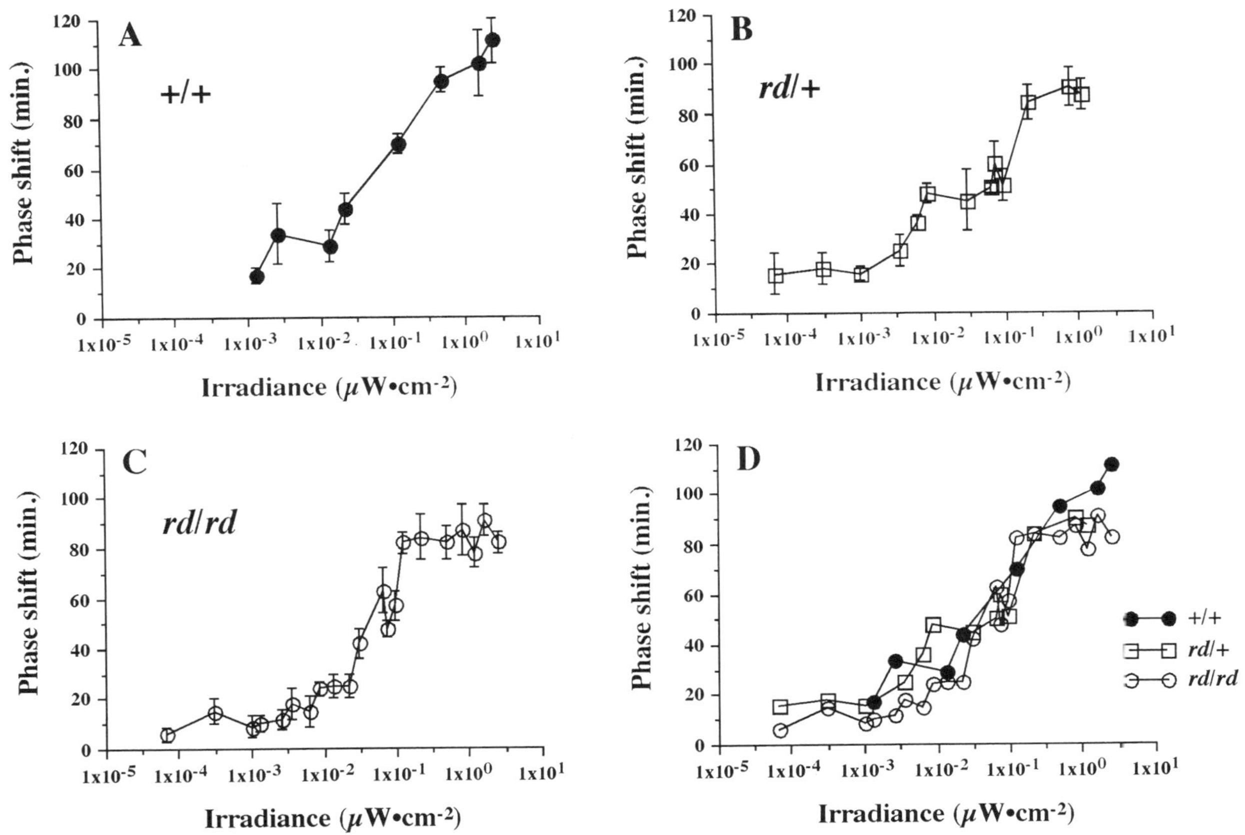

Figure 1 The magnitude of phase shifts produced by a 15 min. light pulse as a function of light intensity in mice with: **A** normal retinas (+/+); **B** normal retinas (*rd*/+); **C** degenerate retinas (*rd/rd*). For comparison the 3 irradiance response curves (**A**,**B**,**C**) have been superimposed in **D**. Despite the massive damage to the visual photoreceptors and loss of visual function in *rd/rd* mice, these animals show circadian responses to light that are indistinguishable from those of mice with normal retinas. Error bars ± SEM. Based on (Foster, Provencio *et al.*, 1991)

tracers have been used to explore the impact of the *rd/rd* mutation on the projections from the retina to the SCN. In detailed studies, no obvious differences in the nature of the projections from the eye to the SCN were identified between +/+ or *rd/rd* mice (Provencio, Cooper *et al.,* 1998). Finally, it should be stressed that the site of circadian photoreception must reside within the eye of *rd/rd* mice because bilateral enucleation of *rd/rd* animals abolishes all circadian responses to light (Foster, Provencio *et al.,* 1991).

Another retinal mutation, *retinal degeneration slow* (*rds/rds*) has provided a second approach to the question of which elements in the eye mediate circadian responses to light. In *rds/rds* mice, the retina undergoes normal development until the first postnatal week, then both rod and cone photoreceptors fail to develop outer segments and then gradually degenerate. In *rds/rds* mice approximately half of all the rods and cones have degenerated by 3 months, and most photoreceptors seem to have degenerated by one year of age. Despite this loss circadian responses to light are identical in *rds/rds, rd/rd* and +/+ genotypes (Argamaso, Froehlich *et al.,* 1995). This provides overwhelming evidence that the photoreceptive elements mediating circadian responses to light do not require an outer segment.

Circadian responses to light have also been investigated in rodless transgenic mice (*rdta*), which have an integrated fusion gene consisting of a 1 kb fragment of the human rhodopsin promoter linked to an attenuated diphtheria toxin gene. Morphological and physiological analysis of the retinae, and the use of reverse transcriptase polymerase chain reaction (RT-PCR) approaches using mouse specific primers, has demonstrated that rod degeneration begins soon after the onset of rhodopsin expression (around postnatal day 6) and that rhodopsin expression is reduced by approximately 80% by postnatal day 28 (McCall, Gregg *et al.,* 1996). Cone cell bodies (with no outer segments), as well as cone opsins, however, persist until at least postnatal day 150. The *rdta* phenotype resembles the *rd/rd* phenotype; the only obvious difference being that rods degenerate earlier in development in the *rdta* mice. However, *rdta* mice show abnormally large responses to light (Lupi, Cooper *et al.,* 1998). At light intensities that produce maximum phase shifts in locomotor rhythms for +/+ mice (around 80 min.), *rdta* mice show shifts around 230 min. The different responses of *rdta* mice compared to *rd/rd* and +/+ mice, could have resulted from changes in the entrainment pathway, and occurs in the *rdta* because rods are affected earlier in postnatal development. Clearly rod photoreceptors are not required for photoentrainment, but the loss of rods early in development does influence clock responses to light. The basis for this difference remains to be determined.

Until the studies on *rd/rd* mice, it had been assumed that rods regulate circadian responses to light in mammals. This belief was based upon an action spectrum for phase shifting of locomotor activity rhythms in the golden hamster (*Mesocricetus auratus*) (Takahashi, DeCoursey *et al.,* 1984). These data show a spectral maximum (λ_{max}) around 500 nm, which correlates well with the absorbance maximum (λ_{max} = 502 nm) of the extractable rod photopigment in this species. More recent studies in rodents, however, have demonstrated the existence of cones with spectral sensitivities near 500 nm (Jacobs, Neitz *et al.*, 1991), including the hamsters, in which a single cone sensitivity has been identified between 505 – 506 nm (Calderone and Jacobs 1995). Because the action spectrum for phase shifting locomotor rhythms cannot resolve differences between sensitivities at 502 nm (rod) or 506 nm (cone), the roles of rods and/or cones, therefore, remains uncertain in this species. Two types of cones have been identified in the normal mouse retina. Electroretinogram (ERG) and behavioural studies have shown two sensitivity maxima; a green-sensitive cone near 510 nm and an ultraviolet sensitive cone near 360 nm (Jacobs, Neitz *et al.*, 1991). Action spectra for phase-shifting in aged (80 – 90 days) *rd/rd* (rodless) and +/+ mice show two spectral sensitivities around 510 nm and 360 nm, corresponding well with the absorption maxima of the two known mouse cone types (Provencio and Foster 1995). By using RT-PCR techniques, followed by cloning and sequencing of the amplified cDNAs, low levels of both the green and UV cone opsin have been isolated from the aged *rd/rd* retina (animals > 2 years). However, rod opsin mRNA was not detected beyond approximately 1 year of age (Argamaso, Froehlich *et al.*, 1995). On the basis of the molecular analysis, and the similarity of the action spectrum results with the spectral sensitivity of the known cones, the cones became strong candidates for circadian regulation. However it is worth stressing that although cone opsins and cell bodies remain in the *rd/rd* retina, most of the cones have been lost and the remaining cones lack outer segments. If the remaining cones do mediate circadian responses to light then one must propose a mechanism that can compensate for massive photoreceptor loss and loss of outer segments. Such a mechanism has been proposed and outlined in Figure 2.

The requirement for cone photoreceptors in photoentrainment has been explicitly examined in very recent experiments that have utilised transgenic mice which contain an integrated fusion gene consisting of a portion of the human red cone opsin promoter (Wang, Macke *et al.*, 1992) linked to an attenuated diphtheria toxin gene. Analysis of the retinae of these mice shows that the rod photoreceptors remain intact, UV cones are reduced, and green cones are eliminated (Freedman, von Schantz *et al.*, 1997). Despite the loss of green cones, the circadian system of these mice show saturating and sub-saturating phase shifts in response to monochromatic green light (λ_{max} 515 nm) that are indistinguishable from congenic +/+ controls (Freedman, Soni, Foster – unpublished). These data suggest that there may be complete redundancy of the photoreceptor inputs to the clock, so that in the absence of the cones, rods mediate photoentrainment while in the absence of the rods, cones regulate the clock. Alternatively, the possibility that neither the rods nor cones are required for photoentrainment cannot be excluded. To test the possibility that there may be novel photoreceptors/ photopigments within the mammalian eye, the rodless *rdta* mice and green

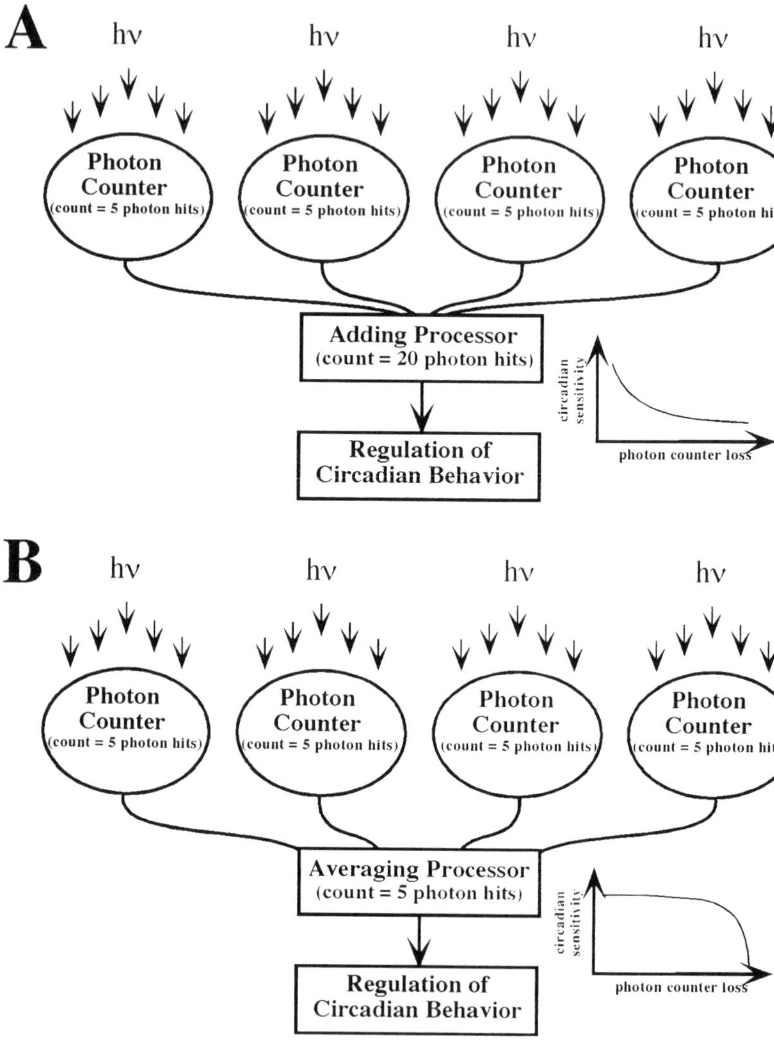

Figure 2 It is possible that relatively few surviving photoreceptors (lacking outer segments) are sufficient to maintain unattenuated circadian responses to light in retinally degenerate mice. This figure shows two possible models for how light could be processed by the circadian system. (**A**) Model showing a direct relationship between photoreceptor number (photon counters) and sensitivity. In this model the 'circadian processor' adds all photic input. As a result, the loss of photoreceptors would be directly proportional to the sensitivity of the system. It seems unlikely that the circadian system processes light using an additive mechanism. (**B**) Model in which there is no relationship between photoreceptor number and sensitivity. Because information from the photoreceptors is averaged by the circadian processor, loss of photoreceptors would not lead directly to a loss in the sensitivity of the circadian system. Only when photoreceptor number decreased to a level when quantal hits failed to occur would a drop in sensitivity be observed. Such a system would be buffered from the degenerative loss of photoreceptive elements in *rd/rd, rds/rds* and *rdta/rdta* retinal mutants (see text for details). Based on (Foster, Argamaso *et al.*, 1993)

coneless transgenic mice were bred and the progeny of these genetic crosses examined. Preliminary results show that green monochromatic light stimuli (λ_{max} 515 nm) are still capable of phase shifting circadian behaviour in these mice (Freedman, Soni, Foster – unpublished). If confirmed, these results provide strong evidence that the mammalian eye contains novel photoreceptors which mediate photoentrainment. Of course the finding of novel circadian photopigments does not mean that the rods and cones play no role in mammalian photoentrainment. The experiments outlined above merely suggest that rods and cones are not required for photoentrainment.

The possibility that the mammalian eye contains novel photopigments has attracted some general interest because of the recent discovery of novel retinal photopigments within the eyes of fish (Soni and Foster 1997; Soni, Philp *et al.*, 1998) and amphibians (Provencio, Jiang *et al.*, 1998) (see Chapter 6). Indeed, the general expectation seems to be that novel photopigments will be found in the mammalian retina. This has prompted both serious research and some rather speculative proposals. For example, in a recent paper a molecule called cryptochrome was claimed to be the mammalian circadian photopigment (Miyamoto and Sancar 1998). This claim was made even though cryptochrome has not been shown to act as a photopigment in any vertebrate, that it has an absorption spectrum which shows no resemblance to the action spectra for mammalian photoentrainment, and that crypotochrome is expressed in almost every organ of the mouse (Miyamoto and Sancar 1998). Further experiments are needed to determine what, if any, role cryptochrome may play in mammalian photoentrainment.

Perhaps the most surprising observation to emerge from the studies outlined above is that mice which lack classical visual responses are still capable of regulating their circadian rhythms by light with unattenuated sensitivities. These types of observation, however, are not restricted to mice. Several studies have demonstrated that clock driven rhythms can be regulated by light in a sub-set of humans that have their eyes but remain 'blind' to visual stimuli (Czeisler, Shanahan *et al.,* 1995). In addition, the blind mole rat (*Spalax ehrenbergi*) provides a further example of a mammal lacking classical visual responses but which is still capable of photoentrainment. In this highly specialised subterranean mammal, the minute eyes (600 μm in diameter) are completely buried beneath the skin and brain structures involved in image analysis are severely reduced or absent. For example, the dorsal lateral geniculate nucleus consists of a narrow sheet of tissue 3 – 5 neurones thick, the superficial layers of the superior colliculus are collapsed into a single layer, and the medial terminal nucleus is absent. By contrast, the SCN is well developed and receives a bilateral projection from the retina, and the absolute size, cytoarchitecture, and distribution of retinal afferents in *Spalax* are comparable with those of other rodents (Cooper, Herbin *et al.,* 1993). Behavioural studies show that *Spalax* lacks any image forming ability but will entrain circadian rhythms of activity to the external light cycle (Nevo, Guttman *et al.,* 1982; Haim, Heth *et al.,* 1983; Pevet, Heth *et al.,* 1984; Rado, Gev *et al.,* 1988; Rado and Terkel *et al.,* 1989; Nevo, Heth *et al.,* 1991), and significantly, removal of the eye blocks all photoentrainment (Pevet, Heth *et al.,* 1984). It appears that during the past 30 million years, the blind mole rat has undergone dramatic changes. Evolutionary processes have effectively disentangled and isolated the photoentrainment pathway of *Spalax* with some components of the visual system being lost, while those components that regulate the biological clock have been retained.

Results in retinally degenerate mice, mole rats and blind humans, have led to the realisation that two functionally distinct systems exist for processing light information in the mammal eye (and perhaps the eye of other vertebrates): the 'image-forming' photoreceptor system, which constructs a representation of the environment (classical vision) and the 'non-image-forming' photoreceptor system, which instead detects changes in the overall quality of light at different times of day. In view of the different sensory demands of image detection and the regulation of biological clocks it is not surprising that two systems for processing light information have evolved. Note that some confusion has arisen recently because references to these data have not been qualified by the explicit statement that enucleation of retinally degenerate mammals will block photoentrainment. For example, a recent paper proposed that mammalian photoentrainment is caused by the illumination of the skin (Campbell and Murphy 1998), and these researchers supported their argument with references to findings in *rd/rd* mice! A full critique of this paper can be found in (Foster 1998).

7.4 The sensory ecology of photoentrainment

The research to understand photoentrainment in the vertebrates has been driven by the search for the photoreceptors and their projections to the clock. By contrast, there has been little consideration of the sensory ecology of the entrainment pathway and the features of the light environment at dawn and dusk that may be important for circadian regulation. During twilight, the quality of light changes in three important respects (see Chapters by Ellis Loew and McFarland *et al.*): 1) the amount of light, 2) the spectral composition of light, 3) and the source or direction of light (i.e. the position of the sun). These photic parameters all change in a systematic way, and could be used by organisms to detect the phase of twilight. The possibility that one or more of these features are utilised by vertebrates to achieve photoentrainment will be considered in this section.

7.4.1 The amount of light

The unique anatomical and physiological features of the photic input to the SCN seem to protect the circadian axis from stimuli incapable of serving as reliable time cues. The circadian and visual systems occupy different regions of a theoretical plane where the x dimension is sensitivity and the y dimension is integration time (Figure 3). It is striking that in those animals studied the threshold for phase shifting circadian rhythms is significantly higher than that required to elicit visual responses. For example, hamsters can recognise optical gratings at a luminance level 200 times less than the level necessary to elicit phase shifts in locomotor rhythms (Emerson 1980). This relative insensitivity of the circadian system may function to filter out 'photic noise' (Nelson and Takahashi 1991). The irradiance of starlight is approximately 9.3×10^8 photons.cm^{-2}.s^{-1} (total irradiance between 400 and 700 nm) while the irradiance of the full moon is approximately 32-fold greater (3×10^{10} photons.cm^{-2}.s^{-1} total irradiance between 400 and 700 nm) (Munz and McFarland 1977). Both of these photic noise sources fall below the threshold for photoentrainment, and therefore cannot interfere with circadian function.

Lightning could also potentially 'confuse' the circadian system. It can be 50 times greater than that of direct sunlight. In addition, shading by vegetation or cloud cover can greatly alter the amount of light falling upon an organism. As a result, any reliable measure of light level (and hence time of day) must

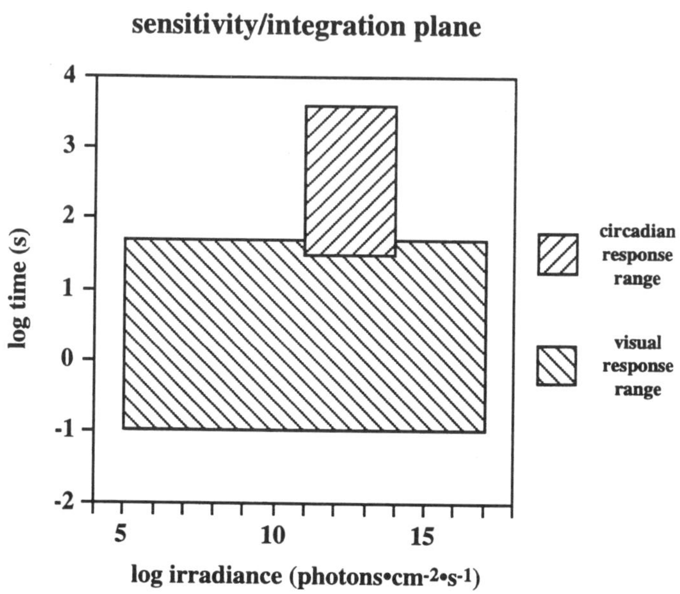

Figure 3 Graphical representation of the response ranges of the visual and circadian systems. The response range of sensitivity (x-axis) and integration time (y-axis) of the circadian and visual systems are contained within their respective rectangles. Note that the circadian system is insensitive to light and requires stimuli of long durations relative to the visual system. The unique anatomical and physiological features of the photic input to the SCN protect the circadian system from stimuli that do not provide reliable time cues. See text for references

compensate for local fluctuations. In view of these considerations, it is not surprising that the circadian system of those animals examined is very insensitive to light stimuli of short durations. For example, the circadian system of the hamster is relatively insensitive to stimulus durations of less than 30 seconds (Nelson and Takahashi 1991).

Finally, the circadian system needs to measure overall light levels in the environment (irradiance) and ignore brightness in particular areas of the sky (radiance). For image detection, the visual system maintains complete retinotopic order. The eye focuses light onto a particular region of the retina, and this radiance information is then mapped to a specific position in the brain. By contrast, the circadian system requires measures of irradiance. Photoreceptors located beneath the skull or in the brain are unable to extract any image information. The overlying tissues scatter light to such an extent that all features are lost. By their very nature the pineal and deep brain photoreceptors of non-mammals are excellent irradiance detectors. But mammals lack extraretinal photoreceptors (why this may have occurred is discussed in Chapter 6) and are 'forced' to use their eyes for photoentrainment. In the blind mole rat the subcutaneous eyes provide irradiance information but in most mammals the eyes are obligate radiance detectors, and the question has been how do mammals attempting to extract irradiance information compensate for a lens? The answer seems to be as follows. The retinal ganglion cells (RGCs) projecting to the SCN, variously called type III in the rat and mouse or W-cells in the cat, are relatively scarce, are spread evenly across the retina, and have extensive dendritic arbours (Moore, Speh *et al.,* 1995; Provencio, Cooper *et al.,* 1998). This reduces spatial resolution and increases the sampling area. In addition, there is an absence of retinotopic order in the RHT. RGCs project randomly to the retinorecipient areas of the SCN, which further blurs any image. These combined effects provide the SCN with irradiance information (Cooper, Tessonneaud *et al.,* 1993). This arrangement is analogous to the use of extraretinal photoreceptors in the non-mammalian vertebrates.

7.4.2 The spectral composition of light

In addition to large changes in irradiance, there are very precise spectral changes associated with twilight (Chapter by McFarland *et al.*). Primarily there is an enrichment of the shorter wavelengths (< 500 nm) relative to the mid-long wavelengths (500 – 650 nm) at twilight. If the circadian system were capable of some form of spectral discrimination, and able to ratio changes in the relative amounts of short and long wavelength radiation, then it could determine the phase of twilight very accurately. The fresh water algae *Chlamydomonas* (Kondo, Johnson *et al.,* 1991), and the dinoflagellate *Gonyaulax polyedra* (Roenneberg and Deng 1997), appear to utilise the spectral changes at dawn and dusk to regulate their circadian clocks, but there is little evidence for this in the vertebrates. We do know that the circadian system of mice is sensitive to both green light (λ_{max} at 511 nm) and near-UV irradiation, and that multiple photopigments with different spectral maxima, exist in many of the extraocular photoreceptors (Chapter 6 – Shand & Foster), but we do not know how the signals from these different spectral channels are utilised by the circadian system. If wavelength discrimination is used by the circadian system for photoentrainment, and based on the assumption that photoentrainment evolved before image detection, then it is possible that multiple photopigments and wavelength discrimination may have evolved originally as a means of detecting changes in twilight. These mechanisms only later became specialised for contrast detection in the image-forming visual systems.

7.4.3 The position of the sun in the sky

The position of the sun in the sky is used by many different animals for time compensated sun-compass orientation (for review see (Wallraff 1981)). Whether this information is also used by organisms to entrain circadian systems remains a mystery. For this task radiance detection, and topographic mapping, would be required to determine the position of the sun. In this way perhaps the 'classical' visual system does contribute to photoentrainment.

7.5 Conclusions

Organisms have evolved to occupy both a spatial and temporal niche within the environment. If an organism is to take full advantage of its temporal niche it needs some sense of time. The possession of a circadian system enables an organism to anticipate the different demands of the day/night cycle and fine-tune physiology and behaviour to exploit the changed circumstances as soon as they occur. To ensure that internal and environmental time coincide, a biological clock must be entrained to local time. For the majority of circadian clocks, photoreceptors detect changes in the light environment at twilight and align circadian time to environmental time. This process is called photoentrainment and is brought about by both ocular and extraocular photoreceptors.

As a species we are dominated by the input of our 'image-forming' photoreceptor system. We take for granted the fact that the vertebrate eye has evolved specialisations to enhance visual contrast and improve sensitivity. This has made us slow to recognise that the eye has also evolved specialisation that enable it to extract time cues from the solar cycle. Our image-forming bias has also led to a disregard of the extraocular photoreceptors of the non-mammals. Extraocular photoreceptors have frequently been dismissed as either vestigial or redundant. The possibility that these photoreceptors might provide light information critical to the survival of the organism has been largely ignored. These views are slowly being replaced. The realisation that the sensory demands of photoentrainment have imposed a unique set of selection pressures, which have led to the evolution of specialised photoreceptor systems, is beginning to be appreciated by both circadian and photoreceptor biologists alike. However, a detailed understanding of the photoentrainment pathway is lacking in any vertebrate (or invertebrate). If we are to address seriously the question of how light is processed to regulate biological time, the expertise of both circadian and photoreceptor biologists will need to be coordinated and brought to bear upon this fascinating problem.

Acknowledgments

We thank Dr. Gianluca Tosini, Dr. Malcolm von Schantz and Elizabeth Downes for their comments on this manuscript.

References

Araki, M. and Taketani, S. (1992) A PCR analysis of rhodopsin gene transcription in rat pineal photoreceptor differentiation. *Developmental Brain Research*, 69, 149 –152.

Argamaso, S. M., Froehlich, A. C., McCall, M. A., Nevo, E., Provencio, I. and Foster, R. G. (1995) Photopigments and circadian systems of vertebrates. *Biophysical Chemistry*, 56, 3 –11.

Aschoff, J. (1981) *Handbook of Behavioural Neurology: Biological Rhythms.* New York & London: Plenum Press.

Benoit, J. (1935a) Le role des yeux dans l'action stimulante de la lumiere sur le developpement testiculaire chez le canard. *Comptes Rendus de Séances de la Société de Physique et de ses Filiales*, 118, 669 – 671.

Benoit, J. (1935b) Stimulation par la lumiere artificielle du developpment testiculaire chez des canards aveugles par section du nerf optique. *Comptes Rendus de Séances de la Société de Physique et de ses Filiales*, 120, 133 –136.

Cahill, G. M. and Besharse, J. (1995) Circadian rhythmicity in vertebrate retinas: Regulation by a photoreceptor oscillator. *Progress in Retinal and Eye Research*, 14, 267– 291.

Calderone, J. B. and Jacobs, G. H. (1995) Photopigments of two types of hamsters. *Investigative Ophthalmology & Visual Science*, 36(4), S276 (abstract 1260).

Campbell, S. S. and Murphy, P. (1998) Extraocular circadian phototransduction in humans. *Science*, 279, 396 – 399.

Chabot, C. and Menaker, M. (1992) Effects of physiological cycles of infused melatonin on circadian rhythmicity in pigeons. *Journal of Comparative Physiology A,* 170(5), 615 – 22.

Chiba, A., Kikuchi, M. and Aoki, K. (1993) The effects of pinealectomy and blinding on the circadian locomotor activity rhythm in the Japanese newt, *Cynops pyrrhogaster. Journal of Comparative Physiology A,* 172, 683 – 691.

Colwell, C. S. and Foster, R. G. (1992) Photic regulation of Fos-like immunoreactivity in the suprachiasmatic nucleus of the mouse. *The Journal of Comparative Neurology*, 324, 135 –142.

Cooper, H. M., Herbin, M. and Nevo, E. (1993a) Visual system of a naturally microphthalmic mammal: The blind mole rat, *Spalax ehrenbergi. The Journal of Comparative Neurology,* 328, 313 – 350.

Cooper, H. M., Tessonneaud, A., Caldani, A., Locatelli, A., Richard, S. and Viguier-Martinez, M. C. (1993) Morphology and distribution of retinal ganglion cells (RGC) projecting to the suprachiasmatic nucleus in the sheep. *Society for Neuroscience Abstracts*, 19, Abstract 701.11.

Czeisler, C. A., Shanahan, T. L., Klerman, E. B., Martens, H., Brotman, D. J., Emens, J. S., Klein, T. and Rizzo III, J. F. (1995) Suppression of melatonin secretion in some blind patients by exposure to bright light. *The New England Journal of Medicine*, 332, 6 –11.

Ebihara, S. and Kawamura, H. (1981) The role of the pineal organ and suprachiasmatic nucleus in the control of circadian locomotor rhythms in the Java sparrow, *Padda oryzivora. Journal of Comparative Physiology A*, 141, 207– 214.

Emerson, V. F. (1980) Grating acuity in the golden hamster: The effects of stimulus orientation and luminance. *Experimental Brain Research*, 38, 43 – 52.

Falcon, J., Guerlotte, J. F., Voison, P. and Collin, J.-P. (1987) Rhythmic melatonin biosynthesis in a photoreceptive pineal organ: A study in the pike. *Neuroendocrinology,* 45, 479 – 486.

Foster, R. G. (1998) Shedding light on the biological clock. *Neuron,* 20(5), 829 – 832.

Foster, R. G., Argamason, S., Coleman, S., Colwell, C. S., Lederman, A. and Provencio, I. (1993) Photoreceptors regulating circadian behavior: A mouse model. *Journal of Biological Rhythms*, 8(supplement), S17– S23.

Foster, R. G. and Follett, B. K. (1985) The involvement of a rhodopsin-like photopigment in the photoperiodic response of the Japanese quail. *Journal of Comparative Physiology A*, 157, 519 – 528.

Foster, R. G., Garcia-Fernandez, J. M., Provencio, I. and DeGrip, W. J. (1993) Opsin localization and chromophore retinoids identified within the basal brain of the lizard *Anolis carolinensis. .ournal of Comparative Physiology A*, 172, 33 – 45.

Foster, R. G., Provencio, I., Hudson, D., Fiske, S., De Grip, W. and Menaker, M. (1991) Circadian photoreception in the retinally degenerate mouse *(rd/rd). Journal of Comparative Physiology A*, 169, 39 – 50.

Foster, R. G., Schalken, J. J., Timmers, A. M. and De Grip, W. J. (1989) A comparison of some photoreceptor characteristics in the pineal and retina: I. The Japanese quail *(Coturnix coturnix). Journal of Comparative Physiology A*, 165, 553 – 563.

Foster, R. G., Timmers, A. M., Schalken, J. J. and De Grip, W. J. (1989) A comparison of some photoreceptor characteristics in the pineal and retina: II. The Djungarian hamster *(Phodopus sungorus). Journal of Comparative Physiology A*, 165, 565 – 572.

Freedman, M. S., von Schantz, M., Soni, B. G. and Foster, R. G. (1997) Molecular dissection of the coneless transgenci mouse retina. *Investigative Ophthalmology and Visual Science*, 38, S322, A –1500.

Frisch, K. v. (1911) Beitrage zur Physiologie der Pigmentzellen in der Fischhaut. *Pfluger's Archv. Gesamte Physiol. Menschen Tiere*, 138, 319 – 387.

Gaston, S. and Menaker, M. (1968) Pineal Function: The biological clock in the sparrow? *Science*, 160, 1125 –1127.

Haim, A. G., Heth, H., Pratt, H. and Nevo, E. (1983) Photoperiodic effects on thermoregulation in a 'blind' subterranean mammal. *Journal of Experimental Biology*, 107, 59 – 64.

Homma, K., Ohta, M. and Sakakibara, Y. (1980) Surface and deep photoreceptors in photoperiodism in birds. *Biological Rhythms in Birds.*Y. Tanake, K. Tanaka and T. Ookawa. Berlin: Springer-Verlag. 149 –156.

Innocenti, A., Minutini, L. and Foa, A. (1993) The pineal and circadian rhythms of temperature selection and locomotion in lizards. *Physiology & Behaviour*, 53(5), 911– 915.

Jacobs, G. H., Neitz, J. and Deegan, J. F. (1991) Retinal receptors in rodents maximally sensitive to ultraviolet light. *Nature*, 353, 655 – 656.

Janik, D. S., Pickard, G. E. and Menaker, M. (1990) Circadian locomotor rhythms in the desert iguana. II. Effects of electrolytic lesions to the hypothalamus. *Journal of Comparative Physiology A*, 166(6), 811– 6.

Kezuka, H., Aida, K. and Hanyu, I. (1989) Melatonin secretion from goldfish pineal gland in organ culture. *Gen. Comp. Endocrinol.*, 75, 217– 221.

Kondo, T., Johnson, C. H. and Hastings, J. W. (1991) Action spectrum for resetting the circadian phototaxis rhythm in the CW15 strain of *Chlamydomonas*. I: cells in darkness. *Plant Physiology,* 95, 197– 205.

Korf, H.-W., Kramm, C. and DeGrip, W. J. (1991) Further analysis of photoreceptor-specific proteins in the rodent pineal organ and retina. *Advances in Pineal Research,* 5, 115 –122.

Kornhauser, J. M., Nelson, D. E., Mayo, K. E. and Takahashi, J. S. (1990) Photic and circadian regulation of c-fos gene expression in the hamster suprachiasmatic nucleus. *Neuron*, 5(2), 127– 34.

Lupi, D., Cooper, H. M., Froehlich, A. C., Stanford, L. R., McCall, M. A. and Foster, R. G. (1999) Transgene ablation of rod photoreceptors alters the circadian phototype of mice. *Neuroscience,* (in press).

Max, M. and Menaker, M. (1992) Regulation of melatonin production by light, darkness, and temperature in the trout pineal. *Journal of Comparative Physiology A*. 170(4), 479 – 489.

McCall, M. A., Gregg, R. G., Merriman, K., Goto, N. S., Peachey, N. S. and Stanford, L. R. (1996) Morphological and physiological consequences of the selective elimination of rod photoreceptors in transgenic mice. *Experimental Eye Research,* 63, 35 – 50.

Menaker, M. and Tosini, G. (1996) The evolution of vertebrate circadian systems. *Sapporo Symposium on Biological Rhythms.* K. Honma. and S. Honma. Sapporo: Hokkaido University Press.

Menaker, M. and Underwood, H. (1976) Extraretinal photoreception in birds. *Photochemistry and*

Photobiology, 23, 299 – 306.
Menaker, M. and Wisner, S. (1983) Temperature-compensated circadian clock in the pineal of *Anolis*. *Proc. Natl. Acad. Sci. USA*, 80, 6119 – 6121.
Miyamoto, Y. and Sancar, A. (1998) Vitamin B_2-based blue-light photoreceptor in the retinohypothalamic track as the photoactive pigments for setting the circadian clock in mammals. *Proceedings of the National Academy of Science*, 95, 6097– 6102.
Moore, R. Y. and Klein, D. C. (1974) Visual pathways and the central neural control of a circadian rhythm in pineal serotonin N-acetyltransferase activity. *Brain Research*, 71, 17– 33.
Moore, R. Y., Speh, J. C. and Card, J. P. (1995) The retinohypothalamic tract originates from a distinct subset of retinal ganglion cells. *The Journal of Comparative Neurology*, 352, 351– 366.
Morita, Y., Tabata, M., Uchida, K. and Samejima, M. (1992) Pineal-dependent locomotor activity of lamprey, Lampetra japonica, measured in relation to LD cycle and circadian rhythmicity. *Journal of Comparative Physiology (A)*, 171, 555 – 562.
Munz, F. W. and McFarland, W. N. (1977) Evolutionary adaptations of fishes to the photic environment. *Handbook of Sensory Physiology*. F. Crescitelli. Berlin: Springer-Verlag. VII/5, 193 – 274.
Nelson, D. E. and Takahashi, J. S. (1991) Comparison of visual sensitivity for suppression of pineal melatonin and circadian phase-shifting in the golden hamster. *Brain Research*, 554(1– 2), 272 –7.
Nelson, D. E. and Takahashi, J. S. (1991) Sensitivity and integration in a visual pathway for circadian entrainment in the hamster (*Mesocricetus auratus*). *Journal of Physiology*, 439(115), 115 – 45.
Nelson, R. J. and Zucker, I. (1981) Absence of extraocular photoreception in diurnal and nocturnal rodents exposed to direct sunlight. *Comparative Biochemistry and Physiology*, 69A, 145 –148.
Nevo, E., Guttman, R., Haber, M. and Erez, E. (1982) Activity patterns of evolving mole rats. *Journal of Mammal*, 63, 453 – 463.
Nevo, E., Heth, G. and Pratt, H. (1991) Seismic communication in a blind subterranean mammal – a major somatosensory mechanism in adaptive evolution underground. *Proceedings of the National Academy y of Science (USA)*, 88, 1256 –1260.
Oliver, J., Jallageas, M. and Bayle, J. D. (1979) Plasma testosterone and LH levels in male quail bearing hypothalamic lesions or radioluminous implants. *Neuroendocrinology*, 28, 114 –122.
Pevet, P., Heth, H., Haim, A. G. and Nevo, E. (1984) Photoperiod perception in the blind mole-rat (*Spalax ehrenbergi*). Involvement of the harderian gland, atrophied eyes and melatonin. *Journal of Experimental Zoology*, 232, 41– 50.
Pickard, G. E., Turek, F. W., Lamperti, A. A. and Silverman, A.-J. (1982) The effect of neonatally administered monosodium glutamate (msg) on the development of retinofugal projections and the entrainment of the circadian locomotor activity. *Behavioural and Neural Biology*, 34, 433 – 444.
Pittendrigh, C. S. (1981) Circadian systems: Entrainment. *Handbook of Behavioral Neurobiology: Biological Rhythms*. J. Aschoff. New York & London: Plenum Press. 4, 95 – 124.
Provencio, I., Cooper, H. M. and Foster, R. G. (1997) Retinal projections in mice with inherited retinal degeneration. *The Journal of Comparative Neurology*, 395, 417– 439.
Provencio, I. and Foster, R. G. (1993) Vitamin A_2-based photopigments within the pineal gland of a fully terrestrial vertebrate. *Neuroscience Letters*, 155, 223 – 226.
Provencio, I. and Foster, R. G. (1995) Circadian rhythms in mice can be regulated by photoreceptors with cone-like characteristics. *Brain Research*, 694, 183 –190.
Provencio, I., Jiang, G., DeGrip, W. J., Hayes, W. P. and Rollag, M. D. (1998) Melanopsin: An opsin in melanophores, brain and eye. *Proceedings of the National Academy of Science*, 95, 340 – 345.
Provencio, I., Wong, S., Lederman, A., Argamaso, S. M. and Foster, R. G. (1994) Visual and circadian responses to light in aged retinally degenerate mice. *Vision Research*, 34, 1799 –1806.
Rado, R. and Terkel, J. (1989) Circadian activity of the blind mole rats, Spalax ehrenbergi monitored by radio-telemetry in seminatural and natural conditions. *Environmental quality and ecosystem stability*. E. Spanier, Y. Steinberger and M. Luria. Jerusalem, 391– 400.
Rado, R. H., Gev, H. and Terkel, J. (1988) The role of light in entraining mole rats' circadian rhythms. *Israel Journal of Zoology*, 35, 105 –106.
Roenneberg, T. and Deng, T.-S. (1997) Photobiology of the *Gonyaulax* circadian system I: Different phase response curves for red and blue light. *Planta*, 202, 494 – 501.
Rusak, B., Robertson, H. A., Wisden, W. and Hunt, S. P. (1990) Light pulses that shift rhythms induce gene expression in the suprachiasmatic nucleus. *Science*, 248(4960), 1237– 40.
Simpson, S. M., Urbanski, H. F. and Robinson, J. E. (1983) The pineal gland and the photoperiodic control of luteinizing hormone secretion in intact and castrated Japanese quail. *Journal of Endocrinology*, 99, 281– 287.

Soni, B. G. and Foster, R. G. (1997) A novel and ancient vertebrate opsin. *FEBS Letters,* 406, 279 – 293.

Soni, B. G., Philp, A., Knox, B. E. and Foster, R. G. (1998) Novel Retinal Photoreceptors. *Nature,* 394, 27– 28.

Takahashi, J. S., DeCoursey, P. J., Bauman, L. and Menaker, M. (1984) Spectral sensitivity of a novel photoreceptive system mediating entrainment of mammalian circadian rhythms. *Nature*, 308, 186 – 8.

Takahashi, J. S. and Menaker, M. (1979) Brain mechanisms in avian circadian systems. *Biological rhythms and their central mechanisms.* M. Suda, H. Osamu and H. Nakagawa. Elsevier, 95 –108.

Takahashi, J. S., Murakami, N., Nikaido, S. S., Pratt, B. L. and Robertson, L. M. (1989) The avian pineal, a vertebrate model system of the circadian oscillator: Cellular regulation of circadian rhythms by light, second messengers, and macromolecular synthesis. *Recent Progress in Hormone Research*, 45, 279 – 352.

Tosini, G. and Menaker, M. (1996) Circadian rhythms in cultured mammalian retina. *Science*, 272, 419 – 421.

Tsin, A. T. C., Philips, T. S. and Reiter, R. J. (1989) An evaluation on the level of retinoids in the bovine pineal body. *Advances in Pineal Research*, 3, 147–150.

Underwood, H. (1977) Circadian organization in lizards: The role of the pineal organ. *Science*, 195, 587– 589.

Underwood, H. (1983) Circadian organization in the lizard *Anlois carolinensis*: A multioscillator system. *The Journal of Comparative Physiology A,* 152, 265 – 274.

Underwood, H. (1986) Circadian rhythms in lizards: phase response curve for melatonin. *Journal of Pineal Research*, 3, 187– 96.

Underwood, H. (1986) Light at night cannot suppress pineal melatonin levels in the lizard Anolis carolinensis. *Comparative Biochemistry & Physiology*, 84(4), 661– 3.

Underwood, H., Barrett, R. K. and Siopes, T. (1990) Melatonin does not link the eyes to the rest of the circadian system in quail: a neural pathway is involved. *Journal of Biological Rhythms*, 5(4), 349 – 61.

Underwood, H. and Menaker, M. (1976) Extraretinal photoreception in lizards. *Journal of Comparative Physiology A*, 83, 187– 222.

Vigh-Teichmann, I. and Vigh, B. (1992) Immunocytochemistry and calcium cytochemistry of the mammalian pineal organ: A comparison with retina and submammalian pineal organs. *Microscopy Research and Technique*, 21, 227– 241.

Wallraff, H. G. (1981) Clock-controlled orientation in space. *Handbook of Behavioural Neurobiology.* J. Aschoff. New York & London: Plenum Press. Biological Rhythms: Chapter 16, 299 – 309.

Wang, Macke and Merbs (1992) A locus control region adjacent to the human red and green visual pigment genes. *Neuron,* 9(3), 429 – 440.

Yokoyama, K., Oksche, A., Darden, T. R. and Farner, D. S. (1978) The sites of encephalic photoreception in the photoperiodic induction of growth of the testes in the white crowned sparrow, *Zonotrichia leucophrys gambelii. Cell and Tissue Research,* 189, 441– 467.

Zhao, X., Haeseleer, F., Fariss, R. N., Huang, J., Baehr, W., Milam A. H. and Palczewski, K. (1997) Molecular cloning and localization of rhodopsin kinase in the mammalian pineal. *Visual Neuroscience,* 14, 225 – 232.

Zimmerman, N. H. and Menaker, M. (1979) The pineal: a pacemaker within the circadian system of the house sparrow. *Proceedings of the National Academy of Science*, 76, 999 –1003.

Section III

Biochemical and Physiological Adaptations

Introduction

S.N. ARCHER and M.B.A. DJAMGOZ

For a sensory system to function optimally, it must be adapted to receiving and responding to specific sets of physical stimuli. If there exists the possibility that the characteristics of the stimulus can change over time, then the sensory system must further adapt to those changes in order to remain functionally expedient. From this type of simple assumption, we can make some basic predictions about the adaptation of a sensory system: 1) Where we observe the widest range of stimuli characteristics we should expect to see the greatest adaptation in the sensory system. 2) Where the range of stimulus characteristics, become restricted or specialised, we should expect to see a corresponding focus of sensory system adaptation. 3) If the characteristics of the sensory stimuli are changing over a period of time, then we should expect to find adaptive mechanisms in the sensory system that are able to respond within the same time scale.

When John Lythgoe first wrote 'The Ecology of Vision' (1979) he was exploring the natural environment for evidence of correlations between the visual system and the behavioural ecology of an animal. He was essentially documenting examples of where the above predictions were obviously in operation. Many of these examples are found in fish that inhabit a wide diversity of spectral environments and are perhaps presented with the greatest set of spectral stimuli as well as extreme light-limiting conditions. As light fades and becomes monochromatic in the deep sea it is not surprising to discover that many deep-living species shift their spectral sensitivity to match the precious narrow band of penetrating light. Although this correlation is easy to understand, the ecological approach to vision often identifies other mechanisms that are much more subtle components of the complex adaptation.

This section of the book looks at how the function of the visual system relates to the ecology of vision and how evolution has been able to work upon biochemical and physiological aspects of function to produce adaptation to the environment. Adaptations of visual pigments to the aquatic environment are discussed in detail by Partridge and Cummings in Chapter 8. This chapter begins by making the important point that past correlations between vision and the environment highlighted by the ecology of vision have been extremely useful in identifying interesting areas for further research but, importantly, have not always addressed specific questions. Thus, it is not enough to observe an interesting correlation but it is also necessary to understand how the particular adaptation benefits the overall biology of the organism. In other words, what evolutionary pressures have been instrumental in selecting the adaptation? For example,

S.N. Archer et al. (eds.), Adaptive Mechanisms in the Ecology of Vision, 247–250

although most deep-sea fishes are maximally sensitive to blue light around 480 – 490 nm, the deep-sea fish *Aristostomias tittmanni* also has a visual pigment that is maximally sensitive to red light (Partridge & Douglas, 1995). In order to understand this apparent rule-breaking adaptation, it is necessary to know that the fish also emits red bioluminescence and is probably using this visual pigment for a private communication channel. This is an example of the visual system focusing on a specialised stimulus. But, as Partridge and Cummings point out, for an explanation such as this to be without doubt we must be sure that the adaptation of red sensitivity has been selected in this fish because it confers to the animal the real advantage of a private communication channel.

In Chapter 9, Marshall, Kent and Cronin review the visual pigment sensitivity adaptation found in the crustaceans. These animals inhabit a similar variety of niches to fish but have not produced such diversity in visual pigment adaptation. Most crustaceans have only a pair of visual pigments in the UV/violet and blue/green regions of the spectrum. These sensitivities are well separated and it is not clear if the pigments interact to provide colour vision. In freshwater crayfish these pigments can become further separated as the sensitivity of the blue/green pigment is pushed up towards the red end of the spectrum. As in fish, this adaptation has been achieved by replacing rhodopsin with porphyropsin pigments (or a variable mixture) and presumably the selection pressure has also been the same (i.e. that freshwater generally transmits more longer wavelength light).

In the crustaceans it is not so easy to find precise correlations between spectral sensitivity and environmental light. For many crustaceans the spectral sensitivity of their visual pigments is not exactly matched to the light available for vision (as in deep-sea crustaceans). This may be due to the fact that they have a limited number of visual pigments available and compromises have to be made as regards matching spectral sensitivity to the underwater light spectrum, matching the spectrum at different times, being receptive to bioluminescent light sources or being tuned for species specific behavioural tasks. Within the crustaceans there are probably more examples where the ecology of vision approach identifies correlations that are composed of many subtle elements. Another overriding pressure may come from the requirement to possess polarised light sensitivity. Sensitivity to polarised light could be achieved by the spectral positioning of a visual pigment either side of the polarisation minimum and this could further influence visual pigment adaptation. Lastly, it should be remembered that the crustaceans also give us the most numerous set of visual pigments yet encountered in an organism. The stomatopods can have up to 16 different visual pigments that show a high degree of potential specialisation in function. In these animals, it appears that selection pressure from behaviour has had a large influence in the adaptation of visual pigments.

In 'The Ecology of Vision', John Lythgoe emphasized mainly the relevance of photoreceptor/photopigment characteristics to visual ecology. However, the last two decades have seen tremendous advances in our understanding of the post-receptoral neuronal wiring of the vertebrate retina, and how this varies in animals with different visual requirements. Furthermore, we now appreciate that the adult retina is remarkably plastic, especially in lower vertebrates (fish, amphibians), and retinal functional organization can be modified significantly by light/dark adaptational mechanisms with distinct neurochemical control (Wagner and Djamgoz, 1993).

The next two chapters (10 and 11) deal with these aspects as regards outer and inner retinal signalling, respectively. Djamgoz, Vallerga and Wagner (Chapter 10) adopt a comparative approach to terrestrial mammals and fish as examples of vertebrates living in markedly different visual environments and review the functional connectivity of photoreceptors, horizontal cells and bipolar cells. Emphasis is placed upon mechanisms of synaptic transfer and spatio – chromatic signalling. The role played by dopamine in mediating light/dark adaptational synaptic plasticity is covered in some detail. Frishman and Robsin (Chapter 11) complement this by a parallel account of the inner retina. In the latter case, however, the mammalian retina is covered in greater detail due to the imbalance in the available information. In fact, much work remains to be done on understanding the functional organization of the inner retinae in fish in relation to visual ecology.

As already noted, for a sensory system to remain efficient it must be capable of responding to changing stimuli, some times within the same time scale. In Chapter 12, Beaudet and Hawryshyn describe ontogenetic mechanisms in fish that enable them to adapt their visual systems rapidly in response to the changing light environments that are encountered during migration. Flatfish larvae occupy bright, surface water and have pure-cone retinae with a single visual pigment. When they metamorphose and become bottom dwellers their visual system adapts to more dim light conditions by developing rods and expressing new visual pigments (in addition to a rather spectacular eye migration). Juvenile salmon migrate from the streams where they were hatched to live in ocean water. During this migration they lose a class of photoreceptors from their retinal mosaic that provided them with UV sensitivity. The UV sensitivity must be beneficial to the juveniles in shallow freshwater but is no longer required when they migrate into deeper sea water. As John Lythgoe (1979) pointed out, a cone that contains a non-functional visual pigment is occupying valuable space in the retinal mosaic. Interestingly, there is also evidence that the UV-sensitive pigment may reappear when the adults return to streams to reproduce. An opposite sense of migration occurs in the European eel where mature adults leave freshwater to return to deep-sea breeding grounds. Here the sensitivity of the rod-dominated retina shift from being green to blue sensitive, presumably as an adaptation to deep sea light conditions. In this adaptation the visual pigments of the rod photoreceptors become entirely composed of rhodopsins and a new opsin is introduced that shifts the maximum spectral sensitivity to the blue region of the spectrum. The switch in opsin has been shown to be brought about by a change in opsin gene expression and, because of the high levels of opsin protein turnover in outer segments, this adaptation can occur very rapidly. All these examples clearly demonstrate further that the visual system has a high degree of in-built plasticity. This plasticity provides the visual system of a particular species with the mechanisms to adapt quickly to changes in external stimuli but must also provide the variation upon which selection pressure can drive evolution of the visual system.

In the final chapter (13) of this section, Bowmaker and Hunt examine in detail the molecular mechanisms that determine the spectral sensitivity of visual pigments. The visual pigments are the molecular interfaces between the outside environment and the visual system. All visually coded information passes through the visual pigments and an understanding of how they encode the information and transduce photons into electrical signals is essential. The molecular dissection of visual pigments has allowed us to understand how spectral sensitivity can be altered by changing opsin structure. From this we can appreciate how evolutionary selection pressure has been able to adapt visual pigment spectral sensitivity and how plasticity during development can be achieved by changes in opsin gene expression. Knowledge of the molecular biology of visual pigments has also allowed us to determine the molecular basis for our own colour vision deficiencies.

References

Lythoe, J. N. (1979) *The Ecology of Vision*. Clarendon Press, Oxford.

Partridge, J. C. and Douglas, R. H. (1995) Far-red sensitivity of dragon fish. *Nature* 375, 21–22.

Wagner, H.-J. and Djamgoz, M. B. A. (1993) Spinules: A case for retinal synaptic plasticity. *Trends in Neuroscience* 16, 201–206.

8
Adaptation of visual pigments to the aquatic environment

J.C. PARTRIDGE and M.E. CUMMINGS*
School of Biological Sciences, University of Bristol, Woodland Road, Bristol BS8 1UG, UK. email: j.c.partridge@bristol.ac.uk
*Department of Ecology, Evolution and Marine Biology, University of California at Santa Barbara, Santa Barbara, CA 93106, USA. email: cummings@lifesci.ucsb.edu

8.1 Introduction

A central theme of John Lythgoe's book 'The Ecology of Vision' (Lythgoe, 1979) was that of adaptation: the matching between an organism's phenotype and its natural environment. In 'The Ecology of Vision', numerous adaptations of animal visual systems were identified, covering diverse aspects of life related to vision including ocular anatomy, visual physiology, biochemistry, behaviour, diet and life style. In this process various striking correlations between the ecology of different species and aspects of their visual systems were described. One of the strongest themes of the book is that of the matching of photoreceptor spectral sensitivities to the visual environment, a theme that was illustrated by numerous examples giving many insights into the evolutionary pressures on visual systems. Indeed, much of 'The Ecology of Vision' is concerned with the way in which evolution has driven different species to have the diverse spectral sensitivities that they have today. Nevertheless, the future of the ecology of vision lies in understanding visual adaptations by studies that explicitly identify the selective pressures acting on visual systems.

The theme of visual pigment adaptation in aquatic animals will be the basis of this chapter. John Lythgoe approached this subject by emphasising correlations between the visual pigments of animals and their ecology, and a limited number of such examples are described in Sections 8.8 and 8.9. For a number of reasons, it is important to appreciate the problems inherent in this approach and to realise that it is best at identifying *questions* rather than supplying *answers*. The first and fundamental reason for this is that, from a neo-Darwinian perspective, adaptations are linked to reproductive success; adaptive traits leading to a greater genetic legacy for an individual. Therefore, to fully assess the adaptive significance of some feature of an animal, one needs to measure the capacity of the adaptation to promote both viability (e.g. growth, foraging ability, longevity etc.) and reproductive success. Such measurements are rarely attempted (but see Endler, 1986, 1987, 1991). Secondly, the uncritical interpretation of supposedly adaptive features presents the danger of *properties* being mistaken for function

S.N. Archer et al. (eds.), Adaptive Mechanisms in the Ecology of Vision, 251–283

and being given undue importance. Visual ecologists must be careful that the visual property selected for any 'adaptive' correlation is actually under significant evolutionary pressure for visual tasks, and is not simply a feature that is correlated with some other property of the visual system. Animals function as the sum of their parts and any given adaptation will never function in isolation from other adaptations, or be unconstrained by physical limitations (see Chapter 1). Furthermore, we must keep in mind the phylogenetic history of the trait, and be aware that the visual properties we are investigating are not merely a function of some historic genetic constraint (Harvey and Pagel, 1991). Once appropriate visual properties have been identified, and proper phylogenetic comparisons made, another persistent problem that adaptive investigations must face is that the identification of correlations between supposedly adaptive features (e.g. visual pigment spectral sensitivities) and environmental variables (e.g. spectral irradiance) will inevitably be based on 'noisy' data. No two species ever live in the same place, at the same time, and in exactly the same way, so the selective pressures operating on them are never identical. Thus, the interpretation of visual adaptation needs specific knowledge of animals' visual ecology, without which generalisations can well be spurious. Lastly, it is a mistake to assume that a putative adaptation should be perfect: natural selection moulds adaptations to a certain level of effectiveness, which may be a far cry from perfection. Sensory systems are, however, generally acknowledged to be under high selective pressure and, therefore, apparent correlations between some aspect of a visual system and some variable of its operating environment are likely, at least, to provide interesting questions (Endler, 1991; Endler and Théry, 1996; Marchetti, 1993; Fleishman, 1992). This chapter will focus on ways in which visual systems operate in the aquatic environment which, we hope, may suggest avenues for further research in the ecology of vision.

8.2 Light underwater

The electromagnetic (EM) spectrum covers radiation with frequencies ranging from 10^3 to 10^{22} Hz (Campbell and Dwek, 1984) but only a tiny fraction, known as light, with frequencies between approximately 7.5×10^{14} and 4.3×10^{14} Hz (corresponding to wavelengths between 400 nm and 700 nm) is visible to humans. In fact, humans can see EM radiation outside this range provided it is of high enough intensity and, when animal vision is considered, it is useful to redefine 'light' as that part of the EM spectrum used by all animals, this covering the range from the ultraviolet (UV) to the far-red, with wavelengths (in a vacuum) of approximately 300 nm to 800 nm. Most light originates from the sun but the spectrum of light arriving at the earth's surface is greatly modified by its passage through the atmosphere in ways which vary greatly with latitude, season, time of day, cloud cover etc. (Henderson, 1970; Gates, 1980; Kirk, 1994).

The spectrum of light available at the earth's surface varies considerably throughout the day. When expressed in irradiance units of photons s^{-1} nm^{-1} m^{-2}, (photon units being appropriate for the consideration of visual systems which are essentially photon counters rather than cells that measure energy; Dartnall, 1975), the spectrum of light varies from a predominance of longwave photons during much of the day, to a predominance of shortwave and longwave photons (and dearth of mid-wave photons) at sunset (Henderson, 1970; Munz and McFarland, 1973; McFarland and Munz, 1976). Because of the phenomenon of colour constancy (Pokorny *et al.*, 1991) the magnitude of diurnal spectral change is rarely appreciated by the human observer. Also, because the fraction of light reflected from the water surface is wavelength dependent and is greatly affected by solar angle, the spectrum of light even immediately under the surface can be radically different from that above, especially at sunrise and sunset (Munz and McFarland, 1973; Munz and McFarland, 1977; for review see also Loew and McFarland, 1990). For these reasons it is not easy to calculate the irradiance spectrum at depth from measurements of above surface light, even if diffuse attenuation coefficients (e.g. Jerlov, 1976) are well known. Measurements of spectral irradiance at depth are, however, difficult to make and relatively few data are available (but see Jerlov, 1976; and Kirk, 1994) and for this reason calculation of underwater spectral irradiance must often be attempted. Such calculations may be based directly on measured spectral diffuse attenuation coefficients (Jerlov, 1976), but Baker and Smith (1982) and Prieur and Sathyendranath (1981) have developed models which can be used to calculate these coefficients from measurements of the concentrations of dissolved organic matter (DOM) and chlorophyll. Such calculations demonstrate that the spectral filtering of light constitutes a major constraint on the light available for vision and not only is the total photon flux rapidly reduced as depth increases, but also the light is quickly restricted in spectral range.

Light underwater is absorbed and scattered in a manner essentially similar to atmospheric attenuation, but with visually significant effects over a few meters path length (Duntley, 1962,1963). This attenuation is strongly dependent on wavelength due to spectral absorption and spectral scattering by: (1) water molecules; (2) chlorophyll in phytoplankton; (3) DOM; and (4) suspended solids. Because the concentrations of 2, 3 and 4 vary widely in different fresh and marine waters, there is wide variation in spectral irradiance at different depths in different waters. Pure water transmits light best at wavelengths close to 460 nm (Morel, 1974; Tyler and Smith, 1970; Jerlov, 1976) and for this reason oligotrophic waters, whether fresh or saline, appear blue. In coastal waters where phytoplankton and run-off from land increases the levels of chlorophyll, DOM, and organic and terrigenous particles, water colour is usually green or yellow. Levels of chlorophyll, DOM and suspended solids vary dramatically within short periods in most inshore waters and plankton blooms and variations in run-off cause unpredictable and extreme variations in spectral transmissions, and hence in the light available for vision. Freshwater is similarly changeable and can vary from extremely clear to green, yellow or brown, although some freshwaters such as Scottish tarns are so heavily stained with DOM that red light penetrates best (Spence *et al.*, 1971) and in the blackwater tributaries of the Amazon far-red or even infra red wavelengths predominate (Muntz, 1978; Muntz and Mouat, 1984). Because of the extreme effect of filtering on the spectrum of light available for vision underwater, and the great variability in this spectral filtering in different waters, the study of the evolution of visual spectral sensitivity of animals living in different waters can be thought of as using the underwater environment as a 'natural laboratory' (Levine, 1982). Nevertheless, to fully understand the data provided from this natural laboratory it is necessary to exactly identify the selective pressures that the underwater light environment imposes on visual systems.

For botanists interested in ascertaining the light available for photosynthesis spectral irradiance at depth can be an adequate measure. For visual ecologists, however, this is rarely sufficient because eyes *image* light with optical structures and thus the *direction* at which the light arrives is of fundamental importance. For this reason the vision ecologist is forced to consider the spectrum of light flux from all directions surrounding an aquatic animal, the three dimensional spectral radiance distribution, which is both complex in structure and difficult to measure. Close to the water surface the spectral radiance distribution is particularly complicated, the entire above surface scene being viewed through a cone, with a solid angle of some 97°, known as Snell's window (Jerlov, 1976; Smith, 1974). When an animal looks upwards towards the water surface Snell's window is seen surrounded by light originating from deeper water that is reflected by the water surface (Lythgoe, 1979; Partridge, 1990). The edge of Snell's window, which may be highly distorted by surface ripples and waves, is marked by a narrow ring of different hues where light from the above water horizon is split into a spectrum due to the way in which the refractive index of water varies with wavelength (see Partridge, 1990). Near the surface the 3-dimensional spectral radiance distribution is highly asymmetric, with a marked drop in intensity at the edge of Snell's window and strong shadows cast by the sun (Smith, 1974; Partridge, 1990). However, as depth increases in open water the spectral radiance rapidly approaches an asymptotic distribution that is symmetrical about the vertical axis (Jerlov, 1976). In contrast, the spectral radiance distribution in shallow water, and close to the bottom or other underwater surfaces, is highly asymmetrical. In such habitats the ambient light thus varies markedly in both intensity and spectral composition depending on direction of view (Munz and McFarland, 1977; Levine and MacNichol, 1979, 1982). Typically, the radiance of shallow water substrates is similar in spectral distribution to downwelling light and is spectrally broader than light originating from open water and viewed horizontally (Munz and McFarland, 1977; Levine and MacNichol, 1982).

Given the potential complexity of the light field underwater it is perhaps surprising that a mathematical framework exists for the description of radiance underwater (e.g. Duntley, 1962, 1963; Preisendorfer, 1964) and has been used successfully as a basis for the understanding of vision underwater in both man and fishes (e.g. Lythgoe, 1968; Lythgoe and Partridge, 1991). At heart the radiance transfer equations emphasise the way in which image-forming light reflected from an underwater object is attenuated by absorption and scattering on its passage to an observer, and how this light is 'veiled' by a 'coloured fog' of light that is scattered into the light path. Figure 1 shows how an underwater visual situation can be reduced to a mathematical description of radiance transfer. In this example, light incident on a small area of a target, (fish 2), originating from all directions surrounding fish 2, but primarily from above, is reflected towards an observing animal (fish 1), which is at a horizontal distance z. This spectral radiance, $R(\lambda)$, which has units of photons s^{-1} m^{-2} nm^{-1} $steradian^{-1}$, is attenuated by absorption and scattering such that the image forming light reaching fish 1 is reduced. This light is also supplemented by non-image-forming 'veiling' light that is scattered into the visual pathway and which reduces the contrast between light from target fish 2, ${}_tR(\lambda,z)$, and that of the background spacelight, ${}_bR(\lambda)$. The apparent radiance, ${}_tR(\lambda,z)$, of the patch of fish 2 seen at horizontal distance z, is given by an equation that depends on the narrow beam spectral attenuation coefficient, $\alpha(\lambda)$:

$${}_tR(\lambda,z) = R(\lambda)e^{-\alpha(\lambda,z)} + {}_bR(\lambda)(1-e^{-\alpha(\lambda,z)}) \qquad [1]$$

The effect of the veiling light is to degrade the contrast of the image in much the same way as atmospheric fog degrades contrast in the terrestrial environment. This has important implications because most visual detection tasks underwater can be reduced to a problem of contrast detection; objects often having inherently low visual contrasts. Because of this fact, and the difficulty that this presents for even the simplest visual task, *a priori* one would expect aquatic animals to have adaptations to their visual systems that would enhance their operational abilities in conditions of low inherent contrast. Contrast is conventionally defined as the proportional difference between the target radiance and the background radiance (Duntley, 1963). In the scenario depicted in Figure 1 the apparent 'whole body' contrast, $C(\lambda,z)$, at a particular wavelength, λ, between fish 2 and the background light at fish 1 is:

$$C(\lambda,z) = ({}_tR(\lambda,z) - {}_bR(\lambda)) \,/\, {}_bR(\lambda) \qquad [2]$$

This definition of contrast, which has values ranging from –1 to infinity, is similar to a Weber-Fechner fraction (Wyszecki and Stiles, 1967) and can thus readily be related to just detectable contrast differences (e.g. Lythgoe and Northmore, 1973). It is evident that contrast is highly dependent on the visual range and if $C(\lambda,0)$ is the inherent contrast of the target against the background (i.e. the contrast when the sighting range, z, is zero) then the apparent contrast at horizontal visual range z is:

$$C(\lambda,z) = C(\lambda,0)e^{-\alpha(\lambda,z)} \qquad [3]$$

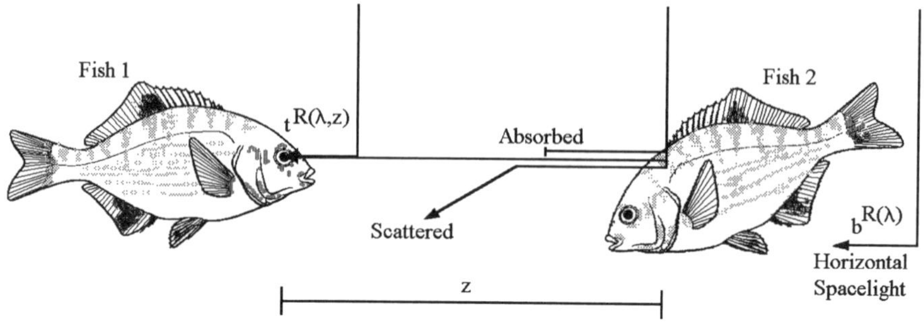

Figure 1 Light incident on fish 2, originating from all directions surrounding fish 2 but primarily from above, is reflected towards an observing fish 1 which is at a horizontal distance z. This radiance, $R(\lambda)$, which has units of photons s^{-1} m^{-2} nm^{-1} $steradian^{-1}$, is attenuated by absorption and scattering such that the image forming light reaching fish 1 is reduced. This image forming light is also degraded by the addition of non-image-forming 'veiling' light that is scattered into the visual pathway and which reduces the contrast between radiance from target fish 2, $_tR(\lambda,z)$, and that of the background spacelight, $_bR(\lambda)$. As explained in Section 8.3, the apparent radiance, $_tR(\lambda,z)$, and the apparent contrast, $C(\lambda,z)$, between the target fish and the background light, are given by equations that depend on the narrow beam attenuation coefficient $\alpha(\lambda)$ and the horizontal visualization distance z.

This measure of contrast is relevant for animals engaged in the detection of targets, such as whole organisms (as shown in Figure 1), against the background spacelight. For some visual tasks, however, the background is not spacelight but a substrate such as the sea floor, or even the water surface. A particularly important set of visual tasks are those associated with mate selection and courtship or other social communication and in these cases the background may be part of the animal itself, the target being a specific region of colour on the animal. With such types of visual task, the specific colour contrasts may be compared with the rest of the animal (Endler, 1986, 1991; Endler and Théry, 1996). The relevant measure of contrast in these circumstances may be termed 'within body contrast' and can be calculated as:

$$C'(\lambda,z) = ({}_{t1}R(\lambda,z) - {}_{t2}R(\lambda,z)) / ({}_{t1}R(\lambda,z) + {}_{t2}R(\lambda,z)) \quad [4]$$

In this equation target spectral radiance of one body area, $_{t1}R(\lambda,z)$, presents a contrast with the spectral radiance of another body area, $_{t2}R(\lambda,z)$. In this case the calculated contrast is the difference between the two radiances normalised to the sum of the two radiances, a measure that has somewhat different properties from whole body contrast given in equation (2), and ranges between –1 and +1.

Both whole body contrast and within body contrast will vary depending on the spectrum of the incident light, the viewing distance, the spectral attenuation of the water, and the spectral reflection of the objects being viewed. Aquatic animals, therefore, are likely to be under direct selection for the possession of both effective detecting systems, including retinal and photoreceptor adaptations, and for the use of readily detectable signals such as colour patterns or behaviour that best transmit information within signal degrading environments. In aquatic environments where images are easily degraded, *a priori* reasoning suggests that maximising contrast, both in production and perception, will be a common theme. Such contrasts may be calculated using radiance transfer mathematics but the future importance of this approach in the understanding of aquatic vision will, without doubt, depend on the identification of visual tasks that are essential to an animal's survival and reproductive success. Because such tasks will depend on the behaviour and ecology of the species being considered the successful identification of these tasks will depend on the ethological interpretations of visual behaviour monitored by direct underwater observation and the assessment of these behaviours in terms of their effect on fitness.

8.3 Spectral reflections and the measurement of target radiance

The consideration of radiance transfer in the context of visual tasks of the type shown in Figure 1 underlines an important point about visual systems: they have evolved to look at *things*, rather than to look at the general light field. In the terrestrial world such a statement would seem trite: we are so used to living in a relatively complicated incident spectral radiance distribution and most visual tasks do not involve contrasts between objects and a background spacelight. In some terrestrial habitats, such as forests, the complexity of the incident spectral radiance distribution is very obvious (Endler, 1993) and has a profound effect on the spectral radiance of viewed objects. Underwater, however, the inherent spectral radiance distribution is, in open water especially, dominated by the underwater spacelight, rather than by the spectral reflections of objects in the environment. (see Section 8.2). Consequently, underwater vision ecologists have traditionally concentrated on the optical properties of the water rather than the ethology of vision. Thus, for instance, correlations have been sought (and found, see Section 8.7) between the spectrum of irradiance at depth and the spectral sensitivities of aquatic animals, often without full consideration of the way in which this is evolutionary significant. Irradiance is a measure of light arriving at a point, either from all directions in the case of scalar irradiance, but also as (e.g.) downwelling, or upwelling vector irradiance. No animal is interested in this *per se* and, instead, the significance of spectral irradiance measurements is that they are a simple measure of the light potentially available for vision. Nevertheless, it is not straightforward to predict the spectral radiance of an underwater target from some measure such as spectral downwelling irradiance because targets are physical objects with particular textures and particular orientation in an underwater light field that is more complex than simple irradiance measurements would suggest (e.g. see Jerlov, 1976). Target radiance is a function of the reflecting properties of the target and the light environment within which it is viewed. Ideally, measurements of the 3-dimensional spectral radiance distribution are required: a Herculean task, and even armed with such data, the calculation of spectral radiance from a target of known spectral reflection is far from straightforward. For this reason there is considerable merit in the direct measurement of spectral radiances from objects in the natural environment.

When *in situ* spectral radiance measurements are not possible, then it is *necessary* to calculate target spectral radiance from measurements of spectral irradiance, target spectral reflections, and attenuation properties of the viewing medium (e.g. Lythgoe, 1966, 1968; Endler, 1990; Lythgoe and Partridge, 1991). Spectral reflections are an important measurement in the determination of target radiance, and one that cannot be assessed by human eyes alone. The use of spectral reflection measurements based on human perception, even relatively quantitative ones such as those provided by colour comparisons with Munsell chips (Munsell, 1976), are simply not appropriate. Comparisons conducted in this way *may* yield data relevant to the vision of the species under consideration; or they may not. The usefulness of human colour assessment for visual adaptation studies depends on the similarity of the species' visual system to that of the human observer. This argument has been made persuasively by Bennett, Cuthill and Norris (1994) and by Cuthill and Bennett (1993) who have likened the use of human vision in the evaluation of animal visual systems to the use of human hearing in the understanding of bat echo location. This comparison is valid: just as we are unaware of bat ultra-sound, so we are unaware of hues and even entire parts of the EM spectrum that may be visually accessible to other species. For instance, animal reflections have UV components that we do not see: some tropical marine fish have blue and yellow reflections without a UV component, others have blue and yellows that are also strongly UV reflecting (N. Justin Marshall, pers. com.). To a fish with UV vision the perceived hues of these different blue or yellow patches may look completely different although they look identical to us.

Because of the limitations of human colour vision, the way in which visualised objects reflect light needs to be objectively quantified by reflection spectrophotometry using a reflection system in which the orientation of the object relative to the directions of the measuring device and the illumination source is carefully controlled. Ideally, the illuminating system should emulate the underwater radiance distribution but for practical reasons this is rarely attempted (but see Denton and Rowe, 1994). Indeed, if the animal is subject to variable light conditions, it is useful to make spectral reflection measurements with a full spectrum light source rather than one that emulates only one specific underwater habitat. These types of reflection measurements provide relative reflection spectra which can be integrated with spectral irradiance measurements to produce first approximations of target radiance under various spectral irradiance and transmission conditions.

8.4 Photoreceptors

The scenario depicted in Figure 1 presents a representation of a visual task reduced to the fundamentals of radiance transfer. To utilise such an approach to understand the adaptation of visual pigments to visual tasks demands the incorporation of the process of photon capture by visual pigments located in photoreceptors. In order to function, a visual system needs its photoreceptors to catch adequate numbers of photons to support the visual tasks in which the animal is engaged. In direct sunlight the irradiance in the animal visible spectrum is extraordinarily high, but many animals, particularly aquatic species, never experience high light levels of full sunlight because the environment in which they live attenuates the available light. Furthermore, even species that sometimes experience full sunlight spend most of their lives at lower light intensities due to atmospheric attenuation and diurnal variations. At wavelengths around 500 nm (i.e. in the middle of animal visible spectrum) full sunlight has an intensity of about 3×10^{16} photons s^{-1} nm^{-1} m^{-2} (McFarland and Munz, 1975a), but at twilight this has fallen by three orders of magnitude, and starlight has less than 10^7 photons s^{-1} nm^{-1} m^{-2} (Munz and McFarland, 1973). Similarly, full sunlight at 1000 m depth in the clearest tropical oceans is attenuated to approximately 10^8 photons s^{-1} nm^{-1} m^{-2} (Dusenbery, 1992).

The brightness of the retinal image depends largely on the optical construction of an eye (Land, 1981, 1990) but there are constraints on eye design (especially maximum size) that cannot be overcome to increase photon catch. As a result, the main problem associated with many visual tasks, even those undertaken in daylight (Land, 1981), is to catch sufficient photons. Both the emission and absorption of photons are stochastic processes and, as a result, the detection of photons and the visual detection of objects in the environment are, at root, statistical problems (Rose, 1973) and the amount of information in a retinal image is fundamentally limited by the number of photons that form it (Buschbaum, 1981). When photon flux is especially low there is no alternative but to increase either the area of the retina over which photon catch is pooled (at the expense of spatial resolution) or to increase the integration time (at the expense of temporal resolution). Both measures will increase the signal to noise (S/N) ratio of the image and the ability of an animal to detect visually large and slow moving objects, but the functional significance of this depends on the animal's visual ecology and the visual tasks that it must undertake in order to survive. Alternatively, animals may increase the amount of visual pigment in a photoreceptor to enhance the cell's efficiency at catching photons, most obviously by increasing the size of the photoreceptor. However, this itself has penalties: firstly, a large photoreceptor may be more energetically expensive, or structurally difficult to maintain; secondly, spurious signals from spontaneously 'activated' visual pigments, or other sources of receptor noise, will increase with photoreceptor volume; and, thirdly, long photoreceptors conflict with a fundamental requirement for colour vision, the need for adequate separation of spectral sensitivity of different photoreceptor classes.

Visual pigments are membrane bound proteins and are situated in specialised photoreceptors in which many membranes are arranged into compact stacks. The density of packing of rhodopsin in vertebrate rods is about 20,000 molecules μm^{-2} of membrane (Krebs and Kühn, 1977) and the fraction of photons absorbed (i.e. the absorptance; P) of a membrane stack of N membranes is, at the wavelength of peak absorption (λ_{max}), given by Dusenbery (1992) as:

$$P_{\lambda_{max}} = 1\text{-}10^{-0.00013N} \qquad [5]$$

Thus a single membrane absorbs only 0.0003 (i.e. 0.03%) of the incoming photons, even when they have a wavelength at which the rhodopsin is most sensitive. For this reason a visual photoreceptor needs many membranes if its sensitivity is to be adequate for any realistic visual task. In vertebrate rod outer segments membrane discs are separated by approximately 30 nm (Knowles and Dartnall, 1977) and there is therefore effectively one membrane per 15 nm, or about 70 membranes per micrometer of outer segment length. A 1 μm length of such an outer segment would thus absorb approximately 0.02 of the incoming photons if they had a wavelength equal to the visual pigment's λ_{max} (i.e. the stack would have an absorbance at the λ_{max} of *ca.* 0.0091). Correspondingly, a 10 μm long outer segment, which is not atypical of rods or cones of shallow living fish (Ali and Anctil, 1976), would absorb approximately 0.19 of the incoming photons (an absorbance of 0.091), and a 100 μm long outer segment (common in deep-sea fish retinas) would absorb approximately 0.88 of the incoming photons, at the λ_{max} (an absorbance of 0.91).

The above calculations disguise the fact that as the length of the outer segment increases the absorptance spectra of photoreceptors become broader, conferring greater sensitivity at all wavelengths, but particularly at those away from the λ_{max} (Knowles and Dartnall, 1977). This broadening of spectral sensitivity also has consequences for the number of different types of photoreceptor that an animal is likely to have because, on theoretical grounds, the number of photoreceptor types that animals 'should' have is intimately linked to the spectral bandwidth of the photoreceptors' absorptance spectra. Simply put, an animal's ability to discriminate radiance spectra from each other will not necessarily be improved by the addition of extra photoreceptors with different spectral sensitivities unless those spectral sensitivities are relatively narrow compared with the extent of the spectrum covered. Barlow (1982) has argued that for sinusoidally modulated ('comb filtered') spectral radiance distributions the possession of three types of visual pigment is probably the optimum solution unless the visible spectrum is extended beyond the human visible wavelengths or the photoreceptors' absorptance spectra are considerably narrowed by intracellular filters (see Douglas and Marshall, in press). Although this conclusion was derived from the consideration of artificial comb filtered spectra similar conclusions have been reached by others (see Osorio and Vorobyev, 1997) with the consideration of additional factors including the spectral reflections of natural objects and the need for colour constancy in colour vision systems.

In the photon limited underwater world there results a particularly acute conflict between sensitivity and the type of colour vision an animal can have. Most fish have more than one type of photoreceptor in their retina: almost all having duplex retinas boasting both rods and cones, and usually several types of the latter (Ali and Anctil, 1976; Bowmaker 1995). In general, rods function at low light levels and cones at higher light levels and fish have complex mechanisms for sensory adaptation to different irradiances, including diurnal retino-motor movements of rods and cones to bring the active photoreceptors into play (Wagner, 1990). All fish with more than one type of cone seem to have the retinal neural machinery to compare the signals from the different cone types, giving the ability to distinguish spectral radiances not just by intensity but also in terms of their spectral composition – i.e. they have the potential for colour vision. For colour vision more than one type of photoreceptor with different spectral sensitivities are required. However, as discussed above, the amount of colour information that can be extracted from a visual scene is limited if the spectral range of environmental light is relatively restricted (Bowmaker, 1983) and the photoreceptor spectral sensitivities are not narrowed by photostable filters (see Douglas and Marshall, in press). In fish, intracellular spectral filtering is relatively undeveloped and, although cone inner segments may in some cases have significant filtering (MacNichol *et al.*, 1978), in most species photoreceptor spectral sensitivities are largely determined by the absorptance spectra of the cells' visual pigments. For this reason it is not surprising that there is good evidence that the number of photoreceptor types present in the retina of different fishes correlates with the spectral bandwidth of the available downwelling light, with the number of spectrally distinct cone types ranging from one to five. For instance, temperate coastal marine fishes living at moderate depths, and experiencing a restricted spectrum of light, have less shortwave and less longwave sensitivity and fewer cone types than tide-pool species living in very shallow water in the same area which experience the full spectrum of daylight (Loew and Lythgoe, 1978; Lythgoe, 1984; Partridge, 1986, 1990). Typically, temperate coastal marine fishes have relatively few types of cones, usually two, with relatively large outer segments and λ_{max} values close to 460 nm and 530 nm (Partridge, 1986, 1990; Lythgoe, 1988; Lythgoe and Partridge, 1991).

For the reasons given above, polychromatic colour vision systems are most likely to be found in shallow living aquatic animals that are exposed to relatively high light levels with a broad spectral range. Indeed, shallow living freshwater fishes generally have at least three types of retinal cone (Levine and MacNichol, 1979; Lythgoe, 1988). Nevertheless, we cannot be sure how the number of photoreceptor types is related to the number of dimensions of an animal's colour vision (Jacobs, 1981). Thus, for instance, we cannot assume that a guppy, which has at least four types of cone (Archer *et al.*, 1987), has tetrachromatic colour vision. Indeed, there may be plasticity in colour vision and the processing of the output from different photoreceptors. For instance, the goldfish, which has four types of cone, has been shown to be tetrachromatic at high light levels but trichromatic when the level of illumination is reduced (Neumeyer, 1992).

8.5 Photoreceptor spectral sensitivity

Photoreceptors of the visual system absorb electromagnetic radiation and amplify the energy content of the arriving photons by the transduction of this photon signal into an ion flow across cell membranes and, ultimately, give rise to a biologically relevant output in the form of neurotransmitter release at a synapse. In this way information about the visual environment is transferred to the internal, physiological environment of an animal's nervous system. The information capacity of visual systems, especially those of vertebrates, is extraordinarily high and the human eye is capable of transferring information to the brain at rates equivalent to 10^6 to 10^8 bits s^{-1} (Dusenbery, 1992) but such rates of information transfer can only be attained if photon capture itself is efficient.

In all photoreceptors involved in vision it is the absorption of light by coloured compounds, the visual pigments, that initiates the transduction process in the photoreceptor and constitutes the first step in the visual process. Fundamental to this is the successful absorption of photons: only those absorbed by visual pigments can play any part in vision and no amount of subsequent neural processing can replace the information lost when photons are not absorbed (for recent review of photoreceptor spectral sensitivity, see Loew, 1995). The absorption of photons is, however, a stochastic process and the ability of a photoreceptor to signal the arrival of a photon is largely determined by the probability of that photon being absorbed. This probability is a function of the frequency of the photon and the probability of photon absorption varies greatly across the spectrum. Arguably spectral absorptions should be presented as a function of frequency (which is a fundamental property of a photon) but this is rarely done. Instead, visual pigment absorption spectra are generally presented as a function of wavelength, tacitly assumed to be that of the photon in a vacuum despite the fact that wavelength varies with the speed of light (and hence the refractive index of the transmitting medium). In this format, all visual pigments have a major (α-band) absorption peak, the absorbance spectrum of which is asymmetrical and bell-shaped with a wavelength of peak absorbance known as the λ_{max}, at which the pigment is most effective at absorbing photons. If intracellular spectral filters (see Douglas and Marshall, in press) are ignored, the way in which a photoreceptor cell responds to arriving photons of different frequencies is almost entirely determined by the spectral absorptance of the visual pigment that it contains. Spectral absorptance is, itself, dependent on both the spectral specific absorbance (i.e. absorbance μm^{-1}) of the visual pigment and the path length (effectively the photoreceptor outersegment length) presented to the incoming photons (Knowles and Dartnall, 1977). Most available evidence suggests that the spectral sensitivity of a photoreceptor, i.e. the way in which it responds to incident photons of different frequencies, is directly proportional to its spectral absorptance over much of its sensitivity spectrum (Dartnall, 1972). For the above reasons, much of the investigation of the adaptation of photoreceptor spectral sensitivity to animals' ecology involves the measurement of visual pigment absorbance spectra. Such measurements may be made by spectrophotometry of detergent extracts of visual pigments (e.g. Knowles and Dartnall, 1977; Douglas *et al.*, 1995) or by microspectrophotometry of visual pigments in individual photoreceptor outersegments (e.g. Partridge and DeGrip, 1991).

8.6 Visual pigment λ_{max} and spectral tuning

Fundamental to the appreciation of adaptations in sensory systems is an understanding of the constraints that may be operating on a given system. Such constraints may be due to the physics of the sensory modality, but may also be due to biological limitations operating at all levels from primary protein structure and cell biochemistry to animal behaviour. In the case of the ecology of vision, an understanding of the way in which visual pigments may be adapted to suit specified visual tasks requires knowledge of how such pigments are spectrally 'tuned' and the constraints that may be acting on this tuning.

Visual pigments are the coloured compounds found in photoreceptors, the principle function of which is to absorb light, become 'activated' and, in this state, initiate a cascade of intracellular reactions that results in a change in the polarisation of the cell membrane and release of neurotransmitter at a synapse. All visual pigments consist of a chromophore, which is a derivative of vitamin A, and a protein, known as opsin. Vitamin A occurs in two natural forms, A1 and A2, and the aldehydes of these (retinal and 3,4-dehydroretinal respectively) are both used as chromophores by vertebrates (Knowles and Dartnall, 1977). Invertebrates also use these chromophores, but insects also employ a hydroxylated derivative of retinal, 3-hydroxyretinal (Vogt, 1983), and a few cephalopods use 4-hydroxyretinal in addition to retinal and 3,4-dehydroretinal (Seidou *et al.*, 1990) . On their own these compounds absorb most strongly in the near UV, retinal having a λ_{max} close to 377 nm (in solution in ethanol; Knowles and Dartnall, 1977), and 3,4-dehydroretinal, which has an extra double bond in the terminal ring of the molecule, having a λ_{max} at about 400 nm. These absorption properties are not visually important, however, because the absorption spectrum of the chromophore is radically altered when combined with an opsin to form a visual pigment.

Opsins are transmembrane proteins with seven alpha helices of mainly hydrophobic amino acids (Hargrave *et al.*, 1984). Opsins belong to a large family of cell receptor proteins collectively known as G-protein linked cell membrane receptors, which includes β-adrenergic receptors, muscarinic acetyl choline receptors, dopamine receptors and others. These receptors clearly arose by diversification from a common ancestral protein and are highly conserved in their primary, secondary and tertiary structure, although modified for radically different purposes and signalling intracellularly via somewhat different G-protein mediated mechanisms. All opsins comprise some 350 amino acids, the amino terminal of the protein being exposed on the external surface of the cell membrane (the luminal surface of the outersegment disks of vertebrate rod photoreceptors), and the carboxyl terminal being on the internal cytoplasmic surface (Hargrave *et al.*, 1984). Each α-helix consists of approximately 26 amino acids with a central 18 or so embedded in the membrane (Baldwin, 1993; Schertler *et al.*, 1993). Within the palisade of seven α-helices is a ligand binding pocket for retinal, which is bound to the protein via a Schiff base linkage to a lysine residue in helix VII. Like virtually all proteins, opsin absorbs maximally at wavelengths below 300 nm but the electrostatic interaction between the opsin and the chromophore shifts the λ_{max} of the latter to other wavelengths in the electromagnetic spectrum. The exact location of the resulting visual pigment's λ_{max} depends on both the chromophore employed and the amino acid sequence of particular opsins but, in fish, which have the widest spread of λ_{max} values among the vertebrates, the range extends from ca. 350 nm to 635 nm. Thus the λ_{max} of the chromophore can be displaced by this 'opsin shift' to either shorter wavelengths (a hypsochromatic shift) or to longer wavelengths (a bathochromatic shift) compared with the free chromophore. If the λ_{max} of a visual pigment is displaced to longer wavelengths this implies that there is a smaller energy difference between the ground and excited states of the molecule (Campbell and Dwek, 1984) and thus low energy, longwave photons are more likely to be absorbed than photons with shorter wavelengths. In the case of longwave absorbing visual pigments the interaction between opsin and the chromophore must result in an increase in the delocalisation of the chromophore's π electrons. In practice this is achieved by charged and polar amino acids positioned in the transmembrane helices of opsin and particularly by such replacements at sites on the inner surfaces of the helix palisade positioned close to the chromophore. There have been numerous studies of the way in which the λ_{max} values of different visual pigments are spectrally 'tuned' by this interaction and work is particularly advanced in the case of vertebrate cones (see Bowmaker and Hunt, in press) in which various key amino acids involved in λ_{max} tuning have now been identified.

Nevertheless, a given opsin is capable of forming visual pigments with a variety of chromophores, both natural and unnatural. This property has been utilised to study various biochemical properties of visual pigments but is also used by animals. In particular many aquatic vertebrates, especially freshwater fish and amphibians, use 3,4-dehydroretinal as a chromophore to form 'porphyropsin' pigments in their rods and cones, whereas most marine species use retinal to form 'rhodopsins' in their photoreceptors. The absorption spectra of visual pigments based on different chromophores differ significantly: a visual pigment based on a given opsin and employing 3,4-dehydroretinal having a λ_{max}

bathochromically (i.e. longwave) displaced compared with a visual pigment based on the same opsin but employing retinal. Not only is the λ_{max} shifted to longer wavelengths but also the normalised absorbance spectrum is generally broader (Knowles and Dartnall, 1977). The relationship between the λ_{max} of analogous pairs of such visual pigments (i.e. based on the same opsin, but employing different chromophores) has been investigated by several workers (e.g. Dartnall and Lythgoe, 1965; Whitmore and Bowmaker, 1989) who have proposed simple equations linking the λ_{max} of porphyropsin visual pigments to the λ_{max} of analogous rhodopsins. Various animals exploit the relationship between rhodopsin and porphyropsin λ_{max} and a number of species, particularly fish living in freshwater, shift their visual pigment chromophores between retinal and 3,4-dehydroretinal either seasonally or during development (see Section 8.9). Although the changes in visual pigment λ_{max} and breadth of spectral absorptance that occur in such cases are well documented, the adaptive features of such shifts are not well understood. For instance, the adaptive value could be related not so much to λ_{max} but to other properties of rhodopsin and porphyropsin pigments such as thermal stability (see Section 8.9, Loew and Archer, in press).

The discussion of the underwater light field (section 8.2) and the spectral reflections of objects (section 8.3) emphasised the fact that the wavelengths available for vision are affected by wavelength dependent functions. In order to calculate how well tuned a given photoreceptor's spectral sensitivity may be to a particular visual task one must integrate the spectral radiance of the arriving light with the spectral absorption of the photoreceptor. In order to do this the fraction of light absorbed by a photoreceptor (i.e. its absorptance) must be calculated for each wavelength considered in the numerical integration. Spectral absorptance calculations require information about the characteristic way in which a pigment absorbs light of different wavelengths, most fundamentally given by its normalised absorbance spectrum (A(λ); Knowles and Dartnall 1977); the length specific absorbance of the visual pigment (s; i.e. the absorbance per micron of photoreceptor, measured at the λ_{max}); and the total path length (l) presented by that structure to incoming photons (e.g. the length of the outer segment in the case of vertebrate rods and cones). The spectral absorptance (P(λ)) is thus given by:

$$P(\lambda) = 1 - 10^{-A(\lambda)sl} \qquad [6]$$

Although A(λ) can be measured by microspectrophotometry of individual photoreceptors (Liebman, 1972; Partridge, 1986; Partridge and De Grip, 1991) or by spectrophotometry of visual pigments extracted in detergent or solvents (Knowles and Dartnall, 1977; Douglas *et al.*, 1995) such measurements are not always necessary, particularly in theoretical studies. This is because the normalised absorbance spectrum of a visual pigment of a given λ_{max} is readily calculated by using visual pigment template curves, transformed appropriately on the spectral axis: i.e. visual pigments have a more or less standard normalised absorbance spectrum if the wavelength axis is transformed in an appropriate way. There have been numerous attempts to describe standard normalised absorbance spectra of different classes of visual pigments (principally those based on different chromophores) and various spectral axis transforms have been suggested (e.g. Dartnall 1953; Partridge and De Grip, 1991; Stavenga *et al.*, 1993). Such templates provide a simple way of calculating the normalised spectral absorbance of visual pigment with a given λ_{max}, but all have limitations: for instance many only consider the principal (α) absorbance band of the visual pigment (Knowles and Dartnall, 1977), although the rhodopsin and porphyropsin templates of Stavenga *et al.* (1993) extend to shortwave β-band and γ-band absorbance. In addition, there are many assumptions underlying the templates: for instance Stavenga *et al.*, (1993) assume that the β-band does not shift its spectral location when the α-band λ_{max} is shifted, although others have suggested relationships between these bands (e.g. Palacios *et al.*, 1996). There is also accumulating evidence that visual pigments with very shortwave λ_{max} values, especially UV sensitive pigments have normalised absorbance spectra that are substantially narrower than is predicted by standard templates (e.g. Palacios *et al.*, 1996). Despite such reservations, however, visual pigment templates provide excellent tools for the analysis of visual pigment spectroscopic measurements (e.g. Partridge and De Grip, 1991). Moreover, in the context of the ecology of vision, visual pigment templates are particularly useful when constructing optimisation models of visual pigment function. Such models are essential to develop hypotheses concerning the spectral tuning and adaptive value of visual pigment spectral sensitivities and have considerable predictive potential (e.g. Lythgoe and Partridge, 1989, 1991; Vorobyev and Osorio, 1998).

As previously noted, the range of known visual pigments is large, with peak sensitivities ranging in wavelength from the UV to longwave red. Nevertheless there is evidence that there are limits on the variability of λ_{max} values. Principle among these data is the discontinuous nature of the distribution of λ_{max} values, some values being far more common than others and some apparently never being found. These data have been gathered principally from measurements of aquatic animals, especially fishes, and such spectral 'clustering' of λ_{max} values has been demonstrated both for porphyropsins (Bridges, 1964) and for rhodopsins (Dartnall and Lythgoe, 1965; Partridge *et al.*, 1989). These data indicate that visual pigment measurements fall into distinct classes with λ_{max} separations of 5 – 8 nm, individual species having visual pigments selected, teleologically speaking, from a finite and discontinuous palette of λ_{max} values. As our understanding of the molecular basis for spectral tuning has increased (see Bowmaker and Hunt, in press; Loew and Archer, in press) an explanation for λ_{max} clustering has been presented: only a limited number of amino acids within the opsin molecule can have an effect on π-electron delocalisation; others are simply too remote from the chromophore to have significant influence on a visual pigment's absorption spectrum. It is also likely that there are other molecular constraints on the spectral location of a visual pigment's λ_{max}. In particular, visual pigments must function in ways other than simply absorbing photons: they must be structurally coherent, stable (i.e. only become activated when a photon has been absorbed), and they need to interact efficiently with the other proteins involved in transduction (Piantanida, 1991). It is entirely likely that these requirements impose irreducible constraints on the range of λ_{max} values that functional visual pigments can have.

As knowledge of the molecular biology of visual pigments has increased an explanation for the empirical observation of spectral clustering of λ_{max} values has been presented as particular amino acids in the opsin molecule have been identified as being responsible for discrete shifts in λ_{max} (see Bowmaker and Hunt, in press; Loew and Archer, in press). Such data reinforce the earlier assumption that λ_{max} clustering is driven by physical limitations associated with opsin-chromophore interactions rather than environmental constraints: a conclusion reached simply because all the variables associated with radiance transfer are continuous rather than discrete. Indeed, the discontinuous distribution of λ_{max} values may well represent the limit to which λ_{max} adaptation by opsin evolution can be achieved in response to selective pressure for the spectral tuning. Thus, natural selection does not *cause* the λ_{max} clusters (which are caused by biophysical or biochemical constraints) but natural selection *will* determine the cluster into which a given photoreceptor visual pigment will fall. Given the potential for fine tuning λ_{max} by using a mixture of different chromophores, not to mention the potential for refinement of absorptance spectra by altering the size of photoreceptor outer segments, the discontinuity in λ_{max} due to opsin sequence may not be particularly restrictive in terms of optimisation of spectral tuning of photoreceptors. Nevertheless, it is likely that the adaptation of λ_{max} in response to selective pressure associated with environmental variables and specific visual tasks cannot be perfect because of the clustering phenomenon.

8.7 Visual pigments of aquatic animals: background

There is now a considerable body of information about the visual pigments of aquatic animals, much of this gleaned from studies inspired by the supposition that the severe limits on the spectrum of light available for vision underwater would be reflected in adaptations of visual sensitivity. Among the invertebrates, most work has centred on the crustacea, species of which have a great diversity of visual pigments, utilising both retinal and 3,4-dehydroretinal as chromophores, and ranging in number from single pigments in many amphibious and deepsea decapods to at least 11 visual pigments in stomatopods (see Marshall, Kent and Cronin, in press). Crustaceans have evolved visual pigments with wavelengths of peak absorbance encompassing UV sensitive rhodopsins to far-red sensitive porphyropsins and, together with great variety in optical design, have possibly the most diverse visual systems of any group (for reviews, see

Marshall, Kent and Cronin, in press; and Land, in press). Within the phylum Mollusca the visual pigments of several cephalopods have been measured, principally by extraction of visual pigments (Seidou, 1990), but also by microspectrophotometry (Michinomae *et al.*, 1994), and several species have been found to have multiple visual pigments separated in different types or parts of rhabdoms (Seidou, 1990). Surprisingly, however, there is no evidence for colour vision in any cephalopod (Hanlon and Messenger, 1996; Marshall and Messenger, 1996), despite their obvious and spectacular ability (to a human observer) to modulate their dermal coloration for intraspecific display and crypsis. Most cephalopods utilise retinal as their visual pigment chromophore (Seidou *et al.*, 1990) but two species, *Watasenia scintillans* and *Bathyteuthis* sp. also use 3,4-dehydroretinal (Matsui *et al.*, 1988; Kito *et al.*, 1992a) and several deepwater species (the squids *Pyroteuthis* sp. and *Pterigioteuthis* sp. and the octopus *Japetella* sp.) also employ 4-hydroxyretinal (Kito *et al.*, 1992a). Visual pigments based on 4-hydroxyretinal (which could be named 'cephalopsins' in view of the class of molluscs in which they were first discovered, although this may be etymologically slovenly) have λ_{max} values at shorter wavelengths than rhodopsin pigments based on the same opsin. For instance the cephalopsin analogue of a rhodopsin having a λ_{max} of 501 nm has a λ_{max} at 486 nm, but otherwise the two classes of pigments have very similar normalised absorbance spectra and very similar extinction coefficients (Kito *et al.*, 1992b). All available evidence suggests that the rhodopsin and cephalopsin visual pigments of deep-sea cephalopods have λ_{max} values spectrally tuned to confer high sensitivity at different depths in the open ocean (in *Watasenia scintillans*, for instance the cephalopsin, rhodopsin and porphyropsin pigments have λ_{max} values of 470, 484 and 500 nm respectively; Matsui *et al.*, 1988; Seidou *et al.*, 1990). Such a range of pigments demonstrates an interesting evolutionary convergence with both deep sea crustaceans (see Marshall, Kent and Cronin, in press) and mesopelagic deep-sea fishes (e.g. Partridge *et al.*, 1988, 1989, 1992) that also have visual pigments with similar λ_{max} values but employing retinal rather than 4-hydroxyretinal as the chromophore.

More work has been conducted on the visual pigments of aquatic vertebrates. Mammals are not a taxon with well developed colour vision and aquatic mammals such as seals and cetaceans have retinae dominated by rods. Little information is available about cone visual pigments in these animals but data on rod visual pigments demonstrate that there is λ_{max} convergence with deep-sea fishes: deeper diving mammals having visual pigments with more shortwave λ_{max} values (Lythgoe, 1972a). Few birds have had their visual pigments measured (Bowmaker *et al.*, 1997) and the only aquatic, or semi-aquatic species so far investigated are the mallard duck (*Anas platyrhynchos*; Jane and Bowmaker, 1988) and the Huboldt penguin (*Spheniscus humboldti*; Bowmaker and Martin, 1985). No firm conclusions can be made about avian visual pigment adaptations to underwater vision based on such limited data, although there is evidence that diving birds may have different proportions of cone types and oil droplets with cut-off wavelengths different from most birds (Partridge, 1989). Unlike birds and mammals, which only use retinal as a chromophore, aquatic reptiles have evolved both rhodopsin and porphyropsin visual pigments, with the latter being more common, apparently, in freshwater species (e.g. the terrapin *Pseudomys scripta elegans*; Liebman and Granda, 1971). The predominance of porphyropsins in freshwater reptiles is mirrored in amphibians which, as tadpoles, tend to have predominantly porphyropsins in their retinas, but at metamorphosis and emergence from water, switch more or less to rhodopsins (see Knowles and Dartnall, 1977; Lythgoe, 1979). Species that remain aquatic, such as *Xenopus laevis*, are exceptions to this rule and retain porphyropsins as adults (Crescitelli, 1973; Witkovsky *et al.*, 1981).

Naturally, the most obvious group of vertebrates to study when searching for adaptations of visual pigments to the underwater world are fishes: they have had a long evolutionary history and in that time have radiated into *ca.* 25000 extant species (Nelson, 1994). Fishes occupy an extraordinary range of aquatic habitats that includes near terrestrial visual worlds (e.g. mudskippers such as *Boleophthalmus*) to habitats that never include the water surface (e.g. most deep-sea fish); from lives in hot soda lakes at 44 °C (e.g. *Tilapia* in East Africa) to sub zero temperatures in the Antarctic (e.g. Nototheneid cod ice fish); from high altitude freshwater lakes (e.g. cyprinodontids in Lake Titicaca, 3812 m above sea level) to the extreme depths of the ocean (e.g. species in the Marianas Trench, 11 km deep); from existences in the clearest oceans to lives in highly turbid estuaries and rivers (Nelson, 1994). Many data are available about the visual pigments of both the rods and cones of many hundreds of species of fishes and these data have been regularly and exhaustively reviewed by numerous authors, usually from a tacitly adaptationist perspective (e.g. Lythgoe 1972b, 1979, 1980, 1984; Loew and Lythgoe, 1978; Ali and Wagner, 1975; Crescitelli *et al.*, 1985; McFarland and Munz, 1975b; Munz and McFarland 1973, 1975, 1977; Knowles and Dartnall, 1977; Levine and MacNichol 1979; Bowmaker 1990, 1991a, 1991b, 1995; Partridge, 1990; Douglas *et al.*, 1995; Partridge *et al.*, 1988, 1989, 1992). In almost all cases these authors have described correlations between the observed visual pigment data and some aspect, or aspects, of the underwater visual environment. In most cases, however, there has been little effort to separate the potentially confusing combination of phylogenetic and ecological information, far less to eliminate this confusion by use of an appropriate statistical method (cf. Harvey and Pagel, 1991).

Nevertheless, two approaches have partly circumvented this problem. The first has been to reduce the potential phylogenetic confusion by concentrating on closely related, sometimes endemic species. Such studies include the work of Muntz (1976) on Lake Malawi cichlids; that of Bowmaker *et al.*, (1994) and Hunt *et al.*, 1996 on Lake Baikal cottoids; and that of Lythgoe *et al.*, 1994) on Great Barrier Reef snappers of the genus *Lutjanus*. The merits of this approach are further discussed in Section 8.8. The second approach is to suggest a hypothesis that defines the 'best' visual pigments for certain visual tasks and to construct models, with predictive values, with which the visual pigment data may be compared.

Such models are not new in the context of underwater vision and one of the first, the Sensitivity Hypothesis, simply predicts that, because of the importance of photon capture in any visual task, visual pigments should have λ_{max} values spectrally located to maximise photon catch during adaptively important visual tasks undertaken in the visual environment in which the animal lives. Underwater, the environmental attenuation of light ensures that light levels are often low and catching as many photons as possible is of paramount importance. The eyes of fishes show numerous adaptations towards the maximisation of photon catch including optics designed to give a bright retinal image (Fernald, 1988) and tapeta which reflect light back through the retinal photoreceptors for a 'second chance' of being absorbed. For similar reasons a number of fishes, including crepuscular predators (Lythgoe, 1988), but more especially deep-sea fishes (Ali and Anctil, 1976), have evolved particularly large photoreceptors in which the unusually long path-length that light must travel in the outer segments, often as much as 100 μm, substantially increases the chances of photon absorption. Deep-sea fishes have also increased the density of visual pigment in the outer segments (Munk, 1966; Partridge *et al.*, 1989) and specific optical densities may be as high as 0.028 μm^{-1} in these animals (Partridge *et al.*, 1989), very much higher than the range 0.010 μm^{-1} to 0.015 μm^{-1} that is typical of shallow living fishes (MacNichol *et al.*, 1973; Knowles and Dartnall, 1977; Partridge, 1986). Given these anatomical adaptations it is surprising that there is very little support for the Sensitivity Hypothesis, except in the case of deep-sea species, and most animals do not have visual pigments, even those subserving vision at scotopic light levels, with λ_{max} values spectrally located to maximise photon catch (Partridge, 1990). One possible reason for this is that 'sensitivity' in visual systems is actually best expressed in terms of the signal to noise ratio (S/N) associated with the visual task. Although S may be increased by spectral tuning of λ_{max} this will only be advantageous if N is increased by a lesser proportion, otherwise S/N may, in fact, decrease. At low light levels the noise, both from the light signal itself (channel, or photon noise) and from the visual system (dark noise), imposes limits on S/N and hence on visual performance. Although visual pigments are extremely stable and rarely signal spontaneously (i.e. without photon absorption) visual pigment molecules are found in retinas in such high numbers that some visually significant spontaneous activity is inevitable. For this reason an animal's ability to detect dim lights is limited by noise in its own visual system, with rhodopsin being, in all probability, the primary noise source (Barlow *et al.*, 1993). Indeed, as the volume of an outersegment essentially determines the number of visual pigment molecules present in a photoreceptor, rhodopsin dark noise may well be an important constraint on photoreceptor size. Moreover, if some visual pigments are less susceptible to spontaneous isomerisations than others then there will be selection pressure for these pigments regardless of $\lambda_{max,}$ especially in photon-limited environments. On thermodynamic grounds visual pigments with longer wavelength λ_{max} values are more susceptible to spontaneous isomerisations due to thermal excitation (Barlow, 1957) and empirical evidence suggests that the most stable visual pigments do indeed have shortwave λ_{max} values (Firsov and Govardovskii, 1990). In photon limited environments noise prone visual pigments will be unacceptable, even if their spectral sensitivities better match the available light than other, less noisy, pigments if their effective S/N is lower.

It is quite possible that the evolution of visual pigments for scotopic vision has been driven largely by the need to maximise S/N by the employment of noise-free visual pigments (see Loew and Archer, in press). Nevertheless, an alternative explanation of why visual pigments may not be spectrally located to maximise sensitivity is provided by the Contrast Hypothesis (Lythgoe, 1966, 1968). Most visual tasks involve the detection of changes in light levels, either with time or in space, and the larger the change the easier the task of detection (Land, 1981). Spatial contrasts may be greater in regions of the spectrum where radiances are less than at other wavelengths and in these cases visual pigments with λ_{max} values 'offset' from the wavelengths of greatest light flux may confer advantages in contrast detection tasks. In the context of the visual task depicted in Figure 1 the contrast between the observed radiance from fish 2 and that of the background can be calculated by computing the photon catch of photoreceptors receiving light from the two sources. The visual pigment that provides the greatest visual contrast (C″) is then given by:

$$C'' = (I_t - I_b)/I_b \qquad [7]$$

Where I_t is the photon catch of a photoreceptor in the retina of fish 1 receiving light from target fish 2 and I_b is the catch of a photoreceptor receiving horizontal background spacelight. Relative photon catches can be approximated as:

$$I_t = \sum_{\lambda=300}^{\lambda=800} P(\lambda)\, T(\lambda)\, {}_tR(\lambda,z) \qquad [8]$$

$$I_b = \sum_{\lambda=300}^{\lambda=800} P(\lambda)\, T(\lambda)\, {}_bR(\lambda) \qquad [9]$$

Where $P(\lambda)$ is the spectral absorptance of the photoreceptors, $T(\lambda)$ is the spectral transmission of intraoccular tissues such as the lens and cornea, and ${}_tR(\lambda)_z$ and ${}_bR(\lambda)$ are spectral radiances from target and background respectively, as shown in Figure 1, although absolute catches require other variables to be taken into account.

Using such methods it can be shown that visual contrasts are often greater for visual pigments with λ_{max} values offset from the wavelengths of maximum irradiance and several authors have discussed the advantages of such 'offset' visual pigments in contrast enhancement (Lythgoe, 1966, 1968, 1979; McFarland and Munz, 1975b; Munz and McFarland, 1977; Levine and MacNichol, 1979; Loew and Lythgoe, 1978). In general, 'offset' visual pigments increase the contrast of light targets seen against a darker background spacelight, but when a dark target is seen against a lighter background a visual pigment most sensitive to I_b will maximise contrast. Once again, however, it is important to appreciate that the contrast signal C'' is subject to noise due to the inherent fluctuations in I_t and I_b due to photon noise. Thus some 'offset' visual pigments may appear to produce a very high contrast but in fact have a photon capture rate that is so low that the contrast signal is too noisy to be visually useful. Because I_t and I_b follow Poisson distributions the noise associated with these signals can be readily calculated (Land, 1981) and thus the S/N of C'' can be computed (see Lythgoe and Partridge, 1991) *as long as* I_t and I_b have been measured in absolute units (photons s^{-1}) not relative ones (see Rodieck 1998).

This approach can be extended to consider multiple visual pigments and has been used by Lythgoe and Partridge (1991) to predict the best pairs of rhodopsins for the discrimination of targets on the basis of luminosity and colour contrast ('chromaticity'). The model, which considers dichromats living in green coastal waters, utilises the statistics of photon capture by cones and predicts that a λ_{max} of 535 nm to 560 nm would be best for brightness discriminations, depending on the objects considered and viewing conditions. For 'chromaticity' discriminations the model predicts that a shortwave sensitive cone (λ_{max} *ca.* 440 nm – 470 nm) should be paired with a longer-wave cone having a λ_{max} of approximately 530 nm. The predictions of this model compare well with actual data from temperate coastal fishes which generally have two cone types: a single shortwave sensitive cone with a λ_{max} between 440 nm and 470 nm, and double cones with λ_{max} values between 520 and 540 nm (Lythgoe, 1984; Lythgoe and Partridge, 1991).

8.8 Case studies

Investigations that seek to understand the ways in which visual systems vary among species living under different optical environments, require that due account be taken of other sources of influence such as phylogenetic constraint and non-environmental sources of selection (e.g. predation avoidance, foraging etc.). In the last decade a few investigations have attempted to control for phylogeny by examining the visual systems of closely related species that live in optically different habitats. Bowmaker *et al.* (1994) conducted a microspectrophotometric survey on an endemic group of 17 Cottoid species in Lake Baikal. Species within this group of fish partition the lake into different habitats, principally by depth range. Thus the most shallow dwelling species (e.g. *Cottus kessleri* and *Paracottus kneri*) live at depths between 1 m and 5 m, whereas the pelagic species (e.g. *Comephorus dybowskii* and *Comephorus baicalensis*) live at various relatively species specific depth ranges between 300 m and 1500 m. The total depth range occupied by these fishes is correlated with dramatic changes in optical habitat, from full spectrum sunlight in the shallow regions to dark and nearly homochromatic in the deepest habitats. The results of the survey by Bowmaker and colleagues illustrate that species living at different depths have different sets of visual pigments, this variation appearing to be related to the wavelength of maximum irradiance at different depths. Shallow dwelling species, living in waters with wavelengths of maximum transmission between 550 nm and 600 nm, have double cones with λ_{max} values between 500 nm and 545 nm. As depth increases, the λ_{max} values of the longwave sensitive double cones found in deeper living species shift to shorter wavelengths, between 500 nm and 521 nm, and the wavelength of maximum transmission of the water at such depths shifts to *ca*. 490 nm. A similar shift occurs in the λ_{max} of the single blue cones, with shallow species (e.g. *Cottus kessleri*) having a λ_{max} around 449 nm, whereas in the deeper species (e.g. *Limnocottus griseus*) the blue cone has a λ_{max} close to 431 nm. There is also a shift to shorter wavelengths in rod λ_{max} values of the deeper dwelling species: shallow dwelling cottoids having rod λ_{max} values ranging from 510 – 520 nm whereas abyssal species have rod λ_{max} values between 480 nm and 500 nm.

The assumption underlying the study of closely related species is that they all share a relatively recent common ancestor, and therefore differences observed among the descendants are likely to be driven by different selection pressures that have operated during and since speciation. The factor most likely to be responsible for the changes in visual sensitivity among the Baikal cottoid fishes is the variation in available light at different depths. The shift in the λ_{max} values of the double cones to shorter wavelengths with increasing depth gives credibility to the selective role that the environment has on photoreceptor sensitivity, since the λ_{max} of the double cones is correlated with a shift in wavelength of maximum irradiance of the environment. Such results lends support to the 'Sensitivity Hypothesis' which suggests that there is likely to be selection pressure operating on photoreceptors to maximise photon capture (see Section 8.7 and Loew and Archer, in press). The shortwave shift in blue cone λ_{max} values, however, appears to contradict the predictions of the Sensitivity Hypothesis for two reasons: (1) the light available at these depths is found in the longer wavelength region of the spectrum, and (2) the total amount of light available at these depths is well below known cone threshold levels. Why, therefore, would blue cone λ_{max} values be shifted to a part of the spectrum where there are fewer photons available in the downwelling spectrum, and why bother with cone photoreceptors at all in so dim an environment? These questions suggest that other selection pressures may be involved in the observed shortwave shifts of rod and cone photoreceptor sensitivities with increasing depth. One possible explanation is that shortwave photoreceptors are likely to have less thermally induced noise (see Section 8.7 and Loew and Archer, in press). In visual conditions where the signal to noise (S/N) ratio is under selection for maximisation but where the inherent signal (S) is unavoidably low, such as in deeper regions of Lake Baikal, there may be high selection pressures for photoreceptor sensitivity that maximises S/N by minimising noise (N). Undoubtedly our understanding of this issue would be greatly enhanced by further measurements of the intrinsic noise (Barlow *et al.*, 1993) of rod and cone photoreceptors containing visual pigments of different λ_{max}.

The study of the visual pigments of Baikal Cottoids is likely to become a classic in the ecology of vision and is considerably strengthened by the associated work on the molecular biology of Cottoid opsins (Hunt *et al.*, 1996) which explains how the visual pigments of these animals are tuned by opsin sequence. Nevertheless, the species examined are relatively diverse in both their taxonomy and in their visual behaviour. This problem can be overcome, at least partly, by the study of very closely related species that exhibit behavioural similarities and have superficially similar visual tasks while living in habitats with contrasting environmental conditions. This approach was taken by Lythgoe *et al.* (1994) who surveyed the visual pigments of 12 species of snappers (genus *Lutjanus*) living in different optical habitats in the Australian Great Barrier Reef. This study attempts to isolate the effect of the environment on photoreceptor sensitivity by focusing on congeners with similar foraging habits and behaviour. The results show that congeners living in different light environments have divergent visual pigment combinations. Lutjanid species that occupy the outer reef environments, in which the irradiance spectrum is dominated by short wavelengths (maximum transmission *ca.* 490 nm), have three cone classes with relatively shorter wavelength λ_{max} values (e.g. *L. kasmira* has cone λ_{max} values at 430 nm, 487 nm, and 518 nm). In contrast, Lutjanids living in the more heavily DOM dominated waters of the near-shore estuaries have cones with relatively long wavelength λ_{max} values (e.g. *L. fulviflamma* has cone λ_{max} values at 505 nm, 534 nm, and 568 nm). The study attempted to test the predictions of the Sensitivity Hypothesis by comparing the λ_{max} data with calculations of the λ_{max} of the rhodopsin visual pigment that would confer the greatest sensitivity in each optical habitat. The λ_{max} measurements reveal that all but one of the 12 Lutjanid species have double cone λ_{max} values spectrally positioned such that they would be more than 90% as effective at capturing photons as the most sensitive possible visual pigment.

This spectral location of double cone λ_{max}, tuned to confer high sensitivity to the spectrum of available light, also appears to be the case for temperate coastal marine fish (Lythgoe and Partridge, 1991) as well as the Lake Baikal cottoids. The placement of the λ_{max} values for the blue sensitive photoreceptors, however, is not as simple to understand. Lythgoe *et al.* (1994) calculated rates of photon capture by shortwave sensitive cones for the 12 Lutjanid species and determined their photon capture rates to be only 20% – 40% of that of the other cone types in any given optical habitat. The suggested explanation for this difference is that, whereas the double cones offer sensitivity to the downwelling spectrum, the blue cones may have a greater role in hue discrimination when light is not limiting. How the placement of blue cone λ_{max} is determined by such a putative need for hue discrimination is not well understood, and any attempts to reveal their role will require detailed knowledge of relevant (i.e. adaptively important) visual tasks.

The identification of such visual tasks, particularly those associated with foraging, predator avoidance, and reproductive behaviour will provide information essential to the further understanding of the spectral tuning of λ_{max}. Implicit in such studies is the identification of visual tasks that confer some fitness advantage. The analysis of such tasks in terms of photon capture by photoreceptors will involve measurement of radiance spectra from the visual targets involved in defined visual tasks, and the use of these data in models that consider such issues as how λ_{max} values vary in the optimisation of sensitivity and contrast (see Sections 8.2 and 8.7). Such data are not easy to obtain, particularly as fitness advantage related to these tasks also needs to be properly quantified, but such an approach is now both timely and essential if further evaluation of the adaptive pressures on visual pigment λ_{max} is to be achieved.

8.9 Visual pigment plasticity

The study of closely related species, such as those detailed above, at least partially eliminates the potentially confusing interaction of phylogeny and ecology in studies focused on the understanding of visual pigment adaptation. Nevertheless, another route offers even greater potential: the study of the adaptive value of visual pigment shifts within individual species. As explained in section 8.6, the λ_{max} of a visual pigment is determined both by the amino acid sequence of its opsin and by the nature of the chromophore employed. Many species of fish exhibit λ_{max} plasticity during their lives, achieving this plasticity both by opsin shift and by chromophore exchange. Often such shifts are correlated with environmental changes, with migration between different photic environments (e.g. catadromous and anadromous species), or with ontogenetic events.

Many species of freshwater teleosts, including the speciose cyprinids, have retinae dominated by porphyropsin visual pigments, but have the potential for the incorporation of rhodopsin pigments. The best studied example is probably the rudd (*Scardinius erythrophthalmus*) in which both rod and cone visual pigments have been shown to be based on mixtures of retinal and 3,4-dehydroretinal chromophore (Whitmore and Bowmaker, 1989; for review see Bowmaker, 1995). In general wild rudd have retinas dominated by porphyropsins, with rod pigments having λ_{max} values close to 543 nm and longwave cones most sensitive at *ca.* 620 nm, but in summer these shift by chromophore exchange to have λ_{max} values close to 510 nm and 565 nm respectively. The adaptive significance of such shifts is, however, far from clear. Although correlations with seasonal shifts in spectral irradiance in rudd habitats have been noted, the λ_{max} shifts are generally contradictory to the observed irradiance shifts, suggesting that sensitivity maximisation is not the primary selection pressure. Also, different populations of rudd exhibit different degrees of chromophore shift, and the degree of annual chromophore shift diminishes with increasing age such that older animals have, more or less, only porphyropsin pigments. The explanation most focused on the concept of adaptation to a vital visual task is that of Muntz and Wainwright (1978) who suggest that the rudd's seasonal visual pigment shift may be due to annual changes in feeding strategy since rudd tend to feed closer to the surface in summer than in winter. However, the validity of this suggestion, and its adaptive significance, remains to be tested.

Shifts in rhodopsin:porphyropsin ratios are also found in migratory fishes and, once again, it is tempting to identify environmental correlates with visual pigment sensitivity. One of the best documented is the European eel (*Anguilla anguilla*) which hatches in the Sargasso Sea, and drifts as a leptocephalus larva to the coastline of Europe where it metamorphoses into a glass elver and enters the river systems. At this stage the rod visual pigment shifts from a rhodopsin with a λ_{max} close to 502 nm, typical of marine fishes, to a porphyropsin with a λ_{max} close to 523 nm typical of freshwater fish (Wood *et al.*, 1992). Eels remain in freshwater as 'yellow eels' for several years before migrating downstream, during which time the chromophore exchange is reversed and is accompanied by the expression of a new opsin, giving a rhodopsin pigment with a λ_{max} of 482 nm (Wood and Partridge, 1993). The new pigment essentially gives the now 'silver eel' the retinal sensitivity of a deep-sea fish (Partridge *et al.*, 1989) and may, in this case, be an adaptation for visual sensitivity during the deep-water migration across the Atlantic.

The new 'deep-sea' opsin of the eel, which has been sequenced and putative spectral tuning sites identified (Archer *et al.*, 1995), has been shown to be incorporated into the outersegments of existing rods (Wood and Partridge, 1993). Thus for some time two different opsins are found in the single photoreceptors. Rod opsin exchange is also reported in the cardinal fish, *Apogon brachygrammus*, (Munz and McFarland, 1973) which, as a larva, has a rhodopsin with a λ_{max} at 482 nm and, as an adult, has a rhodopsin with a λ_{max} of 494 nm. However, it is unknown whether this involves expression of different opsins in the same photoreceptors. In cone photoreceptors such dual expression of opsin genes has been suggested in the longwave sensitive cones of the guppy (Archer and Lythgoe, 1990) and is also implicated in the shortwave cones of the pollack (*Pollachius pollachius*; Shand *et al.*, 1988). In the latter species, which is a coastal marine gadoid, the duplex retina contains rhodopsin pigments both in its rods (λ_{max} 489 nm) and in its two types of cone pigment. The identical paired cones contain a visual pigment with a λ_{max} at 521 nm, but the single cones contain a shortwave pigment which, in small fish has a λ_{max} close to 410 nm, but in larger fish has a λ_{max} close to 460 nm. The shift in shortwave cone λ_{max} occurs during the development of the fish as they move into deeper water, where there is less shortwave light, and change their diet from mainly plankton to mainly benthic crustaceans. Whether it is the change in light environment *per se*, or the change in diet, and thus visual task, that has led to the evolution of this shortwave cone shift has yet to be determined, but clearly there is potential in such a study.

A similar uncertainty surrounds the adaptive significance of the ontogenetic changes in the visual pigments of the tropical reef goat fish, *Upeneus tragula* (Shand, 1993, 1994). As pelagic, surface living larvae this species has double cones with one member containing a longwave sensitive rhodopsin (λ_{max} 580 nm), and the other containing a relatively shorter-wave sensitive pigment (λ_{max} 487 nm). At metamorphosis to a benthic subadult extensive somatic morphological changes are accompanied by a shift in the longwave cone to a λ_{max} at 530 nm, although the other cone and rod pigments do not appear to shift. Once again, the adaptive advantage of this visual pigment shift may be purely due to environmental factors (there is somewhat less longwave light even at the relatively shallow benthic depths of the metamorphosed animals than are found at the surface) or it may be related to the visual tasks associated with the new mode of life.

Similar ontogenetic shifts in opsin expression are known in a number of other fishes, including salmonids in which the UV sensitive pigments appear to be lost and regained at different phases of life (see Beaudet and Hawryshyn, in press); the temperate pomacentrid *Chromis punctipinnis* (McFarland and Loew, 1994) which has a violet-sensitive cone as a juvenile (λ_{max} 417 nm) but loses this in favour of a UV-sensitive cone as an adult (λ_{max} *ca.* 350 nm); and the flounder *Pseudopleuronectes americanus* (Evans *et al.*, 1993), which has a single cone with a λ_{max} at 519 nm as a larval fish, but which develops a duplex retina with both rods and cones as an adult, the latter cells expressing three visual pigments with λ_{max} values at 457, 531 nm and 547 nm. Such examples of visual pigment plasticity, exhibited by species with diverse phylogenies, makes it highly likely that many fish will be found to shift their visual pigments at times of development when the nature of their visual tasks change significantly. The proper investigation of such changes is, arguably, the most useful means by which the adaptive advantage of visual pigment sensitivities can be directly investigated.

8.10 Conclusions

The study of visual pigment adaptation in the aquatic environment has been an active field of research for several decades and in this time most effort has been concentrated on the matching of spectral sensitivity (λ_{max}) to environmental factors, especially spectral irradiance. Such studies have identified numerous correlations between λ_{max} and the underwater light environments, although rigorous statistical methods have not been employed and many apparent correlations may be confounded by phylogenetic factors. This problem has been minimised in some studies which focused on closely related species and has been avoided by the use of modelling, principally to determine what visual pigments should best suit some defined visual task, which has often proved to have useful predictive value. The future understanding of the way in which visual pigments are adapted to the underwater environment will, however, depend on the identification visual tasks that demonstrably affect fitness: i.e. tasks that are essential for survival and reproduction. Such an experimental approach will perhaps be most successful when the visual tasks of individual species, particularly those that experience significant developmental changes, are examined in detail.

Acknowledgements

We thank several people for reading drafts of this paper, including Jim Bowmaker, Julia Shand, Innes Cuthill, Simon Archer, and Jeremy Kent. We are particularly indebted to John Endler for thorough and constructive criticism, and Ron Douglas for other suggestions that are incorporated in this work.

References

Aidley, D. J. (1989) *The physiology of excitable cells.* 3rd Edition. C.U.P., Cambridge.

Ali, .M. A. and Anctil, M. (1976) *Retinas of fishes: an atlas.* Springer-Verlag. Berlin.

Ali, M. A. and Wagner, H.-J. (1975) Visual pigments: phylogeny and ecology. In: *Vision in fishes* (Ed. M. A. Ali), pp. 481– 516. Plenum, New York.

Archer, S. N. and Lythgoe, J. N. (1990) The visual pigment basis for cone polymorphism in the guppy, *Poecilia reticulata. Vision Res.*, 30, 225 – 233.

Archer, S. N., Endler, J. A., Lythgoe, J. N. and Partridge, J. C. (1987) Visual pigment polymorphism in the guppy *Poecilia reticulata. Vision Res.* 27, 1243 –1252.

Archer, S. N., Hope, A. J. and Partridge, J. C. (1995) The molecular basis for the green-blue sensitivity shift in the rod visual pigments of the European eel. *Proc. R. Soc. B.*, 262, 289 – 295

Baker, K. S. and Smith, R. C. (1982) Bio-optical classification of natural waters 2. *Limnol. and Oceanogr.*, 27, 500 – 509.

Baldwin, J. M. (1993) The probable arrangement of the helices in G protein-coupled receptors. *EMBO J.*, 12, 1693 –1703.

Barlow, H. B. (1957) Purkinje shift and retinal noise. *Nature,* 179, 255 – 256.

Barlow, H. B. (1982) What causes trichromacy – a theoretical analysis using comb-filtered spectra. *Vision Res.*, 22, 635 – 643.

Barlow, R. B., Birge, R. R., Kaplan, E. and Tallent, J. R. (1993) On the molecular origin of photoreceptor noise. *Nature,* 366, 64 – 66.

Beaudet, L. and Hawryshyn, C. W. (in press) Ecological aspects of vertebrate visual ontogeny. In: *Adaptive Mechanisms in the Ecology of Vision* (Eds. S. N. Archer, M. B. A. Djamgoz, E. Loew, J. C. Partridge, and S. Vallerga) Chapman & Hall, London.

Bennett, A. T. D., Cuthill, I. C. and Norris, K. J. (1994) Sexual selection and the mismeasure of color. *Am. Nat.*, 144 (5), 848 – 860.

Bowmaker, J. K, & Hunt, D. M. (in press) Molecular biology of photoreceptor spectral sensitivity. In: *Adaptive Mechanisms in the Ecology of Vision* (Eds. S. N. Archer, M. B. A. Djamgoz, E. Loew, J. C. Partridge, and S. Vallerga) Chapman & Hall, London.

Bowmaker, J. K. and Martin, G. R. (1985) Visual pigments and oil droplets in the penguin, *Spheniscus humboldti. J. Comp. Physiol. A.,* 156, 71–77.

Bowmaker, J. K. (1983) Trichromatic colour vision – why only three receptor channels. *Trends in Neuroscience*, 6(2), 41– 43.

Bowmaker, J. K. (1990). Visual pigments of fishes. In: *The visual system of fishes* (Ed. R. H. Douglas and M. B. A. Djamgoz). pp. 81–108.

Bowmaker, J. K. (1991a) Visual pigments, oil droplets and photoreceptors. In: *Vision and Visual Dysfunction Vol. 6* (Ed. P. Gouras) pp. 108 –127. MacMillan, London.

Bowmaker, J. K. (1991b) The evolution of vertebrate visual pigments and photoreceptors. In: *Vision and Visual Dysfunction Vol. 2* (Eds. J. R. Cronley-Dillon, and R. L. Gregory) pp. 63 – 81. MacMillan, London.

Bowmaker, J. K. (1995) The visual pigments of fish. In: *Progress in retinal and eye research* Vol. 15(1):1–31. Elsevier, London.

Bowmaker, J. K., Govardovskii, V. I., Zueva, L. V., Hunt, D. M. and Sideleva, V. (1994) Visual pigments and the photic environment: the cottoid fish of Lake Baikal. *Vision Res.* 34, 591– 605.

Bowmaker, J. K., Heath, L. A., Wilkie, S. E., and Hunt, D. M. (1997) Visual pigments and oil droplets from six classes of photoreceptor in the retinas of birds. *Vision Res.* 37: 2183 – 2194.

Bridges, C. D. B. (1964) Periodicity of absorption properties in pigments based on vitamin A2 from fish retinae. *Nature,* 203, 303 – 304.

Buschbaum, G. (1981) The retinal as a two-dimensional detector array in the context of color vision theories and signal detection theory. *Proceedings of the IEEE*, 69, 772 –785.

Campbell, I. D. and Dwek, R. A. (1984) *Biological Spectroscopy* Benjamin/Cummings Publishing Co., Menlo Park.

Cott, H.B. (1940) *Adaptive coloration in animals.* 508 pp. Methuen, London.

Crescitelli, F. (1990) Adaptations of visual pigments to the photic environment of the deep-sea. *J. Expl. Zool.* 5(Supl.), 66 –75.

Crescitelli, F., (1973) The visual pigment systems of *Xenopus laevis* tadpoles and adults. *Vision Res.* 13, 855 – 865.

Crescitelli, F., McFall-Ngai, M. and Horwitz, J. (1985) The visual pigment sensitivitity hypothesis: further evidence from fishes of varying habitats. *J. Comp. Physiol. A.*, 157, 323 – 333.

Cuthill, I.C. & Bennett, A.T.D. (1993) Mimicry and the eye of the beholder *Proc. R. Soc. Lond. B.*, 253, 203 – 204.

Dartnall, H. J. A. and Lythgoe, J. N. (1965) The spectral clustering of visual pigments. *Vision Res.*, 5, 81–100.

Dartnall, H. J. A. (1953) The interpretation of spectral sensitivity curves. *British Med. Bull.*, 9, 24 – 30.

Dartnall, H. J. A. (1972) Photosensitivity. In: *Handbook of Sensory Physiology*, VII/1. (Ed. Dartnall H. J. A.) pp. 122 –145. Springer Verglag, Berlin.

Dartnall, H. J. A. (1975) Assessing the fitness of visual pigments for their photic environments. In: *Vision in Fishes*, pp. 543 – 563 (Ed. M.A., Lli). Plenum Press, New York.

Denton, E.J., & Rowe, D.M. (1994) Reflective communication between fish, with special reference to the greater sand eel, *Hyperoplus lanceolatus. Phil. Trans. R. Soc. Lond.,* B Vol. 344 No. 1309, 221– 237

Douglas, R. H. and Marshall, N. J. (in press). A review of vertebrate and invertebrate ocular filters. In: *Adaptive Mechanisms in the Ecology of Vision* (Eds. S. N. Archer, M. B. A. Djamgoz, E. Loew, J. C. Partridge and S. Vallerga) Chapman & Hall, London.

Douglas, R. H., Partridge, J. C. and Hope, A. J. (1995) Visual and lenticular pigments of demersal deep-sea fishes. *J. Comp. Physiol. A.* 177, 111–122.

Duntley, S. Q. (1962) Underwater visibility. In: *The Sea,* Vol. 1, *Physical Oceanography* (ed. M. N. Hill), pp. 452 – 455. Interscience, New York.

Duntley, S. Q. (1963) Light in the sea. *J. Opt. Soc. Am.* 53: 214 – 233.

Dusenbery, D. B. (1992) *Sensory Ecology.* Freeman, New York. pp. 558.

Ebrey, T. S. and Honig, B. (1977) New wavelength dependent visual pigment nomograms. *Vision Res.* 17, 147–151.

Endler, J. A. and Théry, M. (1996) Interacting effects of lek placement, display, ambient light and color patterns in three neotropical forest-dwelling birds. *Am. Nat.* 148, 421– 452.

Endler, J. A. (1986) Defence against predation. In: *Predator-Prey Relationships, Perspectives and Approaches from the Study of Lower Vertebrates* (Eds. M. E., Feder and G. E. Lauder), pp. 109 –134. University of Chicago Press, Chicago.

Endler, J.A. (1987) Predation, light intensity and courtship behaviour in *Poecilia reticulata. Animal Behav.*, 35, 1376 –1385.

Endler, J. A. (1990) On the measurement and classification of color in studies of animal color patterns. *Biol. J. Lin. Soc. Lond.*, 41, 315 – 352.

Endler, J. A. (1991) Variation in the appearance of guppy color patterns to guppies and their predators under different visual conditions. *Vision Res.* 31, 587– 608.

Endler, J. A. (1993) The color of light in forests and its implications. *Ecol. Monog.* 63, 1–27.
Evans, B. I., Harosi, F. I. and Fernald, R. D. (1993) Photoreceptor spectral absorbancy in larval and adult winter flounder (*Pseudopleuronectes americanus*). *Visual Neurosci.* 10, 1065 –1071.
Fernald, R. D. (1988) Aquatic adaptations in fish eyes. In *Sensory Biology of Aquatic Animals* (eds. J. Atema *et al.*) pp. 435 – 466. Springer Verlag, New York
Firsov, M. L. and Govardovskii, V. I. (1990) Dark noise of visual pigments with different absorption maxima. *Sensory Systems* 4, 25 – 34 (In Russian)
Fleishman, L. J. (1992) The influence of the sensory system and the environment on motion patterns in the visual-displays of anoline lizards and other vertebrates. *Am. Nat.* 139, 36 – 61
Gates, D. M. (1980) *Biophysical Ecology,* Springer Verlag, New York.
Hanlon R. T. and Messenger, J. B. (1996) *Cephalopod Behaviour.* Cambridge University Press, Cambridge.
Hargrave, P. A., McDowell, J. H., Feldman, R. J., Atkinson, P. H., Rao, J. K. M. and Argos, P. (1984). Rhodopsin's protein and carbohydrate structure: selected aspects. *Vision Res.* 24, 1487–1499.
Harosi, F. I. (1985). Ultraviolet- and violet-absorbing vertebrate visual pigments: dichroic and bleaching properties. In: *The Visual System* (ed. A Fein and J. S. Levine), pp. 41– 55. Liss, New York.
Harvey, P. H. and Pagel, M. D. (1991). *The comparative method in evolutionary biology.* Oxford University Press, pp. 239.
Henderson, S. T. (1970) *Daylight and Its Spectrum,* Elsevier, New York.
Hunt, D. M., Fitzgibbon, J., Slobodyanyuk, S. J. and Bowmaker, J. K. (1996) Spectral tuning and molecular evolution of rod visual pigments in the species flock of cottoid fish in Lake Baikal. *Vision Res.* 36, 1217–1224.
Jacobs, G. H. (1981) *Comparative Color Vision.* Academic Press, New York.
Jane, S. D. and Bowmaker, J. K. (1988) Tetrachromatic colour vision in the duck (*Anas platyrhynchos* L.): microspectrophotometry of visual pigments and oil droplets. *J. Comp. Physiol.* A 162, 225 – 235.
Jerlov, N. G. (1976) *Marine Optics.* Elsevier, Amsterdam. pp. 231.
Kirk, J. T. O. (1994) *Light and Photosynthesis in Aquatic Ecosystems* 2nd Ed. Cambrdge University Press, Cambridge. pp. 509.
Kito, Y., Narita, K., Seidou, M., Michinomae, M., Yoshihara, K., Partridge, J. C. and Herring, P. J. (1992a) A blue sensitive visual pigment based on 4-hydroxyretinal is found widely in mesopelagic cephalopods. In: *Structures and Functions of Retinal Proteins.* (Ed. J. L. Rigaud). Vol. 221, pp. 41– -414. John Libbey Eurotext Ltd.
Kito, Y., Partridge, J. C., Seidou, M., Narita, K., Michiniomae, M. and Yoshihara, K. (1992b). The absorbance spectrum and photosensitivity of a new synthetic 'visual pigment' based on 4-hydroxy-retinal. *Vision Res.* 32, 310.
Knowles, A. and Dartnall, H. J. A. (1977). The photobiology of vision. In: *The Eye* (ed. H. Davson), pp. 1– 689. London, Academic Press.
Krebs, W. and Kühn, H. (1977) Structure of isolated bovine rod outer segment membranes. *Exp. Eye. Res.* 25, 511– 526.
Lamb, T. D. (1995) Photoreceptor spectral sensitivities: common shape in the longwavelength region. *Vision Res.* 35, 3083 – 3091.
Land, M. F. (1981) Optics and vision in invertebrates. In: *Handbook of Sensory Physiology*, vol VII/6B. *Comparative Physiology and Evolution of Vision in Invertebtrates* (ed. H. Autrum), pp. 471– 592. Springer-Verlag, Berlin.
Land, M. F. (1990) Optics of the eyes of marine animals. In: *Light and Life in the Sea* (Eds. Herring, P. J., Campbell, A. K., Whitfield, M. and Maddock, L.) pp. 149 –166. Cambridge University Press, Cambridge.
Land, M. F. (in press) Compound eye structure: matching eye to environment. In: *Adaptive Mechanisms in the Ecology of Vision* (Eds. S. N. Archer, M. B. A. Djamgoz, E. Loew, J. C. Partridge, and S. Vallerga) Chapman & Hall, London.
Levine, J. S. and MacNichol, E. F. Jr. (1979) Visual pigments in teleost fises: effects of habitat, micro-habitat, and behavior on visual system evolution. *Sensory Processes* 3, 95 –131.
Levine, J. S. and MacNichol, E. F. Jr. (1982) Color Vision in fishes. *Sci. Am.* 246, 108 –117.
Levine, J. S. (1982) Water colors: the sea as laboratory for the study of color vision. In: *The Sciences*, pp. 20 – 24. New York Academy of Science.
Liebman, P. A. and Granda, A. M. (1971) Microspectrophotometric measurements of visual pigments in two species of turtle *Pseudomys scripta* and *Chelonia mydas. Vision Res.* 11, 105 –114.
Liebman, P. A. (1972) Microspectrophotometry of photoreceptors. In: *Handbook of Sensory Physiology VI/I* (Photochemistry of Vision). (Ed. H. J. A. Dartnall) pp. 481– 528 Springer-Verlag, Berlin.

Loew, E. R. (1995) Determinants of visual pigment spectral location and photoreceptor cell spectral sensitivity, in *Neurobiology and Clinical aspects of the outer retina*. (Eds. M. B. A. Djamgoz, S. N. Archer and S. Vallerga), Chapman & Hall, London.

Loew, E. R. and Archer, S. N. (in press) Light & Photoreception. In: *Adaptive Mechanisms in the Ecology of Vision* (Eds. S. N. Archer, M. B. A. Djamgoz, E. Loew, J. C. Partridge and S. Vallerga) Chapman & Hall, London.

Loew, E. R. and Lythgoe, J. N. (1978) The ecology of cone pigments in teleost fish. *Vision Res.* 18, 715 –722.

Loew, E. R. and McFarland, W. N. (1990) The underwater visual environment, in *The Visual System of Fish* (Eds R. H. Donjlag and M. B. P. Djamgoz), Chapman & Hall, London.

Lythgoe, J. N, Muntz, W. R. A., Partridge, J. C., Shand, J. and Williams, D. McB. (1994). The ecology of the visual pigments of snappers (*Lutjanidae*) of the Great Barrier Reef. *J. Comp. Physiol.* A. 174, 461– 467

Lythgoe, J. N. and Northmore, D. P. M. (1973) Colours underwater. In: *Colour 73* pp. 77– 98. Adam Hilger, London.

Lythgoe, J. N. and Partridge, J. C. (1991) The modelling of optimal visual pigments of dichromatic teleosts in green coastal waters. *Vision Res.* 31(3), 361– 371.

Lythgoe, J. N. and Shand, J. (1983) Diel changes in the neon tetra *Paracheirondon innesi*. *Environmental Biology of Fishes*, 8, 249 – 254.

Lythgoe, J. N. (1966) Visual pigments and underwater vision. In: *Light as an Ecological Factor.* (Ed. R. Bainbridge, C. C. Evans and O. Rackman), pp. 375 – 392. Blackwell, Oxford.

Lythgoe, J. N. (1968) Visual pigments and visual range underwater. *Vision Res.*, 8, 997–1012

Lythgoe, J. N. (1972a) List of vertebrate visual pigments. In: *Handbook of sensory Physiology. Vol VII/I.* (Ed. H. J. A. Dartnall), pp. 604 – 624. Springer-Verlag, Berlin.

Lythgoe, J. N. (1972b) The adaptation of visual pigments to their photic environment. In: *Handbook of Sensory Physiology. Vol. VII/I* (Ed. H. J. A. Dartnall), pp. 566 – 603. Springer-Verlag, Berlin.

Lythgoe, J. N. (1979) *The Ecology of Vision*. Clarendon Press, Oxford.

Lythgoe, J. N. (1980) Vision in fishes: ecological adaptations. In: *Environmental Physiology of fishes* (Ed. M. A. Ali), pp. 431– 445. Plenum, New York.

Lythgoe, J. N. (1984) Visual pigments and environmental light. *Vision Res.* 24, 1539 –1550.

Lythgoe, J. N. (1988) Light and vision in the aquatic environment. In Sensory Biology of Aquatic Animals (eds. J. Atema *et al.*) pp. 57– 82. Springer Verlag, New York.

Lythgoe, J. N. and Partridge, J. C. (1989). Visual pigments and the acquisition of visual information. *J. Exp. Biol.*, 146, 1– 20.

MacNichol, F. E. Jr., Feinberg, R. and Harosi, F. I. (1973) Colour discrimination processes in the retina. In: *Colour 73*, pp. 191– 251. Adam Hilger, London.

MacNichol, F. E. Jr., Kunz, Y. W., Levine, J. S., Harosi, F. I. and Collins, B. A. (1978) Ellipsosomes: organelles containing cytochrome like pigment in the retinal cones of certain fishes. *Science* 200: 549 – 551.

Marchetti, K. (1993) Dark habitats and bright birds illustrate the role of the environment in species divergence. *Nature*, 362, 149 –152.

Marshall, N. J. and Messenger, J. B. (1996) Colour-blind camouflage. *Nature* 382, 408 – 409.

Marshall, N. J., Kent, J. K, and Cronin, T. C. (in press). Visual adaptations in crustaceans: spectral sensitivity in diverse habitats. In: *Adaptive Mechanisms in the Ecology of Vision* (Eds. S. N. Archer, M. B. A. Djamgoz, E. Loew, J. C. Partridge and S. Vallerga) Chapman & Hall, London.

Matsui, S., Seidou, M., Horiuchi, S., Uchiyama, I. and Kito, Y. (1988) Adaptation of a deep-sea cephalopod to the photic environment. *J. Gen. Physiol.*, 92, 55 – 66

McFarland, W. N. and Lowe, E. R. (1994) Ultraviolet visual pigments in marine fishes of the family Pomacentridae. *Vision Res.*, 34, 1393 –1396.

McFarland, W. N. and Munz, F. W. (1975a). Part II: The photic environment of clear tropical seas during the day. *Vision Res.* 15, 1071–1080.

McFarland, W. N. and Munz, F. W. (1975b). Part III: The evolution of photopic visual pigments in fishes. *Vision Res.* 15, 1071–1080.

McFarland, W. N. and Munz, F. W. (1976). The visible spectrum during twilight and its implications to vision. In *Light as an Ecological Factor*, vol. 2 (ed. G. C Evans et al.) pp. 249 – 270. Blackwell, Oxford.

Michinomae, M., Masuda, H., Seidou, M. and Kito, Y. (1994) Structural basis for wavelength discrimination in the banked retina of the firefly squid *Watasenia scintillans*. *J. Exp. Biol.* 193, 1–12.

Morel, A. (1974) Optical properities of pure water and pure sea water. In: *Optical Aspects of Oceanography* (ed. N. G. Jerlov and E. Steeman Nielsen), pp. 1– 24. Academic Press, London.

Munk, O. (1966) Ocular anatomy of some deep-sea teleosts. *Dana Reports*, 70, 1– 62.

Munsell Color Company. (1976) *Munsell book of color.* Glossy finish ed. 2 vols. Munsell/Macbeth/ Koomorgenh, Baltimore.
Muntz, W. R. A. (1976) Visual pigments of cichlid fishes in Lake Malawi. *Vision Research* 16, 897– 903.
Muntz, W. R. A. (1978) A penetraçáo de luz águas de rias amazônicos. *Acta Amazonica* 8, 613 – 619.
Muntz, W. R. A., and Mouat, G. S. V. (1984) Annual variation in the visual pigments of brown trout inhabiting lochs providing different light environments. *Vision Res.* 24, 1575 –1580.
Muntz, W. R. A. and Wainwright, A. W. (1978) Annual cycles in the light environment and visual mechanisms of fishes. In: *Rhythmic Activity in Fishes* (ed. J. E. Thorpe) pp. 105 –129. Academic Press, London.
Munz, F. W. and McFarland, W. N. (1973) The significance of spectral position in the rhodopsins of tropical marine fishes. *Vision Res.* 13, 1829 –1874.
Munz, F. W. and McFarland, W. N. (1975). Part I: Presumptive cone pigments extracted from tropical marine fish. *Vision Res.* 15, 1045 –1062.
Munz, F. W. and McFarland, W. N. (1977) Evolutionary adaptations of fishes to the photic environment. In: F. Crescitelli (Ed.) *Handbook of sensory physiology, VIII/5: The visual system in vertebrates.* Springer-Verlag, New York.
Munz, F. W. and Schwanzara, S. A. (1967) A nomogram for retinene$_2$-based visual pigments. *Vision Res.* 7, 111–120.
Nelson, J. S. (1994) *Fishes of the World* (3rd edition) Wiley, New York. pp. 600.
Neumeyer, C. (1992) Tetrachromatic color vision in goldfish: evidence from color mixture experiments. *J. Comp. Physiol.* A 171, 639-649.
Osorio, D. and Vorobyev, M. (1997) Sepia tones, stomatopod signals and the uses of colour. *Trends in Ecol. Evol.* 12, 167–168.
Palacios, A. G., Goldsmith, T. H. and Bernard, G. D. (1996) Sensitivity of cones from a cyprinid fish (*Danio aequipinnatus*) to ultraviolet and visible light. *Visual Neurosci.* 13, 411– 421.
Partridge, J. C. (1986) *Microspectrophotometry of vertebrate photoreceptors.* Ph.D. Thesis, University of Bristol.
Partridge, J. C. (1989). The visual ecology of avian cone oil droplets. *J. Comp. Physiol. A* 165, 415 – 426.
Partridge, J. C. (1990) The colour sensitivity and vision of fishes. In: *Light and Life in the Sea* (Eds. Herring, P. J., Campbell, A. K., Whitfield, M. and Maddock, L.) pp. 167–184. Cambridge University Press, Cambridge.
Partridge, J. C. and De Grip, W. J. (1991). A new template for rhodopsin (A1 based) visual pigments. *Vision Res.* 31, 619 – 630.
Partridge, J. C., Archer, S. N. and Lythgoe, (1988). Visual pigments in the individual rods of deep-sea fish. *J. Comp. Physiol. A* 162, 543 – 550.
Partridge, J. C., Archer, S. N. and van Oostrum, J. (1992) Single and multiple visual pigments in deep-sea fishes. *J. Mar. Biol. Assoc. UK* 72, 113 –130.
Partridge, J. C., Shand, J., Archer, S. N., Lythgoe, J. N. and van Groningen-Luyben, W. A. M. (1989). Interspecific variation in the visual pigments of deep-sea fshes. *J. Comp. Physiol. A*, 164, 513 – 529.
Piantanida, T. P. (1991) Molecular biology of colour vision. In: *Vision & Visual Dysfunction Vol. 6* (Ed. P. Gouras) pp. 90 –107.
Pokorny, J., Shevell, S. K., Smith V. C. (1991) Colour appearance and colour constancy. In: *Vision and Visual Dysfunction Vol. 6* (Ed. P. Gouras) pp. 43 – 61. CRC Press, Boca Raton.
Preisendorfer, R.W. (1964) A model for radiance distribution in natural hydrosols. In: *Physical aspects of light in the sea* (ed. J. E. Tyler), pp. 51– 59. University of Hawaii Press.
Prieur, L. and Sathyendranath, S. (1981) An optical classification of coastal and coeanic waters based on the specific spectral absorption curves of phytoplankton pigments, dissolved organic matter and other particulate materials. *Limnmol. Oceanogr.* 26, 671– 688.
Rodieck, R.W. (1998) *The First Steps in Seeing.* Sinauer, Sunderland MA.
Rose, A. (1973). *Vision human and electronic.* Plenum, New York.
Schertler, G. X., Villa, C. and Henderson, R. (1993) Projection structure of rhodopsin. *Nature,* 362, 770 –772.
Seidou, M., Sugahara, M., Uchiyama, H., Hiraki, K., Hamanaka, T., Michiomae, M., Yoshihara, K. and Kito, Y. (1990) On the three visual pigments in the retina of the firefly squid, *Watsenia scintillans. J. Comp. Physiol. A* 166, 769 –773.
Shand, J. (1993) Changes in the spectral absorption of cone visual pigments during the settlement of the goatfish *Upeneus tragula* – the loss of red sensitivity as a benthic existence begins. *J. Comp. Physiol. A.* 173, 115 –121.
Shand, J. (1994) Changes in retinal structure during development and settlement of the goatfish *Upeneus tragula. Brain Behav. Evol.* 43, 51– 60.

Shand, J., Partridge, J. C., Archer, S. N., Potts, G. W. and Lythgoe, J. N. (1988). Spectral absorbance changes in the violetblue sensitive cones of the juvenile pollack, *Pollachius pollachius*. *J. Comp. Physiol. A* 63, 699 –703.

Smith, R. C. (1974) Structure of solar radiation in the upper layers of the sea. In: *Optical Aspects of Oceanography* (ed. N. G. Jerlov and E. Steemann Nielsen), pp. 95 –119. Academic Press, London.

Spence, D. H. N., Campbell, R. M. and Chrystaol, J. (1971) Spectral intensity in some Scottish freshwater lochs. *Freshwater Biol.* 1, 321– 337.

Stavenga, D. G., Smits, R. P. and Hoenders, B. J. (1993) Simple exponential functions describing the absorbance bands of visual pigment spectra. *Vision Res.* 33, 1011–1017.

Tyler, J. E. and Smith, R. C. (1970) *Measurement of Spectral Irradiance Under Water.* Gordon & Breach, New York

Vogt, K. (1983) The chromophores of the visual pigments in some insect orders. *Zeitschrift Naturforschung*, 39c, 196 –197.

Vorobyev, M. and Osorio, D. (1998) Receptor noise as a determinant of colour thresholds. *Proc. R. Soc. Lond., B,* 265: 315 – 358.

Wagner, H.-J. (1990) Retinal structure of fishes. In: *The visual system of fish* (Ed. R. H. Douglas and M. B. A. Djamgoz). pp. 109 –158, Chapman and Hall, London.

Whitmore, A. V. and Bowmaker, J. K. (1989) Seasonal variation in cone sensitivity and shortwave absorbing visual pigments in the rudd, *Scardinius erythrophthalmus*. *J. Comp. Physiol. A*, 166: 103 –115.

Witkovsky, P., Levine, J. S., Engbretson, G. A., Harris, G. and MacNichol, E. F. Jr. (1981) A microspectrophotometric study of normal and artificial visual pigments in the photoreceptors of *Xenopus laevis*. *Vision Res.* 21, 867– 873.

Wood, P. and Partridge, J. C. (1993). New opsin induced in the retinal rods of the eel (*Anguilla anguilla* L.). *Proc. Roy. Soc. B.*, 254, 227– 232

Wood, P., Partridge, J. C. and De Grip, W. J. (1992). Rod visual pigment changes in the elver of the eel *Anguilla anguilla* L. measured by microspectrophotometry. *J. Fish Biol.* 41, 601– 611.

Wyszecki, G. and Stiles, W. S. (1967) *Colour Science.* Wiley, New York. pp. 628.

9
Visual adaptations in crustaceans: Spectral sensitivity in diverse habitats

J. MARSHALL*, J. KENT†, and T. CRONIN‡
*V.T.H.R.C., Physiology and Pharmacology, The University of Queensland, Brisbane, Qld 4072, Australia.
†School of Biological Sciences, University of Bristol, Woodland Road, Bristol BS8 1UG, England.
‡Department of Biological Sciences, University of Maryland Baltimore County, Maryland 21228, USA.

9.1 Introduction

In all aspects of their biology, crustaceans exhibit a remarkable degree of variation. A cursory examination of a barnacle, a giant spider crab and a copepod, for instance, might fail to place these organisms together in the same subphylum. This diversity of form reflects the ecological demands of different lifestyles and habitats and extends to include a great variety of eye designs (Land 1981a, and see Chapter 2). As stated neatly by Land in his review of (1984) 'Although one usually thinks of the Crustacea as linked to the insects by the presence of an exoskeleton and compound eyes, there is actually a much greater diversity of eye types in the Crustacea than in the insects or any other invertebrate group' (Figure 1).

Crustaceans are found in all reaches of the marine environment from the sea shore to the deepest ocean, in fresh water habitats and on the land. They may live in bright tropical sun, or in niches such as caves and the deep ocean where daylight does not penetrate. Not all crustaceans possess eyes but where present they are built for a variety of tasks to be performed at different times of the day. The need for effective vision under these broad ranging conditions has produced several different eye design strategies (Land 1984 and 1990). This chapter addresses only one aspect of crustacean eyes, their spectral sensitivities, and explores the ecological explanations behind trends discovered so far.

As well as the huge range of intensities (10^{-13} photons cm^{-1} sr^{-1} nm^{-1} to less than 10 photons cm^{-1} sr^{-1} nm^{-1}) experienced by crustaceans in different niches, the colour and polarization content of light also varies considerably. Perception of polarized light is outside the boundaries of this chapter but may be even more important to many crustacean species than colour (Waterman 1981, Marshall *et al.,* 1991a and see conclusion for further comments). Crustaceans experience light which varies in its chromatic content from the broad 'white' sunlight of the reef top to the narrow, blue waveband of oceanic mid-water, green of coastal

S.N. Archer et al. (eds.), Adaptive Mechanisms in the Ecology of Vision, 285–327

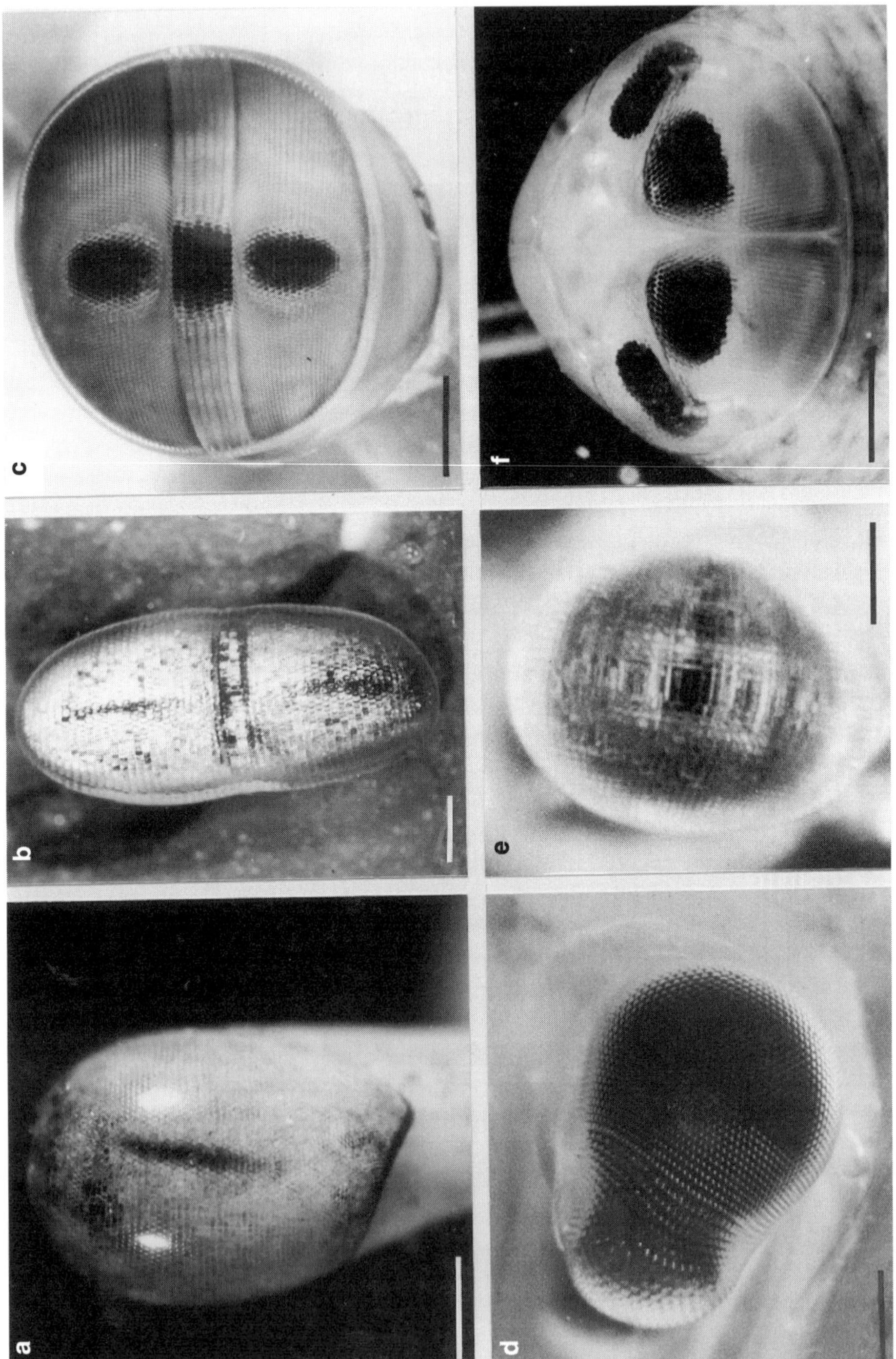
a
b
c
d
e
f

Figure 1 The compound eyes of a variety of crustaceans. **a**) The apposition compound eye of *Uca lactea annulipes* a fiddler crab which spends much of its time out of water in bright sunlight. Scale 0.5 mm. (Photograph by Jochen Zeil). **b**) The eye of *Squilla ensigera* a stomatopod inhabiting deep water. Squilloid stomatopods lack the mid-band rows specialised for colour and polarized light perception and only have 2 rows in their mid-band. The photoreceptors of these rows are similar in design to those of the peripheral regions. Scale 1.0 mm. **c**) The eye of *Odontodactylus scyllarus* a stomatopod with a six-row mid-band containing specialised ommatidia. Scale 1 mm. **d**) A divided eye in the deep-sea euphausid *Stylocheiron maximum*. The smaller 'half' of the eye with enlarged facets usually looks upwards. See Chapter 2 Figure 7d for another example of a divided euphausid eye. Scale 0.5 mm. **e**) The superposition compound eye of *Machrobrachium sp*. a freshwater shrimp which feeds largely at night. Eyes of this design are generally around 1000x more sensitive than apposition eyes. Scale 1.0 mm. **f**) Dorsal view of the head of *Brachyscelus ovoides*, a hyperiid amphipod with eyes divided into dorsal upward looking and ventral downward looking halves, each equipped with its own retina (the 4 dark regions seen within). See Chapter 2 Figure 7 for further examples of divided hyperiid eyes. Scale 1.0 mm water and even predominantly yellow and red in some freshwater environments (Jerlov 1976, Lythgoe 1976). In common with fish, their neighbours in all these habitats, one would expect crustacean spectral sensitivities to be adapted to suit these different requirements. While this is true to some extent, given their diversity of eye design and based on present data, most crustaceans seem to be surprisingly conservative in their spectral sampling capabilities. Many possess only 1 or 2 photoreceptor types, irrespective of habitat, and in a comparison between species, these are restricted to relatively narrow spectral sensitivity ranges. Thus, most have a photoreceptor population maximally sensitive to blue/green light between 480 nm and 540 nm and in some species a secondary set of UV/blue photoreceptors, sensitive near to 400 nm (Figure 2).

As a cautionary note, most of the crustaceans examined to date fall into three large groups: the crabs, stomatopods and deep-sea shrimps. Either in taxonomic or environmental terms this is not a broad cross section and more work is needed on the subphylum. The fish, which will frequently be called upon by way of comparison in this chapter, have received more attention and a number of ecological trends with regard to the tuning of their visual pigments are well documented (Lythgoe 1966, 1972 and 1979, Partridge 1990 and see Chapter 8). At present it seems that fish, which occupy almost precisely the same niches as the crustacea, possess a greater diversity of photoreceptor types (between 1 and 5) with broader ranging peak sensitivities (340 to 560 nm; see Partridge 1990, Bowmaker 1990). Most crustaceans have apparently settled on a basic plan including a single or more frequently two visual pigments, which operate under a number of different conditions. It may be that the reason for this lies, not with limitations in spectral space but in factors concerned with polarisation vision.

A few exceptions to this conservatism are known but by far the most spectacular are the stomatopods (mantis shrimps). These hoplocarid crustaceans appear in the fossil record and have been in their own limb of the evolutionary tree for 400 million years (Caldwell and Dingle 1976). During this time they followed a very different visual strategy to other crustaceans. Species from 2 of the 5 existing superfamilies possess up to 16 different photoreceptor types in their eyes, each with its own spectral sensitivity, and sample light from 312 nm to over 700 nm (Marshall 1988, Cronin and Marshall 1989a and b, Cronin *et al.*, 1994a, Marshall *et al.*, 1994). This diversity of spectral sensitivity classes far exceeds that of any other animal known. The latter part of this chapter examines reasons behind this and compares the habitat-linked variations that exist within the stomatopods. First, the ecology of spectral sensitivities of other crustaceans is reviewed.

It is important not to lose sight of the individuality of animals when trying to draw broad conclusions about visual ecology. Many live in microhabitats, which both in geographical and behavioural terms may not be obvious to us. Furthermore, the spectral sensitivity of photoreceptors is not just a function of environmental factors. It is notable for instance that most crustacean visual

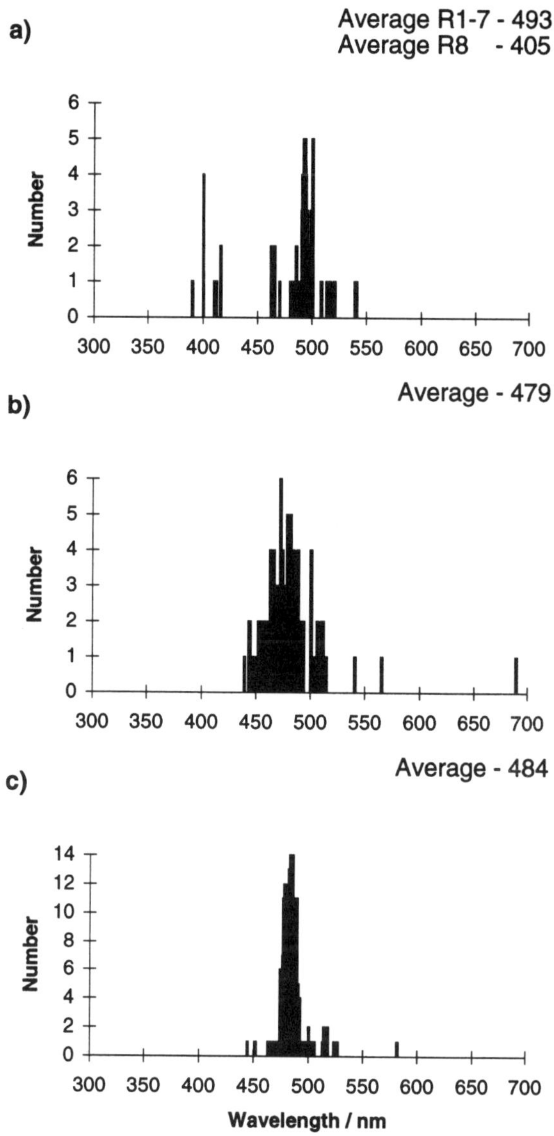

Figure 2 Features of life in the deep-sea. **a**) λ_{max} values for the visual pigments of deep-sea decapod, euphausid and mysid shrimps. Values close to 400 nm are from the R8 cells where present and those close to 500 from R1-7 cells. (For data source see Table 1 also Kent pers.obs.). **b**) The λ_{max} of light emitted from bioluminescent sources in the deep-sea (data from Widder *et al.,* 1983 and Latz *et al.,* 1988). **c**) Visual pigment λ_{max} values of deep-sea fish (data from Partridge *et al.,* 1989 and 1992, Douglas and Partridge 1997 and Douglas and Partridge pers. comm.)

pigments have λ_{max} values below 512 nm and this may be due to a number of factors such as thermal instability of visual pigments or evolutionary inertia (see Goldsmith 1991 for a useful discussion of this). Apparent phylogenetic restrictions to λ_{max} positions are known in crustaceans (Cronin 1990) as they are in fish (Douglas and Partridge 1997). Also, as noted by Dartnall and Lythgoe (1965), visual pigment sensitivity peaks from a variety of animals seem to be clustered around certain regularly spaced modal values (see also Partridge *et al.,* 1989 and Figure 5d). These are probably set by interactions between the molecular components of visual pigments. As a result of all these factors, any interpretation of photoreceptor spectral sensitivity just in terms of ecological factors is fraught with complications. As John Lythgoe has shown us, however, it is lots of fun. Before examining environmental adaptations in the crustacea, some background information is needed.

9.2 Methods of measurement and some theoretical background

Spectral sensitivities have been obtained from at least 228 photoreceptors in a minimum of 121 species of crustaceans using four principal methods: a) Microspectrophotometry (MSP), where a direct measure of the absorbance of the photoreceptors visual pigment is obtained in situ (eg Cronin and Forward 1988, Cronin and Marshall 1989a); b) Behavioural measures, usually involving phototactic action spectra (eg Stearns 1975, Hyatt 1975, Stearns and Forward 1984). These are generally not very precise, making a comparison between behavioural and other methods problematic; c) Electrophysiological responses, either of the whole eye using electroretinogram (ERG – eg Frank and Case 1988a and b), extracellular electrode (eg Smith and Macagno 1990) or intracellular recording from photoreceptors (eg Martin and Mote 1982); and d) Visual pigment extract measured using a spectrophotometer (eg Goldsmith and Bruno 1973). A non-invasive physiological method examining retinal pigment movements in response to different wavelengths of light has also been used (Cronin 1989). It is now recognised that it is hard to extract invertebrate visual pigments without including non-visual pigments so this method has largely been dropped.

In nearly all invertebrates including crustaceans, visual pigment molecules are situated on stacks of tube shaped membranes or microvilli. Microvilli constitute the photoreceptive portion (rhabdom) of the retinal (retinular) cells and are the direct equivalent of the plate-shaped visual pigment bearing membranes of outer segments in vertebrate rods and cones. In common with vertebrates, most crustacean visual pigments consist of the 11-*cis* isomer of the vitamin A derived retinal (A_1-retinal) combined with a protein 'opsin' to make rhodopsin (Goldsmith 1972). Freshwater crayfish also combine 3,4-dehydroretinal (A_2-retinal) with an opsin in order to produce a longer wavelength absorbing visual pigment or porphyropsin (Suzuki *et al.,* 1984), an adaptation also found in freshwater fish (Bowmaker 1990, Partridge 1990 and see Chapter 8) and amphibia (Bridges

1972). The spectral position of the peak absorbance, or λ_{max}, of visual pigment absorbance curves (eg Figure 3) is set by the specific combination of chromophore and opsin. Various opsins combined with retinal will produce a population of visual pigments with different λ_{max} values. The same opsins combined with 3,4-dehydroretinal produce a different population of visual pigments with longer λ_{maxes} (Bridges and Yoshikama 1970, Partridge 1990).

A characteristic of invertebrate visual pigments, which sets them apart from the vertebrates, is the formation of an intermediate, non-dissociated stable state

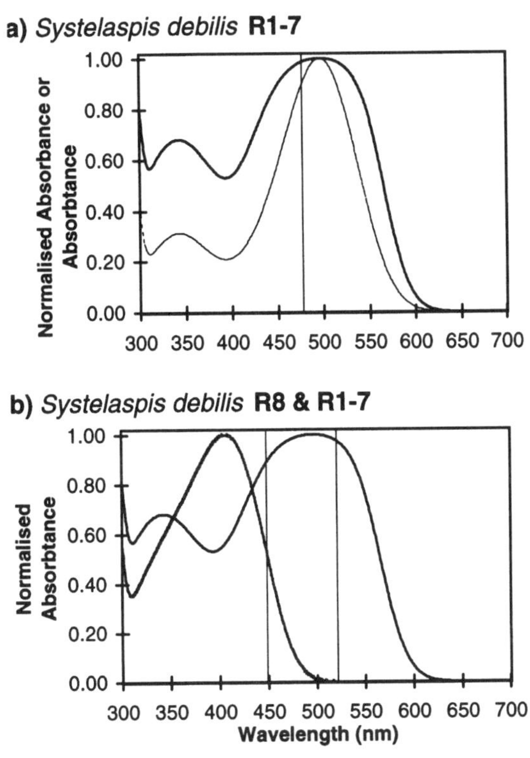

Figure 3 The spectral sensitivities of the deep-sea shrimp *Systellaspis debilis* (λ_{max} values in Cronin and Frank 1997). Nomograms are calculated according to Stavenga *et al.,* (1993) **a**) R1-7 cell visual pigment absorbance nomogram (thin line). Calculated absorptance curve (where for each wavelength absorptance $A = 1\text{-}10^{(-a.l.u)}$ where a is the absorbance of the visual pigment at λ, l is photoreceptor length (R1–7 cells, 150 μm Shelton *et al.,* 1992) and *u* the absorbance of the photoreceptor per unit length measured at 0.01 μm^{-1} by Cronin and Frank (1997). Absorptance is a more realistic estimate of visual sensitivity, than simply using the λ_{max} of the visual pigment or its nomogram. The thin vertical line is at the wavelength of peak transmission (475 nm) in clear oceanic water (Jerlov type I water – Jerlov 1976) **b**) Absorptance curves of both R8 and R1–7 photoreceptor types in S. debilis. To calculate R8 absorptance, a rhabdom length *l* of 40 μm and a unit length absorbance of 0.006 μm^{-1} were used (Shelton *et al.,* 1992). The thin vertical lines enclose the spectral region within which the λ_{max} of most bioluminescent emissions lie (Figure 5)

known as metarhodopsin. This is not visually active but can be reconverted to rhodopsin on the absorbance of photons (Stavenga and Schwemer 1984, see chapter 4 for further details). Metarhodopsin has a similarly shaped absorbance spectrum to the rhodopsin or porphyropsin it is derived from but the λ_{max} is generally displaced to one side (Stavenga and Schwemer 1984). For instance the crab *Gecarcinus lateralis* has a rhodopsin contained in its main rhabdom with λ_{max} at 485 nm while the metarhodopsin λ_{max} is situated close to 491 nm (Cronin and Forward 1988). Active photoreceptors always contain some metarhodopsin and its presence may substantially change their spectral sensitivity by acting as a filter (Stavenga and Schwemer 1984, Goldsmith 1978a).

Coloured photostable pigments are also found in and around the rhabdom which filter the spectrum of light reaching the photoreceptor (Goldsmith 1978a, Marshall *et al.,* 1991b and Chapter 4). This may sharpen or even shift the photoreceptors spectral sensitivity away from the λ_{max} value of the visual pigment alone (Snyder *et al.,* 1973, Goldsmith 1978a, Cronin and Marshall 1989a). Ultimately the spectral sensitivity of a crustacean photoreceptor is dependent on multiple factors: the visual pigment's λ_{max} and density, filtering by a variety of photostable pigments in series or lateral to the photoreceptor, self screening within each photoreceptor by rhodopsin and its photoproducts such as metarhodopsin, screening by the visual pigments of distally placed photoreceptors, reflecting pigments or tapeta and the dimensions of the photoreceptor itself. Filters in eyes and their effect on spectral sensitivity are the subject of chapter 4 in this volume and are therefore not discussed further here, aside from a section on their ecology in stomatopod eyes. It is worth emphasising, however, that λ_{max} values of visual pigments alone and the best fit nomogram curves one often sees illustrating them are not inclusive of important factors. As a result, they may be a misleading representation of the real spectral sensitivity of the photoreceptor considered as a whole. Some peak sensitivity values of stomatopod photoreceptors are shifted over 100 nm to longer wavelengths from the visual pigment λ_{max} by filtering (Cronin and Marshall 1989, Figures 3 and 10).

Long photoreceptors, which are common in invertebrates to increase sensitivity, will tend to result in broadened sensitivity functions (Snyder *et al.,* 1973). Using λ_{max} values only, especially with respect to evolutionary optimisation or ecological tuning, is short-sighted. Unfortunately visual pigment λ_{max} values are often the only information available and therefore form the basis of any speculation. From here on the term λ_{max} is used for photoreceptor peak sensitivity based only on MSP data and S-max is used for data from electrophysiological or behavioural studies, or sensitivity maxima calculated to include filtering.

Reflecting their widespread distribution, the greatest body of knowledge on spectral sensitivities is from crustaceans with compound eyes. Copepod and ostracod crustaceans possess simple eyes with a retina situated behind a single optical element (Land 1984) but little is known of their visual pigments (Stearns and Forward 1984). Compound eyes come in two varieties, superposition or apposition, along with some hybrids between these found in galatheid crabs (Nilsson 1989) and odd examples which seem to be 're-evolving' the simple eye (Nilsson and Modlin 1994, Land *et al.,* 1979). Both main designs generally consist of many hundred or thousand individual units or omatidia (Land 1981a, Nilsson 1989). Largely due to their optical arrangement, superposition eyes are around 1000 times more sensitive than apposition eyes and are found in animals living in dim, scotopic conditions (see Chapter 2 for a summary of their differences). Beneath the dioptric apparatus of both eye types, the rhabdomeric part of the omatidium consists of 8 retinular cells in nearly all crustaceans (Eakin 1972). Seven of these (R1-7) generally form the bulk of the rhabdom while the remaining cell (R8) is smaller and is situated either distally or basally in the rhabdom making the whole photoreceptor a 2-tiered structure. It is the R1-7 cells that contain the blue/green sensitive visual pigments (480 – 540 nm) and R8 cells the UV/violet sensitive visual pigments (near 400 nm). It is not too over speculative to assume this relationship holds for most crustacea with 2 photoreceptor types. Interestingly the axons projecting from R8 cells terminate in a different optic neuropil (the medulla externa) to those of the R1-7 cells (which terminate in the lamina ganglionaris, Martin and Mote 1982, Cummins and Goldsmith 1981), potentially suggesting different functions for these 2 cell populations.

9.3 Adaptations to the deep-sea

In common with deep-sea fish it is curious that, although amongst the hardest to catch, one of the best understood group of crustaceans are the deep-sea shrimps (eg Frank and Case 1988a&b, Shelton *et al.,* 1992). These include the decapod sergestids (eg Hiller-Adams *et al.,* 1988), oplophorids (eg Cronin and Frank 1996) and penaeids (eg Fernandez (1965), the mysid *Gnathophausia ingens* (Frank and Case 1988), and the euphausids (Kampa 1955 and 1965, Denys 1982, Denys and Brown 1982, Widder *et al.,* 1987) (Table 1). Recently one of us has undertaken a major MSP survey of deep-sea shrimps examining over 35 species.

The habitat of these animals extends form a few hundred meters to benthic depths well below 1000 m and should be considered as several niches rather than a single one as it often is. It is also a mistake to treat the mesopelagic depths as 'monochromatic' but this is historically the case, partly because it looks blue to us and partly because it has previously been quantified with relatively insensitive instruments. As elegantly demonstrated by Frank and Widder (1996), there are UV wavelengths, for instance, which penetrate to several hundred meters and thus well into the in the habitat of the deep-sea shrimps. Also, different behaviours and modes of life exist. Some mesopelagic decapods such as *Systellaspis debilis* and *Oplophorus spinosus*, for example, are part of the massive diurnal vertical migration undertaken by deep-sea shrimp, fish and other animals (Frank and Case 1988). Others such as *Notostomus gibbosus* and *Acanthephyra curtirostris* apparently stay at the same depth (Frank and Widder 1994b) or reside

Table 1:

Species	λ_{max} (MSP) S_{max}/nm	Method of study	Author(s)	Year
Subphylum Crustacea				
Class Branchiopoda				
Subclass Diplostraca				
Order Cladocera				
Daphnia magna	348, 434, 525 & 608	EON	Smith and Macagno	1990
Subclass Sarsostraca				
Order Anostraca				
Artemia salina	410	ERG	Hertel	1977
Class Maxillopoda				
Subclass Copepoda				
Acartia tonsa	450 – 520	BP	Stearns & Forward	1984
Subclass Cirripedia				
Order Thoracica				
Suborder Balanomorpha				
Balanus amphitrite	532	ERG	Hillman, Dodge, Hochstein, Knight & Minke	1973
	532	MSP	Minke & Kirschfield	1978
Balanus balanoides (nauplius)	510 – 530	BP	Barnes & Klepal	1972
Balanus eburneus	532	ERG	Hillman, Dodge, Hochstein, Knight & Minke	1973
	532	MSP	Minke & Kirschfield	1978
Class Malacostraca				
Subclass Eumalacostraca				
Superorder Holpocarida				
Order Stomatopoda				
Superfamily Squilloidea				
Cloridopsis dubia	510	MSP	Cronin, Marshall & Caldwell	1994c
Squilla empusa	517	MSP	Cronin, Marshall & Caldwell	1994c
	360 & 500	OP	Cronin	1989
	535 – 555	IC	Schiff	1963

Continues on next page

Table 1: Continued

Species	λ_{max} (MSP) S_{max}/nm	Method of study	Author(s)	Year
Superfamily Gonodactyloidea				
Gonodactylus oerstedii	400 to 551 (11)	MSP	Cronin & Marshall	1989b
	325	MSP	Cronin, Marshall, Quinn, & King	1994b
	400 to 695 (11)	CS	Cronin, Marshall, Caldwell & Shashar	1994a
	312 to 710	IC	Marshall and Oberwinkler	unpubl.
Gonodactylus aloha	400 to 551 (11)	MSP	Cronin, Marshall & Caldwell	1996a
Gonodactylus curacaoensis	400 to 520 (8)	MSP	Cronin, Marshall & Caldwell	1996a
Psuedosquilla ciliata	400 to 539 (11)	MSP	Cronin & Marshall	1989b
	400 to 680 (11)	CS	Cronin, Marshall, Caldwell & Shashar	1994b
Hemisquilla ensigera	414 to 535 (11)	MSP	Cronin, Marshall, Caldwell & Shashar	1994a
	440 to 625 (11)	CS	Cronin, Marshall, Caldwell & Shashar	1994a
	330	MSP	Cronin, Marshall, Quinn, & King	1994b
Odontodactylus brevirostris	402 to 535 (11)	MSP	Cronin, Marshall, Caldwell & Shashar	1994a
	400 to 623 (11)	CS	Cronin, Marshall, Caldwell & Shashar	1994a
Odontodactylus scyllarus	400 to 546 (11)	MSP	Cronin, Marshall & Caldwell	1994c
	425 to 655 (11)	CS	Cronin, Marshall & Caldwell	1994c
Odontodactylus havanensis	407 to 520 (8)	MSP	Cronin, Marshall & Caldwell	1996a
Superfamily Lysiosquilloidea				
Coronis scolopendra	407 to 533 (11)	MSP	Cronin, Marshall, Caldwell & Shasher	1994a
	425 to 620 (11)	CS	Cronin, Marshall, Caldwell & Shasher	1994b
Lysiosquilla maculata	330	MSP	Cronin, Marshall, Quinn, & King	1994a
Lysiosquilla sulcata	397 to 538 (11)	MSP	Cronin, Marshall, Caldwell & Shashar	1994a
	410 to 620 (11)	CS	Cronin, Marshall, Caldwell & Shashar	1994a
Superfamily Eucarida				
Order Euphausiacea				
Eusphausia pacifica	462	EX	Kampa	1955
	483	MSP	Widder, Hiller-Adams & Case	1987
Euphausia superba	462	EX	Denys	1982
	485	EX	Denys & Brown	1982

Continues on next page

Table 1: Continued

Species	λ_{max} (MSP) S_{max}/nm	Method of study	Author(s)	Year
Meganyctiphanes norvegica	460 – 465	EX	Fisher & Goldie	1959
	460 – 515, 490	ERG	Boden, Kampa & Abbott	1961
	488	MSP	Denys & Brown	1982
Nematoscelis megalops	465	EX	Fisher & Goldie	1961
Stylocheiron maximum	470	EX	Fisher & Goldie	1961
Thysanoessa raschii	460 – 465	EX	Fisher & Goldie	1961
Thysanopoda acutifrons	480	EX	Fisher & Goldie	1961
Order Decapoda				
Suborder Dendrobranchiata				
Penaeus duorarum	516	EX	Fernandez	1965
Family Sergestidae				
Sergestes tenuiremis	495	MSP	Hiller-Adams, Widder & Case	1988
Suborder Pleocyemata				
Infraorder Caridea				
Family Oplophoridae				
Acanthephyra curtirostris	510	ERG	Frank & Case	1988
	485	MSP	Hiller-Adams, Widder & Case	1988
Acanthephyra smithi	510	ERG	Frank & Case	1988
	491	MSP	Hiller-Adams, Widder & Case	1988
Janicella spinacauda	400 & 500	ERG	Frank & Case	1988
Notostomus gibbosus	490	ERG	Frank & Case	1988
Notostomus elegans	490	ERG	Frank & Case	1988
Oplophorus gracilirostris	400 & 500	ERG	Frank & Case	1988
Oplophorus spinosus	400 & 500	ERG	Frank & Case	1988
Systellaspis debilis	400 & 500	ERG	Frank & Case	1988
	493	MSP	Hiller-Adams, Widder & Case	1988
	410 & 498	MSP	Cronin & Frank	1996
Family Palaemonidae				
Palaemonetes palludosus	539	EX	Fernandez	1965

Continues on next page

Table 1: Continued

Species	λ_{max} (MSP) S_{max}/nm	Method of study	Author(s)	Year
Palaemonetes vulgaris	390 & 540	ERG	Wald & Seldin	1968
Infraorder Astacidea				
Astacus fluviatilis	530	MSP	Hamacher & Kohl	1981
Astacus leptodactylus	530	MSP	Hamacher & Stieve	1984
Homarus americanus	515	EX	Wald & Hubbard	1957
	515	MSP	Bruno, Barnes & Goldsmith	1977
Orconectes rusticus	530	MSP	Goldsmith	1978b
	535	MSP	Cronin & Goldsmith	1982
Procambarus clarkii	530	MSP	Cronin & Goldsmith	1982
	567 (A2)	MSP	Zeiger & Goldsmith	1989
	440 & 530	IC	Cummins & Goldsmith	1981
Procambarus milleri	522	MSP	Crandell and Cronin	1995
Camberellus schufeldtii	526	MSP	Crandell and Cronin	1995
Cambarus ludovicianus	529	MSP	Crandell and Cronin	1995
Engaeus cunicularius	522	MSP	Crandell and Cronin	1995
Infraorder Palinura				
Panulirus argus	504	EX	Fernandez	1965
Infraorder Anomura				
Pleuroncodes planipes	503	EX	Fernandez	1973
	523	IC	Fernandez	1973
Superfamily Paguridea Family Diogenidae				
Clibanarius vittatus	510	MSP	Cronin & Forward	1988
Dardanus fucosus	511	MSP	Cronin & Forward	1988
Petrochirus diogenes	508	MSP	Cronin & Forward	1988
Family Coenobitidae				
Coenobita clypeatus	508	MSP	Cronin & Forward	1988
Coenobita rugosa	491	MSP	Cronin & Forward	1988

Continues on next page

Table 1: Continued

Species	λ_{max} (MSP) S_{max}/nm	Method of study	Author(s)	Year
Family Paguridae				
Pagurus annulipes	495	MSP	Cronin & Forward	1988
Pagurus longicarpus	515	MSP	Cronin & Forward	1988
Pagurus pollicaris	515	MSP	Cronin & Forward	1988
Superfamily Galatheoidea				
Family Porcellanidae				
Polyonyx gibbesi	–	MSP	Cronin & Forward	1988
Superfamily Hippoidea				
Family Hippoidae				
Emerita talpoioda	–	MSP	Cronin & Forward	1988
Infraorder Brachyura				
Hemigrapsus edwardsii	513	EX	Briggs	1961
Leptograpsus variegtus	513	EX	Briggs	1961
	483	IC	Stowe	1980
Section Oxystomata				
Family Calappidae				
Calappa flammea	486	MSP	Cronin & Forward	1988
Hepatus epheliticus	487	MSP	Cronin & Forward	1988
Section Oxyrhyncha				
Family Majidae				
Libinia dubia	489	MSP	Cronin & Forward	1988
Libinia emarginata	493	MSP	Briggs	1961
	493	MSP	Hays & Goldsmith	1969
Section Cancridea				
Cancer irroratus	496	MSP	Cronin & Forward	1988
Pandalus borealis (nauplius)	475 – 500	ERG	Eaton & Brown	1970
Section Brachyrhyncha				
Family Portunidae				
Arenaeus cribrarius	498	MSP	Cronin & Forward	1988
Callinectes ornatus	501	MSP	Cronin & Forward	1988

Continues on next page

Table 1: Continued

Species	λ_{max} (MSP) S_{max}/nm	Method of study	Author(s)	Year
Callinectes sapidus	477	EX	Bruno & Goldsmith	1974
	440 & 508	IC	Martin & Mote	1982
	503	MSP	Cronin & Forward	1988
Carcinus maenes	508	B	Horridge	1967
	508	MSP	Bruno, Mote & Goldsmith	1973
	440 & 508	IC	Martin & Mote	1982
Ovipales stephensoni	505	MSP	Cronin & Forward	1988
Portunus spinimanus	483	MSP	Cronin & Forward	1988
Family Xanthidae				
Eurypanopeous depressus	490	MSP	Cronin & Forward	1988
Mennippe mercenaria	494	MSP	Cronin & Forward	1988
Panopeus herbstii	493	MSP	Cronin & Forward	1988
Panopeus obesus	493	MSP	Cronin & Forward	1988
Pilumnus sayi	489	MSP	Cronin & Forward	1988
Rhithropanopus harrisii	495	MSP	Cronin & Forward	1988
Family Geryonidae				
Geryon quinquendens	473	MSP	Cronin & Forward	1988
Family Grapsidae				
Sesarma cinereum	492	MSP	Cronin & Forward	1988
Sesarma reitrulatum	508	IC	Scott & Mote	1974
	493	MSP	Cronin & Forward	1988
Family Ocypodidae				
Uca pugilator	508	IC	Scott & Mote	1974
Uca pugnax	508	IC	Scott & Mote	1974
Family Gecarcinidae				
Gecarcinus lateralis	510	ERG	Lall & Cronin	1987
	487	MSP	Cronin & Foward	1988

Continues on next page

Table 1: Continued

Species	λ_{max} (MSP) S_{max}/nm	Method of study	Author(s)	Year
Superorder Peracaridia				
Order Mysia				
Gnathophausia ingens	490 & 520	ERG	Frank & Case	1988
Order Isopoda				
Ligia	330, 460 & 520	ERG	Hariyama, Tsukahara & Meyer-Rochow	1993

λ_{max} from MSP and S_{max} from a variety of methods. For the stomatopoda, both λ_{max} and S_{max} calculated from λ_{max} and intrarhabdomal filtering are given. Abbreviations as follows: MSP – microspectrophotometry, EX – visual pigment extract, OP – optical physiology, ERG – electroereintogram, EON – extracellular/optic nerce, IC – intracellular electrophysiology, B – behavioural, BP – behavioural/phototaxis, CS – calculated sensitivity close to the bottom (eg *Plesiopenaeus armatus*). Compared to other crustaceans however, the habitat of deep-sea shrimps, both in spectral and spatial terms, is a relatively simple one and we have therefore decided to examine this ecosystem first.

As even the clearest sea water filters out all downwelling light from the sun by 1000 m (Jerlov 1976, Denton 1990), animals living at increasing depths must try increasingly hard to catch photons. Several adaptations including superposition optics, tapeta and enlarged eyes bear testament to this effort (Land 1990, Shelton *et al.,* 1992, Gaten *et al.,* 1992). An alternative source of light below 1000 m, and also above this depth, is bioluminescence. Indeed anything living below 1000 m will only see light produced by other animals and the fact that many highly visual animals exist below this depth demonstrates the importance of bioluminescence (Herring 1977).

Along with optical adaptations (Land 1981a & b and 1990) one might also expect the sensitivity of visual pigments to be tuned to the blue light which penetrates best in the ocean (Jerlov 1976). Alternatively their sensitivity may match bioluminescence which is also usually blue in order to match the downwelling light or to carry information over a long distance (Figure 2b, Herring 1977, Latz *et al.,* 1988). This is the sensitivity hypothesis of Clarke (1936, and see Lythgoe eg 1972, and McFarland and Munz 1975a for excellent discussions of this). While true to an extent, in that most λ_{max} values are clustered between 450 nm and 500 nm (Figure 2), it is an oversimplification.

The λ_{max} of deep-sea shrimp photoreceptors, deep-sea bioluminescent emissions and, for comparison, the λ_{max} of deep-sea fish visual pigments are shown in Figure 2. The sensitivity hypothesis is inconsistent with these data in several ways. Firstly, the average λ_{max} of the R1-7 cells is 493 nm (Figure 2a), 18 nm longer than the light at 475 nm, which from past work we presume most abundant at depth in the open oceans (this is based on Jerlov type I water – Jerlov 1976, Lythgoe 1979, McFarland and Munz 1975a, Levine and MacNichol 1982) and therefore is apparently not optimal for maximum sensitivity. Also if crustacean visual pigments are tuned to downwelling light or bioluminescence, which has an average emission at 479 nm (Figure 2b), one would not expect the 460 nm to 540 nm diversity of λ_{max} values. The λ_{max} of fish inhabiting the same photic region averages 484 nm (Figure 2) so, although a little better matched to downwelling light or bioluminescence, the same basic problem applies (Figure 2c). A possible explanation for the crustacea emerges when one considers their sensitivity function, taking into account the length and optical density of photoreceptors, not just their λ_{max}. This is illustrated in Figure 3a for *Systellaspis debilis*, a mesopelagic decapod whose λ_{max} is at 498 nm, but by having a 150 mm long photoreceptor with a relatively high density (0.01 compared to 0.008 in most crustaceans, Cronin and Frank 1996) catches virtually as many photons at 475 nm as it does at 498 nm (Figure 3a and see caption to Figure 3 for further explanation). That is, within a 30 to 40 nm window around 475 nm, the exact

λ_{max} placement is not important. Other mesopelagic crustaceans examined anatomically also possess long photoreceptors (eg Gaten *et al.,* 1992) and perhaps this allowing some 'slack' in the exact position of their λ_{maxes}.

It is also possible that many deep-sea shrimps do not, as has been assumed, inhabit Jerlov type I water, the clearest of the sea-water types classified by Jerlov (Jerlov 1976). One has only to move to Jerlov type III water to find a minimum attenuation at 500 nm and indeed with a close examination of Jerlov's map of the worlds type I water is relatively infrequent. Moreover such a bulk classification of a water column takes no account of vertical heterogeneity or the seasonal variation in phytoplankton which will tend to push minimum attenuation to longer wavelengths (see Chapter 8 for a more in depth discussion of this).

More importantly in this discussion, we are grouping many different species with different microhabitats and different 'micro-adaptations' together. Considering species living in Jerlov I, others in Jerlov II or III, some tuned through evolution to search for bioluminescent prey emitting at 450 nm and others searching for conspecifics emitting at 500 nm, there is more than enough potential visual adaptation to account for the variation seen. One must know which visual stimuli are most relevant to the animal and very little detailed observation of this kind exists (Frank and Widder 1994 a and b, Land *et al.,* 1995). Lythgoe (1972) provides a valuable caveat in a discussion of coastal waters and fish, but this could equally be applied here:

'Any attempt to relate the visual properties of fishes to their photic environment must therefore be tentative. There must also be a certain allowance made for the tendency of authors (this one included)' (*and these ones*) 'to treat a particular set of measurements from a locality as applying equally to neighbouring areas, and all the time'.

One argument put forward by Lythgoe (1972) and championed by McFarland and Munz (1975a and b) to explain non sensitivity matched fish visual pigments is the 'offset' or contrast enhanced visual pigment. Here it is better to view a bright object against a dark background (as often occurs in surface waters) with a visual pigment whose λ_{max} is not matched to the background light. In the depths of the sea many animals are either very black, effectively black (by being red) or good reflectors of the ambient illumination by being silver (Denton 1990) and thus appear dark against the brighter background of downwelling light. Under these conditions, a matched visual pigment is best (McFarland and Munz 1975a). Applying the offset idea to deep-sea shrimp visual pigments is therefore hard.

Animals living in mid-water, either deep or a few meters below the surface, do not experience spectrally or quantally even illumination. Upwelling light is around 200x dimmer than downwelling light (Denton 1990, Land 1990) and contains relatively more shortwave light due to particulate scatter being more effective at the shortwave end of the spectrum (Figure 4 and Jerlov 1976). Essentially the underwater world is divided into the narrow, bright window of downwelling light against which silhouettes might appear and a rapidly darkening violet/blue view beneath, punctuated by occasional bioluminescent flashes. Several mesopelagic animals including fish (Land 1990), polychete worms (Wald and Rayport 1977) and crustaceans, notably certain euphausids (Land *et al.,* 1979) and hyperiid amphipods (Land 1989a), have divided their eyes into regions for looking up and regions for looking down (Figure 1). There are also adaptive optical trends with increasing depth (Land *et al.,* 1979, Land 1989a, Chapter 2), and one might predict different spectral sensitivities in the two eye regions. However it appears that the λ_{max} of visual pigments in dorsal and ventral eye regions of euphausids (Table 1, Denys and Brown 1982, Bowden *et al.,* 1961, Widder *et al.,* 1987) and hyperiids (Cronin unpublished obs) are the same. What may differ are screening pigments to tune photoreceptor sensitivity. The hyperiid *Platyscelus ovoides* for instance has red pigment in its ventral eye and yellow pigment in its dorsal eye (Land *et al.,* 1995 and see Figure 1). Alternatively, if the job of the dorsal eye is to receive downwelling light and that of the ventral eye is to see bioluminescence, which itself is usually blue (Figure 2), the spectral requirements of the dorsal and ventral eye may be the same.

An increasing number of crustaceans, including the deep-sea shrimps, are being shown to possess more than one visual pigment (Cummins and Goldsmith 1981, Martin and Mote 1982, Frank and Case 1988, Cronin and Frank 1996, JK pers obs.). The eyes of *Systellaspis debilis*, a deep-sea oplophorid, contain both R8 and R1-7 photoreceptors containing visual pigments sensitive to 410 nm and 498 nm respectively (Figure 3b, Frank and Case 1988, Cronin and Frank 1996). Although it does not possess externally divided eyes, the relative lengths of the R8 and R1-7 receptors change within the eye, such that, in receptors looking up, the R8 length / R1-7 length ratio is 0.15, whereas for downward looking receptors this ratio is 0.8. This is achieved by both a shortening of R1-7 and elongation of R8 in the ventral retina (Shelton *et al.,* 1992). The tapetum is also more extensive in the downward looking region, presumably to increase sensitivity (by reflecting light back through the photoreceptors) in the part of the eye viewing the darkest part of the visual field (Shelton *et al.,* 1992 and see Chapter 4). A reduced upward- pointing tapetum may also minimise upward eyeshine which is potentially visible to downward-looking predators (Shelton *et al.,* 1992). The eye can therefore be viewed as a ‘matched filter’ to the distribution of light around the animal. Shelton *et al.,* (1992) argue convincingly that the enlarged ventral R8 cells are spectrally adapted to the relatively more abundant UV/violet wavelengths in light scattered back from the depths (Figure 4).

S. debilis produces two spectrally differing bioluminescent emissions in the form of a defensive spew or from ventral photophores for counter-illumination camouflage (Herring 1977, Frank and Case 1988). Many of its cohabitants are also bioluminescent and the UV/violet photoreceptors of *S. debilis* (and others – Table 1) may aid in visualising this source of light or even distinguishing different bioluminescent sources which are quite spectrally diverse (Figure 2b and Frank and Case 1988, Widder *et al.,* 1983 and Latz *et al.,* 1988). This may

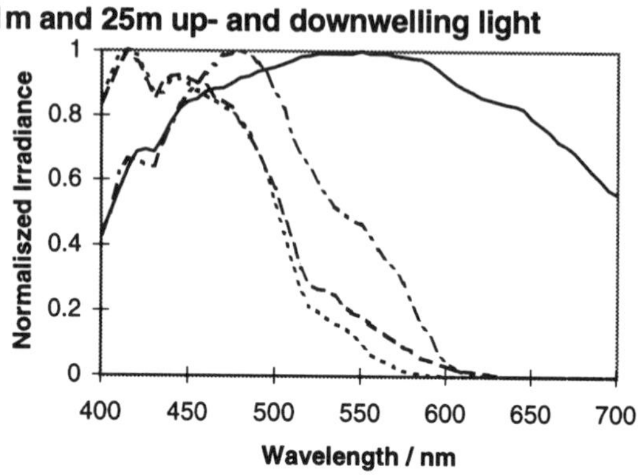

Figure 4 Spectral distribution of light (normalised photons cm^{-1} sec^{-1} nm^{-1}) looking up (zenith) and down (nadir) in clear, deep tropical sea. Solid – 1 m zenith, dashed – 1 m nadir, dotted – 25 m nadir and dot/dash – 25 m zenith. This illustrates the difference in the spectral quality of light from different angles and at different depths. Also note the relatively large amount of short wavelength light from nadir views due to particle scattering, this trend continues to deeper depths also (Gaten *et al.,* 1992). Data from McFarland and Munz (1975b)

be particularly useful in vertically migrating species such as *S. debilis* which undertake this journey in congener swarms (Gaten *et al.,* 1992). That is, being able to see each others' light may help the swarm to stick together, an advantage in the deep-sea that the visual system of *S.debilis* may be designed for. The spectral sensitivities of the R8 and R1-7 photoreceptors of *S. debilis* are optimally placed for maximum contrast (ie the stimulation ratio of the R8 to R1-7 cells) when viewing bioluminescence (Figure 3b, Cronin *et al.,* 1996a). It is not clear at this stage if this putative task is also linked to the dorsal/ventral differences of the eye of *S. debilis*. It is known that the ventral eye of hyperiid amphipods (Figure 1), is used for tracking blue bioluminescent-like sources (Land *et al.,* 1995) while the dorsal eyes are optically adapted for identifying small silhouettes against downwelling light (Land 1989a).

Further uses for dichromacy in the deep-sea are colour vision, for which no proof exists (unless one includes bioluminescence spotting as outlined above), or more simply as a depth gauge. The ratio of UV/violet and blue/green light (ie those wavelengths that the R8 and R1-7 cells are tuned to) changes with depth (and time of day) and may be of use to vertical migrators (Frank and Case 1988). There is clearly an advantage in possessing more than one visual pigment for some deep-sea fish also as out of the 183 species studied 19 are multi-pigment species (Douglas and Partridge 1997).

Rimicaris exoculata is a deep-sea caridean shrimp which lives in large swarms around volcanic hydrothermal vents, at depths in the ocean where no sunlight penetrates. The retinae of *R. exoculata* have migrated from the eye stalk and spread throughout the carapace (van Dover *et al.,* 1989). Reasons for this unusual structural reorganisation are not clear although, intriguingly, it is paralleled in other crustaceans (Land 1989b) and *Ipnops* a deep-sea fish (Marshall 1979). The function of the 'eye', however, may be to warn the shrimp as they cluster close to the super heated plumes of water (305°C) found at such sites. The visual pigment, whose λ_{max} is 500 nm, may just be sensitive enough to see the IR wavelengths produced by heat and, although the eye is non-directional lacking optical components, it could keep the shrimp from being boiled (van Dover *et al.,* 1989, Land 1989b). Very little bioluminescence is reported from the hydrothermal vents, lending support to this unusual claim.

9.5 Crabs, lobsters and vision through day and night in different habitats

Perhaps due to the edibility of some species, the spectral sensitivities of the crabs and lobsters are also quite well known (Figure 5 and Forward *et al.,* 1988, Cronin and Forward 1988, Martin and Mote 1982). Compared to deep-sea crustaceans, crabs dwell in more complex and diverse habitats and experience far more varied light regimes (Figures 4 and 6). It is therefore surprising to find that the λ_{max} of crab visual pigments is even more conservative than those of the deep sea shrimps (compare Figure 5c and 2a). The λ_{max} of their R1-7 cells average around 494 nm and spread no more than 20 nm either side of this. As mentioned in the introduction, it is possible that visual pigment λ_{maxes} do not reflect the true sensitivity and filtering by photostable filters may shift spectral sensitivity and even mediate colour vision (Goldsmith 1978a, Hyatt 1975, Leggett 1979, Kong *et al.,* 1980, Stowe 1980). These claims are, however, difficult to interpret and have yet to be confirmed. Colour vision in the fiddler crabs is a particularly attractive possibility as the males are well known for their claw waving display, often involving colour (Hyatt 1975).

As with some deep-sea shrimp, crabs such as *Carcinus maenas* and *Callinectes sapidus* also possess R8 cells containing UV/violet visual pigments (Martin and Mote 1982). It may be that colour vision in such animals could result from this potential dichromacy but again this has never been examined thoroughly. R8 cells are often small and not easy to quantify with MSP so have frequently been overlooked in the past. Their existence in crabs may be the norm rather than the exception (Lall and Cronin 1987).

Forward *et al.,* (1988) examined 27 crab species (Table 1 and Figure 6) and looked hard for environmental adaptations, examining animals from and measuring the spectral quality of light in marine and estuarine habitats. The variable nature of irradiance at various times of the day, in the sea, estuary, on land and in fresh water (all of which are typical of various crab habitats) are illustrated in Figure 6. Estuarine and the algal and silt-laden coastal waters generally transmit more light at longer, green/yellow, wavelengths than oceanic water (Jerlov 1976,

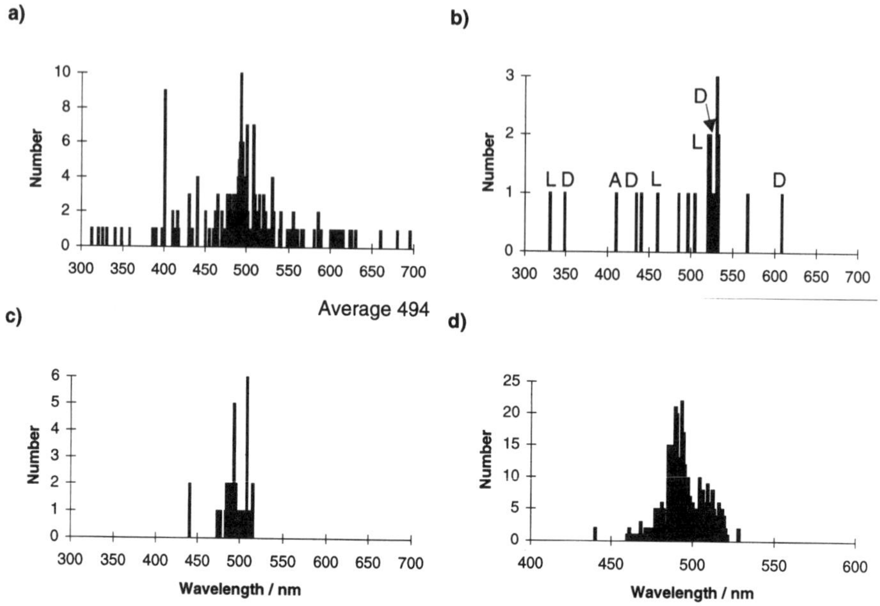

Figure 5 Sensitivity maxima either from S_{max} or λ_{max} data. **a**) All 228 photoreceptor sensitivities from the 121 species included in this study. Many of these are λ_{max} values of individual visual pigments from MSP studies: 112 are from other methods such as electrophysiology, behaviour and in the stomatopods, sensitivity calculations based on retinal anatomy and λ_{max} values (see Table 1 and Cronin and Marshall 1989a for details). The apparent peak at 400 nm is an artefact due to the limitations of MSP for examining photoreceptors with short λ_{max} values (see text for details). These photoreceptors do have sensitivities close to, but not necessarily on 400 nm. **b**) S_{max} values for crustaceans not in the stomatopods, deep-sea shrimps and crabs (Table 1). Sensitivities of *Daphnia magna* are labelled D, *Artemia salina* A and *Ligia exotica* L, all others are form crayfish, copepods or barnacles. **c**) Crab λ_{max} values from MSP only. **d**) λ_{max} of all crab photoreceptors measured. Two clusters either side of 500 nm are apparent in these data

Lythgoe 1976, Figure 6d). This is also true in many fresh water habitats (Jerlov 1976, Munz and McFarland 1977, Loew and McFarland 1990, Figure 6c). The sort of adaptation to the freshwater or estuarine photic environments one finds in fish, where λ-maxes are generally at longer wavelengths than oceans species (Lythgoe 1972 Partridge), are disappointingly absent in crabs. The following λ_{max} ranges are found, arranged here according to Forward *et al.*, (1988) by environment:

Semi-terrestrial	487– 508
Estuarine	489 – 515
Coastal	483 – 515
Continental shelf	473

Only a single species, *Geryon quinquidens* a deep living continental shelf species is well matched to the predominant wavelengths of light (475 nm) in its habitat. It is notable that most species are poorly adapted to the light spectrum at midday, so for this time of day, the sensitivity hypothesis is not well founded in crabs. That is, the visual pigments of estuarine and coastal crabs are not tuned to absorb as many photons as possible at midday.

It is not obvious if crab photoreceptor sensitivities are deliberately offset to maximise contrast. As explained above this only works for bright objects viewed horizontally against dark backgrounds (Lythgoe 1972, McFarland and Munz 1975a, Forward *et al.,* 1988). A crab sitting on the sea bed or sea shore is faced with several different angles of view, from centimetres in front to the open space above, all containing different objects and backgrounds. A matched visual pigment will actually perform better for most visual tasks here (Forward *et al.,* 1988). We must look elsewhere for reasons behind this apparent ecological mismatch. Also, at the risk of using it as a scape-goat it is worth re-emphasising the importance of treating individual species as individuals, not just 'all crabs under the sun'.

Activity period and time of day may provide an explanation. Many crab species seem better adapted to the blue-shifted light of the twilight hours (see Lythgoe 1972 and 1976, McFarland and Munz 1975a for a discussion of this in fish). This is particularly striking in the estuarine habitat (Figure 6d). Of the terrestrial species only one species *Gecarcinus lateralis* has been examined and was found to have visual pigment with a λ_{max} at 485 nm. Though it is not a good match to the light available around sunset, which peaks around 450 nm, as it is active throughout the day (Lall and Cronin 1987) it may be a good compromise between this and the daylight spectrum (Figure 6a). There is some evidence for a broader sensitivity in this species possibly due to the presence of an R8 cell (Lall and Cronin 1987).

The graphs of Figure 6 are plotted as normalised irradiances to emphasise colour differences but this can be misleading. During the day in all habitats more light is available in the 425 – 525 nm range, which encompasses crab photoreceptors, than at twilight, so the light at midday is probably not a powerful limiting factor on λ_{max} position. Also in common with the deep-sea shrimps, many crabs have photoreceptors several hundred microns long (eg Stowe 1980, Arikawa *et al.,* 1987) resulting in broad sensitivity functions. More probably, it is for vision at light-limited times that the photoreceptors of crabs, and indeed many other crustaceans, are optimised.

Crabs lobsters and other crustaceans are often more active during the dawn and dusk periods and during the night also. There is a substantial rhabdom size increase to boost sensitivity occurring every night in some species (Arikawa *et al.,* 1987). Dawn and dusk in reef habitats have been shown to be critical periods with much activity and increased predation occurring (McFarland and Munz 1975a) and this is likely to be the case elsewhere. Being able to see well at this time of the day may therefore be more important than at any other time. The spectral distribution of moonlight in clear tropical water is maximal at 570 nm (Figure 6b). No crustaceans yet examined are well matched to this, although the night active lobsters *Homarus gammarus, Homarus americanus* and *Panulirus argus* all have λ_{maxes} over 500 nm (Table 1). Perhaps during the darkest hours, other sense organs are more important.

Two patterns in crab λ_{max} values apparently not related to ecology are 'clustering' of maximal sensitivities around certain modal points and a loose phylogenetic trend. A plot of all crab photoreceptors examined (Figure 5d) reveals that their visual pigments tend to cluster near 490 nm and 510 nm. The diogenid

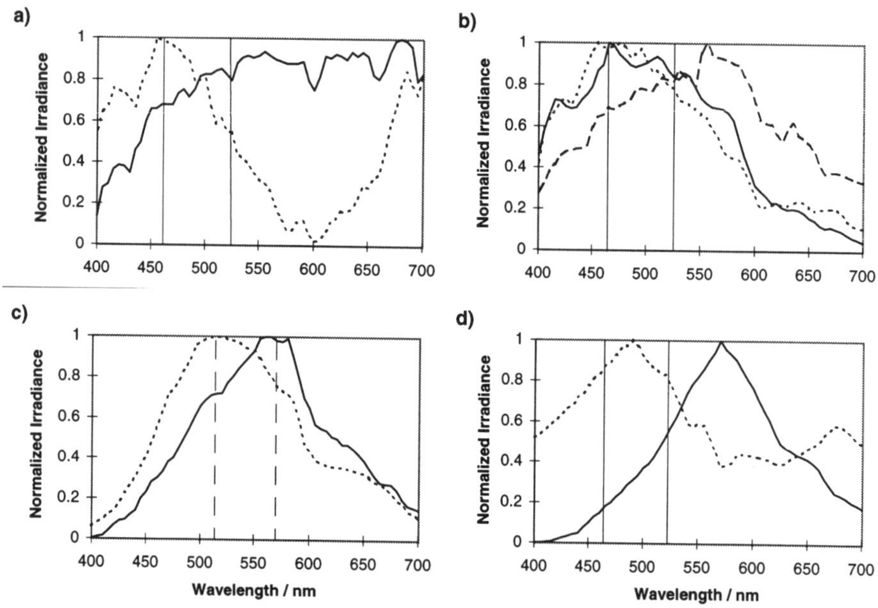

Figure 6 Downwelling light environments in a variety of crustacean habitats (normalised photons cm^{-1} sec^{-1} nm^{-1}). Thin vertical lines in (**a**), (**b**) and (**d**) enclose the spectral region within which the S and λ_{maxes} of 90% of crustacean R1-7 photoreceptor lie, taken from Figure 2(a) and include all crab R1-7 photoreceptor sensitivities. **a**) Light at midday (solid line) and at twilight after sunset, (dotted line) measured in air during the summer at 3 m in Eniwetok Atoll in the Pacific ocean. Note the relative abundance of short- and longwave light at sunset. Data from Munz and McFarland (1973). **b**) The spectral distribution of light at 3 m in Eniwetok Atoll at midday (solid line), twilight (dotted line) and moonlight (dashed line). Note how the long wavelengths present at twilight are rapidly lost in water. Many reef crustaceans are active during crepuscular or nocturnal periods. Data from Munz and McFarland (1973) and McFarland and Munz (1975a). **c**) Normalised irradiance at 3 m in lake Michigan, USA, at midday (solid line) and during twilight (dotted line). This is chosen as a 'typical' chlorophyll-rich freshwater habitat where, compared to sea-water, longer wavelengths predominate. The thin, dashed vertical lines here enclose the known λ_{max} values of crayfish visual pigments (Table 1). Data from Munz and McFarland 1977. **d**) Downwelling light at 0.5 m in the Beaufort Inlet N.Carolina, USA at midday (solid line) and twilight (dotted line)

(hermit) crabs generally fall in the latter group and brachyura the former, both groups including species from diverse habitats (Cronin 1990). Some degree of phylogenetic constraint on λ_{max} values is also known in fish (Douglas and Partridge 1997). Cluster points are noted at approximately 5 nm or 10 nm intervals in deep-sea fish eg at 442, 452, 463, 470, 477, 483, 490, 501, 505, 511, 519, and 524 nm, (Partridge *et al.,* 1989) and other vertebrates (Lythgoe 1976). These imply a certain degree of molecular constraint to the position of λ_{max} values, probably relating to the opsin-chromophore interaction which sets visual pigment sensitivities (Dartnall and Lythgoe 1965, Partridge *et al.,* 1989).

9.6 Ligia, an exception to the dichromat 'rule'

Ligia exotica is a beach dwelling isopod active during the day and therefore exposed to a broad spectrum of light (Figure 6a). Its omatidia contain only 7 retinular cells but include 3 different spectral sensitivities at 340, 460 and 520 (Hariyama *et al.,* 1993, Figure 5b). It is not known how these sensitivities integrate in the life of *Ligia* but the occurrence of a UV photoreceptor lends support to the likelihood that short wavelength receptors are common in the crustacea. Chittka (1996) has recently suggested that trichromacy such as this is the plesiomorphic condition for many crustaceans as well as insects.

9.7 Adaptations to freshwater

A number of crustacean species live in fresh water and of these crayfish and cladoceran spectral sensitivities are known. Structurally crayfish eyes are much like the superposition eyes of marine crustacea. In all species investigated the rhabdoms include an R8 cell and in *Procambarus clarkii* it is known to be UV/violet sensitive with a λ_{max} at 440 nm (Cummins and Goldsmith 1981). As with marine crustacea, few suggestions exist as to the function of these cells. The light environment in freshwater is very variable but in general peak transmission is shifted to longer wavelengths (Figure 6c). This is usually due to dissolved vegetable matter and chlorophyll in the water (Lythgoe 1966 and 1972, Chapter 1 which absorb or scatter short wavelengths preferentially. Many crayfish species are partially terrestrial and are often found in clear water, both being spectral environments where UV/violet is abundant and possibly maintaining the evolutionary pressure for their UV/violet sensitivity.

Conversely, the sensitivities of the R1-7 cells of crayfish are apparently adapted to the relatively long wavelengths which penetrate turbid fresh water (Figure 6c) and as not all shortwave photons are absorbed, the R8 cells may still function in such waters. Freshwater fish and amphibians use 3-4 dehydroretinal, the vitamin A2 derived chromophore, to construct long wavelength sensitive visual pigments, the 'porphyropsins' (Bridges and Yoshikama 1970). Intriguingly, this is also the mechanism used by the crayfish *Procambarus clarkii* (and three other crayfish genera – Suzuki and Eguchi 1987) to produce a visual pigment with λ_{max} at 562 nm (Suzuki *et al.,* 1984, Zeiger and Goldsmith 1989). How the pigment is distributed in retina is not known and it is not found in all crayfish. Nonetheless all λ_{maxes}, weather A_1 or A_2 derived, are above 520.

In common with fish such as the brown trout *Salmo trutta* (Muntz and Mouat 1984, Bowmaker 1990) or the bullfrog (Reuter *et al.,* 1971), there are seasonal variations in the amount of porphyropsin in the eyes of *Procambarus clarkii*, more being present in winter months. This is perhaps an adaptation to increased water turbidity as a result of increased rainfall in winter. Unfortunately, little information exists on the habitats and habits of any crayfish.

The water flea, *Daphnia magna*, has only 22 omatidia in its compound eye, each containing 8 rhabdomeres as found in other crustacea. Since the work of von Frisch (von Frisch and Kupelwieser 1913) followed by Smith and Baylor (1953) and Stearns (1975) a number of species of *Daphnia* have been known to respond in characteristic ways to different wavelengths of light. These 'colour dances' (Smith and Baylor 1953) apparently enable avoidance of damaging short wavelengths and swarming in areas of food (phytoplankton) and may be more akin to the wavelength specific behaviours known for some insects (Menzel 1979). A recent electrophysiological investigation of *Daphnia magna* has unearthed an astonishing 4 spectrally distinct photoreceptors within one omatidium (Schehr 1984, Smith and Macagno 1990), their S-maxes at 348, 434, 525 and 608. *Daphnia* live in the shallow depths of freshwater ponds where the spectral distribution of light may not be very different to air, but with the UV wavelengths decaying most rapidly with depth. Smith and Macagno (1990) in fact suggest the UV photoreceptor in *Daphnia* may enable them to detect proximity to the surface or its position to enable body orientation.

It is hard to believe *Daphnia* would need tetrachromatic vision. A more parsimonius explanation of their spectral sampling diversity being that each photoreceptor type is 'hard wired' into a specific behaviour related to a narrow wavelength range. The early 'colour dance' observations (Smith and Baylor 1953), which recorded stereotyped behaviours, such as clustering or sideways migration, in response to narrow wavelength stimuli, support this hypothesis. Wavelength specific behaviours, where a direct correlation exists between wavelengths of light encountered and individual photoreceptors, should provide a particularly clear-cut relationship between ecological influences and spectral sensitivities. Unfortunately the *Daphnia* micro-environmental and behavioural data needed are unknown.

9.8 The stomatopods

In terms of their spectral sampling abilities, the stomatopods are unsurpassed in the animal kingdom. From MSP we know members of 2 of the existing 5 superfamilies (Manning and Camp 1983) possess at least 11 visual pigments (Cronin and Marshall 1989a and b, Cronin *et al.,* 1994a). Recent electrophysiological data suggests there may be as many as 16 S_{maxes} (one for each morphological type of photoreceptor cell in the retina, Marshall and Oberwinkler, 1993), each probably based on a different visual pigment (Cronin *et al.,* 1994a, Table 1, Figure 7). Photoreceptors can be divided into: colour receptors (12), polarisation receptors (2) and those for spatial vision (2), although there may be some overlap in function. Colour and polarization receptors are found in the mid-band, a specialised zone of omatidia obvious by external examination of the eye (Figure 1). The complexity of the stomatopod eyes means that before their visual ecology is discussed, some background is needed.

The mid-band in gonodactyloid and lysiosquilloid stomatopods contains 6 rows of enlarged omatidia, 4 of which (rows 1 to 4, labelling dorsally downwards in Figure 1c) contain rhabdoms adapted for colour vision, the remaining 2 (rows 5 and 6) containing structural adaptations which suggest they are sensitive to polarized light (Marshall 1988, Cronin and Marshall 1989a, Marshall *et al.,* 1991a). Colour vision has recently been confirmed behaviourally in the gonodactyloid mantis shrimp *Odontodactylus scyllarus* (Marshall *et al.,* 1996). The remainder of the eye, the 'periphery', contains photoreceptors much like those of other crustaceans with a short R8 cell distally placed over a longer R1-7 rhabdom tier. The periphery is for spatial analysis of the world, possibly including monocular stereopsis (Exner 1891, Marshall and Land 1993).

R8 cells are present in all eye areas but have become modified for reception of polarized light in rows 5 and 6 (Marshall *et al.,* 1991a) and for colour vision in rows 1 to 4 (Marshall *et al.,* 1991b, Marshall *et al.,* 1996). The R1-7 cells of rows 1 to 4 are highly modified for their role in colour vision. Adaptations include tiering of photoreceptor segments (beneath the R8 cells into a 3 cell and a 4 cell section made from the R1-7 cells), coloured intrarhabdomal filters in between tiers in rows 2 and 3 (Figure 8, Chapter 4), different visual pigments in each photoreceptor tier and different screening pigments including, in some species, coloured lateral filters used in spectral tuning (Marshall *et al.,* 1991b, Cronin and Marshall 1997 in prep). The intrarhabdomal filters, a unique adaptation amongst the crustacea, consist of densely coloured blocks of varying length (5 –70 μm) made from vesicles containing carotenoid pigment (Marshall 1988, Marshall *et al.,* 1991b, Cronin *et al.,* 1994d). They may be red, orange, yellow, blue or pink and are situated either between the R8 and the top tier of the R1-7 cells (distal filters) or between the two tiers formed by the R1-7 cells (proximal filters, Chapter 4 Plate 1).

MSP analysis of each of the photoreceptor tiers in rows 1 to 4 demonstrated they contain different visual pigments (Cronin and Marshall 1989, Cronin *et al.,* 1994a, Figures 7 and 9). The tiering of photoreceptors and inclusion of coloured filters in rows 2 and 3 mean that as light passes down the rhabdom, through each photoreceptor tier or filter, its spectrum is modified. Each tier, whether colour filter or visual pigment acts as a long-pass filter, passing increasingly longer

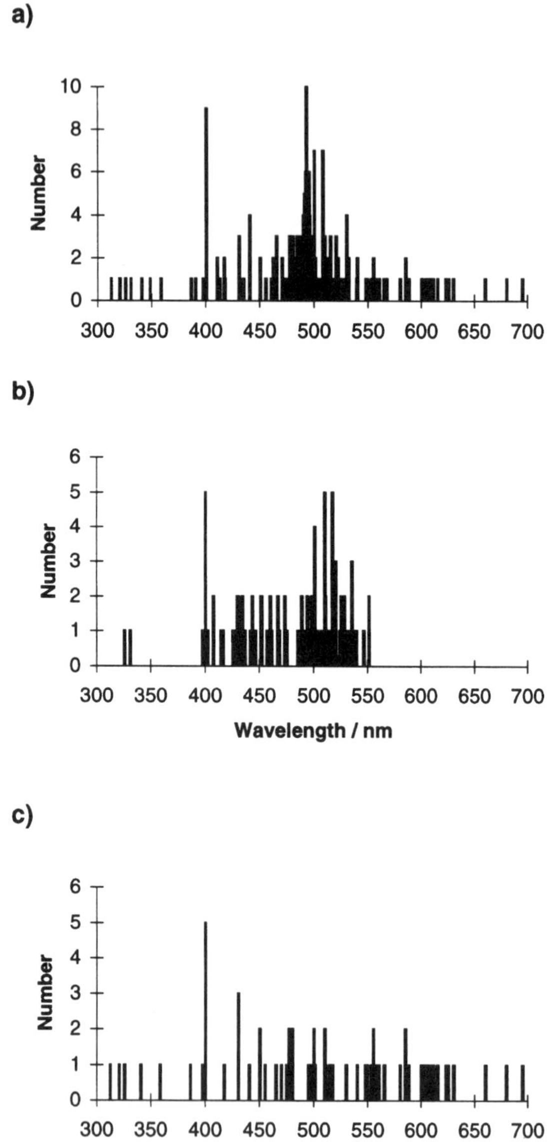

Figure 7 S-max and λ_{max} values of stomatopods compared to other crustaceans (for data sources see Table 1). **a**) The same graph as in Figure 2 (a) with all known crustacean S_{max} and λ_{max} values re-plotted for comparative purposes. **b**) The λ_{max} of all known stomatopod visual pigments. The apparent peak in all these plots at 400 nm is an artefact due to the difficulty of measuring below 400 nm with MSP (Cronin *et al.*, 1994a). It consists mostly of the visual pigment λ_{max} from R1-7 cells of the distal tier of row 1 in the mid-band, a 'blue/UV' photoreceptor whose λ_{max} values would probably lie near, but not at, 400 nm. The two visual pigments at 325 and 330 are best MSP estimates of the λ_{max} values (Cronin *et al.*, 1994b). **c**) The S_{max} of stomatopod photoreceptors calculated using the λ_{max} values in (b), and the photoreceptor dimensions and filtering by photostable filters and tiers of photoreceptor given by Cronin and Marshall 1989a. S_{maxes} below 400 nm are from electrophysiological intracellular recordings (Marshall and Oberwinkler, 1993)

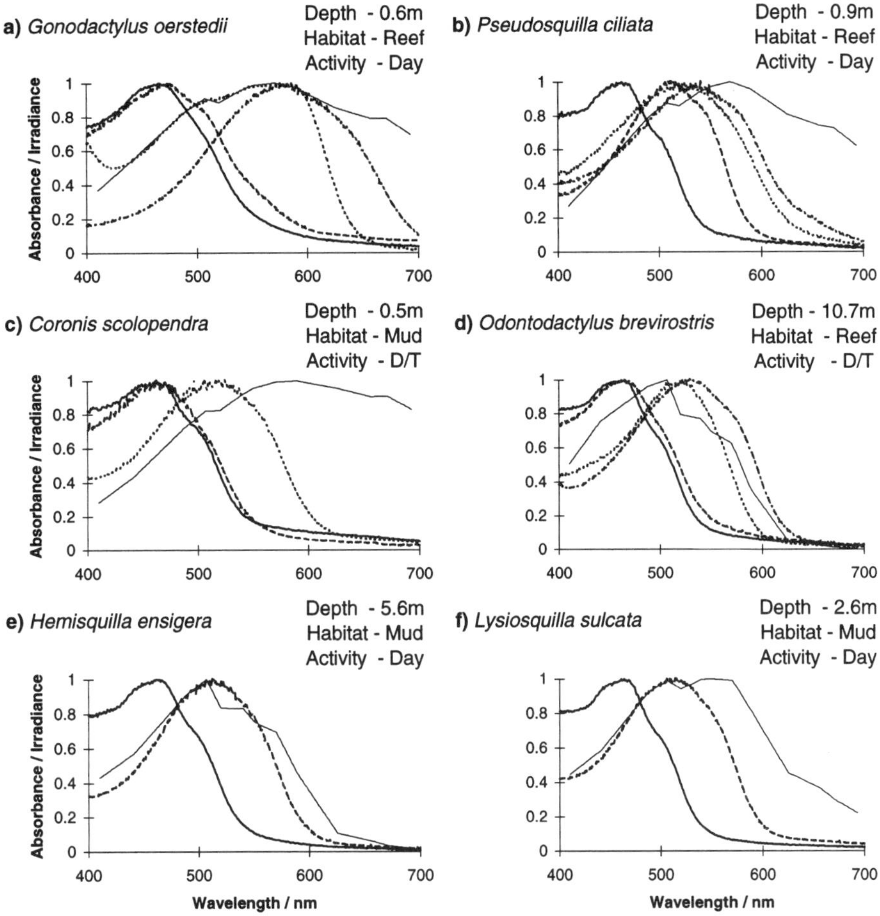

Figure 8 Normalised absorption curves of intrarhabdomal filters in the mid-band of stomatopod species from different habitats. Plotted on top of each set of filter curves are normalised spectral irradiance distributions (thin lines) measured in the habitat of each species and at their usual time of peak activity. Depth of measurements are given at the top right of each graph along with a habitat description and time of activity (D/T – diurnal and twilight active). The more bathychromic the filters long wavelength transmission, the longer the S_{max} of the row 2 and 3 photoreceptors will be (Cronin *et al.,* 1994b). One can see that in many cases, the filters absorption is matched to the habitat irradiance (c) being an exception to this

wavelengths to the tiers below. This results in the sensitivity of each retinal area beneath the first tier being very sharply tuned and shifted from the visual pigment λ_{max} to an S_{max}, in some cases over 100 nm away, towards longer wavelengths (Cronin and Marshall 1989, Cronin *et al.,* 1994a and Figure 9). One can calculate the result of all this filtering and show that the sensitivity of the photoreceptors are spread to sample much of the spectrum (Figure 10 and compare Figure 7 b and c). In all species, row 1 contains the shortest wavelength sensitivities followed by row 4, then 2, then 3 (Figure 10).

At present the R1-7 photoreceptors of this system are understood best and the remainder of this discussion centres mainly around them and their 8 sharply tuned spectral sensitivities. The R8 cells, as in other crustaceans, are sensitive to short UV/violet wavelengths, however, in stomatopods there may be as many as 6 different S_{max} values in the 300 – 400 nm range resulting from these cells. They are not fully characterised as yet but seem to fit the polychromatic tendencies of this remarkable visual system. They are unusually long in mid-band rows 1 and

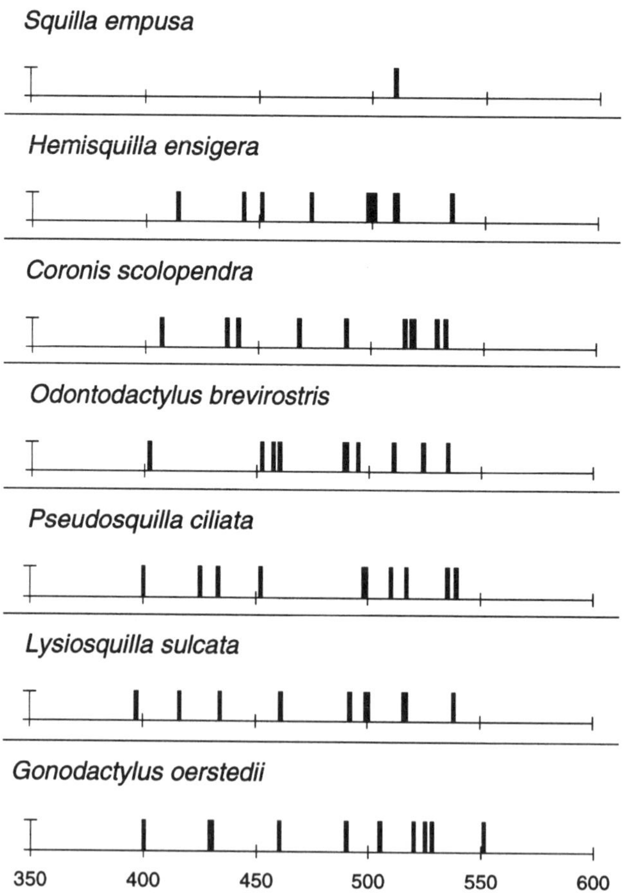

Figure 9 The λ_{max} positions of the visual pigments of stomatopods from various habitats. Squilloid stomatopods live in deep waters and have 'discarded' the spectral sampling capability of a six-row mid-band. Stomatopods with six-row mid-bands living in light-limited habitats squash the λ_{max} of their photoreceptors towards the middle of the spectrum. This helps to limit the S_{max} of the photoreceptors to the available light

4 and will therefore act as UV/violet absorbing filters. This will result in more sharply tuned violet/blue R1-7 sensitivities in these rows than those illustrated in Figure 10 (Marshall *et al.,* 1991a gives an example of this), in this instance sharpening the rising or short wavelength component of the visual pigment spectrum. As we will see shortly, this is important for spectral tuning of this system to the environmental light available to each species.

Eight spectral sensitivities, in R1-7 of rows 1 to 4, are only found in stomatopods living at, or relatively near to the surface. Members of the superfamily Bathysquilloidea are deep water benthic animals with no mid-band and in some cases reduced eyes (Manning *et al.,* 1984). Nothing is known of their spectral sensitivities but they are likely to be tuned close to the wavelengths of down-welling light or bioluminescence, as in other deep-sea animals. Stomatopods from the superfamily Squilloidea dig burrows in muddy turbid areas and may be found at several hundred meters. In such a limited light environment (Figure 10a) a retina with the full spectral coverage of the gonodactyloids and lysiosquilloids would be wasted. As a result the squilloids have 'lost' the 6 row mid-band also. A loss of the mid-band in this group seems likely as they do retain what appears to be a rudimentary 2 row mid-band, these omatidia contain photoreceptors identical to the hemispheres (Figure 1). Squilloid rhabdoms have R8 cells which are uncharacterised due to their small size (Marshall *et al.,* 1991a) and R1-7 cells with λ_{max} values around 500 nm (Table 1, Figure 10a). Their length gives these photoreceptors high sensitivity from 400 – 600 nm (Figure 10a), ideal for their murky situation. Figure 10 also includes spectral irradiance in the habitat of *Squilla empusa*, measured at the time of day the animals are active.

The stomatopods from superfamily Gonodactyloidea, eg *Gonodactylus oerstedii* and *Pseudosquilla ciliata*, are often found inhabiting the top few centimetres of clear tropical waters where they are exposed to high intensity light of all wavelengths. Their typical habitat is in holes in coral rubble on reef tops or in sea-grass areas and they are active during the day. Their R1-7 photoreceptors have S_{maxes} ranging from 400 nm to just below 700 nm (Table 1, Figure 10c and d). Four filters, with widely spaced spectral absorbances, are present in rows 2 and 3 of such species (Figure 8a and b). From the partial classification of R8 cells in *G. oerstedii* (Marshall and Oberwinkler, 1993) it is known that these cells sample a range of light from 400 nm down to 312 nm.

A number of gonodactyloids and more notably species from the lysiosquilloid and erythrosquilliod stomatopod superfamilies, live in light-limited environments. This may be the result of increased habitat depth, water turbidity, being active during light-limited hours, activity over a depth range or a combination of such factors. If these species were to retain a broad sampling range, particularly at the red end of the spectrum which is rapidly attenuated with depth (Figure 4), their photoreceptors would need to become prohibitively large in order to catch enough photons to operate (Land 1981, Cronin *et al.,* 1994c). Mid-band omatidia are around twice the size of those in the periphery for precisely this reason

(Figure 1, Marshall *et al.,* 1991b, Marshall and Land 1993). Serial filtering, especially where it involves the dense intrarhabdomal filters, is costly in terms of loss of photon catch. The sensitivity curves plotted in Figure 10 are normalised to illustrate their spectral position and tend to hide this problem (for non-normalised values see Cronin *et al.,* 1994a and c). It is notable that, as yet, we have not found any species which have discarded one or more of the spectral sensitivities in rows 1 to 4, clearly suggesting the four row colour vision system functions as a unit.

Four classes of adaptations are known in stomatopods living in dim light conditions. All tend to 'squash' S_{max} values in from the ends of the spectrum in order to match the long and short wavelength limited spectrum of available light.

a) Changing filter colour, and thereby tuning its longwave pass, towards shorter wavelengths. Compare for instance row 2 and row 3 filters in *G. oerstedii* and *Odontodactylus brevirostris* (Figure 8a and d). All filters in *O. brevirostris* pass shorter wavelengths than their equivalent in *G. oerstedii.* The proximal filter in row 3 of G. oerstedii passes such long wavelengths that a substantial short wavelength window exists in the absorbance curve (Figure 10a) making this filter appear blue (Plate 1, Chapter 4).

b) Using filters of similar spectral type in one row (eg Figure 8c and d). In a surface dwelling stomatopod, proximal filters pass longer wavelengths than the distal filters. There may be a yellow filter over a red filter in row 2 for instance. By using filters of the same absorbance, this shift is reduced and the sensitivity of the proximal photoreceptors kept shorter (Cronin *et al.,* 1994a and d).

c) 'Loss' of proximal filters. This is typical of lysiosquilloids which are mud-burrow dwellers, often inhabiting rather turbid waters such coastlines with mangrove or estuarine influence. It is also known in 2 gonodactyloid species, *Hemisquilla ensigera* and *Echinosquilla guerini.* The result is to reduce the extent to which the sensitivities of the proximal tier is pushed to long wavelengths.

d) Reduction in the spectral spread of visual pigment λ_{max} positions at either end of the sensitivity range (Figure 9). As the final S_{max} of a photoreceptive tier is dependent on the λ_{max} of its visual pigment and, if it is a proximal tier, the visual pigment in the distal tier above it, this reduces the spread of sensitivities from either end of the spectrum. It is implemented particularly by the short wavelength photoreceptors of row 1 and 4 and to draw the short wavelength sensitivities in towards the centre of the spectrum, away from 400 nm.

Some specific examples of each of these adaptations and their ecological significance now follow.

The small lysiosquilloid *Coronis scolopendra* lives in mud borrows at depths from 0.5 –10 m round estuarine entrances and in coastal water and is often twilight active. In such a photic environment the spectral distribution of light will be reduced at both ends of the spectrum. Comparing the sensitivity curves of *C. scolopendra* to *G. oerstedii* (Figure 10c and e) one can see the 'squashing in' of *C. scolopendra* spectral sensitivities at both ends of the spectrum. This is achieved in row 3 by discarding the proximal filter and shortening the wavelengths passed by the distal filter, in row 2 by having similar filters in the distal and proximal position (Figure 8c) and by drawing the λ_{max} of the visual pigments in at either end of the spectrum (Figure 9). In other words, *C. scolopendra* uses all 4 adaptations to reduce the spectral range of its mid-band.

O. brevirostris is a moderately deep-living (7– 20 m) gonodactyloid. It is most active around dawn and dusk, making effective vision with the mid-band even harder (Figure 10f). No filters are discarded in this species but those in row 2 are similar in absorbance and both row 2 and 3 filters pass relatively short wavelengths (compare Figure 8a and d). Visual pigment λ_{maxes} are squashed in towards the centre of the spectrum (Figure 9). *Echinosquilla guerini*, lives in close proximity to *O. brevirostris* but is active just after dusk and has no proximal filter in row 3 (Cronin *et al.,* 1994c).

H*emisquilla ensigera* unusually for a gonodactyloid lives in a mud burrow in moderately deep turbid temperate water. It is the most light-limited of the species discussed here (Figure 10g). This species has lost both proximal filters, shows shortened long wavelength pass in the remaining filters (Figure 8e) and has substantially drawn in visual pigment λ_{maxes} at both ends of the spectrum (Figure 9).

Lysiosquilla sulcata is a lysiosquilloid that has discarded both proximal filters (Figure 8f), the remaining filters having absorbances very much like *H. ensigera.* It, however, lives in slightly shallower, clearer water in mud or sand close to coral reefs and has relatively well spread visual pigment λ_{maxes} (Figure 9) to take advantage of the short wavelengths present in this habitat (Figure 10h).

Rows 5 and 6 of the mid-band (which in all species possess identical visual pigments) and the peripheral retina also exhibit R1-7 sensitivities correlated to environmental conditions. These photoreceptors are so long that they absorb very efficiently between 400 nm and their long wavelength limb (Figure 10b). It is the position of the long wavelength limb which shows environmental tuning. For example, the reef top species *G. oerstedii* samples light to longer wavelengths than the deep, temperate, turbid mud burrow dweller, *H. ensigera* in both the peripheral retina and rows 5 and 6.

Reasons for designing a colour vision system with multiple sharp tuned sensitivities, rather than 3 broad functions like our own, is largely outside the scope of this chapter. Some possible explanations are presented in Cronin and Marshall (1989), Marshall *et al.,* (1991b), Marshall *et al.,* (1996) and Osorio *et al.,* (1997 in press). Colour, colour signalling and aggression play an important part in the life of many stomatopods. They are from tropical seas and reefs where their surroundings contain much colour detail. Individual species often have a distinguishing colour spot (the meral spot) which is displayed to con- and intraspecific rivals in competition for food, holes in the reef to live in or during sexual encounters (Caldwell and Dingle 1976). Mantis shrimps are so called because they have a raptorial limb similar to that of the preying mantis but capable of far greater force. It is used for prey capture or in battles with other stomatopods. A

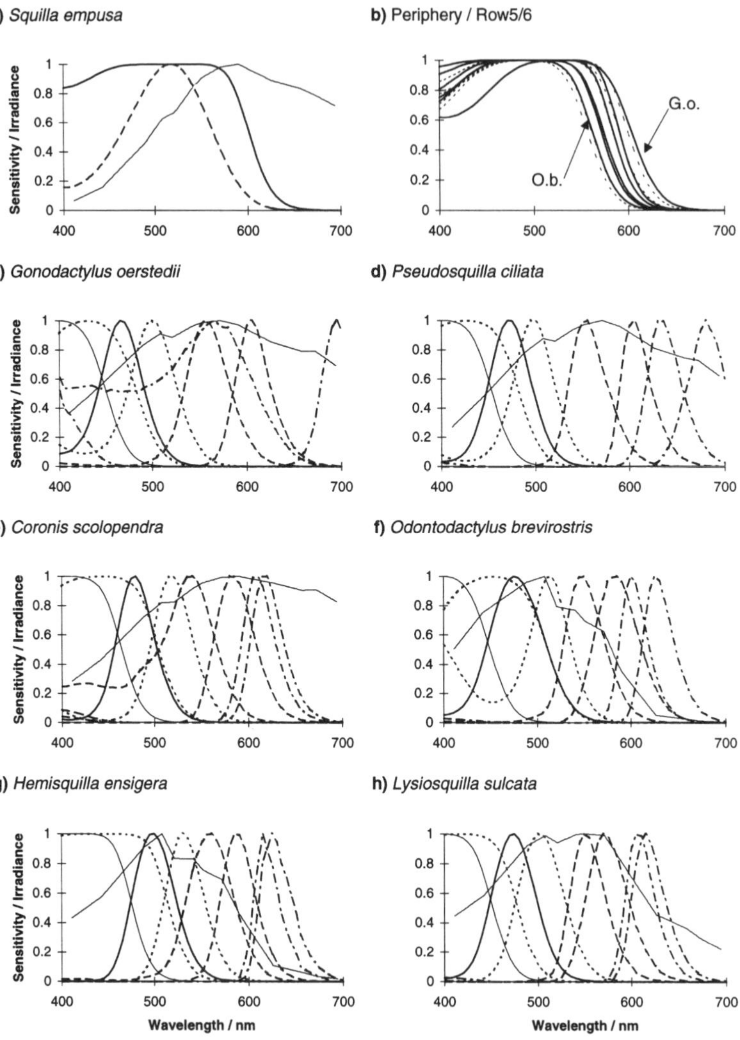

Figure 10 The calculated spectral sensitivities (S_{max}) of the R1-7 cells in stomatopods from different habitats. On each graph (except (b)) the spectral distribution of light at the habitat measured during known activity periods of the species is plotted (thin line). Solid lines – row 1 spectral sensitivities, dotted lines – row 4, dashed lines – row 2, dot/dash lines – row 3. In each case the distal most photoreceptor layer samples shorter wavelengths than the proximal layer. **a**) *Squilla empusa* lacks a six-row mid-band and has only a single visual pigment in its eye with λ_{max} at 517 nm shown as a visual pigment nomogram (dotted line). Absorbtance or sensitivity, solid curve. **b**) Spectral sensitivities of peripheral retinae (solid lines) and mid-band rows 5 & 6 (dotted lines) for the 6 species shown in (**c**) – (**h**). The peripheral retina sensitivity curves of *Odontodactylus brevirostris*, a deep living gonodactyloid, and *Gonodactylus oerstedii*, a shallow clear water gonodactyloid, are labelled. **c** & **d**) R1-7 cell sensitivities in Row 1-4 of the mid-band for stomatopods from shallow clear water. **e** – **h**) R1-7 cell sensitivities in Row 1-4 of the mid-band for 4 stomatopods from habitats which are light limited due to water depth, turbidity or which may be active at dawn and dusk. The population of both *Coronis scolopendra* and *Squilla empusa* chosen were living in relatively shallow (0.5 and 0.8 m) water for these species, hence the high long wavelength component of the irradiance distributions. In deeper habitats sometimes frequented by these animals, this would not be present (*eg* Figure 4)

single blow could kill an opponent (Caldwell and Dingle 1976) and their meral spot colour signals may be a way to communicate important information prior to any potential combat. Colour has therefore become an integral part of the lives of these animals.

Maintaining the effectiveness of their colour signals is therefore important to stomatopods. The perception of body colours will be greatly influenced both by environmental conditions and the efficacy of stomatopod vision. Maintaining colour constancy with increasing depth or other light-limiting factors is hard. To our visual system red at the surface appears muddy green at 10 m and this is an example of colour constancy failure, a phenomenon well known to divers with red snorkels. It is due both to strong absorbance of red light by water (Figure 4) and the relative insensitivity of the human visual system at this end of the spectrum. One way to better maintain colour constancy is to use sharply tuned sensitivities. These give higher colour contrast ratios, with some dependence on the colour being looked at (Osorio *et al.,* 1997). It may be the stomatopod's need to maintain colour constancy in order to interpret colour signals over a variety of ecological conditions that has driven the evolution of this unique visual system. The fine tuning of the retinal sensitivities of the sort seen suggests the mid-band of many species is working at the edge of its capability.

9.9 Ontogenetic changes

A number of fish and amphibia undergo changes in spectral sensitivity during development or maturation, invariably associated with a change in habitat and often, although not always, the result of a migration from fresh to sea water or water to land (Wood *et al.,* 1992, Shand *et al.,* 1993 and see Chapter 12 for a review of this). Many crustaceans also undergo dramatic changes in lifestyle as their larvae leave the plankton to take up adult life on the sea bed, in more defined niches in mid-water or on land. The visual demands of the planktonic versus adult niche are very different and often involve substantial optical reorganisation (Nilsson 1989, Cronin *et al.,* 1995a). As the light environment at the ocean surface, where many larvae are found, is very different to the sea bed, one might expect concomitant spectral sensitivity changes. Anatomically, R8 cells are known in the larval eye of some stomatopods (Cronin *et al.,* 1995a) and these are likely to be UV/violet sensitive, however their spectral sensitivities are unknown. The following summarises what is known of larval to adult changes in R1-7 cell visual pigment λ_{max} values:

Callinectes sapidus	Larva (Meg.)- 504 nm	Adult – 504nm
Habitat	Pelagic	Muddy bottom
	(Cronin and Marshall 1995)	
Squilla empusa	Larva – 512 nm	Adult – 517 nm

Habitat	Pelagic (Cronin and Marshall in prep)	Muddy bottom
Gonodactylus aloha	Larva – 499 nm	Adult – 10+ none at 499 nm Peripheral retina at 510 nm closest
Habitat	Pelagic (Cronin *et al.*, 1995a)	Reef crest
Pullosquilla tomasini	Larva – 447 nm	Adult – 10+ Periphery at 461 nm row 4P at 449 nm closest
Habitat	Pelagic (Jutte, Caldwell and Cronin pers.comm.)	Reef 2 – 40 m
Pullosquilla litoralis	Larva – 446 nm	Adult – 10+ Periphery at 511 nm row 4D at 450 closest
Habitat	Pelagic (Jutte, Caldwell and Cronin pers.comm.)	Reef 2 m

Spectrally at least, it seems that the demands placed on both larvae and adult eyes of *C. sapidus* and *S. empusa* are similar. *C. sapidus* megalopus larvae and adult apparently posses the same rhodopsin while the 5 nm shift to long wavelengths in *S. empusa* may not be enough to assign these visual pigments to different classes (Cronin *et al.*, 1993).

Given the abundance and spectral coverage of adult visual pigments, it would not be surprising to find stomatopod larval visual pigments occurring in one of the spectral classes of the adult. Surprisingly this is not so and suggests that the ecological tuning of larval spectral sensitivity is rather precise. Alternatively, as the adult eye is a radically new structure, replacing the larval eye completely on metamorphosis (Cronin *et al.*, 1995a), the differences in λ_{max} may be due to developmental constraints. It is interesting that the spectral sensitivity of the larvae of *G. aloha* and the two *Pullosquilla* species (lysiosquilloid stomatopods) are over 50 nm different. These larvae are found together in the plankton so their larval sensitivity differences, if ecologically significant, must reflect yet another aspect of crustacean micro-habitats which we don't yet understand. Both these sensitivities, however, fall within the blue/green behavioural sensitivity peak found in a number of crustacean larvae (Forward and Costlow 1974, Forward and Cronin 1979, Forward and Douglas 1989).

9.10 Conclusions and speculations on polarization sensitivity

Influences determining crustacean spectral sensitivity fall into a variety of categories listed below, none of these are mutually exclusive. In other words the final spectral position of photoreceptors may be the result of a variety of factors.

Ecological factors	
Sensitivity hypothesis	Dependent on spectral distribution of light available and behavioural factors such as time of day active and periods of significance eg dawn/dusk.
Offset/contrast hypothesis	Only for viewing bright objects against darker backgrounds so probably limited to the lower half of the visual field in surface dwellers.
Behaviourally significant tasks	Specific events or tasks of importance such as colour communication of aggressive intent, or sexual display, bioluminescence, depth determination.
Phylogenetic trends	Notable in crabs (possibly due to large data set).
Biochemical constraints	
Chromophore/opsin interaction.	λ_{max} clustering and opsin/porphyropsin short and long wave-length families of visual pigments.
Noise/thermodynamic	Reason for lack of long wavelength λ_{max}. Long wavelength S_{max} therefore achieved with filtering (eg stomatopods).

Evolutionary inertia is sometimes used to explain visual pigment λ_{max} positions, specifically the lack of red sensitivities (Goldsmith 1991). In the light of new evidence, however, it seems that visual pigment evolution (Cronin *et al.,* 1996b) and eye evolution (Nilsson and Pelger 1994) can occur rapidly. This implies that the spectral sampling strategies observable today are the 'right solution' for the animal in question. The challenge in visual ecology is to determine what factor or combination of factors generate this solution.

In general conclusion decapod crustaceans usually possess 2 spectral sensitivities, one around 500 nm (blue/green sensitive R1-7 cells) and one around 400 nm (UV/violet sensitive R8 cells). Their small size and short wavelength sensitivity means R8 cells are often overlooked or hard to measure from but, in both surface dwelling and deep-sea crustaceans, are not always present. The euphausids and amphipods, despite often having divided eyes, probably possess only a single visual pigment (as do other crustaceans from the minor groups listed in Table 1). The stomatopods, perhaps the mysids, *Daphnia* and *Ligia* all have more than two spectral sensitivities.

The combination of UV/violet and blue/green sensitivities in different photoreceptor populations is used by a surprisingly diverse assemblage of crustacea in many different habitats. No such basic pattern exists in the fish whose visual pigment number per species and spectral range exceeds the basic crustacean design. This begs some further explanation other than ‘arm waving’ about crustacean phylogeny, biochemical constraint and unknown ecological factors, no matter how real these are. The answer may lie in polarisation rather than spectral sensitivity.

The microvillar design of crustaceans makes them inherently sensitive to the E-vector of polarized light (Waterman 1981). That they are concerned with maintaining this sensitivity is suggested by the regular array of orthogonal microvilli in R1-7 cells and occasional regular arrays in the R8 cells also (Waterman 1981, Krebs and Lietz 1982, Marshall *et al.,* 1991a). There is a significant amount of polarized light above and particularly below the water produced mainly by reflection and particle scattering (Ivanoff and Waterman 1958). Functionally, its overall pattern in the sky is known to be used for navigation (Wehner 1983) and it may enhance resolution in scattering media such as turbid waters (Lythgoe 1976). Behaviourally a number of crustaceans are known to respond to polarized light (eg Goddard and Forward 1991). Also, crustacean and other animal body patterns may selectively reflect polarized light making it a potential signal (Neville and Luke 1971, Marshall *et al.,* 1991a, Cronin *et al.,* 1995b, Shashar *et al.,* 1996). Beneath the retina there is reason to believe that signals from cells sensitive to orthogonal E-vectors are separated until the first neural integration site, the lamina, where differential signals may then be integrated (Nassel 1976, Glantz 1996). In stomatopods, the advanced colour system seems to ‘borrow’ this plesiomorphic wiring system, possibly to achieve dichromatic opponency (Marshall *et al.,* 1991b, Marshall *et al.,* 1996). Stomatopods are also known to discriminate E-vectors (Marshall unpublished results).

The degree to which light is polarized in water is known to be wavelength dependent with a minimum at 450 – 500 nm (Ivanoff and Waterman 1958). That is, the least attenuated light (in blue waters) shows the lowest polarization (Jerlov 1976). It may therefore be that efficient perception of polarized light is also important in the spectral tuning of crustacean photoreceptors, their S_{max} values often lying on either side of this polarization minimum. Similar explanations have been suggested for the spectral positioning of insect photoreceptors known to be used for polarized light perception (Wehner 1983, Seliger *et al.,* 1994). Given the apparent emphasis crustaceans place on this sensory modality, perhaps we should look more closely at polarized light for a functional explanation of their spectral sensitivities.

References

Arikawa, K., Kawamata K., Suzuki, T. and Eguchi, E. (1987) Daily changes of structure, function, and rhodopsin content in the compound eye of the crab *Hemigrapsus sanguineus*. *J. Comp. Physiol. A.*, 161, 161–174.

Barnes, H. and Klepal, W. (1972) Phototaxis in stage I nauplius larvae of two cirripedes. *J. Exp. Mar. Ecol.*, 10, 267–273.

Boden, B. P., Kampa, E. M. and Abott, B. C. (1961) Photoreception of a planktonic crustacean in relation to light penetration in the sea. In *Progress in Photobiology.* (eds. Christensen, B. and Buchamann, B.),

Elsevier North-Holland, Inc., New York. 189 –196.
Bowmaker, J. K. (1990) Visual pigments of fishes, in *The visual system of fishes*. (eds Douglas, R. H. and Djamgoz, M. B. A.), Chapman and Hall, London, pp. 81–107.
Bridges, C. D. B. (1972) The rhodopsin-porphyropsin visual system, in *Handbook of sensory physiology. vol. VII/1 Photochemistry of vision* (ed Dartnall, H. J. A.), Springer-Verlag, Berlin, Heidelberg, New York.
Bridges, C. D. B. and Yoshikama, S. (1970) The rhodopsin-porphyropsin system in fresh-water fishes. *Vision Res.* 10, 1333 –1345.
Briggs, M. H. (1961) Visual pigment of grapsoid crabs. *Nature,* 190, 784 –786.
Bruno, M. S., Barnes, S. N. and Goldsmith, T. H. (1977) The visual pigment and visual cycle of the lobster, *Homarus. J. Comp. Physiol.*, 120, 123 –142.
Bruno, M. S. and Goldsmith, T. H. (1974) Rhodopsin of the blue crab Callinectes: evidence for absorption differences *in vitro* and *in vivo*. *Vision Res.*, 14, 653 – 658.
Bruno, M. S., Mote, M. I. and Goldsmith, T. H. (1973) Spectral absorption and sensitivity measurements in single omatidia of the green crab, *Carcinus. J. Comp. Physiol.*, 82, 151–163.
Caldwell, R. L. and Dingle, H. (1976) Stomatopods. *Scient. Am.* 234, 80 – 89.
Chittka, L. (1996) Does bee colour-vision predate the evolution of flower colour. *Naturwissenschaften* 83, 136 –138.
Clarke, G. L. (1936) On the depth at which fishes can see. *Ecology* 17, 452 – 456.
Crandell, K. A. (1995) Opsin evolution in crayfishes: effects of functional constraints, in *Current topics on molecular evolution* (eds Nei, M. and Takahata, N.). Institute of molecular evolutionary genetics, The Pennsylvania State University. pp. 19 – 27.
Cronin, T. W. (1985) The visual pigment of a stomatopod crustacean, *Squilla empusa. J. Comp. Physiol. A.*, 156, 679 – 687.
Cronin, T. W. (1986) Photoreception in Marine Invertebrates. *Amer. Zool.*, 26, 403 – 415.
Cronin, T. W. (1989) Application of intracellular optical techniques to the study of stomatopod crustacean vision. *J. Comp. Physiol. A.*, 164, 737–749.
Cronin, T. W. (1990) Pigments in crustacean compound eyes, in *Frontiers in crustacean neurobiology*. (eds Wiese, K., Kreuse, W.-D., Tautz, J., Reichert, H. and Mulloney, B.), Birkhäuser, Basel. pp 58 – 65
Cronin, T. W. and Forward, R. B. Jr. (1988) The visual pigments of crabs – I. Spectral characteristics. *J. Comp. Physiol. A.*, 162, 463 – 478.
Cronin, T. W. and Goldsmith, T. H. (1982) Quantum efficiency and photosensitivity of the rhodopsin ⇔ metarhodopsin conversion in crayfish photoreceptors. *Photochem. Photobiol.* 36, 447– 454.
Cronin, T. W. and Marshall, N. J. (1989a) Multiple spectral classes of photoreceptors in the retinaes of gonodactyloid stomatopod crustaceans. *J. Comp. Physiol A*, 166, 267– 275.
Cronin, T. W. and Marshall, N. J. (1989b) A retina with at least ten spectral types of photoreceptors in a stomatopod crustacean. *Nature*, 339, 137–140.
Cronin, T. W., Marshall, N. J., Caldwell, R. L. and Shashar, N. (1994a) Specialization of retinal function in the compound eyes of mantis shrimps. *Vision Res.*, 34, 2639 – 2656.
Cronin, T. W., Marshall, N. J., Quinn, C. A. and King, C. A. (1994b) Ultraviolet photoreception in mantis shrimp. *Vis. Res.*, 34, 1443 –1452.
Cronin, T. W., Marshall, N. J. and Caldwell, R. L. (1994c) The retinas of mantis shrimps from low-light environments (Crustacea, Stomatopoda, Gonodactylidae). *J. Comp. Physiol. A*, 174, 607– 619.
Cronin, T. W., Marshall, N. J. and Caldwell, R. L. (1994d) The intrarhabdomal filters in the retinas of mantis shrimps. *Vision Res.*, 34, 279 – 291.
Cronin. T. W., Marshall, N. J., Caldwell, R. L. and Pales, D. (1995a) Compound eyes and ocular pigments of crustacean larvae (Stomatopoda and Decapod Brachyura). *Mar. Fresh. Behav. Physiol.* 26, 219 – 231.
Cronin T. W., Shashar, N. and Wolff, L. (1995b) Imaging technology reveals the polarized light fields that exist in nature. *Biophotonics Int.* 2, 38 – 41.
Cronin, T. W. and Frank, T. (1996) A short-wavelength photoreceptor class in a deep-sea shrimp. *Proc. R. Soc. Lond. B* 263, 861– 865.
Cronin, T. W., Kent, J., Frank, T., Widder, E., Partridge, J. C., Herring, P., Robinson, P. (1996a) *Visual pigments and photoreceptor classes of deep-sea shrimps*. Society for Neuroscience Abstracts.
Cronin, T. W., Marshall, N. J., Caldwell, R. L. (1996b) Visual pigment diversity in two genera of mantis shrimps implies rapid evolution (Crustacea; Stomatopoda). *J. Comp. Physiol. A* 179, 371– 384.
Cronin, T. W. and Marshall, N. J. (1997) Lateral filtering in photoreceptors of two species of stomatopod crustaceans. In prep.
Cummins, D. and Goldsmith, T. H. (1981) Cellular identification of the violet receptor in the crayfish eye.

J. Comp. Physiol., 142, 199 – 202.

Dartnall, H. J. A. and Lythgoe, J. N. (1965) The spectral clustering of visual pigments. *Vision Res.* 5, 81–100.

Denton, E. J. (1990) Light and vision at depths greater than 200 metres, in *Light and life in the sea* (eds Herring, P. J., Campbell, A. K., Whitfield, M. and Maddock, L.), Cambridge University Press, Cambridge, New York, pp. 127-148.

Denys, C. J. (1982) Ommochrome pigments in the eyes of *Euphausia superba* (Crustacea, Euphausiacea). *Polar Biol.*

Denys, C. J. and Brown P. K. (1982) The rhodopsins of *Euphausia superba* and *Meganyctiphanes novegica* (Crustacea, Euphausiacea). *J. Gen. Physiol.,* 80, 451– 472.

Douglas, R. H. and Partridge, J. C. (1997) On the visual pigments of deep-sea fish. J. *fish biol.* 50, 68 – 85.

van Dover, C. L., Szuts, E. Z., Chamberlain, S. C. and Cann, J. R. (1989) A novel eye in the 'eyeless' shrimp from hydrothermal vents of the Mid-Atlantic Ridge. *Nature*, 337, 458 – 461.

von Frisch, K. and Kupeiwieser, H. (1913) Über den einfluss der lichtfarbe auf die phototaktischen reaktionen niederer krebse. *Biol. Centrabl.* 33, 517– 552.

Eakin, R. M. (1972) Structure of invertebrate photoreceptors, in *Handbook of sensory physiology.* VII/1. (ed. Dartnall, H. J. A.), Springer, Berlin, Heidelberg, New York. pp. 625 – 684.

Eaton, B. P. and Brown, C. M. (1970) Photoreception in the nauplius eye of *Pandalus borealis* Kröger: Decapoda, Crustacea. *Canad. J. Zool.* 48, 119 –121.

Exner, S. (1891) *Die physiologie der facettirten augen von krebsen und insecten.* Liepsig, Wein, Deuticke.

Fernandez, H. R. (1965) *A survey of the visual pigments of decapod crustacea of South Florida.* Ph.D. Thesis, University of Miami, Coral Gables, Florida.

Fernandez, H. R. (1973) Spectral sensitivity and visual pigment of the compound eye of the galatheid crab *Pleuroncodes planipes. Mar. Biol.*, 20, 148 –153.

Fisher, L. R. and Goldie, E. H. (1959) The eye pigments of a euphausiid crustacean, *Meganyctiphanes norvegica* (M. Sars). *Proc. XV Intern. Congr. Zool.*, London. 533 – 535.

Fisher, L. R. and Goldie, E. H. (1961) Pigments of compound eyes, in *Progress in Photobiology.* (eds. Christensen, B. and Buchamann, B.), Elsevier North-Holland, Inc., New York. 153 –154.

Forward, R. B. Jr., Cronin, T. W. (1979) Spectral sensitivity of larvae from intertidal crustaceans. *J. Comp. Physiol.* 133, 311– 315.

Forward, R. B. Jr., Cronin, T. W. and Douglass, J. K. (1988) The visual pigments of crabs – II. Environmental adaptations. *J. Comp. Physiol. A.*, 162, 479 – 490.

Forward, R. B. and Douglas, J. K. (1989) Crustacean larval visual sensitivity during diel vertical migration. *Proc. 21st EMBS* 59 – 66.

Frank, T. M. and Case, J. F. (1988a) Visual spectral sensitivities of bioluminescent deep-sea crustaceans. *Biol. Bull.*, 175, 261– 273.

Frank, T. M. and Case, J. F. (1988b) Visual spectral sensitivities of the bioluminescent deep-sea mysid, *Gnathophausia ingens. Biol. Bull.*, 175, 274 – 283.

Frank, T. M. and Widder, E. A. (1994a) Evidence for behavioural sensitivity to near-UV light in the deep-sea crustacean *Systellaspis debilis. Mar. Biol.*, 118, 279 – 284.

Frank, T. M. and Widder, E. A. (1994b) Comparative study of behavioural-sensitivity thresholds to near UV and blue-green light in the deep-sea crustaceans. *Marine Biology* 121, 229 – 235.

Frank, T. M. and Widder, E. A. (1996) UV light in the deep-sea: *In situ* measurements of downwelling irradiance in relation to the visual threshold sensitivity of UV-sensitive crustaceans. *Mar. Fresh. Behav. Physiol.* 27, 189 –197.

Gaten, E, Shelton, P. M. J. and Herring, P. J. (1992) Regional morphological variations in the compound eyes of certain mesopelagic shrimps in relation to their habitat. *J. mar. biol. Ass. U.K.* 72 ,61–75.

Glantz, R. M. (1996) Polarization sensitivity in Crayfish lamina monopolar neurons. *J. Comp. Physiol. A* 178, 413 – 425.

Goddard, S. M. and Forward, R. B. (1991) The role of underwater polarised light patterns in sun compass navigation of the grass shrimp *Palaemonetes vulgaris J. Comp. Physiol. A* 169, 479 – 491.

Goldsmith, T. H. (1972) The natural history of invertebrate photoreceptors, in *Handbook of Sensory Physiology* Vol. VII, Part 1 (ed. Dartnall, H. J. A.). Springer-Verlag, New York, pp. 685 –719.

Goldsmith, T. H. (1978a) The effects of screening pigments on the spectral sensitivity of some crustacea with scotopic (superposition) eyes. *Vision Res.*, 18, 475 – 482.

Goldsmith, T. H. (1978b) The spectral absorption of crayfish rhabdoms: Pigment, photoproduct, and pH sensitivity. *Vision Res.* 18, 463 – 473.

Goldsmith, T. H. (1991) The evolution of visual pigments and colour vision. in *The perception of colour.*

(ed Gouras, P.) *Vision and visual dysfunction.* vol 6 (ed gen J. Cronly-Dillon), Macmillan, pp. 62 – 89

Goldsmith, T. H. and Bruno, M. S. (1973) Behavior of rhodopsin and metarhodopsin in isolated rhabdoms of crabs and lobster, in *Biochemistry and Physiology of Visual Pigments* (ed. Langer, H.). Springer-Verlag, Berlin, Heidelberg, New York, pp. 147–153.

Hamacher, K. J. and Kohl, K. D. (1981) Spectroscopical studies of the *Astacus* visual pigment. *Biophys. Struct. Mech.*, 7, 338.

Hamacher, H. and Stieve, H. (1984) Spectral properties of the rhodopsin-system of the crayfish *Astacus leptodactylus. Photochem. Photobiol.* 39, 379 – 390.

Hariyama, T, Tsukahara, Y. and Meyer-Rochow, V. B. (1993) Spectral responses including a UV-sensitive cell type in the eye of the isopod *Ligia exotica. Naturwissenschaften* 80, 233 – 235.

Hays, D. and Goldsmith, T. H. (1969) Microspectrophotometry of the visual pigment of the spider crab *Libinia emarginata. Z. vergl. Physiol.* 65, 218 – 232.

Herring, P. J. (1977) Luminescence in cephalopods and fish. *Symp. zool. Soc. Lond.* 38, 127–159.

Hertel, H. (1972) Aspekte zur photorezeption von *Artemia salina. Verh. dtsch. Ges. Zool. Erlangen.*

Hiller-Adams, P., Widder, E. A. and Case, J. F. (1988) The visual pigments of four deep-sea crustacean species. *J. Comp. Physiol. A.*, 163, 63 –72.

Hillman, P., Dodge, F. A., Hochstein, S., Knight, B. W. and Minke, B. (1973) Rapid dark recovery of the invertebrate early receptor potential. *J. Gen. Physiol.* 62, 77– 86.

Horridge, G. A. (1967) Perception of polarization plane, colour, and movement in two dimensions by the crab, *Carcinus. Z. Vergel. Physiol.* 55, 207– 224.

Hyatt, G. W. (1975) Physiological and behavioural evidence for colour discrimination by fiddler crabs (Brachyura, Ocypodidae, Genus, Uca), in *Physiological ecology of estuarine organisms* (ed. Vernberg, J.). University of South Carolina Press, Columbia.

Ivanoff, A. and Waterman, T. H. (1958) Factors, mainly depth and wavelength, affecting the degree of underwater light polarization. *J. Mar. Res.* 16, 283 – 307.

Jerlov, N. G. (1976). *Marine Optics*. Elsevier, Amsterdam.

Kampa, E. M. (1955) Euphausiopsin, a new photosensitive pigment from the eyes of euphausiid crustaceans. *Nature (Lond.)*, 175, 996 – 998.

Kampa, E. M. (1965) The euphausid eye – a re-evaluation. *Vision Res.* 5, 475 – 481.

Kong, K. L., Fung, Y. M. and Wasserman, G. S. (1980) Filter-mediated color vision with one visual pigment. *Science*, 207, 783 –786.

Krebs, W. and Lietz, R. (1982) Apical region of the crayfish retinula. *Cell. Tiss. Res.* 222, 409 – 415.

Lall, A. B. and Cronin, T. W. (1987) Spectral sensitivity of the compound eyes in the purple land crab, *Gecarcinus lateralis. Biol. Bull.*, 173, 398 – 406.

Land, M. F. (1981a) Optics and vision in invertebrates, in *Handbook of sensory physiology.* VII/6B. (ed. Autrum, H.) Springer-Verlag. Berlin, Heidelberg, New York. pp. 471– 594.

Land, M. F. (1981b) Optical mechanisms in the higher Crustacea with a comment on their evolutionary origins, in *Sense organs* (eds. Laverack, M. S. and Cosens, D. J.), Blackie, London, Glasgow, pp. 31– 48.

Land, M. F. (1984) Crustacea, in *Photoreception and vision in invertebrates* (ed. Ali, M. A.), Plenum, pp. 401– 438.

Land, M. F. (1989a) The eyes of hyperiid amphipods: relation so optical structure to depth. *J. Comp. Physiol. A* 164, 751–762.

Land, M. F. (1989b) The sight of deep wet heat. *Nature* 337, 404.

Land, M. F. (1990) Optics of the eyes of marine animals, in *Light and life in the sea* (eds Herring, P. J., Campbell, A. K., Whitfield, M. and Maddock, L.), Cambridge University Press, Cambridge, New York, pp. 149 – 166.

Land, M. F., Burton, F. A. and Meyer-Rochow, V. B. (1979) The optical geometry of euphausid eyes. *J. Comp. Physiol. A* 130, 49 – 62.

Land, M. F., Marshall, N. J. and Diebel, C. (1995) Tracking of blue lights by hyperiid amphipods. *J. mar. biol. Ass. U.K.* 75, 71– 81.

Latz, M. I., Frank, T. M. and Case, J. F. (1988) Spectral composition of bioluminescence of epipelagic organisms from the Sargasso Sea. *Marine Biology* 98, 441– 446.

Leggett, L. M. W. (1979) A retinal substrate for colour discrimination in crabs. *J. Comp. Physiol.*, 133, 159 –166.

Levine and MacNichol (1982) SCI AM.

Loew, E. R. and McFarland, W. N. (1990) The underwater visual environment, in *The visual system of fish.* (eds Douglas, R. H. and Djamgoz, M. B. A.), Chapman and Hall, London. pp. 1– 49.

Lythgoe, J. N. (1966). Visual pigments and underwater vision, in *Light as an Ecological Factor* (ed. Bainbridge, R., Evans, G. C. and Ruckham, O.), pp. 375 – 391. Blackwell, Oxford.
Lythgoe, J. N. (1972). The adaptation of visual pigments to the photic environment, in *The Handbook of Sensory Physiology* (ed. Dartnall, H. J. A.), Vol. VII/1, pp. 566 – 603. Springer-Verlag, Berlin, Heidelberg, New York.
Lythgoe, J. N. (1979) *The ecology of vision*. Clarendon, Oxford.
Manning, R. B. and Camp, D. K. (1983) Erythrosquilloidea, a new superfamily, and tetrasquillidae, a new family of stomatopod crustaceans. *Proc. Biol. Soc. Wash.* 106, 85 – 91.
Manning, R. B., Schiff, H. and Abbott, B. C. (1984) Eye structure and the classification of stomatopod crustacea. *Zoologica Scripta* 13, 41– 44.
Marshall, N. B. (1979) Developments in deep-sea biology. Blandford, Dorset.
Marshall, N. J. (1988) A unique colour and polarization vision system in mantis shrimps. *Nature*, 333, 557– 560.
Marshall N. J., Land M. F., King C. A. and Cronin T. W. (1991a) The compound eyes of mantis shrimps (Crustacea, Hoplocarida, Stomatopoda). I. Compound eye.structure: the detection of polarised light. *Phil.Trans. R. Soc. Lond. B* 334, 33 – 56.
Marshall N. J., Land M. F., King C. A., and Cronin T. W. (1991b) The compound eyes of mantis shrimps (Crustacea, Hoplocarida, Stomatopoda). II Colour. pigments in the eyes of stomatopod crustaceans: polychromatic vision by serial and lateral filtering. *Phil.Trans. R. Soc. Lond. B* 334, 57– 84.
Marshall, N. J. and Oberwinkler (1993) Electrophysiology of Mantis Shrimp retina. *Abstracts of the British Photobiology Society.*
Marshall, N. J. and Land, M. F. (1993) Some optical features of the eyes of stomatopods. I. Eye shape, optical axes and resolution. *J. Comp. Physiol. A* 173, 565 – 582.
Marshall, N. J., Jones, J. P. and Cronin, T. W. (1996) Behavioural evidence for colour vision in stomatopod crustaceans. *J. Comp. Physiol. A* 197, 473 – 481.
Martin, F. G. and Mote, M. I. (1982) Colour receptors in Marine Crustaceans: A second class of retinular cell in the compound eyes of *Callinectes* and *Carcinus. J. Comp. Physiol.*, 145, 549 – 554.
McFarland, W. N. and Munz, F. W. (1975a) Part II: The photic environment of clear tropical seas during the day. *Vision Res.*, 15, 1063 –1070.
McFarland, W. N. and Munz, F. W. (1975b) Part III: The evolution of photopic visual pigments in fishes. *Vision Res.*, 15, 1071–1080.
Minke, B. and Kirschfield, K. (1978) Microspectrophotometric evidence for two photo-interconvertible states of visual pigment in the barnacle lateral eye. *J. Gen. Physiol.*, 71, 37– 45.
Munz, F. W. and McFarland, W. N. (1977) Evolutionary adaptations of fishes to the photic environment, in *The visual system in vertebrates.* (ed. Crescitelli, F.), Springer, Berlin, Heidelberg, New York. pp. 193 – 274.
Muntz, W. R. A. and Mouat, G. S. V. (1984) Annual variations in the visual pigment of brown trout inhabiting lochs providing different light environments. *Vision Res.* 24, 1575 –1580.
Nässel, D. R. (1976) The retina and the retinal projection on the lamina ganglionaris of the crayfish *Pacifasticus leniusculus* (Dana). *J. comp. physiol.* 167, 341– 360.
Neville, A. C. and Luke, B. M. (1971) Form optical activity in crustacean cuticle. *J. Insect Physiol.* 17, 519 – 526.
Nilsson, D.-E. (1989) Optics and evolution of the compound eye, in *Facets of vision* (eds Stavenga, D. G. and Hardie, R. C.) Springer-Verlag, Berlin, Heidelberg, New York. pp. 30 –73.
Nilsson, D.-E. and Modlin, R. F. (1994) A mysid shrimp carrying a pair of binoculars. *J. exp. Biol.,.* 189, 213 – 236.
Nilsson, D.-E. and Pelger, S. (1994) A pessimistic estimateof the time required for an eye to evolve. *Proc. R. Soc. Lond. B* 256, 53 – 58.
Osorio, D., Marshall, N. J. and Cronin, T. W. (1997) Stomatopod photoreceptor spectral tuning as an adaptation for colour constancy in water. *Vision. Res.* In press.
Partridge, J. C. (1990) The colour sensitivity and vision of fishes, in *Light and Life in the Sea* (ed. Herring, P. J., Campbell, A. K., Whitfield, M. and Maddock, L.), pp. 167–184. Cambridge University Press.
Partridge, J. C., Shand, J., Archer, S. N., Lythgoe, J. N. and van Groningen-Luyben, W. A. H. M. (1989) Interspecific variation in the visual pigments of deep-sea fishes. *J. Comp. Physiol. A* 164, 513 – 529
Partridge, J. C., Archer, S. N. and Van Oostrum, J. (1992). Single and multiple visual pigments in deep-sea fishes. *J. mar. biol. Ass. U.K.*, 72, 113 –130.
Schehr, R. S. (1984) Spectral sensitivities of anatomically identified photoreceptors in the compound eye of *Daphnia magna*. PhD thesis, Columbia University.

Schiff, H. (1963) Dim light vision of *Squilla mantis L. Amer. J. Physiol.*, 205, 927– 940.
Scott, S. and Mote, M. I. (1974) Spectral sensitivity in some marine crustacea. *Vision Res.*, 14, 659 – 663.
Shand, J. (1993) Changes in the spectral absorption of cone visual pigments during the settlement of the goatfish *Upeneus tragula*: the loss of red sensitivity as a benthic existence begins. *J. Comp. Physiol. A.*, 173, 115 –121.
Shashar, N., Rutledge, P. S. and Cronin, T. W. (1996) Polarization vision in cuttlefish – a concealed communication channel? *J. Exp. Biol.* 199, 2077– 2084.
Shelton, P. M. J., Gaten, E. and Herring, P. J. (1992) Adaptations of tapeta in the eyes of mesopelagic decapod shrimps to match the oceanic irradiance distribution. *J. mar. biol. Ass. U.K.* 72, 77– 88.
Smith, F. E. and Baylor, E. R. (1953) Colour responses in the cladocera and their ecological significance. *Am. Nat.* 87, 49 – 55.
Smith, K. C. and Macagno, E. R. (1990) UV photoreceptors in the compound eye of Daphnia magna (Crustacea, Branchiopoda). A fourth spectral class in a single omatidium. *J. Comp. Physiol. A* 166, 597– 606.
Snyder, A. W., Menzel, R. and Laughlin, S. B. (1973) Structure and function of the fused rhabdom. *J. Comp. Physiol. A* 87, 99 –135.
Stavenga, D. G. and Schwemer, J. (1984) Visual pigments of invertebrates, in *Photoreception and Vision in Invertebrates* (ed. Ali, M. A.), Plenum Press, New York and London.
Stavenga, D. G., Smits, R. P. and Hoenders, B. J. (1993) Simple exponential functions describing the absorbance bands of visual pigment spectra. *Vision Res.*, 33, 1011–1017.
Stearns, S. C. (1975) Light responses of *Daphnia pulex. Limnol. Oceanogr.* 20, 564 – 570.
Stearns, D. E. and Forward, R. B. Jr. (1984) Photosensitivity of the calanoid copepod, *Acartia tonsa. Mar. Biol.*, 82, 85 – 90.
Stowe, S. (1980) Spectral sensitivity and retinal pigment movement in the crab *Leptograspsus variegatus* (Fabricius). *J. Exp. Biol.*, 87, 73 – 98.
Suzuki, T., Makino-Tasaka, M. and Eguchi, E. (1984) 3-dehydroretinal (vitamin A_2 aldehyde) in crayfish eye. *Vision Res.*, 24, 783 –787.
Suzuki, T. and Eguchi, E. (1987) A survey of 3-hydroretinal as a visual pigment chromophore in various species of crayfish and other crustaceans. *Experientia* 43, 1111–1113.
Wald, G. and Hubbard, R. (1957) Visual pigment of a decapod crustacean: the lobster. *Nature (Lond.),* 180, 278 – 280.
Wald, G. and Seldin, E. B. (1968) Spectral sensitivity of the common prawn, *Palaemonetes vulgaris. J. Gen. Physiol.*, 51, 694 –700.
Wald, G. and Rayport, S. (1977) Vision in annelid worms. *Science* 196, 1434 –1439.
Waterman, T. H. (1981) Polarization sensitivity, in *Handbook of sensory physiology*. VII/6C (ed. Autrum, H.), Springer Verlag, Berlin, Heidelberg, New York. pp. 281– 469.
Wehner, R. (1983) The perception of polarized light. in *The biology of photoreception* (eds Cosens, D. J. and Vince-Prue, D.), Symposia of the Society for Experimental Biology. Symposium XXXVI. Cambridge University Press, Cambridge, London, New York. pp. 331– 369.
Widder, E. A., Hiller-Adams, P. and Case, J. F. (1987) A multichannel microspectrophotometer for visual pigment investigations. *Vision Res.,* 27, 1047–1055.
Widder, E. A., Latz, M. I. and Case, J. F. (1983) Marine bioluminescence spectra measured with an optical multichannel detection system. *Biol. Bull.* 165, 791– 810.
Wood, P., Partridge, J. C. & De Grip, W. J. (1992) Rod visual pigment changes in the elver of the eel *Anguilla anguilla* L. measured by microspectrophotometry. *J. Fish Biol.* 41, 601-611.
Zeiger, J. and Goldsmith, T. H. (1989) Spectral properties of porphyropsin from an invertebrate. *Vision Res.* 29, 519 – 527.

10
Functional organization of the outer retina in aquatic and terrestrial vertebrates: comparative aspects and possible significance to the ecology of vision

M.B.A. DJAMGOZ[1,2]*, S. VALLERGA[2,3] and H-J. WAGNER[4]
[1]Department of Biology, Neurobiology Group, Imperial College of Science, Technology and Medicine, London SW7 2AZ, U.K.
[2]International Marine Centre and [3]Ecology of Vision Section, Institute of Biophysics and Cybernetics – CNR, 09072 Torregrande, Oristano, Italy.
[4]Institute of Anatomy, Eberhard-Karls University of Tuebingen, 72 –74 Tuebingen, Germany.
*Corresponding author

10.1 Introduction

The visual world is extremely diverse in several different respects. For example, the intensity of terrestrial illumination can vary by 10^{10}–10^{11} times, even during the course of one day. Visual conditions are made even more complex after light enters under water due to the additional effects of scattering. Thus, depending upon the type of water (e.g. freshwater, brackish vs. ocean), the quality and quantity of suspended material, depth, season and viewing angle, visual including chromatic factors can change significantly (e.g. Loew & McFarland, 1990). Migratory aquatic animals may be subject to an even more dynamic situation as they move vertically or horizontally through their environment. It would follow, therefore, that the visual system, including the retina, must adapt, on varying time scales, in order to retain optimal functional efficiency.

Visual information in the optical image focussed on the retina is extracted by photoreceptor cells and processed further by the post-receptoral neural retina in both serial and parallel pathways. In the outer retina, these pathways comprise, on the one hand, interactions amongst the photoreceptor cells (rods and cones) and the second-order neurones (horizontal and bipolar cells), and on the other, the modulatory role of the dopaminergic interplexiform cells and possibly other neuromodulatory mechanisms.

This Chapter deals with the functional organization of the vertebrate outer retina in relation to visual ecology. It covers morphological and physiological aspects of the outer plexiform layer (OPL) in a comparative approach to terrestrial *vs.* aquatic animals (mainly mammals and fish, respectively). Some species comparisons are also made within each class, where interesting. Particular emphasis has been placed upon the plasticity of the functional organization of the OPL in relation to changes in visual conditions, and the role played by dopamine

S.N. Archer et al. (eds.), Adaptive Mechanisms in the Ecology of Vision, 329–382

in this respect. Extensive reviews of basic neurobiological aspects of the vertebrate (fish and mammalian) retinae were published earlier (Douglas & Djamgoz, 1990; Djamgoz *et al.*, 1995a). A complementary account of the inner retina is given in the next Chapter.

10.2 General aspects of cellular and synaptic organization of the outer plexiform layer

10.2.1 Overview

The OPL constitutes the first synaptic layer of the retina. In the 'through' pathway, photoreceptor cells (PCs) are linked to bipolar cells (BCs) in the centripetal feed-forward projection. The classic intrinsic neurone of this layer is the horizontal cell (HC) which is thought to influence significantly the flow of signals from photoreceptors to BCs. In addition, there are neuromodulatory neurones, interplexiform cells (IPCs) which form a diffuse centrifugal pathway transmitting information from the inner plexiform layer (IPL) back to the OPL, in some cases as the final element of a neural chain originating in other parts of the central nervous system (CNS). Synaptic activity in the OPL is complicated by two additional factors: First, each of the main types of neurone can have several sub-types based on their morphology, physiology and/or neurochemical signature; this implies considerable potential for different degrees of specificity in their connectivities. Second, there are not only synapses linking one population of neurone to another, but in addition, multiple, and in general highly specific, interactions or couplings occur among cell types of each population.

This section considers the general cellular organization of the outer retina, the synaptic connectivity and the functional architecture of the OPL. Figures 1 and 2 highlight the main differences between schematised retinae of teleost fish (as an example of adaptation to the diversity of aquatic habitats) and mammal (adapted to the constraints of terrestrial environment).

10.2.2 Spatial arrangements of retinal neurones and synapses

Regular, mosaic-like patterning of neurones (sensory, motor etc.) is a typical feature of all nervous systems resulting in partitioning of cells into functional modules. In the retina, this general principle can be observed through all layers, from PCs to ganglion cells (GCs) (Wässle & Rieman, 1978; Marc, 1986). It is intuitively obvious that the patterning of photoreceptors determines the quality and resolution of the neural rendering of the optical image falling onto the retina, whilst the arrangement of ganglion cells limits the transmission between the eye and the brain. It was Helmholtz (1896) who originally suggested that the spatial frequency of photoreceptors represented the anatomical resolution limit of the eye. Modern information science has introduced the 'Sampling Theorem' stating that a regular sampling array with an inter-receptor spacing of d will allow an unambiguous reconstruction of signals band-limited to frequencies of $d/2$ ('Nyquist limit'; Yellott *et al.,* 1984). Signals above the Nyquist limit result in sampling artifacts known as 'aliasing'.

In the case of the primate fovea, cones form a regular triangular mosaic with a minimal centre-to-centre spacing of $2-3\ \mu m$. The optical qualities of the human eye transmit spatial frequencies of 1 min/arc, corresponding well to the Nyquist limit, and ensuring aliasing-free sampling (Williams, 1986; Boycott & Wässle, 1991). Outside the fovea, cone density decreases more rapidly than the optical qualities of the eye; the potential danger of aliasing in this region is counteracted by a less regular arrangement of cones (Wässle & Riemann, 1978; Yellott, 1982). The arrangement and spacing of HCs has been studied in several mammals and found to be in the form of a regular mosaic (Wässle & Riemann, 1978). With respect to the array of synapses in the OPL, no systematic data are available; however, it may be assumed that they follow the basic arrangement of cones and HCs, but it is unclear whether they contribute to the problem of aliasing.

In many fish retinae, photoreceptors show a mosaic-like arrangement which differs from that in mammals in that it may not be statistically significant, although the appearance is highly ordered and repetitive with crystal-like arrays reminiscent of arthropod ommatidia (Ryder, 1895; Figure 1). The most common type is the square pattern consisting of double and single cones. Since, again unlike mammals, teleost cones with different chromatic sensitivities can be discriminated morphologically, a typical square mosaic is composed of (i) double cones, the principal member of which is long wavelength-sensitive (LWS) and the accessory member middle wavelength-sensitive (MWS); (ii) a central single cone (occupying the space where the double cone clefts would intersect) which is short wavelength-sensitive (SWS); and (iii) an ultraviolet-sensitive (UVS) miniature single cone located in the corner of the square (Levine & MacNichol, 1979). This type of square mosaic can be found in many visually oriented teleosts such as salmonids and percids, including cichlids (review: Wagner, 1972). It arises from a row pattern of alternating double and single cones found at the peripheral growth zone of the retina. Molecular biological techniques suggest that SWS or UVS cones are the first to be committed and that these somehow induce the differentiation of the LWS and MWS double cones (Raymond, 1995), thus emphasising the role of SWS/UVS cones in maintaining a regular pattern and spacing even when the arrangement of the other cone types is more irregular (e.g. goldfish and other cyprinids: Stell & Lightfoot, 1975; H-J. Wagner, unpublished observations; Figure 9C lower right – tangential view of H3 HC cone-connectivity; primates: Wikler & Rakic, 1990). The UVS cone is interesting also because, in retinae of many teleosts (e.g. trout, cyprinids), it is present only in juvenile specimens and lost during maturation (Avery *et al.,* 1983; Bowmaker & Kunz, 1987; Bowmaker *et al.,* 1991).

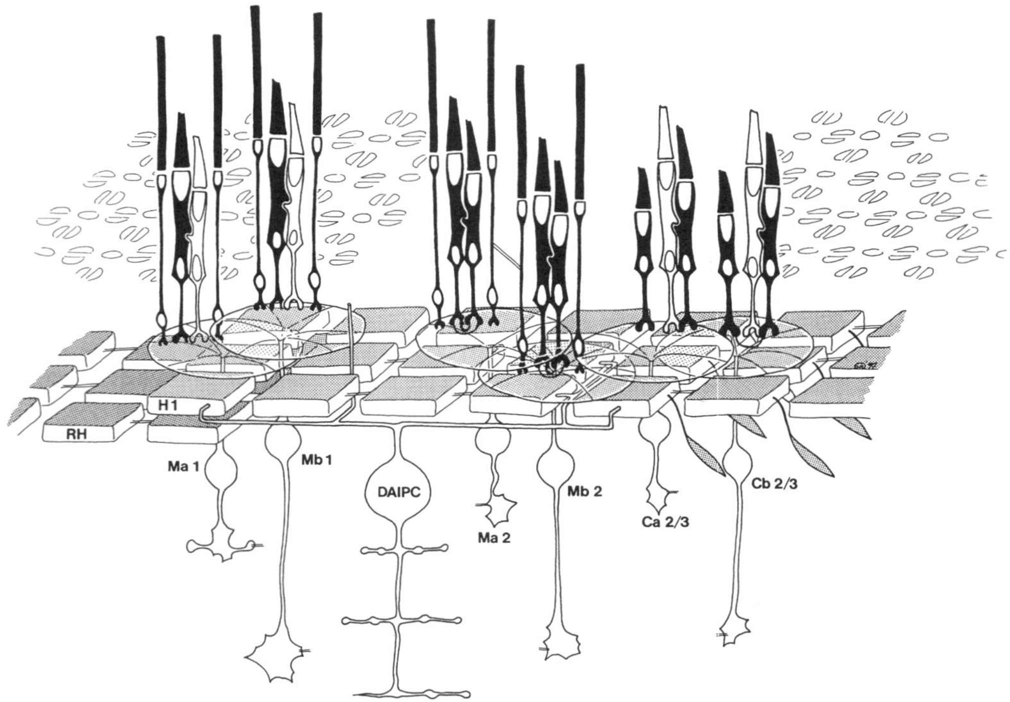

Figure 1 Highly schematical diagram of the outer plexiform layer and the participating cells of a (hypothetical) typical teleost with a mosaic retina in a combination of a lateral view and a perspectival representation demonstrating the regular, tile-like arrangement of horizontal cells and a square-type cone mosaic composed of double cones (longer member: long wave sensitive [LWS] and shorter member: medium wave sensitive [MWS]) and one type of single cone (short wave sensitive [SWS]). A single row of cone horizontal cells (H1) is shown, with axons and axon terminals included only at the right hand side for greater clarity; rod horizontal cells are shown on the left hand side; they carry no axon. Bipolar cells are depicted as ON/OFF pairs (based on Scholes, 1975). Ma/Mb1 cells are mixed bipolar cells receiving input from LWS and rods; Ma/Mb2 bipolar cells receive input from LWS and MWS cones and occasional rods; Ca/b2/3 bipolar cells are cone specific and contact either MWS or SWS cones. In addition, a dopaminergic interplexiform cell (DAIPC) is involved in the synaptic interactions of the outer plexiform layer. Double bars indicate presence of (gap junctional) coupling

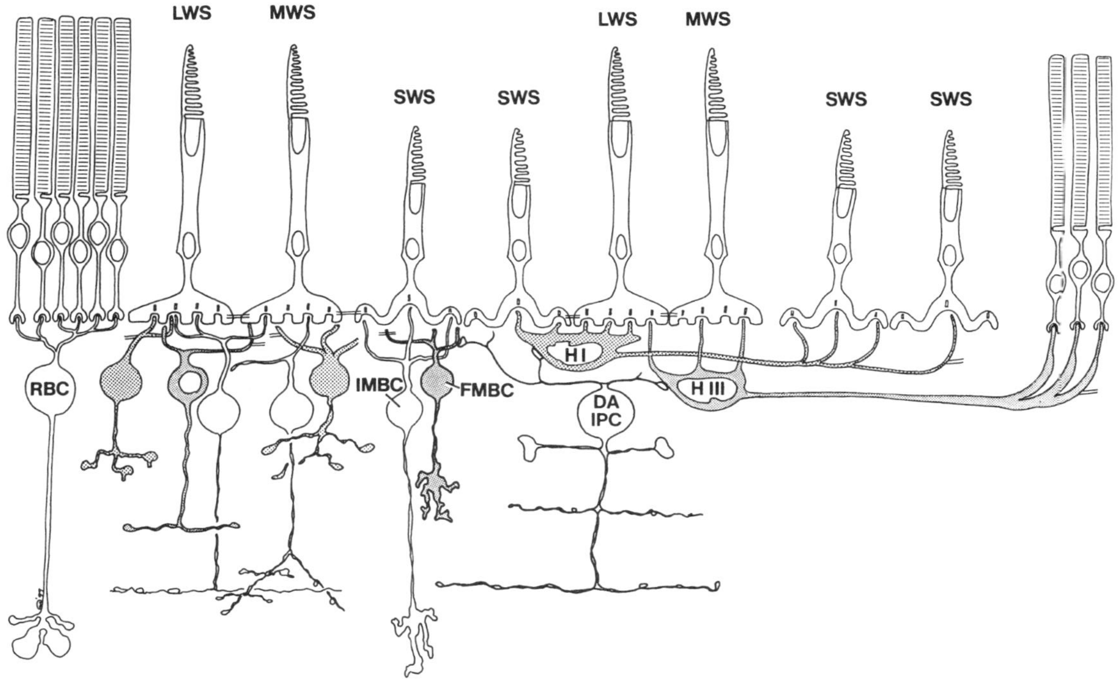

Figure 2 Highly schematical diagram of the outer plexiform layer and the participating cells of a (hypothetical) typical mammal in lateral view. Rod photoreceptors shown on the left and right; they are linked to rod bipolar cells (RBC) and axon terminals of B-type horizontal cells (H1); they are not linked by gap junctions (double bars). Large cone photoreceptors are LWS or MWS; smaller cones are SWS. They are presynaptic to various types of bipolar cells: midget cells contacting only a single cone (IMBC, FMBC), and several other bipolar cell types; they all occur as ON/OFF centre pairs; their dendrites are linked by gap junctions. Gap junctions are also present between rods and cones. Horizontal cells have perikaryal dendrites linked to cones. Horizontal cells are coupled by gap junctions (HI and HIII represent horizontal cells of the human retina; Ahnelt & Kolb, 1994). DAIPC dopaminergic interplexiform cell. Large cones are LWS and MWS; small cones are SWS

The biological significance of row or square mosaics has been the subject of numerous speculations, yet little direct experimental evidence is available to show that mosaic retinae might have superior visual capabilities compared to non-mosaic retinae. The one conclusive case concerns the perception of polarized light, which has been shown electrophysiologically and behaviourally in rainbow trout (Novales-Flamarique & Hawryshyn, 1997) and green sunfish (Cameron & Pugh, 1991). It has thus been argued that the interaction of LWS/MWS double cones and UVS single cones forms the basis of a W-shaped polarization sensitivity function (Parkyn & Hawryshyn, 1993).

It is relevant to note that cells of the inner nuclear layer (INL) can also participate in a crystal-like lattice arrangement. This is especially true for the most distal (H1) type HCs, the perikarya of which are localised under the central single cones of the square mosaic (Wagner, 1972, 1978) and BCs (Hibbard, 1971; Haesendonck & Missotten, 1983b). In this type of mosaic retina, the regular arrangement also includes the cone pedicles and their synapses (Wagner, 1978). These retinae are ideal, therefore, for connectivity studies (Braun *et al.*, 1997). However, the functional (as well as any ecological) significance of such modular arrangements remain to be elucidated.

10.3 Photoreceptor cells

10.3.1 Basic morphological and physiological aspects

All vertebrate PCs (rods and cones) have the same basic cellular architecture comprising the following functional compartments: (1) An outer segment incorporating some 1000 membranous discs which contain the photopigment molecules. This is where the phototransduction machinery is located. (2) An ellipsoid region containing a high density of mitochondria (one of the highest in the CNS) which is essentially the cell's metabolic 'factory'. (3) An inner segment and cell body. (4) A bulbous synaptic terminal where photoreceptor signals are transmitted to HCs and BCs. It is well known that rods are involved in dim-light vision and, consequently, are relatively more abundant in nocturnal mammals and deep-sea fish. On the other hand, cones mediate bright-light including colour vision. The degree of the latter varies depending upon the presence of different chromatic (LWS, MWS, SWS or UVS) types of cone. In a few species (e.g. frogs) rods also appear to have limited chromatic differentiation (e.g. Liebman & Entine, 1968). Some freshwater fish (e.g. cyprinids) have unusually wide spectral sensitivities, ranging from some 380 nm to 700 nm (Avery *et al.,* 1983; Bowmaker, 1984; Downing *et al.,* 1986), probably reflecting the extent and dynamic nature of the chromatic conditions in shallow water.

In fishes living in murky waters, and/or depths which are below the photic zone of the ocean (*ca.* 200 m), photoreceptors display numerous specializations adapting these retinae to the constraints of the photic environment. On the whole, such morphological adaptations include (i) reduction of cones and dominance of the rod system up to the the occurrence of all-rod retinae; (ii) elongation of rod outer segments and/or arrangement of rods in multiple banks; and (iii) grouping of rods and/or cones into larger bundles (reviews: Locket, 1977; Wagner & Ali, 1978; Wagner, 1990). In multi-bank retinae, rod synapses are also arranged in several tiers. As for retinae with bundled photoreceptors, two observations suggest that each bundle acts as a macro-receptor: Pigment epithelial cells and their processes form pits to accommodate each bundle (while isolating it from its neighbours), and reflective material occurs within each bundle presumably to increase photon catch (Locket, 1977).

The mechanism of phototransduction is essentially the same in all vertebrate PCs and has been reviewed extensively (e.g. Kawamura, 1995). In essence, in the dark, cyclic nucleotide-gated (CNG) cation channels present in the plasma membrane of the outer segment are kept open by the allosteric action of cyclic guanosine monophosphate (cGMP) molecules acting as second messengers. Consequently, cations (mainly Na^+ and Ca^{2+}) rush in and cause the plasma membrane to depolarize to –30 to –40 mV. During illumination, photolysed rhodopsin (a G-protein-coupled receptor) activates transducin (G-protein) which, in turn, stimulates a phosphodiesterase (PDE) enzyme. The latter hydrolyzes cGMP and thus leads to closure of the CNG channels, thereby generating an outward current and hyperpolarizing the plasma membrane (Figure 3). This is a 'slow', graded response which is transmitted synaptically to BCs and HCs. Rods are significantly more sensitive to light than cones due to differences in the intrinsic amplification of the transduction cascades (Lagnado & Baylor, 1992). Furthermore, synaptic gain associated with rod-driven signal transmission may be higher, also adding to the sensitivity of the rod system (Falk & Shiells, 1995). The neurotransmitter of photoreceptors is glutamate which is released in the dark when the cells are depolarized; release of glutamate is suppressed during the light-evoked response when the cells hyperpolarize (Copenhagen & Jahr, 1989; Trifonov, 1968; Kaneko & Shimazaki, 1975).

The dark-evoked Ca^{2+} influx through the CNG channels has an important role in *adaptation* of the photoreceptors: Ca^{2+} inhibits guanylate cyclase (the enzyme synthesizing cGMP) and thus tends to induce a light-adapted state. Conversely, reduction of Ca^{2+} influx during the light-evoked response disinhibits the cyclase and exerts a dark adaptation effect (review: Kawamura, 1995).

10.3.2 Functional organization of photoreceptor synaptic terminals

Photoreceptor terminals belong to the more complex synaptic arrays found in the vertebrate CNS and may contain up to about 100 dendritic processes from the OPL engaging in functional contact with a single PC. For this purpose, PCs have enlarged vitread ends, small and spherical in rods (hence the term 'spherule'), and more conical, with a flattened base in cones (hence the term 'pedicle')

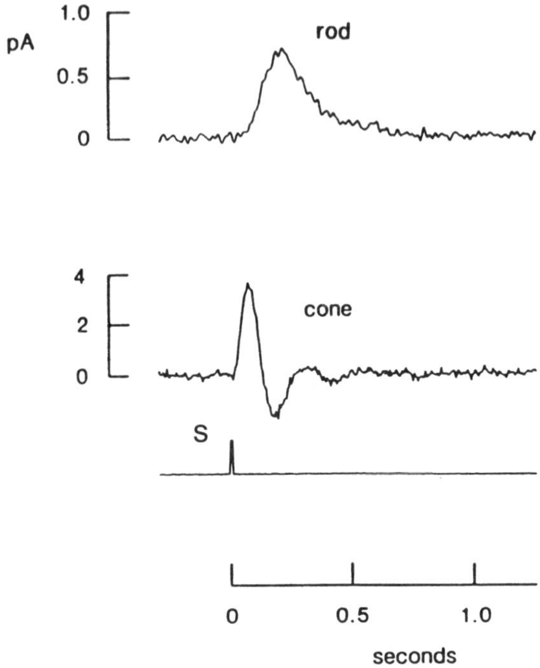

Figure 3 Kinetics and amplification of rod and LWS-cone responses to weak light stimuli(s) recorded from the macaque retina using the extracellular suction electrode method. Changes in outer segment membrane current, averaged from multiple trials, is plotted as a function of time indicated on the lowermost trace. For the rod, the flash caused an average of one isomerization per trial; at the peak of the response, 3.3% of the channels in the outer segment closed. For the cone, the flash caused an average of 35 isomerizations per trial, and one isomerization caused an estimated 0.36% of the channels to close. Modified from Lagnado & Baylor (1992)

(Figure 4). While the overall morphology of rod spherules is highly conserved throughout all vertebrates, the shape and complexity of cone pedicles vary considerably depending on species and spectral identity. Teleost cone pedicles are bell-shaped, containing a single cavity, in which most of the synaptic interactions take place (Stell, 1976); in mammals, the base is rather flat and invaginated at several independent sites (Figure 4; Dowling & Boycott, 1966; Kolb, 1970, 1977).

The most conspicuous structure common to all vertebrate PC terminals is a plate- or crescent-shaped structure, called 'synaptic ribbon' (from its ultrastructural appearance when viewed in cross section; Sjöstrand, 1953). Ribbons play an important part in determining the pattern of synaptic connectivity within PC terminals, as well as providing a cytoskeletal scaffold designed to facilitate exocytosis of synaptic vesicles at tonically active synapses (review: Wagner, 1997). In the 'classic' arrangement, a photoreceptor ribbon synapse consists of a

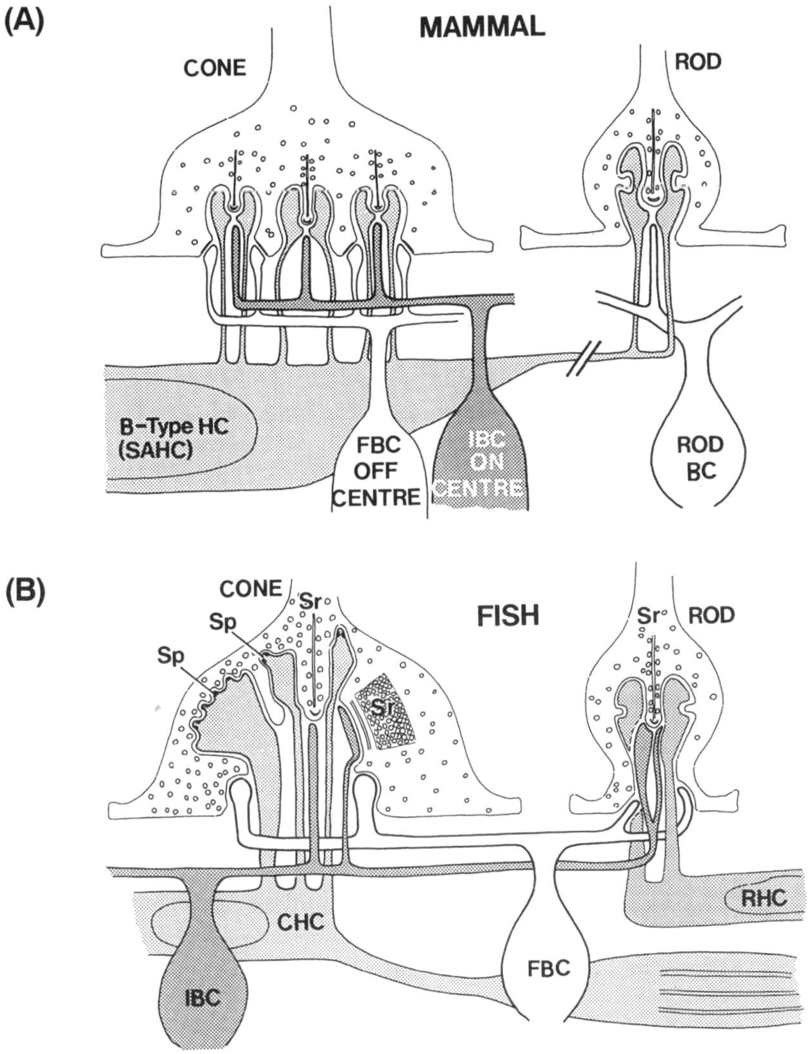

Figure 4 Schematic diagram of the basic organisation of photoreceptor terminals and their synaptic contacts with bipolar and horizontal cells. In mammals (A), cone pedicles have flat bases and several invaginations, each containing a ribbon synapse. Some bipolar cell dendrites make invaginating contacts as central elements of the ribbon synapse (invaginating bipolar cells, IBC) mediating sign inverting transmission (ON-centre responses), while other bipolar cell dendrites terminate more superficially, at flat contacts (flat bipolar cells, FBC), mediating sign conserving transmission (OFF centre responses). Additional contacts not shown. Horizontal cell dendrites (from a B-type, SAHC) form the lateral elements at the ribbon synapse. Rod spherules typically contain a single ribbon synapse, two horizontal, and a bipolar cell (ROD BC) dendrites. During dark adaptation, finger-like extensions of the rod spherule invaginate into the horizontal cell dendrites. In teleosts (B), cone pedicles contain a single cavity with numerous ribbon synapses. Bipolar cell dendrites are found at invaginating, and flat positions. Horizontal cell processes form lateral elements of ribbons synapses, but may also occur as central elements (not shown). During light adaptation, the lateral processes of horizontal cells carry finger-like extensions, called spinules (Sp), and are thought to meditate feedback transmission. CHC cone horizontal cell; RH rod horizontal cell; IBC invaginating bipolar cell; FBC flat bipolar cell; Sr synaptic ribbon

triad of post-synaptic elements: Two flanking ('lateral') HC dendrites, and a BC terminal as 'central' component beneath the inner edge of the ribbon. It should be noted, however, that additional patterns of connectivity are associated with synaptic ribbons, especially in fish retinae (Stell, 1976); such patterns are characteristic for the different spectral cone types (e.g. Downing & Djamgoz, 1989). In addition to ribbon synapses, there are other contact sites at the photoreceptor synapse, where dendrites do not invaginate as deeply into the pedicle, but engage instead at or near the base of the pedicle.

Rod spherules have been used as a model system for detailed studies of the processes taking place at ribbon synapses. Compared to cones, the morphology of the spherule is simple, typically consisting of a single synaptic ribbon, two HC and two BC processes (goldfish: Stell, 1967; cat: Rao-Mirotznik *et al.,* 1995). In mammals, physiology of rod spherules is characterised by a very short integration time (*ca.* 250 ms) and a sensitivity in the scotopic range which is less than one photon per integration time (Sterling *et al.,* 1987). Thus, the basic operation of the rod synapse is binary, signalling the presence of either 0 or 1 molecule of bleached photopigment. This signal can be picked up at the GC level as a burst of 2 – 3 spikes (Barlow *et al.,* 1971). Since the absorption of light leads to a reduction of the rate of transmitter release, the faithful rendering of a photon absorption at the level of the rod synapse critically depends on the capability of the synaptic terminal to ensure tonic transmitter release. The latter occurs through an uninterrupted sequence of vesicle exocytosis, thus enabling any interruption to be interpreted as the consequence of rhodopsin bleaching. The synaptic ribbon plays a special role in this process: It has a surface area of some 0.77 μm^2 and the edge directed towards the OPL is some 2.2 μm long. This edge corresponds to the active zone next to which exocytosis of vesicles occurs. It has been calculated that about 130 vesicles can be accommodated ready for release along the active zone of the ribbon edge. In the absence of any photic stimulus, at a membrane potential of –40 mV, at least 40 vesicles/second (v/s) are released (Rao-Mirotznik *et al.*, 1994); given the 130 docking sites, the turnover time would be 3.25 v/s. The total number of vesicles attached to the plate-like synaptic ribbon is about 770, and the special role of the ribbon would be to ensure an uninterrupted flow of vesicles towards the active zone by 'catching' those free-floating in the cytoplasm and forwarding them via actin dependent mechanisms to the site of release (review: Wagner, 1997).

10.3.3 Photoreceptor coupling

Many PCs extend lateral, fin-like projections near their soma at the level of the external limiting membrane. In addition, there are often short, lateral processes emerging from the synaptic terminal in the OPL. These were first noted by Cajal (1892) and have later been shown to interconnect rods, rods and cones, and cones in a cell-type- and species-specific manner (Figures 1 & 2). *Rod-rod* coupling occurs via both lateral fins and telodendria. In fish and amphibia, the interdigitating fins are linked by gap junctions (Witkovsky *et al.,* 1974; Gold & Dowling, 1979; R.H. Douglas and H-J Wagner, unpublished observations). Compared to other gap junctions in the CNS, the photoreceptor gap junction channels are smaller by a factor of ten and, thus may be expected to have different conductance properties (Gold, 1979). In amphibia, the electrical coupling between rods is specific in that only MWS rods are not linked to LWS rods, but LWS rods are strongly coupled among themselves (Gold & Dowling, 1979). In mammals such as rabbit, monkey (Raviola & Gilula, 1975) and cat (Kolb, 1977), rods lack lateral fins and no gap junctions have been observed. Morphological studies on functional *cone-cone* contacts have been performed in lower vertebrates and mammals. In turtles and teleost fish, there are several telodendria originating from the edges of the cone pedicles and invaginating the cavities of neighbouring cone pedicles. In turtles, they terminate as basal junctions (Ohtsuka & Kawamata, 1990), while in carp and goldfish, the telodendria can form components of the ribbon synapses (Stell, 1980). The telodendritic processes linking the cone pedicles have a strict chromatic specificity. In cyprinid teleosts, MWS cones send pedicles to LWS cones, which, in turn project back onto MWS cone pedicles. Furthermore, SWS cones contact MWS cones but not LWS cones, and, in goldfish, are contacted by LWS and MWS cones (Scholes, 1975; Stell, 1980; Lockhart & Stell, 1979).

Except for mammals, cones are often found associated in pairs, triplets or even quadruplets. In the case of double cones, some are twin cones with identical visual pigments, but in most cases, they consist of a MWS and a LWS partner. Because of the close physical association of the inner segments and, more specifically, extensive submembraneous cisterns of smooth endoplasmic reticulum along the plasmalemmata facing each other, it has been speculated that double cones might act as macroreceptors (Baldridge *et al.,* 1987). However, no morphological evidence of gap junctions has been found at this level, and no indication of dye-coupling observed between the members of double cones (Burkhardt, 1986; Baldridge *et al.,* 1987). In mammals, in contrast to lower vertebrates, cone-cone junctions show no chromatic selectivity (Cohen 1965; Raviola & Gilula, 1973). Quantitative analysis demonstrated that among the 150 or so cone pedicles located about 300 μm from the centre of the fovea, each makes some 5 junctions involving almost all of the neighbouring cones indiscriminately (Tsukamoto *et al.,* 1992). These junctions consisted of a desmosome-like contact and 1-6 focal gap junctions each comprising about 10 connexons. *Rod-cone* contacts have been found in amphibia and mammals; by contrast, they seem to be absent in teleosts and reptiles (Stell & Lightfoot, 1975; Stell *et al.,* 1982; Othsuka & Kawamata, 1990). Gap junctions are present between LWS rods and cones in toads (Gold & Dowling, 1979). Furthermore, morphological evidence for gap junctional coupling between rods and cones has been presented in cat (Kolb & Famiglietti, 1976; Sterling, 1983) and monkey

(Raviola & Gilula, 1975). In the area centralis of cats, each spherule is coupled to 2.4 cone pedicles, on average, and each cone pedicle is linked to about 48 rod spherules (Smith *et al.,* 1986).

The functional aspects and consequences of PC coupling have been reviewed by Barnes (1995) and Djamgoz and Petruv (1995). In the case of *rod-rod* coupling, Owen and Copenhagen (1977) first confirmed in the turtle retina that this is electrotonic. Although coupling would reduce the signal-to-noise ratio of an individual rod (by dissipating the signal within the coupled network), it would allow the network of rods to average the individual responses and smoothen the fluctuations (thus maintaining an even adaptational state) before transmitting the signal to second-order neurones (Lamb & Simon, 1976; Barnes, 1995). The consequent increased input to BCs can, therefore, be considered to be a 'coincidence detector' – coincidence occurring in the rod network, the BC acting as the detector. Thus, a large post-synaptic response may be generated by small signals transmitted with high gain at numerous synaptic sites (Barnes, 1995). Another characteristic of the rod network, first reported in the turtle retina is 'negative propagation velocity' i.e. the peak response of a distant rod precedes that of the rod in which the signal was generated (Detwiler *et al.,* 1978). One functional consequence of this is to enable higher order neurones to 'focus their attention' upon the centre of the stimulus. The rod network can also be viewed as a 'spatial frequency filter' which suppresses the high frequencies. Thus, signalling of precise spatial location is lost when it is more important to capture what little visual information may be available at low light levels. High-acuity visual signalling is handled by the cone system at photopic levels when ample visual information enters the retina. Accordingly, in the turtle retina, the spatial extent of rod-rod interaction is nearly three times that of cone-cone interaction.

As regards *cone-cone* coupling, intracellular recordings in retinae of pikeperch (Burkhardt, 1977) and turtle (Baylor *et al.,* 1971) have shown that cones of a given spectral type summate their light-evoked responses over retinal areas of 50 – 250 μm. Electrical coupling via gap junctions between cones has been reported in the carp retina (Witkovsky *et al.,* 1974; Stell, 1980). Computer modelling of the efficiency of cone-cone coupling in walleye and salamander has indicated that bidirectional information transfer along the teleodendria could occur (Kraft & Burkhardt, 1986; Merchant & Barnes, 1991). Furthermore, although the cellular mechanisms remain unclear, there is evidence of signal transfer at the inner segment level of double and twin cones in teleost fish (Marchiafava, 1985; Miller & Korenbrot, 1993) suggesting that double cones may act as luminosity rather than strict chromaticity receptors. Weak *rod-cone* coupling also occurs and can contribute to photoreceptor spectral sensitivity in several retinae e.g. salamander (Wu & Yang, 1988; Hare & Owen 1992), turtle (Schwartz, 1975) and cat (Nelson, 1977).

10.4 Bipolar cells

Bipolar cells form the centripetal link between the first and second/final synaptic layer, conveying the electrical signals generated by the photoreceptors onto the ganglion cells (GCs). There are two main principles that govern the connectivity between photoreceptors and BCs. First, there is the functional dichotomy into ON-centre (i.e. centre-depolarizing) and OFF-centre (i.e. centre-hyperpolarizing) BCs, which is reflected structurally at the level of termination and ramification of the axon terminals in the inner plexiform layer (IPL). As demonstrated initially by Famiglietti and Kolb (1976) and Famiglietti *et al.,* (1977), the inner/proximal sublamina (*b*) of the IPL processes signals in the ON-centre pathway, and thus receives inputs from the long-axon BCs, whereas the outer/distal sublamina (*a*) deals with signal processing in the OFF-centre pathway and contains the synaptic terminals of the short-axon BCs (Figures 5 & 6; Schiller, 1992). Second, the functional organisation of BCs is determined by their rod/cone connectivity and by the chromatic sub-type(s) of the latter. In lower vertebrates, on the whole, mixed BCs combine inputs from rods and cones, whilst cones have separate pathways (Figure 5; Scholes, 1975; Stell *et al.,* 1977; Ishida *et al.,* 1980; Saito *et al.,* 1985). In contrast, in mammals, there is strict separation of rod and cone pathways at the BC level (Figure 6; Boycott & Kolb, 1973a; Boycott & Wässle, 1991; Wässle *et al.,* 1991).

10.4.1 Bipolar cell types and photoreceptor connectivities in teleost fishes

Cyprinid fishes have been studied most intensively with respect to the organization of BCs (Scholes, 1975; Saito *et al.*, 1985; Stell *et al.*, 1977; Ishida *et al.*, 1980; Sherry & Yazulla, 1993). A summary diagram of the current views is shown in Figure 5. In the first major systematic study, Scholes (1975) demonstated in the rudd the existence of two major populations of BCs – those connected to cones exclusively, and those with mixed rod/cone input. Selective cone-BCs were found to be delicate neurones with wide dendritic trees which contact MWS and SWS cones specifically (no BC was found which was linked exclusively to LWS cones). These findings have been confirmed for the goldfish where C2b BCs receive inputs from MWS cones, and C3b BCs from SWS cones, respectively (Ishida *et al.*, 1980). In later studies, 9 different types of cone-specific BC have been identified, taking into account also the pattern of axonal termination in the IPL (Sherry & Yazulla, 1993). In some of these BCs, the complementary pattern of *a/b* stratification of the axon terminals could be demonstrated, whereas in others, a more complex situation was found with synaptic swelling in several sublaminae of the IPL.

The mixed BCs of the cyprinid retina are robust cells with wide dendritic fields and prominent axon terminals (Figure 5). They were further subdivided into Ma/b1 cells, receiving input from rods and LWS cones, and Ma/b2 and Ma3 cells which contact to rods and MWS cones (Ishida *et al.*, 1980).

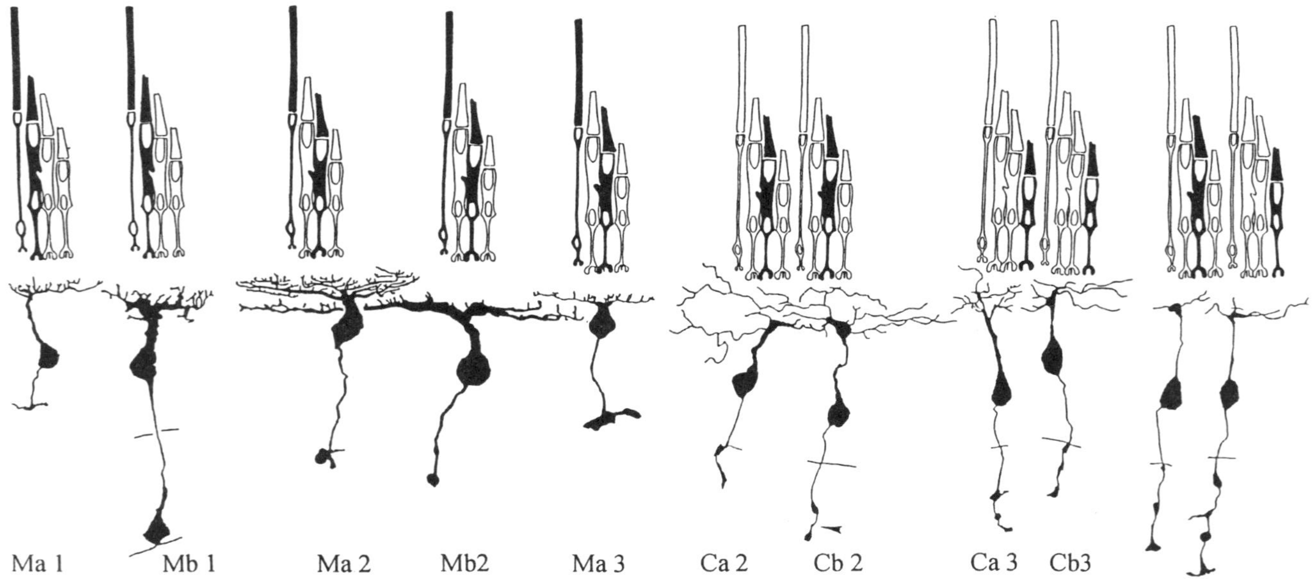

Figure 5 Bipolar cell types of the teleost retina in lateral view. In the retinae of cyprinid fish, there are several types of mixed bipolar cells receiving input from rods and cones (Ma/b1, Ma/b2, Ma3) and several types of cone-specific bipolar cells (Ca/b2; Ca/b3). In this figure, ON-and OFF-centre bipolar cells have been grouped as pairs. The filled photoreceptor symbols indicate the chromatic cone types and/or rods contacted by the respective types of bipolar cells (same as in Figure 1a): Principal double cone members are LWS; accessory double cone members are MWS; single cones are SWS. This illustration is based on figures by Scholes (1975), Ishida et al. (1980) and Sherry and Yazulla (1993)

10.4.2 Bipolar cell types and photoreceptor connectivities in mammals including primates

Mammalian retinae are characterised by separate rod and cone inputs to the BCs. A great physiological and morphological variety of cone-BCs is in marked contrast to the single type of rod-BC; the general characteristics of the latter are well conserved throughout mammals.

Rod-BCs are distinguished by a bushy dendritic tree and delicate dendritic branches. Depending on retinal eccentricity, these cells contact 15 – 40 rod spherules in cat and monkey (Figure 6; Boycott & Kolb, 1973a; Polyak, 1941), and up to 100 rods in rabbit (Dacheux & Raviola, 1986). Rod-BCs are immunoreactive to protein kinase C (Greferath *et al.,* 1990) and have their perikarya in the outer portion of the INL. Their axons terminate as large lobulated complexes exclusively in the innermost sublamina of the IPL, where, unlike other BCs which also project onto GCs, their output is directed via chemical synapses towards two types of amacrine cells (ACs) (Kolb & Famiglietti, 1976 McGuire *et al.,* 1986; for details, see next Chapter).

Most of the cone-specific BCs exhibit the classic OFF-ON/ *a-b* dichotomy; in addition, they can be subdivided based on their morphology and connectivity with different chromatic types of cone. In many species, BCs with dendritic fields matching roughly the dimensions of a single cone pedicle, termed 'midget' bipolars, are distinguished from other cell types with wider dendritic fields, 'diffuse' BCs (Polyak, 1941). In the following, the situation in the rod-dominated retina of the nocturnal rat will be compared to that of the primate retina: In rat, there are 9 different types of cone-BCs, constituting at least 50% of the entire BC population (Figure 6; Euler & Wässle, 1995). These belong to the population of 'diffuse' BCs and are distinguished on the basis of their dendritic trees and pattern of termination of the IPL. The fact that 4 of them resemble strongly diffuse type BCs of the primate retina may be taken as an indication that the morphological characteristics of BCs are well conserved among mammals. In macaque monkey, there are 2 types of 'midget' BCs, the ON-midget immunoreactive for cholecystokinin, and the OFF-midget immunoreactive for recoverin, which form the first distinct link in the neural chain of the parvocellular pathway (Wässle *et al.,* 1994). Up to a retinal eccentricity of 10 mm, the density of the OFF-midget BCs matches the cone density; in the periphery, however, several cones converge onto one OFF-midget BC (Wässle *et al.,* 1994). As for the diffuse BCs, 6 different types have been identified in the rhesus monkey (Figure 6; Boycott & Wässle, 1991). Apart from the IPL level(s) of termination, these cells differ in their dendritic branching pattern and contact 5 –10 cones with no apparent chromatic selectivity (Boycott & Wässle, 1991; Kolb *et al.,* 1992). In addition, there are other cone-specific BCs which contact SWS cones exclusively. Such BCs are characterised by large, diffuse dendritic trees, and occur as ON/ OFF pairs (Mariani, 1984; Kolb *et al.,* 1992). Finally, one or several types

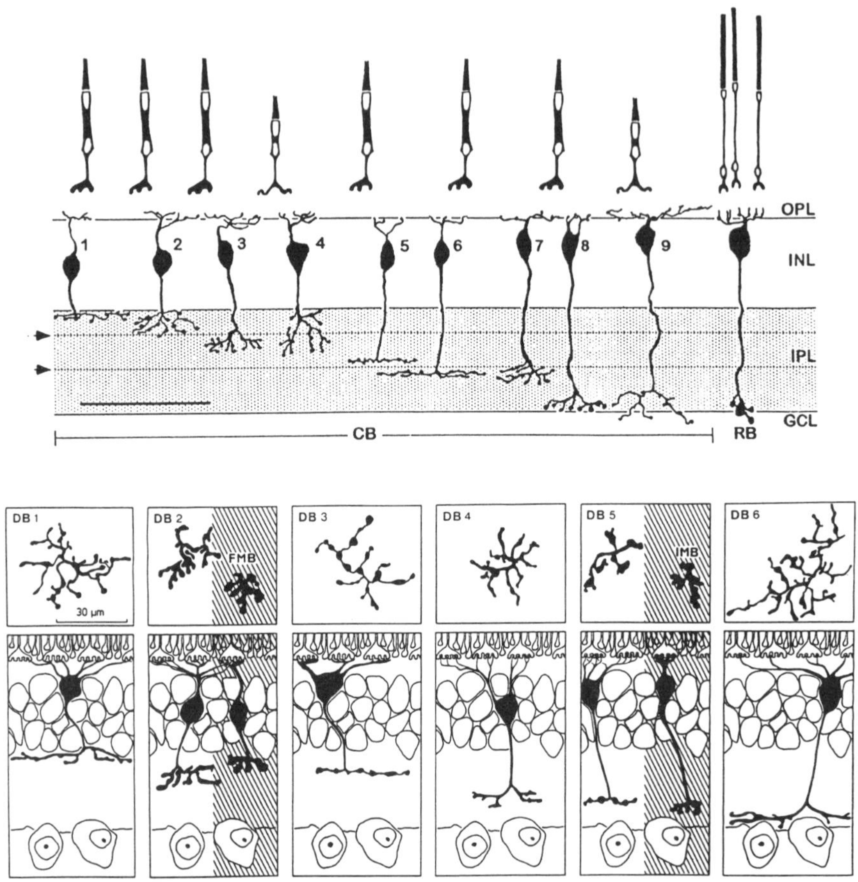

Figure 6 Bipolar cells of the mammal retina in lateral view. Upper panel: bipolar cells of the rat retina (from Euler & Wässle, 1995); types 1-8: diffuse cone bipolar cells (CB); type 9: rod bipolar (RB); OPL outer plexiform layer; INL inner nuclear layer; IPL inner plexiform layer; GCL ganglion cell layer; the chromatic identity of the cones types contacted by these cone-specific bipolar cells has not been determined; cone symbols identical to Figure 2. Lower panel: Cone bipolar cells of the macaque retina (from Boycott & Wässle, 1991) in lateral view and in wholemount view of the dendritic field; midget BCs have been highlighted by shading; DB1-6 diffuse BC; fmb flat midget BC; imb invaginating midget BCs; not shown in this graph are BCs selectively contacting SWS cones and giant bistratified BCs

of giant, bistratified BCs occur in the retina of many mammals (Mariani, 1983; Boycott & Wässle, 1991; Kolb *et al.,* 1992; Linberg *et al.,* 1996). Their dendritic trees are more than 100 μm across and appear to contact the overlying cones indiscriminately; their axons have terminal specialisations at two different levels of the IPL.

10.4.3 Photoreceptor – bipolar cell synaptic transmission

As already noted, there are two main physiological types of BC: ON-centre and OFF-centre (Werblin & Dowling, 1969; Kaneko, 1970, 1973). This difference is basically due to the cells receiving the photoreceptor signals via qualitatively different glutamate receptors expressed upon their dendrites within the photoreceptor terminals (review: Shiells, 1995). OFF-centre BCs possess ionotropic receptors where binding of glutamate in the dark directly opens cation channels thereby depolarizing the cells; during illumination, when the photoreceptor neurotransmitter is absent, the cells hyperpolarize. The post-synaptic signalling mechanism involving the ON-centre BCs is more complex. The rod input is mediated most commonly via metabotropic (i.e. G-protein coupled, mGluR6 type) glutamate receptors for which 2-amino-4-phosphonobutyrate (APB) is a specific agonist (Slaughter & Miller, 1981). Accordingly, binding of glutamate in the dark activates a signalling mechanism (rather similar to the transduction cascade of photoreceptors) whereby (i) a PDE is stimulated; (ii) cGMP molecules are hydrolysed; and (iii) cationic CNG channels are kept closed, thereby hyperpolarizing the membrane. During illumination, the membrane depolarizes due to the absence of glutamate. It would follow, therefore, that the ON-BCs are intrinsically capable of amplifying the visual signals. In fact, in the all-rod retina of dogfish, it has been estimated that synaptic transmission to ON-BCs has a gain of 100 – 200, compared to a gain of only some 10 in the OFF-BC/HC pathway (review: Shiells, 1995). The cone input(s) to ON-BCs involves another type of glutamate receptor that is not APB sensitive (Grant & Dowling, 1996) and may not be metabotropic (Grant & Dowling, 1995). The essential electrochemical characteristic of the cone input to teleost ON-BCs is that the ion(s) involved has a reversal potential more negative than the resting potential (Saito *et al.,* 1979) and could be Cl^- (Grant & Dowling, 1995, 1996) and/or K^+ (Nawy & Jahr, 1990; Hirano & MacLeish, 1991). Accordingly, these cells depolarize when transmitter release from PCs is reduced by light. The situation in the mammalian (cat) retina may be simpler with rod and cone inputs to ON-BCs sharing the same cGMP-dependent ionic mechanism (de la Villa *et al.,* 1995). However, much more work remains to be done to generalise these findings to other fish and mammalian species.

As regards the ultrastructural correlates of the different modes of PC → BC transmission, it can be stated to a first approximation (i) that BC dendrites making wide-cleft, basal junctions at some distance from the synaptic ribbons mediate sign-conserving, i.e. OFF-centre responses [BCs involved in this type of junction have traditionally been called 'flat' BCs]; and (ii) that BC dendrites engaged in narrow-cleft, invaginating contacts close to the ribbons mediate sign-inverting (i.e. ON-centre) responses; BCs with this type of contact have been referred to as 'invaginating' BC (Figure 4). However, there are other BC types in many vertebrate species where this simple dichotomy does not apply (Kolb &

Nelson, 1995). As a rule, the proximity to the synaptic ribbon appears to be less important for the transmission characteristics of the BC synapse than the detailed ultrastructure, such as the width of the synaptic cleft, the presence of extracellular material in the cleft, and the pattern of A- or B-face particles in the synaptic membranes (Raviola & Gilula, 1975).

10.4.4 Bipolar cell receptive field

In addition to the central zone, which is driven by direct PC input(s) to the cells' dendritic trees, BCs have a peripheral ('surround') receptive field where the response sign is opposite to that in the corresponding centre, i.e. BC receptive fields are centre-surround antagonistic (Figure 7). The surround region can be several mm's in extent and is due to HC input (either direct or indirect via negative feed-back to photoreceptors). The primary role of the surround field is to control the efficacy of PC → BC synaptic transmission in the through pathway, and thus greatly influence the signals reaching the GCs (Werblin, 1979). Since the strength of HC coupling varies depending upon the dark-light adaptational state of the retina (section 10.8.2), the antagonistic balance in the BC receptive fields can change with ambient light intensity.

In aquatic or semi-aquatic vertebrates (e.g. fish and turtles), there is evidence that the spatial organization of the cone input(s) to BC receptive fields has spectral selectivity (Kaneko, 1973; Kaneko & Tachibana, 1981, 1983; Yazulla, 1976). On the whole, two types of colour-coded BCs have been found (Figure 8): (i) *Single-colour-opponent BCs*. These receive centre responses from LWS cones (and rods), and opponent inputs from LWS- and MWS-cones in the periphery (Figure 8a). The peripheral responses may be driven by biphasic/chromaticity (H2) type HCs. The two sub-types of this kind of receptive field are red-ON-

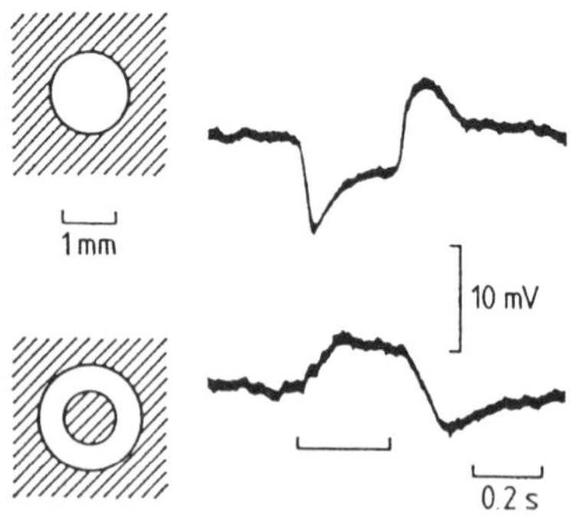

Figure 7 Basic light-evoked response characteristics of the bipolar cells, recorded in the carp retina. The responses are from an OFF-centre BC and demonstrate the centre-surround antagonistic receptive field. The unlabelled bar denotes presentation of stimuli: spot (above); annulus (below). Modified from Kaneko (1970)

centre/red- and green-OFF-surround; and red-OFF-centre/red- and green-ON-surround. (ii) *Double-colour-opponent BCs*. These are colour-opponent in both the centre and surround regions of the receptive fields (Figure 8b,c). One sub-type has a centre response comprising hyperpolarization to red light and depolarization to blue/green light. The other main sub-type is the exact mirror image for both centre and surround. These BCs are thought to be the origin of double-colour-opponent GC receptive fields (Daw, 1968).

The basic centre-surround spatial antagonistic organization of BC receptive fields, and its dependence upon variable HC coupling constitutes a powerful

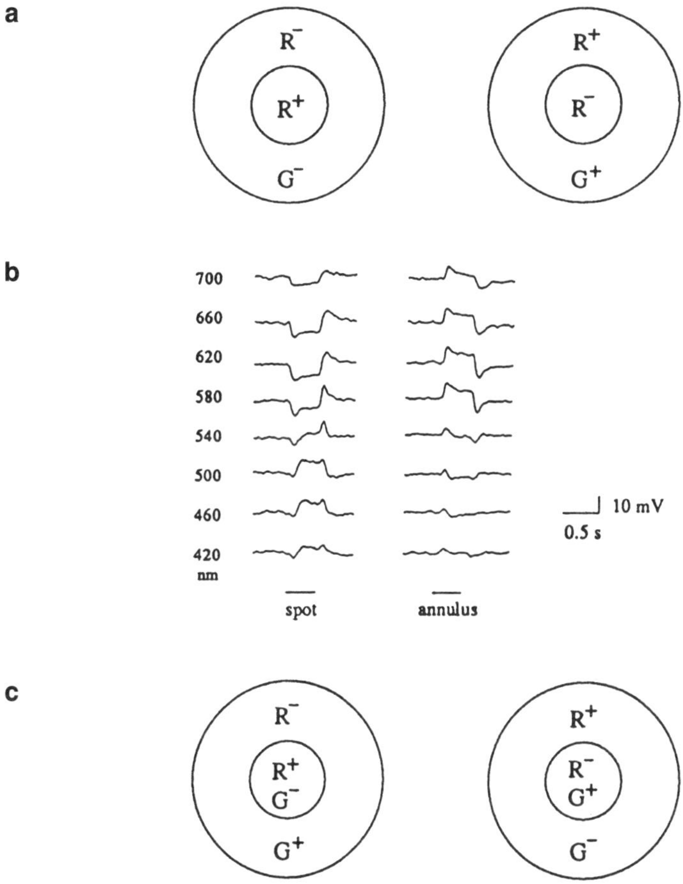

Figure 8 Chromatic aspects of bipolar cell receptive fields in the cyprinid fish retina. (**a**) Single-colour-opponent receptive fields; mirror-symmetric sub-types. (**b**) Double-colour-opponent light-evoked responses in the spectral range 420 –700 nm showing spectral opponency both for spot and annular stimuli. (**c**) Mirror-symmetric double-colour-opponent receptive fields. The receptive field organization shown on the right corresponds to the responses shown in (**b**). Modified from Djamgoz & Petruv (1995); original data shown in (**b**) from Kaneko & Tachibana (1981)

mechanism that most probably plays a crucial, early role in the way visual signalling in the outer retina is adjusted in response to changing illumination conditions. Furthermore, since BC receptive field also have chromatic organization, signalling by these cells could be influenced by the spatio-chromatic characteristics as well as the illumination level of the visual environment.

10.4.5 Bipolar cell coupling

Since BC receptive field centres in lower vertebrates appear to be about ten times larger than the cells' dendritic trees (tiger salamander: Borges & Wilson, 1987; carp: Saito & Kujiraoka, 1982, 1988) it was concluded that BCs are coupled. Indeed, prominent gap junctions were found between the telodendria of the large, mixed BCs in salamander (Wong-Riley, 1974), dogfish (Witkovsky & Stell, 1973) and goldfish (Marc *et al.,* 1988). While in these species the coupling appeared to be homologous, involving BCs of the same type, studies of the dragonet BC terminals showed evidence also for heterologous coupling, comprising mixed chemical and electrical synapses (van Haesendonck & Missotten, 1983a). Since the dendrites of pairs of electrically coupled BCs showed considerable overlap, it was suggested that coupling was also present at the OPL level (Saito & Kujiraoka, 1988). This was also found in goldfish BCs microinjected with neurobiotin (A.K. Ball unpublished data, cited by Vaney, 1994). Finally, Yamada and Saito (1997) confirmed by electrophysiological recording and modelling that the spatial summation profile by carp BCs has two components corresponding to coupling of dendritic fields and axon terminals, the latter being relatively (*ca.* 100-fold) stronger.

There is direct evidence for occurrence of gap junctions between BC dendrites also in mammalian retinae. Thus, Raviola and Gilula (1975) found gap junctions in both ON- and OFF-centre BC dendrites, probably making heterologous couplings with as yet unidentified partner BCs. Heterologous gap junctional coupling has also been found between BC axon terminals of primate, rabbit and cat on the basis of serial section and freeze fracture studies (Reale *et al.,* 1978; Kolb, 1979; Raviola & Raviola, 1982). In the case of the rabbit BCs, Vaney (1994) has raised the possibility that such coupling may be effective only in one direction, based on observations with neurobiotin-injected cells.

10.5 Horizontal cells

Horizontal cells (HCs) can be regarded as the 'intrinsic' neurones of the OPL. Their cell bodies form the inner border of this layer and are easily identified because of their perpendicular orientation relative to the centripetal mainstream pathway from photoreceptors to BCs. Especially in lower vertebrates, HCs are so numerous and so densely arranged that they appear to form 'fenestrated membranes' leaving room only for the radial processes of BCs and Müller cells. This observation from early morphological studies has been confirmed at the physiological level where HCs were found to be strongly coupled electrically forming the so-called 'S-space' (Naka & Rushton, 1967).

Attempts to distinguish different types of HCs have to take into account (i) the number of layers next to the OPL (outer, middle, internal etc.); (ii) the presence or absence of a short axon, giving short-axon HCs (SAHC) and axonless HCs (ALHCs; Gallego, 1985); and (iii) the pattern of connectivity with PCs (rod-specific; cone-specific, with possible chromatic selectivity etc.).

10.5.1 Sub-types, photoreceptor connectivities and light-evoked response patterns in teleost fishes

In many teleost fishes with highly developed colour vision, such as many cyprinids, three different types of SAHCs are found in addition to a single type of ALHC (Figure 9; Stell, 1967). The perikaryon of the latter is situated farthest away from the OPL in the INL; its stout dendrites run laterally at first, before turning radially and tapering to contact rod spherules exclusively, i.e. it is a rod-HC.

The cyprinid SAHCs are arranged in three tiers, usually referred to as H1, H2 and H3, and receive PC input from cones only (Figure 9A). Their delicate axons run horizontally for several hundreds of microns before increasing in diameter and forming 5 –10 μm thick cylindrical structures of about another 100 μm length which may or may not give off short lateral branches. These 'nematode'-like axon terminals course with a slight 'dip' and rest among AC and BC perikarya close to the IPL (Figure 9B). Conventional chemical synapses are formed upon BC somata or dendrites (60%) and interplexiform cell and AC perikarya (Sakai & Naka, 1986; Marshak & Dowling, 1987).

The pattern of connectivity of SAHCs with chromatic sub-types of cone has been studied most thoroughly in tetrachromatic cyprinids: goldfish (Stell *et al.*, 1975; Stell, 1980; Stell *et al.*, 1982, 1994); carp (Weiler & Zettler, 1979; Kamermans *et al.*, 1989a, b); and roach (Djamgoz & Downing, 1988; Downing & Djamgoz, 1989; Kirsch *et al.*, 1990; Greenstreet & Djamgoz, 1994a; Djamgoz & Greenstreet, 1996). The available evidence supports the following hierarchical connectivity scheme: H1 HCs contact all the chromatic types of cone within their dendritic trees; H2 HCs invaginate the pedicles of MWS, SWS and UVS cones; and H3 HCs interact with SWS and UVS cones (Figure 9C).

In fish species with fewer spectral cone types, there are correspondingly fewer types of cone HCs: In cichlid fish (*Aequidens pulcher*) with three spectral types cones, there are only two types of HCs – H1 cells link to all cone types while H2 cells contact MWS and SWS cones (Braun *et al.,* 1997). In the monochromatic catfish (having only LWS cones), there is only one type of cone-HC (Sakai & Naka, 1986). Detailed analyses and modelling of the presumed synaptic feed-forward and feed-back HC pathways within the cone pedicles have revealed an intricate cascade model of connectivity (widely referred to as the 'Stell model').

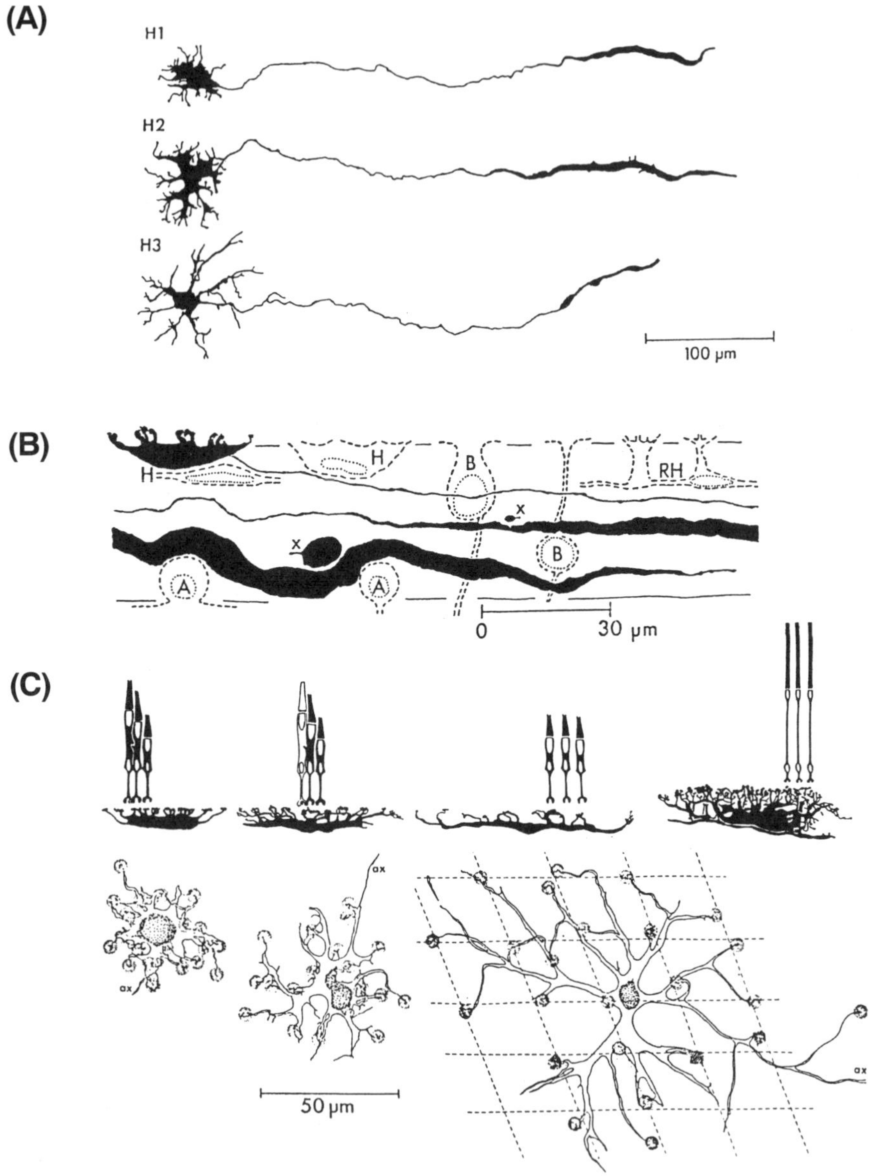

Figure 9 Horizontal cell types (H1-H3) and their relationship to various kinds of photoreceptor cells in the cyprinid retina. (A) Whole-mount view of cell bodies, axons and axon terminals. (B) Lateral view of the inner nuclear layer with the localisation of a H1 perikaryon, its axon and terminal (from Stell, 1975). (C) Connectivities of horizontal cells and photoreceptors, in lateral and tangential view: H1 contact all spectral cone types, H2 contact MWS and SWS cones and H3 contact SWS and UVS cones. RH rod horizontal cells are axonless. (from Stell & Lightfoot, 1975; photoreceptor symbols as in Figure 1)

This provides a convincing morphological basis for explaining the chromatic characteristics of the light-evoked electrophysiological signals recorded from the different types of cone-HCs. Accordingly, intracellular recording and correlative staining studies have shown the following (Figure 10):

1. H1 HCs generate generally hyperpolarizing ('luminosity' type) responses with a spectral sensitivity function that matches closely (but not exactly) the action spectrum of the LWS cone pigment. This is consistent with these HCs receiving their predominant cone input from LWS cones, about ten times stronger than other cones. H1 HCs provide negative feed-back to cones, most strongly to LWS but, significantly, also to MWS cones (Figure 11).
2. H2 HCs generate biphasic, red-depolarizing/green-hyperpolarizing ('chromaticity' type) responses. This is consistent with H2 cells being driven

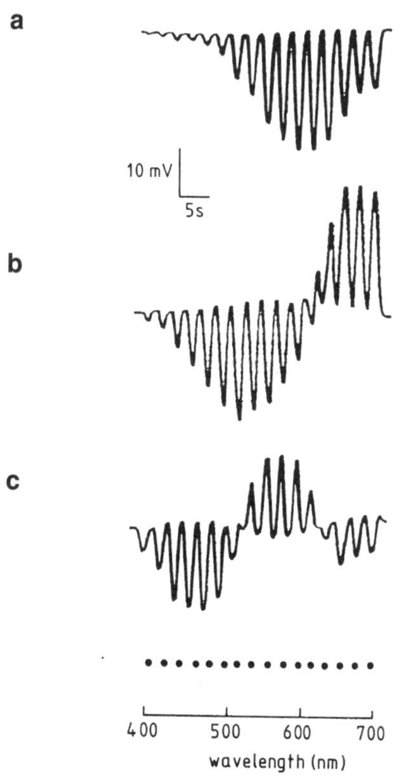

Figure 10 Chromatic aspects of light-evoked horizontal cell responses recorded in the cyprinid fish (roach) retina. (**a**) Long-wavelength-sensitive, luminosity (H1) type responses. (**b**) Biphasic, chromaticity (H2) type responses. (**c**) Triphasic, chromaticity (H3) type responses. The stimuli comprised sixteen different spectral flashes of near-equal quantum content in the range 400 to 700 nm (inclusive; 20 nm intervals). Modified from Djamgoz & Yamada (1990)

predominantly directly by MWS cones and indirectly by LWS cones (via H1 → MWS cone negative feed-back). In turn, the H2 HCs feed back strongly onto SWS and UVS cones (Figure 11).

3. H3 HCs generate triphasic, chromaticity responses (red-hyperpolarizing/green-depolarizing/blue-hyperpolarizing). This response pattern is due to the cells being post-synaptic just to SWS cones, the other chromatic response components being driven by indirect, presynaptic effects (Figure 11).

A further (tetrachromatic/H4) HC type receiving direct UVS cone input, also mediating indirect H3 HC input (via negative feed-back) could also occur (Stell, 1980; Hashimoto *et al.,* 1988). An alternative model of cone-HC connectivity has been proposed by Kamermans *et al.* (1989a, b). This involves non-selective cone contacts, ascribing the different chromatic patterns of response instead largely to different intrinsic properties of the HCs. It is, however, based on ambiguous morphological evidence and largely fails to explain the characteristics of the H3 cells (Stell *et al.,* 1994).

10.5.2 Sub-types, photoreceptor connectivities and light-evoked response patterns in mammals

ALHC and SAHC types of HC are also found in mammals (Figure 12; reviews: Djamgoz *et al.,* 1995; Sandermann *et al.,* 1996a b). Traditionally, HCs in mammals have been termed A-cells (ALHCs), and B- or C-cells in the case of SAHCs. In the cat, which was the first mammal to be studied in detail with ultrastructural methods, there is a single type of A cells, contacting cones exclusively; and one type of B-cell which contacted cones with the dendrites originating from the cell body, and rods with the elaborate axonal telodendria (Kolb, 1974; Fisher & Boycott, 1974). This basic organisation of HCs which is found in most non-primate mammals is complicated by a further differentiation of the B-cells into several subtypes, and also by the fact that humans and other primates appear to lack A-cells altogether (Boycott & Kolb, 1973b; Ahnelt & Kolb, 1994).

The differentiation of B-type HCs is based on the observation that some of these cells show preference in their cone contacts, especially with regard to SWS cones. This was first realised in rabbits (Famiglietti, 1990), confirmed in horses (Sandemann *et al.,* 1996a) and also demonstrated in humans (Ahnelt & Kolb, 1994). In the latter, somatic dendrites of B-cells (H1 cells) contact predominantly LWS and MWS cones (90%) and only occasionally SWS cones (10%); axonal telodendria contact rods. C-cells (H2-cells) receive preferential input from SWS cones via both their perikaryal dendrites and axonal terminals. Finally, H3 cell dendrites avoid SWS cones altogether and their telodendria contact rod spherules (Figure 12).

The surprising morphological diversity and apparent confusion in the distribution of type A (ALHC) and B (SAHC) HCs have recently been discussed in the light of new findings in retinae of ox, sheep and pig (Sandemann *et al.,*

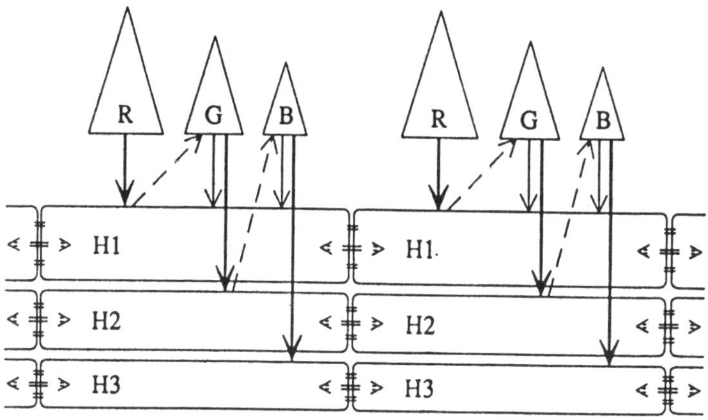

Figure 11 Schematic diagram showing the hierarchical, cascade model of spatio-chromatic organization of horizontal cells in cyprinid fish retinae, based mainly on the findings of Stell *et al.* (1975); Downing & Djamgoz (1989); and Djamgoz & Greenstreet (1996). H1 type horizontal cells receive a major excitatory input (as indicated by the thick solid arrows) from red-sensitive (R-) cones and minor inputs (thin solid arrows) from green (G-) and blue (B-) sensitive cones. H1 cells form an inhibitory feed-back pathway (as shown by the dashed arrows) mainly onto G- cones, but also B-cones and back onto R-cones (not illustrated). H2 horizontal cells receive major excitatory input from G-cones and feed-back onto B-cones. In turn, the latter forms the major feed-forward cone input to H3 cells. All three types of horizontal cells are coupled by gap junctions (indicated by the dashed double-headed arrows); each cell type forms a separate syncytium across the outer retina. Modified from Djamgoz & Petruv (1995)

1996b). B-type cells are common to all mammals and are characterised by the presence of an axon and an axon terminal system based on an enlargement of the axon into thicker processes giving off delicate branchlets contacting rod spherules. It is above all the distribution of A-type cells which shows the greatest variability among mammals. These cells have no obvious axon and their dendritic trees are always larger than those of B-cells. However, the dendritic tree is often asymmetric up to a point where some of the dendrites may resemble B-cell axonal telodendria, whilst the terminal branches of the A-type cells are thin and delicate throughout. In the artiodactyl retina, Sandemann *et al.* (1996b) have identified A-type cells with intermediate axonless and axon-bearing morphologies and have argued that the primate H2 cell may be regarded as an A-type cell – an interpretation further corroborated by the fact that this particular cell type contacts cones with the perikaryal dendrites as well as with the processes traditionally regarded as axon terminals. If one further considers the primate H3 cell as a B-cell variant, the primate HCs can easily be integrated into the basic pattern of A/B (ALHC/SAHC) type HCs of mammals in general. In the light of the evidence available, the basic A/B-type dichotomy is also independent of the rod/cone ratios in different mammalian species (Sandemann *et al.,* 1996b).

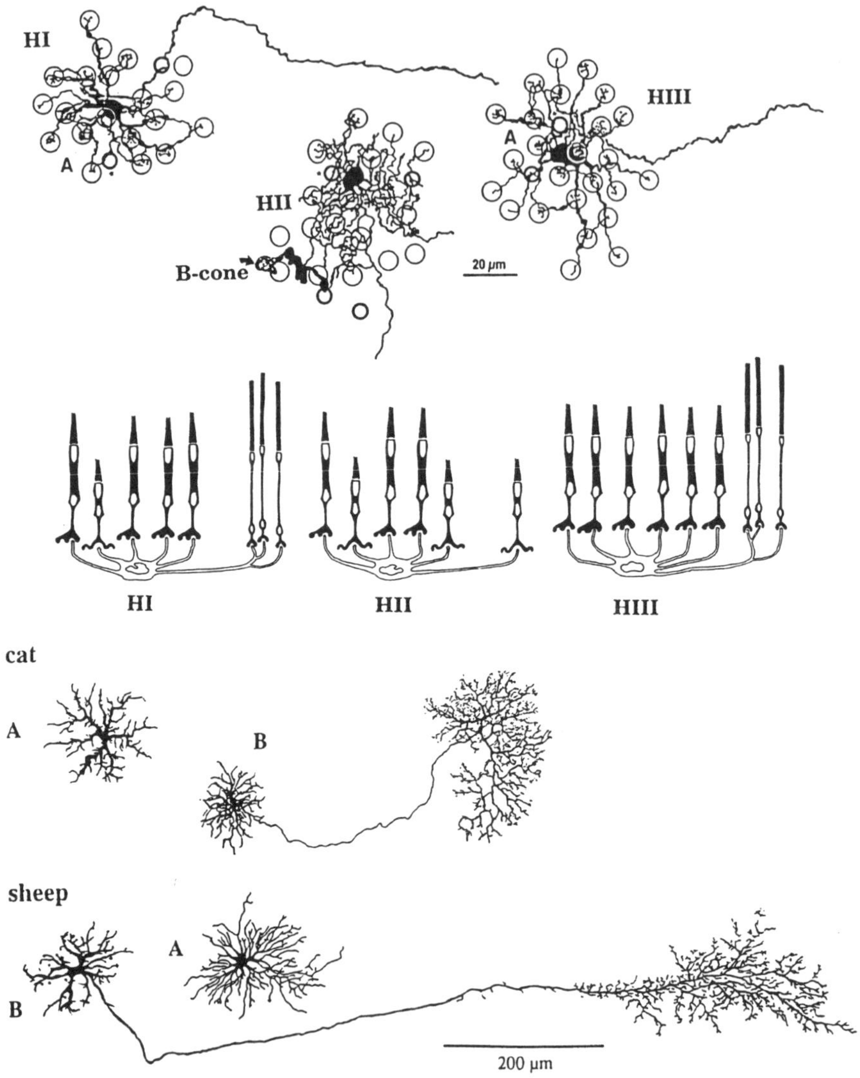

Figure 12 Horizontal cell types of the mammalian retinae. Upper panel: horizontal cell types of the human retina (from Ahnelt & Kolb, 1994). Top: Cell bodies of all three types of axon-bearing cells in tangential view (circles indicate cone pedicles contacted, B-cone indicates SWS cone); axon terminals not shown; below: pattern of connectivity with photoreceptors in lateral view; symbols identical to Figure 2: large cones: LWS or MWS; small cones: SWS. Lower pane: axonless (A-type) and axon-bearing (B-type) horizontal cells of cat and sheep in wholemount view (from Sandemann *et al.,* 1996b) These authors argued that the human HII cell correspond to the general mammal A-type cell; it contacts cones exclusively; HI and HIII cells correspond to the general mammal B-type cell which contacts cones with its perikaryal dendrites and rods with its axonal telodendria

Light-evoked response characteristics of mammalian HCs are illustrated in Figures 13 and 14. Importantly, all responses recorded to date are hyperpolarizing. ALHCs demonstrate mixed cone- rod input, whilst the perikaryon and axon terminal of SAHCs generate selective cone- and rod-driven activity, respectively (Figure 13) (Nelson *et al.,* 1975; Nelson, 1977, 1985; Gouras & Zrenner, 1981; Bloomfield & Miller, 1982; Dacheux & Raviola, 1982). Chromatic aspects of mammalian (primate) HCs have been studied by Dacey *et al.* (1996). All HCs were found to receive sign-conserving inputs from LWS- and MWS-cones (Figure 14). In addition, H2 (but not H1) HCs also received a similar input from SWS-cones. Thus, the cells' spectral inputs were consistent with their cone connectivity. Importantly, however, there was no evidence for any colour-opponent activity (Dacey, 1996).

10.5.3 Feed-forward and feed-back mechanisms

Photoreceptor to HC synaptic transmission occurs as in the case of the OFF-centre BCs, i.e. in the dark, glutamate depolarizes the HCs by acting on ionotropic receptors and allowing Na^+ influx (Kaneko & Shimazaki, 1975). Thus, HCs receiving direct cone input(s) mainly hyperpolarize during illumination. This mechanism explains LWS, MWS and SWS cone inputs to H1, H2 and H3 HCs, respectively. Other response components may involve feed-back interactions or different receptor mechanisms.

As already noted, there is general consensus that HCs are responsible for the

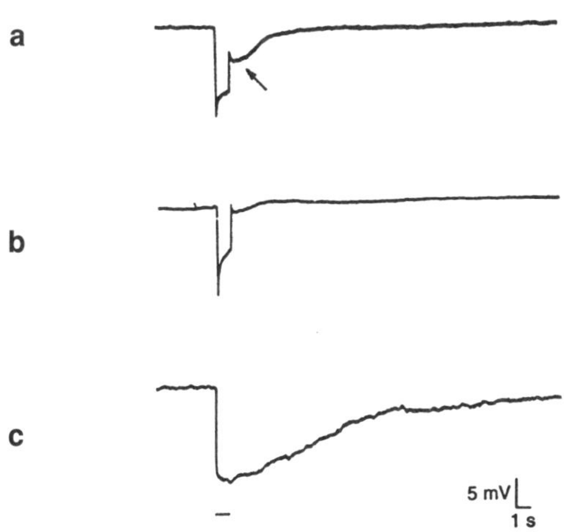

Figure 13 Horizontal cell responses in the mammalian (rabbit) retina evoked by high-intensity white light indicated by short bar under bottom trace. (**a**) ALHC response showing mixed cone- and rod-driven activity; transition indicated by small arrow. (**b** & **c**) Perikaryon and axon terminal responses of SAHC showing mainly cone- and rod-driven activity, respectively. Modified from Dacheux & Raviola (1982)

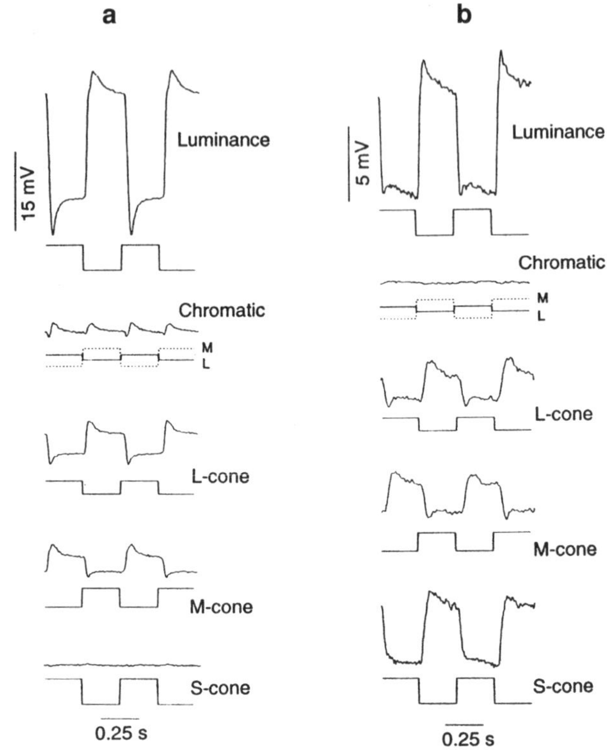

Figure 14 Lack of chromatic opponency in light-evoked horizontal cell responses of the primate retina. (**a**) H1 cells. (**b**) H2 cells. In each case, the top trace illustrates that the cell gives a sustained hyperpolarizing response to luminance increments. Equiluminant counterphase modulation of LW and MW stimulators (responses marked 'chromatic') elicits a small amplitude hyperpolarization (**a**) or cancels the cell response (**b**). Selective modulation of LWS or MWS cones also elicits a hyperpolarizing response that is in phase with the modulated cone type (traces marked 'L-cone' and 'M-cone'). Selective stimulation of SWS-cones elicits no response in H1 cells (**a**) but hyperpolarizes the H2 cells (**b**). Modified from Dacey *et al.* (1996)

generation of the antagonistic surround of BC receptive fields. At the synaptic level, this requires, in addition to sign-conserving feed-forward transmission from photoreceptors to HCs, a negative feed-back pathway from HCs to photoreceptors (and/or to BCs). In mammals, HC dendrites occupy the typical position lateral to the synaptic ribbons. Near the ribbon apex, there is an accumulation of A-face particles on both the synaptic ridge and the HC membrane (Raviola & Gilula, 1975). It is here that pre-synaptic (Ω-shaped) exocytotic membrane profiles, L-type Ca^{2+} channels and the molecular machinery for fusion and exocytsis of synaptic vesicles (e.g. neurexin, SNAP 25) and post-synaptic features [e.g. α-amino-3-hydroxy-5-methyl-4-isoxazolepropionate (AMPA)-type glutamate receptor proteins] have been observed (Wagner, 1997). It is generally agreed that this is the site where sign-conserving feed-forward transmission from photoreceptors to HCs occurs. In teleosts, HC dendrites are also found as central elements underneath the ribbons, and from their similar appearance in freeze-fracture micrographs (Stell *et al.,* 1982), it is assumed that they, too, are post-synaptic to photoreceptors.

The synaptic mechanism(s) underlying the feed-back from HCs to photoreceptors has been considerably more elusive. Gamma-aminobutryic acid (GABA; the inhibitory neurotransmitter of some HCs) and/or the GABA-synthesizing enzyme glutamine acid decarboxylase have been localised in HCs (mammals: Vardi *et al.,* 1994; teleosts: Marc *et al.,* 1978; reviewed by Yazulla, 1995), and GABA receptors have been attributed to BC and cone synaptic membranes (Vardi *et al.,* 1992; Tachibana & Kaneko, 1984; Yazulla *et al.,* 1989; review: Djamgoz, 1995), suggesting that the molecular machinery required for a sign-inverting feed-back synapse is present in HCs. There is good evidence to suggest that GABA release from HC (at least at those sites which are thought to be involved in negative feed-back to photoreceptors) occurs through a Na^+-dependent, Ca^{2+}-independent (i.e. non-vesicular) carrier (Yazulla, 1985, 1995). However, this does not apply to the few sites where HCs synapse onto BCs (catfish: Sakai & Naka, 1986; mammals: Fisher & Boycott, 1974; Mangel, 1991; Smith & Sterling, 1990). In mammals, in fact, HC pathways in the OPL appear to have limited role in forming the surround field of BCs, suggesting that instead interactions involving ACs may be important for shaping BC receptive fields (Kolb & Nelson, 1995).

In teleosts, there is good evidence that in addition to the formation of BC receptive fields, HC feed-back mechanisms play a major part in the generation of multi-phasic, chromatic signals of H2 and H3 HCs (Djamgoz & Ruddock, 1979; Murakami *et al.,* 1978, 1982). Importantly, these experiments provided evidence that HC negative feed-back in fish is directed primarily towards cone pedicles rather than BCs. After establishing the cone-HC connectivity in cyprinids, the first morphological interpretation of the synaptic arrays in the pedicles of the various chromatic cone types concluded that ‘central’ HC dendrites located under the edge of the synaptic ribbon mediated feed-forward transmission, whereas ‘lateral’ HC dendrites would be involved in the feed-back process (Stell *et al.,* 1975). This idea was extended following the observation that the presence and absence of small, finger-like extensions from the lateral HC processes – ‘spinules’ (Wagner, 1980; Figure 15) correlated with the presence and absence of the biphasic chromaticity response in H2 cells (Weiler & Wagner, 1984; Djamgoz *et al.,* 1988; Wagner & Djamgoz, 1993). These spinules are characterised by (i) conspicuous membrane densities, which react positively with ethanolic phosphotungstic acid, a non-specific, general label for synaptic membrane densities; (ii) an absence of typical clear synaptic vesicles, consistent with the idea of GABA being liberated by a carrier mechanism, not by exocytosis (Yazulla, 1985, 1995); (iii) intramembranous particles on the inner face of the outer lipid monolayer at the spinule tip being not noticeably specialised (in marked contrast to the part of the HC dendrite membrane facing the synaptic ribbon) in conjunction with an absence of intracleft material near the apex of the spinules (Stell *et al.,* 1982). On the basis of strong parallels between spinule dynamics and amplitude of the depolarizing component of biphasic/H2 HCs in cyprinid fish, observed in a variety of morphological, pharmacological and physiological experiments, it was argued that spinules are the sites for the negative feed-back transmission from HC to cones in teleosts (Wagner & Djamgoz, 1993). Furthermore, compartmental model calculations based on serial electron microscopic sections of HC spines entering amphibian rods and cones have indicated that these can act as structural entities independent from the perikaryon, well suited for mediating local feed-back circuits (Winslow *et al.,* 1989).

More recently, immunocytochemical staining demonstrated that the lateral HC dendrites contain ionotropic glutamate receptors (Nakanishi, 1992; Wenthold *et al.,* 1992) at those parts of the membrane facing the synaptic ribbon (Stell *et al.,* 1994) indicating that these sites are postsynaptic to photoreceptors. This is consistent with the view that, in teleosts, lateral HC dendrites serve a dual function, mediating (i) feed-forward transmission next to the ribbon, and (ii) feed-back by means of the spinules, as previously proposed by quantitative and correlative evaluation of cone-HC connectivity and the HC response patterns (Downing & Djamgoz, 1989; Kirsch *et al.,* 1990). In short, the lateral HC processes constitute a reciprocal synapse and a micro-circuit in themselves (Figure 15).

From the above account of the synaptic organisation of the photoreceptor, and especially cone pedicle, two points emerge: (i) The synaptic interactions occur at highly specialised sites, most of which bear little resemblance in structural and molecular organisation to the conventional (e.g. Gray Type I and II) synapses found in the IPL and elsewhere in the CNS. (ii) Cone pedicles are an exquisite, complex structure where numerous delicate dendritic processes are arranged next to each other within the cavity without any isolation by glial lamellae. These considerations have also led to an alternative, 'electrical' theory of information transfer within the cone pedicle, whereby the extracellular current flow generated by HCs in response to chemical synaptic input from the photoreceptor would be sufficient to directly influence neighbouring cone or BC membranes via an extracellular field effect (Byzov *et al.,* 1977; Byzov & Shura-Bura, 1986). This type of interaction is all the more plausible since the cone pedicle cavities often have a very narrow opening towards the OPL, effectively restricting the exchange of ions and other metabolites. However, there is no direct evidence for this type of signal transfer.

10.5.4 Coupling among horizontal cells

Due to the large size of HC perikarya and axon terminals, coupling is particularly well characterised in this population of retinal neurones (reviews: Vaney, 1994;

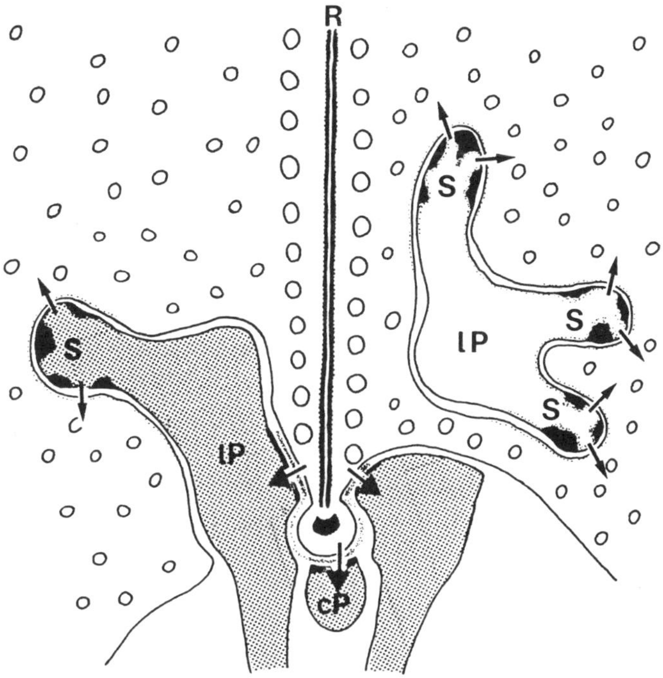

Figure 15 Semi-schematic drawing of a ribbon (R) synaptic complex in green-sensitive cone pedicle in cyprinid fish (roach) retina. According to the model, the pedicle is contacted by a H1 horizontal cell 'lateral' process (lP; clear, unstained cytoplasm), and 'central' (cP) and 'lateral' (lP) H2 cell processes (shaded; the latter was initially intracellularly stained with horseradish peroxidase). Arrows indicate the direction of signal transmission. Information from LWS-H1 cells is transmitted sign-inverted (small arrows) via spinules (S) to the cone pedicle. This feed-back input is then integrated with the activity of the pedicle and fed sign-conserved (large arrows) to the H2 cell dendrites near the tip of the synaptic ridge. Modified from Kirsch *et al.* (1990)

Hankins, 1995). In fish and other lower vertebrates, there is ample ultrastructural evidence for the presence of numerous gap junctions between cell bodies in each layer of HCs indicating extensive homologous coupling (Yamada & Ishikawa, 1965; Witkovsky & Dowling, 1969; Marc *et al.,* 1978). This has been confirmed in many physiological and pharmacological and dye-coupling experiments for cone-HCs (SAHCs) (Vaney, 1994, and below). However, for the teleost rod ALHCs, the situation is less certain; while obvious gap junctions have not been observed (Marc *et al.,* 1978) and dye-coupling is weak (Kaneko & Stuart, 1984), electrotonic transmission results in a receptive field about 20 times the size of the dendritic field (Tornqvist *et al.,* 1988). In teleosts, coupling has also been observed between homologous axon terminals of cone-HCs (Kouyama & Watanabe, 1986; Kurz-Isler & Wolburg, 1986). Interestingly, the coupling of HC axon terminals in teleost fish was found to be stronger than the corresponding somata (Kaneko, 1970; Djamgoz *et al.,* 1996a).

In mammals, numerous studies have demonstrated the occurrence of gap junctions linking homologous HCs (Fisher & Boycott, 1974; Raviola & Gilula, 1975; Kolb, 1977). The dendrites of A-type cells are permeable to Lucifer Yellow and coupled electrically (e.g. rabbit: Dacheux & Raviola, 1982; Bloomfield & Miller, 1982). The case for dendritic coupling in B-type HCs was more ambiguous since no gap junctions were detected (Kolb, 1977), and dye coupling with Lucifer Yellow could not be shown (Dacheux & Raviola, 1982). However, coupling was inferred from the fact that the receptive fields were several times larger than the dendritic trees; dye-coupling was subsequently demonstrated by using neurobiotin or biocytin as a low molecular weight tracer (Vaney, 1991, 1993; Bloomfield, 1993). These studies also found that axon terminals of mammalian B-type HCs too are dye-coupled. Interestingly, the strength of the dye coupling of dendrites and axon terminals in rabbit B-type HCs were inversely related (Vaney, 1993) suggesting the existence of different quality and/or quantity of connexin proteins, and/or subtle modulatory mechanisms.

The main functional role of HC coupling is to integrate in 'space' the intensity distribution in the visual image and use this to set the gain of PC → BC transmission (Werblin, 1979). There is also some evidence that this function may vary depending upon the spectral content of the image. In particular, the receptive field of cyprinid fish H1 HC syncytium appears wider for signals generated by long-wavelength stimuli compared with short-wavelengths (Teranishi *et al.,* 1982; Lipetz & Kaneko, 1984; Djamgoz *et al.,* 1998). Thus, fish HC receptive fields are capable of integrating spatial and spectral information in parallel. Accordingly, control of the receptive field size in the H1 HC system would be expected to depend not only on the light/dark adaptational state of the retina (section 10.8.2) but also on the spectral content of the visual scene. It is not known if a comparable mechanism operates in mammals.

10.6 Comparative aspects of bipolar and horizontal cells in fish and terrestrial mammals: an overview

10.6.1 Bipolar cells

The most salient feature shared by many BCs throughout vertebrates is the dichotomy into ON- and OFF-centre cells. As regards fish *vs.* mammals, there are two major differences between the functional organization of BCs: (i) the autonomy of the rod pathway; and (ii) the differentiation of the cone pathways. In fish, the rod signal is grafted onto the pathways of both LWS and MWS cones. Most notably, only LWS-specific cone pathway is always accompanied by rods; in teleosts, this is represented by the largest of all ('giant') BCs, which occur as a canonical pair. In view of the fact that, during the ontogeny of the retina, cones and their associated neurones precede the differentiation of rods, which then impinge upon pre-existing pathways, the situation in teleosts has been interpreted as being more primitive. In mammals, by contrast, rods have a single 'private' BC channel conveying their signals to the inner retina. Only at that level is rod-driven activity taken up by a special (AII) type of AC and transmitted to ON- and OFF-GCs (Kolb & Famiglietti, 1976; Wässle & Boycott, 1991; for details see next Chapter). The importance of the cone pathway for the processing of rod-related information in mammals is best demonstrated by the fact that even in species (e.g. rat) with rods comprising ≥ 95% of the photoreceptors, about 50% of the BC population are cone-BCs.

Although there are several types of cone-specific BCs in teleosts, no evidence has been found to suggest that there are 'midget' BCs in addition to the more widely distributed diffuse cells. This suggests that segregation into parvocellular (P) and magnocellular (M) pathways is not yet present in fish retinae, at least at the BC level. Furthermore, there seems to be a 'privileged' pathway for SWS cones in mammals, whilst it is unclear whether there are separate channels for LWS and MWS, or how this information is segregated in BCs. In fish, on the other hand, there are discrete chromatic channels for MWS and SWS cones at this level.

10.6.2 Horizontal cells

The basic morphology of HCs including the dichotomy into ALHCs and SAHCs appears to be well conserved throughout vertebrates. This is certainly true with respect to electrical coupling which seems to be more extensive in HCs than in any other class of retinal neurone. A closer look indicates, however, profound differences in the pattern of connectivity with rods and cones, and the organisation of the axon terminals, as follows: (i) ALHCs are the typical rod-HCs in teleosts, whereas in mammals they contact cones exclusively; (ii) the axons of teleost SAHCs are directed towards the inner retina and synapse onto BCs and ACs; in mammals, by contrast, the axonal telodendria, like the perikaryal dendrites are confined to the OPL, where the latter contact cones and the former rods; and finally (iii) SAHCs in teleosts are characterised by a high degree of specificity in the way they contact the chromatic cone types; in mammals, and especially primates, only limited selectivity in SWS cone contacts have been observed.

Whilst in mammals there is a separate rod-BC pathway and a mixed rod/cone system occurs at least in B-type HCs, in teleosts, rod and cone pathways are separated at the HC level and mixed in the BC pathway. Furthermore, it is important to comment on the different degrees of selectivity with which HCs contact spectral cone types in mammals and teleosts. In teleosts, there are cone-HCs (H2 and H3) coding for chromaticity, which show high spectral specificity in the way they contact cones. In mammals, on the other hand, two types of HCs are only biased towards the SWS or the LWS/MWS cones. Furthermore, while in teleosts the 'Stell' model and its amendments (Downing & Djamgoz, 1989; Wagner & Djamgoz, 1993) provide a firm basis for the explanation of the bi- and triphasic chromatic responses, its hard to see in mammals how a similar degree of chromatic preprocessing could be afforded at the HC level.

In conclusion, therefore, due to the substantial differences in the types of BCs and HCs, and their connectivities with the photoreceptors, fish and mammalian retinae are markedly different in their mechanisms of both scotopic and photopic information processing, as well as the initial stages of the mechanism mediating colour vision.

10.7 Interplexiform cells

Interplexiform cells (IPCs) are characterised by perikarya in the INL, and extensive, diffuse or multilayered neurites – dendrites or axons – in both plexiform layers. Cajal (1892) observed such structures originally in perciform fishes and recognised their special nature; Gallego (1971) described these cells in mammals and classified them as 'interplexiform cells'. Subsequent application of histochemical and neurochemical methods revealed the existence of up to four different subtypes of IPC with specific morphologies and neurochemical signatures, confirming them as an essential neuronal component of all vertebrate retinae (Marc, 1995). The following sections highlight features (morphology, neurochemistry and 'connectivity') of IPCs which may be of relevance to retinal neuromodulation and ecology of vision.

10.7.1 Interplexiform cells in fish retinae

Fish retinae contain the greatest diversity of IPCs (Figure 16). Based on their content of transmitter (or its synthesising enzyme), there are dopaminergic, GABAergic, glycinergic (GLY), serotonergic (5-HT), and possibly, adrenergic subtypes of IPC. This variety underlines the fact that IPCs play a major, if in many cases yet poorly understood, role in retinal signal processing.

Dopaminergic IPCs (DA-IPCs) are among the most prominent and best characterized. In general, they have three tiers of ramification in the IPL (most prominent in layer 1, and additionally in layers 3 and 5). From the most proximal stratum (layer 1), several radial processes originate which cross the INL and form an extensive, dense network of varicose fibres at the level of the external (H1) HCs with occasional branches penetrating the layer of photoreceptor terminals (Figure 16; Dowling & Ehinger, 1975). These cells receive synaptic input in the IPL from GABAergic ACs, some BCs and other DA-IPCs (Yazulla & Zucker, 1988). Interestingly, their perikara are post-synaptic to GnRH-containing efferent fibres reaching the retina from the olfacto-retinalis system (Zucker & Dowling, 1987). In the OPL, DA-IPCs direct more than 90% of their output towards H1 HCs; some HC axon terminals are also 'contacted' (Haesendonck *et al.*, 1993; Wagner & Wulle, 1990, 1992). A great deal of morphological variability has been observed in the IPC systems of fishes. There are marked differences in the density of the telodendrial plexus in the OPL; many advanced diurnal species possess a well developed plexus, whereas in retinae which are less cone dominated, the OPL fibres are considerably sparser (Wagner & Wulle, 1990; de Miguel & Wagner, 1990; Marc, 1995). Dopaminergic ACs, which completely lack processes in the OPL, are most common in deep-sea fish with all-rod retinae (Fröhlich *et al.*, 1995) and more primitive fishes like holostean gars and *Polypterus* (Marc, 1995). These examples indicate that phylogenetic factors play an important role in determining the differentiation of the dopaminergic cells, but that constraints related to adaptation to the visual environment may also be effective. Dopamine has been recognised as an important agent involved in the control of adaptational processes in the transition from scotopic to photopic vision (reviews: Negishi *et al.*, 1990; Witkovsky & Dearry, 1991; Djamgoz & Wagner, 1992).

Glycinergic IPCs (GLY-IPCs), next best characterized in teleost retinae, are multipolar neurones having several processes running from the cell body towards the OPL where they ramify profusely distal of the H1 HC layer (Figure 16). In the IPL, the lateral branches are most concentrated at the inner border of the INL; other processes reach sublayers 2 and 5 of IPL (Marc & Lam, 1981; Kalloniatis & Marc, 1990). GLY-IPCs are post-synaptic to 5-10 GABAergic H1-HCs and pre-synaptic in the IPL to GABAergic ACs which, in turn, synapse onto BCs and GCs, thus copying the activity of the S-space to the inner retina without passing through the high spatial filter of the BCs (Kalloniatis & Marc, 1990; Marc, 1995). It is also apparent that the direction of information flow in the GLY-IPCs is opposite to that of DA-IPCs. The connectivity of the other IPCs in teleost retinae remains to be determined.

10.7.2 Interplexiform cells in mammalian retinae

In mammals, DA-IPCs and GABA-IPCs have been identified unequivocally; the existence of additional types cannot be excluded. As in fish, most work has been done on DA-IPCs which occur as a heterogenous group. At one extreme is the common type of dopaminergic AC which is characterized by a prominent axonal system (Dacey, 1989, 1990). In many cases, there is only a sparse arborization of fibres in the OPL, but in the platyrhine Cebus monkey, this almost equals that found in most teleosts (Ehinger & Falck, 1969). Occasional pre-synapses in the OPL are directed towards HCs and BCs (Dowling & Ehinger, 1975). However, since even closely related primates lack comparable fibre plexus in the outer retina, Marc (1995) argued that the DA-IPCs in mammals and teleosts must have evolved independently. This is further supported by (i) the morphological differences concerning the radial processes in the INL, which arise from the

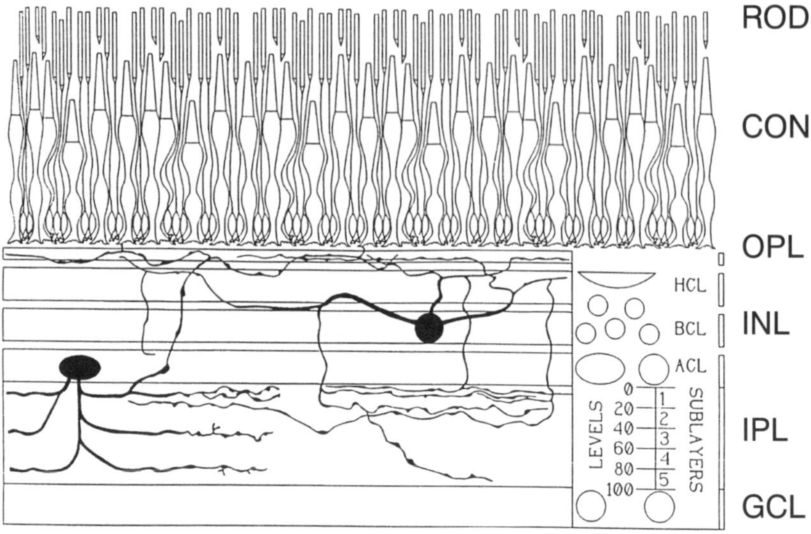

Figure 16 Schematic diagram of a dopaminergic (left) and a glycinergic interplexiform cell of the teleost retina in lateral view. Dopaminergic cells have large cell bodies in the inner part of the inner nuclear layer (INL) scattered among the perikarya of amacrine cells (ACL amacrine cell layer), and dendritic arborisations in sublayers 1, 3 and 5 of the inner plexiform layer (IPL), which receive input predominantly from GABAergic amacrine cells. Axonal processes arise from the dendrites, cross the INL and form a dense telodendrial plexus of varicose branches which contact horizontal cells in the horizontal cell layer (HCL), and occasional photoreceptor terminals. Glycinergic interplexiform cells have cell bodies situated among the bipolar cell layer (BCL). Their dendrites ramify extensively in the horizontal cell layer and receive input from GABAergic H1 cells. The dendrites give off axonal processes which descend to the outer half of the IPL (sublayers 1-3) and are presynaptic to GABAergic amacrine cells. Slightly modified from Marc (1995)

perikarya in mammals but from the fibre plexus in teleosts; and (ii) the fact that DA-IPCs in mammals, but not in teleosts, also contain GABA (Wulle & Wagner, 1990). Ultrastructural and morphometric data in the primate retina suggested that there is also a separate population of IPCs which contain only GABA (Nakamura *et al.,* 1980; Kolb *et al.,* 1990; Ryan & Hendrickson, 1987). There is no evidence for existence of GLY-IPCs in mammals (Marc, 1989, 1995).

In conclusion, IPCs of the vertebrate retina form a heterogenous group of neurones which differ as much in their morphology as in their transmitter content. From the established connectivity, it would appear in fish that DA-IPCs relay information from the IPL to the OPL, whilst GLY-IPC map the functional state of the OPL to the IPL. On the basis of the transmitters used and the receptor mechanisms involved, Marc (1995) proposed that the glycinergic system is relatively faster. In both cases, however, the wide diameter of the dendritic or axonal trees would suggest that the actions of these cells are more modulatory in nature, rather than being involved primarily in the synaptic structuring of receptive fields. In mammals, by contrast, the main site of action of DA-IPCs would appear to be the IPL (see next Chapter). Yet, like its counterpart in fish, dopamine seems to be involved in the control of light adaptive mechanisms in the outer retina. Finally, in view of the major differences in HCs and BCs between fish and mammals regarding basic morphology, connectivity and thus typology, it does not come as a surprise that a similar or even greater degree of diversity is found in the case of IPCs. Unfortunately, much less is known about the functional role of most of them; the diversity, therefore, is all the more intriguing.

10.7.3 Control of dopamine release

It is generally accepted that there is some basal release of retinal DA in the dark and that this is enhanced significantly by light adaptation (reviews: Witkovsky & Dearry, 1991; Djamgoz & Wagner, 1992; Besharse & Iuvone, 1992). Interestingly, prolonged dark adaptation could also increase DA release but only at night (Weiler *et al.,* 1997). Both rods and cones may contribute to the light adaptive release process (Besharse & Witkovsky, 1992; Myhr *et al.,* 1994), the former acting through APB-sensitive ON-centre BCs (Besharse, 1992). Release of DA is Ca^{2+}-dependent (i.e. vesicular) and controlled by several different neurochemical pathways (review: Djamgoz *et al.,* 1997). For example, GABA, glycine, melatonin, nitric oxide, enkephalins and prostaglandins inhibit the DA release, whilst glutamate, serotonin, noradrenaline, substance P and thyrotropin-releasing hormone are excitatory. Importantly, release of DA may depend upon the pattern of the light stimulus. For example, in teleost fish retina, flickering stimulus is more effective than steady in releasing endogenous DA (Kirsch & Wagner, 1989). On the other hand, both patterns of stimulus release DA equally in the amphibian retina (Boatright *et al.,* 1989). Consequently, DA normally brings about a variety of light adaptive effects on retinal neurones, and accordingly, visual signal processing would be expected to depend upon the dynamics of environmental light.

10.8 Plasticity of synaptic activity in the outer retina

10.8.1 Light-dark adaptation dependent morphological changes in photoreceptor terminals

The synaptic activity of photoreceptors and second-order neurones is greatly influenced by ambient lighting conditions. Consequently, adaptation-dependent changes in the morphology of the synaptic complexes of the OPL have been examined in some detail. In the retinae of lower vertebrates, morphological changes at synapses occur in addition to gross cellular changes such as photomechanical responses which bring rods and, alternatively, cones into an optimal position near the external limiting membrane to catch the incoming light (Ali,

1975; Burnside & Nagle, 1983; Wagner, 1990). At the synaptic level, the following parameters have been studied in rod and cone terminals: The density of synaptic vesicles, shape, size or numbers of synaptic ribbons, and the morphology of the HC dendrite/photoreceptor interface away from the synaptic ribbon.

During physiological light/dark adaptation conditions, no statistically significant change was found in the density of synaptic vesicles (Mountford, 1963; Cragg, 1972; Rao-Mirotznik *et al.,* 1994). Only, when the tissue was cooled artificially, darkness led to a decrease in vesicle density, accompanied by a swelling of the entire terminal, indicating impairment of membrane retrieval and vesicle recycling mechanisms (Schaeffer and Raviola, 1978).

Synaptic ribbons have been shown to be dynamic structures in many, but not all, vertebrates studied; in addition to retinal rod and cone synapses, ribbons in pineal photoreceptors also show impressive adaptation-related changes (review: Vollrath & Spiwoks-Becker, 1996). In mammals, as well as changing their shape (to spheres, sticks, or clubs), the overall dimensions also show various degrees of change, most notably in the direction perpendicular to the pre-synaptic membrane (Vollrath *et al.,* 1989). In more severe cases, the ribbons may become so short that they all but disappear. This has been observed in many teleost cones after dark adaptation (Wagner, 1973; Wagner *et al.,* 1992). In albino rats, the number of rod ribbons changes between light- and dark-adapted states by a factor of three (Spadaro *et al.,* 1978; Hermes *et al.,* 1992). In turtle, cone ribbons are unchanged, but rod ribbons start to proliferate during the second half of the night and the first half of the day to form aggregates; subsequently, the numbers decrease, until a single ribbon is left in the usual position (Abe & Yamamoto, 1984). Apart from light and darkness, endogenous circadian factors have been shown to be involved in the control of ribbon dynamics, and Ca^{2+} has been identified as an intracellular messenger (Wagner, 1997). Interestingly, in hibernating squirrels, ribbons leave their normal place in the synaptic ridge and move some 5.5 μm into the photoreceptor cytoplasm (Remé & Young, 1977). It may be assumed that the major role of synaptic ribbons is to catch and bind an appropriate number of vesicles and to guide them to the docking sites on the presynaptic membrane in order to ensure uninterrupted transmitter release in photoreceptors. Accordingly, one might expect the ribbons to be largest (and/or most numerous) during, or at the beginning of the dark phase, since under these conditions, the membrane potential is maximally depolarized and the highest transmitter release occurs. This situation has been found in rod spherules of rats (Hermes *et al.,* 1992), and to a certain degree in turtles (Abe & Yamamoto, 1984). The opposite behaviour observed in teleost cones may be due to the fact that the ribbon synapses are not functional during the dark, since cone outer and inner segments migrate towards the back of the eye, becoming shielded from and possibly not corresponding to any incoming light.

While in the region of the synaptic ridge, the contact zone between photoreceptors and HC dendrites is characterized by flat, parallel membranes, the situation is more complex in those regions facing away from the synaptic ribbon. In light-adapted rod spherules, this contour is smooth and cross-sections of HC terminals are regular and ovoid; upon dark adaptation, however, finger-like protrusions of photoreceptor cytoplasm extend into the HC cytoplasm on either side of the ribbon (rat and goldfish: Brandon & Lam, 1983). Although the functional significance of these morphological changes in the rod/HC dendrite contact zone is elusive, it is noteworthy that the formation of such protrusions results in an increased area of contact between the two cells and that this occurs at a time when rods would be most active. The morphological changes noted above are the exact reciprocal of the situation in teleost cone pedicles, where HC spinules are present during the day, and absent at night. Formation and degradation of HC spinules take 30 – 40 min. Both processes are initiated before the actual change in the lighting conditions, indicating the influence of an endogenous circadian oscillator (Wagner *et al.,* 1992). The major photic signalling system controlling spinule formation involves dopamine released from DA-IPCs, acting via D1 type DA receptors with cAMP as the second messenger (Wagner & Djamgoz, 1993). The 'dark' adaptive signal, on the other hand, is mainly glutamate released from cones and acting upon AMPA type receptors, gating Ca^{2+} influx, to reduce the number of spinules (Weiler & Schultz, 1993).

10.8.2 Physiological and possible ecological aspects

Light adaptation of the retina (and/or application of exogenous DA) is accompanied by a number of physiological changes. Such changes have been studied mostly at the level of the HCs in fish retinae. The available evidence, which may have ecological significance, can be summarized as follows:

i. Photoreceptor → HC synaptic transmissions

Dark ↔ light adaptation of varying extents has been found to have a number of complex effects upon photoreceptor → HC synaptic transmission in vertebrate retinae; most work has been done on H1 (luminosity – type) HCs. Cone → HC transmission is suppressed in the dark and potentiated by background light adaptation (Yang *et al.,* 1994; Yang & Wu, 1996). As regards chromatic effects, prolonged darkness caused the spectral peak of the H1 HC responses to shift towards the green, possibly due to a rod input becoming apparent (Yang *et al.,* 1994). On the other hand, light adaptation from moderately dark-adapted state was found to sharpen the LW spectral peak of the H1 HCs by 'truncating' SW sensitivity (Figure 17); this effect was partially mimicked by DA (Djamgoz & Yamada, 1992; Djamgoz *et al.,* 1996b).

In amphibian HCs which receive mixed rod-cone input, application of DA enhanced the cone input whilst inhibiting rod → HC transmission (Witkovsky *et al.,* 1989; Krizaj *et al.,* 1994). Krizaj *et al.* (1994) also studied effects of gluta-

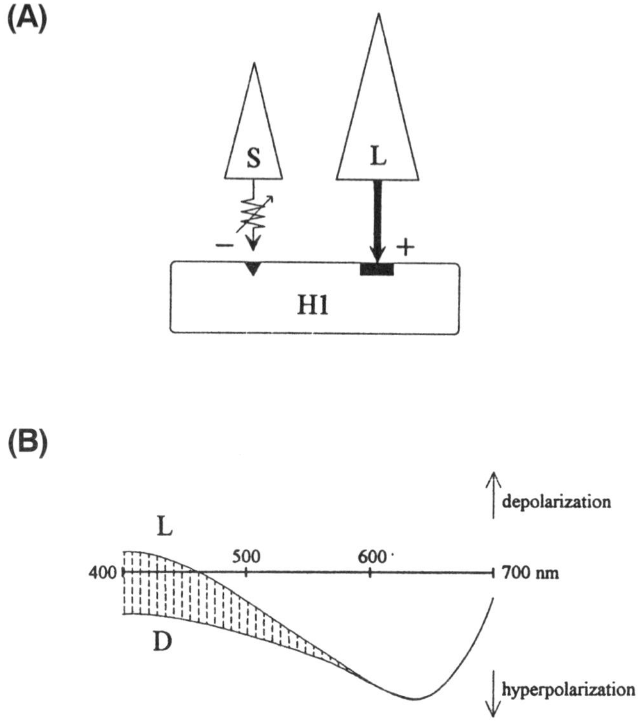

Figure 17 A simple connectivity model of the H1 horizontal cell's dynamic spectral sensitivity in the carp retina. (A) The cell receives two sets of cone inputs: The main LWS-cone input (L), indicated by the thick arrow, has a conventional, excitatory post-synaptic effect associated with a receptor protein where the influence of the photoreceptor transmitter (released in the dark) is to increase conductance (+). The minor, SWS-cone input (S) indicated by the think arrow line is dynamic (denoted by the variable resistor) and associated with a synaptic receptor protein mechanism where the photoreceptor transmitter decreases conductance (−). (B) The effect of LW stimuli is to hyperpolarize whilst the effect of SW light is to depolarize the H1 cell. The latter input is suppressed in the dark-adapted retina (D) whereby the spectral sensitivity becomes broader and mainly follows the action spectrum of the LWS cone pigment. The SWS cone input is activated by light adaptation and 'truncates' the spectral profile of the cell. Thus, the cell's spectral response is dynamic and can demonstrate considerable variability as indicated by the shading in the SW region of the spectrum. Modified from Djamgoz *et al.* (1996b)

mate and its agonists on HCs as a function of the adaptational state of the retina. Glutamate elicited desensitizing responses under light-adapted conditions but larger and more sustained responses when the retina was dark adapted. These observations extend the original findings of Knapp and Dowling (1987) that modulation of glutamate receptors by DA could underlie some aspects of light ↔ dark adaptation effects upon HCs. The basic effect of DA (and possibly light adaptation) on mammalian HCs may be the same since Hankins and Ikeda (1991b, 1993) have found that exogenous DA depolarized the resting potential and reduced the amplitude of response to full-field illumination in the rat, as in the teleost, retina.

ii. HC coupling

Light adaptation reduces the electrotonic coupling of HCs thereby enhancing their ability to signal spatial contrast. In the dark, HCs are strongly coupled so as to maximize their chance for signalling photon capture at the expense of loss of spatial information. The effect of light adaptation involves release of DA and accumulation within the HCs of cAMP via activation of D1 receptors. As a result, PKA phosphorylates the connexins and thus reduces their open probability by some four-fold (McMahon *et al.,* 1989). This uncoupling effect of DA seems general in vertebrates and can be inferred to occur even in humans during Parkinson's disease where there is loss of retinal DA. The change in the coupling efficacy of the HCs would be expected to have profound effects in the balance of the centre-surround antagonism in BC and/or GC receptive fields, and hence the visual signals leaving the retina. This aspect is reviewed in the next Chapter.

Spatio-chromatic signalling in the H1 HC network (section 10.5.4) is also dependent upon light ↔ dark adaptation. In essence, this characteristic is apparent only in the light-adapted state of the retina; during dark adaptation, the HC receptive field (for both SW and LW stimuli) gets broader (as discussed generally above) and the chromatic difference disappears (Djamgoz *et al.*, 1998). Furthermore, DA partially mimics the effect of light adaptation (Djamgoz *et al.,* 1998). The light adaptive spatio-chromatic signalling by the H1 HC network would imply that spatial integration of visual information would be affected by the chromatic content of the visual scene. This is probably a reflection of the complexity of the visual spatio-chromatic conditions under water, as argued in the Introduction.

iii. HC → Cone negative feed-back

A clear case of light ↔ dark adaptation – induced change in HC signalling is seen in the H2 type HC responses of cyprinid fish. In light-adapted retinae, these cells generate well-balanced biphasic/chromaticity type responses (Figures 10 & 18). During dark adaptation however, the LW-sensitive depolarizing component is suppressed and the cells' spectral response becomes monophasic, peaking in the MW part of the spectrum (Figure 18). This effect is not due to suppression of LWS-cone activity since H1 HC response amplitudes appear normal. Instead, it is H1 HC → LWS-cone negative feed-back that is suppressed during dark adaptation. The loss of LW-MW contrast in the H2 HC system leads to a behavioural loss of red-green discrimination under dark-adapted condition (Neumeyer, 1986). This mechanism of plasticity is controlled (at least partially) by DA since in DA-depleted or D1-receptor antagonist-treated retinae, LW-MW discrimination is also suppressed (Djamgoz *et al.,* 1989; Kirsch *et al.,* 1991). Furthermore,

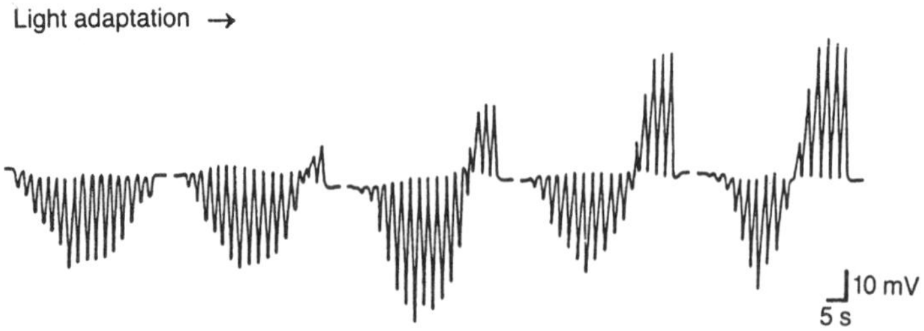

Figure 18 Dark → light adaptation-dependent plasticity of the spectral opponency in H2 type horizontal cell responses in the cyprinid fish (roach) retina. In the dark-adapted state, the cells' spectral response is monophasic, peaking in the MW region of the spectrum. The cell's response gradually develops the LWS depolarizing component during steady light adaptation (from left to right, as indicated by the arrow). In the fully light-adapted retina (righthand-most response trace), the cell's spectral response is well-balanced between the LWS depolarizing and SWS hyperpolarizing components. Spectral stimuli as in Figure 10 legend. Modified from Djamgoz *et al.* (1988)

the dynamics of HC spinules correlates strongly with the change in the balance of the LW-MW components in H2 HC responses (Kirsch *et al.,* 1990, 1991).

From these observations two major points have emerged: (a) Spinules could be the site of HC → cone negative feed-back (Wagner & Djamgoz, 1993); and (b) such feed-back is 'plastic' in teleost retinae so as to generate a wide variety of spectral responses, again presumably, to match the dynamic chromatic conditions under water. In contrast, there are no such mechanisms operating in mammalian HCs.

10.9 A novel light adaptative neurotransmitter/modulator: nitric oxide

Increasing evidence suggest that retinal light adaptation can occur, at least partially, independently of DA (e.g. Baldridge & Ball, 1991; Douglas *et al.,* 1992; Djamgoz *et al.,* 1996b). Interestingly, several independent recent studies indicate that nitric oxide (NO) is a novel light adaptive neurotransmitter/modulator in vertebrate retinae. Thus, Greenstreet and Djamgoz (1994b) showed originally that application of NO donors to dark-adapted carp retinae produced light adaptive photomechanical movements of cones and formation of HC spinules. NO donors or cGMP analogues caused uncoupling of fish HCs, again similar to the effect of light adaptation (Miyachi *et al.,* 1990; Lu & McMahon, 1997). Furukawa *et al.* (1997) found that NO and light adaptation both enhanced SW-sensitive synaptic input to carp H1 HCs. Finally, direct evidence for light-induced NO production in mammalian and teleost retinae has been presented by Donati *et al.* (1995), Neal *et al.* (1998), and Sekaran *et al.* (1998).

As a signalling molecule, NO is unusual in several different respects. First, it is produced in neurones upon demand (mainly following a rise in intracellular Ca^{2+}), rather than being synthesized and stored. Second, it can diffuse and act upon target cells within 100s of μm for 100s of ms. Third, its receptor is an enzyme (soluble guanylate cyclase) activation of which leads to accumulation of cGMP molecules within the target cells. In fact, NO does not fit readily into the general scheme of first or second messenger.

Many aspects of the role of NO in retinal processing of visual information, especially in relation to dark ↔ light adaptation, remain to be elucidated. One particular important aspect of retinal NO is its interaction with DA. Thus ON-BCs/APB receptors, which control DA release in the retina, are also involved in NO production (rabbit: Neal *et al.,* 1998). On the other hand, NO has been found to suppress depolarization – induced DA release in fish and mammalian retinae (Bugnon *et al.,* 1994; Djamgoz *et al.,* 1995b; Pottek *et al.,* 1997). In conclusion, light adaptation of the retinae involves release of both DA and NO which are interactive. Such multiplicity and interaction of light adaptive neurochemical mechanisms probably reflect the complexity of the essential light adaptation processes.

10.10 Conclusions and future perspectives

Comparative neurobiology of the outer retinae clearly suggests that there are marked differences in the synaptic connectivities of PCs, HCs, BCs and IPCs between terrestrial and aquatic vertebrates, represented in this review by mammals and fish, respectively. On the other hand, some fundamental aspects of retinal organization (especially segregation of visual signals into 'ON' and 'OFF' pathways) are clearly the same. Another theme that is maintained is the role of DA-IPCs in controlling lateral interactions (HC coupling) in the outer retina (Djamgoz & Wagner, 1992; Djamgoz *et al.,* 1997).

The many differences in the neurobiology of the outer retinae that have been noted occur probably so that the networks can process qualitatively and quantitatively different kinds of visual information. A related, remarkable property of the outer retina of teleost fish is its synaptic plasticity. It is not yet known whether the PC – HC – BC systems of terrestrial mammals are equally plastic, as much less work has been done in this area due to obvious technical difficulties. Another important, emerging (but not so surprising) concept is the multiplicity of light ↔ dark adaptive neuromodulators. Whilst DA has firmly been established as a light adaptive retinal neuromodulator, an increasing body of experimental evidence suggests that NO is a novel modulator involved in the light adaptive cellular and synaptic plasticity of the outer retina of teleost fish. Clearly much work remains to be done in several broad areas (i) to enhance our comparative understanding of the neurobiology of the outer retina; (ii) to determine the capacity of outer retinal synaptic connectivity to undergo modulatory changes in accordance with visual conditions in the environment; (iii) to characterize the cellular and subcellular signalling mechanisms involved in synaptic interactions and modulation; and (iv) to relate all these to the ecology of vision!

Acknowledgements

We would like to thank Mrs Wendy McGovern and Miss Sumathi Sekaran for help in preparing the manuscript.

References

Abe, H. and Yamamoto, T. Y. (1984) Diurnal changes in synaptic ribbons of rod cells of the turtle. *Journal of Ultrastructure Research* 86, 246 – 251.

Ahnelt, P. and Kolb, H. (1994) Horizontal cells and cone photoreceptors in human retina: a Golgi-EM study of spectral connectivity. *Journal of Comparative Neurology* 343, 406 – 427.

Ali, M. A. (1975) Retinomotor responses. In *Vision in Fishes: New Approaches in Research* (Ali, M. A. Ed.). Plenum Press, New York. pp. 313 – 356.

Avery, J. A., Bowmaker, J. K., Djamgoz, M. B. A. and Downing, J. E. G. (1983) Ultraviolet receptors in a freshwater fish. *Journal of Physiology* 334, 23 – 24P.

Baldridge, W. H. and Ball, A. K. (1991) Background illumination reduces horizontal cell receptive field size in both normal and 6-hydroxydopamine-lesioned goldfish retinas. *Visual Neuroscience* 7, 441– 450.

Baldrige, W. H. and Ball, A. K. (1993) A new type of interplexiform cells in the goldfish retina is PNMT immunoreactive. *NeuroReport* 4, 1015 –1018.

Baldridge, W. H., Marchiafava, P. L., Miller, R. G. and Strettoi, E. (1987) Coupling in the absence of gap junctions in the fish double cone: a dye diffusion and freeze fracture study. *Journal of Submicroscopic Cytology* 19, 545 – 554.

Barlow, H. B., Levick, W. R. and Yoon, M. (1971) Response to single quanta of light in retinal ganglion cells of the cat. *Vision Research* S3, 87–101.

Barnes, S. (1995) Photoreceptor synaptic output: neurotransmitter release and photoreceptor coupling. In *Neurobiology and Clinical Aspects of the Outer Retina* (Djamgoz, M. B. A., Archer, S. N. & Vallerga, S. Eds.). Chapman & Hall, London. pp. 133 –153.

Baylor, D. A., Fuortes, M. G. F. and O'Bryan, P. M. (1971) Receptive fields of cones in the retina of the turtle. *Journal of Physiology (London)* 214, 265 – 294.

Besharse, J. C. (1992) The 'ON'-bipolar agonist, L-2-amino-4-phosphonobutyrate, blocks light-evoked cone contraction in *Xenopus* eye cups. *Neurochemical Research* 17, 75 – 80.

Besharse, J. C. and Witkovsky, P. (1992) Light evoked contraction of red absorbing cones in the *Xenopus* retina is maximally sensitive to green light. *Visual Neuroscience* 8, 243 – 249.

Besharse, J. C. and Iuvone, P. M. (1992) Is dopamine a light adaptive or dark-adaptive modulator in retina? *Neurochemistry International* 20, 193 –199.

Bloomfield, S. A. (1993) A comparison of receptive field and tracer coupling size of horizontal cells in the rabbit retina. *Investigative Ophthalmoloy and Visual Science* 34, 1153.

Bloomfield, S. A. and Miller, R. F. (1982) A physiological and morphological study of the horizontal cell types of the rabbit retina. *Journal of Comparative Neurology* 208, 288 – 303.

Boatright, J. H., Hoel, M. J. and Iuvone, P. M. (1989) Stimulation of endogenous dopamine release and metabolism in amphibian retina by light- and K^+-evoked depolarization. *Brain Research* 482, 164 –168.

Borges, S. and Wilson, M. (1987) Structure of the receptive fields of bipolar cells in the salamander retina. *Journal of Neurophysiology* 58, 1275 –1291.

Bowmaker, J. K. (1984) Microspectrophotometry of vertebrate photoreceptors: a brief review. *Vision Research* 24, 1641–1650.

Bowmaker, J. K. and Kunz, Y. W. (1987) Ultraviolet receptors, tetrachromatic colour vision and retinal mosaics in the brown trout (*Salmo trutta*), age-dependent changes. *Vision Research* 27, 2101– 2108.

Bowmaker, J. K., Thorpe, A. and Douglas, R. H. (1991) Ultraviolet sensitive cones in the goldfish. *Vision Research* 31, 349 – 352.

Boycott, B. B. and Kolb, H. (1973a) The connection between the bipolar cells and photoreceptors in the retina of the domestic cat. *Journal of Comparative Neurology* 148, 91–114.

Boycott, B. B. and Kolb, H. (1973b) The horizontal cells of the rhesus monkey retina. *Journal of Comparative Neurology* 148, 115 –139.

Boycott, B. B. and Wässle, H. (1991) Morphological classification of bipolar cells of the primate retina. *European Journal of Neuroscience*, 3, 1069 –1088.

Brandon, C. and Lam, D. M.-K. (1983) The ultrastructure of rat rod synaptic terminals: Effects of dark adaptation. *Journal of Comparative Neurology* 217, 167–175.

Braun, S. C., Kröger, R. H. H. and Wagner, H.-J. (1997) Connectivity patterns of cone horizontal cells in blue acara (*Aequidens pulcher*, Cichlidae) reared in different light regimes. *Neuroscience Letters* 235, 145 –148.

Bugnon, O., Schaad, N. C. and Schorderet, M. (1994) Nitric oxide modulates endogenous dopamine release in bovine retina. *Neuro Report* 5, 401– 40.

Burkhardt, D. (1977) Responses and receptive field organization of cones in perch retina. *Journal of Neurophysiology* 40, 53 – 62.

Burkhardt, D. (1986) Functional properties of twin and double cones. *Neuroscience Research, Supplement,* 4, S554 – S558.

Burnside, B. and Nagle, B. (1983) Retinomotor movements of photoreceptors and retinal pigment epithelium: Mechanisms and regulation. *Progress in Retinal Research* 2, 67–109.

Byzov, A. L. and Shura-Bura, T. M. (1986) Electrical feedback mechanism in the processing of signals in the outer plexiform layer of the retina. *Vision Research* 26, 33 – 44.

Byzov, A. L., Golbtzov, K. V. and Trifonov, Y. (1977) The model of mechanisms of feedback between horizontal cells and photoreceptors in the vertebrate retina. In *Vertebrate Photoreceptors* (Barlow, H. B. & Fatt, P. Eds.). Academic Press, New York. pp. 265 – 274.

Cajal, S. R. y (1892) La rétine des vertébrés. *La cellule* 9, 17– 257.

Cameron, D. A. and Pugh, E. N. Jr. (1991) Double cones as a basis for a new type of polarization vision in vertebrates. *Nature* 353, 161–164.

Cohen, A. I. (1965) Some electron microscopic observations on inter-receptor contacts in the human and macaque retinae. *Journal of Anatomy* 99, 595 – 610.

Copenhagen, D. R. and Jahr, C. E. (1989) Release of endogenous excitatory amino acids from turtle photoreceptors. *Nature* 341, 5365 – 539.

Cragg, B. G. (1972) Plasticity of synapses. In Bourne, G. H. ed. *The Structure and Function of Nervous Tissue*, New York, Academic Press, Vol. IV, pp. 1– 60.

Dacey, D. M. (1989) Axon-bearing amacrine cells of the macaque monkey retina. *Journal of Comparative Neurology* 284, 175 –193.

Dacey, D. M. (1990) The dopaminergic amacrine cell. *Journal of Comparative Neurology* 301, 461– 489.

Dacey, D. M. (1996) Circuitry for color coding in the primate retina. *Proceedings of the National Academy of Science (U.S.A.)* 93, 582 – 588.

Dacey, D. M., Lee, B. B., Stafford, D. K., Pokorny, J. and Smith, V. C. (1996) Horizontal cells of the primate retina: cone specificity without spectral opponency. *Science* 271, 656 – 659.

Dacheux, R. F. and Raviola, E. (1982) Horizontal cells in the retina of the rabbit. *Journal of Neuroscience* 2, 1486 –1493.

Dacheux, R. F. and Raviola, E. (1986) The rod pathway in the rabbit retina: a depolarizing bipolar and amacrine cell. *Journal of Neuroscience* 6, 331– 345.

Daw, N. W. (1968) Colour-coded ganglion cells in the goldfish retina. Extension of their receptive fields by means of new stimuli. *Journal of Physiology (London)* 197, 567– 592.

Detwiler, P. B., Hodgkin, A. L. and McNaughton, P. A. (1978) A surprising property of electrical spread in the network of rods in the turtle's retina. *Nature* 274, 562 – 565.

Djamgoz, M. B. A. and Ruddock, K. H. (1979) Effects of picrotoxin and strychnine on fish retinal S-potentials: evidence for inhibitory control of depolarizing responses. *Neuroscience Letters* 12, 329 – 334.

Djamgoz, M. B. A. (1995) Diversity of GABA receptors in the vertebrate outer retina. *Trends in Neurosciences* 18, 118 –120.

Djamgoz, M. B. A., Petruv, R., Yasui, S., Furukawa, T. and Yamada (1998) Modulation of chromatic difference in receptive field size of H1 horizontal cells in carp retina: dopamine-and APB-sensitive mechanisms. *Neuroscience Research.* 30, 13 – 24.

Djamgoz, M. B. A. and Downing, J. E. G. (1988) A horizontal cell selectively contacts blue-sensitive cones in cyprinid fish retina: intracellular staining with horseradish peroxidase. *Proceedings of the Royal Society of London B* 235, 281– 287.

Djamgoz, M. B. A. and Greenstreet, E. H. (1996) Quantitative analysis of triphasic (H3) horizontal cell – cone connectivity in the cyprinid fish (roach) retina. *Vision Research* 36, 4007– 4014.

Djamgoz, M. B. A. and Petruv, R. (1995) Spatio-chromatic signalling in the vertebrate retina. In *Basic and Clinical Perspectives in Vision Research* (Robbins, J., Djamgoz, M. B. A. and Taylor, A. Eds.). Plenum Press, New York. pp. 77– 94.

Djamgoz, M. B. A. and Wagner, H.-J. (1992) Localization and function of dopamine in the adult vertebrate retina. *Neurochemistry International,* 20, 139 –191.

Djamgoz, M. B. A. and Yamada, M. (1990) Electrophysiological characteristics of retinal neurones: synaptic interactions and functional outputs. In *The Visual System of Fish* (Douglas, R. and Djamgoz, M. B. A., Eds.). Chapman & Hall, London. pp. 159 – 210.

Djamgoz, M. B. A. and Yamada, M. (1992) Dopamine and light adaptation sharpen the spectral response of H1 horizontal cells in carp retina. *Neuroscience Research Communications* 10, 149 –153.

Djamgoz, M. B. A., Downing, J. E. G., Kirsch, M., Prince, D. J. and Wagner, H.-J. (1988) Plasticity of cone horizontal cell functions in cyprinid fish retina: Effects of background illumination of moderate intensity. *Journal of Neurocytology* 17, 701–710.

Djamgoz, M. B. A., Kirsch, M. and Wagner, H.-J. (1989) Haloperidol suppresses light-induced spinule formation and biphasic responses of horizontal cells in fish (roach) retina. *Neuroscience Letters* 107, 200 – 204.

Djamgoz, M. B. A., Archer, S. N. and Vallerga, S. (1995a) *Neurobiology and Clinical Aspects of the Outer Retina*. Chapman & Hall, London.

Djamgoz, M. B. A., Cunningham, J. R., Davenport, S. L. and Neal, M. J. (1995b) Nitric oxide inhibits depolarization-induced release of endogenous dopamine in the rabbit retina. *Neuroscience Letters* 198, 33 – 36.

Djamgoz, M. B. A., Evans-Capp, A. J. and Wagner, H.-J. (1996a) Intra-vitreal injection of substance P antibodies as an antagonist in the vertebrate (fish) retina. *Journal of Neuroscience Methods* 64, 237– 243.

Djamgoz, M. B. A., Fitzgerald, E. M. and Yamada, M. (1996b) Spectral plasticity of H1 horizontal cells in carp retina: Independent modulation by dopamine and light adaptation. *European Journal of Neuroscience* 8, 1571–1579.

Djamgoz, M. B. A., Hankins, M. W., Hirano, J. and Archer, S. N. (1997) Neurobiology of retinal dopamine in relation to degenerative states of the tissue. *Vision Research* 37, 3509 – 3529.

Donati, G., Pournaras, C. J., Munoz, J.-L., Poitry, S., Poitry-Yamate, C. L. and Tsacopoulos, M. (1995) Nitric oxide controls arteriolar tone in the retina of the miniature pig. *Investigate Ophthalmology and Visual Science* 36, 2228 – 2237.

Douglas, R. H. & Djamgoz, M. B. A. (1990) *The Visual System of Fish.* Chapman & Hall, London.

Douglas, R. H., Wagner, H-J., Zaunreiter, M., Behrens, U. D. and Djamgoz, M. B. A. (1992) The effect of dopamine depletion on light-evoked and circadian retinomotor movements in the teleost retina. *Visual Neuroscience,* 9, 335 – 343.

Dowling, J. E. and Boycott, B. B. (1966) Organization of the primate retina: electron microscopy. *Proceedings of the Royal Society of London, B*, 116, 80 –111.

Dowling, J. E. and Ehinger, B. (1975) Synaptic organisation of the amine-containing interplexion cells of the goldfish and Cebus monkey retinas. *Science* 188, 270 – 273.

Downing, J. E. G. and Djamgoz, M. B. A. (1989) Quantitative analysis of cone photoreceptor-horizontal cell connectivity patterns in the retina of a cyprinid fish: Electron microscopy of functionally-identified and HRP-labelled horizontal cells. *Journal of Comparative Neurology* 289, 537– 553.

Downing, J. E. G., Djamgoz, M. B. A. and Bowmaker, J. K. (1986) Photoreceptors of a cyprinid fish, the roach: morphological and spectral characteristics. *Journal of Comparative Physiology* 159, 859 – 868.

Ehinger, B. and Falck, B. (1969) Adrenergic retinal neurons of some new world monkeys. *Zeitschrift für Zellforschung und mikroskopische Anatomie* 100, 364 – 375.

Euler, T. and Wässle, H. (1995) Immunocytochemical identification of cone bipolar cells in the rat retina. *Journal of Comparative Neurology* 361, 461– 478.

Falk, G. and Shiells, R. A. (1995) On-biopolar cells, visual sensitivity and the b-wave. In *Basic and Clinical Perspectives in Vision Research* (Robbins, J., Djamgoz, M. B. A. and Taylor, A. Eds.). Plenum Press, New York. pp. 95 –102.

Famiglietti, E. V. Jr (1990) A new type of wide-field horizontal cell, presumably linked to blue cones, in rabbit retina. *Brain Research* 535, 174 –179.

Famiglietti, E. V. Jr and Kolb, H. (1976) Structural basis for ON- and OFF-centre responses in the retinal ganglion cells. *Science* 194, 193 –195.

Famiglietti, E. V. Jr., Kaneko, A. and Tachibana, M. (1977) Neuronal architecture of on and off pathways to ganglion cells of the carp retina. *Science* 198, 1267–1269.

Fisher, S. K. and Boycott, B. B. (1974) Synaptic connections made by horizontal cells within the outer plexiform layer of the retina of the cat and the rabbit. *Proceedings of the Royal Society of London B* 186, 317– 331.

Fröhlich, E., Negishi, K. and Wagner, H.-J. (1995) The occurrence of dopaminergic interplexiform cells correlates with the presence of cones in the retinae of fish. *Visual Neuroscience* 12, 359 – 369.

Furukawa, T., Yamada, M., Petruv, R., Djamgoz, M. B. A. and Yasui, S. (1997) Nitric oxide, 2-amino-4-phosphonobutyric acid and light/dark adaptation modulate short-wavelength-sensitive synaptic transmission to retinal horizontal cells. *Neuroscience Research* 27, 65 –74.

Gallego, A. (1971) Horizontal and amacrine cells in the mammal's retina. *Vision Research* 3, 33 – 50.

Gallego, A. (1985) Advances in horizontal cell terminology since Cajal. In *Neurocircuitry of the Retina, A Cajal Memorial* (Gallego, A. and Gouras, P. Eds.). Elsevier, Amsterdam. pp. 122 –140.

Gold, G. H. (1979) Photoreceptor coupling in the retina of the toad, *Bufo marinus* II. Physiology. *Journal of Neurophysiology* 42, 311– 328.

Gold, G. H. and Dowling, J. E. (1979) Photoreceptor coupling in the retina of the toad, *Bufo marinus*. I. Anatomy. *Journal of Neurophysiology* 42, 292 – 310.

Gouras, P. and Zrenner, E. (1981) Color coding in primate retina. *Vision Research* 21, 1591–1598.

Grant, G. B. and Dowling, J. E. (1995) A glutamate-activated chloride current in cone-driven ON bipolar cells of the white perch retina. *Journal of Neuroscience* 15, 3852 – 3862.

Grant G. B. and Dowling, J. E. (1996) ON bipolar cell responses in the teleost retina are generated by two distinct mechanisms. *Journal of Neurophysiology* 76, 3842 – 3849.

Greenstreet, E. H. and Djamgoz, M. B. A. (1994a) Triphasic chromaticity-type horizontal cells selectivity contact short wavelength-sensitive cone photoreceptors in the retina of a cyprinid fish *Rutilus rutilus*. *Proceedings of the Royal Society of London B* 256, 227– 230.

Greenstreet, E. H. and Djamgoz, M. B. A. (1994b) Nitric oxide induces light-adaptive morphological changes in retinal neurones. *NeuroReport* 6, 109 –112.

Greferath, U., Grünert, U. and Wässle, H. (1990) Rod bipolar cells in the mammalian retina show protein kinase C-like immunoreactivity. *Journal of Comparative Neurology* 301, 433 – 442.

Haesendonck, E. van and Missotten, L. (1983a) Interbipolar contacts in the dorsal inner plexiform layer in the retina of *Callionymus lyra L. Journal of Ultrastructure Research* 83, 303 – 311.

Haesendonck, E. van and Missotten, L. (1983b) Stratification and square pattern of arrangements in the dorsal inner plexiform layer in the retina of *Callionymus lyra L. Journal of Ultrastructure Research* 83, 296 – 302.

Haesendonck, E. Van, Marc, R. E. and Missotten, L. (1993) New aspects of dopaminergic interplexiform cell organisation in the goldfish retina. *Journal of Comparative Neurology* 333, 503 – 518.

Hankins, M. W. (1995) Horizontal cell coupling and its regulation. In *Neurobiology and Clinical Aspects of the Outer Retina* (Djamgoz, M. B. A., Archer, S. N. and Vallerga, S. Eds.). Chapman & Hall, London. pp. 195 – 220.

Hankins, M. W. and Ikeda, H. (1991a) Non-NMDA type excitatory amino acid receptors mediate rod input to horizontal cells in the isolated rat retina. *Vision Research* 31, 609 – 617.

Hankins, M. W. and Ikeda, H. (1991b) The role of dopaminergic pathways at the outer plexiform layer of the mammalian retina. *Clinical Vision Sciences* 31, 87– 93.

Hankins, M. W. and Ikeda, H. (1993) Dopamine pathways in the outer mammalian retina and abnormalities expressed in hereditary retinal dystrophy. *Zdravinski vestnik* 62 Supp. 1, 33 – 38.

Hare, W. A. and Owen, W. G. (1992) Effects of 2-amino-4-phosphonobutyric acid on cells in the distal layers of the tiger salamander retina. *Journal of Physiology (London)* 445, 741–757.

Hashimoto, Y., Haros, F. I., Ueki, K. and Fukurotani, I-I. (1988) Ultraviolet-sensitive cones in the color-coding systems of cyprinid retinas. *Neuroscience Research, Suppl.* 8, S81– S95.

Helmholtz, H. v. (1896) Handbuch der Physiologischen Optik. Hamburg & Leipzig, Leopold Voss.

Hermes, B., Reuss, S. and Vollrath, L. (1992) Synaptic ribbons, spheres and intermediate structures in the developing retina. *International Journal of Developmental Neuroscience* 10, 215 – 223.

Hibbard, E. (1971) Grid patterns in the retinal organization of the cichlid fish *Astronotus ocellatus. Experimental Eye Research* 12, 175 –180.

Hirano, A. A. and MacLeish, P. R. (1991) Glutamate and 2-amino-4-phosphonobutyrate evoke an increase in potassium conductance in retinal bipolar cells. *Proceedings of the National Academy of Science (U.S.A.)* 88, 805 – 809.

Ishida, A. T., Stell, W. K. and Lightfoot, D. O. (1980) Rod and cone inputs to bipolar cells in goldfish retina. *Journal of Comparative Neurology* 191, 315 – 335.

Kalloniatis, M. K. and Marc, R. E. (1990) Interplexiform cells of the goldfish retina. *Journal of Comparative Neurology* 297, 340 – 358.

Kamermans, M., van Dijk, B. W. and Spekreijse, H. (1989a) Lateral feedback from monophasic horizontal cells in carp retina. I. Experiments. *Journal of General Physiology* 97, 819 – 843.

Kamermans, M., van Dijk, B. W., Spekreijse, H. and Zweypfenning, R. C. V. J. (1986b) Lateral feedback from monophasic horizontal cells in carp retina. II. A quantitative model. *Journal of General Physiology* 93, 681– 694.

Kaneko, A. (1970) Physiological and morphological identification of horizontal, bipolar and amacrine cells in goldfish retina. *Journal of Physiology (London)* 207, 623 – 633.

Kaneko, A. (1973) Receptive field organization of bipolar and amarcrine cells in the goldfish retina. *Journal of Physiology (London)* 235, 133 –153.

Kaneko, A. and Stuart, A. E. (1984) Coupling between horizontal cells in the carp retina revealed by diffusion of Lucifer Yellow. *Neuroscience Letters* 47, 1– 4.

Kaneko, A. and Shimazaki, H. (1975) Effects of external ions on the synaptic transmission from photoreceptors to horizontal cells in the carp retina. *Journal of Physiology (London)* 252, 509 – 522.

Kaneko, A. and Tachibana, M. (1981) Retinal bipolar cells with double colour-opponent receptive fields. *Nature* 293, 220 – 222.

Kaneko, A. and Tachibana, M. (1983) Double colour opponent receptive fields of carp bipolar cells. *Vision Research* 23, 381– 388.

Kawamura, S. (1995) Phototransduction, excitation and adaptation. In *Neurobiology and Clinical Aspects of the Outer Retina* (Djamgoz, M. B. A., Archer, S. N. and Vallerga, S. Eds.). Chapman & Hall, London. pp. 105 –131.

Kirsch, M., Djamgoz, M. B. A. and Wagner, H.-J. (1990) Correlation of spinule dynamics and plasticity of the horizontal cell spectral response in cyprinid fish retina: quantitative analysis. *Cell Tissue Research* 260, 123 –130.

Kirsch, M. and Wagner H.-J. (1989) Release pattern of endogenous dopamine in teleost retinae during light adaptation and pharmacological stimulation. *Vision Research* 29, 147–154.

Kirsch, M., Wagner, H.-J. and Djamgoz, M. B. A. (1991) Dopamine and plasticity of horizontal cell function in the teleost retina: regulation of a spectral mechanism through D1 receptors. *Vision Research* 31, 401– 412.

Knapp, A. G. and Dowling, J. E. (1987) Dopamine enhances excitatory amino acid-gated conductances in horizontal cells. *Nature*, 325, 437– 439.

Kolb, H. (1970) Organization of the outer plexiform layer of the primate retina: electron microscopy of Golgi impregnated cells. *Philosophical Transactions of the Royal Society of London B* 258, 261– 283.

Kolb, H. (1974) The connections between horizontal cells and photoreceptors in the retina of the cat. Electron microscopy of Golgi preparations. *Journal of Comparative Neurology* 155, 1–14.

Kolb, H. (1977) The organization of the outer plexiform layer in the retina of the cat. *Investigate Ophthalmology and Visual Science* 6, 131–153.

Kolb, H. (1979) The inner plexiform layer of the retina of the cat. Electron microscopic observations. *Journal of Neurocytology* 8, 295 – 329.

Kolb, H. and Famiglietti, E. V. (1976) Rod and cone pathways in the retina of the cat. *Investigative Ophthalmology and Visual Science* 15, 935 – 946.

Kolb, H. and Nelson, R. (1995) The organization of photoreceptor to bipolar synapses in the outer plexiform layer. In *Neurobiology and Clinical Aspects of the Outer Retina* (Djamgoz, M. B. A., Archer, S. N. and Vallerga, S. Eds.). Chapman & Hall, London. pp. 273 – 296.

Kolb, H. Cuenca, N., Wang, H.-H. and Decorver, L. (1990) The synaptic organization of the dopaminergic amacrine cell in the cat retina. *Journal of Neurocytology* 19, 343 – 366.

Kolb, H., Linberg, K. A. and Fisher, S. K. (1992) Neurons of the human retina: A Golgi study. *Journal of Comparative Neurology* 318, 147–187.

Kouyama, N. and Watanabe, K. (1986) Gap-junctional contacts of luminosity-type horizontal cells in the carp retina: a novel pathway of signal conduction from the cell body to the axon terminal. *Journal of Comparative Neurology* 249, 404 – 410.

Kraft, T. W. and Burkhardt, D. A. (1986) Telodendrites of cone photoreceptors: structure and probable function. *Journal of Comparative Neurology* 249, 13 – 27.

Krizaj, D., Akopian A. and Witkovsky, P. (1994) The effects of L-glutamate, AMPA, quisqualate and kainate on retinal horizontal cells depend on adaptational state: implications for rod-cone interactions. *Journal of Neuroscience* 14, 5661– 5671.

Kurz-Isler, G. and Wolburg. H. (1986) Gap junctions between horizontal cells in the cyprinid fish alter rapidly their structure during light and dark adaptation. *Neuroscience Letters* 67, 7–12.

Lagnado, L. and Baylor, D. (1992) Signal flow in visual transduction. *Neuron* 8, 995 –1002.

Lamb, T. D. and Simon, E. J. (1976) The relation between intercellular coupling and electrical noise in turtle photoreceptors. *Journal of Physiology (London)* 263, 257– 286.
Lasansky, A. (1973) Organization of the outer synaptic layer in the retina of the larval tiger salamander. *Philosophical Transactions of the Royal Society of London B* 265, 471– 489.
Levine, J. S. and MacNichol, E. F. (1979) Visual pigments in teleost fishes: effects of habitat, microhabitat and behaviour on visual system evolution. *Sensory Processes* 3, 95 – 130.
Liebman, P. A. and Entine, G. (1968) Visual pigments of frog and tadpole *(Rana pipiens.) Vision Research* 8, 761–775.
Linberg, K. A., Suemune, S. and Fisher, S. K. (1996) Retinal neurons of the California ground squirrel, Spermophilus beecheyi: A. Golgi study. *Journal of Comparative Neurology* 365, 173 – 216.
Lipetz, L. E. and Kaneko, A. (1984) Receptive field properties of the photopic luminosity horizontal cells of carp retina. *Vision Research* 24, 1947–1950.
Locket, N. A. (1977) Adaptations to the deep-sea environment In: Crescitelli, F. ed. *The Visual System of Vertebrates; Handbook of Sensory Physiology,* Volume VII/5, Berlin, Heidelberg, Springer verlag, pp. 67–192.
Lockhart M. E. and Stell, W. K. (1979) Invaginating telodendria: a pathway for color-specific interconnections between goldfish cones. *Investigative Ophthalmology and Visual Science (Suppl.),* 19, 92.
Loew, E. R. and McFarland, W. N. (1990) The underwater visual environment. In *The Visual System of Fish* (Douglas, R. H. and Djamgoz, M. B. A. Eds.). Chapman & Hall, London. pp. 1– 43.
Lu, C. and McMahon, D. G. (1997) Modulation of hybrid bass retinal gap junctional channel gating by nitric oxide. *Journal of Physiology (London)* 499, 689 – 699.
MacNichol, E. F. and Svaetichin, G. (1958) Electrical responses from the isolated retinas of fishes: *American Journal of Ophthalmology* 46, 26 – 40.
Mangel, S. C. (1991) Analysis of horizontal cell contribution to the receptive field surround of ganglion cells in the rabbit retina. *Journal of Physiology (London)* 442, 211– 234.
Marc, R. E. (1986) The development of retinal networks. In *The Retina: a Model for Cell Biology* (Adler, R. and Farber, D. Eds.). Academic Press, New York. pp. 17– 65.
Marc, R. E. (1989) Evolution of retinal circuits. In *Neural Mechanisms of Behaviour* (Erber, Menzel, R., Pflüger, H.-J. and Todt, D. Eds.). Thieme, Stuttgart, pp. 146 –147.
Marc, R. E. (1992) The structure of GABAergic circuits in ectotherm retinas. In *GABA in the Retina and the Central Visual System* (Mize, R., Marc, R. E. and Sillito, A. Eds.). Elsevier, Amersterdam, pp. 61– 92.
Marc, R. E. (1995) Interplexiform cell connectivity in the outer retina. In *Neurobiology and Clinical Aspects of the Outer Retina* (Djamgoz, M. B. A., Archer, S. N. and Vallerga, S. Eds.). Chapman & Hall, London. pp. 369 – 393.
Marc, R. E. and Lam, D. M. K. (1981) Glycinergic pathways in the goldfish retina. *Journal of Neuroscience* 1, 152 –165.
Marc, R. E., Stell, W. K., Bok, D. and Lam, D. M. K. (1978) GABAergic pathways in the goldfish retina. *Journal of Comparative Neurology* 182, 221– 246.
Marc, R. E., Liu, W.-L.S. and Muller, J. F. (1988) Gap junctions in the inner plexiform layer of the goldfish retina. *Vision Research* 28, 9 – 24.
Marchiafava, P. L. (1985) Cell coupling in double cones of the fish retina. *Proceedings of the Royal Society of London B* 226, 211– 215.
Mariani, A. P. (1983) Giant bistratified bipolar cells in the monkey retina. *Anatomical Record* 206, 215 – 220.
Mariani, A. P. (1984) Bipolar cells in the monkey retina selective for cones likely to be blue sensitive. *Nature* 308, 184 –186.
Marshak, D. W. and Dowling, J. E. (1987) Synapses of cone horizontal cell axons in goldfish retina. *Journal of Comparative Neurology* 256, 430 – 443.
McGuire, B. A., Stevens, J. K. and Sterling, P. (1984) Microcircuitry of bipolar cells in cat retina. *Journal of Neuroscience* 4, 2920 – 2938.
McGuire, B. A., Stevens. J. K. and Sterling, P. (1986) Microcircuitry of beta ganglion cells in cat retina. *Journal of Neuroscience* 6, 907– 918.
McMahon, D. G., Knapp, A. G. and Dowling, J. E. (1989) Horizontal cell gap junctions: single-channel conductance and modulation by dopamine. *Proceedings of the National Academy of Science (U.S.A.)* 86, 7639 –7643.
Merchant, V. and Barnes, S. (1991) Lateral sign propagation via cone telodendrites in tiger salamander retina. *Society for Neuroscience Abstrates* 17, 298.

Miguel, E. de, and Wagner, H-.J. (1990) Tyrosine hydroxylase immunoreactive interplexiform cells in the lamprey retina. *Neuroscience Letters* 113, 151–155.
Miller, J. L. and Korenbrot, J. I. (1993) Phototransduction and adaptation in rods, single cones and twin cones of the striped bass retina: a comparative study. *Visual Neuroscience* 10, 653 – 667.
Miyachi, E-I., Murakami, M. and Nakaki, T. (1990) Arginine blocks gap junctions between retinal horizontal cells. *NeuroReport* 1, 107–110.
Mountford, S. (1963) Effects of light and dark adaptation on the vesicle population of receptor-bipolar synapses. *Journal of Ultrastructure Research* 9, 403 – 419.
Murakami, M., Shiomoda, Y. and Nakatani, K. (1978) Effects of GABA on neuronal activities in the distal retina of the carp. *Sensory Processes* 2, 334 – 338.
Murakami, M., Shimoda, Y., Nakatani, K., Miyachi, E. and Watanabe, S. (1982) GABA-mediated negative feedback from horizontal cells to cones in carp retina. *Japanese Journal of Physiology* 32, 911– 935.
Myhr, K. L., Dong, C-J. and McReynolds, J. S. (1994) Cones contribute to light-evoked, dopamine-mediated uncoupling of horizontal cells in the mudpuppy retina. *Journal of Neurophysiology* 72, 56 – 62.
Naka, K. I. and Rushton, W. A. H. (1967) The generation and spread of S-potentials in fish *(Cyprinidae). Journal of Physiology (London)* 192, 437– 461.
Nakamura, Y., McGuire, B. A. and Sterling, P. (1980) Interplexiform cell in cat retina: identification by uptake of gamma-(3H)-aminobutyric acid serial reconstruction. *Proceedings of the National Academy of Sciences (U.S.A.)* 77, 658 – 661.
Nakanishi, S. (1992) Molecular diversity of glutamate receptors and implications for brain function. *Science* 258, 597– 603.
Nawy, S. and Jahr, C. E. (1990) Suppression by glutamate of cGMP-activated conductance in retinal bipolar cells. *Nature* 346, 269 – 271.
Neal, M. J., Cunningham, J. R. and Matthews, K. (1998) Selective release of nitric oxide from retinal amacrine and bipolar cells. *Investigate Ophthalmology and Visual Science* 39, 850 – 853.
Negishi, K., Teanishi, T. and Kato, S. (1990) The dopamine system of the teleost fish retina. *Progress in Retinal Research* 9, 1– 48.
Nelson, R. (1977) Cat cones have rod input: a comparison of the response properties of cones and horizontal cell bodies in the retina of the cat. *Journal of Comparative Neurology* 172, 109 –136.
Nelson, R., Lützow, A. S. C., Kolb, H. and Gouras, P. (1975) Horizontal cells in the cat retina with independent dendritic systems. *Science* 189, 137–139.
Nelson, R. (1985) Spectral properties of cat horizontal cells. *Neuroscience Research Suppl.* 2, S167– S183.
Neumeyer, C. (1986) Wavelength discrimination in the goldfish. *Journal of Comparative. Physiology A* 158, 203 – 213.
Novales Flamarique, I. and Hawryshyn, C. W. (1997) Is the use of underwater polarized light by fish restricted to crepuscular time periods? *Vision Research* 37, 975 – 989.
Ohtsuka, T. and Kawamata, K. (1990) Telodendrial contact of HRP-filled photoreceptors in the turtle retina: pathways of photoreceptors coupling. *Journal of Comparative Neurology* 292, 599 – 613.
Owen, W. G. and Copenhagen, D. R. (1977) Characteristics of the electrical coupling between rods in the turtle retina. In *Vertebrate Photoreception* (Barlow, H. B. and Fatt, P. Eds.). Academic Press, New York. pp. 169 –192.
Parkyn, D. C. and Hawryshyn, C. W. (1993) Polarized light sensitivity in rainbow trout *(Oncorynchus mykiss):* characterization from multiunit ganglion cell responses in the optic nerve fibres. *Journal of Comparative Physiology A* 172, 493 – 500.
Peichl, L. (1991) Catecholaminergic amacrine cells in the dog and wolf retina. *Visual Neuroscience* 7, 575 – 587.
Polyak, S. L. (1941) *The Retina.* Chicago University Press, Chicago.
Pottek, M., Schultz, K. and Weiler, R. (1997) Effects of nitric oxide on the horizontal cell network and dopamine release in the carp retina. *Vision Research* 37, 1091–1102.
Rao-Mirotznik, R., Buchsbaum, G. and Sterling, P. (1994) Rate of quantal transmitter release at a mammalian rod synapse. *Biophysical Journal* 67, 57– 63.
Rao-Mirotznik, R., Harkins, A. B., Buchsbaum, G. and Sterling, P. (1995) Mammalian rod terminal: architecture of a binary synapse. *Neuron* 14, 561– 569.
Raviola, E. and Gilula, N. B. (1973) Gap junctions between photoreceptor cells in the vertebrate retina. *Proceedings of the National Academy of Sciences (U.S.A.)* 70, 1677– 81.
Raviola, E. and Gilula, N. B. (1975) Intramembrane organization of specialized contacts in the outer plexiform layer on the retina. A freeze-fracture study of monkeys and rabbits. *Journal of Cell Biology* 65, 192.

Raviola, E. and Raviola, G. (1982) Structure of the synaptic membranes in the inner plexiform layer of the retina: freeze-fracture study in monkeys and rabbits. *Journal of Comparative Neurology* 209, 233 – 248.

Raymond, P. A. (1995) Development and morphological organization of photoreceptors. In *Neurobiology and Clinical Aspects of the Outer Retina* (Djamgoz, M. B. A., Archer, S. N. and Vallerga, S. Eds.). Chapman & Hall, London. pp. 1– 23.

Reale, E., Luciano, L. and Spitznas, M. (1978) Communicating junctions of the human sensory retina. *Albrecht von Gaefes Archiv für experimentelle Ophthamologie* 208, 77– 92.

Remé, C. E. and Young, R. W. (1977) The effects of hibernation on cone visual cells in the ground squirrel. *Investigate Ophthalmology and Visual Science* 16, 815 – 840

Ryan, M. K. and Hendrickson, A. E. (1987) Interplexiform cells in macaque monkey retina. *Experimental Eye Research* 45, 57– 66.

Ryder, J. A. (1895) An arrangement of retinal cells in the eyes of fishes partially simulating compound eyes. *Proceedings of the National Academy of Science (U.S.A.)* 161–166.

Saito, T. and Kujiraoka, T. (1982) Physiological and morphological identification of two types of on-centre bipolar cells in the carp retina. *Journal of Comparative Neurology* 205, 161–170.

Saito, T. and Kujiraoka, T. (1988) Characteristics of bipolar-bipolar coupling in the carp retina. *Journal of General Physiology* 91, 275 – 287.

Saito, T., Kujiraoka, T., Yonaha, T. and Chino, Y. (1985) Re-examination of photoreceptor-bipolar connectivity patterns in carp retina: HRP-EM and Golgi-EM studies. *Journal of Comparative Neurology* 236, 141–160.

Saito, T., Kondo, H. and Toyodo, J-I. (1979) Ionic mechanisms of two types of on-center bipolar cells in the carp retina. I. The responses to central illumination. *Journal of General Physiology* 73, 73 – 90.

Sakai, H. M. and Naka, K. I. (1986) The synaptic organization of the cone horizontal cells in the catfish retina. *Journal of Comparative Neurology* 245, 141–160.

Sandermann, D., Boycott, B. B. and Peichl, L. (1996a) Blue-cone horizontal cells in the retinae of horses and other equidae. *Journal of Neuroscience* 16, 3381– 3396.

Sandermann, D., Boycott, B. B. and Peichl, L. (1996b) The horizontal cells of artiodactyl retinae: A comparison with Cajal's description. *Visual Neuroscience* 13, 735 –746.

Schaeffer, S. F. and Raviola, E. (1978) Membrane recycling in the cone cell endings of the turtle retina. *Journal of Cell Biology* 79, 802 – 825.

Schiller, P. H. (1992) The ON and OFF channels of the visual system. *Trends in Neuroscience* 15, 86 – 92.

Scholes, J. H. (1975) Colour receptors and their synaptic connexions in the retina of a cyprinid fish. *Philosophical Transactions of the Royal Society of London B* 270, 61–118.

Schwartz, E. A. (1975) Cones excite rods in the retina of the turtle. *Journal of Physiology (London)* 246, 639 – 651.

Sekaran, S., Matthews, K. L., Cunningham, J. R., Neal, M. J. and Djamgoz, M. B. A. (1998) Nitric oxide release during light adaptation of the carp retina. *Journal of Physiology*. In press.

Sherry, D. M. and Yazulla, S. (1993) Goldfish bipolar cells and axon terminal patterns: A Golgi study. *Journal of Comparative Neurology* 329, 188 – 200.

Shiells, R. A. (1995) Photoreceptor-bipolar cell transmission. In *Neurobiology and Clinical Aspects of the Outer Retina* (Djamgoz, M. B. A., Archer, S. N. and Vallerga, S. Eds.). Chapman & Hall, London. pp. 297– 324.

Sjöstrand, F. S. (1953) Ultrastructure of retinal rod synapses of the guinea pig eye, as revealed by three dimensional reconstruction of serial sections. *Journal of Ultrastructure Research* 5, 13 –17.

Slaughter, M. M. and Miller, R. F. (1981) 2-amino-4-phosphonobutyric acid: a new pharmacological tool for retina research. *Science* 211, 182 –184.

Smith, R. G. and Sterling, P. (1990) Cone receptive field in cat retina computed from microcircuitry. *Visual Neuroscience* 5, 453 – 461.

Smith, R. G., Freed, M. A. and Sterling, P. (1986) Microcircuitry of the dark adapted cat retina: functional architecture of the rod-cone network. *Journal of Neuroscience* 6, 3505 – 3517.

Spadaro, A, de simone, I. and Puzzolo, D. (1978) Ultrastructural data and chronobiological patterns of the synaptic ribbons in the outer plexiform layer of albino rats. *Acta anantomica* 102, 365 – 373.

Stell, W. K. (1967) The structure and relationships of horizontal cells and photoreceptor-bipolar synaptic complexes in goldfish retina. *American Journal of Anatomy* 120, 401– 423.

Stell, W. K. (1975) Horizontal cell axons and axon terminal in goldfish retina. *Journal of Comparative Neurology* 159, 503 – 520.

Stell, W. K. (1976) Functional polarization of horizontal cells dendrites in goldfish retina. *Investigative Ophthalmology and Visual Science* 15, 895 – 908.

Stell, W. K. (1980) Photoreceptor-specific synaptic pathways in goldfish retina: a world of colour, a wealth of connections. In *Colour Vision Deficiencies V* (Verriest, G. Ed.). A. Hilger, Bristol. pp. 1–14.

Stell, W. K. and Lightfoot, D. O. (1975) Color-specific interconnections of cones and horizontal cells in the retina of the goldfish. *Journal of Comparative Neurology* 159, 473 – 502.

Stell, W. K. and Lightfoot, D. O. (1979) Computer-aided reconstruction and analysis of goldfish rod synapses. *Seitai no Kagaku* 30, 173 –177.

Stell, W. K., Ishida, A. T. and Lightfoot, D. O. (1977) Goldfish retina: structural basis for ON- and OFF-centre responses in retinal bipolar cells. *Science* 198, 1269 –1271.

Stell, W. K., Kretz, R. and Lightfoot, D. O. (1982) Horizontal cell connectivity in goldfish. In *The S-Potential* (Drujan, B. D. and Laufer, M. Eds.). Liss, A. R. New York. pp. 51–75.

Stell, W. K., Lightfoot, D. O., Wheeler, T. G. and Leeper, H. F. (1975) Goldfish retina: functional polarization of cone horizontal cell dendrites and synapses. *Science* 190, 989 – 890.

Stell, W. K., Barton, L., Ohtsuka, T. and Hirano, J. (1994) Chromatic and neurochemical correlates of synapses between cones and horizontal cells. In *Great Basin Visual Science Symposium* (Olson, R. J., and Lasater, E. M. Eds.). University of Utah, Salt Lake City. pp. 41– 48.

Sterling, P. (1983) Microcircuitry of the cat retina. *Annual Reviews of Neuroscience* 6, 149 –185.

Sterling, P., Cohen, E., Freed, M. A. and Smith, R. G. (1987) Microcircuitry of the on-beta ganglion cell in daylight, twilight and starlight. *Neuroscience Research Supplement* 6, S269 – 285.

Tachibana, M. and Kaneko, A. (1984) γ-aminobutyric acid acts at axon terminals of turtle photoreceptors: difference in sensitivity among cell types. *Proceedings of the National Academy of Science (U.S.A.)* 81, 7961–7964.

Teranishi, T., Kato, S. and Negishi, K. (1982) Lateral spread of S-potential components in the carp retina. *Experimental Eye Research* 34, 389 – 399.

Tornqvist, K., Yang, X.-L. and Dowling, J. E. (1988) Modulation of the cone horizontal cell activity in the teleost fish retina. III. Effects of prolonged darkness and dopamine on electrical coupling between horizontal cells. *Journal of Neuroscience* 8, 2279 – 2288.

Trifonov, Y. A. (1968) Study of synaptic transmission between photoreceptors and horizontal cells by means of electrical stimulation in the retina. *Biofizika* 13, 809 – 817.

Tsukamoto, Y., Masarachia, P., Schein, S. J. and Sterling, P. (1992) Gap junctions between the pedicles of macaque foveal cones. *Vision Research* 32, 1809 –1815.

Vaney, D. I. (1990) The mosaic of amacrine cells in the mammalian retina. *Progress in Retinal Research* 9, 49 –100.

Vaney, D. I. (1991) Many diverse types of retinal neurons show tracer coupling when injected with biocytin or neurobiotin. *Neuroscience Letters* 125, 187–190.

Vaney, D. I. (1993) The coupling pattern of axon-bearing horizontal cells in the mammalian retina. *Proceedings of the Royal Society of London B* 252, 93 –101.

Vaney, D. I. (1994) Patterns of neuronal coupling in the retina. *Progress of Retinal Research* 13, 301– 355.

Vardi, N., Masarachia, P. and Sterling, P. (1992) Immunoreactivity to $GABA_A$ receptor in the outer plexiform layer of the cat retina. *Journal of Comparative Neurology* 320, 394 – 397.

Vardi, N., Kaufman, D. I. and Sterling, P. (1994) Horizontal cells in cat and monkey retina express different isoforms of glutamic acid decarboxylase. *Visual Neuroscience* 11, 135 –142.

Villa de la, P., Kurahashi, T. and Kaneto, A. (1995) L-Glutamate-induced responses and cGMP-activated channels in three subtypes of retinal bipolar cells dissociated from the cat. *Journal of Neuroscience* 15, 3571– 3582.

Voigt, T. and Wässle, H. (1987) Dopaminergic innervation of AII amacrine cells in mammalian retina. *Journal of Neuroscience* 7, 4115 – 4128.

Vollrath, L., Meyer, A. and Buschmann, F. (1989) Ribbon synapses of the mammalian retina contain two types of synaptic bodies – ribbons and spheres. *Journal of Neurocytology* 18, 115 –120.

Vollrath, L. and Spiwoks-Becker, I. (1996) Plasticity of retinal ribbon synapses. *Microscopy Research and Technique* 35, 472 – 487.

Wagner, H.-J. (1972) Vergleichende Untersuchungen über das Muster der Sehzellen und Horizontalen in der Teleostier-Retina (Pisces). *Zeitschrift für Morphologie der Tiere* 72, 77–130.

Wagner, H.-J. (1973) Darkness-induced reduction in the number of synaptic ribbons in fish retina. *Nature New Biology* 246, 53 – 55.

Wagner, H.-J. (1978) Cell types and connectivity patterns in mosaic retinae. *Advances in Anatomy and Embryology* 55, 1– 81.

Wagner, H.-J. (1980) Light-dependent plasticity of the morphology of horizontal cell terminals in cone pedicles of fish retina. *Journal of Neurocytology* 9, 573 – 590.

Wagner, H.-J. (1990) Retinal structure in fishes. In *The Visual System of Fish* (Douglas, R. H. and Djamgoz, M. B. A. Eds.). Chapman & Hall, London. pp. 109 –158.

Wagner, H.-.J. (1997) Presynaptic bodies ('ribbons'): from ultrastructural observations to molecular perspectives. *Cell and Tissue Research* 287, 435 – 446.

Wagner, H.-J. and Ali, M. A. (1978) Retinal organization in goldeye and mooneye. *Revue Canadienne de Biologie* 37, 65 – 84.

Wagner, H.-J. and Wulle, I. (1990) Dopaminergic interplexiform cells contact photoreceptor terminals in catfish retina. Cell and Tissue Research 261, 359 – 365.

Wagner, H.-J. and Wulle, I. (1992) Contacts of dopaminergic interplexiform cells in the outer retina of the blue acara. *Visual Neuroscience* 9, 325 – 333.

Wagner, H.-J. and Djamgoz, M. B. A. (1993) Spinules: a case for retinal synaptic plasticity. *Trends in Neurosciences* 16, 201– 206.

Wagner, H.-J., Behrens, U. D., Zaunreiter, M. and Douglas, R. H. (1992) The circadian component of spinule dynamics in teleost retinal horizontal cells is dependent on the dopaminergic system. *Visual Neuroscience* 9, 345 – 351.

Wässle, H. and Riemann, H. J. (1978) The mosaic of nerve cells in the mammalian retina. *Proceedings of the Royal Society of London B* 200, 441– 461.

Wässle, H. and Boycott, B. B. (1991) Functional architecture of the mammalian retina. *Physiological Reviews* 71, 447– 480.

Wässle, H., Grünert, U., Martin, P. R. and Boycott, B. B. (1994) Immunocytochemical characterisation and spatial distribution of midget bipolar cells in the macaque monkey retina. *Vision Research* 34, 561– 579.

Wässle, H., Yamashita, M., Greferath, U., Grünert, U. and Müller, F. (1991) The rod bipolar cell of the mammalian retina. *Visual Neuroscience* 7, 99 –112.

Weiler, R. and Zettler, F. (1979) The axon bearing horizontal cells in the teleost retina are functional as well as structural units. *Vision Research* 19, 1261–1268.

Weiler, R. and Wagner, H.-J. (1984) Light dependent change in cone-horizontal cell interactions in carp retina. *Brain Research* 298, 1– 9.

Weiler, R. and Schultz, K. (1993) Ionotropic non-N-methyl-D-aspartate agonists induce retraction of dendritic spinules from retinal horizontal cells. *Proceedings of the National Academy of Science (U.S.A.)* 90, 6533 – 6537.

Weiler, R., Baldridge, W. H., Mangel, S. C. and Dowling, J. E. (1997) Modulation of endogenous dopamine release in the fish retina by light and prolonged darkness. *Visual Neuroscience* 14, 351– 356.

Wenthold, R. J., Yokotani, N., Doe, K. and Wada, K. (1992) Immunochemical characterization of the non-NMDA glutamate receptor using subunit-specific antibodies. Evidence for a heterooligomeric structure in rat brain. *Journal of Biological Chemistry* 267, 501– 507.

Werblin, F. S. (1979) Integrative pathways in local circuits between slow-potential cells in the retina. In *The Neurosciences, Fourth Study Program* (Schmitt, F. O. and Worden, F. G. Eds.). The MIT Press, Cambridge MA. PP. 193 – 211.

Werblin, F. S. and Dowling, J. E. (1969) Organization of the retina of the mudpuppy, *Necturus maculosus*: II Intracellular recording. *Journal of Neurophysiology,* 40, 23 – 62.

Wikler, K. C. and Rakic, P. (1990) Distribution of photoreceptor subtypes in the retina of diurnal and nocturnal primates. *Journal of Neuroscience* 10, 3390 – 4401.

Williams, D. R. (1986) Seeing through the photoreceptor mosaic. *Trends in Neurosciences* 9, 193 –198.

Winslow, R. L., Miller, R. F. and Ogden, T. E. (1989) Functional role of spines of the retinal horizontal cell network. *Proceedings of the National Academy of Science (U.S.A.)* 86, 387– 391.

Witkovsky, P. and Dowling, J. E. (1969) Synaptic relationships in the plexiform layers of the carp retina. *Zeitschrift für Zellforschung und Mikroskopische Anatomie* 100, 60 – 82.

Witkovsky, P. and Stell, W. K. (1973) Retinal structure in the smooth dogfish *Mustelus canis*: electron microscopy of serially sectioned bipolar cell synaptic terminals. *Journal of Comparative Neurology* 150, 147–168.

Witkovsky, P., Stone, S. and Tranchina, D. (1989) Photoreceptor to horizontal cell synaptic transfer in the Xenopus retina: modulation by dopamine ligands and a circuit model for interaction of rod and cone inputs. *Journal of Neurophysiology* 62, 864 – 881.

Witkovsky, P. and Dearry, A. (1991) Functional roles of dopamine in the vertebrate retina. *Progress in Retinal Research* 11, 247– 292.

Witkovsky, P., Shakib, M. and Ripps, H. (1974) Interreceptoral junctions in the teleost retina. *Investigative Ophthalmology and Visual Science* 13, 996 –1009.

Wong-Riley, M. T. T. (1974) Synaptic organization of the inner plexiform layer in the retina of the tiger salamander. *Journal of Neurocytology* 3, 1– 33.

Wu, S. M. and Yang, X.-L. (1988) Electrical coupling between rods and cones in the tiger salamander retina. *Proceedings of the National Academy of Science (U.S.A.)* 85, 275 – 278.

Wulle, I. and Wagner, H.-J. (1990) GABA and tyrosine hydroxylase immunocytochemistry reveal different patterns of colocalization in retinal neurons of various vertebrates. *Journal of Comparative Neurology* 196, 173 –178.

Yamada, M. and Saito, T. (1997) Dual component in receptive field centre of bipolar cells in carp retina. *Vision Research* 37, 2331– 2338.

Yamada, E. and Ishikawa, T. (1965) The fine structure of the horizontal cells in some vertebrate retinas. *Cold Spring Harbour Symposia on Quantitative Biology* 30, 383 – 392.

Yang, X-L., Fan, T.-X. and Shen, W. (1994) Effects of prolonged darkness on light responsiveness and spectral sensitivity of cone horizontal cells in carp retina *in vivo*. *Journal of Neuroscience* 14, 326 – 334.

Yang, X-L. and Wu, S. M. (1996) Response sensitivity and voltage gain of the rods- and cone-horizontal cell synapses in dark- and light-adapted tiger salamander retina. *Journal of Neurophysiology* 76, 3863 – 3874.

Yazulla, S. (1976) Cone input to bipolar cells in the turtle retina. *Vision Research* 16, 737–744.

Yazulla, S. (1985) Factors controlling the release of GABA from goldfish retinal horizontal cells. *Neuroscience Research Suppl.* 2, S147–165.

Yazulla, S. (1995) Neurotransmitter release from horizontal cells. In *Neurobiology and Clinical Aspects of the Outer Retina* (Djamgoz, M. B. A., Archer, S. N. and Vallerga, S. Eds.). Chapman & Hall, London. pp. 249 – 271.

Yazulla, S., Studholme, K. M., Victoria, J. and de Blas, A. L. (1989) Immunocytochemical localization of $GABA_A$ receptors in goldfish and chicken retinas. *Journal of Comparative Neurology* 280, 15 – 26.

Yazulla, S. and Zucker, C. L. (1988) Synaptic organization of dopaminergic interplexiform cells in the goldfish retina. *Visual Neuroscience* 1, 13 – 29.

Yellott, J. I. Jr. (1982) Spectral analysis of spatial sampling by photoreceptors: topological disorders prevent aliasing. *Vision Research* 22, 1205 –1210.

Yellott, J. I. Jr., Wandell, B. A. and Cornsweet, T. N. (1984) In *Handbook of Physiology, Section I.* American Physiological Society, Bethesdy, Md, pp. 257– 316.

Zucker, C. L. and Dowling, J. E. (1987) Centrifugal fibres synapse on dopaminergic interplexiform cells in the teleost retina. *Nature*, 330, 166 –168.

11
Inner retinal signal processing: adaptation to environmental light

L.J. FRISHMAN and J.G. ROBSON
College of Optometry, University of Houston, Houston, TX, USA

11.1 Introduction

11.1.1 The range of visual operation

The human visual system operates over an enormous range of ambient illumination. From starlight to bright sunlight, the mean level of illumination increases well over a billion times, covering a range of about 10 log units. Over much of this range, due to various mechanisms of light adaptation, visual sensitivity remains fairly high for small light increments or decrements from the steady level of illumination. The maintenance of a relatively high contrast sensitivity over such a large range have been the subject of numerous psychophysical and non-invasive physiological studies in humans, as well as physiological and some behavioral studies in animal models with similar capabilities. A comparison of the psychophysical results with recordings from retinal ganglion cells in cats and primates indicates that most of the light adaptation reported by human subjects can be observed in individual retinal ganglion cells. This means that most of the light adaptation occurs in the retina before the visual signals travel to the brain for further processing. The question then is where in the retina, and by what mechanisms, the adaptation occurs. The previous chapter of this book mainly examined mechanisms of adaptation in the vertebrate outer retina. Special emphasis was placed on lower vertebrates and specifically the teleost retina where light-related morphological plasticity in the outer retina is well documented and its relation to light adaptation has been studied extensively. Mechanisms of light-adaptation in the inner retina of fish have not been as thoroughly studied. Although adaptation-dependent morphological plasticity in goldfish inner retina of Mb (Yazulla and Studholme, 1992) and Ma (Behrens and Wagner, 1996) bipolar cell terminals has been described, the physiological consequences of these changes are not yet clear. Nor is there a large body of work examining possible roles of inner *vs* outer retina on retinal output in this species, though effects of dark-adaptation on color-opponency in ganglion cells would appear to originate distally (e.g. Raynauld *et al.*, 1979.) In amphibia it also appears that inner retina is not a major site of adaptation, although some occurs (e.g. Donner *et al.*, 1990), whereas in mammals, for whom more information is available, there is strong

S.N. Archer et al. (eds.), Adaptive Mechanisms in the Ecology of Vision, 383–412

evidence of a prominent role for inner retinal mechanisms, at least at low light levels. Therefore, rather than focusing on lower vertebrates, the present chapter will concentrate primarily on comparing the role of inner and outer retinal neurons in light adaptation in mammals where morphological plasticity, aside from modulation of gap junctions (see section 11.3 on switching from rod to cone circuitry) has not been implicated in light adaptation. However, before turning our attention to specific sites and neuronal mechanisms, we will review briefly, the structural features of the eye, as well as morphological features of the mammalian retina that are integrally involved in the process of light adaptation.

11.1.2 Structural and morphological features affecting light adaptation

To some extent, the amount of light that reaches the mammalian retina is regulated by the pupil, which changes in size as a function of the ambient level of illumination. Under fully dark-adapted conditions, such as in starlight, the natural pupil diameter dilates to as much as 7 – 8 mm in young humans, maximizing the possibility of catching photons. Any reduction of the pupil area reduces the amount of light proportionally. But under normal conditions, the pupil diameter only reduces by about 1 mm from its dark-adapted maximum at light levels of around 1 photopic troland, (phot td; a troland is defined one candle per meter^2 viewed through a one mm^2 pupil), where cone vision becomes of consequence. Thus changes in pupil area are of very little importance in adjusting scotopic sensitivity. Over the photopic range, after taking the Stiles-Crawford effect into account, the pupil, when maximally constricted (to about 2 – 3 mm in diameter) at 10^4 phot td, reduces the effective collection of light by the cones by a factor of about 6 which helps to protect against extremes of cone photoreceptor pigment bleaching (Woodhouse and Campbell, 1975), and reduces the time for recovery following exposure to very high luminances.

Morphologically, most mammalian retinas have two receptor systems with different ranges of operation and absolute sensitivities. In starlight which provides a retinal illuminance less than about 10^{-2} scotopic trolands (scot td), the rods, which mediate scotopic vision, can signal the absorption of single photons, whereas cones cannot. In primates, the full range over which rods signal individual photons is from absolute threshold, (about 10^{-4} scot td), to almost 1 scot td, a level which, in humans, results in 8.6 photoisomerations per second per rod (Rh*.sec^{-1}rod^{-1}) (derived using Wyszecki and Styles, 1982); at higher illuminances, the signal in the rods becomes continuous. It should be noted that visual neurons proximal to the rods can only signal single photons over a smaller range because their signals become effectively continuous at a lower illuminance as a result of the convergence onto them of many rod signals. The rods also dominate over much of the mesopic range of vision at the upper end of which they saturate, rod-mediated vision giving way completely to cone-mediated photopic vision.

In order to relay signals from the two receptor systems to the inner retinal neurons, in mammals there are different rod and cone pathways through the retina. A specialized rod circuit is active at the very low light levels, (i.e. starlight levels), at which photon-signaling occurs. This circuit, illustrated in Figure 1,

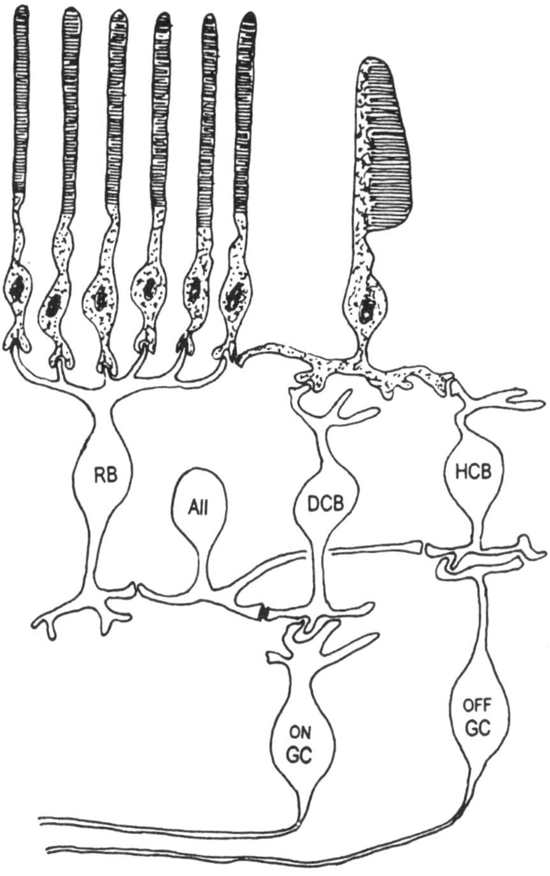

Figure 1 The specialized rod circuit in mammalian retina. Under 'starlight' conditions when photons are discretely signaled, the rod signals travel via rod bipolar cells to the inner retina where the bipolar cells contact AII amacrine cells. The AII amacrine cells are bistratisfied; more distally, in sublamina A, they form chemical (glycinergic) synapses with Off cone bipolar cells; more proximally, in sublamina B, they form gap junctions with On cone bipolar cells. The rod signals ultimately reach the On and Off retinal ganglion cells via the chemical synapses (ionotropic glutamatergic) that the On and Off bipolar cells make with the ganglion cells. Under mesopic conditions, the rod signal which passes via a gap junction into the cone, is transmitted via cone bipolar cells to the retinal ganglion cells. Horizontal cells which contact the photoreceptors in the outer plexiform layer have been omitted from the figure because they do not form part of the straight-through signal pathway. See Daw et al., 1990, and Wässle and Boycott, 1991 for reviews of this circuitry.

includes a single type of rod bipolar cell, which has L-2-amino-4-phosphonobutyrate- (L-APB)-sensitive metabotropic glutamate receptors (Nakajima *et al.*, 1993) and is known to be depolarizing in the rabbit (Dacheux and Raviola, 1986), cat (de la Villa *et al.*, 1995), and rat (Yamashita and Wässle 1991), and almost certainly in primates as well (Dolan and Schiller, 1989). In the inner retina, the rod signals converge onto a distinctive type of narrow field amacrine cell, the AII amacrine cell (Kolb and Famiglietti, 1974; Strettoi *et al.*, 1992; Wässle *et al.*, 1995). The AII amacrine cell, a depolarizing neuron that is glycinergic (Pourcho, 1980; Pourcho and Goebel, 1985; Hendrickson *et al.*, 1988), contacts On cone bipolar cells via gap junctions and Off cone bipolar cells via chemical synapses; these relay the signals to On and Off ganglion cells respectively (Kolb and Famiglietti, 1974; Famiglietti and Kolb, 1975; 1976; Strettoi *et al,* 1992). At mesopic light levels when there are sufficient photons available for the signal in the rods to become continuous, the rod bipolar cells' responses saturate. Several lines of evidence, morphological (Raviola and Gilula, 1973; Kolb, 1977), physiological (Nelson, 1977; Nelson and Kolb, 1985; Schneeweis and Schnapf, 1995; Stockman *et al.*, 1995; De Vries and Baylor, 1995), psychophysical (Conner and MacLeod, 1977; Conner, 1982; Sharpe *et al.*, 1989; Stockman *et al.*, 1991) as well as neural models (Smith *et al.*, 1986) indicate that under these conditions the rod signals travel via gap-junctions to the medium and long wavelength cones, and then to inner retina via the midget and diffuse bipolar cells that the cones contact. At light levels at which the rod photoreceptors themselves have saturated, only the cones, including the short-wave length cones whose signals travel in private circuitry, are available to vision. The transition from rod- to cone-mediated vision also effects a change from achromatic to chromatic vision, and an increase in spatial resolution, temporal resolution and contrast sensitivity.

11.1.3 Bleaching of visual pigment

Another important factor involving a physical change rather than neuronal processing, that must be considered when discussing light adaptation, is pigment bleaching. For rods, the electrical response is fully saturated at light levels that bleach only a very small portion of the visual pigment. Thus, although the slow removal of products of pigment bleaching appear to prolong the recovery of sensitivity of the rod after exposure to large amounts of light, a reduction in the amount of available pigment does not itself play a direct functional role in reducing the sensitivity of the rod. For cones the situation is quite different. Cone-mediated vision continues for ambient illuminations up to 10^6 phot td when 99% of the pigment is bleached in monkeys (Valeton and Van Norren, 1983). About 10% of the pigment will be bleached by 3000 phot td, and about half will be bleached at 20000 phot td (Rushton & Henry, 1968). At these high levels, a substantial reduction in sensitivity can clearly be attributed to a reduction in the quantum-catching capability of the cones.

11.2 Adaptation of rod-mediated vision

11.2.1 A comparison of psychophysics and physiology

Adaptation of rod-mediated vision has been studied more thoroughly than that of cone-mediated vision, and the retinal sites of adaptation are better defined. Over a considerable range of scotopic background intensities, too weak to affect appreciably the rod photoreceptor response (Baylor *et al.*, 1984; Tamura *et al.*, 1989; Kraft *et al.*, 1993), the human psychophysical threshold increases as background illumination is increased. This increase occurs with a slope of between 0.5 and 1.0 (Weber's Law) on logarithmic coordinates (e.g. Aguilar and Stiles, 1954 ; Sharpe, 1990 for review). Figure 2 illustrates the classic study of Aguilar and Stiles (1954) of the effect of adapting backgrounds on the threshold for rod-mediated vision with average data from four human subjects. The limited role played by the rod photoreceptors themselves in the reduction of sensitivity due to light adaptation is also illustrated in this figure. Curves fitted to data measured *in vitro* from single rod outer segments of humans (rightmost curve, Kraft *et al.*, 1993) and macaque monkeys (dashed curve, Tamura *et al.*, 1991) are shown on the right side of the figure. Very similar effects of light adaptation on rod outer segment photocurrents have been found for cats (Tamura *et al.*, 1989), rabbits

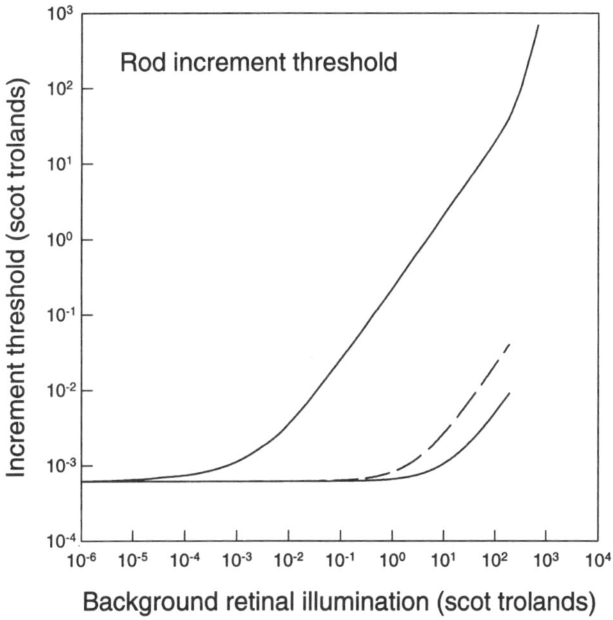

Figure 2 Increment threshold and inverse sensitivity curves. The curve on the left is the average result of the classical psychophysical study of four human subjects by Aguilar and Stiles (1954). The two curves on the right are fits of Weber's Law to recordings from isolated rod outer segments of humans (solid lines; Kraft *et al.*, 1993) and macaque monkeys (dashed line; Tamura *et al.*, 1991).

and other nonprimates (Nakatani *et al.*, 1991). It is clear that although the psychophysical response is well into the Weber region by 10^{-2} td, the photoreceptor response is only affected similarly by illumination levels at least 3 logs units higher. Experiments on single photoreceptors in several laboratories (Torre *et al.*, 1986; Matthews *et al.*, 1988; Nakatani and Yau, 1988; reviewed by Koutalos and Yau, 1996) have demonstrated that intracellular Ca^{2+} (via feedback pathways) is involved in the photoreceptor-mediated adaptation. The reduction of intracellular Ca^{2+} concentration induced by light increases synthesis of cGMP by modulating the enzyme guanylate cyclase (Koch and Stryer, 1988) as well as affecting other elements of the transduction cascade (e.g. Kawamura and Murakami, 1991; Koutalos *et al.*, 1995; reviewed by Koutalos and Yau, 1996).

Although the rod photoreceptors in mammalian retina are not the main site of light adaptation of rod-mediated vision, as already pointed out in the introduction, most light adaptation of rod-derived signals is accomplished before they leave the retina. Incremental sensitivity measures derived from recordings of action potentials from individual (cat) retinal ganglion cells, the last cells to process the visual signal before it travels to the brain, are quite similar to those measured psychophysically in humans (thoroughly reviewed by Shapley & Enroth-Cugell, 1984). While many studies have examined the issue of where in the retina this postreceptoral 'field' (Rushton, 1965) or 'network' adaptation occurs, and general loci have been demonstrated, the neuronal mechanisms are not well understood. We now will review the evidence that much of the adaptation of the rod signal at very low light levels occurs in the innermost retina, where amacrine and ganglion cells process visual signals. Much of the evidence that we will cite comes from studies using the electroretinogram (ERG), a signal that can be measured at the cornea of the intact eye and that represents the summed activity of retinal neurons.

11.2.2 Studies of adaptation using the scotopic electroretinogram

11.2.2.1 Adaptation of the rod circuit

The dark-adapted ERG elicited by a weak diffuse flash is primarily a cornea-negative signal in humans (Sieving and Nino, 1988; Sieving and Wakabayashi, 1991), and several other species including cats (Sieving *et al.*, 1986), monkeys (Wakabayashi *et al.*, 1988) and rats (Bush and Reme, 1992; Hubbard and Naarendrop, 1994). This negative signal, called the scotopic threshold response (STR) by Sieving, Frishman and Steinberg (1986) first appears in the ERG when the stimulus is no more than a few quanta.deg^{-2} (about 1 Rh* per 10^4 rods), an intensity roughly one hundred times less than that eliciting a discernible cornea-positive 'b-wave' (Granit's PII). These features of the cat ERG can be seen in Figure 3 which shows an intensity series for the cat ERG measured over a 6 log unit range. The small and relatively insensitive a-wave of the ERG that arises from photoreceptor currents (Granit, 1933; Penn and Hagins, 1972), is just

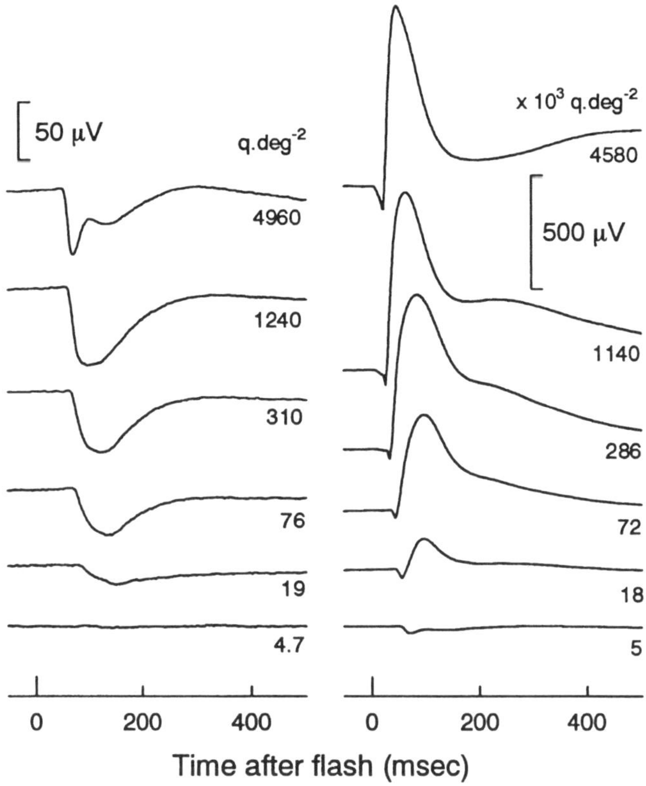

Figure 3 The ERG of a dark-adapted cat in response to brief ganzfeld flashes of increasing intensity. The ganzfeld was provided by an array of green LEDs behind a diffuser 35 mm in diameter close to the eye. Stimulus intensity increased from bottom to top and left to right, ranging from 4.7 to 4.6×10^6 q.deg^{-2} (507 nm). A brief stimulus was flashed at zero time for each trace in this and subsequent figures of similar format. A detailed description of the recording procedure can be found in Robson and Frishman (1995). Briefly, recordings were made in urethane-anesthetized cats that were immobilized with pancuronium bromide. ERGs were recorded between an intravitreal silver wire and a silver plate in the orbit behind the eye (measurements from cat number 25). (From Robson and Frishman, 1995, courtesy of Visual Neuroscience)

visible at the highest intensity, and is best-studied with stimuli of even higher intensity. The figure shows that the amplitude of the negative STR which is present at the lowest intensities saturates around the intensity that the b-wave first appears, while the less sensitive b-wave continues to grow until the intensity is about one hundred times higher still. Studies, mainly in cats, have indicated that whereas the b-wave (PII) component of the ERG originates from bipolar cells (Newman, 1980; Newman and Odette, 1984; Stockton and Slaughter, 1989; Gurevich and Slaughter, 1993; Robson and Frishman, 1995), the STR arises more proximally from amacrine and ganglion cells (Sieving *et al*., 1986; Naarendrop and Sieving, 1991; Sieving, 1991; Frishman *et al.,* 1993; Robson *et al*., 1993;

Frishman *et al.*, 1996B). At the light intensities used to elicit the STR and rod b-wave, it is likely that all ERG signals arise from the specialized rod circuit found in mammals that is illustrated in Figure 1. In addition to post-synaptic currents of the neurons involved, K^+ currents in Müller cells mediate the STR (Frishman *et al.*, 1988; Frishman and Steinberg, 1989), and contribute, but to a more limited extent, to the b-wave (Newman, and Odette, 1984; Wen and Oakley; 1990; Xu and Karwoski, 1994; Robson and Frishman, 1995, Hanitzsch *et al.*, 1996).

The separation of the ERG into components that arise from bipolar cells, and others that arise more proximally, has provided an opportunity to study light adaptation at different stages of processing in the retina. Recent studies in cats and primates have shown that the STR sensitivity is much reduced by weak backgrounds that hardly affect the b-wave sensitivity, indicating that adaptation occurs in stages in the retina, with adaptation to weak backgrounds occurring more proximally than the site at which higher backgrounds affect more distally generated signals (Frishman and Sieving, 1995; also see Green and Powers, 1982). This finding is illustrated in Figure 4 which shows the intensity needed to elicit a small criterion response in the ganzfeld ERG as a function of background illumination for a cat, a macaque and a human subject (Frishman and Sieving, 1995). Desensitization of the STR occurs for weaker backgrounds than those affecting PII. From human ERG recordings (Hood and Birch, 1993A), we know that photoreceptor-generated a-wave is not reduced appreciably by steady backgrounds until they are at least 32 scot td, near the end of the range illustrated in Figure 4.

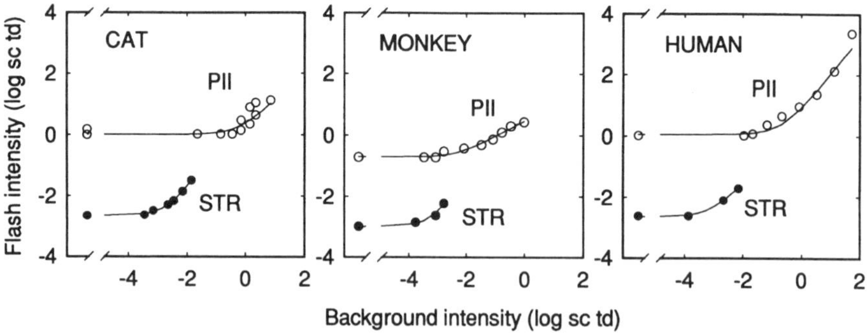

Figure 4 Plots of 1/sensitivity vs background illumination for a cat, a monkey and a human subject. The Y axis is the test flash intensity required at each adapting background intensity (X axis) to evoke a response of 5 μV for the STR (filled circles) of the cat and the monkey, 4 μV for the human subject. The criterion for the b-wave (PII) (open circles) was 40 μV for all three plots. Lines in this figure were fitted to the STR using equation 3 (see text). Recordings were made using Burian Allen electrodes and the cats and monkeys were anesthetized. (From Frishman and Sieving, 1995, courtesy of Vision Research.)

In the ERG study illustrated above, although b-waves and STRs were measured, no effort was made to separate overlapping components from different stages of processing in order to examine more carefully the roles of individual retinal neurons. It has been known since the early studies of Granit (1933), that the ERG represents the sum of the various positive and negative going contributions from the retina. In recent work in cats, we have used pharmacological agents to dissect the components of the dark-adapted ERG (Robson and Frishman, 1995; 1996). This allows us to be more specific about the cells involved, to model their component contributions to the ERG, and to study adaptation of the components. We can model the way in which the amplitude of the cat ganzfeld ERG at a particular time after the stimulus flash depends upon stimulus intensity by assuming, as did Granit (1933), that the response is the sum of the individual positive and negative components. We also assume that the contribution of each component increases in proportion to the stimulus intensity at low levels, but then saturates at a level and in a way that is characteristic of the particular cells involved. This assumption is justified by observations from several stages of retinal processing. In the intensity range that we are considering, both the stimulus intensity and background illuminance are such that the individual photoreceptors will rarely be stimulated by more than one photon at a time. Stimulus response functions of individual photoreceptors have an initially linear response range when stimulated by one or only a few photons, which then, at higher levels, show a nearly exponential form of saturation. More pertinent for modeling of inner retinal activity, is the demonstration by Barlow, Levick and Yoon (1971) that cat retinal ganglion cells can signal the activation by a single photon of a single photoreceptor, and that their responses to very weak flashes are directly proportional to stimulus strength up to a level corresponding to the activation of the cell by several photons. This finding by Barlow *et al.* indicates that all of the stages in the retina prior to the ganglion cells operate linearly over this stimulus range.

We find further support for the assumption of linear behavior of retinal neurons in the behavior of a pharmacologically isolated PII component of the cat ERG that we believe to reflect the activity of rod bipolar cells (Robson and Frishman, 1995). Isolation of PII is achieved using the glutamate agonist *n*- methyl-DL-aspartate (NMDLA) to suppress responses from inner retinal neurons, kainate (KA) and α-amino-3-hydroxy-5-methyl-4-isoxazolepropionate (AMPA) receptor antagonists to suppress Off bipolar and horizontal cells, and APB to isolate photoreceptoral activity, which then can be subtracted from the NMDLA-isolated response. We found that the isolated PII (rod bipolar cell response) measured at times both before and after its peak is initially proportional to stimulus intensity over a wide range of stimulus intensities and then saturates according to a hyperbolic Michaelis-Menten function. The initial linearity of isolated PII is illustrated in Figure 5. The figure also shows lack of initial linearity for the control b-wave.

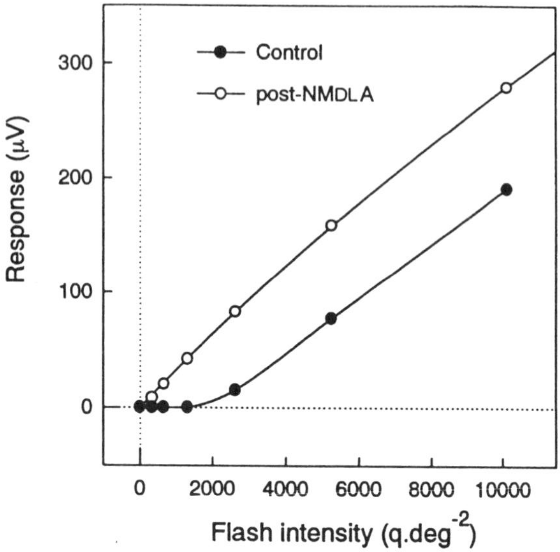

Figure 5 Plot of the peak amplitude of the cat b-wave (trough to peak) vs intensity before (filled symbols) and after (open symbols) intravitreal injection of NMDLA (concentration of 3.8 mM). The methods were as described for Figure 3 (measurements from cat no. 25). (From Robson and Frishman, 1995, courtesy of *Visual Neuroscience*)

The proportionality of the response to stimulus intensity for retinal ganglion cells (Barlow *et al.*, 1971) and the pharmacologically isolated rod bipolar cell response (Figure 5) also provides evidence that there is no quantum-coincidence requirement (i.e. no threshold non-linearity) operating for these retinal neurons. If the components of the ERG do not show threshold non-linearities, then neither should the ERG. However, both the negative STR and the positive b-wave in ERG traces in Figure 3, as well as the control b-wave amplitude in Figure 5 appear to show threshold-like behavior rather than rising gradually with intensity. The nonlinear rise of the STR and b-wave (PII) can also be seen in Figure 6A which is a plot, from another experiment, of ERG amplitude measured at 140 ms after the flash when the STR was maximal. If we are to retain the notion that all ERG components initially increase in proportion to stimulus intensity, we must assume that the 'negative' STR in fact conceals a positive component. (Robson *et al.*, 1993; Frishman *et al.*, 1996A). ERG recordings from cats consistently show similar apparent threshold-like behavior for the STR, and in roughly 1 in 10 individuals the ERG actually shows a small positive peak for very weak stimuli. We have called this small positive response in the cat ERG the positive scotopic threshold response, or pSTR (Frishman *et al.*, 1993; Robson *et al.*, 1993; Frishman *et al.,* 1996A), to distinguish it from the negative STR, now labeled, nSTR. When a positive peak is overtly present (e.g. inset to Figure 6A), it always saturates at very low intensities and amplitudes, reaching no more

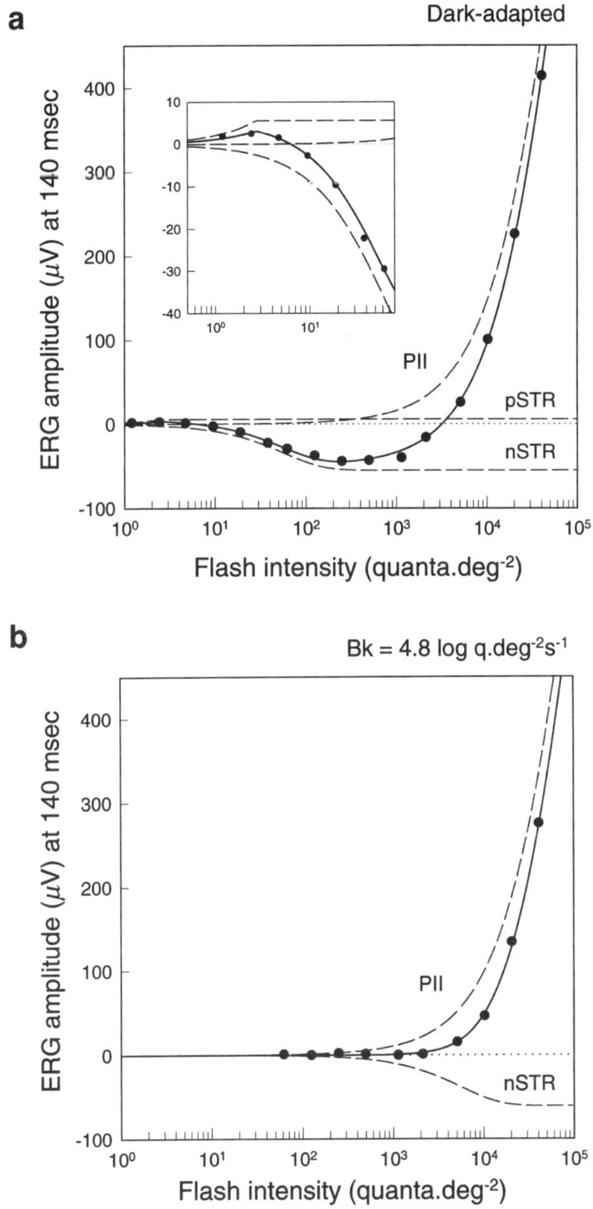

Figure 6 Amplitudes of cat ERG responses measured 140 ms after a flash at the maximum negativity of the nSTR. A) Dark-adapted ERG. Dashed lines show model curves for the pSTR, nSTR and PII. The inset shows the pSTR and nSTR at low light levels. The solid line shows the prediction of the complete model. Model curves for the responses assume an abrupt saturation for the pSTR, an exponential saturation for the nSTR and hyperbolic saturation for PII. B) ERG amplitudes in the presence of an adapting background of 4.8 log q.s^{-1}.deg^{-2} (0.63 scot td) with model component curves (dashed) and the complete model (solid) (measurements from cat no. 27)

than 1/10 the maximal amplitude of the nSTR. We find that a pSTR is necessary for describing the apparent threshold nonlinearity shown by all cat (and human) ERG responses to weak stimuli, even when there is no overt positive response to be seen in the ERG traces.

The manner in which the pSTR and nSTR in cats (and humans, see below) are affected by adapting backgrounds and the similarity of their sensitivities coupled with the relatively lower V_{max} for the pSTR are consistent with the pSTR reflecting activity of Off-center ganglion (or possibly amacrine) cells, and the nSTR reflecting that of On-center cells at the same level of retinal processing. The very low V_{max} of the positive response to a light flash likely reflects a reduction of maintained activity, typical of an Off response, that 'bottoms out' as it reduces the maintained activity of the inner retinal neuron to zero. While there are well-described depolarizing amacrine cells, AII and A17, in the rod circuit that pool signals over many rods, less is known about the Off-amacrine cells, A13 and A2/3 (in the cat) that receive major rod input (Nelson, 1982; Nelson *et al.*, 1985; Kolb and Nelson, 1984, 1985, 1993; Sterling *et al.*, 1987). On the other hand, On and Off ganglion cells with similar receptive field center sizes are known to have similar absolute sensitivities (Harding and Enroth-Cugell, 1978), and ganglion cell contributions to the dark-adapted ganzfeld ERG are at least plausible. Although we know that some portion of the nSTR survives after section of the optic nerve in cats and after degeneration of retinal ganglion cells in cats and humans (Sieving, 1991), it seems that it is a slower portion that is removed by retinal ganglion cell loss. In cats, the pSTR, as well a slow portion of the nSTR, are suppressed by intravitreal injection of tetrodotoxin (TTX; Robson *et al.*, 1993) which blocks the activity of spiking cells (all ganglion cells, but at least some amacrine cells as well). In macaques, the STR also disappears after longstanding ocular hypertension has destroyed ganglion cells. (Frishman *et al.*, 1996B) .

Figure 6A shows model curves (dashed lines) for the amplitudes of the components of the cat ERG at low light levels. As shown in the figure, the responses to weak stimuli are assumed to be the sums of (1) a cornea-positive component (pSTR) that saturates abruptly at some intensity, (2) a slightly less sensitive cornea-negative component (STR) that saturates as an exponential function at a significantly higher intensity, and (3), at higher light levels, a third, less sensitive positive component, PII, that is assumed to saturate as a hyperbolic function at a much higher intensity (though the exact way in which PII saturates is of little consequence in fitting most of the data here because of its limited range of high intensities). The solid line in Figure 6A shows the predicted outcome for the ERG, and it fits the data well.

Explicitly for our modeling, the pSTR rises as a linear function that saturates abruptly at V_{max}, while the exponential saturation for the nSTR is defined by:

$$V = V_{max} (1-\exp(-I/I_o)) = S\ I_o (1-\exp(-I/Io)) \qquad [1]$$

where V_{max} is the maximum (saturated) amplitude, S is the sensitivity and I_o is the intensity for an amplitude of $(1\text{-}1/e)V_{max}$, while the hyperbolic relation used for PII is defined by:

$$V = V_{max}\ I/(I + I_o) = S\ I_o\ I/(I + I_o) \qquad [2]$$

where V_{max} and S have the same meaning but I_o is the intensity at which the amplitude is $V_{max}/2$. For both equations, the response amplitude is approximately proportional to intensity so long as the response is sufficiently small (less than about 10 – 20% of the maximum). Within this range V = SI. The nature of the saturation for the STR components is not determined from independent observations but chosen on the basis of providing the best description of the present data.

Having separated the components of the dark-adapted ERG, it is possible to model the three components at each of several steady levels of background illumination . As an example, model curves for the components when a steady background of 0.63 scot td is present are shown in Figure 6B. Although the nSTR is not overtly seen in the records, it is still necessary to include it in the model at this higher background, in order to describe the rather abrupt initial rise of the b-wave.

Figure 7A shows data and the fitted model curves from another experiment. The figure illustrates well the reduction in sensitivity of STR over a range of background illuminations that hardly affect the sensitivity of the b-wave portion of the curves. A plot of 1/sensitivity (the slope of the initially linear 10 – 20% of the pSTR and nSTR model curves) versus background illumination for the same data is shown in part B of the figure. The curves in Figure 7B and subsequent plots of 1/sensitivity vs background in Figure 8 for humans are fitted with the equation:

$$1/S = K(I + I_D) \qquad [3]$$

where I_D is the intensity at which sensitivity is reduced by a factor of 2. I_D also indicates the intensity below which backgrounds are not shifting the sensitivity an appreciable amount. I_D, calculated for psychophysical data is often termed 'dark light' when determined with respect to absolute threshold. The dark light for both the pSTR and nSTR components in this experiment was about 2.0 quanta.deg^{-2}.sec^{-1}, equivalent to 4.5×10^{-3} scot td. The plot also contains a measure of V_{max} for the nSTR from the exponential curve fits. Even though the sensitivity of the nSTR is reduced substantially, V_{max} hardly varies across backgrounds.

Guided by our analyses in cat, for a few human subjects we also have analyzed the amplitude of the dark-adapted ERG at low light levels and in the presence of steady backgrounds (Frishman *et al.*, 1996A). In addition, for some of the human subjects we have measured the effects of steady backgrounds on psychophysical responses in the same experimental setup (Frishman *et al.*, 1996A). Commensurate with previous observations, a measurable STR in dark-adapted human subjects

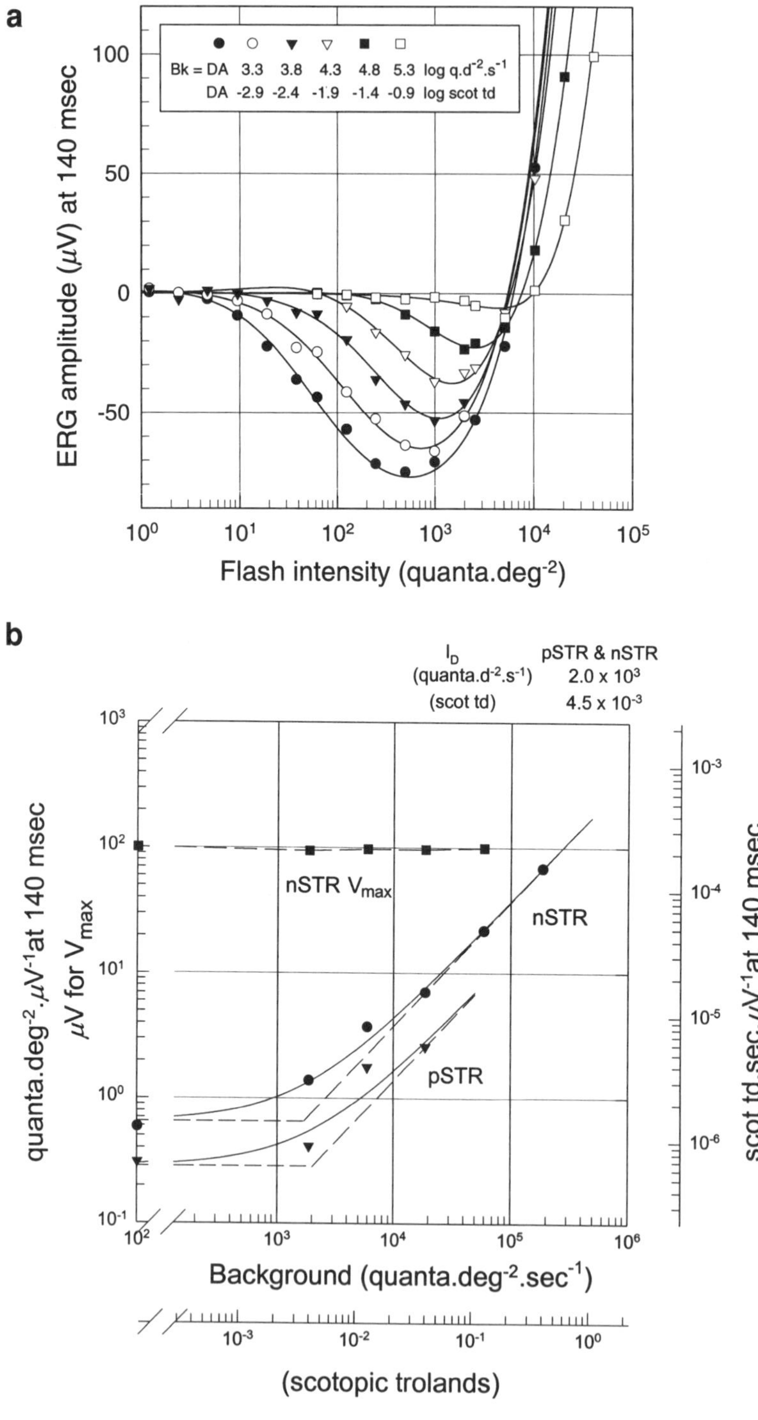
a
Bk = DA 3.3 3.8 4.3 4.8 5.3 log q.d^{-2}.s^{-1}
DA -2.9 -2.4 -1.9 -1.4 -0.9 log scot td
ERG amplitude (μV) at 140 msec
Flash intensity (quanta.deg^{-2})
b
I_D
(quanta.d^{-2}.s^{-1})
(scot td)
pSTR & nSTR
2.0 x 10^3
4.5 x 10^{-3}
nSTR V$_{max}$
nSTR
pSTR
quanta.deg^{-2}.μV^{-1}at 140 msec
μV for V$_{max}$
scot td.sec.μV^{-1}at 140 msec
Background (quanta.deg^{-2}.sec^{-1})
(scotopic trolands)

Figure 7 (opposite) Effect of steady backgrounds on the amplitude of cat ERG responses and components at 140 msec after the flash. A) Amplitudes measured under dark-adapted conditions and in the presence of 3 different steady backgrounds. Solid lines show complete model fits: See legend to Figure 6 for details of the individual component fits. B) Plot of 1/sensitivity and V_{max} vs background intensity for the pSTR and nSTR based on the curves shown in part A. The solid curve is a fit of equation 3 to the data from which we can derive I_D. I_D also is illustrated graphically as the intersection of the dashed lines of slope zero and slope one. The slope zero line passes through the fully dark-adapted value for 1/sensitivity, a slope one line is fit to the light-adapted values at backgrounds for which the data followed Weber's Law (measurements from cat no. 28)

occurs within about 0.7 log units of the psychophysical threshold for the same ganzfeld stimulus (Sieving and Nino, 1988). Our ERG results for the human subjects are similar to those obtained for the cats. For example, the initial apparent threshold-like behavior of the negative STR amplitude is seen in the records of all subjects. However, assuming, as discussed above, that there is no threshold non-linearity in the retinal pathway, this finding can be taken to indicate the existence of a sensitive positive component that, at stimulus levels below those at which the negative STR becomes evident, is a mirror image of the negative response and effectively cancels it out in the ERG. Therefore the human data measured at 200 msec after the flash when the nSTR is maximal in amplitude are modeled using the three components found for the cat, the pSTR, the nSTR, and PII. The 1/sensitivity functions for the negative STR components are shown in the lower portion of Figure 8A, and the psychophysical increment-threshold results in Figure 8B. As observed for the cat (Figure 7B), V_{max} for the nSTR in the human also remains fairly constant with increasing background illumination. The increment threshold (1/sensitivity) function for PII in the upper portion of Figure 8A is measured at 110 ms after the flash when PII is maximal. In the human ERG, responses measured at 110 msec after the flash did not include components other than PII whose amplitude was fitted well by a hyperbolic function. The results for the human subject support the notion that effects of weak background lights on psychophysical increment sensitivity for stimuli close to psychophysical threshold, i.e. under conditions where the ERG is dominated by the STR, largely reflect adaptation at a site downstream of the bipolar cells, a site that involves amacrine or ganglion cells. However, as background illuminance is increased, more distally located sites of adaptation affecting PII become involved in reducing the sensitivity. As summarized in Table 1, and reported originally by Biersdorf *et al.* (1965), the human b-wave is not desensitized by adapting backgrounds until they are about 1000 times above visual threshold.

Although the sites of adaptation are different for the STR and PII, the adaptation of both components occurs very rapidly. As an indication of the time required for adaptation of the nSTR and PII, we have observed effects of backgrounds pulsed briefly at various times with respect to brief stimulus flashes.

a

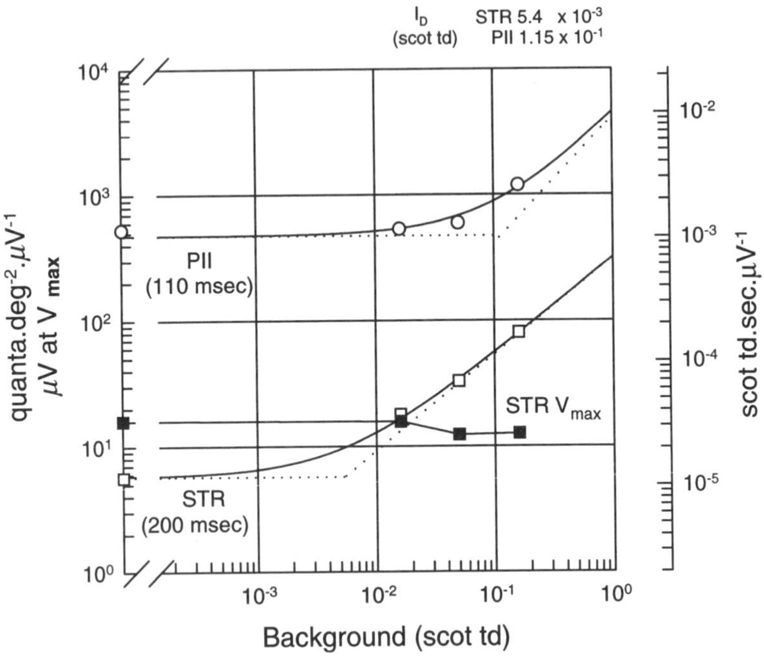
I_D (scot td) STR 5.4 x 10^-3 PII 1.15 x 10^-1
quanta.deg^-2.µV^-1 µV at V_max
scot td.sec.µV^-1
PII (110 msec)
STR (200 msec)
STR V_max
Background (scot td)

b

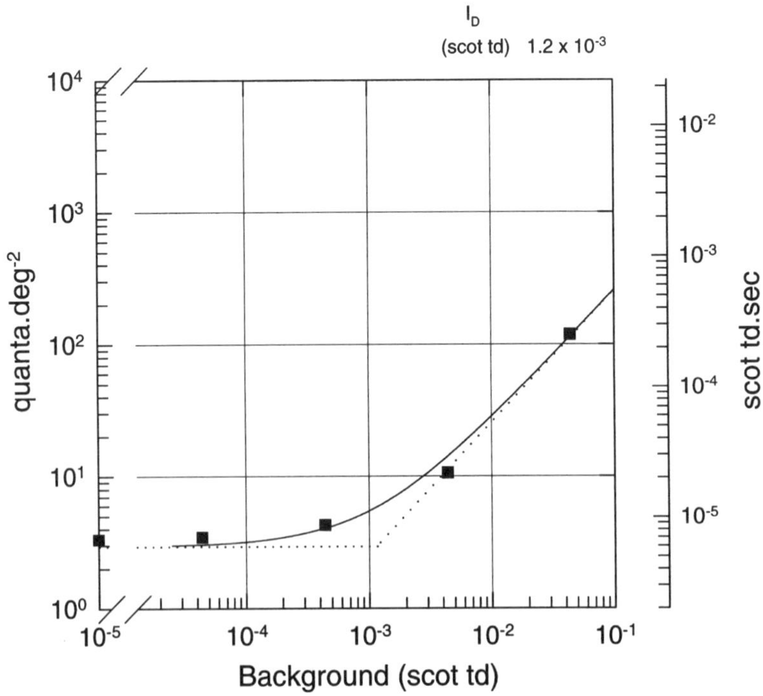
I_D (scot td) 1.2 x 10^-3
quanta.deg^-2
scot td.sec
Background (scot td)

Figure 8 (opposite) Effect of steady adapting backgrounds on the ERG and psychophysical threshold for the same human subject (SR). A) Plot of 1/sensitivity and V_{max} (filled squares) vs background intensity for the nSTR (open squares, measurements made at 200 msec when the nSTR amplitude was maximal), and 1/sensitivity for PII (open circles, at 110 msec when the b-wave amplitude was maximal). The pSTR is too small to derive sensitivity (measurements from subject SR). Corneal ERGs in humans were recorded, using a DTL fiber electrode (Dawson *et al.*, 1979) referenced either to the ipsilateral temple, or another DTL fiber placed in the cul-de-sac of the fellow eye). The solid curve was fit with the equation 3, and I_D also is illustrated as the intersection of the dashed lines of slope zero and slope one (see legend to Figure 7B). (From Frishman *et al.*, 1996A, courtesy of Journal of the Optical Society of America A.) B) The psychophysical threshold vs background intensity (TVI) function derived from probability of seeing functions. The solid curve was fit with the equation 3 (see text), and I_D is illustrated as the intersection of the dashed lines of slope zero and slope one (see legend to Figure 7B). Probability-of-seeing functions were determined in the dark and at four different steady background illuminations for the same stimulus conditions as those used for recording STRs, after an hour of dark-adaptation using a method of constant stimuli, with 10 randomly ordered repetitions of each of 10 stimuli. A Weibull function was fit to each set of frequency-of-seeing data and psychophysical threshold was computed as the intensity for 50% seen. False-positive rates were less than 2%. (From Frishman *et al.*, 1996A, courtesy of Journal of the Optical Society of America A.)

Table 1 The table compares values for I_D for humans from the study of Frishman *et al.*, 1996A and previous studies. Values are compared for psychophysical measures, and for electrophysiological measures where possible

Level	Source	I_D (quanta.deg^{-2}.sec^{-1})[#]	I_D (scotopic trolands)
Psychophysics	Frishman *et al.* (1996A)	200 – 900	$5.0 \times 10^{-4} - 2.0 \times 10^{-3}$
	Barlow (1957)	900	2.0×10^{-3}
	Sharpe *et al.* (1993)	110 – 450	$2.5 \times 10^{-4} - 1.0 \times 10^{-3}$
STR (scotopic threshold response)	Frishman *et al.* (1996A)	1200 – 4400	$3.2 \times 10^{-3} - 0.01$
	Frishman and Sieving (1995)	446	1.0×10^{-3}
b-wave	Frishman *et al.* (1996A)	$(2.8 - 5.6) \times 10^4$	0.06 – 0.13
	Fulton and Rushton (1978)	$(1.4 - 4.4) \times 10^4$	0.03 – 0.1
	Frishman and Sieving (1995)	2.6×10^4	0.06
Rods	Kraft *et al.* (1993)	5.6×10^6	12.6
Rod *a*-wave	Hood and Birch (1993A)*	1.4×10^7	31.6

[#]At the cornea

*Lowest background intensity where a-wave showed Weber behavior

Figure 9 shows some results from the human. Both for the nSTR and PII adaptation and recovery from adaptation are complete within a second. In our studies of anesthetized cats, we can observe these changes at 500 msec and even less. Such rapid adaptation has been described previously in mammalian retina. For example, Shapley and Enroth-Cugell (1973) recording from cat retinal ganglion cells under scotopic and mesopic conditions, and Yeh *et al.* (1996) recording under photopic conditions from macaque ganglion cells, observed adaptation that occurred within 100 msec. of the onset of an adapting light. In

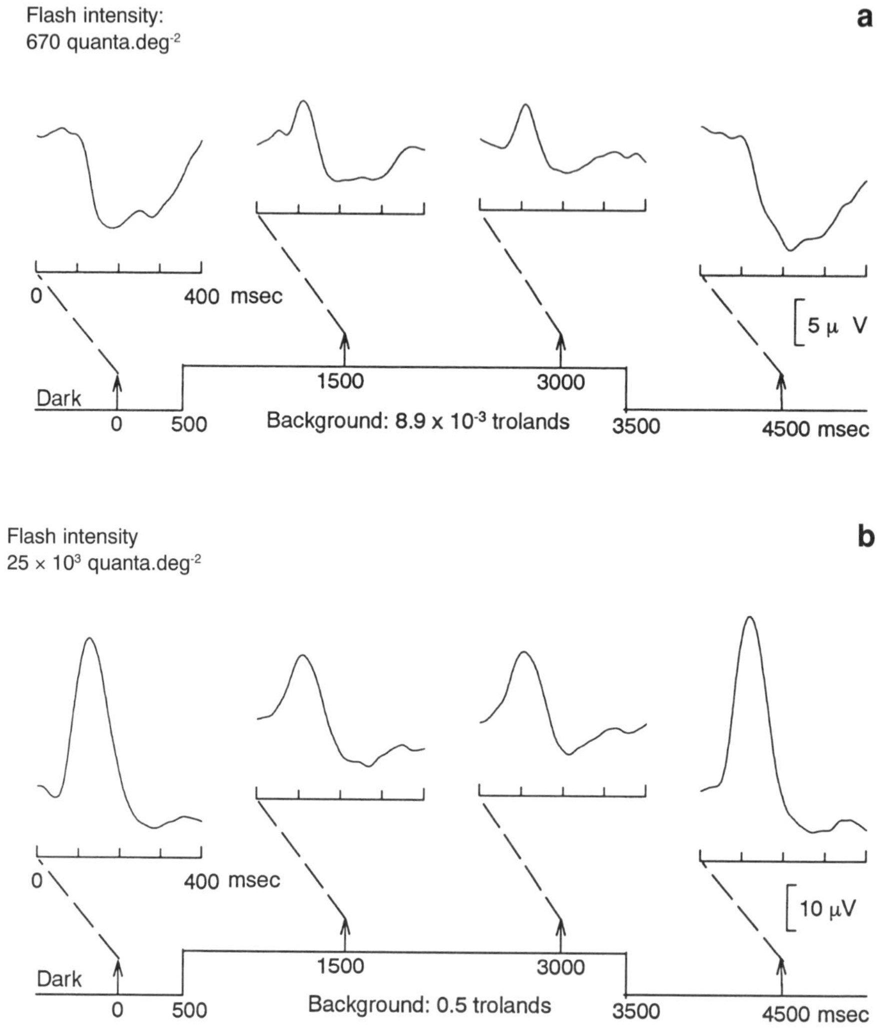

Figure 9 Timecourse of adaptation. Brief stimulus flashes of constant intensity were presented before, during and after (arrows) a background (the pedestal above the baseline) that was pulsed on for 3 seconds. The interval between the onset of the background, the offset of the background, and the subsequent flashes was one second. A) nSTR: The flash intensity was 670 q.deg^{-2}, and the background intensity was 8.9×10^{-3} scot td. B) b-wave: The flash intensity was 25×10^3 q.deg^{-2}, and the background intensity was 0.5 scot td (subject LD). (From Frishman *et al.*, 1996A courtesy of Journal of the Optical Society of America A. 13: 601–612, 1996)

goldfish, under photopic conditions recordings from ganglion cells also showed an early (i.e. rapid) loss of sensitivity at onset of an adapting light (Afanador and Adams, 1974).

A comparison of 1/sensitivity vs. background level for the human ERG (Figure 8A) with the psychophysical increment threshold over the same range (Figure 8B) indicates that the psychophysical absolute threshold is adapted at even lower background levels than the STR. A way of quantifying the difference is to compare I_D calculated used equation 3 for the two sets of data. This 'dark-light' sets the illuminance limit, below which imposing backgrounds has essentially no effect on sensitivity; it represents an equivalent background light that may be assumed to be present even in the absence of an imposed background, due at least in part, to spontaneous isomerizations in the photoreceptors (Barlow, 1956; 1957; Sharpe, 1990 for review). On average, the estimated psychophysical dark-light (I_D) was about 1/5 the I_D of the STR (see Table I). The psychophysical dark-light (I_D) for the subject whose thresholds shown in Figure 8 was 1.2×10^{-3} scot td, the range was 0.5×10^{-3} to 2.1×10^{-3} scot td for the six subjects (see Table 1), which is similar to the range found in other psychophysical studies. The similarity among studies may be related to the observation that the dark-light for a given subject is very stable (unlike the threshold) when stimulus parameters are varied (Sharpe *et al.*, 1993).

11.2.2.2 The nature of the adaptation: shifting of the operating curve, or response compression.

Our modeling of ERG data from cats and humans shows that V_{max} of the nSTR is unchanged by backgrounds thereby indicating that the whole operating curve of the nSTR is shifted with changes in adapting illumination (Figures 5 and 6). This maintenance of V_{max} with changes in background is reminiscent of results reported many years ago by Sakmann and Creutzfeld (1969) for the adapting effects of backgrounds on cat retinal ganglion cell responses. They showed that the entire response vs log intensity curve from the smallest response to saturation was shifted essentially unaltered along the intensity axis as mean luminance was increased, thus providing strong motivation for describing retinal adaptation as for automatic gain control. The gain control referred to here is a control of response to luminance steps which is set by background illuminances, and should be distinguished from contrast gain control which is set by an initial contrast level (e.g. Shapley and Victor, 1981). Since Sakmann and Creutzfeld measured some of the ganglion cell responses at mesopic levels of illumination, it is likely that the gain control occurs for cone-mediated as well as rod-mediated responses. Similar evidence for gain control under photopic conditions has been provided for ganglion cells in the goldfish retina (Afanador and Adams, 1974). It is not clear where in the retina this special type of adaptation first develops, but, in mammals it appears to be an inner retinal phenomenon, associated with the amacrine or ganglion cells. It is not seen in photoreceptor current recordings (Baylor *et al.*, 1984; Tamura *et al.*, 1989, Tamura *et al.*, 1991; Nakatani *et al.*, 1991; Kraft *et al.*, 1993), or in studies of a-wave adaptation (Hood and Birch,

1993A) nor does it occur in the cat horizontal cell recordings (Steinberg, 1971; Lankheet *et al.*, 1991), or in recordings of the rat or cat rod b-wave (Green and Powers 1982; and our unpublished observations). In these recordings response vs intensity functions at each higher background show a reduced maximum amplitude (i.e. reduced V_{max}). Even though there is some adjustment as sensitivity for photoreceptors, horizontal cells and the b-wave, V_{max} declines when the steady background illumination is increased. The reduction in V_{max} indicates the presence of response compression. The available response range for the neurons is reduced by the steady polarization level produced by the background. Only the range between the level of polarization and saturation remains available. The incremental responsivity of photoreceptors, horizontal cells and the b-wave are not totally predicted by compression, mechanisms for adjusting sensitivity also must be present, even though they do not lead to a preservation of the entire operating range. In the case of the photoreceptor, as noted above, calcium feedback is involved, in the case of the other neurons, the mechanism is unknown.

11.2.2.3 Role of convergence in adaptation of the rod signal in the retina

Convergence of the rod signals is an important factor in determining the sensitivity, saturation, and I_D or dark-light at the various stages of scotopic vision that we have examined. With respect to sensitivity, the convergence of light-driven quantal events from individual rod photoreceptors onto a given postreceptoral neuron determine the size of the signal in that cell when the signal is generated as a response to a uniformly illuminated field of large area. In turn, assuming that the unadapted postsynaptic effect of a single Rh* is the same at each level (1– 2 mV), the size of the signal, larger when convergence is greater, will set the operating range of the neuron from its smallest response to saturation. A neuron located in inner retina receiving inputs from many rods will saturate at a lower luminance than a neuron located in outer retina with less convergence of rod signal.

With respect to I_D, or dark-light, the sensitivities of the most sensitive components of the ERG appear to be reduced by backgrounds slightly higher than those required to reduce psychophysical sensitivity. The value that we find for the psychophysical dark-light, a value similar to that found by previous investigators (see Table 1) is 10^{-3} scot td. Assuming that 1 scot td produces 8.6 Rh* per sec per rod (derived from Wyszecki and Stiles, 1982), the dark-light corresponds to 1 Rh* per 116 sec per rod. The minimum possible value for this dark-light, would be set by the rate of spontaneous isomerizations in the photoreceptors. This rate according, to Baylor *et al.*(1984) for isolated macaque rod outer segments, is about 1 Rh* per 160 sec per rod. These values are close, suggesting that much of the limitation for adaptation of the psychophysical threshold is set by spontaneous isomerizations. The value that we find psychophysically is only about 1/5 that required for adaptation of the STR indicating that the dark-adapted sensitivity of the STR must be set by something other than spontaneous isomerizations.

Adaptation of the nSTR has its complete effect in less than a second (Figure 9) which suggests that the adapting signal is already effectively continuous on this time scale. In order for the signal to be continuous despite the discrete nature of quantal absorptions, there must be a sufficiently large input signal, and this must be orchestrated by convergence of inputs. Consider for example, our finding that I_D for the STR is about $2.5*10^3$ $q.deg^{-2}.sec^{-1}$, or with a quantum efficiency of 20%, about 500 $Rh^*.deg^{-2}.sec^{-1}$ at the retina, i.e. about 5 Rh* per 100 $rods.sec^{-1}$. Because each bipolar cell receives input from about 40 rods (Kolb, Linberg and Fisher, 1992), there are only 2 Rh* per sec for each bipolar cell for this I_D level. Judging from the isolated PII in cat, the photon signals in the bipolar cell have a duration of a little more than 100 msec (Robson and Frishman 1995), and 2 Rh* per sec would not produce an even roughly continuous signal. To obtain an approximately continuous signal then would require perhaps 10 Rh* per sec, 5 times more input signal than that calculated for individual bipolar cells. This means that it is very unlikely that the reduced STR is brought about by an effect in bipolar cells or in their effectiveness in exciting amacrine cells. However, the signal at the next stage of convergence, i.e. at the AII amacrine cell, could be effectively continuous, given a similar response duration (i.e. 100 msec), if there is convergence of 5 or more bipolar cells. The factor of 20 for the difference in I_D between STR and PII that we found in this study may be a better indicator of the real situation.

The psychophysical threshold was adapted by backgrounds 5 times lower than those that adapted the STR. While it is possible that this change in threshold resulted from changes of sensitivity in retinal mechanisms that adapt at the same very low levels but do not contribute significantly to the ERG, and which we therefore could not observe, it is also possible that the effects of the weakest adapting fields depend upon the increased noise associated with quantal absorptions from the background rather than from a change in responsiveness of the cells involved (Barlow and Levick, 1969). In this case, we may suppose that the increased noise takes effect in reducing sensitivity at some site more central than the retina (possibly visual cortex).

11.3 Switching from rod-circuits to cone-circuits

11.3.1 Modulation of gap junctions

The switch-over from rod to cone circuits in the mammalian retina is effected not only by the differential sensitivity and saturation of the photoreceptor types, but also, as noted above, probably by saturation of the rod circuit that is specialized to transmit information about single photons (Figure 1). In addition, it has been suggested that release of dopamine may affect the rod circuit. Because dopamine-containing A18 amacrine cells, whose processes surround the AII amacrine cells (Voigt & Wässle, 1987; Dacey, 1990), receive their input from cone bipolar cells (Mariani *et al.*, 1984), it has been hypothesized that dopamine is involved in switching from rod to cone circuits (Witkovsky and Dearry, 1992 for review). While there is evidence that exogeneously applied dopamine (via the second messenger cAMP) reduces the coupling through gap junctions between AII amacrine cells in the rod circuit of rabbit (Hampson *et al.*, 1992; Mills and Massey, 1995; Bloomfield *et al.*, 1997), there is little evidence that dopamine affects the gap-junction coupling of AII to cone bipolar cells that forms the essential link between rod bipolar cells and On ganglion cells. Partial uncoupling of the AII and cone bipolar cells has been achieved by pharmacologically-induced release of NO which increases cGMP in the bipolar cell (Mills and Massey, 1995). Although direct evidence that local release of dopamine and NO can have an effect on coupling is limited, Bloomfield *et al.*(1997) recently have shown that weak adapting backgrounds reduce the tracer-coupling of AII amacrine cells in the dark-adapted rabbit retina eyecup, and lead to an increase in the size of the physiologically-determined receptive field. However Bloomfield *et al.* (1997) did not observe uncoupling of the AII from the cone bipolar cells for the background illuminations that they tested. Recent work of Dacey (1996) recording from AII amacrine cells in the monkey eyecup also suggests that the junctions between AII and cone bipolar cells may remain functional under light-adapted conditions. Cone-driven signals, presumably received via the junction with the cone bipolar cells were recorded in the AII at photopic light levels. The switching between circuits in inner retina is clearly an area that requires further investigation.

11.3.2 Size of retinal ganglion cell receptive fields

Another issue related to adaptation that has stimulated considerable interest over the years, is the relation between mean illumination and the size and sensitivity of the center and surround of retinal ganglion cell receptive fields. A number of studies in cats have shown little or no change in receptive field center size (e.g. Barlow *et al.*, 1957; Cleland and Enroth-Cugell, 1968; Enroth-Cugell *et al.*, 1977; Derrington and Lennie, 1982), with changes from scotopic to photopic luminance. Although Andrews and Hammond (1970) reported a two-fold increase in center diameter when measuring with a spot under scotopic conditions. However, it is likely, from more recent observations that use gratings and fit a Gaussian-center surround model to the results, that the center increases by only about 20% on average (Troy *et al.*, 1993; 1997). Some of these measures were made in the scotopic range of illuminations where the rod circuit would be operative, so differences simply due to a switch in postreceptoral circuitry between rod and cone pathways is unlikely. Ahmed (1981), using more traditional light spots, found similar increases in receptive field centers for a large sample of cat retinal ganglion cells when changing from photopic to scotopic lighting.

A change in surround sensitivity relative to that of the center could explain the larger increases in apparent center size reported by Andrews and Hammond (1970). Barlow *et al.* (1957) reported a large reduction of retinal ganglion cell surround sensitivity at very low light levels in cat retina, and others have since reported reduced sensitivity, or a frank loss of surround in cat (e.g. Rodieck and Stone, 1965; Enroth-Cugell and Robson, 1966; Maffei *et al.*, 1971; Cleland *et al.*, 1973; Enroth-Cugell & Lennie, 1975; Chan *et al.*, 1992) and rabbit (Jensen, 1991; Muller & Dacheux, 1997). It is important to note that the loss of surround sensitivity occurs only for a small portion of the range over which visual sensitivity is adjusted, and for most of the range, surrounds are present.

It seems likely that in cats, where rods outnumber cones even in the area centralis, the small (e.g. 20%) changes in receptive field center size related to a transition from rod-dominated to cone-dominated vision might depend upon the distribution of rod and cone inputs to a particular cell. In primates with relatively more cones, and a specialized cone-dominated foveal region, the relative rod vs cone input probably is an important determinant of scotopic vs photopic receptive field size of retinal ganglion cells that project to parvocellular layers of the lateral geniculate (Wässle & Boycott., 1991 for review).

11.4 Adaptation in cone pathways in the primate

Less has been established about the sites of adaptation of cone signals than rod signals. Similar to the case of rod-mediated vision, a comparison of human psychophysical results and studies in retinal ganglion cells (macaque) indicate that most of the adaptation occurs in the retina. Although the extent to which that adaptation occurs in the cone photoreceptors themselves rather than more proximally in the retina is still not entirely resolved, most physiological measures are consistent with the existence of substantial postreceptoral adaptation. Recordings from single primate cones (Schnapf *et al.*, 1990), and the human ERG (Hood and Birch, 1993B), have indicated that cone sensitivity is not measurably reduced until the retinal illumination reaches between 10^3 and 10^4 phot td. This is illustrated in Figure 10, taken from Hood and Birch (1993B). In contrast with the cones themselves, the human psychophysical sensitivity enters the Weber region when the background illumination is roughly 3 orders of magnitude lower. Similarly, in recordings from ganglion cells in macaque retina, Purpura *et al.* (1990) find that some light-adapted M ganglion cells also show behavior conforming to Weber's Law when the illumination is at least three orders of magnitude lower than that at which individual cones (Figure 10) were desensitized. P-cells enter the Weber region at slightly higher intensities, but still well below the range at which individual photoreceptors are desensitized.

Although the studies just cited indicate postreceptoral adaptation of cone signals, other studies have been used as evidence that most adaptation of cone signals occurs in the cone photoreceptors. For example, Boynton and Whitten

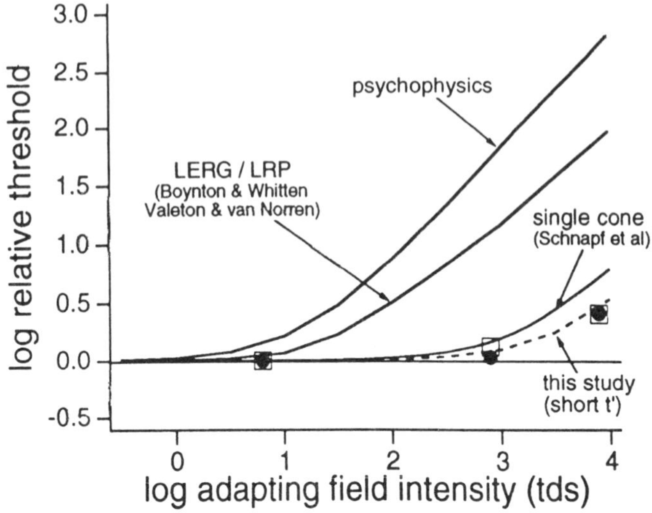

Figure 10 Sensitivity of cone-mediated vision. The figure compares psychophysical results (left) with results from (left to right) intraretinal late receptor potential recordings (Boynton and Whitten, 1970; Valeton and Van Norren, 1983), single rod outer segment recordings (Schnapf *et al.*, 1990) and the human photopic ERG. (From Hood and Birch, 1993B, courtesy of *Visual Neuroscience*, 10, 857– 872, 1993)

(1970), and Valeton and Van Norren (1983) find that the intraretinally recorded late receptor potential in macaque adapts at levels that approach that found in psychophysical studies (Figure 10). The discrepancy between this and the observations on single cones and the ERG data have not been resolved. Furthermore, results of some psychophysical studies in humans (MacLeod *et al.*, 1992; Chaparro *et al.*, 1995) and physiological studies in macaque ganglion cells (Yeh *et al.*, 1996) have been interpreted to indicate both an early, as well as a late or 'second' site of adaptation. The studies show that at the early site of adaptation, signals from individual M or L cones are adapted, while at the second site the adaptation takes place after color-opponent interactions have occurred. The early adaptation is thought to occur either within the photoreceptor themselves or perhaps at the point of transmission to bipolar cells, because of its spatial and spectral restriction to a single cone. However, it is known that in foveal regions of primate retina signals from single cones are transmitted via midget bipolar cells to midget ganglion cells, whereas it has not been established that color opponent interactions occur in (all) bipolar cells. If opponent interactions do not occur before the signals reach the ganglion cells, then the 'early' site of adaptation could be as late as the site of interaction within ganglion cells of signals from single cones. Clarification of the sites of adaptation under photopic conditions awaits further knowledge of the cone circuitry.

11.5 Concluding remarks

There is very strong evidence that there are postreceptoral mechanisms for light adaptation for rod-mediated signals in the retina. The case is weaker for cone-mediated signals, but it is clear that at least some of the adaptation is post-receptoral. Although we know little about all the neuronal mechanisms involved, we do know adaptation occurs at multiple sites, and via several different mechanisms. Perhaps the least understood aspect of postreceptoral adaptation in the mammalian retina are the mechanisms for adjusting of sensitivity of bipolar cells, and for automatic gain control of inner retinal neurons. It is likely that mechanisms in addition to compression and saturation are involved in both cases. The rapidity with which postreceptoral adaptation occurs should help to constrain hypotheses about possible mechanisms. For example, the rapid onset and offset of the adaptation would preclude the slower process of morphological plasticity of the type described in teleost retina.

At present we have little solid data on the mechanisms of sensitivity adjustment and automatic gain control. We don't know the extent of involvement of feedback or feedforward mechanisms, presynaptic or postsynaptic mechanisms, neuromodulatory mechanisms, second messenger systems, or specific types of ion channels. With respect to feedback, it could be intrinsic to the cell, as it is for regulation of photoreceptor sensitivity. However, while intracellular calcium feedback regulation conceivably could occur in the OPL in the dendrites of On bipolar cells that have metabotropic glutamate receptors and a G-protein cascade, such feedback regulation of cation channels would not apply for the amacrine or ganglion cell in the IPL where glutamatergic transmission via ionotropic glutamate receptors, although local modulation of intracellular calcium concentration could occur via NMDA receptors or metabotropic GABA receptors (e.g. work of Zhang, Shen, and Slaughter, 1997 in tiger salamander). Intercellular feedback (or feed forward) could involve GABAergic or glycinergic synapses. Evidence for this kind of feedback might be detected in surrounds. This feedback could be fine-tuned by metabotropic glutamate receptors in the IPL (or OPL) modulating activity of GABAergic or glycinergic inhibitory feedback (Koulen *et al.*, 1997). Such feedback could regulate the amount of glutamate released by photoreceptors in the OPL and bipolar cells in the IPL and hence the sensitivity or gain in the next neurons. With respect to automatic gain control, another specialization, the presence of voltage-gated sodium channels, and the production of action potentials might also be related. Although preliminary experiments with TTX measuring scotopic ERGs in cat retina suggests that gain control can exist in the absence of sodium spikes (Robson *et al.*, 1993), other related specializations of the cells might be involved in the process. We could speculate further that the great diversity of amacrine cells in the IPL (in excess of 25 types) with specialized contacts and a range of neurotransmitters and receptors and ion channels could provide further mechanisms for the processing necessary for automatic gain control. With the rapid advances in understanding of cellular mechanisms and retinal circuitry this mystery may be solved in the not too distant future.

Acknowledgments

Supported by NEI grant RO1 EY06671, NEI core grant P30 EY07751, and LGIA funds from the University of Houston.

References

Afanador, A.J. and Adams, A.J. (1974) Early visual adaptation in goldfish retinal ganglion cells. *Nature*, 250, 346 – 348.

Ahmed, B. (1981) The size and shape of rod and cone centers of cat retinal ganglion cells. *Experimental Brain Res.,* 43, 422 – 428.

Aguilar, M. and Stiles, W. S. (1954) Saturation of the rod mechanism at high levels of stimulation. *Optica Acta.*, 1, 59 – 65.

Andrews, D. P. and Hammond, P. (1970) Suprathreshold spectral properties of single optic tract fibres in cat, under mesopic adaptation: cone-rod interaction. *J. Physiol. (London)*, 209, 83 –103.

Barlow, H. B. (1956) Retinal noise and absolute threshold. *J. Opt. Soc. Am.,* 46, 634 – 639.

Barlow, H. B. (1957) Increment thresholds at low intensities considered as signal/noise discriminations. *J. Physiol. (London)*, 136, 469 – 488.

Barlow, H. B., Fitzhugh, R. and Kuffler, S. W. (1957) Change of organization in the receptive fields of the cat's retina during dark adaptation. *J. Physiol. (London)*, 136, 469 – 488.

Barlow, H. B. and Levick, W. R. (1969) Three factors limiting the reliable detection of light by retinal ganglion cells of the cat. *J. Physiol. (London)*, 200, 1– 24.

Barlow, H. B., Levick,W. R. and Yoon, M. (1971) Responses to single quanta of light in retinal ganglion cells of the cat. *Vis. Res.*, 11, Suppl. 3, 87–102.

Baylor, D. A., Nunn, B. J. and Schnapf, J. L. (1984) The photocurrent, noise and spectral sensitivity of rods of the monkey Macaca Fascicularis. *J. Physiol. (London)*, 357, 575 – 607.

Behers, U. D. and Wagner, H. J. (1996) Adaptation-dependent changes in bipolar cell terminals in the fish retina: Effects of overall morphology and spinule formation in Ma and Mb cells. *Vis. Res.*, 36, 3901– 3911.

Biersdorf, W. R., Granda, A. M. and Lawson, H. F. (1965) Electrical measurement of incremental thresholds in the human eye. *J. Opt. Soc. Am.*, 55, 454 – 455.

Bloomfield, S.A., Xin, D. and Osburne, T. (1997) Light-induced modulation of coupling between AII amacrine cells in the rabbit retina. *Vis. Neurosci.*, 14, 565 – 576.

Boynton, R. M. and Whitten, D. N. (1970) Visual adaptation in monkey cones: recording of late receptor potentials. *Science*, 170, 1423 –1426.

Bush, R. and Remé, C. E. (1992) Chronic lithium treatment induces reversible and irreversible changes in the rat ERG *in vivo*. *Clin. Vis. Sci.*, 5, 393 – 401.

Chan, L.-H., Freeman, A. W., and Cleland, B. G. (1992) The rod-cone shift and its effects on ganglion cells in the cat's retina. *Vis. Res.*, 32, 2209 – 2219.

Chaparro, A., Stromeyer, C. F. III, Chen, G. and Kronauer, R. E. (1995) Human cones appear to adapt to low light levels: measurements on the red-green detection mechanism *Vis. Res.*, 35, 3103 – 3118

Cleland, B. and Enroth-Cugell, Ch. (1968) Quantitative aspects of sensitivity and summation in the cat retina. *J. Physiol. (London)*, 198, 17– 38.

Cleland, B., Levick, W. R. and Sanderson, K. J. (1973) Properties of sustained and transient ganglion cells in the cat retina. *J. Physiol. (London)*, 228, 649 – 680.

Conner, J. D. (1982) The temporal properties of rod vision. *J. Physiol. (London)*, 332, 139 –155.

Conner, J. D. and MacLeod, D. I. A. (1977) Rod photoreceptors detect rapid flicker. *Science,* 195, 698 – 699.

Dacey, D. M. (1990) The dopaminergic amacrine cell. *J. Comp. Neurol.*, 301, 461– 489.

Dacey, D. M. (1996) Morphology and physiology of the AII amcrine cell network in the macaque monkey retina. *Soc. Neurosci. Abs.*, 22, 492.

Dacheux, R. F. and Raviola E. (1986) The rod pathway in the rabbit retina: a depolarizing bipolar and amacrine cell. *J. Neurosci.*, 6, 331– 345.

Daw, N. W, Jensen, R. J. and Brunken, W. J. (1990) Rod pathways in mammalian retinae. *Trends in Neurosci.* 13, 110 –115.

Dawson, W. W., Trick, G. L. and Litzkow, C. A. (1979) Improved electrode for electroretinography. *Invest. Ophthalmol. Vis. Sci.* 18, 988 – 991.

de la Villa, P., Kurahashi T. and Kaneko, A. (1995) L-Glutamate-induced responses and cGMP- activated channels in three subtypes of retinal bipolar cells dissociated from the cat. *J. Neurosci*, 15, 3571– 3582.

Derrington, A. and Lennie, P. (1982) The influence of temporal frequency and adaptation level on receptive field organization of retinal ganglion cells in the cat. *J. Physiol. (London),* 333, 343 – 366.

DeVries, S. H. and Baylor, D. A. (1995) An alternative pathway for signal flow from rod photoreceptors to ganglion cells in mammalian retina. *Proc. Natl. Acad. Sci.*, 92, 10658 –10662.

Dolan, R. P. and Schiller, P. H. (1989) Evidence for only depolarizing rod bipolar cells in the primate retina. *Vis. Neurosci.* 2, 421– 424.

Donner, K., Copenhagen, D. R. and Reuter, T (1990). Weber and noise adaptation in the retina of the toad Bufo marinus. *J. Gen Physiol.,* 95, 733 – 53.

Enroth-Cugell, Ch., Hertz, B. G., Lennie, P. (1977) Convergence of rod and cone signals in the cat's retina. *J. Physiol., (London),* 269, 297– 318.

Enroth-Cugell, Ch. and Lennie, P. (1975) The control of retinal ganglion cell discharge by receptive field surrounds. *J. Physiol., (London),* 247, 551– 578.

Enroth-Cugell Ch. and Robson J. G. (1966) The contrast sensitivity of retinal ganglion cells of the cat. *J. Physiol. (London),* 187, 517– 552.

Famiglietti, E. V. and Kolb, H. (1975) A bistratified amacrine cell and synaptic circuitry in the inner plexiform layer of the retina. *Brain Res.*, 84, 293 – 300.

Famiglietti, E.V. and Kolb, H. (1976) Structural basis for on- and off- center responses in retinal ganglion cells. *Science*, 194, 193 –195.

Frishman, L. J., Reddy, M. G. and Robson, J. G. (1996A) Effects of background light on the human dark-adapted ERG and psychophysical threshold. *J. Opt. Soc. Am. A.*, 13, 601– 612.

Frishman, L. J., Robson, J. G. and Du, L. (1993) Contributions of the positive and negative scotopic threshold responses to the scotopic cat ERG. *Invest. Ophthalmol. Vis. Sci.* Suppl. 34, 1273.

Frishman, L. J., Shen, F., Du, L., Robson, J. G., Harwerth, R. S., Smith III, E. L., Carter-Dawson, L. and Crawford, M. L. J. (1996B) Scotopic ERG of macaque monkey after ganglion cell loss due to experimental glaucoma. *Invest. Ophthalmol. & Vis. Sci.*, 37, 125 –141.

Frishman, L. J. and Sieving, P. A. (1995) Evidence for two sites of adaptation affecting the dark- adapted ERG of cats and primates. *Vis. Res.*, 35, 435 – 442.

Frishman L. J., Sieving, P. A. and Steinberg, R. H. (1988) Contributions to the electroretinogram of currents originating in proximal retina. *Vis. Neurosci.*, 1, 307–15.

Frishman, L. J. and Steinberg, R. H. (1989) Light evoked increases in [K^+] in proximal portion of the dark adapted cat retina. *J. Neurophysiol.,* 61, 1233 –1243.

Fulton, A. B. and Rushton, W. A. H. (1978) The human rod ERG: Correlation with psychophysical responses in light and dark adaptation. *Vis. Res.*, 18, 793 – 800.

Granit, R. (1933) The components of the retinal action potential in mammals and their relation to the discharge in the optic nerve. *J. Physiol. (London)*, 77, 207– 239.

Green, D. G. and Powers, M. K. (1982) Mechanisms of light adaptation in rat retina. *Vis. Res.*, 22, 209 – 216.

Gurevich, L. and Slaughter, M. M. (1993) Comparison of the waveforms of the ON bipolar neuron and the *b*-wave of the electroretinogram. *Vis. Res.*, 33, 2431– 2435.

Hampson, E. C. G. M., Vancy, D. I. and Weiler, R. (1992) Dopaminergic modulation of gap junction permeability brtween amacrine cell in Mammalian retina. *J. Neurosci.,* 12, 4911– 4922.

Hanitzsch, R., Lichtenberger R. Th. and Mättig, W-U. (1996) The influence of $MgCl_2$ and APB on the light-induced potassium changes and the ERG b-wave of the isolated superfused rat retina. *Vis. Res.* 35, 499 – 507.

Harding, T. H. and Enroth-Cugell, C. (1978). Absolute dark sensitivity and center size in cat retinal ganglion cells. *Brain Res.*, 153, 157–162.

Hecht, S., Shlaer, S. and Pirenne, M. H. (1942) Energy, quanta, and vision. *J. Gen. Physiol.*, 25, 819 – 840.

Hendrickson, A. E., Koontz, M. .A., Pourcho, R. G., Sarthy, P. V. and Goebel, D. J. (1988) Localization of glycine-containing neurons in the Macaca Monkey retina. *J. Comp. Neurol.*, 273, 473 – 487.

Hood, D. C. and Birch , D. G. (1993A) Light adaptation of human rod receptors: the leading edge of the human *a*-wave and models of rod receptor activity. *Vis. Res*,. 33, 1605 –1618.

Hood, D. C. and Birch, D. G. (1993B) Human cone receptor activity: the leading edge of the human *a*-wave and models of receptor activity. *Vis. Neurosci.*, 10, 857– 872.

Hubbard, N. P. and Naarendorp, F. (1994) Effects of dim background lights on the scotopic threshold response of the rat ERG. *Invest. Ophthalmol. Vis. Sci.* Suppl., 35, 2047.

Jensen, R. J. (1991) Involvement of glycinergic neurons in the diminished surround activity of ganglion cells in the dark-adapted rabbit retina. *Visual Neurosci.* 6, 43 – 53.

Kawamura, S. and Murakami, M. (1991) Calcium dependent regulation of cGMP phophodiesterase by a protein from frog retinal rods. *Nature,* 349, 420 – 423.

Koch, K.-W. and Stryer, L. (1988) Highly cooperative feedback control of retinal rod guanylate cyclase by calcium ions. *Nature*, 334, 64 – 66.

Kolb, H. (1977) The organization of the outer plexiform layer of the cat: electron microscopic observations. *J. Neurocytol.*, 6, 131–153.

Kolb, H. and Famiglietti, E. V. (1974) Rod and cone pathways in the inner plexform layer of the retina. *Science*, 186, 47– 49.

Kolb, H, Linberg, K. A. and Fisher, S. K. (1992) Neurons of the human retina: A Golgi study. *J Comp Neurol.,* 318, 147–187.

Kolb, H. and Nelson, R. (1984) Neural architecture of the cat retina. *Prog. Ret. Res.* 3, 21– 60.

Kolb, H. and Nelson, R. (1985) Functional neurocircuitry of amacrine cells in the cat retina. In: *Neurocircuitry of the Retina, A Cajal Memorial*, edited by Gallego, A. and Gouras, P. New York: Elsevier, p. 215 – 232.

Kolb, H. and Nelson, R. (1993) OFF-Alpha and OFF-beta ganglion cells in cat retina: II. Neural circuitry as revealed by electron microscopy of HRP stains. 329, 85 –110.

Koulen, P., Kuhn, R., Wässle, H. and Branstätter, J H. (1997) Group l metabotropic glutamate receptors mGluRla and mGluR5a: localization in both synaptic layers of the rat retina *J. Neurosci.*, 17, 2200 – 2211

Koutalos, Y. Nakatani, K. and Yau, K-W. (1995) The cGMP-phosphodiesterase and its contribution to sensitivity regulation in retinal rods *J. Gen. Physiol.*, 106, 891– 921.

Koutalos, Y. and Yau, K-W. (1996) Regulation of sensitivity in vertebrate rod photoreceptors by calcium. *Trends in Neurosci.*, 73 – 81.

Kraft, T. W. Schneeweis, D. M. and Schnapf, J. L. (1993) Visual transduction in human rod photoreceptors. *J. Physiol. (London),* 464, 747–765.

Lamb, T. D., McNaughton, P. A. and Yau, K.-W. (1981) Spatial spread of activation and background desensitization in toad rod outer segments. *J. Physiol. (London)* 319, 463 – 496.

Lankheet, M. J. M., van Wezel, R. J. A. and van de Grind, W. A. (1991) Effects of background illumination on cat horizontal cell responses, Vis. Res., 31, 919 – 932.

MacLeod, D. I. A., Williams, D. R. and Makous, W. (1992) A visual nonlinearity fed by single cones. *Vis Res.* 32, 347– 363

Maffei, L., Fiorentini, A. and Cervetto, L. (1971) Homeostasis in retinal receptive fields. *J. Neurophysiol.,* 34, 579 – 587.

Mariani, A. P., Kolb, H., and Nelson R. (1984) Dopamine-containing amacrine cells of rhesus monkey retina parallel rods. *Brain Res.* 322: 1–7.

Mathews, H. R., Fain, G., Murphy, R. L. W. and Lamb, T. D.(1988) Photoreceptor adaptation is mediated by cytoplasmic calcium concentration. *Nature,* 334, 67– 69.

Mills, S. L., and Massey, S. C. (1995) Differential properties of two gap junctional pathways made by AII amacrine cells. *Nature*, 377, 734 –737.

Muller, J. F. and Dacheux, R. F. (1997) Alpha ganglion cells of the rabbit retina lose antagonistic surround responses under dark-adaptation. *Vis. Neurosci.*, 14, 395 – 401.

Naarendorp, F. and Sieving, P. A. (1991) The scotopic threshold response of the cat ERG is suppressed selectively by GABA and glycine. *Vis. Res.*, 31, 1–15.

Nakajima, Y., Iwakabe, H., Akazawa, C., Nawa, H., Shigemoto, R, Mizuno, N. and Nakaishi, S., (1993) Molecular characterization of a novel retinal metabotropic glutamate receptor mGluR6 with a high agonist selectivity for L-2-amino-4-phosphonobutyrate. *J Biol. Chem.*, 268, 11868 –11873.

Nakatani, K., Tamura, T. and Yau, K.-W. (1991) Light adaptation in retinal rods of the rabbit and two other nonprimate mammals. *J. Gen. Physiol.*, 97, 413 – 435.

Nakatani, K., and Yau, K.-W. (1988) Calcium and light adaptation in retinal rods and cones. *Nature,* 334, 69 –71.

Nelson, R. (1977) Cat cones have rod input: a comparsion of the response properties of cones and horizontal cell bodies in the retina of the cat. *J. Comp. Neurol.*, 172, 109 –136.

Nelson, R. (1982) AII amacrine cells quicken time course of rod signals in the cat retina. *J. Neurophysiol.*, 47, 928 – 947.

Nelson, R. and Kolb, H. (1985) A17: a broad field amacrine cell in the rod system of the cat retina. *J. Neurophysiol.*, 54, 592 – 614.

Nelson, R., Lynn, T., Dickinson-Nelson, A. and Kolb, H. (1985) Spectral mechanisms in cat horizontal cells. Neurocircuitry of the retina: A Cajal Memorial. 109 –121.

Newman, E. A. (1980) Current source density analysis of the *b*-wave of frog retina. *J. Neurophysiol.*, 43, 1355 –1366.

Newman, E. A. and. Odette, L. L. (1984) Model of electroretinogram *b*-wave generation: a test of the K^+ hypothesis. *J. Neurophysiol.*, 51, 164 –182.

Penn, R. D. and Hagins, W. A. (1972) Kinetics of the photocurrent of retinal rods. *Biophys. J.*, 12, 1073 –1094.

Pourcho, R. (1980) Uptake of [^{3}H]glycine and [^{3}H]GABA by amacrine cells in the cat retina. *Brain Res.*, 198, 333 – 346.

Pourcho R. G. and Goebel, D. J. (1985) A combined Golgi and autoradiographic study of [^{3}H]glycine-containing amacrine cells in the cat retina. *J. Comp. Neurol.*, 233, 473 – 480.

Purpura, K., Tranchina, D., Kaplan, E. and Shapley, R. M. (1990) Light adaptation in primate retina: analysis of changes in gain and dynamics of monkey retinal ganglion cells. *Vis. Neurosci.*, 4, 75 – 83.

Raviola, E. and Gilula, N. B. (1973) Gap junctions between photoreceptor cells in the vertebrate retina. *Proc. Nat. Acad. Sci.*, 6, 1677–1681.

Raynauld, J-P., Laviolette, J. R. and Wagner, H-J. (1979) Goldfish retina: A correlate between cone activity and morphology of the horizontal cell in cone pedicules. *Science*, 204, 1436 –1438.

Robson, J. G. and Frishman, L. J. (1995) Response linearity and dynamics of the cat retina: the bipolar cell component of the dark-adapted ERG. *Vis. Neurosci.*, 12, 837– 850.

Robson, J. G. and Frishman, L. J. (1996) Photoreceptor and bipolar-cell contributions to the cat electroretinogram: a kinetic model for the early part of the flash response. *J. Opt. Soc. Am. A.*, 13, 613 – 622.

Robson, J. G. , Frishman, L. J. and Du, L. (1993) Component potentials of the dark-adapted cat ERG. *Soc. Neurosci. Abstracts*, 1413.

Rodieck, R. W. and Stone, J. (1965) Analysis of receptive fields of cat retinal ganglion cells. *J. Neurophysiol.*, 28, 833 – 849.

Rushton, W. A. H. (1965) The Ferrier Lecture 1962. Visual adaptation. *Proc. Roy. Soc. Lond B.*, 162, 20 – 46.

Rushton, W. A. H. and Henry, G. H (1968) Bleaching and regeneration of cone pigments in man. *Vis. Res.*, 8, 617– 631.

Sakmann, B. and Creutzfeldt, O. D. (1969) Scotopic and mesopic light adaptation in the cat's retina. *Pflugers Arch.*, 313, 168 –185.

Schnapf, J. L. , Nunn, B. J., Meister, M. and Baylor, D. A (1990) Visual transaction in cones of the monkey, Macaca Fascicularis. *J. Physiol. (London)*, 427, 618 –713.

Schneeweis, D. M. and Schnapf, J. L. (1995) Photovoltage of rods and cones in the macaque retina. *Science*, 268, 1053 –1056.

Shapley, R. M. and Enroth-Cugell, Ch. (1973) Adaptation and dynamics of cat retinal ganglion cells. *J. Physiol. (London)*, 233, 271– 309.

Shapley, R. M. and Enroth-Cugell, Ch. (1984) Visual adaptation and retinal gain controls. *Prog. Ret. Res.*, 3, 263 – 346.

Shapley, R. M. and Victor J. D. (1981) How the contrast gain control modifies the frequency responses of cat retinal ganglion cells. *J. Physiol. (London)* 318, 161–79.

Sharpe, L. T. (1990) The light adaptation of the human rod visual system, in *Night vision: basic, clinical and applied aspects*, eds. Hess, R. F. , Sharpe, L. T. and Nordby, K. Cambridge University Press, 49 –124.

Sharpe, L. T., Stockman, A., Fach, C. C. and Markstahler, U. (1993) Temporal and spatial summation in the human rod visual system. *J. Physiol. (London)*, 461, 325 – 348.

Sharpe, L. T., Stockman, A. and MacLeod, D. I. A. (1989) Rod flicker perception: scotopic duality, phase lags, and destructive interference. *Vis. Res.*, 1539 –1559.

Sieving, P. A. (1991) Retinal ganglion cell loss does not abolish the STR of the cat and human ERG. *Clin. Vis. Sci.*, 6, 149 –158.

Sieving, P. A. and Nino, C. (1988) Scotopic threshold response (STR) of the human electroretinogram. *Invest. Ophthalmol. Vis. Sci.*, 29, 1608 –1614.

Sieving, P. A. and Wakabayashi, K. (1991) Comparison of rod threshold ERG from monkey, cat and human. *Clin. Vis. Sci.*, 6, 171–179.

Sieving, P. A., Frishman, L. J. and Steinberg, R. H. (1986) Scotopic threshold response of proximal retina in cat. *J. Neurophysiol.*, 56, 1049 –1061.

Zhang, J., Shen, W. and Slaughter, M. M. (1997) Two metabotropic γ-aminobutyric acid receptors differentially modulate calcium currents in retinal ganglion cells. *J. Gen. Physiol.*, 110, 45 – 58.

Smith, R. G., Freed, M. A. and Sterling, P. (1986) Microcircuitry of the dark-adapted cat retina: functional architecture of the rod-cone network. *J. Neurosci.*, 6: 3505 – 3517.

Steinberg, R. H. (1971) Incremental responses to light recorded from pigment epithelial cells and horizontal cells of the cat retina. *J. Physiol. (London)*, 217, 93 –110.

Sterling, P., Cohen, E., Freed, M. A. and Smith, R. G. (1987) Microcircuitry of the on-beta ganglion cell in daylight, twilight and star light. *Neurosci. Res.*, 6, s269 – s285.

Stockman, A., Sharpe L. T., Zrenner, E. and Nordby, K. (1991) Slow and fast pathways in the human rod visual system: ERG and psychophysics. *J. Opt. Soc. Am. A.*, 8, 1657–1665.

Stockman, A., Sharpe, L. T., Rüther, K. and Nordby, K. (1995) Two signals in the human rod visual system: a model based on electrophysiological data. *Vis. Neurosci.*, 12, 951– 970.

Stockton, R. A. and Slaughter, M. M. (1989) *B*-Wave of the electroretinogram: a reflection of ON bipolar cell activity. *J. Gen. Physiol.*, 93, 101–122.

Strettoi, E., Raviola, E. and Dacheux, R. F. (1992) Synaptic connections of the narrow-field, bistratified rod amacrine cell (AII) in the rabbit retina. *J. Comp. Neurol.*, 325, 125 –168.

Tamura, T. Nakatani, K. and Yau, K.-W. (1989). Light adaptation in cat retinal rods. *Science*, 245, 755 –758.

Tamura, T., Nakatani, K. and Yau, K.-W. (1991) Calcium feedback and sensitivity regulation in primate rods. *J. Gen. Physiol.,* 98, 95 –130.

Torre, V., Mathews, H. R. and Lamb, T. (1986) Role of calcium in regulating the cyclic GMP cascade of phototransduction in retinal rods. *Proc. Natl. Acad. Sci. USA*, 83, 7109 –7113.

Troy, J. B., Oh, J. K. and Enroth-Cugell, Ch. (1993) Effect of ambient illumination on the spatial properties of the center and surround of Y-cell receptive fields. *Vis. Neurosci.*, 10, 753 –764.

Troy, J. B., Bohnsak, D. I., Diller, L. C., Enroth-Cugell, Ch. and Zhan, X. J. (1997) Spatial filtering by X cells as a function of light level. *Invest. Ophthalmol. Vis. Sci.* Suppl. 38, S52.

Valeton, J. M. and van Norren, D. (1983) Light adaptation of primate cones: an analysis based on extracellular data. *Vision Res.*, 23, 1539 –1547.

Voigt, J. T., and Wässle, H. (1987) Dopaminergic innervation of AII amacrine cell in mammalian retina. *J. Neurosci.*, 7, 4115 – 4128.

Wakabayashi, K., Gieser, J. and Sieving, P. A. (1988) Aspartate separation of the scotopic threshold response (STR) from the photoreceptor *a*-wave of the cat and monkey ERG. *Invest. Ophthalmol. Vis. Sci.*, 29, 1615 –1622.

Wässle, H. and Boycott, B. B. (1991) Functional architecture of the mammalian retina. *Physiol. Reviews,* 71, 447– 480.

Wässle, H., Grünert, U., Chun, M.-H. and Boycott, B. B. (1995) The rod pathway of the macaque retina; identification of AII-amacrine cells with antibodies against calretinin. *J. Comp. Neurol.*, 361, 537– 551.

Wen, R. and Oakley, B. (1990) $K(^+)$-evoked Müller cell depolarization generates b-wave of electroretinogram in toad retina. *Proc. Nat. Acad. Sci. (USA)*, 1990, 87, 2117– 2121.

Witkovsky, P. and Dearry, A. (1992) Functional roles of dopamine in the vertebrate retina. *Prog. Ret. Res.*, 10, 247– 292.

Woodhouse, J. M. and Campbell, F. W. (1975) The role of the pupil light reflex in aiding adaptation to the dark. *Vis. Res.*, 15, 649 – 653.

Wyszecki, G. and Stiles W. S. (1982) *Color Science, Concepts and Methods, Quantitative Data and Formulae*, Wiley, New York.

Xu, X. J. and Karwoski, C. J. (1994) Current source density analysis of retinal field potentials. 2. Pharmacological analysis of the *b*-wave and *m*-wave. *J. Neurophysiol.*, 72, 96 –105.

Yamashita, M. and Wässle, H. (1991) Responses of rod bipolar cells isolated from the rat retina to the glutamate agonist 2-amino-4-phosphonobutyric acid (APB). *J. Neurosci.*, 11, 2372 – 2382.

Yazulla, S. and Studholme, K. M. (1992) Light-dependent plasticity of the synaptic terminals of Mb bipolar cells in the goldfish retina. *J. Comp. Neurol.* 320, 521– 530.

Yeh, T., Lee, B. B. and Kremers, J. (1996) The time course of adaptation in macaque retinal ganglion cells. *Vis. Res.*, 36, 913 – 931.

12
Ecological aspects of vertebrate visual ontogeny

L. BEAUDET and C.W. HAWRYSHYN
*Department of Biology, University of Victoria, P. O. Box 3020, Victoria, B. C., Canada V8W 3N5

12.1 Introduction

The life history of many species is comprised of several stages characterized by distinct anatomical, physiological, behavioral and ecological adaptations. Through evolution, some animals have come to select different habitats for each of the stages of their life history. This is because some environments provide conditions favorable to early growth and development, but may be limiting for the juvenile and adult stages. To take advantage of conditions more favorable to the later stages of their life history, animals often move to another environment, at some point in their ontogeny. Movement from one habitat to another, while providing benefits to the animal, also subjects it to new environmental conditions, some of which may affect the visual system's performance. Through ontogenetic adaptation processes, the visual system of such species is modified, presumably to increase its efficiency in each of the new environments it encounters.

In this chapter, we review adaptations of the visual system that illustrate how various species respond to shifts in habitat during ontogeny. Because the changes in visual environment that have a noticeable effect on the visual system occur mostly in fishes and amphibians, we will focus our attention on these two groups of vertebrates.

As we will show, evolution has brought about several means by which the visual system copes with changing environmental conditions. In this chapter, we will consider the transformations of the visual system at ontogenetic 'turning points' during the life of animals that undergo such behavioral and environmental changes.

12.2. Changes associated with continual development and growth of the visual system

12.2.1 Development of the eye: a brief overview

Before discussing in detail the changes that occur at specific times during ontogeny, we will examine those associated with gradual development and growth of the visual system. In all vertebrates, the developing retina emerges from an evagination of the diencephalic walls of the developing brain. The lens and

S.N. Archer et al. (eds.), Adaptive Mechanisms in the Ecology of Vision, 413–437

cornea originate from the optic placodes, and the extraocular muscles and sclera develop from neural crest cells (Browder *et al.,* 1991). Consisting initially of a double epithelial layer, the retina quickly differentiates into two distinct layers, the actual neural retina and the pigmented epithelium. The diverse cell types that compose the mature neural retina originate from multipotent precursors; retinal ganglion cells (RGC) and cones appear first, followed by the cells of the inner nuclear layer and finally by the rods. Although this developmental sequence shows some variation between species (for review, see Raymond, 1991a), rods typically differentiate and mature last (Johns and Easter, 1977; Johns and Fernald, 1981). Concomitant with the addition of new cells and the establishment of neuronal connections, is the elimination of non-functional connections and the deletion of cells that are either not functioning adequately, are redundant, or play only a transitory role during development.

The retina of fish and some amphibians is set apart from that of the other vertebrates in that retinal genesis is not restricted to the early ontogeny, but lasts throughout the animals' life (Straznicky and Gaze, 1971; Johns and Easter, 1977; Wetts *et al.,* 1989). Furthermore, fish and amphibian retinae retain the ability to regenerate following injury throughout life (Raymond, 1991b). The implications of this are discussed later in this chapter.

12.2.2 Development of visual acuity and photopic sensitivity

Visual acuity refers to the ability to see small objects or objects that are far away. Visual acuity is a function of resolving power, which, to a first approximation and ignoring lenticular diffraction varies according to the formula:

$$\text{resolving power} = f\,(2p)$$

where f is the focal length of the lens, and p the shortest distance between two adjacent photoreceptors. Because growth of the eye is associated with an increase in retinal surface, changes in visual cell densities and size, and modifications of the optical properties of the dioptric apparatus (cornea and lens), visual acuity changes during development. As the larvae of several vertebrate species forage actively, changes in visual acuity associated with eye and retinal development are likely to have an impact on the visual ecology of various animals, during the early stages of their life history (see below).

As mentioned earlier, fish and amphibian eyes keep growing throughout life, causing continual changes in the optical properties of the eye. To understand how eye growth affects visual acuity, it is important to consider the opposing action of two sets of processes. First, as eye diameter increases, the density of cones tends to diminish and cone size to increase, which would tend to reduce visual acuity. At the same time, lens diameter and focal distance increase, causing an increase in the size of the image projected onto the retina. The opposing action of decreasing cell density and lenticular growth should theoretically result in a life-long increase in visual acuity (Johns and Easter, 1977).

Visual acuity, as measured with behavioral techniques, has indeed been shown to increase with development and growth in several species of fish (Rahmann *et al.,* 1979; Breck and Gitter, 1983; Neave, 1984; Li *et al.,* 1985; Zaunreiter *et al.,* 1991; Browman and O'Brien, 1992; Pankhurst *et al.,* 1993; van der Meer, 1994, 1995 and others). For example, food perception distance has been shown to increase with growth in the larvae of several species (for review, see Blaxter, 1986), and attack distance (the distance between the larva and the prey item at the moment of the attack) has been shown to increase with size (Browman and O'Brien, 1992). The net effect of changes in visual acuity is believed to be an overall increase in foraging efficiency.

Visual acuity also depends on factors other than lens diameter and cone spacing, as estimates of visual acuity based solely on such histological measurements normally differ from those estimated behaviorally (Neave, 1984; Browman *et al.,* 1990; Pankhurst *et al.,* 1993; van der Meer, 1995). In most studies, visual acuity estimated histologically was higher than that measured behaviorally, and this discrepancy was especially large during development and early growth (Neave, 1984; van der Meer, 1995). As a result, the rate of increase in visual acuity as measured behaviorally is larger than that of histological estimates. The relatively larger increase in behavioral estimates of visual acuity has been attributed to the developmental maturation of intercellular connections and the dioptric apparatus (Pankhurst *et al.,* 1993). In addition, when such factors as the divergence of cone signals onto neurons of higher order (bipolar cells) are taken into consideration, histological estimates can be corrected to better reflect behavioral measurements (van der Meer, 1995). It follows that to understand the physiology and anatomy behind ontogenetic changes in visual function, the intricate nature of the organization of the visual system has to be considered.

Another parameter, the ontogenetic increase in cone size, has been related to an increase in absolute photopic sensitivity, that is the sensitivity to light during daylight conditions (when only the cone photoreceptors are active). Using a behavioral technique, van der Meer (1994) found an increase in photopic sensitivity with size, in the cichlid fish *Haplochromis sauvagei.* This was the first study to directly address the effect of eye growth and cone size on photopic sensitivity.

Development and maturation of the cone photoreceptors may be associated with changes in color vision, which depends on the interaction of spectrally distinct cone types within the retina. Differences in the differentiation times of these cone types may result in modifications of color vision as new spectral inputs are added. In goldfish, the ability to discriminate colors in the short-wavelength (violet-blue) part of the visual spectrum is delayed (Ohnishi, 1993). Similarly, a deficit in sensitivity to short wavelengths has been found in human infants (see Brown, 1990 for review). The cause and ecological significance of this slower maturation of the short-wavelength sensitive cone mechanism, if any, are not known.

12.2.3 Development of scotopic sensitivity

So far, we have discussed the changes that affect cone photoreceptors, visual acuity and photopic sensitivity. Changes in eye size, retinal area, cell size and density also influence the developmental properties of scotopic vision, active at night and mediated by the rod photoreceptors. According to Fernald (1990), retinal illuminance for an extended light source (i.e. a non-point light source) does not increase during eye growth but remains constant. Therefore, if rod numbers and size were to remain constant, growth of the eye would result in a decrease in absolute scotopic sensitivity, as less of the retinal surface would be covered by these photoreceptors. It has been shown experimentally, however, that during eye growth in goldfish, rods are added to the retina, resulting in a slight increase in rod density (Johns and Easter, 1977). It follows that during eye growth, absolute scotopic sensitivity should remain constant or increase slightly. Powers *et al* (1988a, b), found a small but statistically significant correlation between scotopic visual sensitivity, as determined psychophysically, and size in goldfish. More recently, however, van der Meer (1994) found a more dramatic increase in scotopic sensitivity with size in *Haplochromis sauvagei.* Specifically, van der Meer found a 2.13 log unit (approx. 100 times) increase in scotopic sensitivity with an approximate five-fold increase in lens diameter.

12.2.4 Development of central visual pathways

While the retina develops and grows, so do the retinal projections to the brain. During gradual and continuous development, the pattern of retinofugal tracts typical of the adult form becomes established early in the ontogenetic sequence of events. With further growth, fibers are added to the existing tracts. In addition to the increase in the number of fibers that compose the complement of retinal projections, a gradual shift in the topographical organization of the projections' terminal fields occurs, most notably at the level of the optic tectum (Chung *et al.*, 1974; Gaze *et al.*, 1974; Gaze *et al.*, 1979). This shift is required to maintain the topographical representation of the visual field on the optic tectum, as the retina grows concentrically but the optic tectum grows longitudinally (see Udin, 1990 for review). In some species, however, the development of the retinofugal projections is not gradual and the establishment of the adult pattern is delayed until major morphological reorganizations, such as those associated with metamorphosis, take place. This is the case in lamprey and some amphibians, which will be discussed later in this chapter.

12.3 Changes that occur during critical periods of development

In some species, several of the changes that the visual system undergoes during ontogeny do not only occur progressively as described earlier but also occur more abruptly, at times when animals experience shifts in their visual behavior or their visual environment. These changes are abrupt when compared to the entire life span of the animal and are accompanied by substantial modifications of the animal's morphology and physiology. The most spectacular of these changes are those that animals such as fishes and frogs undergo at metamorphosis.

The changes that affect the visual system during metamorphosis or comparable events occur at different levels of organization and thus vary in their magnitude. They may affect large structures, such as the eye and associated musculature, alter the cellular composition of the retina, modify the morphology of existing retinal cells, or simply change some chemical properties of retinal or other cells. Changes may be accomplished through one or more of the following processes: cell addition, cell deletion, morphological transformations of existing cells, and changes in the expression of specific genes, resulting in new physiological properties of photoreceptors (e.g. expression of a new visual pigment protein). Subtle changes can also affect the chemical composition of the visual pigment's chromophore and result in significant shifts in the spectral absorption properties of the visual pigment. In the following sections, we discuss a few examples of ontogenetic events that illustrate the above transformation processes, and relate them to the visual ecology of animals.

12.4 Migration and the visual system of fishes

Migration from one habitat to another often exposes organisms to light environments that differ in spectral quality and intensity. As such, it may represent the most significant source of change in the visual environment of animals. Changes in the visual system of animals associated with migratory events are presumed to improve foraging abilities, and provide increased predator avoidance capabilities, but this has never been directly and convincingly tested.

Although complex and varied in nature, migratory events can, for the purpose of our discussion, be classified in general as either vertical or horizontal. Vertical migrations are experienced by fish that spend part of their life as pelagic larvae or juveniles and eventually settle to deeper layers of the water column or to the benthos, as exemplified by flatfishes and some reef fishes. In this case, movement to greater depth is the major source of environmental change impacting the visual system.

Horizontal migrations can be classified as anadromous or catadromous. Anadromy is characteristic of salmon and lamprey and involves a larval or early juvenile stage in freshwater, followed by a movement to the marine environment, where most of the adult life is spent. Catadromy, conversely, involves a larval stage in the marine environment followed by juvenile and adult phases in freshwater, and is characteristic of the eel life history. Migration in fishes is normally preceded by metamorphosis or, in salmonid fishes, by a similar process called smoltification (Hoar, 1988).

12.4.1 Metamorphosis and vertical migration in flatfishes

Larval phase Flatfishes (flounder, plaice, sole etc.) provide a good example of the visual adaptations that can be associated with vertical migrations. The life history of all species of flatfishes is characterized by two discrete phases: the larval, pelagic phase, and the juvenile/adult, benthic phase (Figure 1, right-hand side). In some species, the intervening metamorphosis may also be considered a phase as it may last several months (Markle *et al.,* 1992).

The larval, pelagic phase is believed to favor dispersal and increase foraging opportunities. Effective foraging in these animals is crucial not only to ensure proper development, but also because the larva often has to store energy in planning for metamorphosis, during which it ceases to feed. The distribution of flatfish larvae is heavily influenced by that of their zooplanktonic prey, which are most abundant in the top layers of the water column, the euphotic zone. Thus, the pelagic, predatory larva forages in a bright light environment, and is characterized by positive phototaxis and negative geotaxis (Champalbert *et al.,* 1991).

The larvae of flatfish are bilaterally symmetrical with eyes located on either side of the head. Consistent with the pelagic lifestyle described above, the retina is typically adapted to vision in bright light environments, as its photoreceptor layer contains exclusively single cones and shows no evidence of retinomotor movements (Blaxter, 1968; Evans and Fernald, 1993; Evans *et al.,* 1993; Pankhurst and Butler, 1996). In larvae of the winter flounder, Evans *et al.,* (1993) found a single visual pigment, indicating that these larvae are not capable of color vision and thus probably rely on brightness contrast to locate prey. According to these authors, the peak absorbance (λ_{max} 519 nm) of this cone photopigment is offset from the spectral distribution of the ambient light in which the larvae live, thus providing increased contrast for the prey. Finally, the visual system of flatfish larvae is characterized by a gradual increase in visual acuity which, coupled with an increase in swimming speed, probably improves foraging ability (Neave, 1984).

Metamorphosis Metamorphosis in flatfishes begins when the larva reaches a certain size or age, which differs from species to species. Metamorphosis brings about adaptations for the benthic life style of the juveniles and adults which affect the general body plan, physiology and behavior of the animals. Whereas the larva was a swimming and active zooplanktivore, the juvenile and adult forms are more sedentary, spending most of their time on or near the ocean floor, feeding on benthic invertebrates. In addition, juvenile and adult flatfish lie flat on their side, an adaptation accompanied by a loss of bilateral body symmetry that is reflected in jaw shape and dentition, body coloration and eye position. The most obvious effect on the visual system is the migration of one eye to the contralateral side of the head, adjacent to the other eye, such that in postmetamorphic animals both eyes are on the exposed side of the body and scan the environment located above the plane in which the animal lies.

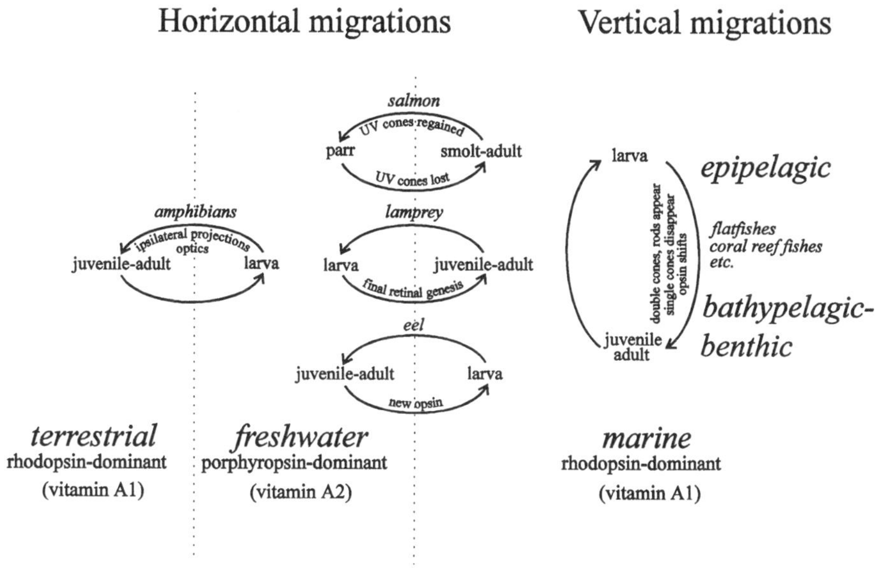

Figure 1 Diagram summarizing the main visual adaptations discussed in this chapter and their associations with the different types of environmental change. Although, for reasons of clarity, exceptions to the processes illustrated here have been omitted, the reader should be mindful of the complexity that exists amongst ontogenetic trajectories

This transition from a pelagic to a benthic habitat is associated with a corresponding decrease in the intensity of ambient light, and a spectral shift towards shorter wavelengths. At the retinal level, the transformations that have been best described occur in the photoreceptor layer and involve the addition of new cell types, and changes in opsin gene expression. Whereas the larval retina contains only single cones, at metamorphosis, rod photoreceptors are added, mostly to the ventral retina, in keeping with the lower light intensities found at greater depths (Blaxter, 1968, Neave, 1984, Evans and Fernald, 1993). Double or twin cones (λ_{max} 531 and 547 nm) are added, mostly to the dorsal retina (Evans and Fernald, 1993; Pankhurst and Butler, 1996). Thus, the juvenile and adult flatfish retina contains mostly cones in the dorsal and rods in the ventral half. Evans and Fernald (1993) suggest that the rod-dominated area might serve as 'an effective visual detector for viewing objects silhouetted against the reduced [down-welling] incident light' at twilight. There is yet no evidence that this would be the case, however. On the contrary, it would seem that the scotopic system of fish is most sensitive to light increments, and thus designed to respond to brighter targets that appear over a darker background (see Wheeler, 1979). The exact manner in which the rod-dominated retina is used by the organism therefore remains to be determined. The dorsal cone-dominated area, conversely, is probably used to scan the surface of the substrate in search of prey during daytime.

With metamorphosis, the peak absorbance of the photopigment contained in the single cones also shifts, towards shorter wavelengths (from 519 nm to 457 nm in winter flounder, Evans *et al.,* 1993). This, presumbly, represents an adaptation to the blue-shifted light environment that prevails at the depths at which adults live. In winter flounder, this shift in the spectral sensitivity of the single cones probably reflects the expression of a new gene, resulting in a different opsin with new spectral properties (Evans *et al.,* 1993). Therefore, whereas in the larval winter flounder the visual pigment contained within the single cones was likely offset with respect to the ambient light spectral composition, in the juvenile and adult, these cones contain a pigment that is probably matched to this ambient light.

The above changes in retinal structure and pigment content are accompanied by several other changes in the flatfish visual system, including the appearance of retinomotor movements, development of photonegative behavior, and increased visual acuity, although the latter appears to be the continuation of a process initiated earlier in the larval stage (Blaxter, 1968, Neave, 1984). In addition, an increase in absolute sensitivity occurs; in plaice, the threshold for dark adaptation falls by two orders of magnitude during metamorphosis (Neave, 1984). The appearance of rods and double cones adapts the visual system to a dimmer environment. Furthermore, the appearance of additional spectral classes of cones indicates the potential for color vision, which could greatly improve visual ability, allowing discrimination not only on the basis of brightness but also of hue.

Whether flatfish possess color vision, however, is still unknown. Color discrimination experiments needed to address this question are still missing. Other questions also remain to be answered regarding changes in the visual system of flatfishes at metamorphosis. For example, the appearance of new classes of cones in the retina probably affects the chromatic organization of RGCs fields; electrophysiology experiments are needed to determine the mechanism by which these new structures become integrated into existing neuronal circuits while maintaining the functionality of the system. In addition, the origin of the new cones added to the retina is not known.

12.4.2 Vertical migration in other species

Transformations associated with vertical migration are also encountered in other species of fish, albeit not as spectacular as those found in flatfishes. In some species, such as *Sebastes diploproa* (Boehlert, 1979) and *Gempylus serpens* (Munk, 1990), single cones disappear from the retina during migration to greater depths and only double cones and rods remain. In yellow perch, single cones sensitive to violet light are lost during vertical migration (Loew and Wahl, 1991) whereas in pollack, these cones are not lost, but become sensitive to longer wavelengths through a change in visual pigment content (Shand *et al.,* 1988).

Changes in photoreceptor density and spectral sensitivity have also been related to settlement into the reef environment in coral reef fishes (Shand, 1994, 1997). This transition from the pelagic to the reef environment is marked by a change in feeding behavior, a decrease in ambient light levels and a shift towards shorter wavelengths (McFarland, 1991). The timing of settlement differs between species, and the timing of changes in the visual system differs accordingly. Shand (1994, 1997) found that an increase in rod density and a decrease in cone density occurred at a smaller size in early-settling species than in species with a prolonged pelagic phase. The final photoreceptor densities depended on the lifestyle of the adults. Fish active exclusively during daytime had lower rod densities than those whose activity extended into the crepuscular periods. Using microspectrophotometry, Shand (1993) also found a loss of red cones in goatfish which she related to the movement towards deep zones of the coral reef after settlement. Whereas larvae possessed double cones with two visual pigments (λ_{max} 487 and 580 nm), settled goatfish lacked the long-wavelength sensitive photopigment (λ_{max} 580 nm) which had been replaced with a combination of 515 and 530 nm λ_{max} photopigments.

12.4.3 Lamprey metamorphosis and migration

In contrast to most species that undergo metamorphosis, lamprey are characterized by an unusually long larval phase, spent in streams (Potter, 1980). The lamprey larva, called ammocoete, hatches from eggs buried in the substrate during spawning (Potter, 1980). Emergence of the ammocoetes occurs a few weeks later and gives rise to larval dispersal, normally over relatively short distances, mostly at night. After dispersing, the ammocoetes bury themselves in the substrate in areas where current and mechanical disturbances are minimal (Just *et al.,* 1981). They remain buried in the substrate or exhibit very limited movements for the remainder of their larval life, which can last up to 8 years depending on the environmental conditions and the species (average of 6 years for most anadromous species). During their larval life, lamprey are filter feeders (Just *et al.,* 1981).

Ammocoete visual system The behavior of the ammocoete, which remains mostly inactive, points to a reduced role for vision during this phase of its life cycle. It is commonly held that the ammocoete eye, which remains subdermal until metamorphosis, is non-functional. Thus, ammocoete are normally referred to as blind. Although there is no electrophysiological evidence that the ammocoete eye is sensitive to light, some anatomical and behavioral observations suggest that it may be. Studnicka (1912) reports that the ammocoete eye contains a patch of differentiated retina that contains all the retinal layers, including fully differentiated photoreceptors. Kennedy and Rubinson (1977) and De Miguel *et al.,* (1990) report the presence of retinofugal projections to the diencephalon, and retinopetal projections from the mid-brain tegmentum, in young larvae. Finally, the nocturnal dispersal and movement patterns of the ammocoete have been regarded as displays of photonegative behaviors (Potter, 1980). Although it has been proposed that extra-retinal photoreceptors may be responsible for the response to light in the ammocoete, the possibility of a retinal involvement warrants closer consideration.

Metamorphosis and migration Lamprey metamorphosis is associated with increased activity levels and movement from areas of low to higher water flow (Potter, 1980). This movement is followed by downstream migration and the adoption of a parasitic lifestyle (Potter, 1980; Just *et al.,* 1981). Downstream migration primarily occurs at night and, depending on the species, may lead to the ocean for marine (anadromous) species, or to a lake environment in freshwater species (Just *et al.,* 1981). Although changes in water current and temperature appear to trigger the lamprey's postmetamorphic migration, light cues play a role in restricting migration to nighttime, which may represent an adaptation to reduce predation (Potter, 1980).

In most fishes, the eye of the premetamorphic larva is fully differentiated, and only a few populations of cells are modified at metamorphosis. In lamprey, the changes that occur at metamorphosis affect the eye as a whole, including the retinal projections to the brain. During metamorphosis the eyes emerge from under the skin, the extraocular muscles increase in size, the retina differentiates and adopts an appearance in many respects similar to that of teleosts. Metamorphosis also results in the final maturation of the retinal projections into a pattern similar to that of teleost fishes (De Miguel *et al.,* 1990).

The metamorphosed lamprey retina contains two morphological classes of photoreceptors, one with a short tapered outer segment and the other longer and more slender (Ohman, 1976). Based on their ultrastructure, Ohman (1976) concluded that both classes of photoreceptors of the metamorphosed lamprey were 'rod-like' although they also possessed some characteristics of cones. Govardovskii and Lychakov (1984), however, based on electrophysiological recordings, claim the retina of *Lampetra fluvialis* is duplex, containing both rods and cones. Based on microspectrophotometric measurements, Harosi and Kleinschmidt (1993) drew similar conclusions, detecting both 'rod-like' and 'cone-like' photopigments in the retina of adult *Petromizon marinus*. Although the data seem to point to the presence of both rods and cones in lamprey, they also highlight some of the ambiguities in defining these structures.

Sea lamprey, in addition to shifting from a sedentary to a more active lifestyle following metamorphosis, also move from an environment dominated by longer wavelengths (freshwater) to one richer in middle to short wavelengths (marine environment). The ability of the lamprey visual system to adapt to this change may lie in the spectral properties that visual pigments can exhibit depending on the type of chromophore they are attached to. The visual system of sea lamprey contains both vitamin A_1-(rhodopsin) and vitamin A_2-based (porphyropsin) photopigments (Crescitelli, 1972). It is not clear whether these two types of photopigment are associated with different stages of the life history of lamprey. One theory, put forth by Wald (1945) attributes vitamin A_2-based photopigments to the freshwater part of the life cycle in sea lamprey and vitamin A_1-based photopigments to the marine phase (Figure 1). Conversely, Crescitelli (1958), found only vitamin A_1-based photopigments in both downstream- and upstream-migrating sea lamprey and concludes that this theory may not apply to all cases. Depending on the timing of the measurements and the stage of the migration, the A_2-based photopigments may not have had time to become expressed in the upstream migrating animals studied by Crescitelli (1958). A recent study by Harosi and Kleinschmidt (1993) found only vitamin A_2-based photopigments in the upstream migrants, providing further support to Wald's (1945) contention that it is the photopigment associated with freshwater.

12.4.4 Salmon migration and smoltification: the loss of ultraviolet light sensitivity

Salmonid migration is associated with both the deletion of a class of photoreceptors and their possible reincorporation into the retina at a later ontogenetic stage. As in lamprey, the general life history of salmon is characterized by two migrations. The first one occurs after smoltification, a process analogous to metamorphosis in other fishes. This involves major physiological and morphological transformations, which in many cases adapt the freshwater juvenile to the marine environment (Hoar, 1988). During this migration, the juvenile leaves the stream where it has hatched and swims to the ocean (anadromous salmon and trout), or to deeper strata in lakes (several trout and kokanee salmon), where it spends most of its life. The adult salmonid later returns to its natal stream to reproduce, and, in the case of Pacific salmon, to die. The visual characteristics of the freshwater and marine environments differ substantially. Prior to smoltification juveniles live in a broad-spectrum light environment that characterizes shallow freshwater. The light environment experienced by adults is probably narrower in its spectrum and shifted to shorter wavelengths for animals that migrate to the marine environment, or longer wavelengths for animals that migrate to greater depths in lakes (Novales Flamarique *et al.,* 1992). Also, the feeding habits of juvenile salmon differ from those of the adult, with young salmonids feeding mostly on small zooplankton and larger animals foraging on large zooplankton and fish. As a result, the visual system of the freshwater juvenile differs appreciably from that of the larger, smolted fish.

The freshwater juvenile The retina of young salmonids (parr stage) is characterized by a relatively high cone to rod ratio. It contains both single and double cones, organized in a square pattern (Figure 2): four double cones (DC) form the sides of a square and surround a central single cone (SC). The pattern is completed by the presence of accessory single cones at the four corners of this square arrangement (A). The parr retina contains four cone and one rod visual pigments (brown trout, Bowmaker and Kunz, 1987, rainbow trout, Hawryshyn and Harosi, 1994). The cone photopigments are sensitive to ultraviolet (UV), short, middle and long wavelengths, respectively, and the rod pigment to middle wavelengths.

Smoltification and UV vision At smoltification, the physiology and structure of the retina are transformed noticeably. Evidence suggests that sensitivity to UV light is affected at the time of smoltification (Bowmaker and Kunz, 1987; Hawryshyn *et al.,* 1989). These studies have indicated that sensitivity to UV light disappears at the time when the fish undergo their first migration, and that this change in spectral sensitivity is accompanied by the loss of a morphological class of photoreceptors. Histological studies have shown that a class of cones, the

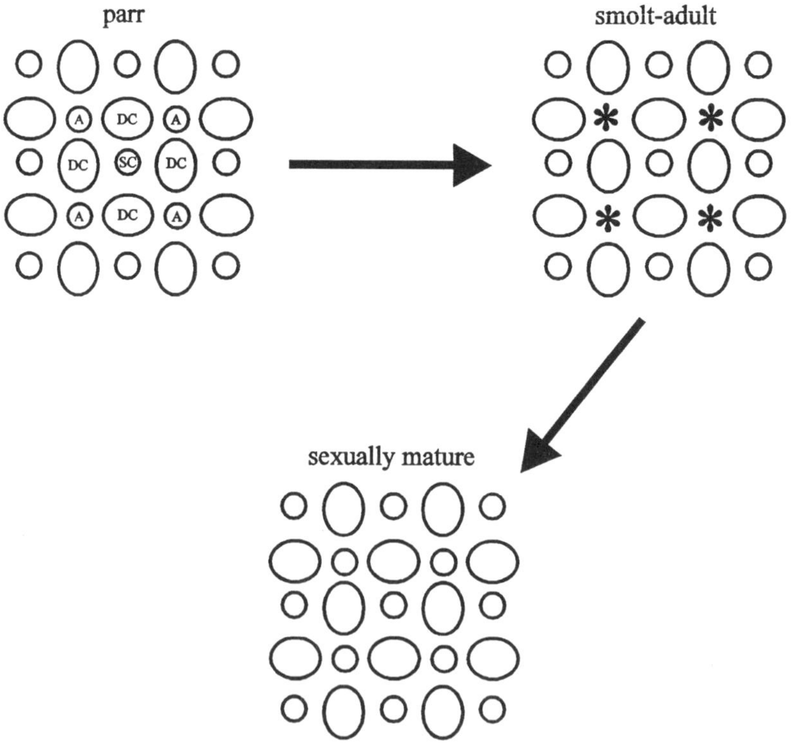

Figure 2 Schematic representation of the ontogenetic sequence of disappearance and reappearance of UV cones in the retina of Pacific salmonids. When sectioned tangentially at the level of the photoreceptors, the fish retina normally exhibits arrays of cone cells organized in a crystal-like fashion. At the time of sea-ward migration, the accessory corner cone (A) is lost from the retina of parr and only double (DC) and central single cones (SC) remain. This process is reversed later in life, at the time of sexual maturation and return to freshwater

accessory corner cones, present in the parr stage is absent in larger juveniles or adults (brown trout: Lyall, 1957; brown trout and Atlantic salmon: Ahlbert, 1976; rainbow trout: Bathelt, 1970; Beaudet *et al.,* 1993). This provided the first evidence that this cone type is the UV cone in salmonids. In addition, the number of rods per unit mosaic almost doubles during this period, possibly reflecting an increase in visual sensitivity associated with migration to darker environments.

Consistent with the contention that UV cones are lost during smoltification, Hawryshyn *et al.,* (1989) were able to show a progressive loss of sensitivity to ultraviolet light with increasing size in rainbow trout. More recently, the loss of putative UV cones was associated with the disappearance of UV inputs to the ON response as recorded electrophysiologically from the optic nerve of rainbow trout (Beaudet *et al.,* 1993), and sockeye salmon (Novales Flamarique and Hawryshyn, 1996). Similarly, although the retina was not examined, Coughlin and Hawryshyn (1994) found a reduction in the number of single units with UV inputs in the *torus semicircularis* of the midbrain in rainbow trout with increasing size. The mechanism by which the UV cones disappear from the retinal cone mosaic is still unknown. However, Browman and Hawryshyn (1992) have shown that exogenous thyroxine causes premature disappearance of putative UV cones and UV sensitivity in rainbow trout, suggesting an involvement of the endocrine system.

Disappearance of the UV cones at the time of sea-ward migration raised the question of whether they might reappear during the return migration, when the animals are once again found within their natal, freshwater environment. Recent evidence supports this suggestion. Browman and Hawryshyn (1994) have shown that UV sensitivity and UV cones are regained in larger rainbow trout, with exogenous thyroxine treatment. These authors, however, did not thoroughly investigate the retinal cone topography in either thyroxine-treated or control animals. Thus, one could not, at the time, exclude the possibility that some of the putative UV cones found in thyroxine-treated fish were part of a small residual population of putative UV cones in the retina (Beaudet *et al.,* 1993; Coughlin and Hawryshyn, 1994).

Beaudet *et al.,* (1997), however, have provided support to the contention that UV cones reappear in the retina of large, sexually mature salmonids, by showing that returning Pacific salmon possess putative UV cones over extensive areas of the retina. These authors examined the topography of cone photoreceptors in the retina of sexually mature Pacific salmonids, at the time of spawning. Putative UV cones were found in the dorso-temporal retina in the four species examined, over an area that represented approximately 20% of the total retinal surface, one order of magnitude larger than the area over which the ‘resident’ population had been previously found. These results suggest that UV cones and their associated UV sensitivity are regained when salmon are ready to either initiate their return migration or, at least, to re-enter the freshwater environment. The mechanism by which putative UV cones are regained in the retina of larger salmonid fishes remains unknown, but it is suspected that thyroxine may play a role, as suggested in Browman and Hawryshyn (1994). Hence, in salmonids, plasticity of the UV light detection mechanism appears to be associated with movements from freshwater to the marine environment and back. The ecological significance of this sensitivity to UV light has not been established.

Smoltification is also associated with changes in the photopigments chemical composition. It is possible that vitamin A_2-based photopigments dominate the parr retina while vitamin A_1-based photopigments dominate after smoltification. The most recent evidence for this comes from Alexander *et al.*, (1994) who found a decrease in the amount of vitamin A_2-based photopigment in small coho salmon, coinciding with the appearance of smolt characteristics such as hypo-osmoregulation. This increase in vitamin A_1-based photopigment would probably increase the sensitivity of the animals in the marine environment, which is generally richer in shorter wavelengths than the freshwater environment. In sockeye salmon (*Oncorhynchus nerka*), however, both anadromous and landlocked forms favor vitamin A_1-based photopigments over vitamin A_2-based photopigments, even while in freshwater (Beatty, 1966).

12.4.5 Eel visual system

American and European eels reproduce in the Atlantic ocean. After a larval stage that lasts up to three years, metamorphosis occurs, the juvenile swims upstream and remains in freshwater until sexual maturation. Following maturation, a second metamorphosis occurs and the eel migrates back to the ocean to reproduce. This catadromous pattern of migration is the reverse of that observed in salmon and lamprey, hence the changes undergone by the eel visual system are the reverse of some of those of salmon and lamprey.

While in the marine environment, the retina of the larval eel contains only 'rod-like' photoreceptors (Pankhurst, 1984) that contain a mixture of vitamin A_1- (λ_{max} 501 nm) and A_2- (λ_{max} 523 nm) based photopigments, with the A_1-based photopigment dominating (Wood *et al.*, 1992). This implies that in eel, rods appear before cones, which is not usually the case in vertebrates. However, since the photoreceptors of the larval eel have been identified as rods only on morphological grounds, it is too early to draw definitive conclusions regarding their identity. Following metamorphosis and migration to freshwater, the eel retina contains both rods and cones (Pankhurst, 1984). Furthermore, it is dominated by a vitamin A_2-based photopigment (λ_{max} 523 nm) (Beatty, 1975; Wood *et al.*, 1992) in agreement with the view that this is an adaptation to environments richer in longer wavelengths (see Beatty, 1984).

Migration back to the marine environment is associated with changes in retinal structure and visual pigment content, some of which are the reverse of those observed during the migration to freshwater. There is an increase in the size of the eye and the number of rods, whose density thus remains constant (Pankhurst,

1984). Cone density decreases as a result of eye growth and possibly also due in part to cellular degeneration (Pankhurst, 1984). The change in visual pigment content is two-fold. First, there is a shift from a dominance of vitamin A_2-based photopigments to that of A_1-based photopigments (Beatty, 1975; Wood and Partridge, 1993). Second, while the 523 nm vitamin A_2-based photopigment disappears and is replaced by the 501 nm A_1-based photopigments, a new opsin, a 482 nm A_1-based photopigment emerges and eventually becomes the main visual pigment of the sexually mature animal (Beatty, 1975; Wood and Partridge, 1993; Archer *et al.,* 1995). Interestingly, during this opsin substitution, the outer segment of individual photoreceptors contains both opsins simultaneously (Wood and Partridge, 1993). This shift towards a preponderance of vitamin A_1-based photopigments and the eventual appearance of a new opsin with its λ_{max} at shorter wavelengths, is probably an adaptation to the deep marine environment in which eel reproduce.

12.5 Metamorphosis in amphibians

The movement from the larval, aquatic environment to the juvenile or adult, terrestrial environment in some amphibian species is akin to the migratory events described in the previous sections. These two environments differ in their optical properties and are likely to require specific morphological, physiological and behavioral adaptations. The different visual requirements of the aquatic and terrestrial environments are illustrated by the transformation the visual system of amphibians undergoes at metamorphosis. Larval amphibians are generally aquatic animals. They possess a fish-like body form, and features such as external gills, a lateral line and an escape motor system involving Mauthner cells. In addition, larval amphibians are mostly herbivorous, feeding on algae and detritus. At metamorphosis, there is a transition to the terrestrial environment, and amphibians take on a carnivorous, predatory lifestyle. Metamorphosis leads to the loss of larval characters, and the appearance of adaptations to the new physical characteristics of the terrestrial environment and to the changes in behavior of the animals. This process also includes changes to the eyes and the central visual pathways.

The aquatic and terrestrial environments differ in many aspects, all of which have an impact on the visual system. First, the desiccant effect of air in the terrestrial environment requires the presence of protective integuments on the eye such as nictitating membranes (Kaltenbach, 1953). Second, because of the differing density indices of water and air, in water the cornea is more or less optically inactive and most of the refractive properties of the eye lie in the lens; the reverse is true in the terrestrial environment (Sivak and Warburg, 1983). Third, the spectral distribution of light is much narrower in the aquatic environment and, in freshwater where amphibians are found, contains longer wavelength radiation, i.e. yellow to red colors (Reuter, 1969). Fourth, the predatory lifestyle of the terrestrial juvenile and adult requires good stereoscopic vision and the necessary ipsilateral visual projections associated with it. All of the changes in the visual environment and behavior that occur at the time of metamorphosis produce changes in the optics, spectral sensitivity and pattern of central retinal connections.

Optics The larval eye of amphibians is slightly myopic or emmetropic in water and strongly myopic in air (Sivak and Warburg, 1980, 1983; Mathis *et al.,* 1988). The adult eye, on the other hand, is strongly hyperopic in water and emmetropic or slightly myopic in air. This indicates that changes in the refractive state of the eye occur at metamorphosis. Using infrared refractometry to measure the refractive state of the eye in air, Sivak and Warburg (1980, 1983) and Mathis *et al.*, (1988) showed a progressive shift from strong myopia to emmetropia or slight hyperopia after metamorphosis. The transformations are dramatic not only because of their amplitude, but also because of the speed at which they occur; in *Bufo sp*, Mathis *et al.,* (1988) found an hourly change of approximately 10 dioptres during metamorphosis. As a result, the transformations are completed within a few days. The change in refractive state of the eye is likely to result from an increase in the refractive power of the cornea and a decrease in that of the lens. The increase in the cornea's refractive effect is achieved through a change in curvature and probably protein composition (Sivak and Warburg, 1980, 1983; Mathis *et al.,* 1988). The decrease in the refractive power of the lens is accomplished by a relative flattening and a change in protein composition (Polansky and Bennett, 1973).

Amphibian retinal photoreceptors The anuran visual system differs from that of most vertebrates in that it contains two classes of rods that differ histologically and spectrally. These two classes are generally referred to misleadingly as 'green' and 'red' rods, even though they contain short- (λ_{max} 430 – 440 nm) and middle- (λ_{max} 500 – 525 nm) wavelength sensitive visual pigments respectively (Reuter, 1969). The outer segment of the short-wavelength sensitive rod is relatively short, which is typical of the outer segment of short-wavelength sensitive photoreceptors found in other vertebrates. Similarly, this class of rods accounts for only a small proportion of the total number of rods. More typical of rods found in other vertebrates, the more numerous middle-wavelength sensitive rod possesses a long, slender outer segment. The anuran retina also contains single and double cones. The single cone contains a long-wavelength sensitive visual pigment, whereas the double cone contains a long-wavelength sensitive photopigment in its principal member and a middle-wavelength sensitive pigment in its accessory member. Interestingly, it is believed that the anuran double cone originates from the fusion between a single cone and a middle-wavelength sensitive rod during development (see Reuter, 1969 for a discussion). This possible relationship between anuran rods and double cones further suggests the arbitrariness of classifying vertebrate photoreceptors simply as rods or cones.

Retinal physiology Metamorphosis in anurans is associated with distinct patterns of expression of rhodopsin and porphyropsin. Generally speaking, tadpole photoreceptors contain a larger proportion of porphyropsin (Wald, 1945; Crescitelli; 1958; Wilt, 1959; Reuter, 1969). Reuter (1969) found a 'red' rod porphyropsin with maximal absorbance at 523 nm in *Rana temporaria* tadpoles. At metamorphosis, the relative abundance of porphyropsin decreased, concomitantly with a shift in spectral absorbance of the photopigment extracts towards shorter wavelengths (Figure 3, modified from Reuter, 1969). The shift was abrupt at first and continued until the photoreceptors contained mostly the rhodopsin, which corresponded to a wavelength of peak absorbance of 502 nm, characteristic of the adult. Interestingly, rods in the retina of *Bufo bufo* tadpoles, contain rhodopsin at the larval stage and retain it in the adult stage (Muntz and Reuter, 1966).

Retinal ganglion cells, retinal projections and binocular vision With metamorphosis, some changes occur in the physiology and morphology of the retinal ganglion cell (RGC) layer of tadpoles. Using single unit recordings, Reuter (1969) discovered the emergence of new RGC receptive field properties at metamorphosis. A physiological type of RGC, responding to the presence of convex

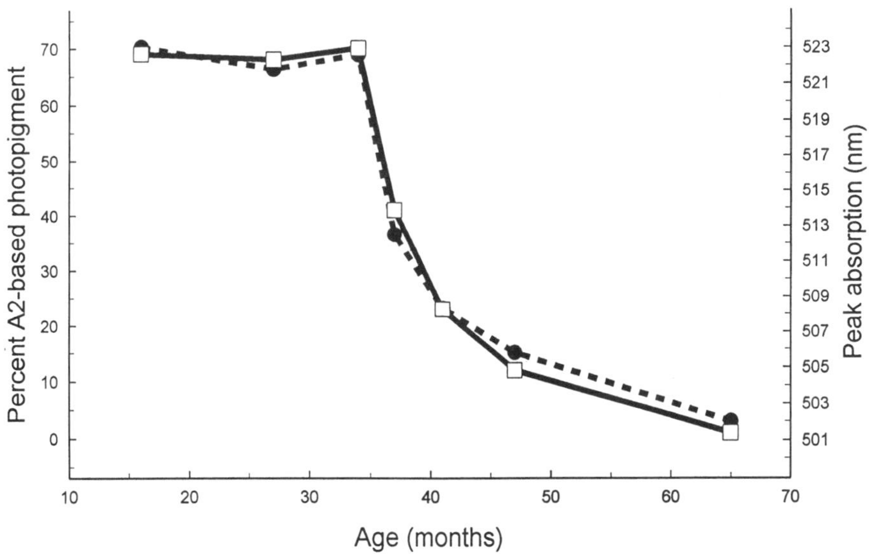

Figure 3 Wavelength of peak absorption of photopigment extracts and percent of vitamin A_2-based photopigment as a function of age in *Rana temporaria* tadpoles. As the relative amount of vitamin A_2-based photopigment decreases, the peak absorbance gradually shifts towards shorter wavelengths. Open squares = percent of vitamin A_2 photopigment and filled circles = wavelength of peak spectral absorption of photopigment extracts as determined from difference spectra. Adapted from Figure 1 and Table of Reuter (1969). Error bars were not provided in Reuter (1969)

edges within its receptive field and which was absent from the tadpole retina was prominent in the adult. Reuter (1969) presumed this type of RGC would be most sensitive to small objects, such as insects, and believed it could 'transmit the information which elicits the feeding behavior of the adult frog [...]'.

Reuter (1969) further suggested that the appearance of this new receptive field type could be the result of the maturation of RGC already present in the tadpole. Findings by Pomeranz (1972), using Golgi silver impregnation, appeared to contradict this, suggesting the appearance of a new morphological class of RGC at metamorphosis. Using a more sensitive technique (HRP labeling), however, Frank and Hollyfield (1987) later confirmed that all the morphological RGC classes present in adult *Rana sp.* (7 classes in total) are also present in tadpoles. What is the anatomical basis, if any, for the difference in RGC receptive field properties? Frank and Hollyfield found a delayed dendritic development of certain classes of RGC, coinciding with metamorphosis, and which they associated with the emergence of the new type of receptive field. This development was characterized by an increase in the complexity of the dendritic arbor. Interestingly, in *Xenopus laevis* which retains an aquatic lifestyle throughout its life, there is no such maturation of the RGC dendritic arbors (Dunlop and Beazley, 1984). It is still not clear how the emergence of new receptive field properties relates to the visual ecology of the animals.

Maturation of dendritic arbors is not the only change that occurs within the RGC layer during metamorphosis. There is also an alteration of the topographic distribution of RGC densities, resulting in two zones of higher cell density, in the nasal and temporal retina (Dunlop and Beazley, 1981; Nguyen and Straznicky, 1989). These two areas eventually contribute to the formation of an equatorial streak of higher cell density, except in aquatic adult forms like *Xenopus* which retain only nasal and temporal zones of high density (Nguyen and Straznicky, 1989; Straznicky and Gaze, 1971). The appearance of a horizontal streak could be associated with that of a 'horizon' in the visual field of the animals when they become terrestrial. This zone would confer higher visual acuity and facilitate the detection of visual targets moving at or near the ground.

In anurans the transition from aquatic to terrestrial life is accompanied by a displacement of the eyes from a lateral towards a more frontal location, permitting stereoscopic vision. Stereoscopic vision requires comparison of the signals that originate from those parts of both retinae that sample the same region of the visual field. This comparison depends on the presence of optic fibers that project to both sides of the brain: contralaterally, to the opposite side of the brain and ipsilaterally, to the same side as the eye. In tadpoles, all retinal fibers project contralaterally. Hoskins (1990) reports that a new class of ipsilaterally projecting retinal ganglion cells is produced at the time of metamorphosis in anurans. These cells appear late in larval development, from the temporal retina, corresponding to the binocular visual field. The development of these fibers follows the increase in overlap between the visual field of the two eyes.

12.6 Seasonal changes in photopigment characteristics

In addition to the major ontogenetic changes accompanying migration, organisms may be exposed to significant seasonal variations in their light environment. Seasons are associated with changes in a number of factors, including light intensity, photoperiod, spectral characteristics and temperature. Some environments are highly seasonal, some less so. Organisms have developed visual adaptations to changing seasonal conditions, such as alterations in the spectral characteristics of their visual pigments.

As we have seen, the retina of certain groups of vertebrates possesses both vitamin A_1- and vitamin A_2-based photopigments i.e. a paired-pigment visual system. In 1961, Dartnall *et al.,* reported that in the rudd (*Scardinius erythrophthalmus*) relative proportions of rhodopsin and porphyropsin vary depending on the light regime under which the animals are kept. Under conditions of constant lighting, rhodopsin becomes dominant in the retina, whereas under constant darkness, porphyropsin becomes dominant. Rhodopsin was thus associated with the summer months and porphyropsin with winter. This lability of the paired-pigment visual system of fish in response to photoperiod has since been observed in several other species and the reader is referred to Beatty (1984) for a detailed review.

The interplay of factors contributing to a change in the proportion of the two types of pigment is complex, and the response to similar light regimes is not consistent from one species to the next. Some animals respond to constant light or darkness in a manner opposite to that observed for the rudd. Hence, McFarland and Allen (1977) coined the expressions 'rudd' and 'anti-rudd' to refer to the two types of response.

Even within a species, seasonal variations in the behavior of the paired-pigment visual system may be manifested differently in different habitats. For example, Muntz and Mouat (1984) measured annual changes in the light characteristics of three bodies of water and the visual pigment content in brown trout that inhabit them. In all sites, they found the retina to be dominated by porphyropsin for most of the year, with rhodopsin appearing for a few months in the summer. The increase in rhodopsin content during summer was best correlated with an increase in temperature and light intensity. In trout inhabiting the most dystrophic, reddish body of water, however, they only found a reduced amount of rhodopsin. They concluded that in such an environment dominated by longer wavelengths, porphyropsin probably increases the sensitivity of the visual system.

As we have just seen, light alone is not the only determinant of the relative proportions of rhodopsin and porphyropsin (Muntz and Mouat, 1984; Beatty, 1984). Moreover, when artificial lighting conditions were used to simulate the change of seasons, the response of cutthroat trout was opposite to that observed based on normal seasonal changes (Allen *et al.,* 1973). This further indicates that factors other than the spectral characteristics of the environment must be considered to account for the variation in visual pigment. Temperature is one factor that has been singled out as determining behavior of the rhodopsin-porphyropsin system, even overriding the effect of light (McFarland and Allen, 1977). However, failure to address *increase* and *decrease* in photoperiod may have lead to an underestimation of the role light plays in the behavior of the paired-pigment visual system in response to seasonal changes. Most studies of the effect of light on the paired-pigment system, have used static light conditions: complete light, complete darkness or 12D:12L (as in McFarland and Allen, 1977). An *increase* or a *decrease* in photoperiod, may represent a more biologically relevant light stimulus for these kinds of experiments.

12.7 Substrate for developmental adaptation

The transformations of the visual system described in this chapter suggest that the visual system must retain plasticity during ontogeny. The addition of retinal ganglion cells in the frog, or cone photoreceptors in winter flounder, at the time of metamorphosis suggests that the visual system of these animals retains undifferentiated precursors during their larval development, within portions of the retina that are, for the most part, functional and differentiated. Potential for plasticity late in life has also been revealed in regeneration experiments. In fish and amphibians, partially destroyed retina (or any part of the CNS for that matter) can regenerate. In goldfish, the potential for retinal regeneration has been demonstrated using ouabain injections (Maier and Wolburg, 1979), and more recently both chemical and laser ablation techniques (Braisted *et al.,* 1994; Raymond, 1991b). Although it was initially believed that stem cells named rod precursors were responsible for the generation of all cell types within the regenerated retina (Raymond *et al.,* 1988), recent evidence indicates that rod precursors may actually be giving rise only to rods and that the other cell types originate from yet unidentified precursors (Braisted *et al.,* 1994). Local interactions are probably responsible for the induction of cellular differentiation leading to retinal regeneration (Raymond, 1991a).

The pathways of cellular differentiation involved in retinal regeneration may be the same ones involved in the ontogenetic changes in the visual system discussed in this chapter. In this case, however, local cell interactions may be replaced or supplemented by hormonal actions. Hormones have been shown to affect the survival of cells, both within the visual system and in the rest of the body (Browman and Hawryshyn, 1992, 1994; Kaltenbach, 1953; Hoskins, 1990). Thyroxine, for example, has been implicated in the disappearance and the possible reappearance of the UV cones in rainbow trout (Browman and Hawryshyn, 1992). Recent evidence from Kelley *et al.,* (1995) suggests thyroxine and retinoic acid induce the differentiation of embryonic retinal precursors into photoreceptors, *in vitro.*

This raises the interesting question of whether or how the plasticity associated with ontogenetic changes in the visual system may be related to the ability of the visual system to regenerate. In other words, is the evolution of regeneration in some way linked to the plasticity that is required by visual systems that have to adapt to changing environmental conditions? Results by Kurz-Isler and Wolburg (1982) at least question the dependence of ontogenetic plasticity on the presence of regenerative abilities. These authors found that the rainbow trout retina, despite the ontogenetic plasticity identified in various studies, possesses limited regenerative abilities. More research is obviously needed to clarify any potential relationship between these two factors.

12.8 Conclusions

In this chapter, we have described changes in the visual system that exemplify the diversity of adaptive mechanisms used by animals in response to changes in their environment (summarized in Figure 1). The link that has been established between these changes in the visual system, and changes in the visual environment is, for the most part, qualitative in nature. However important a role qualitative observations may play in the formulation and development of theories, they do not carry much predictive power. There is much value associated with hypotheses such as the Sensitivity and Contrast (or offset pigment) hypotheses (see Douglas *et al.,* 1995 for a recent review). They have provided an initial framework within which variations in spectral characteristics of visual systems could begin to be explained. Without quantitatively set boundaries, however, a theory based on the Sensitivity-Contrast hypotheses may be incapable of revealing some of the driving forces behind the evolution of spectral sensitivity for it encompasses every possible situation. No matter what the spectral characteristics of a given visual system are, they can be explained in terms of either the Sensitivity or the Contrast hypotheses. It is not surprising that research paradigms inspired by these hypotheses may have difficulty identifying situations in which forces other than those represented by these two hypotheses may be driving the ontogenetic changes described in this chapter. We believe a new, quantitative model of the evolution of visual systems and visual adaptations is needed and will be welcomed. A quantitative approach such as that adopted by Lythgoe and Partridge (1991) is likely to lead to new understandings of the relationships between visual environments and the ontogeny of the vertebrate visual system.

Acknowledgments

We thank Julia Shand for sending us an unpublished manuscript. We would also like to thank Lisa Hartman for providing several insightful comments on the manuscript, Joy Macknight for providing editorial help and Samantha Petticrew for helping with the bibliography search. LB was supported by NSERC and Fonds FCAR postgraduate scholarships and CWH by an NSERC operating grant (#OPG0106102).

References

Ahlbert, I.-B. (1976) Organization of the cone cells in the retinae of salmon (*Salmo salar*) and trout (*Salmo trutta trutta*) in relation to their feeding habits. *Acta Zoologica* (Stockholm), 57, 13 – 35.

Alexander, G., Sweeting, R. and McKeown, B. (1994) The shift in visual pigment dominance in the retinae of juvenile coho salmon (*Oncorhynchus kisutch*): An indicator of smolt status. *Journal of Experimental Biology*, 195, 185 —197.

Allen, D. M., McFarland, W. N., Munz, F. W. and Poston, H. A. (1973) Changes in the visual pigments of trout. *Canadian Journal of Zoology*, 51, 901– 914.

Archer, S., Hope, A. and Partridge, J. C. (1995) The molecular basis for the green-blue sensitivity shift in the rod visual pigments of the European eel. *Proceedings of the Royal Society of London Series B-Biological Sciences*, 262, 289 – 295.

Bathelt, V. D. (1970) Experimentelle und vergleichend morphologische Untersuchungen am visuellen System von Teleostiern. *Zoologische Jahrbucher Abteilung fur Anatomie und Ontogenie der Tiere*, 87, 402 – 470.

Beatty, D. D. (1966) A study of the succession of visual pigments in Pacific salmon (*Oncorhynchus*). *Canadian Journal of Zoology*, 44, 429 – 453.

Beatty, D. D. (1975) Visual pigments of the American eel *Anguilla rostrata*. *Vision Research*, 15, 771–776.

Beatty, D. D. (1984) Visual pigments and the labile scotopic visual system of fish. *Vision Research*, 24, 1563 –1573.

Beaudet, L., Browman, H. I. and Hawryshyn, C. W. (1993) Optic nerve response and retinal structure in rainbow trout of different sizes. *Vision Research*, 33, 1739 –1746.

Beaudet, L., Novales Flamarique, I. and Hawryshyn, C. W. (1997) Cone photoreceptor topography in the retina of sexually mature Pacific salmonid fishes. *Journal of Comparative Neurology*, 383, 49 – 59.

Blaxter J. H. S. (1968) Light intensity, vision and feeding in young plaice. *Journal of Experimental Marine Biology and Ecology*, 2, 293 – 307.

Blaxter, J. H. S. (1986) Development of sense organs and behavior of teleost larvae with special reference to feeding and predator avoidance. *Transactions of the American Fisheries Society*, 115, 98 –114.

Boehlert, G. W. (1979) Retinal development in postlarval through juvenile *Sebastes diploproa*: adaptions to a changing photic environment. *Revue Canadienne de Biologie*, 38, 265 – 280.

Bowmaker, J. K. and Kunz, Y. W. (1987) Ultraviolet receptors, tetrachromatic colour vision and retinal mosaics in the brown trout (*salmo trutta*): age dependent changes. *Vision Research*, 27(12), 2101 –2108.

Braisted, J. E., Essman, T. F. and Raymond, P. A. (1994) Selective regeneration of photoreceptors in goldfish retina. *Development*, 120, 2409 – 2419.

Breck, J. E. and Gitter, M. J. (1983) Effect of fish size on the reactive distance of bluegill (*Lepomis macrochirus*) sunfish. *Canadian Journal of Fisheries and Aquatic Sciences*, 40, 162 –167.

Browder, L. W., Erickson, C. A. and Jeffery, W. R. (1991) *Developmental Biology,* Saunders College, Philadelphia.

Browman, H. I. and Hawryshyn, C. W. (1992) Thyroxine induces a precocial loss of ultraviolet photosensitivity in rainbow trout (*Oncorhynchus mykiss*, Teleostei). *Vision Research*, 32, 2303 – 2312.

Browman, H. I. and Hawryshyn, C. W. (1994) The developmental trajectory of ultraviolet photosensitivity in rainbow trout is altered by thyroxine. *Vision Research*, 11, 1397–1406.

Browman, H. I. and O'Brien, W. J. (1992) The ontogeny of search behavior in the white crappie, *Pomoxis annularis*. *Environmental Biology of Fishes*, 34, 181–195.

Browman, H. I., Gordon, W. C., Evans, B. I. and O'Brien, W. J. (1990) Correlation between histological and behavioral measures of visual acuity in a zooplanktivorous fish, the white crappie (*Pomoxis annularis*). *Brain Behavior and Evolution*, 35, 85 – 97.

Brown, A.M. (1990) Development of visual sensitivity to light and color vision in human infants: a critical review. *Vision Research*, 30, 1159 –1188.

Champalbert, G., Macquart-Moulin, C., Patriti, G. and Chiki, D. (1991) Ontogenic variations in the phototaxis of larval and juvenile sole *Solea solea* L. *Journal of Experimental Marine Biology and Ecology*, 149, 207– 225.

Chung, S. H., Keating, M. J. and Bliss, T. V. P. (1974) Functional synaptic relationships during the development of the retino-tectal projection in amphibians. *Proceedings of the Royal Society of London Series B: Biological Sciences*, 187, 449 – 459.

Coughlin, D. J. and Hawryshyn, C. W. (1994) Ultraviolet sensitivity in the torus semicircularis of juvenile rainbow trout (*Oncorhynchus mykiss*). *Vision Research*, 34, 1407–1413.

Crescitelli, F. (1958) The natural history of photopigments, in *Photobiology*, (ed Newburgh), Colloquium, Oregon State College, Corvallis, pp. 30 – 51.

Crescitelli, F. (1972) The visual cells and visual pigments of the vertebrate eye, in *Textbook of Sensory Physiology,* (ed Dartnall, H. J. A.), Springer-Verlag, Berlin, pp. 246 – 363.

Dartnall, H. J. A., Lander, M. R and Munz, F. W. (1961) Periodic changes in the visual pigment of a fish, in *Progress in Photobiology*, (eds Christiansen, B. and Buchmann, B.), Elsevier, Amsterdam, pp. 203 – 213.

De Miguel, E., Rodicio, M. C. and Anadon, R. (1990) Organization of the visual system in larval lampreys: An HRP study. *Journal of Comparative Neurology*, 302, 529 – 542.

Douglas, R. H., Partridge, J. C. and Hope, A. J. (1995) Visual and lenticular pigments in the eyes of demersal deep-sea fishes. *Journal of Comparative Physiology* [A], 177, 111–122.

Dunlop, S. A. and Beazley, L. D. (1981) Changing retinal ganglion cell distribution in the frog *Heleiloporus eyrei*. *Journal of Comparative Neurology*, 202, 221– 236.

Dunlop, S. A. and Beazley, L. D. (1984) A morphometric study of the retinal ganglion cell layer and optic nerve from metamorphosis in *Xenopus laevis*. *Vision Research*, 24, 417– 427.

Evans, B. I. and Fernald, R. D. (1990) Metamorphosis and fish vision. *Journal of Neurobiology*, 21, 1037–1052.

Evans, B. I. and Fernald, R. D. (1993) Retinal transformation at metamorphosis in the winter flounder (*Pseudopleuronectes americanus*). *Visual Neuroscience*, 10, 1055 – 64.

Evans, B. I., Harosi, F. I. and Fernald, R. D. (1993) Photoreceptor spectral absorbance in larval and adult winter flounder (*Pseudopleuronectes americanus*). *Visual Neuroscience*, 10, 1065 –1071.

Fernald, R. D. (1990) Visual pigments of fishes, in *The Visual System of Fish*, (eds Douglas, R. H. and Djamgoz, M. B. A.), Chapman & Hall, London, pp. 81–107.

Frank, B. D. and Hollyfield, J. G. (1987) Retina of the tadpole and frog: Delayed dendritic development in a subpopulation of ganglion cells coincident with metamorphosis. *Journal of Comparative Neurology*, 266, 435 – 444.

Gaze, R. M., Keating, M. J., Ostberg, A. and Chung, S. H. (1979) The relationship between retinal and tectal growth in larval Xenopus: implications for the development of the retino-tectal projection. *Journal of Embryology and Experimental Morphology*, 53, 103 –143.

Gaze. R. M., Keating, M. J. and Chung, S. H. (1974) The evolution of the retinotectal map during development in Xenopus. *Proceedings of the Royal Society of London Series B: Biological Sciences*, 185, 301– 330.

Govardovskii, V .I. and Lychakov, D. V. (1984) Visual cells and visual pigments of the lamprey, *Lampetra fluvialis*. *Journal of Comparative Physiology* [A], 154, 279 – 286.

Harosi, F. I. and Kleinschmidt, J. (1993) Visual pigments in the sea lamprey, *Petromizon marinus*. *Visual Neuroscience*, 10, 711–715.

Hawryshyn, C. W. and Harosi, F. I. (1994) Spectral characteristics of visual pigments in rainbow trout *Oncorhynchus mykiss*. *Vision Research*, 34, 1385 –1392.

Hawryshyn, C. W., Arnold, M. G., Chiasson, D. J. and Martin, P. C. (1989) The ontogeny of ultraviolet sensitivity in rainbow trout *Salmo gairdneri*. *Visual Neuroscience*, 2, 247– 254.

Hoar, W. S. (1988) The physiology of smolting salmonids. *Fish Physiology*, XIB, 275 – 3443.

Hoskins, S. G. (1990) Metamorphosis of the amphibian eye. *Journal of Neurobiology*, 21, 970 – 989.

Johns, P. R. and Easter, S. S. Jr. (1977) Growth of the adult goldfish eye. II. Increase in retinal cell numbers. *Journal of Comparative Neurology*, 176, 331– 342.

Johns, P. R. and Fernald, R. D. (1981) Genesis of rods in teleost fish retina. *Nature* (London), 293, 178 –198.

Just, J. J., Kraus-Just, J. and Check, D. A. (1981) Survey of chordate metamorphosis, in *Metamorphosis*, (eds Gilbert, L. I. and Frieden, E.), Plenum Press, New York, pp. 265 – 326.

Kaltenbach, J. C. (1953) Local action of thyroxin on amphibian metamorphosis. II. Development of the eyelids, nictitating membrane, cornea, and extrinsic ocular muscles in *Rana pipiens* larvae effected by thyroxin-cholesterol implants. *Journal of Experimental Zoology*, 122, 41– 50.

Kelley, M. W., Turner, J. K. and Reh, T. A. (1995) Ligands of steroid/thyroid receptors induce cone photoreceptors in vertebrate retina. *Development*, 121, 3777– 3785.

Kennedy, M. C. and Rubinson, K. (1977) Retinal projections in larval, transforming and adult sea lamprey, *Petromyzon marinus. Journal of Comparative Neurology*, 171, 465 – 480.
Kurz-Isler, G. and Wolburg, H. (1982) Morphological study on the regeneration of the retina in rainbow trout after ouabain-induced damage: evidence of dedifferentiation of photoreceptor cells. *Cell and Tissue Research*, 225, 165 –178.
Li, K. T., Wetterer, J. K. and Hairston, N. G. Jr. (1985) Fish size, visual resolution, and prey selectivity. *Ecology*, 66, 1729 –1735.
Loew, E. R. and Wahl, C. M. (1991) A short-wavelength sensitive cone mechanism in juvenile yellow perch, *Perca flavescens. Vision Research*, 31, 353 – 360.
Lyall, A. H. (1957) The growth of trout retina. *Quarterly Journal of Microscopical Science*, 98, 101–110.
Lythgoe, J. N. and Partridge, J. C. (1991) The modelling of optimal visual pigments of dichromatic teleosts in green coastal waters. *Vision Research*, 31, 361– 371.
Maier, W. and Wolburg, H. (1979) Regeneration of the goldfish retina after exposure to different doses of ouabain. *Cell and Tissue Research*, 202, 99 –118.
Markle, D. F., Harris, P. M. and Toole, C. L. (1992) Metamorphosis and an overview of early-life-history stages in Dover sole *Microstomus pacificus. Fishery Bulletin*, 90, 285 – 301.
Mathis, U., Schaeffel, F. and Howland, H. C. (1988) Visual optics in toads (*Bufo americanus*). *Journal of Comparative Physiology* [A], 163, 201– 213.
McFarland, W. N. (1991) The visual world of coral reef fishes, in *The Ecology of Fishes on Coral Reefs*, (ed Sale, P. F.), Academic Press, San Diego, pp. 16 – 38.
McFarland, W. N. and Allen, D. M. (1977) The effect of extrinsic factors on two distinctive rhodopsin-porphyropsin systems. *Canadian Journal of Zoology*, 55, 1000 –1009.
Munk, O. (1990) Changes in the visual cell layer of the duplex retina during growth of the eye of a deep-sea teleost, *Gempylus serpens* Cuvier, 1829. *Acta Zoologica*, 71, 89 – 95.
Muntz, W. A and Reuter, T. (1966). Visual pigments and spectral sensitivity in *Rana temporaria* and other European tadpoles. *Vision Research*, 6, 601– 618.
Muntz, W. A. and Mouat, G. S. V. (1984) Annual variations in the visual pigments of brown trout inhabiting lochs providing different light environments. *Vision Research*, 24, 1575 –1580.
Neave, D. A. (1984) The development of visual acuity in larval plaice (*Pleuronectes platessa* L.) and turbot (*Scophthalmus maximus* L.). *Journal of Experimental Marine Biology and Ecology*, 78, 167–175.
Nguyen, V. S. and Straznicky, C. (1989) The development and the topographic organization of the retinal ganglion cell layer in *Bufo marinus. Experimental Brain Research*, 75, 345 – 353.
Novales Flamarique, I. and Hawryshyn, C. W. (1996) Retinal development and visual sensitivity of young Pacific sockeye salmon *Oncorhynchus nerka. Journal of Experimental Biology*, 199, 869 – 882.
Ohman, P. (1976) Fine structure of photoreceptors and associated neurons in the retina of *Lampetra fluviatilis* (Cyclostomi). *Vision Research*, 16, 659 – 662.
Ohnishi, K. (1993) Development of color vision in goldfish: Selective delayed maturation of blue vision. *Vision Research*, 33, 1665 –1672.
Pankhurst, N. W. (1984) Retinal development in larval and juvenile European eel, *Anguilla anguilla* (L.). *Canadian Journal of Zoology*, 62, 335 – 343.
Pankhurst, P. M. and Butler, P. (1996) Development of the sensory organs in the greenback flounder, *Rhombosolea tapirina. Marine and Freshwater Behaviour and Physiology*, 28, 55 –73.
Pankhurst, P. M., Pankhurst, N. W. and Montgomery, J. C. (1993) Comparison of behavioral and morphological measures of visual acuity during ontogeny in a teleost fish, *Forsterygion varium*, Tripterygiidae (Forster, 1801). *Brain Behavior and Evolution*, 42, 178 –188.
Polansky, J. R. and Bennett, T. P. (1973) Alterations in physical parameters and proteins of lens from *Rana catesbeiana* during development. *Developmental Biology*, 33, 380 – 408.
Pomeranz, B. (1972) Metamorphosis of frog vision: changes in ganglion cell physiology and anatomy. *Experimental Neurology*, 82, 187–199.
Potter, I. C. (1980) Ecology of larval and metamorphosing lampreys. *Canadian Journal of Fisheries and Aquatic Sciences*, 37, 1641–1657.
Powers, M. K., Bassi, C. J., Rone, L. A. and Raymond, P. A. (1988a) Lighting conditions and retinal development in goldfish: absolute visual sensitivity. *Investigative Ophthalmology and Visual Science Supplement*, 29, 37– 43.
Powers, M. K., Bassi, C. J., Rone, L. A. and Raymond, P. A. (1988b) Visual detection by the rod system in goldfish of different sizes. *Vision Research*, 28, 211– 221.
Rahmann, H., Jeserich, G. and Zeutzius, I. (1979) Ontogeny of visual acuity of rainbow trout under normal conditions and light deprivation. *Behaviour*, 68, 315 – 322.

Raymond, P. A. (1991a) Cell determination and positional cues in the teleost retina: development of photoreceptors and horizontal cells, in *Development of the Visual System*, (eds Shatz, C. and Lam, D. M. K.), M.I.T. Press, Boston. (Retina Research Foundation Symposium) vol. 3, pp. 59 –78.

Raymond, P. A. (1991b) Retinal regeneration in teleost fish, in *Regeneration of Vertebrate Sensory Cells*, (ed Rubel, E.), Ciba Foundation Symposium, Chichester, Wiley, Vol. 160, pp. 171–191.

Raymond, P. A., Reifler, M. J. and Rivlin, P. K. (1988) Regeneration of goldfish retina: Rod precursors are a likely source of regenerated cells. *Journal of Neurobiology*, 19, 431– 463.

Reuter, T. (1969) Visual pigments and ganglion cell activity in the retinae of tadpoles and adult frogs (*Rana Temporaria* L.). *Acta Zoologica Fennica*, 122, 1– 64.

Shand, J. (1993) Changes in the spectral absorption of cone visual pigments during the settlement of the goatfish *Upeneus tragula*: the loss of red sensitivity as a benthic existence begins. *Journal of Comparative Physiology* [A], 173, 115 –121.

Shand, J. (1994) Changes in retinal structure during development and settlement of the goatfish *Upeneus tragula. Brain Behavior and Evolution*, 43, 51– 60.

Shand, J. (1997) Ontogenetic changes in retinal structure and visual acuity: a comparative study of coral-reef teleosts with differing post-settlement lifestyles. *Environmental Biology of Fishes*, 49, 307– 322.

Shand, J., Partridge, J. C., Archer, S. N., Potts, G. W. and Lythgoe, J. N. (1988) Spectral absorbance changes in the violet/blue sensitive cones of the juvenile pollack, *Pollachius pollachius. Journal of Comparative Physiology* [A], 163, 699 –703.

Sivak, J. G. and Warburg, M. R. (1980) Optical metamorphosis of the eye of *Salamandra salamandra. Canadian Journal of Zoology*, 58, 2059 – 2064.

Sivak, J. G. and Warburg, M. R. (1983) Changes in optical properties of the eye during metamorphosis of an anuran, *Pelobates syriacus. Journal of Comparative Physiology* [A], 150, 329 – 332.

Straznicky, K. and Gaze, R. M. (1971) The growth of the retina in *Xenopus laevis*: an autoradiographic study. *Journal of Embryology and Experimental Morphology,* 26, 67–79.

Studnicka, F. K. (1912) Ueber die Entwicklung und die Bedeutung der Seitenaugen von Ammocoetes. *Anatomischer Anzeiger*, 41, 561– 578.

Udin, S. B. (1990) Development of orderly connections in the retinotectal system, in *Science of Vision*, (ed Leibovic, K. N.), Springer-Verlag, New York, pp. 125 –150.

van der Meer, H. J. (1994) Ontogenetic change of visual thresholds in the cichlid fish *Haplochromis sauvagei. Brain Behavior and Evolution*, 44, 40 – 49.

van der Meer, H. J. (1995) Visual resolution during growth in a cichlid fish: A morphological and behavioral case study. *Brain Behavior and Evolution*, 45, 25 – 33.

Wald, G. (1945) The chemical evolution of vision. *Harvey Lectures*, 41, 117–160.

Wetts, R., Serbedzija, G. N. and Fraser, S. E. (1989) Cell lineage analysis reveals multipotent precursors in the ciliary margin of the frog retina. *Developmental Biology*, 136, 254 – 263.

Wheeler, T. G. (1979) Retinal red sensitivity under dark adapted conditions. *Brain Research*, 175, 140 –144.

Wilt, F. H. (1959) The differentiation of visual pigments in metamorphosing larvae of *Rana catesbeiana. Developmental Biology*, 1, 199 – 233.

Wood, P. and Partridge, J. C. (1993) Opsin substitution induced in retinal rods of the eel (*Anguilla anguilla* (L.)): a model for G-protein-linked receptors. *Proceedings of the Royal Society of London Series B: Biological Sciences,* 254, 227– 232.

Wood, P., Partridge, J. C. and DeGrip, W. J. (1992) Rod visual pigment changes in the elver of the eel *Anguilla anguilla* L. measured by microspectrophotometry. *Journal of Fish Biology*, 41, 601– 611.

Zaunreiter, M., Junger, H. and Kotrschal, K. (1991) Retinal morphology of cyprinid fishes: A quantitative histological study of ontogenic changes and interspecific variation. *Vision Research*, 31, 383-394.

13
Molecular biology of photoreceptor spectral sensitivity

J.K. BOWMAKER[1] and D.M. HUNT[2]
[1]Department of Visual Science; [2]Department of Molecular Genetics, Institute of Ophthalmology, University College London, Bath Street, London WC1V 9EL

13.1 Introduction

The wavelengths of spectral sensitivities of the visual systems of animals extend from about 300 nm in the near ultraviolet to above 800 nm in the far red. Some species utilize almost this entire spectral range, whereas others are restricted to narrower ranges within these broad limits (Bowmaker, 1991a). The spectral range of 'visible' light is limited by two factors. First, because of the filtering effects of the atmosphere, most (about 80%) of the spectral radiation at the earth's surface is restricted to wavelengths between about 300 and 1100 nm with the energy distribution having a maximum at about 480 nm and the quantum distribution maximal at about 555 nm. Secondly, the quantal energy of photons above about 850 nm is too low to lead to photoisomerization of organic molecules, whereas photons below about 300 nm have sufficiently high energies to be destructive to proteins (see Knowles and Dartnall, 1977).

Even though visual pigments have broad spectral bands extending over 300 to 400 nm, a single pigment cannot give sensitivity over the entire 'visible' spectrum. Further, since the sensitivity function of a pigment is approximately bell-shaped, a single pigment cannot give unambiguous information about wavelength. For colour vision, at least two spectral classes of photoreceptor are required that have overlapping spectral sensitivities (dichromacy). In this case, two light stimuli of different wavelengths that elicit identical sensations from one class of photoreceptor will invoke different responses from the other class and may be chromatically differentiated. To extend the spectral range over which colours can be determined and to increase the number of distinct colours that can be discriminated, additional classes of photoreceptor need to be added. Humans are trichromatic, but many vertebrates, including a number of birds and shallow-water fish have four spectral classes of cone and are tetrachromatic. The fourth class of cone is normally sensitive at short wavelengths and extends the vision of these animals into the near ultraviolet. Even pentachromacy has been suggested in the pigeon (for a recent discussion, see Bowmaker *et al.*, 1997) and there is some debate as to the maximum number of spectrally distinct photoreceptors that can give a species useful additional visual information (Barlow, 1982; Bowmaker, 1983).

S.N. Archer et al. (eds.), Adaptive Mechanisms in the Ecology of Vision, 439–462

This chapter discusses the molecular structure of visual pigments with specific reference to their evolution within the different vertebrate classes and to the molecular basis underlying the tuning of visual pigments to specific spectral locations.

13.2 The structure of opsin

The organic molecules, visual pigments, that have evolved to subserve vision, belong to a very large family of structurally similar transmembrane proteins that act as receptor molecules in a wide range of different cell types: all function through the activation of a G protein that binds guanosine diphosphate (GDP) (for a brief review, see North, 1989). These receptor molecules consist of a single polypeptide chain normally made up of between 340 and 500 amino acids that form seven hydrophobic regions spanning the membrane, linked by extra-membrane hydrophilic loops (Figure 1). The transmembrane segments, comprising about 50% of the molecule, are thought to be α-helices each composed of 26 amino acids of which only the central 18 are embedded in the membrane (Baldwin, 1993; Schertler *et al.*, 1993), whereas the remaining hydrophilic regions, constituting about 25% of the molecule on each side of the membrane, are presumed to be straight chains (Hargrave *et al.*, 1984). The seven helices form a bundle or palisade within the membrane creating a cavity on the extracellular side that acts as a ligand-binding pocket. Although opsin polypeptides vary in length, the transmembrane helices are highly conserved, as are the cytoplasmic loops; for easy comparisons between different visual pigments the numbering of the amino acids used here is based on bovine rod opsin, but where helpful the equivalent primate long- and middle-wave-sensitive (LWS and MWS) cone opsin numbers are added in parentheses. The different opsins share a number of features (Figure 1). There are two glycosylation sites near the amino terminus on the extracellular side (luminal side in the closed discs of rod outer segments) and a series of phosphorylation sites near the carboxyl terminus on the cytoplasmic side (Hargrave, 1982). Also, regions of the cytoplasmic loops linking the helices are involved in interacting with the membrane-bound G protein (Hargrave, 1982; Applebury and Hargrave, 1986; Kînig *et al.*, 1989; Franke *et al.*, 1992). In addition, there are two highly conserved cysteine residues in the first and second extracellular loops that form a disulphide bridge essential for the correct folding and formation of the molecule (Karnik *et al.*, 1988) and two further cysteines in the carboxyl tail that are palmitoylated, anchoring the tail and forming a fourth cytoplasmic loop (Ovchinnikov *et al.*, 1988).

In visual pigments the ligand, retinal, is bound into the molecular pocket and must be activated by light, whereas in other membrane receptors, the binding of the ligand itself (for example a neurotransmitter or hormone) activates the receptor molecule. To be photosensitive, the molecule must absorb light and this is achieved by the interaction of the protein, opsin, with the prosthetic group,

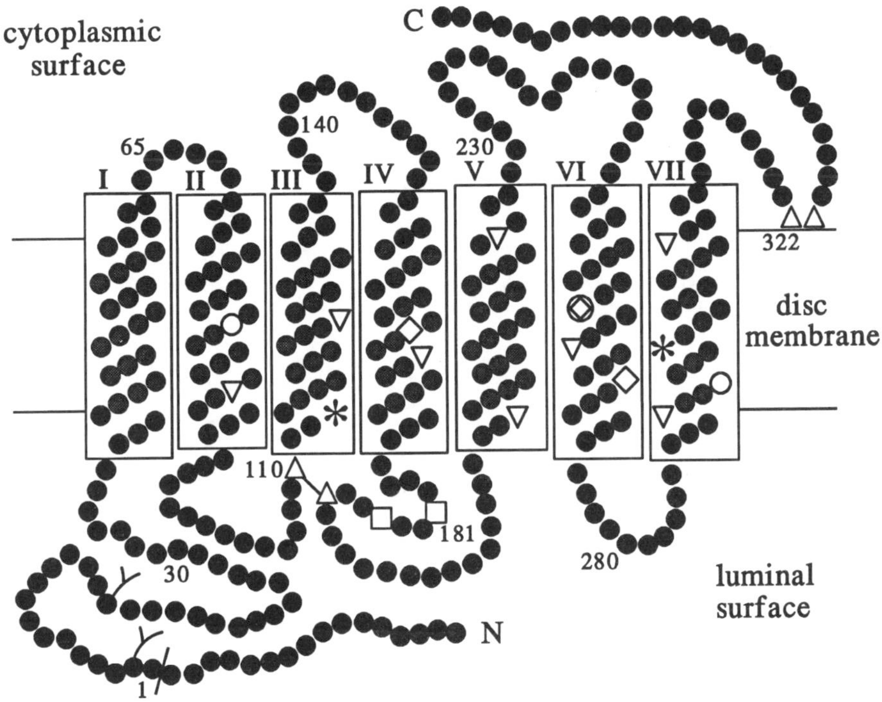

Figure 1 Schematic diagram of the organisation of vertebrate opsin (based on bovine rod opsin) in the lipid bilayer of the disc membrane. Mammalian rod opsins consist of 348 amino acids, whereas the opsins of the primate LWS and MWS cones have an additional 16 amino acids at the amino terminus (N) as indicated by the dividing line at amino acid 1. There are two glycosylation sites (Asparagine 2 and 15) near the amino terminus and a series of phosphorylation sites near the carboxyl terminus (C). The three cytoplasmic loops are involved in protein interactions with transducin and arrestin. The two cysteines at 110 and 187 (△ near the luminal end of helix III) form a disulphide bridge essential for correct folding of the molecule and the two cysteines 322 and 323 (△) on the cytoplasmic surface of the membrane are palmitoylated so forming a fourth cytoplasmic loop. Retinal is attached via a protonated Schiff's base to lysine 296 (* in helix VII) which has a counterion, glutamate at 113 (* in helix III). In the long-wave group of opsins (L opsins), sites 181 and 183 (□ in the second luminal loop) are positively-charged histidine and lysine respectively, forming a chloride-binding site which displaces the λ_{max} of the pigments to longer wavelengths. Sites 164, 261 and 269 (◇ in helix IV and VI) are hydroxyl/non-hydroxyl amino acid exchanges involved in the spectral tuning of primate LWS and MWS cone pigments (L opsins). Sites 83 (helix II), 261 (helix VI) and 292 (helix VII) (o) are thought to be involved in the spectral tuning of rod visual pigments. A number of sites (▽) are proposed as tuning sites responsible for SWS cone pigments (S opsins)

retinal. The absorption of light by the chromophore is a consequence of the π electron shell of the conjugated chain of single and double bonds within the chromophore (Figure 2). Retinal absorbs maximally at about 380 nm, but when bound to an amine, the protonated Schiff's base has a peak sensitivity (λ_{max}) at

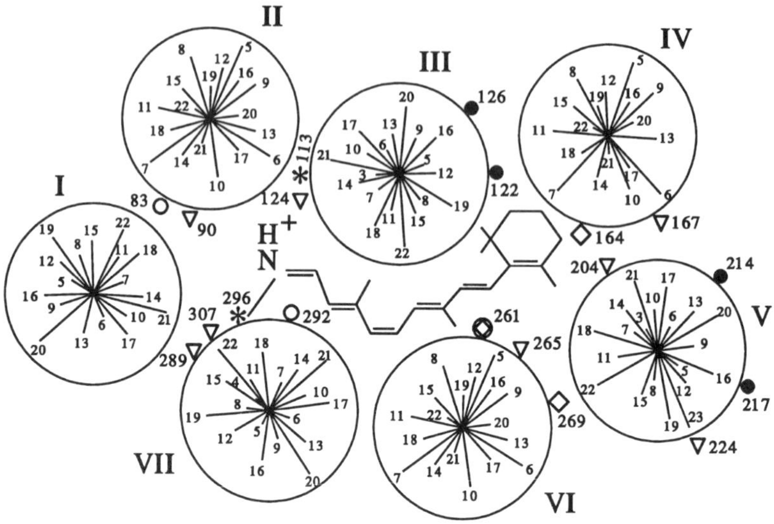

Figure 2 A three dimensional model of a visual pigment showing the relative positions of the seven α-helical transmembrane regions (based on the model of Baldwin, 1993) and the location of retinal within the helical palisade. Each circle represents an α-helix containing 26 amino acids and each radiating line within the circle represents an amino acid. The position of each line indicates the location of the amino acid around the circumference of the helix whereas the length of the line indicates the depth of the amino acid within the helix. Helix III is somewhat buried within the palisade. Retinal is bound via a protonated Schiff's base (NH^+) to lysine 296 (*), position 11 in helix VII, which is stabilized by the counterion, glutamate 113 (*), position 3 in helix III. The three sites (◇) principally involved in tuning L_R and L_G cone pigments are 164, position 14 in helix IV, and 261 and 269, positions 11 and 19 in helix VI. Sites 83 (helix II), 261 (helix VI) and 292 (helix VII) (o) are thought to be involved in the spectral tuning of rod visual pigments and a number of sites (▽) are proposed as tuning sites responsible for SWS cone pigments (S opsins). Further sites (●) have also been proposed as possible tuning sites

about 440 nm (Knowles and Dartnall, 1977). However, visual pigments have λ_{max} ranging from about 350 nm to 570 nm (or 635 nm if based on 3-dehydroretinal) and the precise λ_{max} of a given pigment is determined by interactions between the chromophore and the particular opsin protein. To displace an absorbance maximum to longer wavelengths, for instance, requires that there is a smaller energy difference between the ground and excited states of the molecule: the protein-chromophore interaction must increase delocalisation in the π electron shell of the chromophore.

In all visual pigments, retinal is bound to opsin via a lysine residue in the centre of helix VII, position 296 of bovine rod opsin (Bownds, 1967; Wang *et al.*, 1980), and a negatively charged counterion is required to stabilise the protonated Schiff's base. In most visual pigments, this will cause a red-shift to a λ_{max} close to 500 nm, as seen in a majority of rod visual pigments. Site directed mutagenesis of bovine rod opsin has clearly indicated that the counterion is glutamate at site 113, close to the extracellular side of helix III (Figures 1 and 2) (Sakmar *et al.*, 1989; Zhukovsky and Oprian, 1989; Nathans, 1990a). Replacement of this glutamate with glutamine (E113Q) causes a massive spectral shift to 380 nm that is sensitive to both pH and solvent anions. In non-acidic or chloride-free environments, the Schiff's base becomes unprotonated, resulting in a blue shift to 380 nm. However, chloride in the solvent can act as a counterion so that in acidic conditions, the Schiff's base becomes protonated, displacing the λ_{max} back to about 495 nm. Replacements of the other carboxyl side chain residues in the membrane region of opsin have much smaller spectral effects; the greatest, about a 20-nm blue shift, is caused by the substitution of glutamine for the glutamate at position 122 (E122Q) in helix III. However, the more extreme change to isoleucine or leucine at this site yields only about a 2-nm blue shift, suggesting that site 122 is not intimately involved in spectral tuning (Nathans, 1990b; Decaluwé *et al.*, 1995) (Figures 1 and 2).

These findings imply that visual pigments with λ_{max} longer than 380 nm should all have a protonated Schiff's base stabilized by a counterion equivalent to E113, but that ultraviolet-sensitive pigments with λ_{max} below 380 nm may have an unprotonated Schiff's base and lack a carboxyl side chain residue at site 113. However, with one exception, all vertebrate opsins that have been sequenced have glutamate at site 113, including the putative short-wave-sensitive opsin of the mouse (Chiu *et al.*, 1994) that is thought to be expressed as an ultraviolet-sensitive (UVS) cone pigment with λ_{max} at about 365 nm (Jacobs *et al.*, 1991) and the UVS pigments of the goldfish (Hisatomi *et al.*, 1996) and budgerigar (Wilkie *et al.*, 1997). The significant amino acids responsible for tuning these opsins to an ultraviolet spectral location are thus not immediately apparent (see below), but the position should become clearer as the sequences of more ultraviolet opsins from fish and birds become available. Recently, the sequence of a putative violet-sensitive cone opsin from *Xenopus laevis* has been added to the data base (Knox, 1995) that apparently has glutamate 113 replaced by aspartate. Site directed mutagenesis replacing aspartate for glutamate at 113 in bovine rod opsin has little effect on the λ_{max} of the pigment, causing only a small red shift of about 5 nm (Sakmar *et al.*, 1989; Zhukovsky and Oprian, 1989).

The situation is somewhat different in invertebrates where the opsins, excluding the ultraviolet-sensitive, have a polar (hydroxyl-bearing) residue, tyrosine, in the equivalent position of site 113. However, site 124 which is close to the counterion, is occupied in invertebrate pigments by another polar residue, serine. Thus, unlike vertebrate opsins which have a single carboxyl side chain acting as the Schiff's base counterion, it has been suggested that invertebrate opsins may have two slightly negatively charged polar residues (Chang *et al.*, 1995). However, resonance Raman spectra of octopus visual pigment give no evidence for a tyrosinate, excluding the possibility that the tyrosine serves as the counterion (Hashimoto *et al.*, 1996). It has also been suggested that the Schiff's base in the ultraviolet-sensitive pigments of the fruit fly *Drosophila* is unprotonated, since the polar residues at sites 113 and 124 are replaced by the non-polar amino acids, phenylalanine and alanine respectively (Cowman *et al.*, 1986; Fryxell and Meyerowitz, 1987; Zuker *et al.*, 1987; Montell *et al.*, 1987). Both Bellingham *et al.* (1997) and Towner *et al.* (1997) speculate on additional sites in insect invertebrate opsins that could be involved in spectral tuning into the ultraviolet, but agree on only one position in transmembrane helix VII. In the longer wave pigments, site 302 (position 17 in the helix) is occupied by asparagine, a neutral amino acid with a polar side chain, but is replaced in the UVS opsins by aspartate or glutamate, both negatively charged amino acids (see below).

In vertebrates, the protonated Schiff's base and its counterion can account for the visual pigments with λ_{max} up to around 500 nm. This includes the rod opsins and the middle-wave-sensitive (green) cone pigments of many non-mammalian vertebrates. It cannot account however for the many visual pigments with λ_{max} in excess of 500 nm, as seen for example in the primate red/green-sensitive cone pigments where the λ_{max} ranges from about 530 nm to 565 nm. Crescitelli (1977) reported that the longer-wave-sensitive visual pigment of the gecko (*Gekko gekko*) is sensitive to the ionic environment and that the λ_{max} can be blue-shifted from 521 nm to about 490 nm by lowering the chloride ion concentration. A similar blue-shift was also reported for the chicken LWS cone pigment (Knowles, 1976; Fager and Fager, 1979; Knowles, 1980) and has since been found in the LWS cone pigments of teleosts, amphibians and primates (Novitsky *et al.*, 1989; Kleinschmidt and Hárosi, 1992; Wang *et al.*, 1993). Presumably, chloride ions must bind to a hydrophilic domain on the extracellular region of opsin and perturb the molecule such that it further delocalizes the π electron shell of the chromophore. Using site-directed mutagenesis to substitute 18 different positively-charged amino acids in the human LWS/MWS cone pigments, Wang *et al.* (1993) have identified two residues, histidine at site 181 and lysine at site 184 (bovine rod opsin numbering), as candidates for the chloride-binding site. These sites are in the extracellular loop connecting helices IV and V, but close to the cysteine 187-cysteine 110 disulphide bond linked to the luminal end of helix III, and thus close to glutamate 113 (Figure 1). The two sites are strictly conserved in all LWS cone pigments (including the gecko long-wave 'rod' pigment) but are substituted by glutamate and glutamine respectively in all rod opsins and shorter-wave cone opsins.

13.3 Phylogeny of opsins

From the considerations discussed above and from sequence comparisons of all of the vertebrate opsin sequences available, it is possible to generate detailed phylogenetic trees of visual pigments based on gene duplication and subsequent sequence divergence (Okano *et al.*, 1992; Johnson *et al.*, 1993; Hisatomi *et al.*, 1994; Yokoyama, 1994; Chang *et al.*, 1995). Although the phylogenetic trees proposed by Okano *et al.* (1992), Johnson *et al.* (1993) and Chang *et al.* (1995)

are slightly different from those of Hisatomi *et al.* (1994) and Yokoyama (1994), all trees show the same basic features (Figure 3). The ancestral vertebrate visual pigment would appear to have been cone-like with a λ_{max} below about 520 nm, perhaps close to 500 nm, and based on an opsin with a protonated Schiff's base, but insensitive to chloride ions. Very early in vertebrate evolution, about 500 mya (million years ago), a longer-wave pigment class diverged that contained a chloride ion binding site (Kleinschmidt and Hárosi, 1992; Wang *et al.*, 1993) responsible for tuning the λ_{max} of the pigment to wavelengths in the range from about 530 to 570 nm (Figure 3). Such a division would have formed the basis for a dichromatic system of colour vision. This division was followed rapidly (in evolutionary terms) by the multiplication of the shorter-wave group into at least four major classes of opsin that includes a middle-wave-sensitive group with λ_{max} around 470 – 520 nm and a short-wave-sensitive group with λ_{max} shorter than about 460 nm. The early evolution of four cone opsin groups provides the potential for tetrachromatic colour vision in all of the major vertebrate classes and all four are retained in diurnal teleosts, reptiles and birds (see Bowmaker, 1991a).

Okano *et al.* (1992), Johnson *et al.* (1993) and Chang *et al.* (1995) suggest that violet-sensitive pigments (λ_{max} from about 355 to 430 nm) diverged as a separate

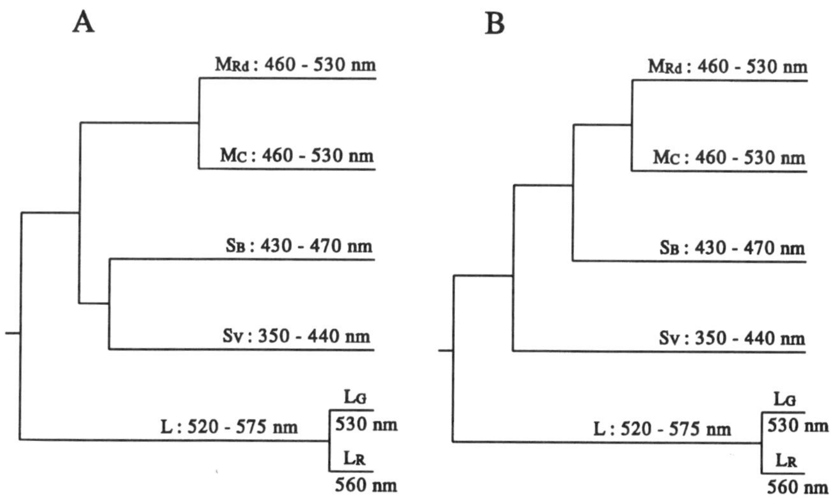

Figure 3 Simplified phylogenetic trees illustrating the evolution of five groups of opsins. See text for details. Both trees show an early gene duplication separating the L opsins (chloride-sensitive) from the remaining groups (chloride-insensitive). Scheme A (e.g. Yokoyama, 1994) suggests that a subsequent gene duplication produced the S and M opsin groups which then divided to produce the S_V and S_B, and the M_C and M_{Rd} groups. Scheme B (e.g. Okano *et al.*, 1992) differs in that the first gene duplication in the chloride-insensitive opsins separated the S_V opsins followed by subsequent duplications separating the S_B, M_C and M_{Rd} groups. The appearance of all five opsin groups occurred early in vertebrate evolution, probably by about 450 mya. The gene duplication within the L opsins separating the L_R and L_G opsins occurred recently, probably about 45 mya, but independently within different vertebrate classes

class before the appearance of the remaining middle-wave (λ_{max} from about 470 to 520 nm) and short-wave-sensitive pigments (λ_{max} from about 430 – 460 nm), with the middle-wave-sensitive group subsequently dividing to give middle-wave-sensitive cone pigments and rod pigments. In contrast, Hisatomi *et al.* (1994) and Yokoyama (1994) propose that the ancestral gene duplicated to produce two families, a shorter-wave group that gave rise to both the short-wave-sensitive and violet/ultraviolet-sensitive pigments and a middle-wave group that gave rise to the middle-wave-sensitive cone pigments and to rod pigments. What is clear is that the divergence into four spectral groups of cone pigments and the rod pigments occurred early in vertebrate evolution, before the divergence of the main vertebrate classes, with the long-wave cone pigments separating first and the rod pigments evolving at a later stage (Figure 3). These phylogenetic trees are based however on relatively little data and only when increased numbers of full sequences from a broader cross section of vertebrates become available will a more precise evolutionary pattern be forthcoming.

One major confusion that becomes apparent from these phylogenetic trees and which has been a problem in the field of visual pigments for many years is that of nomenclature: how to name visual pigments. Using either colour names such as 'red'- or 'green'-sensitive or wavelength regions such as 'long'- or 'middle'-wave becomes highly ambiguous in comparative studies. In humans the LWS/red ($P563_1$) and MWS/green pigments ($P535_1$) belong to the same evolutionary group, whereas the green-sensitive pigments of cyprinids (say a $P509_1$) belong to an entirely different evolutionary group. Similar confusions arise at the short-wave end of the spectrum: the human SWS/blue cone pigment ($P420_1$) belongs to the same evolutionary group as chicken 'violet' ($P418_1$), but is different from chicken 'blue' ($P455_1$) and cyprinid blue-sensitive pigments (say a $P440_1$). The problem has been confounded by some of the terms used to describe the phylogenetic branches, especially such labels as 'protope', 'deuterope' and 'tritope' introduced by Hisatomi *et al.* (1994) which can be confused with 'protanope', 'deuteranope' and 'tritanope' from human colour vision that have entirely different connotations. This is not a trivial problem since deuteranopia in humans refers to problems with the human MWS cones that contain a visual pigment belonging to the long-wave group (a 'protope' pigment), whereas the middle-wave 'deuterope' group is composed of rod visual pigments that are not normally involved in colour vision. The scheme that we will introduce here combines both wavelength and colour names, and seems to be intuitively more helpful. The five divisions become L (long-wave, chloride-sensitive), M_C (middle-wave cone opsins that are chloride-insensitive), S_B (short-wave, blue-sensitive), S_V (short-wave, violet- and ultraviolet-sensitive) and M_{Rd} (middle-wave-sensitive rod opsins). Although four spectral groups of cone pigments appeared at a very early stage in the evolution of vertebrates, some of the groups have been either lost or undergone further spectral diversification. In mammals, for instance, it would appear that only the ancestral shortest-wave and longest-wave groups (S_V and L) have been retained so that most mammals are dichromats (Jacobs, 1993). Trichromacy has only been achieved in primates by a relatively recent (45 Mya) divergence of the longer-wave group into two spectrally distinct classes, the LWS/red (L_R) and MWS/green (L_G) cones (e.g. Nathans *et al.*, 1986b; Ibbotson *et al.*, 1992; Deeb *et al.*, 1994; Dulai *et al.*, 1994). Divergence of the L group has also occurred independently in some teleosts: the cavefish, *Astyanax fasciatus*, apparently has at least three opsins in the group, two that code for green-sensitive pigments (L_{G1} and L_{G2}) and one for a red-sensitive pigment (L_R) (Yokoyama and Yokoyama, 1990).

With this background, it is now possible to identify either particular amino acid substitutions or a combination of substitutions that may be involved in distinguishing the four spectral classes of opsins and of 'fine tuning' the λ_{max} of pigments within each group.

13.4 Long-wave-sensitive (L) cone pigments (λ_{max} 520 – 575 nm)

Nathans *et al.* (1986a) isolated and sequenced the genes coding for the three classes of human cone pigments (λ_{max} at about 420, 530 and 562 nm). The deduced amino acid sequences show that the L_R and L_G cone opsins are 96% homologous and clearly belong to the same phylogenetic group of opsins. Both consist of 364 amino acids and differ at only 15 sites; the 30-nm spectral difference between these two visual pigments must be attributable to substitution at some or all of these sites. Of these, substitutions located within the membrane helices and in close proximity to the chromophore are most likely to have major effects on the spectral tuning of the pigment. Further, only replacements of charged for non-charged or polar for non-polar amino acids are likely to have significant effects on retinal. Nathans *et al.* (1986a) suggested three sites as the most likely candidates; site 164 (180) in helix IV and sites 261 (277) and 269 (285) in helix VI (Figures 1 and 2). In the shorter-wave P530, all three sites are occupied by non-polar residues (alanine, phenylalanine and alanine respectively), whereas in the longer-wave P562 the three sites are occupied by hydroxyl-bearing polar residues with a small negative charge (serine, tyrosine and threonine respectively). These polar residues will further delocalise the π electron shell of the chromophore, lowering the energy difference between the ground and excited states of the molecule and thereby shifting the λ_{max} to longer wavelengths. However, it is not possible from these data alone to establish which if any of the three sites are more or less important.

New World monkeys offer a natural experiment from which to determine the role of each of the three sites. Unlike Old World monkeys and man, most New World monkeys exhibit a marked polymorphism of their red/green-sensitive cones. As a consequence of possessing only a single gene coding for a cone opsin in this spectral region, all males are dichromats, but because each species has three allelic forms of the gene, females heterozygous for the locus are trichromats (for reviews, see Bowmaker, 1991b; Mollon, 1991; Jacobs, 1993). Within the cebids (e.g. capuchins and squirrel monkeys), the three alleles code for cone pigments with λ_{max} at about 535, 549 and 562 nm (e.g. Mollon *et al.*, 1984; Jacobs and Neitz, 1987a; Jacobs and Neitz, 1987b), whereas in the callitrichids (marmosets and tamarins), the alleles code for pigments with λ_{max} at about 543, 556 and 562 nm (Jacobs *et al.*, 1987; Travis *et al.*, 1988; Tovée *et al.*, 1992). These species therefore provide three additional spectral locations for cones absorbing between about 530 and 565 nm from which to extract information about opsin sequences and spectral tuning.

Sequence data from three species of New World monkey, *Saimiri sciureus*, *Callithrix jacchus* and *Saguinus fusicollis*, confirm that the three sites (164, 261 and 269) are involved in spectral tuning, and from a first approximation, can be used in a simple additive pattern (Neitz *et al.*, 1991; Williams *et al.*, 1992; Jacobs *et al.*, 1993). The major spectral shift, about 15 nm, is caused by the threonine/alanine substitution at site 269 (285), followed by a 7 – 9 nm shift from the tyrosine/phenylalanine switch at site 261 (277). The serine/alanine replacement at site 164 (180) causes a smaller 4 – 6 nm shift (Table 1). These conclusions neglect the contribution of any other amino acid differences between the opsins, but sequence comparisons (e.g. Williams *et al.*, 1992) have also implicated sites 214 (230, isoleucine/threonine) and 217 (233, alanine/serine) and recent site-directed mutagenesis studies of the human opsins have confirmed that two additional sites, 100 and 293 (116 and 309), have small spectral effects of the order of about 1 nm (Merbs and Nathans, 1993; Asenjo *et al.*, 1994).

A model for the three-dimensional arrangement of the seven α-helices, based on comparisons of the amino acid sequence of 204 G-protein receptors, including 32 visual pigments, and the two-dimensional crystal structure of rhodopsin (Schertler *et al.*, 1993), has recently been proposed by Baldwin (1993). In this model, each α-helix is composed of 26 residues, although only the central 18 or

Table 1 Summary of deduced amino acid differences at key sites in primate L cone opsins. cal, callitrichid and ceb, cebid New World monkeys; owm, Old World monkeys and Great Apes; man[1], common L_R (LWS) and L_G (MWS) pigments; man[2], less common polymorphic forms. λ_{max} as determined from microspectrophotometry, except for 538-nm pigment. λ_{max} recorded from expressed pigments in solution may be displaced a few nm to shorter wavelength

λ_{max} (nm)	563	556	549	543	538	535
Amino acid site						
164 (180)	Ser^{OH}	Ala	Ala	Ala	Ser^{OH}	Ala
261 (277)	Tyr^{OH}	Tyr^{OH}	Phe	Tyr^{OH}	Phe	Phe
269 (285)	Thr^{OH}	Thr^{OH}	Thr^{OH}	Ala	Ala	Ala
Species	cal	cal		cal		
	ceb		ceb			ceb
	owm					owm
	man[1]	man[2]			man[2]	man[1]

Spectral shift:- 164: Ser^{OH} to Ala, 2 – 7 nm
261: Tyr^{OH} to Phe, 5 – 7 nm
269: Thr^{OH} to Ala, 10 – 14 nm

so residues (5 – 22) are thought to be embedded in the membrane. As shown in Figure 2, the primary tuning sites discussed above, when placed on to the model, face towards either the chromophore or an adjacent helix.

The genes coding for the L_R and L_G opsins are composed of six exons and in humans and Old World primates, the genes lie in a tandem array on the long arm of the X chromosome. This head-to-tail arrangement of the L_R and L_G genes and their close sequence similarity means that this chromosome region is susceptible to mispairing during meiosis, leading to unequal crossing-over between the gene arrays. If the crossover occurs between genes, this will result in the deletion of a gene from one chromosome and its addition to the other, whereas a crossover within a gene will lead to the production of a hybrid or chimeric gene that combines regions of the L_R and L_G genes into a single gene. Such hybrid genes are thought to be responsible for anomalous colour vision in humans (Nathans *et al.*, 1986a).

Further evidence for the role of specific amino acid sites in spectral tuning has come from the correlation of colour vision variations and deficiencies in humans with the molecular genetics of their L_R/L_G opsin genes (Nathans *et al.*, 1986a; Neitz *et al.*, 1989; Deeb *et al.*, 1992; Winderickx *et al.*, 1992; Neitz *et al.*, 1993; Sanocki *et al.*, 1993; Neitz *et al.*, 1995) and the expression of normal, hybrid and mutated genes (Oprian *et al.*, 1991; Merbs and Nathans, 1992b; Merbs and Nathans, 1992a; Merbs and Nathans, 1993). The codons for the three amino acids primarily involved in spectral tuning are located in exon 3 (site 164) and exon 5 (sites 261 and 269). Thus, in hybrid genes composed of a mixture of L_R and L_G exons, sites 261 and 269 will remain together and the λ_{max} of the visual pigment will be largely determined by whether exon 5 is from an L_G gene or an L_R gene. Hybrid genes that contain the L_R exon 5 are expressed as pigments with λ_{max} between about 545 and 557 nm (as determined from difference spectra of the pigments in solution), whereas pigments expressed from genes containing the L_G exon 5 have λ_{max} between about 530 and 536 nm (Merbs and Nathans, 1992a).

The degree of spectral shift resulting from single amino acid substitutions may vary depending on the specific opsin involved. Asenjo *et al.* (1994) suggest that substitution of serine for alanine at site 164 (180) causes a 2-nm red shift in the L_G cone pigment, but a 7-nm red shift in the L_R cone pigment, whereas Merbs and Nathans (1992b) suggest that the substitution leads to about a 4 – 5 nm red shift in the pigments expressed from all of the hybrid genes. Site 100 (116), although located in the first extracellular loop, is also of interest since substitution of tyrosine for serine appears to cause a small blue-shift in the L_R cone pigment (Asenjo *et al.*, 1994). This implies that sites not in the immediate environment of the chromophore can have indirect effects on λ_{max}. The chloride binding site in these L_R/L_G pigments is another such example.

The same three primary sites, 164, 261 and 269, are also involved in spectral tuning in chloride-sensitive L_R cone pigments in non-mammalian vertebrates. Those pigments with λ_{max} close to 560 nm, P563 in the cave fish, *Astyanax fasciatus* (Yokoyama and Yokoyama, 1990; Yokoyama, 1995), P571 in chicken, *Gallus gallus* (Kuwata *et al.*, 1990), P570 in goldfish, *Carassius auratus* (Tsin *et al.*, 1981; Johnson *et al.*, 1993) and P561 in the chameleon, *Anolis carolinensis* (Provencio *et al.*, 1992; Kawamura and Yokoyama, 1993), are all similar to the primate L_R pigments in having the hydroxyl-bearing polar amino acids at all three sites. Both goldfish and *Anolis* have visual pigments with 3-dehydroretinal as their chromophore and have natural L_R cone pigments with λ_{max} greater than about 620 nm. For purposes of comparison, replacement of 3-dehydroretinal by retinal is estimated to blue-shift the λ_{max} to about 560 – 570 nm (e.g. see Bowmaker, 1990). However, the opsin identified as the L_R opsin from the goldfish, when expressed with retinal, gave a λ_{max} at about 525 nm and not at about 565 nm as expected (Tsin *et al.*, 1981; Johnson *et al.*, 1993). The authors suggest that this putative red cone opsin may encode a pigment that is a polymorphic variant of the goldfish L_R pigment.

The shorter-wavelength pigments in this group, the L_G cone pigment, P533, from the cave fish, *Astyanax fasciatus* (Yokoyama and Yokoyama, 1990; Yokoyama, 1995) and the longer-wave 'rod' pigment, P521, of *Gekko gekko* (Kojima *et al.*, 1992) are similar to the L_G primate cone pigments in having the non-polar amino acids at all three tuning sites. Although the gekko has a pure rod retina, the P521 is biochemically cone-like (Crescitelli, 1977) and phylogenetically can be classed with the L cone pigments (Kojima *et al.*, 1992).

The major factors in determining the spectral tuning of the L phylogenetic group of pigments throughout the vertebrates are now relatively clear. First, the protonated Schiff's base of the chromophore and its counterion, glu113, can yield visual pigments with λ_{max} ranging up to about 520 nm. The peak sensitivity is then further shifted bathochromically by a chloride-binding region that includes his181 and lys184. A further red-shift up to the apparent limits of $retinal_1$-based pigments of about 575 nm can be achieved by incorporating hydroxyl-bearing polar amino acids, ser164, tyr261 and thr269. At least within the primates, spectral shifts between about 530 and 565 nm are largely achieved by combinations of polar and non-polar amino acids at these three sites and presumably similar combinations will be found in the other mammalian L cone pigments that lie within this spectral range (Jacobs, 1993). It would also appear that the evolution of this tuning mechanism has occurred independently within each vertebrate class (Okano *et al.*, 1992; Johnson *et al.*, 1993; Hisatomi *et al.*, 1994; Yokoyama, 1994; Chang *et al.*, 1995). This is perhaps not surprising since non-homologous substitutions at other sites may lead to non-functional opsins that either will not bind retinal, or will be much less efficient in activating transducin (see below).

An interesting additional opsin has recently been sequenced from the pineal of the chicken (Okano *et al.*, 1994; Max *et al.*, 1995) which appears to represent a sixth class of opsins. Phylogenetically, the opsin appears to be at least as ancient as the early division between the chloride-sensitive and chloride-insensitive opsins and has been alternatively grouped with either the L opsins (Okano *et al.*, 1994) or the chloride-insensitive opsin groups (Max *et al.*, 1995). The pigment expressed from pineal opsin has a λ_{max} at about 470 nm (Okano *et al.*, 1994) which would suggest that it is chloride-insensitive. This is supported by the apparent absence of the chloride binding region in extracellular loop 2 which contains glutamate at site 181 and arginine at site 184 (rod opsin numbering) as opposed to his181 and lys184 in L pigments.

13.5 Rod (M_{Rd}) and middle-wave-sensitive (M_C) cone pigments (λ_{max} 460 – 530 nm)

Spectral tuning can now be considered in the two pigment groups that have λ_{max} close to 500 nm: the M_C pigments of many non-mammalian vertebrates and the closely-related rod opsins (M_{Rd}) of all the vertebrate groups. Overall, the sequence homology of these pigments with the L pigments (discussed above) and S pigments is only about 40%. Nevertheless, the question can be asked, are the same three sites involved in the 'fine' tuning of these pigments, or are novel sites exploited? Perhaps the most studied opsin is that of the bovine rod pigment (λ_{max} 498 nm) and the role of almost all of the polar and charged amino acids within this opsin have been investigated by site directed mutagenesis (e.g. Sakmar *et al.*, 1989; Zhukovsky and Oprian, 1989; Nathans, 1990b; Nathans, 1990a; Chan *et al.*, 1992). However, although these studies indicate which residues may have an influence on the λ_{max}, they do not necessarily identify the sites that are utilized in native rod opsins.

We have addressed this question by analysing the rod visual pigments from a unique species flock of cottoid fish from Lake Baikal in Siberia. Lake Baikal is the deepest lake in the world (more than 1600 m) and the remarkably clear water is oxygenated even in the deepest regions where oxygen levels are maintained at about 75 – 80% of the surface levels (Weiss *et al.*, 1991). As a consequence, endemic cottoids are found distributed from the surface to the bottom and offer a singular opportunity for examining within a number of closely-related species, the relationships with depth and changing light environment of visual pigments. The rod visual pigments of these fish exhibit a spectral shift to shorter wavelengths correlated with increasing depth, with λ_{max} shifting from about 516 nm in surface species to 484 nm in abyssal species (Bowmaker *et al.*, 1994).

We have derived the amino acid sequences of the rod opsins from 11 species (Table 2) with λ_{max} locations at about 516, 505, 495, 490 and 484 nm (Hunt *et al.*, 1996). The level of homology in the sequences across these species is very high, with an average greater than 93%. Of the amino acid differences between the opsins, twelve changes involve a non-conserved change or the gain or loss of an hydroxyl group, but only three of these sites face into the interior of the binding pocket and are thought likely to be candidate sites for determining the spectral differences between the λ_{max} cluster points. Site 83 in helix III is an aspartate/

asparagine exchange, site 261 in helix VI is a tyrosine/phenylalanine exchange and site 292 in helix VII is an alanine/serine exchange (Figures 1 and 2).

Since the substitution of tyrosine with phenylalanine at site 261 is already known to cause a 6 – 10 nm blue shift in hybrid primate L_G and L_R cone pigments (Merbs and Nathans, 1993; Asenjo *et al.*, 1994), the presence of tyrosine at this site in *Cottus kessleri* and *Paracottus kneri* (λ_{max} 516 nm) and its replacement in all other species by phenylalanine, identifies the site as a strong candidate for the spectral shift between these two species and those with λ_{max} at 505 nm (Figure 4). Yokoyama *et al.* (1995) report that the replacement by site-directed mutagenesis of tyrosine by phenylalanine at this site in the rod opsin of the epigeneous cavefish *Astyanax fasciatus*, produces a blue-shift from the wild-type λ_{max} of 504 nm to 496 nm, and site-directed mutagenesis of this site in bovine rod opsin produces about a 10-nm red shift when phenylalanine is again replaced by tyrosine (Chan *et al.*, 1992). In addition to its role in vertebrate visual pigments, substitution at the equivalent site to 261 has also been proposed as a tuning site in cephalopod visual pigments where two species of squid, *Alloteuthis subulata* and *Loligo forbesi* have pigments with λ_{max} separated by about 5 nm, the longer-wavelength pigment having serine substituted for phenylalanine (Morris *et al.*, 1993).

The two pelagic sub-littoral species of cottoid from Lake Baikal, *Cottocomephorus grewingki* and *Cottocomephorus inermis*, have rod pigments that differ by about 10 nm; the most likely substitution for this spectral shift is an aspartate/asparagine exchange at site 83, since it faces the interior of the chromophore binding pocket (Figure 2). Again, mutagenesis experiments on bovine rod opsin have shown that substitution of aspartate with asparagine at site 83 leads to a short-wave displacement of about 9 nm (Nathans, 1990b; DeCaluwé *et al.*, 1995).

The remaining supra-abyssal species with λ_{max} at about 490 nm lack the substitution at site 83, but all show an alanine to serine change at site 292, as do

Table 2 The λ_{max} of rod visual pigments and the depth distribution of Baikal cottoids.

Spectral class	λ_{max}	Species	Family	Depth habitat (depth)
1	516 nm	*Cottus kessleri*	Cottidae	Littoral (1– 5 m)
		Paracottus kneri	Cottidae	Littoral (1– 5 m)
2	505 nm	*Procottus jettelesi*	Abyssocottidae	Sub-littoral (1–120 m)
		Cottocomephorus grewingki	Cottidae	Sub-littoral, pelagic (1– 300 m)
3	495 nm	*Cottocomephorus inermis*	Cottidae	Supra-abyssal, pelagic (50 – 450 m)
4	490 nm	*Batrachocottus multiradiatus*	Cottidae	Supra-abyssal (100– 500 m)
		Batrachocottus nicolskii	Cottidae	Abyssal (300 –1000 m)
		Limnocottus bergianus	Abyssocottidae	Abyssal (100 –1000 m)
		Limnocottus pallidus	Abyssocottidae	Abyssal (100 –1000 m)
5	484 nm	*Abyssocottus korotneffi*	Abyssocottidae	Abyssal (400 –1500 m)
		Cottinella boulengeri	Abyssocottidae	Abyssal (400 –1500 m)

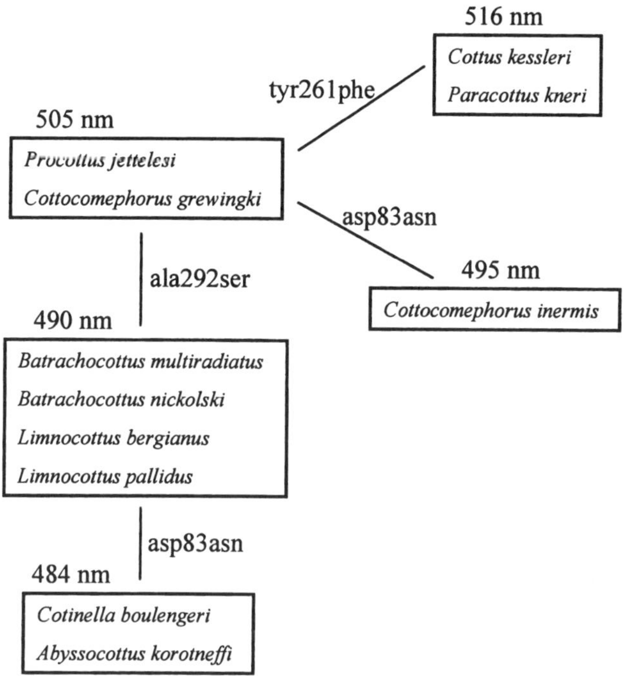

Figure 4 Interrelationships between the five spectral classes of rod opsins in Baikal cottoids. The amino acid substitutions that determine the spectral shifts between classes are shown

the abyssal species with λ_{max} at 484 nm. It would appear therefore that the 15-nm blue shift from 505 to 490 nm is achieved by this substitution, and that the final 8-nm shift to 482 nm in *Abyssocottus korotneffi* and *Cottinella boulengeri* is again the result of an aspartate/asparagine replacement at site 83 (Figure 4). Site 292 has also been proposed as a tuning site in primate S pigments (Hunt *et al.*, 1995a,b; Bowmaker *et al.*, 1996) where the identical change of alanine to serine is associated with a blue shift of around 11 nm. Evidence that substitution at this site can result in a spectral shift comes from mutagenesis experiments on bovine rod opsin, where the replacement of alanine by aspartate produces a 10-nm blue shift (Nakayama and Khorana, 1991). The tuning substitutions in the Baikal cottoid rod opsins are summarised in Figure 4. Sites 83 and 292 also appear to be involved in the shifts to shorter wavelengths of rod opsins of deep-sea fish (Hope *et al.*, 1997).

No similar data are available for spectral tuning within the M_C visual pigments with λ_{max} close to 500 nm since few such opsins have been sequenced. However, since these pigments are closely-related to the M_{Rd} opsins, it will be interesting to see whether a similar set of residues is used. Two M_C opsins from goldfish have been expressed with retinal$_1$ that yield photosensitive pigments with λ_{max} at about 505 and 511 nm (Johnson *et al.*, 1993). A comparison of the amino acid sequences of these two opsins suggests that a candidate site for the 6-nm spectral difference is site 217 (233) where threonine is replaced by alanine in the longer-wave pigment. This is similar to the situation in primate L_R and L_G pigments (Ibbotson *et al.*, 1992; Dulai *et al.*, 1994) where serine is replaced by alanine in the L_R pigments, though in primates the spectral effect is small (see above).

Another animal group that can provide valuable information on spectral tuning of visual pigments in this spectral range is the birds. We have recently re-examined by microspectrophotometry the λ_{max} of M_{Rd} and M_C pigments in a number of avian species and shown that in all cases, the M_{Rd} and the M_C opsins have very similar, if not identical, λ_{max} values, e.g. in the chicken, both the rod and green-sensitive cone visual pigments have λ_{max} at 506 nm (Bowmaker *et al.*, 1997). As might be predicted, the amino acid sites that have already been shown to affect spectral tuning in other species are occupied by the same amino acids in the two classes of pigment, except for site 122 (Okano *et al.*, 1992). In all vertebrate M_{Rd} opsins sequenced to date, site 122 is occupied by glutamate, but in the avian M_C pigment, it is replaced by glutamine. Site directed mutagenesis at this site (E122Q) in bovine M_{Rd} opsin produces a 20-nm blue shift, but the spectral similarity of the two pigments in the avian system suggests that site 122 is not significant in spectral tuning. This interpretation is perhaps supported by the sequence data available for the same classes of pigments in teleosts, where both the M_{Rd} and M_C opsins have glutamate at site 122 (Yokoyama and Yokoyama, 1990; Johnson *et al.*, 1993; Yokoyama *et al.*, 1995).

13.6 Short-wave-sensitive (S_B and S_V) cone pigments (λ_{max} 430 – 470 nm and 350 – 440 nm)

The opsins of these short-wave cone pigments are significantly different from the other classes of opsins, showing only about 40% homology with the M and L opsin groups. Since the displacement of λ_{max} to wavelengths below about 440 nm cannot be attributed to deprotonation of the Schiff's base, the blue shift is most probably due to amino acid substitutions close to either lys296 in helix VII or glu113 in helix III. Nevertheless, it is not immediately apparent from sequence comparisons which specific amino acids are responsible for the blue-shift from 500 nm. The reason for this is perhaps that substitutions at a number of amino acid sites may be responsible for the blue-shifts and too few sequences are available at present to allow a clear identification of these sites by conservation across species.

From a comparison of a number of published opsin sequences, Chang *et al.* (1995) suggested that four amino acid residues may be involved in the blue shift; sites 289, 292 and 307 all close to the lys296 in helix VII and site 124, close to glu113 in helix III (Figures 1 and 2). In general these sites are occupied by non-polar residues in long-wave pigments and by polar residues in short-wave pigments (Table 3). However, their comparisons do not include the sequences for bovine and murine S opsins (Chiu *et al.*, 1994), some additional primate S_V opsins (Hunt *et al.*, 1995a) and the M_{Rd} opsins from the cottoids of Lake Baikal (Hunt *et al.*, 1996), all of which have amino acids at certain sites that contradict some of their proposals. Nevertheless, in all vertebrate opsins except the S_V cone class, site 307 has non-polar residues of valine (cone opsins) or isoleucine (rod opsins), but in the S_V class the valine is replaced by cysteine. Similarly, site 289 is occupied by polar residues threonine and serine in the S_B and S_V cone classes, but by non-polar alanine in the L pigments. This simple division is not however straightforward since M_{Rd} opsins have threonine at 289 whereas the M_C opsins with similar λ_{max} have either alanine or serine (Table 3).

Also contrary to the suggestion of Chang *et al.* (1995) are a number of observations that indicate that site 292 does not solely contribute to the differentiation of the two S opsin classes from the longer-wave opsins. They suggest that all opsins other than the S classes have non-polar alanine at site 292 whereas the S_B and S_V classes have serine. However, as detailed above for the cottoid fish of Lake Baikal, site 292 can be occupied by either serine or alanine in M_{Rd} opsins and it is proposed that the presence of serine at this site results in a blue-shift of about 16 nm. Furthermore, human S_V opsin also has alanine at site 292, and it has been proposed that serine/alanine substitutions at this site may act to fine tune the primate S_V cone pigments (Hunt*et al.*, 1995b; Bowmaker *et al.*, 1996).

Table 3: Summary of possible tuning sites within opsins. Sites include only those within the transmembrane helices. Amino acids are represented by their single letter code. Significant substitutions are indicated by * and the end column summarises the possible tuning function.

	position in helix	amino acid site	S_V	S_B	M_{Rd}	M_C	L_G	L_R	Tuning function
helix II	14	83	G	N	D/N*	D/G	D	D	within M_{Rd} (7 –10 nm)
	21	90	S*	G	G	G	A	A	S_V from other opsins
helix III	3	113	E	E	E	E	E	E	counter ion
	14	124	T*	T/S/G*	A/G	A	A/G	A/G	S_V & S_B from other opsins
helix IV	14	164	G	G/A	A	A	A	S*	L_G from L_R (2 – 7 nm)
	17	167	V*	A*	C	C	W	W	S_V & S_B from other opsins
helix V	3	204	T*	V	V	V	M	M	S_V from other opsins
	16	217	S	A/G	I/V	V/A/T*	S	A/G*	L_G from L_R & within M_C (1 – 6 nm)
	23	224	T/S/A*	G	G	G	L/I	L/I	S_V from other opsins
helix VI	12	261	F	F	F/Y*	F F	Y*	LG	from L_R & within M_{Rd} (6 – 11 nm)
	16	265	Y*	W	W	W	W	W	S_V from other opsins
	20	269	A	A/T	A	A	A	T*	L_G from L_R (15 nm)
helix VII	4	289	T	S/T	T	S/A	A	A	S_V & S_B from L
	7	292	S/A*	S	S/A*	A	A	A	within M_{Rd} & S_V (10 – 16 nm)
	11	296	K	K	K	K	K	K	retinal binding site
	22	307	C*	V	I	V	V	V	S_V from other opsins

The fourth important tuning site proposed by Chang *et al.* (1995) is site 124 that lies close to the Schiff's base counterion in helix III. This site is again non-polar (alanine or glycine) in all opsins except the two S classes where it is replaced in the S_V class by polar threonine and in the S_B class by either threonine or serine. The one exception is goldfish S_B that has glycine at 124 (Table 3).

Another study of S, M and L opsins (Yokoyama, 1995) identifies at least 50 substitutions that could potentially be involved in the blue shift that distinguishes the S classes from the M and L opsins, but no attempt is made to further subdivide these sites. However, if the sites are restricted to those exhibiting polar/non-polar exchanges within the membrane helices, then only five additional sites have to be added to the list of candidates: 90 in helix II, 167 in helix IV, 204 and 224 in helix V and 265 in helix VI. Site 167, located about one turn of the helix above site 164 and therefore in close proximity to the imine ring of retinal, is occupied by tryptophan in L opsins and by cysteine in M_C and M_{Rd} opsins, whereas in the two S opsin classes the site is occupied by alanine in the S_B opsins and valine in the S_V opsins. However, this residue is some distance from the Schiff's base (Figure 2) and so perhaps unlikely to cause dramatic spectral shifts. The remaining four sites are potential candidates for differentiating the S_V class of opsins. Site 90, which is located in helix II about two turns of the helix above site 83 and close to both retinal and the Schiff's base, contains polar serine in the S_V opsins, but glycine (S_B, M_C and M_{Rd} groups) or alanine (L group) in all other vertebrate opsins. Site 265, located in helix VI between sites 261 and 269 and again in close proximity to retinal, has a polar tyrosine in the S_V class of opsins, but tryptophan in all the other opsins. Site directed mutagenesis of tyrosine for tryptophan at this site in bovine rhodopsin leads to a blue shift of about 15 nm (Nakayama and Khorana, 1991) and Hisatomi *et al.* (1994) have suggested that this site may contribute to the blue shift in the S_V opsins. Finally, the two sites in helix V are perhaps the least likely candidates. Site 204, although facing the chromophore pocket, is located near the extracellular end of the helix. This site is occupied by polar threonine in the S_V opsins, but by valine (S_B, M_C and M_{Rd} groups) or methionine (L group) in the other opsins. Perhaps the weakest candidate is site 224 which is located on the outer face of the helix of transmembrane region V facing away from the chromophore pocket. This site is occupied by polar threonine or serine in all S_V opsins, but by glycine in the S_B, M_C and M_{Rd} groups and by leucine or isoleucine in the L opsins. The residues at this site are not so clear cut however, since the S_V opsin of the marmoset has alanine and not serine at site 224 (Hunt *et al.*, 1995a). These substitutions are summarised in Table 3 and their locations in opsin indicated in Figures 1 and 2.

In addition to the difficulties in identifying tuning sites that are involved in separating the L, M and S pigments, is the more specific question of determining which sites are responsible for the 50 – 70 nm difference found within the S_V family between the longer-wave violet-sensitive pigments with λ_{max} to close to 420 nm and the true UVS pigments with λ_{max} around 360 nm. Sequence comparisons between goldfish, avian and mammalian UV opsins with avian and mammalian violet-sensitive opsins are not helpful, perhaps because phylogenetic analysis suggests that the division of the S_V family has occurred separately within the different vertebrate classes (Wilkie *et al.*, 1998). If the analysis is limited solely to avian opsins, then it is possible to identify potential sites responsible for the spectral difference. In a comparison of the opsin sequences of chicken violet (Okano *et al.*, 1992) with the UVS opsins of budgerigar (Wilkie *et al.*, 1998) and canary (Das *et al.*, in prep), five candidate sites have been proposed where non-conserved substitutions occur, A86S, T93V, A118T, T119H and S298A (UVS/chicken violet respectively). Of these, sites 86 and 93 are the more likely contenders since they are situated on the inner surface of helix II in the vicinity of the glutamate counterion (glu113) to the Schiff's base.

13.7 Conclusions

From the above, it is evident that only about 18 amino acid sites have been clearly implicated in the spectral tuning of natural visual pigments and, of these, only about 6 produce substantial spectral shifts (Table 3). One of the reasons for this is that the α-helical structure of the transmembrane regions of the molecule limits the number of amino acid sites that can interact directly with the chromophore. Tuning substitutions appear to involve either the exchange of a polar/non-polar or charged/non-charged amino acid. Changes of this type, as well as more conservative changes, at sites throughout human M_{Rd} opsin are known to cause autosomal dominant retinitis pigmentosa (Al-Maghtheh *et al.*, 1993), indicating that such changes are only tolerated at a few sites. In addition, many of the single amino acid substitutions produced in bovine M_{Rd} opsin by site-directed mutagenesis show reduced regeneration with 11-*cis*-retinal and/or activation of transducin (Nakayama and Khorana, 1991).

This limitation on the sites that can cause spectral shifts is consistent with the presence of convergent evolution of tuning sites in a number of different opsins. Sites 164, 261 and 269 have all been separately used in primates and fish in the evolution of the L_R and L_G opsins (Nathans et al, 1986a; Neitz *et al.*, 1991; Yokoyama and Yokoyama, 1990; Williams *et al.*, 1992; Hunt *et al.*, 1993), and site 261 is also used in the M_{Rd} opsins of fish (Hunt *et al.*, 1996) and in the visual pigment of cephalopods (Morris *et al.*, 1993). Similarly, site 292 has been proposed as a tuning site in primate S_V (Hunt *et al.*, 1995b) and fish M_{Rd} opsins (Hunt *et al.*, 1996; Hope *et al.*, 1997).

A long-running debate in primate colour vision is the origin of trichromacy in Old and New World monkeys (see Chapter 20). As described above, the three cone pigments in Old World primates are each coded by a separate gene, the S_V opsin gene on chromosome 7 (Fitzgibbon *et al.*, 1994) and two members of the L group (L_G and L_R) on the X chromosome (Nathans *et al.*, 1986b). In contrast, most New World primates have only two genes, an autosomal S_V opsin gene, as in Old World species, and a polymorphic L gene on the X chromosome. Trichromacy in this case is restricted to those females that inherit different allelic forms of this X-linked gene (Tovée *et al.*, 1992). With the demonstration that the spectral shifts in the L pigments of both primate groups involve sites 164, 261 and 269 (Neitz *et al.*, 1991; Williams *et al.*, 1992; Jacobs *et al.*, 1993), the separate evolution of both systems seemed less likely than the origin of the Old World system from a pre-existing polymorphism; an unequal crossover that placed two different forms of the polymorphic gene on to the same chromosome would generate the Old World trichromacy. This explanation would predict that the polymorphism seen in present-day New World monkeys pre-dates the separation of the New and Old World lineages around 40 Mya (Hunt *et al.*, 1995a). This interpretation has been questioned however by a recent report by Shyue *et al.* (1995) that the divergence of intron 4 of the three allelic forms of the L gene in two New World primates, the squirrel monkey from the Cebidae, and the marmoset from the Callitrichidae, indicates a much more recent evolution of the L gene polymorphism, perhaps as recently as 5 – 10 Mya. The evolution of trichromacy within primates has become even more complex with the recent finding that at least two species of New World monkeys, the howler monkeys *Alouatta caraya* and *A. seniculus*, have a trichromatic system similar to that of Old World monkeys with two separate genes for the L_G and L_R pigments on the X chromosome (Jacobs *et al.*, 1996). It would appear therefore that the trichromatic colour systems of Old World and New World primates evolved quite separately, with the trichromacy in the howler monkey arising more recently by gene duplication from the pre-existing New World polymorphism (Hunt *et al.*, 1998). The use of sites 164, 261 and 269 in the Old World and New World systems is another example of convergent evolution. Since it is accepted that primate trichromacy is an adaptation to foraging for yellow and red fruits against a green foliage, it is perhaps not surprising that substitution at these three sites has been selected to achieve a similar spectral shift between pigments in the three groups of primate.

References

Al-Maghtheh, M., Gregory, C., Inglehearn, C., Hardcastle, A. and Bhattacharya, S. (1993) Rhodopsin mutations in autosomal dominant retinitis pigmentosa. *Human Mutation*, 2, 249 – 255.

Applebury, M. L. and Hargrave, P. A. (1986) Molecular biology of visual pigments. *Vision Research*, 26, 1881–1885.

Asenjo, A. B., Rim, J. and Oprian, D. D. (1994) Molecular determinants of human red/green color discrimination. *Neuron*, 12, 1131–1138.

Baldwin, J. M. (1993) The probable arrangement of the helices in G protein-coupled receptors. *European Molecular Biology Organisation Journal*, 12, 1693 –1703.

Barlow, H. B. (1982) What causes trichromacy? A theoretical analysis using comb-filtered spectra. *Vision Research*, 22, 635-643.

Bellingham, J., Wilkie, S. E., Morris, A. G., Bowmaker, J. K. and Hunt, D. M. (1997) Characterisation of the ultraviolet-sensitive opsin gene in the honey bee, *Apis mellifera. European Journal of Biochemistry*, 243, 775 – 781.

Bowmaker, J. K. (1983) Trichromatic colour vision: why only three receptor channels? *Trends in Neuro-*

science, 6, 41– 43.

Bowmaker, J. K. (1990) Visual pigments of fishes, in *The Visual System of Fish* (eds R. H. Douglas and M. B. A. Djamgoz), Chapman and Hall, London, pp. 81–107.

Bowmaker, J. K. (1991a) Evolution of visual pigments and photoreceptors, in *Vision and Visual Dysfunction, 2, Evolution of the Eye and Visual System* (eds R.L. Gregory and J.R. Cronly-Dillon), Macmillan, London, pp. 63 – 81.

Bowmaker, J. K. (1991b) Visual pigments and colour vision in primates, in *From Pigments to Perception* (eds A. Valberg and B.B. Lee), Plenum, New York, pp. 1– 9.

Bowmaker, J. K., Govardovskii, V. I., Shukolyukov, S. A., Zueva, L. V., Hunt, D. M., Sideleva, V. G. and Smirnova, O. G. (1994) Visual pigments and the photic environment: the cottoid fish of lake Baikal. *Vision Research*, 34, 591– 605.

Bowmaker, J. K., Heath, L. A., Wilkie, S. E. and Hunt, D. M. (1997) Visual pigments and oil droplets from six classes of photoreceptor in the retinas of birds. *Vision Research,* 37, 2183 – 2194.

Bowmaker, J. K., Hunt, D. M. and Mollon, J. D. (1997) Primate visual pigments: their spectral tuning and evolution, in *John Dalton's Colour Vision Legacy* (eds C. M. Dickinson, I. J. Murray and D. Carden), Taylor and Francis, London, pp. 37– 46.

Bownds, D. (1967) Site of attachment of retinal in rhodopsin. *Nature*, 216, 1178 –1181.

Chan, T., Lee, M. and Sakmar, T. P. (1992) Introduction of hydroxyl-bearing amino acids causes bathochromic spectral shifts in rhodopsin. Amino acid substitutions responsible for red-green color pigment spectral tuning. *Journal of Biological Chemistry*, 267, 9478 – 9480.

Chang, B. S. W., Crandall, K. A., Carulli, J. P. and Hartl, D. L. (1995) Opsin phylogeny and evolution: a model for blue shifts in wavelength regulation. *Molecular Phylogenetics and Evolution*, 4, 33 – 43.

Chiu, M. I., Zack, D. J., Wang, Y. and Nathans, J. (1994) Murine and bovine blue cone pigment genes: Cloning and characterization of two new members of the S family of visual pigments. *Genomics*, 21, 440 – 443.

Cowman, A. F., Zuker, C. S. and Rubin, G. M. (1986) An opsin gene expressed in only one photoreceptor cell type of the Drosophila eye. *Cell,* 44, 705 –710.

Crescitelli, F. (1977) The visual pigments of geckos and other vertebrates: an essay in comparative biology, in *Handbook of Sensory Physiology, VII/5, The Visual System in Vertebrates* (ed F. Crescitelli), Springer, Berlin, pp. 391– 449.

DeCaluwé, G. L. J., Bovee-Geurts, P. H. M., Rath, P., Rothschild, K. J. and de Grip, W. J. (1995) Effect of carboxyl mutations on functional properties of bovine rhodopsin. *Biophysical Chemistry*, 56, 79 – 87.

Deeb, S.S., Jorgensen, A.L., Battisti, L., Iwasaki, L. and Motulsky, A.G. (1994) Sequence divergence of the red and green visual pigments in great apes and humans. *Proceedings of the National Academy of Sciences*, 91, 7262 –7266.

Deeb, S. S., Lindsey, D. T., Hibiya, Y., Sanocki, E., Winderickx, J., Teller, D. Y. and Motulsky, A. G. (1992) Genotype-phenotype relationships in human red/green color-vision defects: molecular and psychophysical studies. *American Journal of Human Genetics*, 51, 687–700.

Dulai, K. S., Bowmaker, J. K., Mollon, J. D. and Hunt, D. M. (1994) Sequence divergence, polymorphism and evolution of the middle-wave and long-wave visual pigment genes of Great apes and Old World monkeys. *Vision Research*, 34, 2483 – 2491.

Fager, L. Y. and Fager, R. S. (1979) Halide control of colour of the chicken cone pigment iodopsin. *Experimental Eye Research*, 29, 401– 408.

Fitzgibbon, J., Appukuttan B., Gayther, S., Wells, D., Delhanty, J. and Hunt, D. M. (1994) Localisation of the human blue cone pigment gene to chromosome band 7q31.3-32. *Human Genetics*, 93, 79 – 80.

Franke, R. R., Sakmar, T. P., Graham, R. M. and Khorana, H. G. (1992) Structure and function in rhodopsin: studies of the interaction between the rhodopsin cytoplasmic domain and transducin. *Journal of Biological Chemistry*, 267, 14767–14774.

Fryxell, K. L. and Meyerowitz, E. M. (1987) An opsin gene that is expressed only in the R7 photoreceptor cell of *Drosophila. European Molecular Biology Organisation Journal*, 6, 443 – 451.

Hargrave, P. A. (1982) Rhodopsin chemistry, structure and topology. *Progress in Retinal Research*, 1, 1– 51.

Hargrave, P. A., McDowell, J. H., Feldmann, R. J., Atkinson, P. H., Rao, J. K. M. and Argos, P. (1984) Rhodopsin's protein and carbohydrate structure: selected aspects. *Vision Research*, 24, 1487–1499.

Hashimoto, S., Takeuchi, H., Nakagawa, M. and Tsuda, M. (1996) Ultraviolet resonance Raman evidence for the absence of tyrosinate in octopus rhodopsin and the participation of trp residues in the transition to acid metarhodopsin. *Febs Letters*, 398, 239 – 242.

Hisatomi, O., Kayada, S., Aoki, Y., Iwasa, T. and Tokunaga, F. (1994) Phylogenetic relationships among vertebrate visual pigments. *Vision Research*, 34, 3097– 3102.

Hisatomi, O., Satoh, T., Barthel, L. K., Stenkamp, D. L., Raymond, P. A. and Tokunaga, F. (1996) Molecular-cloning and characterization of the putative ultraviolet-sensitive visual pigment of goldfish. *Vision Research*, 36, 933 – 939.

Hope, A. J., Partridge, J. C., Dulai, K. S. and Hunt, D. M. (1997) Mechanisms of wavelength tuning in the rod opsins of deep-sea fishes. *Proceedings of the Royal Society of London B*, 264, 155 –163.

Hunt, D. M., Cowing, J. A., Patel, R., Appukuttan, B., Bowmaker, J. K. and Mollon, J. D. (1995a) Sequence and evolution of the blue cone pigment gene in Old and New World primates. *Genomics*, 27, 535 – 538.

Hunt, D. M., Cowing, J. A., Patel, R., Appukuttan, B., Bowmaker, J. K. and Mollon, J. D. (1995b) Sequence and spectral tuning of blue cone photopigments in Old and New World primates. *Investigative Ophthalmology & Visual Science*, 36, 889.

Hunt, D. M., Dulai, K. S., Cowing, J. A., Julliot, C., Mollon, J. D., Bowmaker, J. K., Li, W.-H. and Hewett-Emmett, D. (1998) Molecular evolution of trichromacy in primates. *Vision Research*, 38, 3299 – 3306.

Hunt, D. M., Fitzgibbon, J., Slobodyanyuk, S. and Bowmaker, J. K. (1996) Spectral tuning and molecular evolution of rod visual pigments in the species flock of cottoid fish in Lake Baikal. *Vision Research*, 36, 1217–1224.

Hunt, D. M., Williams, A. J., Bowmaker, J. K. and Mollon, J. D. (1993) Structure and evolution of the polymorphic photopigment gene of the marmoset. *Vision Research*, 33, 147–154.

Ibbotson, R. E., Hunt, D. M., Bowmaker, J. K. and Mollon, J. D. (1992) Sequence divergence and copy number of the middle-wave and long-wave photopigment genes in Old World monkeys. *Proceedings of the Royal Society of London B*, 247, 145 –154.

Jacobs, G. H. (1993) The distribution and nature of colour vision among the mammals. *Biological Reviews*, 68, 413 – 471.

Jacobs, G. H. and Neitz, J. (1987a) Inheritance of color vision in a New World monkey (*Saimiri sciureus*). *Proceedings of the National Academy of Sciences*, 84, 2545 – 2549.

Jacobs, G. H. and Neitz, J. (1987b) Polymorphism of the middle wavelength cone in two species of South American monkey: *Cebus apella* and *Callicebus moloch*. *Vision Research*, 27, 1263 –1268.

Jacobs, G. H., Neitz, J. and Crognale, M. A. (1987) Color vision polymorphism and its photopigment basis in a callitrichid monkey (*Saguinus fusicollis*). *Vision Research*, 27, 2089 – 2100.

Jacobs, G. H., Neitz, J. and Deegan II, J. F. (1991) Retinal receptors in rodents maximally sensitive to ultraviolet light. *Nature*, 353, 655 – 656.

Jacobs, G. H., Neitz, J. and Neitz, M. (1993) Genetic basis of polymorphism in the color vision of platyrrhine monkeys. *Vision Research*, 33, 269 – 274.

Jacobs, G. H., Neitz, M. S., Deegan II, J. F. and Neitz, J. (1996) Trichromatic colour vision in New World monkeys. *Nature*, 382, 156 –158

Johnson, R. L., Grant, K. B., Zankel, T. C., Boehm, M. F., Merbs, S. L., Nathans, J. and Nakanishi, K. (1993) Cloning and expression of goldfish opsin sequences. *Biochemistry*, 32, 208 – 214.

Karnik, S. S., Sakmar, T. P., Chen, H. B. and Khorana, H. G. (1988) Cysteine residue-110 and residue-187 are essential for the formation of correct structure in bovine rhodopsin. *Proceedings of The National Academy of Sciences*, 85, 8459 – 8463.

Kawamura, S. and Yokoyama, S. (1993) Molecular characterization of the red visual pigment gene of the American chameleon (*Anolis carolinensis*). *Febs Letters*, 323, 247– 251.

Kleinschmidt, J. and Hárosi, F. I. (1992) Anion sensitivity and spectral tuning of cone visual pigments *in situ*. *Proceedings of the National Academy of Sciences*, 89, 9181– 9185.

Knowles, A. (1976) The effects of chloride ions upon chicken visual pigments. *Biochemical and Biophysical Research Communications*, 73, 56 – 62.

Knowles, A. (1980) The chloride effect in chicken red cone receptors. *Vision Research*, 20, 475 – 483.

Knowles, A. and Dartnall, H. J. A. (1977) The Photobiology of Vision, in *The Eye*, *2B*, (ed H. Davson), Academic Press, New York, pp. 1– 689.

Knox, B. E. (1995) *Xenopus laevis* violet cone opsin mRNA, complete coding sequence. GenBank, Accession U23463.

Kojima, D., Okano, T., Fukada, Y., Shichida, Y., Yoshizawa, T. and Ebrey, T .G. (1992) Cone visual pigments are present in gecko rod cells. *Proceedings of the National Academy of Sciences*, 89, 6841– 6845.

König, B., Arendt, A., McDowell, J. H., Kahlert, M., Hargrave, P. A. and Hofmann, K. P. (1989) 3 cytoplasmic loops of rhodopsin interact with transducin. *Proceedings of The National Academy of Sciences*, 86, 6878 – 6882.

Kuwata, O., Imamoto, Y., Okano, T., Kokame, K., Kojima, D., Matsumoto, H., Morodome, A., Fukada, Y., Shichida, Y., Yasuda, K., Shimura, Y. and Yoshizawa, T. (1990) The primary structure of iodopsin, a chicken red-sensitive cone pigment. *Febs Letters*, 272, 128 –132.

Max, M., McKinnon, P. J., Seidenman, K. J., Barrett, R. K., Applebury, M. L., Takahashi, J. S. and Margolskee, R. F. (1995) Pineal opsin: a nonvisual opsin expressed in chick pineal. *Science,* 267, 1502 –1506.

Merbs, S. L. and Nathans, J. (1992a) Absorption spectra of human cone pigments. *Nature*, 356, 433 – 435.

Merbs, S. L. and Nathans, J. (1992b) Absorption spectra of the hybrid pigments responsible for anomalous color vision. *Science*, 258, 464 – 466.

Merbs, S. L. and Nathans, J. (1993) Role of hydroxyl-bearing amino acids in differentially tuning the absorption spectra of the human red and green cone pigments. *Photochemistry and Photobiology*, 58, 706 –710.

Mollon, J. D. (1991) The uses and evolutionary origins of primate colour vision, in *Vision and Visual Dysfunction, 2, Evolution of the Eye and Visual System* (eds R. L. Gregory and J. R. Cronly-Dillon), Macmillan, London, pp. 306 – 319.

Mollon, J. D., Bowmaker, J. K. and Jacobs, G. H. (1984) Variations of colour vision in a New World Primate can be explained by polymorphism of retinal photopigments. *Proceedings of the Royal Society of London B,* 222, 373 – 399.

Montell, C., Jones, K., Zuker, C. S. and Rubin, G. A. (1987) A second opsin gene expressed in the ultraviolet-sensitive R7 photoreceptor cells of *Drosophila melanogaster. Journal of Neuroscience*, 7, 1558 –1566.

Morris, A., Bowmaker, J. K. and Hunt, D. M. (1993) The molecular basis of a spectral shift in the rhodopsins of two species of squid from different photic environments. *Proceedings of the Royal Society of London B*, 254, 233 – 240.

Nakayama, T. A. and Khorana, H. G. (1991) Mapping of the amino acids in membrane-embedded helices that interact with the retinal chromophore in bovine rhodopsin. *Journal of Biological Chemistry*, 266, 4269 – 4275.

Nathans, J. (1990a) Determinations of visual pigment absorbance: identification of the retinylidene Schiff's base counterion in bovine rhodopsin. *Biochemistry*, 29, 9746 – 9752.

Nathans, J. (1990b) Determinations of visual pigment absorbance: role of charged amino acids in the putative transmembrane segments. *Biochemistry*, 29, 937– 942.

Nathans, J., Piantanida, T. P., Eddy, R. L., Shows, T. B. and Hogness, D. S. (1986a) Molecular genetics of inherited variations in human color vision. *Science*, 232, 203 – 210.

Nathans, J., Thomas, D. and Hogness, D. S. (1986b) Molecular genetics of human color vision: the genes encoding blue, green and red pigments. *Science*, 232, 193 – 203.

Neitz, J., Neitz, M. and Jacobs, G. H. (1989) Analysis of fusion gene and encoded photopigment of colour-blind humans. *Nature*, 342, 679 – 682.

Neitz, J., Neitz, M. and Jacobs, G. H. (1993) More than three different cone pigments among people with normal color vision. *Vision Research*, 33, 117–122.

Neitz, M., Neitz, J. and Jacobs, G. H. (1991) Spectral tuning of pigments underlying red-green color vision. *Science*, 252, 971– 974.

Neitz, M., Neitz, J. and Jacobs, G. H. (1995) Genetic basis of photopigment variations in human dichromats. *Vision Research*, 35, 2095 – 2103.

North, R. A. (1989) Neurotransmitters and their receptors: from the clone to the clinic. *Seminars in the Neurosciences*, 1, 81– 90.

Novitsky, I. Y., Zak, P. P. and Ostrovsky, M. A. (1989) The effects of anions on absorption spectrum of the longwavelength retinal-containing pigment iodopsin in native frog cones (a microspectrophotometric study). *Bioorganic Chemistry*, 15, 1037–1043, (in Russian).

Okano, T., Kojima, D., Fukada, Y., Shichida, Y. and Yoshizawa, T. (1992) Primary structures of chicken cone visual pigments: vertebrate rhodopsins have evolved out of cone visual pigments. *Proceedings of the National Academy of Sciences*, 89, 5932 – 5936.

Okano, T., Yoshizawa, T. and Fukada, Y. (1994) Pinopsin is a chicken pineal photoreceptive molecule. *Nature*, 372, 94 – 97.

Oprian, D. D., Asenjo, A. B., Lee, N. and Pelletier, S. L. (1991) Design, chemical synthesis, and expression of genes for the three human color vision pigments. *Biochemistry*, 30, 11367–11372.

Ovchinnikov, Y. A., Abdulaev, N. G. and Bogachuk, A. S. (1988) Two adjacent cysteine residues in the C-terminal cytoplasmic fragment of bovine rhodopsin are palmitylated. *Febs Letters*, 230, 1– 5.

Provencio, I., Loew, E. R. and Foster, R. G. (1992) Vitamin A2-based visual pigments in fully terrestrial vertebrates. *Vision Research*, 32, 2201– 2208.

Sakmar, T. P., Franke, R. R. and Khorana, G. H. (1989) Glutamic acid-113 serves as the retinylidene Schiff base counterion in bovine rhodopsin. *Proceedings of the National Academy of Science*, 86, 8309 – 8313.

Sanocki, E., Lindsey, D. T., Winderickx, J., Teller, D. Y., Deeb, S. S. and Motulsky, A. G. (1993) Serine/alanine amino acid polymorphism of the L and M cone pigments: effects on Rayleigh matches among

deuteranopes, protanopes and color normal observers. *Vision Research*, 33, 2139 – 2152.
Schertler, G. X., Villa, C. and Henderson, R. (1993) Projection structure of rhodopsin. *Nature*, 362, 770 –772.
Shyue, S-K., Hewett-Emmett, D., Sperling, H. G., Hunt, D. M., Bowmaker, J. K., Mollon, J. D., and Li, W-H. (1995) Adaptive Evolution of Color Vision Genes in Higher Primates. *Science* 269, 1265 –1267.
Tovée, M. J., Bowmaker, J. K. and Mollon, J. D. (1992) The relationship between cone pigments and behavioural sensitivity in a New World monkey (*Callithrix jacchus jacchus*). *Vision Research*, 32, 867– 878.
Towner, P., Harris, P., Wolstenholme, A. J., Hill, C., Worm, K. and Gärtner, W. (1997) Primary structure of locust opsins: a speculative model which may account for ultraviolet wavelength light detection. *Vision Research*, 37, 495 – 503.
Travis, D. S., Bowmaker, J. K. and Mollon, J. D. (1988) Polymorphism of visual pigments in a callitrichid monkey. *Vision Research*, 28, 481– 490.
Tsin, A. T. C., Liebman, P. A., Beatty, D. D. and Drzymala, R. (1981) Rod and cone visual pigments in the goldfish. *Vision Research*, 21, 943 – 946.
Wang, J. K., McDowell, J. H. and Hargrave, P. A. (1980) Site of attachment of 11-*cis* retinal in bovine rhodopsin. *Biochemistry*, 19, 5111– 5117.
Wang, Z., Asenjo, A. B. and Oprian, D. D. (1993) Identification of the Cl^--binding site in the human red and green color vision pigments. *Biochemistry*, 32, 2125 – 2130.
Weiss, R. E., Carmack, E. C. and Koropalov, V. H. (1991) Deep water renewal and biological productivity in Lake Baikal. *Nature,* 349, 665 – 669.
Wilkie, S. E., Vissers, P. M. A. M., Das, D., DeGrip, W. J., Bowmaker, J. K. and Hunt, D. M. (1998) The molecular basis for UV vision in birds: spectral characteristics, cDNA sequence and retinal localization of the UV-sensitive visual pigment of the budgerigar (*Melopsittacus undulatus*). *Biochemical Journal*, 330, 541– 547.
Williams, A. J., Hunt, D. M., Bowmaker, J. K. and Mollon, J. D. (1992) The polymorphic photopigments of the marmoset: spectral tuning and genetic basis. *European Molecular Biology Organisation Journal*, 11, 2039 – 2045.
Winderickx, J., Lindsey, D. T., Sanocki, E., Teller, D. Y., Motulsky, A. G. and Deeb, S. S. (1992) Polymorphism in red photopigment underlies variation in colour matching. *Nature*, 356, 431– 433.
Yokoyama, R., Knox, B. E. and Yokoyama, S. (1995) Rhodopsin from the fish, *Astyanax:* role of tyrosine-261 in the red shift. *Investigative Ophthalmology & Visual Science*, 36, 939 – 945.
Yokoyama, R. and Yokoyama, S. (1990) Convergent evolution of the red- and green-like visual pigment genes in fish, *Astyanax fasciatus*, and human. *Proceedings of the National Academy of Sciences*, 87, 9315 – 9318.
Yokoyama, S. (1994) Gene duplications and evolution of the short wavelength-sensitive visual pigments in vertebrates. *Molecular Biology and Evolution*, 11, 32 – 39.
Yokoyama, S. (1995) Amino acid replacements and wavelength absorption of visual pigments in vertebrates. *Molecular Biology and Evolution*, 12, 53 – 61.
Zhukovsky, E. A. and Oprian, D. D. (1989) The effect of carboxylic acid side chains on the absorption maximum of visual pigments. *Science*, 246, 928 – 930.
Zuker, C. S., Montell, C., Jones, K. Laverty, T. and Rubin G. M. (1987) A rhodopsin gene expressed in photoreceptor cell R7 of the *Drosophila* eye: homology with other signal-transducing molecules. *Journal of Neuroscience*, 7, 1550 –1557.

Section IV

Behaviour and Communication

Introduction

E.R. LOEW

What is it that distinguishes the visual ecologist from other visual specialists? While the work may involve questions of how the visual system functions and is put together from the molecular on up, the visual ecologist always relates the results back to the broader question of why the system is as it is and how it is adaptive to the animal. This approach ultimately ends with questions of phenotype and behaviour as it is here that natural selection acts to mold the visual system. The elements that need to be integrated in these ecological speculations are: 1) the radiometric and optical characteristics of the environment, 2) the physical nature of relevant targets and how they transfer information about themselves onto the surrounding electromagnetic field, 3) the operating characteristics of the visual system that allow information about the environment to be extracted from the three-dimensional electromagnetic field, and 4) what the animal does with the information. The first element is the independent variable in this ecological equation with the second and third being interdependent to a degree. The last element is often the easiest to examine, but the hardest to understand as it means delving into questions of perception, psychology and meaning – a difficult task when working within our own species let along trying to make sense out of the actions of another species. The best that can be hoped for is that by observing the visual behaviour of animals and correlating these with the other elements we can arrive at some 'universal truths' about visual system capabilities and evolution. The following seven chapters are the ingredients for this speculative stew.

Chapter 14 examines the visual systems and behaviours of cephalopod molluscs. As a group these have well-developed eyes based on the 'camera' design found in vertebrates – an excellent example of convergent evolution. Of particular interest are the primitive pinhole eye of the chambered nautilus and how its simple design and limited performance correlate with the life-style of this interesting animal. The chapter is rounded out with the discussion on the interplay between visual system design and intraspecific signaling.

Birds offer fertile ground for the visual ecologist. The importance of vision to their existence is well documented and their various colourations and complex visual behaviours offer a wealth of opportunity for correlation with underlying structural and functional features. In chapter 15 the focus is on ocular structure and the visual fields of birds. Using a comparative approach these are correlated with visual performance in foraging and in interspecific communication.

S.N. Archer et al. (eds.), Adaptive Mechanisms in the Ecology of Vision, 465–466

Structure and function are sides of the same coin. Therefore, it is not surprising to find that many visual behaviours and capabilities can be correlated with anatomic specializations. Chapter 16 uses a phylogenetic approach to examine how the two-dimensional distribution of retinal neurons and the presence of areas and foveae relate to the visual tasks an organism must perform in the three-dimensional world.

As mentioned above, there is the potential for coevolution between the visual system, the 'detector' and the signal. Nowhere has this been better studied than in the system involving bees and their floral hosts. The sole purpose of the flower as a visual signal is to advertise its presence and present a unique signal that can be quickly identified by potential pollinators. To this end the flower should adopt shapes, patterns and colourations that maximize the probability that they will be seen and identified. Conversely, the visual system of the pollinator should be adapted to operate with maximum efficiency on the target characteristic displayed by the flower. While the role of flower colouration in attracting bee pollinators has been studied qualitatively for many years, only recently have quantitative models been presented to formalize the relationship between bee colour space and flower colourations. One such model is presented in chapter 17.

In a similar vein bioluminescence offers another chance to explore the coevolution of detectors and signals. Chapter 18 concentrates on the aquatic environment where bioluminescence may function as a means of concealment, intraspecific signaling and illumination. Not lost in this treatment is the necessity of taking into account the special optical properties of water and the directional nature of the solar light field.

Chapter 19 brings it all together by looking at the role of light and vision at the community level – in this case coral reefs. Not only do the various species show visual system specializations for photic and target conditions, but the special role of light in establishing the temporality of behaviours is discussed. Central to this discussion is the special case of twilight when the ambient intensity is changing very rapidly along with the spectral quality of sunlight. How this has influenced diurnality and nocturnality at the community level make for an intriguing story.

Lastly, we are presented with an excellent review of primate vision from the molecular to the behavioural. Of particular interest is the question of why dichromacy and trichromacy are distributed as they are. The possible relevance of visual pigment polymorphism to visual success is also discussed.

Taken together this group of chapters presents ample justification for the ecological approach to visual problems. They exemplify the benefits to be had by not being afraid to ask why things are the way they are even if the answers may be qualitative or merge into the presently unpopular world of personal opinion or human perception.

14
Visual systems, behaviour, and environment in cephalopods

W.R.A. MUNTZ
Department of Ecology and Evolutionary Biology, Monash University, Clayton, Victoria 3168, Australia.

14.1 Introduction

The vision of cephalopods is interesting for many reasons. One is the existence of the primitive cephalopod *Nautilus,* which has apparently changed little over the past 150 million years or so. The modern coleoid cephalopods are probably derived from ancestors that resembled *Nautilus* quite closely (Naeff, 1923), and its eye is among its many apparently primitive features, giving an indication of how the visual system of modern cephalopods developed. A second reason is that the eyes of cephalopods provide a striking example of convergent evolution with those of fishes. Cephalopods face similar visual problems to fishes, and superficially the eyes of the two groups are remarkably similar. However, their development and detailed structure are quite different, and fishes and cephalopods have no common ancestors which had camera eyes: they have achieved similar solutions to many problems, but by very different routes.

Eyes are basically of two types: camera eyes, found for instance in vertebrates, molluscs, and arachnids, and compound eyes found only in invertebrates, particularly the crustacea. Kirschfeld (1974, 1976) has considered the relative merits of camera and compound eyes. When the eye is large, camera eyes have potentially better resolution than compound eyes of the same size. Kirschfeld calculates for example that the smallest compound eye that could have the same resolution as a human eye would need to be over one metre in diameter! In small eyes compound eyes have, however, some advantages, in, for example, having a wider field of view. Kirschfeld points out that compound eyes occur in small animals, and camera eyes in larger animals that can have bigger eyes. Cephalopods are among the few invertebrates of sufficient size for camera eyes to be an advantage. Kirschfeld also suggests that small animals have different environmental problems to large ones, in particular in that, because they move relatively slowly, the objects of interest to them will usually be closer to them. Even though their visual resolution is less in absolute terms, at close ranges they can still resolve the same objects as larger animals can at longer ranges, so that their 'subjective resolution' will still be adequate for their needs.

This chapter considers the vision of cephalopods in relation to their behaviour and environment. We will initially consider the primitive camera eye of *Nautilus,*

S.N. Archer et al. (eds.), Adaptive Mechanisms in the Ecology of Vision, 467–483

and discuss how the limited performance possible with it, is nevertheless probably sufficient for the restricted visual environment in which these animals live. We will then consider the advantages of the more sophisticated lens-bearing camera eyes of the coleoid cephalopods, concentrating on the detection of food objects, predators and the like. More complex visual abilities, such as shape discrimination and movement detection, will not be covered.

14.2 Vision in *Nautilus*, a 'primitive' cephalopod

14.2.1 Introduction

The cephalopod *Nautilus* can be considered a 'living fossil': species essentially identical to those found today have existed for 150 million years, and very similar ancestral forms for a further 250 million years or so before that. The eye is one of the features found in *Nautilus* that appears primitive, its structure in a number of respects similar to that found in the embryos of the coleoid cephalopods (Muntz and Wentworth, 1987). Particularly striking is the lack of any dioptric apparatus (lens or cornea): the eye acts as a pin-hole camera, and the interior of the eye is in direct contact with the surrounding sea through the contractile pupil.

Nautilus is still abundant and successful within its localised habitat, and shows limited but reliable visual behaviour (see below). To assess how well the *Nautilus* visual system can cope with its requirements we need to know the visual characteristics of its environment, the capabilities of its primitive visual system, and the needs that *Nautilus* has for vision in its normal life.

Nautilus is found round the west side of the Pacific, between about 30° south and 20° north, on the steep oceanic slopes of fringing or barrier reefs at depths between about 150 m and 600 m. Buoyancy control is through gas contained in the rigid compartments of its shell, which cannot be replenished below about 300 m, so greater depths can only be visited for relatively short periods. The absolute lower depth limit is determined by the depth at which the shell implodes, which is about 800 m. The upper limit is probably controlled by temperature, and where the surface waters are relatively cool, as in Noumea, they are occasionally caught at the surface (see Saunders and Ward (1987) for a summary).

The depth range is mostly below the range of living coral. The water is clear, probably Jerlov type IA or IB (Jerlov, 1968), and the maximum transmission will be around 475 nm. Bioluminescence will be a major source of light. The bottom slopes steeply, and in the range over which the animals are found, at least off Fiji, underwater photography has shown that it is largely featureless (Hyasaka, 1988).

14.2.2 Visual capabilities

Pin-hole cameras, such as the *Nautilus* eye, are simple, do not need focussing, and are not subject to chromatic or spherical aberration. However adequate sensitivity and resolution cannot be obtained simultaneously: obtaining good performance in one results in a catastrophic loss of performance in the other. For example, the theoretical resolving power of an optical system depends only on its aperture, if the dimensions of the other elements of the system are adjusted appropriately. The cost when no lens is present is, however, very great. For a human pupil, say 0.5 mm diameter, Lord Raleigh calculated that a 'perfect' image (limited only by diffraction) would be obtained if the distance from the pupil to the image was 50 m (Raleigh, 1891). Quite apart from the impracticality of an eye this size, the brightness of the image becomes very low: in photographic terms the F-number would be 4,000!

The pupil size necessary to get the best possible resolution in a pin-hole camera of a given size is a trade-off between the wave and particle (ray) properties of light; as the pupil gets smaller the size of the blur circle, as defined by ray optics, gets smaller, but eventually diffraction effects become important and the blur circle starts to get larger again. For a pupil-image distance of 10 mm (as in *Nautilus*) and $\lambda = 470$ nm (in air) the optimal pupil diameter is about 0.14 mm (Young, 1972). The pupil of *Nautilus* is some 3 orders of magnitude bigger than this. This retains some sensitivity for the system, but the potential resolution is very poor, and sensitivity is still reduced compared to that obtainable by lens bearing eyes. Table 1 gives approximate F numbers for *Nautilus* and some lens bearing eyes.

Nautilus can be kept readily in the laboratory, and reliably shows two types of visual behaviour which can be used to assess its visual capabilities. First, when placed in a rotating striped drum the animals will follow the stripes quite precisely, provided the speed of rotation is slow (the optomotor response). Using this it has been shown that the minimum separable angle of *Nautilus* is about 7° of arc, which agrees roughly with expectation for a pin-hole camera with the dimensions of the *Nautilus* eye (Muntz and Raj, 1984).

A second piece of visual behaviour that can be used under laboratory conditions is positive phototaxis. The animals move towards the brighter of two lights presented simultaneously, and if tested in a suitable Y-maze, in which the two arms are differentially illuminated, will choose the brighter arm almost 100% of the time. This response has been used to measure spatial summation by measuring intensity thresholds to stimuli of different sizes. *Nautilus* shows almost complete summation over at least 13°, compared to a summation area of say 5′ for the goldfish (Muntz, 1994a; Northmore, 1977). Along with its very low visual acuity, this confirms that the visual system of *Nautilus* retains adequate sensitivity at the expense of resolution.

Spectral sensitivity has also been determined using this technique (Muntz, 1986). It is found to be maximal at about 470 nm, which agrees with the absorption spectrum of the extractable visual pigment (Muntz, 1987; Kishigami *et al.*, 1988) and is appropriate to the spectral quality of the water in which the animal lives. Finally, it has been used to assess absolute sensitivity to tungsten light

Table 1 Some approximate F numbers for camera eyes

	F- number
Nautilus	
Largest pupil, diam. 2.25 mm	4.4
Smallest pupil, diam. 0.7 mm	14
['optimal' pupil, diam. 0.14 mm]	71
Fish, most cephalopods	
Assuming Matthiessen's ratio 2.5	1.25
Man	
Dark adapted, pupil diam. 7mm	2.8

(Muntz, 1987). While this is of course spectrally unnatural for *Nautilus*, by making appropriate assumptions about the spectral characteristics of daylight and the transmission of the water, an estimate of the greatest depth at which any vision involving downwelling surface light would be possible can be made. This comes out at about 800 m, as we would expect less than for deep sea fish where similar calculations yield an estimate of about 1,000 m (Clarke and Denton, 1962), but nevertheless enough for daylight to have some effect over the depth ranges at which the animals occur (Muntz, 1987).

14.2.3. Nautilus vision in its natural environment

There are at least three visual behaviours that may be important to *Nautilus* in its natural environment. Firstly, they show diurnal activity rhythms being most active in the evening and at night under laboratory conditions (Zahn, 1984). In their natural environment also they often show circadian vertical migrations, although these may not occur in all populations (Saunders and Ward, 1987; O'Dor *et al.*, 1993). The animals are benthic in habit, and with the dark steeply sloping substrate on which they live, light will not always be a reliable indicator of the upward direction. However the animals have statocysts that will give the direction of gravity, and are capable of detecting pressure changes (Jordan *et al.*, 1988), and in combination with the detection of changes in light level this probably controls their vertical migration.

Secondly, their environment is characterised by strong currents. The optomotor response could be involved in holding station, although tracking experiments indicate that they often drift over long distances, presumably carried passively by the currents (Saunders and Ward, 1987).

Finally, bioluminescence is important at the depths at which *Nautilus* lives. Bioluminescent sources can usually be considered as point sources, and their detection depends on catching and pooling as many quanta as possible. Provided the pupil is large and spatial summation sufficient the presence of a lens will not improve detection. Summation is high in *Nautilus*, and the pupil size is comparable to at least some fish.

Nicol (1958) has estimated the greatest distance at which an animal could see various bioluminescent sources in waters with 90% and 99% transmission per metre, assuming that the animal can see a radiant flux of $1.6 \times 10^{-10} \mu W.cm^{-2}$ entering the eye. These estimates are based on experimental determinations of the spectral composition of the bioluminescence of different animals, and on Grundfest's (1932) behavioural data for the absolute threshold of the sunfish, *Lepomis*, assuming a pupil area of 10 mm^2. It is not possible to extrapolate Nicol's estimates directly to *Nautilus* since the *Nautilus* pupil, even when fully dilated, is rather smaller (about 4 mm^2), and we do not know whether the neural sensitivity of the retina and CNS of *Nautilus* is the same as that of fish. Nevertheless, it is likely that the sensitivity is similar, and since the radiance transmitted by water falls off exponentially with distance, small differences will have little effect on visual range.

Two of the sources considered by Nicol are relevant in the environment of *Nautilus*. The shrimp *Acanthephyra pelagica*, in common with many other decapod crustacea, produces a luminous secretion when disturbed, and is similar in this to the decapods *Heterocarpus* and *Aristaemorpha* which are found in association with *Nautilus*, Nicol also presents calculations for the radiolarians *Cytocladus* and *Aulophaera*. The maximum ranges at which *Aristaemorpha* could be seen come out at 7 m and 5 m for the clearer and less clear waters respectively, and at 6 m and 4 m for the radiolarians. Other decapods, such as *Euphasia pacifica*, are brighter and could be seen up to 32 m and 16 m. *Nautilus* can therefore probably see bioluminescent sources in its environment at substantial and useful distances. Since the animals swim slowly, vision at great ranges is probably of no real interest to them.

Decaying food matter, the principle diet of *Nautilus,* may also be associated with bioluminescence. Decaying animal matter itself is often bioluminescent, and other animals in the environment, particularly the shrimp, are also attracted to carrion and will bioluminesce in the activity associated with scavenging. Positive phototaxis might therefore attract *Nautilus* to its sources of food. It has been shown that baited traps catch more *Nautilus* if they are illuminated (Muntz, 1994b).

14.3 Vision in coleiod cephalopods

14.3.1 Introduction

The nautiloids, while originally widespread, are now restricted to one genus, a few species, and to a limited environment for which it is likely that their simple visual system is adequate. On the other hand there are about a thousand species of coleoid cephalopods (Voss, 1977), which are found throughout the oceans, from turbid inshore waters to the clear waters and depths of the open sea. Objects at considerable distances are important to them. *Octopus vulgaris,* for example, lives in crannies in the rocks, occupying one such 'home' for extended periods of time. The territories over which they forage from these are typically oval in shape and between 120 m^2 and 200 m^2 in area (Mather, 1991a), and navigation within them is at least partly based on visual landmarks (Mather, 1991b). Most coleoids swim much faster than *Nautilus* (up to 0.4 $m.s^{-1}$ for sustained swimming in the squid *Loligo opalescens*, and 2 $m.s^{-1}$ for a burst (O'Dor, 1982), which also means that objects that are further away will be more important to them than objects that are of interest to *Nautilus*, requiring better resolution. Shapes and colours are also likely to be more important to them in shallower waters.

Coleoid cephalopods are also well known for their rich behavioural repertoire, Octopuses in particular can be readily trained and will even learn by imitation (Fiorito and Scotto, 1992). Decapods such as the cuttlefish and squid are more difficult to train, partly because they are difficult to keep in good condition in the laboratory, but are also clearly capable of learning (Saunders and Young, 1940; Allen *et al*., 1985). A great deal of behavioural work has been done on cephalopod vision (see Messenger (1981) for a review). This has mostly been on 'higher' visual functions, such as shape discrimination or the effects of brain lesions on visual behaviour. Rather little work has been done on more basic visual abilities, such as sensitivity or visual acuity.

14.3.2 Fundamental visual mechanisms in coleoids

(a) Optics Underwater the cornea is usually optically ineffective and focussing depends entirely on the lens. The lenses of cephalopods are superficially very similar to those of fishes, being roughly spherical, almost free of spherical aberration, and with focal lengths of about 2.5 lens radii. In both groups freedom from spherical aberration is achieved by the grading of the refractive index of the lens material between centre and periphery.

Cephalopods (including *Nautilus*), unlike most fishes, have contractile pupils. A contractile pupil not only affects the amount of light reaching the retina, but also alters the physical characteristics of the image for different directions of view, by determining whether the light rays pass through the centre or through more peripheral parts of the lens. Light rays passing through the centre of the lens (axial rays) will impinge on the retina perpendicularly, but rays passing through more peripheral parts (paraxial rays), while they will be focussed at the same point on the retina as the axial rays, will impinge on it at an angle: in the case of fishes and cephalopods light passing through the edge of the lens will in fact reach the retina at an angle of over 40° from the vertical. Since cephalopod retinal receptors can be over 200 μ long, this would result in the paraxial rays producing a substantial blur circle as they pass through the thickness of the retina, unless there is some mechanism to prevent the spread of light between adjacent receptors.

In some cases there appears to be an interrelationship between pupil shape and structures in the retina that should reduce this blur (Muntz, 1977), related to where the animals live and the directions of view that are most important to them. In *Octopus vulgaris*, for example, while there is no area of specialisation corresponding to the vertebrate fovea, there is a horizontal strip where the receptors are longer, and the screening pigment between them less pronounced, so that the possibility of light passing between adjacent receptors with resulting blur is increased. The horizontal strip, however, matches the shape of the pupil, which when contracted forms a horizontal slit. As a result, when the pupil is contracted, light from the horizontal plane will reach the retina perpendicularly. Light from above or below the horizon will pass through more peripheral parts of the lens and reach the retina at an angle, but in these retinal areas the receptors are shorter, and the screening pigment between them more pronounced, so that the spread of light between adjacent receptors will be reduced. It is likely that the horizontal strip and pupil shape means that the visual system is better adapted for vision anywhere around the horizon, which is appropriate for a benthic animal which attacks its prey monocularly, swimming sideways.

Cuttlefish (*Sepia officianalis*) not only have a horizontal strip in the retina, but also two further areas of specialisation within it, one anterior and one posterior. Their pupils also differ from those of octopus in having roughly the form of a letter 'W', with a horizontal section and two vertical sections at either end. When the pupil is contracted the two vertical sections can act as two separate pupils, and this again results in the light reaching the retina perpendicularly at the two areas of specialisation (Figure 1). In *Sepia* these structures are probably adaptations to living on the bottom, swimming directly backwards, and preying on animals directly in front.

(b) Resolution of spatial detail The minimum separable visual angle of *Octopus australis* and *O. pallidus*, measured behaviourally by their ability to discriminate gratings, is about 7′ (Muntz and Gwyther, 1988), which is about 60 times better than *Nautilus* and comparable to other aquatic animals with lens-bearing camera eyes (Table 2). Most bodies of water, however, transmit much finer detail than this (Jagger and Muntz, 1993), raising the question why the acuity of aquatic animals is not higher: terrestrial animals achieve much better levels of resolution. The answer may lie in part in the rapid reduction of contrast with distance that occurs underwater, which means that objects often disappear from lack of contrast before their details become unresolvable. This, and the generally lower speed at which animals move underwater may, in Kirschfeld's terminology, mean that the 'subjective resolution' allowed by an acuity of 7′ is all that is required in the aquatic environment.

Grating acuity measures the highest spatial frequency at which a square wave grating of black and white bars (a contrast of one) can be discriminated. With coarser gratings below this frequency lower contrasts become discriminable. However, the improvement with lower spatial frequency does not continue indefinitely, and for all animals that have been tested there is an optimal spatial

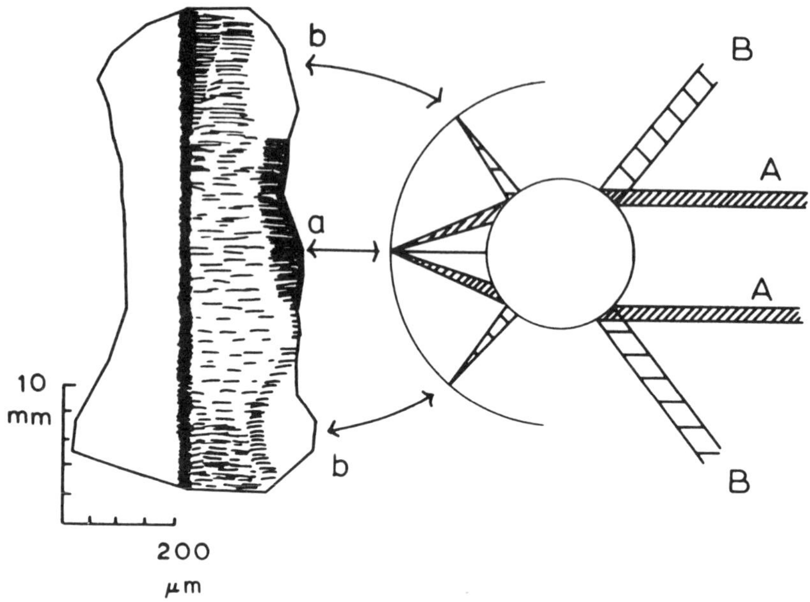

Figure 1 Relationship between pupil shape and retinal anatomy of *Sepia*. The left-hand half of the figure is after Young (1963), and shows the distribution of screening pigment and rhabdome lengths in a horizontal section of a light-adapted *Sepia* retina. The right-hand half of the picture represents a horizontal section through the *Sepia* eye and shows how pencils of light through the anterior and posterior vertical sections of the pupil (B,b) will reach the retina perpendicularly, where the receptors are longer and the screening pigment is less pronounced, whereas light from other directions entering the eye through these vertical sections (A,a) will impinge on the retian at a considerable angle from the vertical

Table 2 Underwater minimum separable visual angles of aquatic animals, determined behaviourally using gratings

Cephalopods		
Nautilus pompilius	7°	Muntz & Raj (1984)
Octopus australis & *O. pallidus*	5′	Muntz & Gwyther (1988)
Fishes		
15 teleost species	3.7′ – 20′	Douglas & Hawryshyn (1990)
Reptiles		
Mud turtle, *Emys orbicularis*	2.9′	Dudziak (1955)
Mammals		
Harbour seal, *Phoca vitulina*	8.3′	Schusterman & Balliett (1970)
Stellar sea lion	7.1′	Schusterman & Balliett (1970)
Bottlenose dolphin, *Turiops truncatus*	8′ –13′	Herman *et al.* (1975)

frequency at which the contrast needed for detection is at a minimum. The goldfish, for example, can detect gratings at 0.3 cycles per degree (cpd) at lower contrasts than gratings of either higher or lower frequency (Northmore and Dvorak, 1979). This optimum in the contrast sensitivity function is a consequence of lateral inhibitory mechanisms in the retina or elsewhere in the visual system.

Contrast sensitivity functions have not been determined for any cephalopod, and it is not known whether such an optimal spatial frequency occurs for them, but it seems almost certain that this must be so. Lateral inhibition has not been directly demonstrated either, but again almost certainly exists. The photoreceptors are known to have collaterals that link neighbouring receptors by gap junctions (e.g. Yamamoto, 1984), and there are ample opportunities for lateral inhibition in the optic lobes (Young, 1971). Electrophysiology is difficult with cephalopods, but Lettvin and Pitts (1962) have reported indications of lateral inhibition in the optic neurones. The question is important in view of the prevalence of stripes on the bodies of underwater animals, which it has been suggested are matched to the visual systems of other animals and result in their being conspicuous (see below).

(c) Sensitivity There appear to have been no behavioural determinations of adaptation or absolute visual sensitivity in cephalopods. Observations by divers have shown that although *Octopus vulgaris* are most active at night, being away from their 'homes' more often between sunset and sunrise than during the day, such foraging activity is by no means restricted to the hours of darkness (Kayes, 1974). Adaptation to different light levels must therefore be important. Some control of sensitivity will be achieved through mechanical means. Changes in pupil size, for example, will alter sensitivity, although probably by only about 0.7 log units (Muntz, 1977). The screening pigment of the retina also shows photomechanical movements, moving out between the receptors in the light and retracting in the dark. It has been shown that the sensitivity change caused by these movements is to the order of 0.6 log units (Daw and Pearlman, 1974). The effect on sensitivity of pupil and screening pigment movements between them are therefore quite small compared to the diurnal range of intensities to which the animals are exposed, and presumably neural and photochemical mechanisms must be involved as well. These are currently little understood.

(d) Spectral sensitivity and colour vision Many cephalopods show very striking and variable body patterns which are often highly coloured (to humans), and which among other things may have a cryptic effect making the animals difficult to discriminate from their backgrounds. In view of this, it is surprising that in most of the species studied there appears to be only a single visual pigment in the retina, and no colour vision. While it is impossible to prove a negative statement such as this, attempts to train *Octopus vulgaris* to discriminate colours have been unsuccessful (Messenger *et al.*, 1973), and other experiments involving the

electroretinogram, pupil response, or optomotor response (Flores, 1983), in a variety of cephalopods, are most economically interpreted by assuming that they are colour-blind (see Wells (1978) for a review). Visual pigment extracts usually reveal only a single pigment and ERG studies of spectral sensitivity, while they sometimes have a number of small humps or irregularities on them, are compatible with expectation on the basis of a single visual pigment. There have been two behavioural studies of spectral sensitivity, both involving phototactic behaviour. White (1924) showed that the spectral sensitivity of larval squid (*Loligo pealii*) followed a typical bell-shaped function, compatible with the absorption spectrum of a single visual pigment, and as we have seen the spectral sensitivity of *Nautilus* agrees with the absorption spectrum of its extractable visual pigment. Recently Marshall and Messenger (1996) have demonstrated that when cuttlefish (*Sepia officianalis*) are placed on different artificially coloured and patterned backgrounds, their body patterns alter to match these in intensity only, without regard to spectral composition.

Among the mesopelagic cephalopods there are, however, exceptions to this general picture. The firefly squid, *Watasenia scintillans*, has three visual pigments in all, one based on retinal (A_1), with λ_{max} 484 nm; a second based on 3 – 4 dehydroretinal (A_2), with λ_{max} 500 nm; and the third based on 4-hydroxyretinal, with λ_{max} 470 nm (Matsui *et al.*, 1988). A similar situation was subsequently found for a further four species (Kito *et al.*, 1992). In *Watasenia* the A_1-based pigment is restricted to the dorsal retina, but the A_2-based pigment and the 4-hydroxyretinal based pigments both occur over the same area of the ventral retina, in the proximal and distal parts respectively of the same receptors, Michinomae *et al.* (1994) discuss how this distribution might underlie colour vision. The photophores of these squid also produce spectrally different light outputs, to which the variety of visual pigments may be related (see below). Colour vision has not, however, been demonstrated directly in any of these species as yet.

(e) Polarised light The importance of polarised light in the underwater environment is becoming increasingly apparent (e.g. Loew and McFarland, 1990). Both fishes and cephalopods are able to respond to polarised light. The data for fishes have been summarised by Douglas and Hawryshyn (1990). Some species exhibit bracycardia in response to the plane of polarised light, and some have a tendency to orientate along or perpendiculary to the e-vector of the light field. Training experiments have further shown that under appropriate circumstances some fish can learn to discriminate lights of different polarisation. Among the cephalopods Jander *et al.* (1963) have shown that innate orientation with respect to the e-vector can also occur in this group, and Moody and Parriss (1961) that octopuses (*Octopus vulgaris*) can be trained to discriminate the plane of polarisation of different lights. Recently Shasher and Cronin (1996) have demonstrated that octopuses (*O. vulgaris and O. briareus*) can discriminate targets on the basis of the presence or absence of a pattern produced by a 90° polarisation contrast

within the target, showing that this ability could be used for object detection or identification. The oriented microvillar arrangement of the photophores of cephalopods presumably underlies their ability to discriminate the plane of polarised light. The mechanism whereby this is done in fishes is, however, still unknown.

14.3.3 Cephalopods as visual stimuli

(a) Introduction Cephalopods respond to a wide variety of visual stimuli. They are also themselves a major visual stimulus, both to other cephalopods and to other groups of animals, such as fishes and birds. We should therefore expect that the coloration and body patterns of cephalopods will be adapted either for camouflage, or to make them as conspicuous as possible.

The relationships between vision, body patterns and environment have been reviewed by Cott (1940) and Lythgoe (1979) among others, and specifically for fishes by Muntz (1990). The cephalopods are particularly rich in the body patterns they can display, and the rapidity with which they can alter them, as can be seen, for example, in the coloured photographs of Hanlon (1988) and Hanlon and Messenger (1988).

(b) Conspicuousness It is often adaptive for animals to reduce their visibility as much as possible, to avoid predation. It is also, however, often adaptive to be as conspicuous as possible, to maximise, for example, the distance over which they can be seen by conspecifics, or to frighten off predators through aposematic displays. To achieve this the animal's form and colour should provide maximal contrast with the background, be transmitted optimally by the water body, to be as effective as possible in stimulating the eye.

Lateral inhibition, which is probably the most ubiquitous spatial mechanism found in the eye (see above), will result in spots and repeating stripes being among the most visible stimuli that there are. Therefore, it is not surprising that eyespots and bars are amongst the commonest body patterns found in animals. Stripes are often considered a feature for concealment, helping to blend the animal with its background, but under many circumstances they in fact make the animal conspicuous (Cott, 1940; Godfrey *et al.*, 1987), and striped fishes underwater are often particularly visible.

The great variety of body patterns shown by cephalopods have been described in detail by Packard and Saunders (1971), Moynihan (1985), and Hanlon and Messenger (1988), among others. The patterns are derived from a number of basic components, frequently consisting of bars or stripes on the arms or body, which are put together under central nervous control to achieve the various final patterns the animals display. Packard and Saunders (1971) describe this for *Octopus vulgaris*, where display patterns, such as the zebra crouch or the dymantic (frightening) display typically depend on stripes and contrasting spots (Figure 2). In *Sepia officianalis*, which has a richer display, Hanlon and

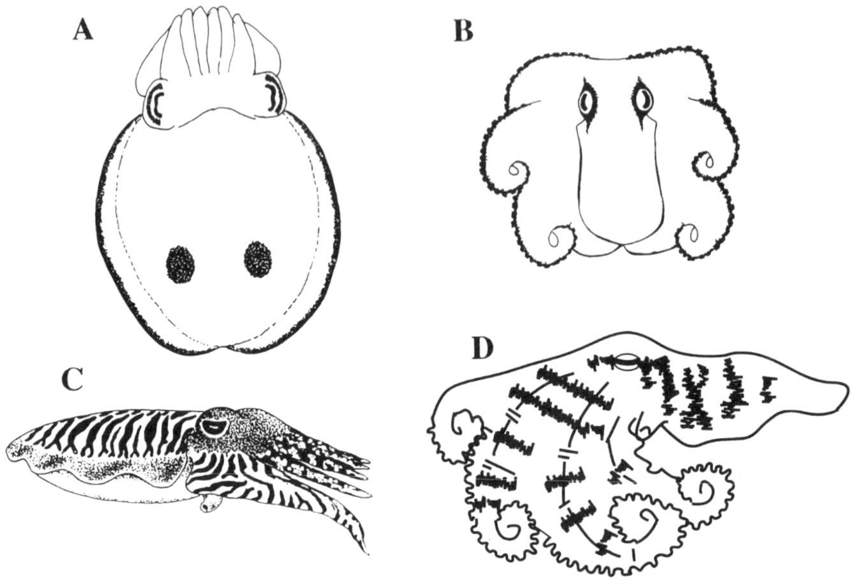

Figure 2 Cephalopod displays. A. *Sepia* dymantic display, after Holmes (1940). B. *Octopus* dymantic display, from a photograph. C. *Sepia zebra* display, from Tinbergen (1939). D. *Octopus* zebra display, after Packard and Saunders (1971)

Messenger (1988) describe no less than 54 pattern components, which can be combined into at least 13 overall patterns. Many of the components, such as stripes, arm bars, and the 'eye spots' of the dymantic pattern, are the same as are found in octopus. The use of 'eye spots' to deflect or deter predators is widespread in the animal kingdom (Cott, 1940), and must have evolved independently on a number of occasions: there is experimental evidence in some cases that it is effective in deterring predators (Blest, 1957).

Cephalopods as a group are conservative in these matters, with the same elements occurring repeatedly in different species (Moynihan, 1985). This conservatism among distinct groups has led Moynihan to suggest that cephalopod body patterns are not only adapted to increase or decrease their visibility to other cephalopods, but also to affect the visual systems of other animals altogether, in particular the fishes, who are often their predators. This raises the interesting possibility that the wide variety of colour patterns they display are adaptive because many of their predators have colour vision, even though the cephalopods themselves may be colour blind (Moynihan,1985). The problem remains as to how, if this is so, they usually achieve the appropriate colour match with their backgrounds. Since however body patterns are achieved through the release of various 'pre-packaged' components, it may be that coloured pattern components have evolved that can be released under natural circumstances without the animal making a detailed spectral comparison with its background: as Marshall and Messenger (1966) have shown, the match achieved is often bad on unnaturally coloured backgrounds.

(c) Camouflage Perfect camouflage of an object would be achieved if the light reaching the observer's eye were identical whether or not the object was present. Cephalopods approach this ideal by matching their environment through changing the albedo, colour, and degree of mottle of their skin. They can also change their shape, by flattening themselves against the substrate, or by disruptive effects in which skin papillae are accentuated, the mantle shape attenuated, and the arms twisted, so that no coherence remains in the body's outline. In its full form this is known as the flamboyant pattern (Packard and Saunders, 1971), and it is particularly effective against a seaweed background. The flamboyant pattern occur in distantly related cephalopods, such as octopuses and cuttlefish. Equivalent patterns also occur in fishes where, however, they are fixed. The general appearance of frog fishes, pipe fishes, and some sea-horses, that live among seaweed, is very reminiscent of the flamboyant pattern of cephalopods (Cott, 1940).

Countershading (or obliterative shading) is another mechanism of camouflage that is very widely spread in the animal kingdom (Cott, 1940). The downwelling light, both on land and underwater, is much more intense than the upwelling light. As a consequence a uniformly coloured animal viewed against a uniform background will be highly visible, with the back appearing brighter and the ventral surface darker than the background. To counteract this in many animals the dorsal surface is darker than the ventral surface. The cephalopods are no exception. Furthermore in *Sepia officianalis*, where the chromatophores are under neural control, the countershading is altered appropriately if the body position is changed, so that if the animal is turned upside down the ventral chromatophores expand and the dorsal ones contract. Turning the animal on its side similarly results in appropriate changes to the two halves of the body, and the reflex also works for orientation changes in the pitching plane (Ferguson and Messenger, 1991; Ferguson *et al.*, 1994).

Countershading cannot help camouflage when an animal is viewed from below, since the upwelling light is less intense than the downwelling light, so that even if the ventral surface had an albedo of one it would still appear darker than the background. However, many midwater marine animals, including cephalopods, have photophores arranged on their ventral surfaces, and it is likely that their function is to generate a light intensity that matches the downwelling light from the surface (Young, 1973). A further sophistication in some cephalopods is the possession of more than one class of photophore, with different spectral outputs, which may vary the spectral quality of the ventral counterillumination to match the changes in the spectral composition of the downwelling light that occur with depth during the animals' diurnal vertical migrations (Young and Mencher, 1980).

14.3.4. Conclusions

Cephalopods are unique, in that a 'primitive' ancestor, *Nautilus*, still exists, with a simple visual system and simple visual behaviour, probably adequate for the limited environment to which it is now restricted. Modern cephalopods have developed from an ancestor probably much like *Nautilus*. They live in varied environments where they compete more or less directly with the fishes, and face the same visual problems. The degree of convergent evolution that has occurred is remarkable. The eyes superficially appear very similar, and many mechanisms of great complexity have appeared independently in the two groups, such as the use of bars and 'eye spots' for conspicuousness, and cryptic patterns such as the flamboyant, countershading, and ventral bioluminescent counterillumination. Visual performance, as measured by acuity, is much the same in the two groups. Within the cephalopods themselves convergent evolution appears to have occurred. Cuttlefish and octopuses show, for example, complex patterns that are similar, such as the dymantic and the flamboyant, yet they are only distantly related.

The many striking similarities that occur between cephalopods and fishes led Packard (1972) to suggest that, functionally, cephalopods should be considered as fishes. Although cephalopods also show differences from fishes in their life cycles, physiology and biochemistry, which may reduce the degree to which they can compete directly with them (O'Dor and Webber, 1986), the remarkably similar solutions to common visual problems achieved by the two groups in many cases suggest that these particular environmental constraints must be especially powerful. Some of these environmental pressures are clear: countershading and ventral counterillumination are related to the distribution of light around the vertical plane, spectral sensitivity, colour vision, and colour matching through photophores, at least in part to the spectral quality of the water body; and body patterns such as stripes and 'eye spots' to basic visual mechanisms such as lateral inhibition.

References

Allen, A. Michels, J. and Young, J. Z. (1985) Memory and visual discrimination by squids. *Marine Behaviour and Physiology*, 11, 271–282.

Blest, A. D. (1957) The function of eyespot patterns in the Lepidoptera. *Behaviour*, 11, 209 – 256.

Clarke, G. L. and Denton, E. J. (1962) Light and animal life. In *The Sea*, Vol. 1, ed. Hill, M. N. John Wiley, New York, pp. 456 – 458.

Cott, H. B. (1940) *Adaptive coloration in animals*. Methuen, London.

Daw, N. W. and Pearlman, A. L. (1974) Pigment migration and adaptation in the eye of the squid, *Loligo pealei*. *Journal of General Physiology*, 63, 22 – 36.

Douglas, R. H. and Hawryshyn, C. W. (1990) Behavioural studies of fish vision: an analysis of visual capabilities. In: *The Visual System of Fish*, ed. Douglas, R. H. and Djamgoz M. B. A. Chapman and Hall, London, pp. 373 – 418.

Dudziak, J. (1955) Ostrosc widzenia u zolwia blotnego (*Emys orbicularis* L.) przypatrznym i wodnym. *Folia Biologica*, 3, 205 – 228.

Ferguson, G. P. and Messenger, J. B. (1991) A countershading reflex in cephalopods. *Proceedings of the Royal Society, London, B.*, 243, 63 – 67.

Ferguson, G. P. Messenger, J. B. and Budelman, B. U. (1994) Gravity and light influence the countershading reflexes of the cuttlefish *Sepia officinalis*. *Journal of Experimental Biology*, 191, 247– 256.

Fiorito, G. and Scotto, P. (1992) Observational learning in *Octopus vulgaris. Science*, 256, 545 – 547.

Flores, E. E. C. (1983) Visual discrimination testing in the squid *Todarodes pacificus*: experimental evidence for lack of color vision. *Memoirs of the Natural Museum of Victoria*, 44, 213 – 227.

Godfrey, D. Lythgoe, J. N. and Rumball, D. A. (1987) Zebra stripes and tiger stripes: the spatial frequency distribution of the pattern compared to that of the background is significant in display and crypsis. *Biological Journal of the Linnean Society,* 32, 427– 433.

Grundfest, H. (1932) The sensitivity of the sun-fish, *Lepomis*, to monochromatic radiation of low intensities. *Journal of General Physiology*, 15, 307– 328.

Hanlon, R. T. (1988) Behavioural and body patterning characters useful in taxonomy and field identification of cephalopods. *Malacologia*, 29, 247– 264.

Hanlon, R. T. and Messenger, J. B. (1988) Adaptive colouration in young cuttlefish (*Sepia officinalis* L.): the morphology and development of body patterns and their relation to behaviour. *Philosophical Transactions of the Royal Society, London,* 320, 437– 487.

Herman, L. M., Peacock, M. F., Yunker, M. P. and Madsen, C. J. (1975) Bottlenosed dolphin: Double-slit pupil yields equivalent aerial and underwater diurnal acuity. *Science*, 189, 650 – 652.

Holmes, W. (1940) The colour changes and colour patterns of *Sepia officinalis* L. *Proceedings of the Zoological Society, London, A*, 110, 17– 36.

Hyasaka, S. D. (ed.) (1988) *Marine ecological study on the habitat of Nautilus pompilius in Fiji (the second operation).* Kogoshima University Research Center for the South Pacific (KUSP), Occasional Papers No. 15.

Jagger, W. S. and Muntz, W. R. A. (1993) Aquatic vision and the modulation transfer properties of unlighted and diffusely lighted natural waters. *Vision Research*, 33, 1755 –1763.

Jander, R. Daumer, K. and Waterman, T. H. (1963) Polarised light orientation by two Hawaiian cephalopods. *Zeitschrift für vergleichen Physiologie*, 46, 383 – 394.

Jordan, M., Chamberlain, J. A. and Chamberlain, R. B. (1988) Response of *Nautilus* to variation in ambient pressure. *Journal of experimental Biology*, 137, 175 –189.

Jerlov, N. G. (1968) *Optical Oceanography*. Elsevier, Amsterdam, London, New York.

Kayes, R. J. (1974) The daily activity pattern of *Octopus vulgaris* in a natural habitat. *Marine Behaviour and Physiology*, 2, 337– 343.

Kirschfeld, K. (1974) The absolute sensitivity of lens and compound eyes. *Zeitschrift für Naturforschung*, 29C, 592 – 596.

Kirschfeld, K. (1976) The resolution of lens and compound eyes. In *Natural processing in visual systems*, ed. Zettler, F. and Weiler, R. Springer-Verlag, Berlin.

Kishigami, A. Hara, R. and Hara, T. (1988) Photopigments in the retina of *Nautilus pompilius*. *Proceedings of the Yamada Conference* XXI, 373 – 374.

Kito, Y., Narita, K., Seidou, M., Michinomae, M., Yoshihara, K., Partridge, J. C. and Herring, P. J. (1992). A blue-sensitive visual pigment based on 4-Hydroxyretinal is found widely in mesopelagic cephalopods. In *Structures and functions of retinal proteins*. Ed. Rigand, J. L. Cooloque INSERM/John Libby Eurotext Ltd., pp. 411– 414.

Lettvin, J. Y. and Pitts, W. H. (1962) Neapolitan studies. *Quarterly Progress Report, Research Laboratory of Electronics, Massachussetts Institute of Technology*, 27, 288 – 290.

Loew, E. R. and McFarland, W. N. (1990) The underwater light environment. In: *The Visual System of Fish,* ed. Douglas, R. H. and Djamgoz M. B. A. Chapman and Hall, London, pp. 1– 43.

Lythgoe, J. N. (1979) *The ecology of vision*. Clarendon Press, Oxford.

Marshall, N. J. and Messenger, J. B. (1996) Colour-blind camouflage. *Nature*, 382, 409.

Mather, J. A. (1991a) Foraging, feeding and prey remains in middens of juvenile *Octopus vulgaris* (Mollusca: Cephalopoda). *Journal of the Zoological Society, London,* 224, 27– 39.

Mather, J. A. (1991b) Navigation by spatial memory and use of visual landmarks in octopuses. *Journal of Comparative Physiology A*, 168, 491– 497.

Matsui, S., Seidou, M., Horiuchi, S., Uchiyama, I. and Kito, Y. (1988). Adaptation of a deep-sea cephalopod to the photic environment. *Journal of General Physiology*, 92, 55 – 66.

Messenger, J. B. (1981) Comparative physiology of vision in molluscs. In *Handbook of Sensory Physiology, Vol. VII/6C, Comparative Physiology and Evolution of Vision in Invertebrates*, ed. H. Autrum. Springer-Verlag, Berling, Heidelberg, pp. 3–200.

Messenger, J. B., Wilson, A. P. and Hedge, A. (1973). Some evidence for colour blindness in *Octopus*. *Journal of Experimental Biology*, 59, 77– 94.

Michinomae, M., Masuda, H., Seidou, M. and Kito, Y. (1994) Structural basis for wavelength discrimination in the banked retina of the firefly squid *Watasenia scintillans. Journal of Experimental Biology,* 193, 1–12.

Moody, M. F. and Parriss, J. R. (1961) The discrimination of polarised light by *Octopus*; a behavioural and morphological study. *Zeitschrift für Vergleichen Physiologie*, 44, 268 – 291.

Moynihan, M. (1985) *Communication and noncommunication in cephalopods*. Indiana University Press, Bloomington.

Muntz, W. R. A. (1977) Pupillary responses of cephalopods. In: *Symposium on the Biology of Cephalopods*, ed. Nixon, M. and Messenger, J. B. *Symposia of the Zoological Society of London,* 38, 277– 285.

Muntz, W. R. A. (1986) The spectral sensitivity of *Nautilus pompilius. Journal of Experimental Biology*, 126, 513 –7.

Muntz, W. R. A. (1987) Visual behaviour and visual sensitivity of *Nautilus pompilius*. In: *Nautilus: The Biology and Paleobiology of a Living Fossil*, ed. Saunders, W. B. and Landman, N. H., Plenum Press, N.Y., pp. 231– 244.

Muntz, W. R. A. (1990) Stimulus, environment and vision in fishes. In: *The Visual System of Fish*, ed. Douglas, R. H. and Djamgoz, M. B. A. Chapman and Hall, London. pp. 491– 511.

Muntz, W. R. A. (1994a) Spatial summation in the photoactic behaviour of *Nautilus pompilius. Marine Behaviour and Physiology,* 24, 183 –187.

Muntz, W. R. A. (1994b) Effects of light on the efficacy of traps for *Nautilus pompilius. Marine Behaviour and Physiology,* 24, 189 –193.

Muntz, W. R. A. and Gwyther, J. (1988) Visual acuity in *Octopus pallidus and Octopus australis. Journal of experimental Biology*, 134, 119 – 29.

Muntz, W. R. A. and Raj, U. (1984) On the visual system of *Nautilus pompilius*. *Journal of Experimental Biology,* 109, 253 – 263.

Muntz, W. R. A. and Wentworth, S. L. (1987) An anatomical study of the retina of *Nautilus pompilius*. *Biological Bulletin*, 173, 387– 97.

Naeff, A. (1923) Die Cephalopoden. *Fauna e Flore del Golfo di Napoli,* 35, 1– 863.

Nicol, J. A. C. (1958) Observations on luminescence in pelagic animals. *Journal of the Marine Biological Association, U.K.* 37, 705 –752.

Northmore, D. P. M. (1977) Spatial summation and light adaptation in the goldfish visual system. *Nature, London,* 268, 450 – 451.

Northmore, D. P. M. and Dvorak, C. A. (1979) Contrast sensitivity and acuity of the goldfish. *Vision Research*, 19, 255 – 261.

O'Dor, R. K. (1982) Respiratory metabolism and swimming performance of the squid, *Loligo opalescens. Canadian Journal of Fisheries and Aquatic Science*, 39, 580 – 587.

O'Dor, R. K. and Webber, D. M. (1986) The constraints on cephalopods: why squid aren't fish. *Canadian Journal of Zoology,* 64, 1591–1605.

O'Dor, R. K., Forsythe, J., Webber, D. M., Wells, J. and Wells, M. J. (1993) Activity levels in *Nautilus* in the wild. *Nature London*, 362, 626 – 628.

Packard, A. (1972) Cephalopods and fish: the limits of convergence. *Biological Reviews*, 47, 241– 307.

Packard, A. and Saunders, G. D. (1971) Body patterns of *Octopus vulgaris* and maturation of the response to disturbance. *Animal Behaviour*, 19, 780 –790.

Raleigh, Lord. (1891). On pin-hole photography. *Philosophical Magazine*, 31, 87– 99.

Sanders, F. K. and Young, J. Z. (1940) Learning and functions of the higher learning centres of *Sepia. Journal of Neurophysiology*, 3, 501– 526.

Saunders, W. B. and Ward, P. D. (1987) Ecology, distribution, and population characteristics of *Nautilus*. In: *Nautilus*: *The Biology and Paleobiology of a Living Fossil,* ed. Saunders, W. B. and Landman, N. H., Plenum Press, N.Y., pp. 137–162.

Schusterman, R. J. and Balliet, R. F. (1970). Visual acuity of the harbour seal and the stellar sea lion underwater. *Nature London*, 226, 563 – 564.

Shasher, N. and Cronin, T. W. (1996) Polarisation contrast vision in *Octopus. Journal of Experimental Biology,* 199, 999 –1004.

Tinbergen, L. (1939) Zur Fortpflanzungsethologie von *Sepia officinalis L. Archive Néerland Zoologie*, 3, 323 – 364.

Voss, G. L. (1977) Present status and new trends in cephalopod systematics. In: *Symposium on the Biology of Cephalopods*, Ed. Nixon, M. and Messenger, J. B., *Symposia of the Zoological Society of London,* 38, 49 – 60.

Wells, M. J. (1978) *Octopus: physiology and behaviour of an advanced invertebrate*. Chapman and Hall, London.

White, G. M. (1924) Reactions of the larvae of the shrimp, *Palaemonetes vulgaris*, and the squid, *Loligo pealii*, to monochromatic light. *Biological Bulletin*, 47, 265 – 273.

Yamamoto, M. (1984) Photoreceptor collaterals in the cuttlefish retina. *Zoological Science*, 1, 501– 503.
Young, J. Z. (1963) Light- and dark-adaptation in the eyes of some cephalopods. *Proceedings of the Zoological Society, London*, 140, 255 – 270.
Young J. Z. (1971) *The anatomy of the nervous system of Octopus vulgaris*. Clarendon Press, Oxford.
Young, M. (1972) Pinhole imagery. *American Journal of Physics,* 40, 715 –720.
Young, R. E. (1973) Information feedback from photophores and ventral countershading in mid-water squids. *Pacific Science*, 27, 1–7.
Young, R. E. and Mencher, F. M. (1980). Bioluminescence in mesopelagic squid: diel color change during counterillumination. *Science N.Y.*, 208, 1286 –1288.
Zahn, L. P. (1984) The rhythmic activity of *Nautilus pompillus* with notes on its behaviour and ecology in Fiji. *The Veliger*, 27, 19 – 28.

15
Optical structure and visual fields in birds: their relationship with foraging behaviour and ecology

G. MARTIN
School of Biology and School of Continuing Studies, The University of Birmingham, Edgbaston, Birmingham B15 2TT, U.K.
Email: G.R.Martin@bham.ac.uk

15.1 Why birds? Why eye structure?

The number of species of birds in the world is trivial compared with the number of other organisms. Current estimates suggest that there may be more than 12.5 million species extant on the earth, and most of these (c. 8 million) are insects (World Conservation Monitoring Centre, 1992). Compared with this, a total of about 10,000 (Sibley and Monroe, 1991; Howard and Moore, 1991) species of extant birds is extremely small. Moreover, compared with other vertebrate taxa, birds are very uniform in their anatomy, morphology, and physiology (Baumel, 1993; Proctor and Lynch, 1993; Welty and Baptista, 1988; Gill, 1990). Indeed it is frequently argued that the diversity of birds is relatively superficial compared with that of other vertebrate taxa. This uniformity results primarily from constraints imposed by the demands of powered flight (Gill, 1990).

While birds may lack the exciting structural and physiological diversity of other animal groups, they do, however, have the advantage of being particularly well known. Precisely because there are so few bird species, a great deal is known about a large number of them and birds have become a 'flagship' taxon for ecological and behavioural research (Lawton, 1996). Ornithologists have been able to study individual species in a way that is impossible for, say, entomologists. Indeed, the amount of detailed information about topics such as breeding and foraging behaviour, diet and distribution, may at times seem overwhelming. Witness, for example, the volume and detail of information which has been condensed into the 7000 pages of *The Handbook of the Birds of Europe the Middle East and North Africa* (1977–1994, Oxford University Press). It is precisely this detail that allows investigations into the sensory ecology of birds to be conducted with some confidence. Such detailed information means that both general and specific relationships between sensory systems and the behavioural and ecological conditions in which they are employed should be relatively amenable to investigation.

This chapter looks at two interrelated areas of investigation. It attempts to account in terms of general ecology and behaviour for variations in, (i) the

S.N. Archer et al. (eds.), Adaptive Mechanisms in the Ecology of Vision, 485–508

optical structure of birds' eyes and, (ii) their associated visual fields. Eye structure and visual fields are particularly suitable for study in avian sensory ecology because of the widely acknowledged importance of vision in the control of avian behaviour which is readily observed. Indeed the maxim of Rochon-Duvigneaud (1943) that 'a bird is wing guided by an eye' is for many species quite appropriate. The behaviour of many birds does appear, at first analysis, to be dominated by the sense of vision with the prime parameters of this sense determined by the information about the animal's world which is projected onto its retina.

A comparative approach is employed in which similar techniques are used to investigate eye structure and visual fields in a range of species which differ in their phylogeny and ecology. This comparative approach increases the likelihood of successfully separating the influences of phylogenetic relatedness and ecological convergence on the evolution of sensory systems. A broad and diverse data base is required for the successful application of this approach. Although the present data base is rather small it does permit some general and specific relationships to be discerned and hypotheses can be proposed which are to some extent testable by the addition of similar data from further species.

15.2 Optical structure

Optically a bird's eye performs a simple function in a simple way. It projects onto the retina a two dimensional image of the bird's environment using a system which consists of just two principal refractive components – lens and cornea. Between them lies the iris which controls pupil aperture and hence image brightness. This general design is shown in Figure 1. For a more general account of avian eye anatomy see Evans and Martin (1993).

Although of simple design, this optical system embodies a number of degrees of freedom capable of producing eyes which have markedly different optical performances. These variations in optical design can involve: (1) the shape and relative positions of all principal refractive surfaces, (2) the effective refractive index of the lens, (3) the relative and absolute sizes of optical components, and (4) the size of the pupil aperture. The outcome of variations in these components is that when viewing the same scene, retinal images in different eyes can have quite different characteristics. In particular, retinal images may differ in absolute size and brightness and the angular extent of the portion of space imaged upon the retina (the monocular retinal visual field).

15.2.1 Descriptions of optical structure

A useful basis for describing the general optical structure of an eye is the construction of a schematic eye model (Hughes, 1986; Martin, 1983, 1994a). This is a mathematical description of the optical system and is usually

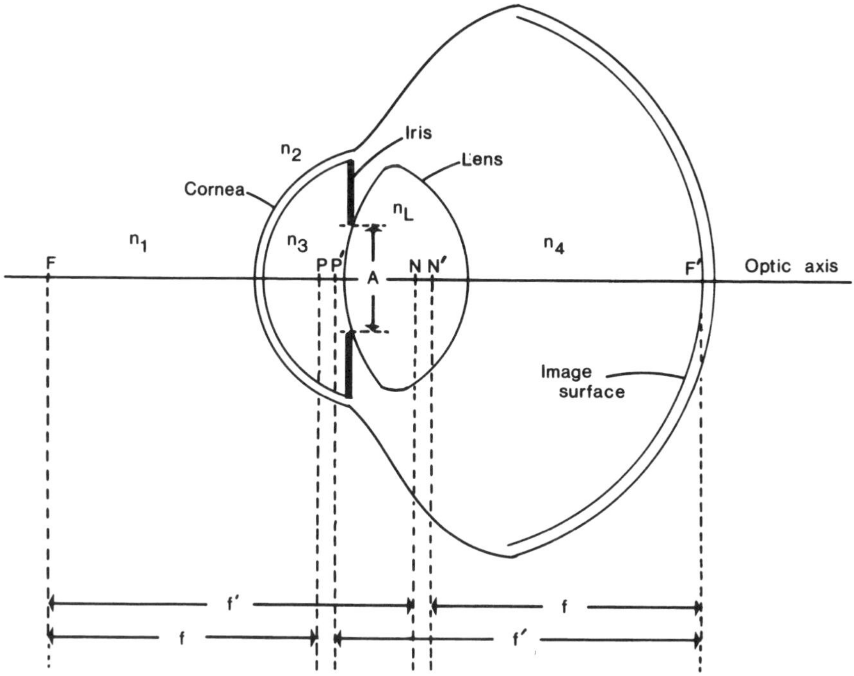

Figure 1 The general optical design of birds' eyes showing the main parameters defined in a schematic eye model. These parameters form the bases of interspecific comparisons as discussed in the text and Table 1. N, N′ are the nodal points, P, P′ are the principal points and F, F′ the focal points of the optical system. These points are mathematical constructs which allow the prediction of image position and size. Strictly they refer to an unaccommodated emmetropic eye imaging an object at infinity. They describe the points in a perfect optical system, of which the schematic eye is a model, through which a bundle of light rays would appear to pass and be focused. They are shown positioned along an optic axis which is an assumed axis of symmetry for the optical system. The distances *f* and *f′* between the nodal and focal points are the anterior and posterior focal lengths of the optical system. It should be noted that the reason why there are two focal lengths for an eye (compared with the single value quoted for a camera lens) is that the object and image points occur in media of different refractive index, n_1 (air) and n_4 (vitreous humour). n_2 and n_3 are the refractive indices of the cornea and aqueous humour, n_L is the calculated equivalent bulk refractive index of the lens. A is the diameter of the real pupil defined by the margins of the iris. A full specification of the eye's optical system also requires the radii of curvature of the two lens surfaces and of the two corneal surfaces. A model such as this will only strictly apply to paraxial rays, i.e. rays which are close to the optic axis, it will not accurately describe optical parameters for peripheral rays. From the schematic eye parameters the refractive power of the cornea F_C, lens *FL*, and whole eye F_E are calculated and these are important tools in interspecific comparisons as shown in Table 1. From Martin (1994a)

accompanied by a schematic diagram. This schematic model represents an hypothetical average eye for the species and examples of four schematic models are shown in Figure 2. While all such schematic models have limitations, the ability to reduce the description of each species' eye to a small number of key

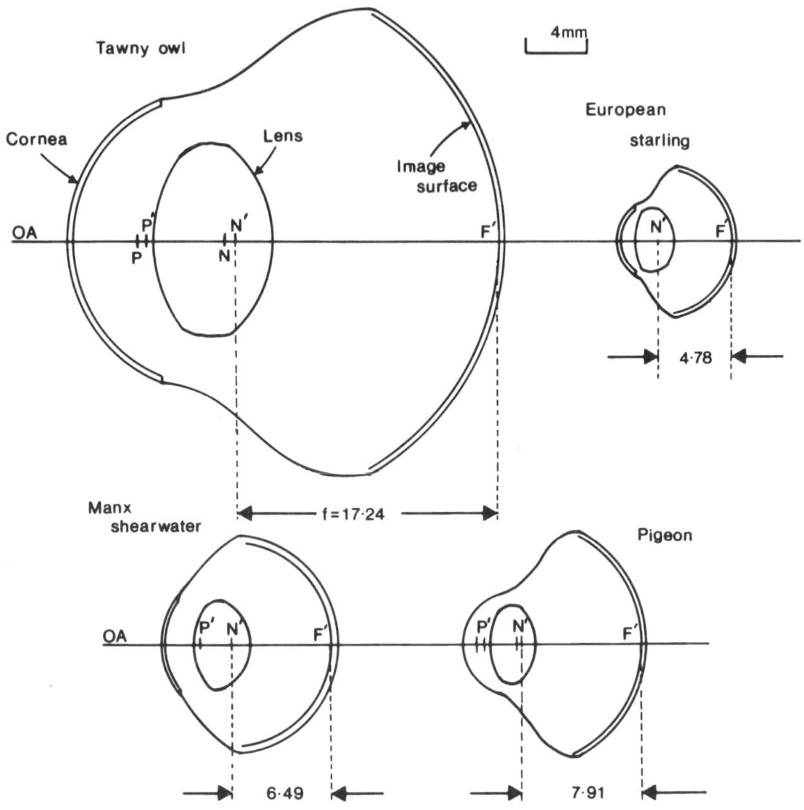

Figure 2 Scaled diagrams of the schematic eye models of four bird species. Also shown is the overall shape of the eye in an approximately horizontal plane. Symbols as in Figure 1. All numerical values in mm. Tawny Owl (Martin, 1982), European Starling (Martin, 1986b), Pigeon (Marshall *et al.*, 1973) and Martin and Brooke, 1991, Manx Shearwater (Martin and Brooke, 1991). From Martin (1993)

variables, allows reasonably robust interspecific comparisons of optical structure and performance to be made. These in turn allow correlations with key environmental variables and hence an exploration of ideas in visual ecology.

15.2.1.1 Eye size

In absolute terms, bird eyes show a wide interspecific variation in size. Figure 2 shows a group of eyes whose axial length varies between 8 and 28.5 mm. The largest eyes of any land vertebrates are thought to be those of ostriches *Struthio camelus* with an axial length of approx 39 mm (Martin and Katzir, 1995). Optically the most important parameter associated with eye size is the anterior focal length (or posterior nodal distance, PND) since it determines the size of the retinal area over which the image of an object is spread. Miller (1979) showed that a large PND is essential if the eye is to achieve the maximum theoretical limit of visual resolution. This is because there exists a finite limit on the minimum size and packing density of the retinal photoreceptors that can sample a retinal image, and this in turn will determine the maximum amount of information which can be extracted from an image (Snyder *et al.*, 1977). The process of increasing PND and hence spreading the image of a given object over more photoreceptors in order to detect more detail cannot, however, be continued *ad infinitum* since diffraction effects and aberrations within the optical system will set limits to image resolution (Miller, 1979). Thus, it is a simple prediction that eyes with the highest resolving power should be absolutely large and indeed this does seem to be the case with respect to diurnally active predatory birds such as the Falconiformes (Reymond, 1985). However, it has also been argued (Martin, 1982) that a large PND, and hence large eye, is also essential if an eye is to function adequately throughout the complete range of naturally occurring luminance levels, from night-time to bright midday sunshine.

In terrestrial habitats light levels can range naturally over 11 orders of magnitude with much greater variation in light levels occurring post sunset than during the day (Martin, 1990a). The argument here is that the optimal extraction of information from the retinal image at low light levels requires large and widely spaced receptors. However, at high light levels the same image will need to be analyzed by small, densely packed receptors if information extraction is to be maximised (Snyder *et al.*, 1977). Thus for an eye to function adequately over a wide luminance range [as might be expected in the case of nocturnally or crepuscularly active species, such as owls (Strigiformes)], both the effective size and density of photoreceptors must alter with luminance level to maximize the amount of information which can be extracted from the retinal image. Snyder *et al.*, (1977) suggest that this flexibility of image sampling can be achieved by pooling photoreceptor (mainly, or exclusively, rods) outputs at the retinal bipolar cells to achieve increasingly larger 'effective receptors' as light levels decrease within the crepuscular-nocturnal range. The degree of flexibility that potentially can be achieved by pooling receptor outputs will be greater if the image is spread over many photoreceptors whose outputs can be summed in various ways by the nervous system.

Thus, the large eyes of owls can perhaps be regarded not simply as nocturnal eyes (designed to function only within a narrow range of lower light levels), but as eyes of arrhythmic species which function adequately (or optimally) throughout the full range of naturally occurring day-time, twilight, and night-time light levels (Martin, 1982, 1990a). Indeed there is psychophysical evidence that although tawny owls *Strix aluco* usually adopt a nocturnal life-style (Martin, 1990a), their visual system functions adequately by both day and night, yielding acuity at day-light levels comparable with that of the strictly diurnal Rock Pigeon *Columba livia* (Martin, 1986c).

Bird species with small eyes, such as many of the smaller Passeriformes, which are strictly diurnal [except perhaps on migration (Martin, 1990b)], may be restricted to their life of diurnal activity by a combination of small eye size and a retina dominated by cone photoreceptors whose outputs are not pooled by bipolar cells. Within the natural day-time luminance range the required variation in receptor pooling for maximizing the extraction of information from the image is relatively small compared with the pooling required in the nocturnal (post-sunset) range. Therefore, it seems that as light levels fall, these smaller eyed birds must roost since their eye design and visual system prohibits the extraction of sufficient information from the visual environment to mediate flight safely under low-light conditions. During nocturnal migratory flights, such birds may collide with even well lit objects such as lighthouses, oil platforms and office blocks, suggesting a high degree of perceptual confusion at low light levels (Martin, 1990b).

For species living in habitats with only very low light levels, there appears to be little selective advantage in the possession of large eyes. This is because at such light levels only low spatial resolution is possible and this can be adequately achieved with a small image analyzed by large effective receptors (Snyder *et al.*, 1977). Thus, although the eyes of nocturnal, flightless kiwis (Apterygidae) are relatively small (axial length approx. 8 mm) they may be regarded, like those of rats (axial length approx 5 mm), as designed to function optimally in the narrow range of low light levels that occur at night below a forest canopy (Reid and Williams, 1975, Martin, 1990a). Indeed, Kiwis seem to depend in their foraging on a sense other than vision. Among birds, these species are unusually reliant upon olfaction (Bang and Wenzel, 1985) and this could be partly attributable to the low spatial resolution which is possible at the nocturnal light levels of forest floors.

15.2.1.2 Optical design

Figure 2 indicates that optical systems within eyes of different sizes are not simply just allometrically scaled models of each other. Furthermore, in eyes of the same size marked interspecific differences in optical structure are also evident. This section deals with functional interpretations of these differences in optical design.

One shorthand way of comparing optical design is to examine the relative contributions of the lens and cornea to the total refractive power of the eye (F_E). Refractive power of a cornea (F_C) is determined principally by its radius of curvature, while the refractive power of the lens (F_L) is a function of surface curvatures, thickness and refractive index gradient. The latter is particularly important since two lenses with identical physical dimensions could differ in refractive power if their internal structures are different (Martin, 1983; Hughes, 1986). The marked ways in which optical designs differ between species are made clear in Table 1 which summarises key parameters taken from schematic eye models of the four species shown in Figure 2.

Table 1: Comparison of eye size (axial length) and certain schematic eye parameters in European Starlings, Manx Shearwaters, Pigeons, and Tawny Owls. Schematic eye models for these species are shown diagrammatically in Figure 2. All data from the same sources as Figure 2. Focal length is the anterior focal length (PND). Refractive powers are in dioptres (D)

	F_{eye} (D)	F_{lens} (D)	F_{cornea} (D)	$\frac{F_{lens}}{F_{cornea}}$	axial length (mm)	focal length (mm)
Starling	209	112.5	124.6	0.90	7.92	4.78
Shearwater	154	108.6	68.1	1.60	11.82	6.49
Pigeon	126	38.7	95.9	0.40	11.62	7.91
Tawny Owl	58	29.9	35.7	0.84	28.50	17.24

A number of differences are readily apparent in Table 1. Thus, although the Manx shearwater and Pigeon eyes are of almost identical axial lengths, the ratio F_L:F_C is quite different. In pigeons the cornea does most of the refraction, contributing about 2.5 times the refractive power of the lens, while in shearwaters the situation is almost reversed, with the lens contributing more than the cornea to total refraction. By contrast, comparison between Starlings and Tawny Owls shows that although these two eyes differ in absolute size by a factor of 3.6, the F_L:F_C ratio is similar, thus indicating that their optical systems are virtually scaled versions of each other.

These differences in optical structure are not functionally trivial. Thus the total refractive power of the Pigeon and Shearwater eyes differ markedly such that the focal length of Shearwaters is less than that of Pigeons. In Pigeons the cornea contributes most of the total refraction and this brings the nodal point of the eye more anterior giving Pigeons a longer focal length than Shearwaters. This has two functional consequences. If these two species were viewing the same scene, pigeons would, by virtue of a larger image, have the potential for higher spatial resolution. On the other hand shearwaters, by virtue of a smaller and hence brighter image, have the potential for greater sensitivity.

15.2.2 Interpretations of optical design

It has long been assumed that at least some of these kinds of interspecific differences in optical design can be attributed to, or correlated with, differences in the sensory problems posed by particular ecological conditions and life-style (e.g. Walls, 1942; Tansley, 1965; Lythgoe, 1979). Two factors which have been considered to be particularly important in this regard are ambient light levels and an amphibious life-style. In analyses of how these factors might be correlated with eye design, data from mammalian as well as avian species have been used. This is based upon the assumption that if differences in eye design are correlated with fundamental ecological or life-style parameters then the same general rules might apply to the optical design of the eyes of both vertebrate groups.

15.2.2.1 Ambient light levels and optical design

It has often been convenient to describe a species' habits or life-style as 'nocturnal', 'diurnal', or 'crepuscular'. However, many 'nocturnal' species may be active over the full range of light levels (typically 7 orders of magnitude) which can occur naturally after sunset. Diurnally active species are typically active over a smaller range (4 orders of magnitude) of higher light levels.

Figure 3 shows the ratio of lens to corneal refractive power F_L:F_C as a function of eye size in four bird and six mammal species which have been allotted to either nocturnal or diurnally active categories. The straight line is the linear regression of the data points for the nocturnal species. It indicates that among nocturnal forms lens power becomes relatively greater as eye size decreases. A similar relationship is not apparent for diurnal species (but sample size is small) and the F_L:F_C appears relatively constant across a three-fold range of eye size.

In these diurnal eyes the cornea always contributes the greater part of total refractive power. This brings the nodal point forward within the eye and maximises focal length. In the nocturnal forms the F_L:F_C ratio is always higher than in a diurnal eye of the same size. These simple general features of eye

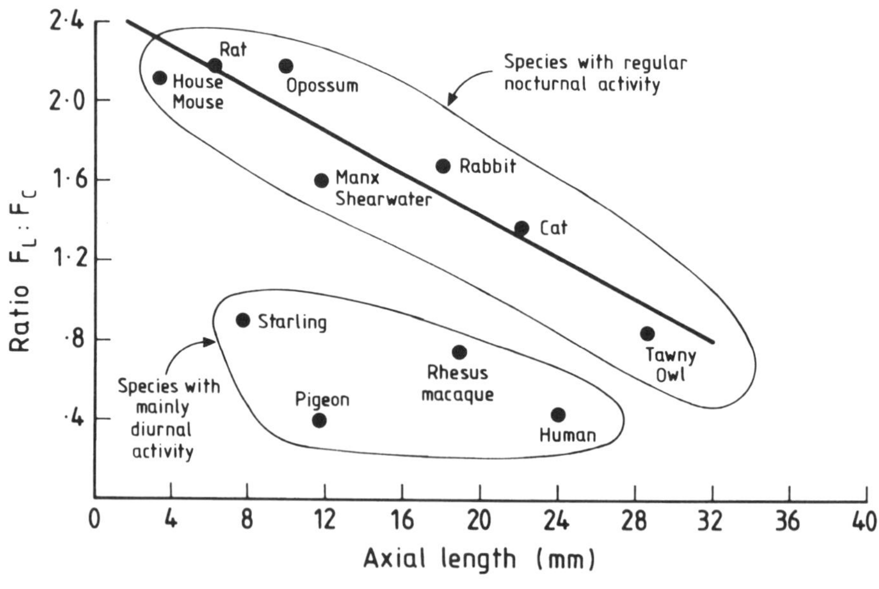

Figure 3 Comparison of schematic eye parameters (ratio of lens : cornea powers (F_L:F_C) as a function of eye axial length) in seven species of mammals and four species of birds which have been allotted to nocturnally and diurnally active categories. The straight line is the linear regression of the data points for the nocturnally active species. Schematic eye data from the same sources as Figure 2 plus, Rat (Hughes, 1979), House Mouse (Remtulla and Hallet, 1985), Opossum (Oswaldo-Cruz *et al.*, 1979), Cat (Vakkur and Bishop, 1963), Human (Hughes, 1977), Rabbit (Hughes, 1972), Rhesus Macaque (Hughes, 1977). From Martin and Brooke (1991)

design indicate that in eyes of the same axial length, diurnal forms, by virtue of their larger image size, have the potential for high visual acuity, while nocturnal eyes produce a brighter retinal image at a given pupil size.

15.2.2.2 Amphibious habits and optical design

Species that employ vision in both air and under water are faced with a specific optical problem. Upon immersion the refractive power of the cornea is abolished because the corneal surface now separates media of near identical refractive index. If the eye of an amphibious animal is to produce a well-focused image in both air and water this loss of corneal refractive power must be compensated for, probably by an accommodative increase in the refractive power of the lens. If the eye is emmetropic in air then the loss of the corneal power upon immersion must be compensated for by an accommodative increase in the power of the lens. If the eye is emmetropic in water the gain in corneal refractive power upon entering air must be compensated for by a loss of refractive power. Table 1 indicates that should an aerial eye like that of the starling be immersed in water, its lens would need to provide 124 Dioptres (D) of additional refractive power in order to produce a clearly focused image.

These considerations led Sivak (1976) to propose that eyes designed for well focused vision in both air and water would benefit from a cornea which is relatively flat, and hence of low refractive power. Such an optical design is found in the eyes of penguins (Sphenisciformes) (Sivak, 1976). In Humboldt Penguins (*Spheniscus humboldti*) the cornea has a refractive power of only 29 D (Martin and Young, 1984), the loss of which upon immersion can perhaps be compensated for by an accommodative mechanism (Sivak *et al.*, 1987). However, Martin and Young (1984) found evidence that an unaccommodated penguin eye is emmetropic in water and that the lens must lose refractive power upon entering air. However, evidence of such an accommodative power loss in any bird eye is lacking.

Analysis of the optical design of Manx Shearwater and penguin eyes presents a further interesting problem in the interpretations of eye design which indicates the difficulty of separating ecological influences. Thus, Shearwaters can be described as semi-nocturnal and amphibious. They catch prey at the sea surface and in shallow dives and there is evidence that vision is used to locate their nest burrows which they enter only under the cover of darkness (Brooke, 1990). Penguins are active mainly during the day but many species dive to great depths (Davis and Darby, 1990) where light levels are strongly attenuated and typically reach levels within the nocturnal range. Both habits predict an eye with short focal length in which the bulk of refraction is produced by the lens rather than the cornea. The amphibious habit predicts a cornea of low power to overcome the problems of immersion, the nocturnal habit predicts a powerful lens to produce a short focal length and hence a brighter retinal image. That the possible amphibious and nocturnal features of these eyes cannot be easily separated indicates the complexity of factors which may underlie optical design in bird eyes and clearly indicates the need for a wide comparative base including detailed analyses of both life style and optical structure.

15.2.3 The role of the iris and pupil

The iris performs functions which are of central importance to the visual performance of the vertebrate eye. The iris regulates pupil size which in turn influences image brightness and image quality and the depth of focus of the optical system. All of these functions are well understood in general terms and will be familiar to the photographer who can skilfully manipulate iris size to produce photographic effects. How iris size can influence parameters of optical performance have been summarised by Land (1981) and Martin (1994a), but there have been few studies of iris function specifically in bird eyes. As well as altering parameters of the image there is evidence that the iris may also serve an accommodatory function by directly influencing the refractive power of the lens in certain amphibious bird species (Schaeffel, 1994).

15.2.3.1 Pupil size and image brightness

When viewing a point source of light (such as a star in the night sky) the characteristics of the optical system does not influence image brightness. This is related only to the diameter of the entrance pupil. However, when viewing an extended light source (the majority of natural stimuli) image brightness does depend both upon the characteristics of the optical system and on the diameter of the pupil, such that image brightness is inversely related to the square of the *f*-number [the anterior focal length/entrance pupil diameter]. Thus the smaller the *f*-number, the relatively brighter the image (Kirschfeld, 1974; Land ,1981). It might be expected therefore that the eyes of nocturnally active species would have lower minimum *f*-numbers (higher maximum image brightness) than diurnal species. Figure 4 shows that this is the case in a range of species of birds and mammals. However, the average difference in image brightness between animals with eyes of equal size which are primarily nocturnal or diurnal, equals only four- or five-fold. Figure 4 also shows that maximum retinal image brightness increases as eye size decreases, such that larger eyed nocturnal owls have lower maximum image brightness than some of the small eyed rodents which they might prey upon.

These differences in maximum image brightness are very small compared with the actual differences in light levels which occur by night and by day (Martin, 1990a). Therefore differences in image brightness cannot be viewed as compensating directly for differences in nocturnal and day-time luminance levels. The small differences in image brightness between humans and owls (approximately

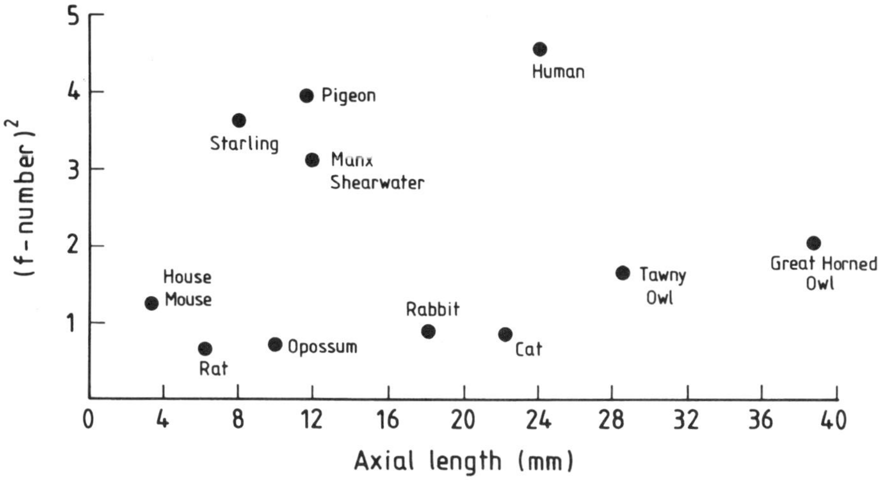

Figure 4 Comparison of schematic eye parameters (relative image brightness (*f*-number)^2 as a function of eye axial length) in six species of mammals and five species of birds. Same sources as Figure 2 and 3 plus Great Horned Owl (Murphy *et al.*, 1985). From Martin (1994a)

2.5 fold) are nevertheless functional. Thus, it is possible to account for the differences in psychophysically measured absolute visual threshold in these species by the difference in the maximum retinal image brightness of the two eyes (Martin, 1977; 1982). Tawny owls have a lower absolute threshold (greater sensitivity) because their retinal images are brighter, not because of differences in the sensitivity of their retinas, which in both species are thought to be close to a theoretical maximum limit (Martin, 1982).

The range over which a mobile circular pupil can alter retinal image brightness is insufficient to equalize image brightness over other than a small part of the natural range of light levels. Within a particular habitat, however, natural light levels usually change by only a small amount. Even thick clouds quickly covering the sun change illuminance levels by a maximum of only ten-fold. A variable aperture pupil, such as that of Pigeons which can alter retinal image brightness over a 16-fold range (Marshall *et al.*, 1973), could therefore serve to equalize retinal image illumination as ambient light levels change throughout the day, or as the animal moves in and out of cover within a particular habitat.

It has been proposed that one function of a variable pupil is in maintaining the retina in a partially dark adapted state (Woodhouse and Campbell, 1975). The adaptive significance of this is that retinal mechanisms of dark adaptation are relatively slow compared with the speed of the pupillary response. Without a mobile pupil a rapid change in ambient light level could result in the eye being inadequately adapted to a change in light level as an animal moved about its habitat between areas of light and shade. A partially closed mobile pupil can be rapidly varied in size and thus maintain retinal adaptation near the optimum as ambient light level changes within a restricted natural range.

15.2.3.2 Pupil size and image quality

Images formed by any optical system are never perfect in that all light from an independent object point is not brought exactly together into a single image point. Generally it is rays which pass through more peripheral parts of an optical system that are the principal source of image imperfection. Therefore, image quality can be enhanced by removal of these peripheral rays by the iris. Such a mechanism has been demonstrated in humans and in birds. Thus, in humans, retinal image quality is high as long as pupil diameter is below 2.4 mm. Beyond that diameter image quality deteriorates as increasingly aberrant rays come to form part of the image (Campbell and Gregory, 1960; Campbell and Gubisch, 1966). However, it is theoretically understood that in a perfect optical system the larger the pupil aperture, the finer the detail that can be resolved in the image (Miller, 1979; Laughlin, 1992). Thus optical systems of higher quality than the human eye may become limited in their optical performance at larger pupil apertures. This has been demonstrated in the eyes of some diurnal raptors (Falconiformes). For example, in Congo Snake Eagles (*Dryotriorchis spectabilis*), image quality does not deteriorate until a pupil diameter exceeds 6.25 mm, and at this diameter image quality is superior to that of human eyes (Shlaer, 1972). Miller (1979) reported that in eight raptor species stable pupil diameter at daylight levels was greater than in human eyes and concluded that all of these birds' eyes have superior optical quality to that of humans.

15.2.3.3 Pupil size and depth of field

The retina of the vertebrate eye is in effect a fixed surface of finite depth onto which images are projected for analysis by the nervous system. At any particular accommodative state only objects at one distance from the eye are sharply focused. Images of objects both closer and further away from the object in focus are defocused to varying degrees. Whether these defocused images result in a reduction in the amount of information that the visual system can extract from them (or even whether they are discerned as out of focus) depends upon properties of both the optical system and of the nervous system that analyzes the image.

The depth of field is the distance that a focused object can be moved towards or away from the optical system without appreciably decreasing the amount of information that can be extracted from its image. In experimental terms the criterion used in determining depth of field of an eye is the distance that an object can be moved without causing perceptible defocus of its image (Campbell, 1957). However, the degree of perceptible defocus is not fixed, but is a function of the analysing system. Furthermore, it will depend upon the resolution of the image analysing system at a particular luminance. The poorer the resolution of the visual system the greater will be the depth of field. The depth of focus of an optical system is the conjugate parameter of the depth of field in the image plane. Formulae permit comparison of depth of focus in eyes with different optical characteristics (Green *et al.*, 1980; Martin, 1994a). These formulae show that depth of focus and hence of field, are related both to the focal length and to the pupil diameter of the eye, as well as resolution of the visual system. The influence of pupil diameter upon depth of focus, and consequently upon depth of field, is far from trivial and it is convenient to consider this as a function of *f*-number in eyes of different focal lengths (Martin, 1994a). It should be noted that depth of focus is often expressed in dioptres because its dioptric equivalent is constant for all viewing distances (Schaeffel, 1994). However, this masks significant effects which occur in the depth of field of eyes of different sizes when viewing objects at different distances. These effects are likely to be significant in behavioural and ecological interpretations of eye design.

The following general points about depth of field can be made which will, perhaps, be familiar to photographers. They are described in more detail in (Martin, 1994a).

1. At a large pupil aperture (low *f*-number, brighter image), depth of field is small. As pupil aperture decreases (*f*-number increases), depth of field expands rapidly, and this happens more quickly the smaller the eye (the shorter the focal length).
2. The smaller the focal length (smaller eye) the wider the depth of field at any given *f*-number. For example at *f*-number = 13, depth of field of an eye with a focal length of 5 mm (approximately that of European Starling *Sturnus vulgaris*) and focused on an object at 500 mm, stretches from 250 mm to infinity, while in an eye with a focal length of 17 mm (approximately that of Tawny Owls *Strix aluco*) only objects lying between 460 mm and 545 mm from the eye are in focus.
3. The distance from the eye to the object in focus has an important influence upon depth of field. Around close objects the depth of field is smaller at a given *f*-number than around more distant objects. This effect is larger in smaller eyes.

These marked changes in depth of field with *f*-number are produced by relatively small changes in absolute pupil diameter. Also, the influence of the pupil aperture becomes increasingly marked as eye size decreases. For example, in a Starling eye a three-fold change in *f*-number (which would produce a 9-fold change in image brightness) is produced by altering the diameter of the pupil by only 0.67 mm (Martin, 1994a). This small change in pupil diameter would produce a marked widening of the depth of field such that around a focused object at 500 mm from the eye, depth of field increases from 460 mm to infinity. Even in a Tawny Owl's eye the same three-fold change in *f*-number is produced by only a 2 mm alteration in pupil diameter but the associated change in depth of field is much smaller.

It is clear that in any eye very small changes in pupil aperture can have marked influence on the performance of the whole visual system. Large eyes will perhaps need more rapid and active focusing mechanisms than small eyes since the images of objects moving towards or away from the eye will appear defocused more readily in the larger eye. This is certainly born out in the case of the larger owl eyes where very rapid accommodative changes occur (Schaeffel, 1994).

It is clear that a full understanding of the factors which have led to the evolution of any particular eye design must take account of the way that pupil size can influence visual performance.

15.3 Visual fields

The visual field of an animal defines that portion of space about it in which visual control of behaviour may be possible. Head movements can allow a bird to scan all segments of space but the visual field defines the portion of space that can be viewed instantaneously without head movements. The visual fields of birds show a high degree of complexity and interspecific variability. This results from difference in the optical structure of eyes, their placement in the skull and the ways in which they can be moved. As in the case of optical structure, functional interpretations of visual fields are dependent upon detailed analyses of only a few bird species. However, these differ in ecology, life-style and evolutionary origins, and this has permitted the postulation of some general functional principles.

The functional (or retinal) visual field is not identical with the optical field of view of the eye. The optical field is the angular extent of space that is imaged by the optical system. It is defined operationally by the angular limit from which the pupil can be seen. The functional field is that portion of the optical field that is served by the retina and is also referred to as the retinal field. It defines the segment of space in which stimuli must lie if they are to elicit a behavioural response. In birds the difference between the boundaries of the optical and retinal fields may be quite large (up to 20°). This observation is important since an estimate of visual field extent based solely upon observation of the pupil could lead to a considerable overestimate especially of the frontal binocular field.

15.3.1 Monocular fields

It is possible to calculate the maximum extent of an eye's optical field from knowledge of its optical structure (Vakkur and Bishop, 1963; Martin, 1983). Such calculations assume that optical fields are symmetrical about the optic axis. While this assumption is probably valid it seems likely that the functional retinal field of an eye is highly asymmetric with respect to the optic axis. This has been described most clearly in Tawny owls (Martin, 1984). Furthermore, in all other bird species investigated to date, marked asymmetries in retinal field in an approximately horizontal plane have been found. In all species the nasal hemi-field (the portion of the visual field between the optic axis and the direction of the bill) is considerably smaller than the temporal hemi-field (See Table 2). In all species studied to date, including Herons, Ciconiiformes, Owls, Ostriches, Woodcocks *Scolopax rusticola*, Albatrosses *Diomedea spp.* and an Eagle *Circaetus gallicus*, (Martin, 1984, 1994b; Martin and Katzir, 1994a,b, 1995, and Martin, 1998) full use is made of the available optical field in the temporal quadrant but not in the nasal quadrant.

The function of this asymmetry is not clear. One consequence is that the frontal binocular field is considerably narrower than it could be if full use were made of the nasal portions of each monocular optical field. This suggests that binocular fields are not maximised in width (see discussions on binocular fields below). In some species binocular fields could be doubled in size if full use was made of the available optical field.

One possible explanation of this failure to use the full potential width of the nasal optical field may be the need for a high quality optical image in frontal vision, which is invariably used to guide flight and foraging. It must be recognised that due to the lateral placement of the eyes in the avian skull, frontal vision, which includes the region of binocular overlap, is always mediated by peripheral portions of the eyes' optics. The assumption is that image quality deteriorates with increasing eccentricity from the optical axis and hence not employing the most extreme portions of the available optical field serves to

Table 2: Retinal visual field widths, axial eye length and separation of eyes in the skull (distance between the nodal points of the two eyes within the skull) in eight bird species. Human data for comparison. All visual fields data refer to visual fields in an approximately horizontal plane when the eyes have adopted a relaxed position and the head is in a typical posture for the species. See Figures 5 and 6 for an explanation of the retinal fields in the horizontal plane and head posture. Note that the first five species all forage by taking individual food items directly in the bill under visual guidance. Foraging in Woodcocks and Mallards is guided by tactile cues from the bill and Tawny Owls employ auditory and visual cues to detect prey which are taken in the feet. All data from the same sources as cited in Figures 5 and 6 and in the text

	Retinal field (degrees)			Eye linear dimensions (mm)	
	Binocular	Monocular	Cyclopean	Axial length	Nodal separation
Ostrich	20	155	290	39.0	82.0
Cattle Egret	22	167	321	13.0	22.0
Night Heron	22	171	320	17.7	29.0
Stone-curlew	18	181	345	20.5	32.0
Pigeon	22	169	316	11.6	16.0
Woodcock	4.5	182	359	16.9	20.0
Mallard	8	183	360	13.5	18.0
Tawny Owl	48	124	201	29.0	40.0
Human	140	170	200	24.0	61.2

maintain high image quality for frontal vision. However, the assumption of poor peripheral image quality may not hold for vertebrate eyes although data on image quality throughout the entire visual field in a bird is not available. In the rabbit eye, however, high quality images are produced in the far periphery (Hughes and Vaney, 1978) and there is some evidence that this may be the case in some owl species (Murphy and Howland, 1983).

Comparison of monocular retinal field widths in tawny owls and starlings (Table 2) shows how eye shape and the extent of the retina restrict the functional (retinal visual field). Thus, although these two eyes differ in size, their optical designs are nearly identical and so calculations show that they are likely to have optical fields of the same size. The Table shows, however, that the owl's monocular retinal field is nearly 40° narrower than that of the starlings. This difference can be attributed to the tubular shape of the owl eyes, compared to the flatter eye of the starling, which result in a less extensive retina in the owl in comparison with the optical field.

15.3.2 Binocular and panoramic fields

No species of bird has been described in which the eyes face forward, parallel to each other, as in primates. Even in species commonly regarded as having frontally placed eyes, for example owls and ostriches, the optic axes diverge by 55°. In some species, such as woodcocks and Mallard Ducks *Anas platyrhynchos*, the eyes may diverge by nearly 180°. The position and movements of eyes in the skull determine the extent of the binocular and panoramic visual fields. These fields are complex three dimensional structures which are difficult to present in illustrations on a flat surface. Marked differences in these three dimensional fields are clearly evident in interspecific comparisons and it is possible to suggest some functional explanations of these differences, especially in relation to the foraging techniques of the species and the extent to which the bird may employ visually guided foraging, in which bill position has to be controlled with respect to individual objects.

15.3.2.1 Binocular fields, foraging and predation

In many species the frontal binocular field is long and narrow and positioned such that the bill lies at its approximate centre. Such an arrangement is found in species of diverse phylogeny and ecology including Ostriches, Starlings, Pigeon, Herons, Stone-curlew *Burhinus oedicnemus*, Manx Shearwater *Puffinus puffinus*, Humboldt Penguin, Black-browed and Grey-headed Albatrosses *Diomedea spp*. Moreover, in all of these species, despite marked differences in eye size and position, and in monocular and panoramic field widths, the width of the binocular field in the region of the bill is approximately 20° (Table 2).

These comparisons suggest that a narrow binocular field about the bill can probably be regarded as a convergent feature of species which feed by visual guidance of the bill directly towards individual objects (Martin and Katzir, 1995). Even Ostriches, whose eyes are amongst the largest of all land vertebrates and the most widely separated of all birds, have a narrow frontal binocular field which is only 20° wide and is very similar to that of such species as Stone-curlews and pigeons, which are very much smaller species (Figure 5). However, all three species feed by the accurate pecking of small individual food items which are taken in single pecks. These pecks occur in the direction of the plane of the bill. In Ostriches, this pecking seems to be entirely under visual control.

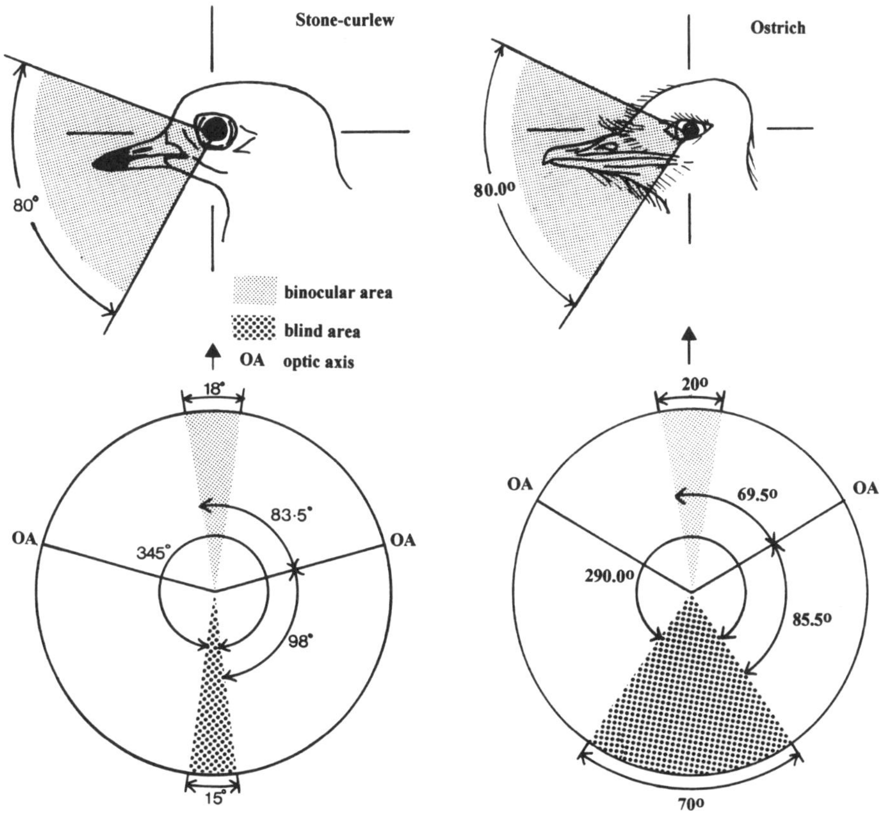

Figure 5 Comparison of visual field parameters in Ostrich and Stone-curlew. Top, the shaded area shows the position relative to the bill and the vertical extent in the median sagittal plane of the area of binocular overlap. Bottom, section through the visual fields in the horizontal plane (as indicated in the top diagrams) showing the width of the area of binocular overlap, the blind area behind the head and the extent of the monocular field (Divided between nasal and temporal sectors). OA is the direction of the optic axes. The drawings of the heads are not to scale but each is shown in a typical resting posture relative to the horizontal. Data from Martin and Katzir (1994a,b)

Each peck, as in pigeons and herons, shows a relatively slow approach and/or static phase, followed by a rapid ballistic movement during which the bill begins to open and the eyelids or nictitating membrane start to close (Zeigler *et al.*, 1993; Katzir, 1994).

The similarity in binocular field width and shape about the bill in this diverse array of species which feed by accurate visually controlled bill-pecking is quite striking (Table 2). However, why there should be convergence upon a resting binocular field width about the bill of about 20° in the plane of the bill is not clear. In all of these species binocular field width could be double that recorded if full use were made of the available optical field in frontal vision (see above). It is thought that in birds, accommodation rather than stereopsis, is the primary cue used to guide absolute distance perception in pecking (McFadden, 1994; Schaeffel, 1994). In pigeons pecking accuracy is not diminished by monocular as opposed to binocular viewing (Jäger and Zeigler, 1991). Thus, binocularity itself may be less important than the fact that each monocular field extends contralaterally. In the pecking species this equals only approximately 10°. It may be that this relatively small (10°) contralateral extent of the visual field of each eye is sufficient to provide visual flow field cues about the pecking target. Two eyes with similar flow field information may be sufficient to locate and maintain accurate fixation upon a target during the actual peck [see, for example, discussions of pecking behaviour and of flow fields in McFadden (1994) and Lee (1994)].

That this apparent convergence of binocular field widths results from the visual demands of accurate visually guided pecking, is further supported by the marked differences in visual field topography in species which do not forage in this way, and by information on how visual fields may be altered by eye movements.

In Tawny Owls the bill lies completely outside the visual field (Figure 6) and the bird has a binocular field width of 48° (Table 2). This field topography can be correlated with the owl's habit of using its feet to catch prey. In its 'perch and pounce' hunting technique, the feet are swung forward to lie within the region of binocular vision just prior to prey capture. All prey is passed to the bill by the feet and is rarely taken directly in the bill.

In Woodcocks the bill also falls outside the visual field (Figure 6) and in Mallard Ducks it falls within its very periphery (Martin, 1986a; Martin, 1994b). This can be correlated with evidence showing that foraging in these birds can be guided exclusively by tactile and taste cues rather than vision (Gottschaldt, 1985; Martin 1990a). In these species eye movements are of small amplitude and the eyes placed so high in the skull that binocular overlap extends from just above the bill through approximately 180° to directly behind the head. Thus, these birds have no blind area above or behind them (except that produced by the visual projections of the pectens in each eye), and can see the whole of the celestial hemisphere. This can be visually scanned for the presence of predators whilst foraging is guided by tactile cues.

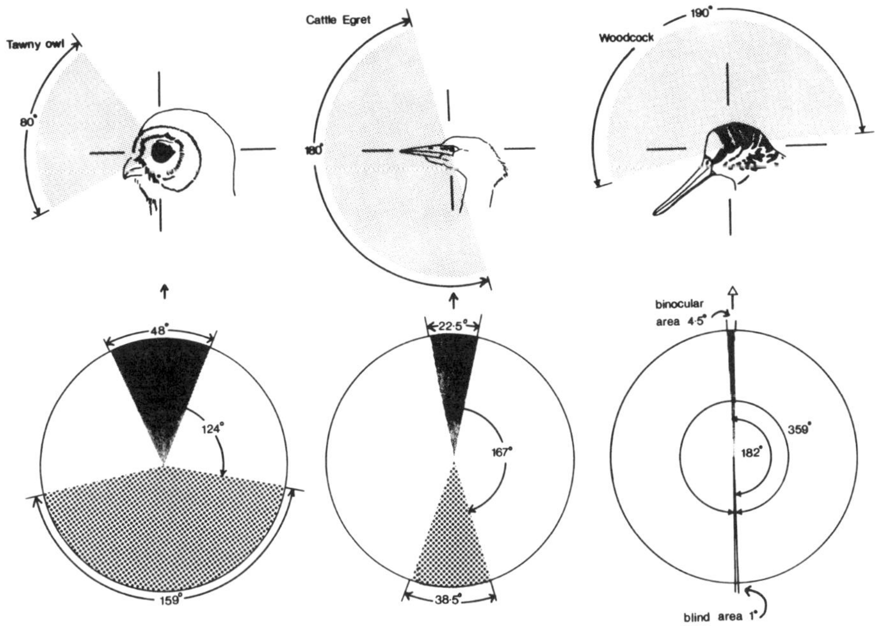

Figure 6 Comparison of visual field parameters in Woodcock, Tawny Owl and Cattle Egret. Top, the shaded area shows the position relative to the bill and the vertical extent of the area of binocular overlap. Bottom, section through the visual fields in the horizontal plane (as indicated in the top diagrams) showing the width of the area of binocular overlap, the blind area behind the head and the extent of the monocular field. The drawings of the heads are not to scale but each is shown in a typical resting posture relative to the horizontal. Data from Martin (1984), Martin and Katzir (1994a,b)

Such comprehensive visual coverage is potentially important in many birds, especially ground feeding species which may be particularly vulnerable to predation by birds of prey or mammals. However, it is perhaps only achieved by those species such as Woodcocks and Ducks which feed by probing into soft substrates for invertebrates, by grazing or by dabbling, and do not need to monitor their bill position visually during foraging. In visually guided ground feeding species, such as Starlings and Pigeons, some degree of comprehensive vision must be sacrificed and the eyes are placed sufficiently forwards for the bill to fall within the frontal binocular field. The nasal:temporal asymmetry of the monocular field which arises due to full use being made of the optical field in the temporal sector of the visual field, but not in the nasal sector, also suggests that maximising panoramic coverage of the space above and behind the head, rather than the maximising of binocular vision, has been an important influence upon the evolution of visual fields in birds.

In many species the degree of binocular overlap can be altered with eye movements. Binocular overlap can be increased as well as decreased from the static resting position of the eyes. In Starlings binocular overlap very similar to that of Tawny Owls can be achieved. Many species [Starlings, Herons, Albatrosses and Tawny Frogmouths (*Podargus strigoides*), Caprimulgiformes (Wallman and Pettigrew, 1985)] have also been shown capable of completely abolishing their frontal binocular fields. Presumably the function of such abandonment of frontal binocular vision is to achieve greater coverage of the lateral and celestial fields of view thus facilitating the search for predators and conspecifics. It should also be noted that eye movements in birds are non-conjugate. This can have the result that one eye may remain in a forward position and achieve a small region of contralateral field about the bill, while the other is moved more laterally and so abolishes both a contralateral and binocular field of view. How this variability in binocular field extent is coped with neurally is unknown.

Although there appears to be convergence in the maximum horizontal width of the frontal binocular field in species employing visual guidance of bill position during foraging, there are marked differences in the vertical extent of the frontal binocular field in these same species. Thus in Ostriches and Stone-curlews the vertical extent of the binocular region is approximately 80°, while in the herons it may extend to 180° (Martin and Katzir, 1994a,b, 1995). In the herons this can be correlated with their prey catching technique which involves the stealthy approach to highly mobile prey items which exhibit rapid escape responses (Katzir and Intrator, 1987). Thus, these birds typically have a single attempt to successfully secure an individual prey item with their bill. Binocular vision 90° below the bill means that they can observe prey items close to their feet while standing motionless with the bill at or slightly below the horizontal. Presumably the birds are able to track such prey without moving themselves until the prey comes within the region of a successful bill strike (Figure 7).

15.3.2.2 Binocular fields and nocturnal habits

The data presented above bring into question the presumed general association of large binocular fields with a predominantly nocturnal life-style (Walls, 1942; Tansley, 1965). Foraging technique may be the most important correlate of visual field topography and it should be noted that of the species listed in Table 2, Night Herons, Stone-curlews, Woodcock and Mallard, as well as Tawny Owls, are all known to forage nocturnally, yet their visual fields differ markedly in topography. Pettigrew (1979) raised the possibility that the more frontal placement of the eyes in both owls and diurnal birds of prey is another example of ecological convergence. He suggested that increased binocular field width is the result of selection to bring more central portions of the eye's optics to point in the forward direction. However, as argued above it cannot be assumed that optical quality in peripheral parts of the visual field is necessarily inferior to that close to the optic axis.

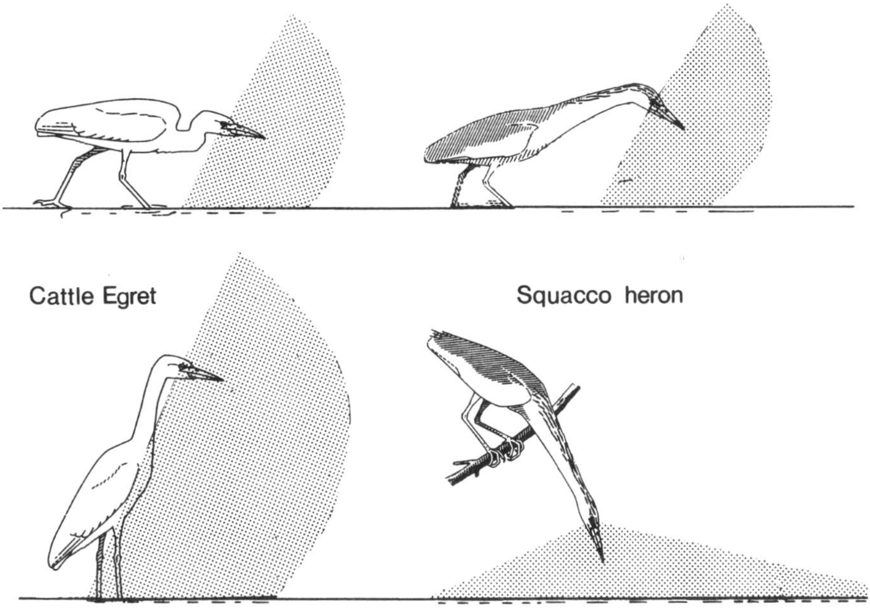

Figure 7 Lateral views of Squacco Herons and Cattle Egrets in typical foraging postures. The stippled areas indicate the vertical extent of the region in the median sagittal plane in which binocular overlap is possible. From Martin and Katzir (1994a)

15.3.3 Visual fields and amphibious habits

One further example of how ecological factors may influence visual field topography is provided by amphibious species. Upon immersion, the refractive power of the cornea is abolished. This not only reduces the refractive power of the optical system but it also narrows the monocular field width. Reducing the width of the monocular fields also reduces the width of the binocular and panoramic fields. It has been demonstrated in penguins (Martin and Young, 1984) that the visual fields are such that upon immersion the binocular field, although reduced, is maintained thus avoiding the production of a blind area in frontal vision. This suggests that the resting visual field of these birds is such as to provide visual guidance of the bill underwater while at the same time maintaining the panoramic field. Penguins are vulnerable to predation by a number of marine mammals and fish and have elaborate predator avoidance behaviours (Davis and Darby, 1990). Thus a broad visual field may be particularly important when underwater and this may have been the principal determinant of visual field topography in these species.

15.4 Conclusions

The above discussions give some indication of the complex factors which must be considered in attempting to examine the optical structure of bird eyes from a sensory ecology perspective. Clearly anatomical and optical constraints provide a set of limitations upon eye structures but within these a great deal of flexibility in design is possible. It seems that evolutionary processes have taken advantage of this flexibility to produce eyes and visual fields which are closely associated with the visual problems of specific life-styles and ecology. Although the case was made that birds are well suited to this kind of comparative approach, it is clear that this approach could be applied to other animal groups. However, all such studies will require an ever broader comparative data base before definitive answers in this branch of sensory ecology can be proposed with confidence.

References

Bang, B. G. and Wenzel, B. M. (1985) Nasal cavity and olfactory system, in *Form and Function in Birds*, vol. 3 (ed King, A. S. and McLelland, J.), Academic press, London and New York, pp. 195 – 225.

Baumel, J. J. (ed) (1993) *Handbook of avian anatomy: Nomina Anatomica Avium*, Nuttall Ornithological Club, Harvard, Mass.

Brooke, M. (1990) *The Manx Shearwater*. Poyser, 1990.

Campbell, F. W. (1957) The depth of field of the human eye. *Optica Acta*, 4, 157–164.

Campbell, F. W and Gregory, A. H. (1960) Effect of size of pupil on visual acuity. *Nature (Lond.)* 187, 1121–1123.

Campbell, F. W. and Gubisch, R. W. (1966) Optical quality of the human eye. *Journal of Physiology London*, 186, 558 – 578.

Davis, L. S. and Darby, J. T. (eds) (1990) P*enguin Biology*, Academic Press, London.

Evans, H. E. and Martin, G. R. (1993) Organa sensum, in *Handbook of avian anatomy: Nomina Anatomica Avium*, (ed. Baumel, J. J.), Nuttall Ornithological Club, Harvard,Mass. pp. 585 – 612.

Gill, F. B. (1990) *Ornithology*, Freeman, New York.

Gottschaldt, K-M. (1985) Structure and function of somatosensory receptors, in *Form and Function in Birds*, vol. 3 (ed King, A. S. and McLelland, J.), Academic press, London and New York, pp. 375 – 461.

Green, D. G., Powers, M. K. and Banks, M. S. (1980) Depth of focus, eye size, and visual acuity. *Vision Research*, 20, 827– 836.

Howard, R. and Moore, A. (1991) *A complete checklist of the birds of the world*, 2nd edn Academic Press, London.

Hughes, A. (1972) A schematic eye for the rabbit. *Vision Research*, 12, 123 –138.

Hughes, A. (1977) The topography of vision in mammals of contrasting life style: comparative optics and retinal organization, in *Handbook of Sensory Physiology*, Vol VII/5 (ed Cresitelli, F.) Springer, Berling, Heidelberg, New York, pp. 613 –756.

Hughes, A. (1979) A schematic eye for the rat. *Vision Research*, 19, 569 – 588.

Hughes, A. (1986) The schematic eye comes of age, in *Visual Neuroscience* (eds Pettigrew, J., Sanderson, K. J. and Levick, W. R.) Cambridge University Press, Cambridge, pp. 60 – 89.

Hughes, A. and Vaney, D. I. (1978) The refractive state of the rabbit eye: variation with eccentricity and correction for oblique astigmatism. *Vision Research*, 18, 1351–1355,

Jäger, R. and Zeigler, H. P. (1991) Visual fields and peck localization in the pigeon, *Behaviour and Brain Research*, 45, 65 – 69.

Katzir, G. (1994) Tuning of visuomotor coordination during prey capture in water birds, in *Perception and Motor Control in Birds* (eds Green, P. and Davies, M.), Springer, Berlin, Heidelberg, New York, pp. 315 – 338.

Katzir, G. and N. Intrator (1987) Striking of underwater prey by a reef heron, *Egretta gularis schistacea. Journal of Comparative Physiology A*, 160, 517– 523.

Kirschfeld, K. (1974) Absolute sensitivity of lens and compound eyes. *Zeitschrift fur Naturforsch*, 29c, 592 – 596.

Land, M. F. (1981) Optics and vision in invertebrates. In: *Handbook of Sensory Physiology*, vol VII/6B (ed. Dartnall, H. J. A.) Springer, Berlin, Heidelberg, New York, pp. 471– 594.

Laughlin, S. B. (1992) Retinal information capacity and the function of the pupil. *Ophthalmology and Physiological Optics*, 12, 161–164.

Lawton, J. H. (1996) Population abundances, geographic ranges and conservation. *Bird Study*, 43, 3 –19.

Lee, D. N. (1994) An eye or ear for flying, in *Perception and Motor Control in Birds* (eds Green, P. and Davies, M.), Springer, Berlin, Heidelberg, New York, pp. 270 – 291.

Lythgoe, J. N. (1979) *The Ecology of Vision*, Clarendon, Oxford.

Marshall, J., Mellerio, J. and Palmer, D. A. (1973) A schematic eye for the pigeon (*Columba livia*). *Vision Research*, 13, 2449 – 2453.

Martin, G. R. (1977) Absolute visual threshold and scotopic spectral sensitivity in the tawny owl, *Strix aluco. Nature (Lond.)* 268, 636 – 638.

Martin, (1982) An owl's eye: schematic optics and visual performance in *Strix aluco* L. *Journal of Comparative Physiology A*, 145, 341– 349.

Martin, G. R. (1983) Schematic eye models in vertebrates, in *Progress in Sensory Physiology*, vol 4 (ed Ottoson, D.), Springer, Berlin, Heidelberg, New York, pp. 43 – 81.

Martin, G. R. (1984) The visual fields of the tawny owl (*Strix aluco*). *Vision Research*, 24, 1739 –1751.

Martin, G. R. (1986a) Total panoramic vision in the mallard duck, *Anas platyrhynchos. Vision Research*, 26, 1301–1305.

Martin, G. R. (1986b) The eye of a passeriform bird, the European starling (*Sturnus vulgaris*): eye movement amplitude, visual fields and schematic optics. *Journal of Comparative Physiology*, 159, 545 – 557.

Martin, G. R. (1986c) Sensory capacities and the nocturnal habit in owls. *Ibis*, 128, 266 – 277.

Martin, G. R. (1990a) *Birds by Night*, Poyser, London.

Martin, G. R. (1990b) The visual problems of nocturnal migration, in *Bird Migration: Physiology and Ecophysiology* (ed Gwinner, E.) Springer, Berlin, Heidelberg, New York, pp. 185 –197.

Martin, G. R. (1993) Producing the image, in *Vision, Brain, and Behaviour in Birds*, (eds Zeigler, H. P. and Bischof, H.-J.), MIT press, Cambridge, Mass, pp. 5 – 24.

Martin, G. R. (1994a) Form and function in the optical structure of bird eyes, in *Perception and Motor Control in Birds* (eds Green, P. and Davies, M.), Springer, Berlin, Heidelberg, New York, pp. 5 – 34.

Martin, G. R. (1994b) Visual fields in woodcocks *Scolopax rusticola* (Scolopacidae: Charadriiformes). *Journal of Comparative Physiology A*, 174, 787–794.

Martin, G. R. (1998) Eye structure and amphibians foraging in albatrosses. *Proceedings of The Royal Society B,* 265, 1–7.

Martin, G. R. and Brooke, M. (1991) The eye of a Procellariiform seabird, the Manx shearwater, *Puffinus puffinus*: visual fields and optical structure. *Brain, Behaviour and Evolution*, 37, 65 –78.

Martin, G. R. and Katzir, G. (1994a) Visual fields and eye movements in herons (Ardeidae). *Brain, Behaviour, and Evolution*, 44, 74 – 85.

Martin, G. R. and Katzir, G. (1994b) Visual fields in Stone curlews *Burhinus oedicnemus*, *Ibis*, 136, 448 – 453.

Martin, G. R. and Katzir, G. (1995) Visual fields in ostriches, *Nature (Lond.)*, 374, 19 – 20.

Martin, G. R. and Young, S. R. (1983) The retinal binocular field of the pigeon (*Columba livia*): English racing homer. *Vision Research*, 23: 911– 915.

Martin, G. R. and Young, S. R. (1984) The eye of the Humboldt penguin *Spheniscus humboldti*: visual fields and schematic optics. *Proceedings of The Royal Society B*, 223, 197– 222.

McFadden, S. A. (1994) Binocular depth perception, in *Perception and Motor Control in Birds* (eds Green, P. and Davies, M.), Springer, Berlin, Heidelberg, New York, pp. 54 –73.

Miller, W. H. (1979) Ocular optical filtering, in *Handbook of Sensory Physiology,* vol VII/6A (ed by Autrum, H.), Springer, Berlin, Heidelberg, New York 69 –143.

Murphy, C. H. and Howland, H. C. (1983) Owl eyes: Accommodation, corneal curvature and refractive state. *Journal of Comparative Physiology A*, 151, 277– 284.

Murphy, C. H., Evans, H. E. and Howland, H. C. (1985) Towards a schematic eye for the great horned owl. *Fortsch. Zool* 30, 703 –706.

Oswaldo-Cruz, E., Hokoc, J. N. and Sousa, A. P. B. (1979) A schematic eye for the opossum. *Vision Research*, 19, 263 – 278.

Pettigrew, J. D. (1979) Comparison of retinotopic organization of the visual wulst in nocturnal and diurnal raptors, with a note on the evolution of frontal vision, in *Frontiers of Visual Science* (eds Cool, S. J. and Smith, E. L.), Springer, Berlin, Heidelberg, New York, pp. 328 – 225.

Proctor, N. S. and Lynch, P. J. (1993) *Manual of ornithology: avian structure and function*, Yale University

Press, New Haven and London.
Reid, B. and Williams, G. R. (1975) The Kiwi, in *Biogeography and ecology in New Zealand* (ed Kuschell, G.), Junk, The Hague.
Remtulla, S. and Hallet, P. E. (1985) A schematic eye for the mouse, and comparison with the rat. *Vision Research*, 25, 21–31.
Reymond, L. (1985) Spatial visual acuity of the eagle *Aquila audax*: a behavioural. optical and anatomical investigation. *Vision Research*, 25, 1477–1491.
Rochon-Duvigneaud, A. (1943) Les yeux et la vision des vertébrés. Masson, Paris.
Schaeffel, F. (1994) Functional accommodation in birds, in *Perception and Motor Control in Birds* (eds Green, P. and Davies, M.), Springer, Berlin, Heidelberg, New York, pp. 35 – 53.
Shlaer, R. (1972) An eagle's eye: quality of retinal image. *Science (Wash DC)*, 176, 920 – 922.
Sibley, C. G. and Monroe, B. L. (1991) *Distribution and taxonomy of birds of the world*, Yale University Press, New Haven.
Sivak, J. G. (1976) The role of the flat cornea in the amphibious behaviour of the blackfoot penguin (*Spheniscus demersus*). *Canadian Journal of Zoology*, 54, 1341–1346.
Sivak, J. G., Howland, H. C. and McGill-Harelstad, P. (1987) Vision of the Humboldt penguin (*Spheniscus humboldti*) in air and water. *Proceedings of the Royal Society B*, 229, 467– 472.
Snyder, A. W., Laughlin, S. B. and Stavenga, D. G. (1977) Information capacity of eyes. *Vision Research*, 17, 1163 –1175.
Tansley, K. (1965) V*ision in Vertebrates*. Chapman & Hall, London.
Vakkur, G. J. and Bishop, P. O. (1963) The schematic eye in the cat. *Vision Research*, 13, 889 – 891.
Wallman, J. and Pettigrew, J. (1985) Conjugate and conjunctive saccades in two avian species with contrasting oculomotor strategies, *Journal of Neuroscience,* 5, 1418 –1428.
Walls, G.L. (1942) *The Vertebrate Eye and its Adaptive Radiation*. Cranbrook Institute of Science, Bloomfield Hills, Michigan.
Welty, J. C. and Baptista, L. F. (1988) *The Life of Birds*, 4th edn, W.B.Saunders, New York.
Woodhouse, J. M. and Campbell, F. W. (1975) The role of the pupil light reflex in aiding adaptation to the dark. *Vision Research*, 15, 649 – 653.
World Conservation Monitoring Centre (1992) *Global Biodiversity: Status of the Earth's Living Resources*, Chapman & Hall, London.
Zeigler, H. P., Jäger, R. and Palacios, A. G. (1993) Sensorimotor mechanisms and pecking in the pigeon, in *Vision, Brain, and Behaviour in Birds*, (eds Zeigler, H. P. and Bischof, H.-J.), MIT press, Cambridge, Mass, pp. 265 – 283.

16
Behavioural ecology and retinal cell topography

S.P. COLLIN
QE II Research Fellow, Marine Neurobiology Laboratory, Department of Zoology, University of Western Australia, Nedlands, 6907, Western Australia, Australia

16.1 Introduction

The vertebrate retina is an extension of the brain, a hemisphere of neural tissue upon which is mapped an image of a particular species visual environment. Each point in visual space is subtended by a corresponding point on the neural retina which in turn is retinotopically mapped onto the visual centres of the brain. Light energy or the ‘optical image’ is transformed into electrical energy or a ‘neural image’ by the photoreceptors and, via a number of interneurons (bipolar, amacrine and horizontal cells), these signals reach the ganglion cells each of which possess an axon carrying information to the central nervous system via the optic nerve. Although processing at the level of the photoreceptors may not necessarily change the neural image, due to the over abundance of photoreceptors relative to the number of ganglion cells, it is the ganglion cells which ultimately define the perception of a species’ environment received by the central nervous system.

The distribution of retinal cells has provided a wealth of information of the importance each species places on observing objects in specific regions of their visual field with higher spatial resolving power. The distribution of retinal cells for each species is unique and closely reflects the symmetry of each species’ perceived world, providing information on the role vision plays in the animals’ behavioural ecology. This chapter attempts to summarise the distribution or topography of retinal cells and the various adaptive strategies employed by a range of vertebrates from various life-styles and habitats to optimise sampling of an image in specific regions of their visual field (Table 1).

In the descriptions that follow, a concentric increase in retinal cell density will be defined as an area centralis and an elongated increase in cell density across the retinal meridian as a horizontal streak. These retinal specialisations may also be associated with deep (convexiclivate) and shallow (concaviclivate) pit-like structures termed foveae. Although examination of retinal specialisations show an increase in central density of all cell populations relative to peripheral densities, most studies have concentrated on either photoreceptor or ganglion cell populations. This is due in part to their locations on either side of the retina allowing an undistorted view in wholemount, and also because of the difficulties in identifying the various populations of interneurons. The photoreceptor array

S.N. Archer et al. (eds.), Adaptive Mechanisms in the Ecology of Vision, 509–535

Table 1: Summary of studies on retinal topography in various species of vertebrates. In species that frequent an open environment (depicted by an O) or an uninterrupted view of their horizon, there is a band-shaped increase in retinal cell density across the horizontal meridian or horizontal streak (HS). The peak retinal ganglion cell densities are × 10^3 cells per mm^2. The peak photoreceptor densities are × 10^3 receptors per mm^2. AC, area centralis; C, central; D, dorsal; dc, double cones; DN, dorso-nasal; DT, dorso-temporal; F, fovea; juv, juvenile stage; lsc, large single cones; met, metamorphic stage; N, nasal; NC, naso-central; ND, naso-dorsal; P, peripheral; pdc, principal double cones; RI, radial isotropy; sc, single cones; T, temporal; tad, tadpole stage; TC, temporo-central; TD, temporo-dorsal; TV, temporo-ventral; V, ventral; VS, vertical streak; *density represents only subpopulation of ganglion cells retrogradely labelled from the medio-rostral pole of the optic tectum (see text)

Species	Class (Order)	Habitat	Peak retinal ganglion cell density	Peak photoreceptor density	Retinal specialisation and retinal location	Reference
Ichthyomyzon unicuspis	Agnatha	Parasitic on fish	3.28; 2.98		AC-TD; AC-ND	Fritzsch and Collin (1990)
Hydrolagus collei	Chondrichthyes	Benthopelagic (O)	0.35*		HS-T	This study
Heterodontus francisci	Chondrichthyes	Benthopelagic (O)	0.20; 0.50		AC-T; HS-C	Peterson and Rowe (1980)
Carcharhinus melanopterus	Chondrichthyes	Benthopelagic (O)	2.00		HS-C	This study
Hemiscyllium ocellatum	Chondrichthyes	Benthic scavenger (O)	1.90		HS-C	This study
Brachaelurus waddi	Chondrichthyes	Benthic feeder (O)	1.75; 1.50		AC-T; HS-N	This study
Mustelus canis	Chondrichthyes	Benthopelagic (O)	1.60		HS-D	This study
Negaprion brevirostris	Chondrichthyes	Pelagic (O)	1.60	6.5 cones	HS-C	Heuter (1991)
Lepisosteus platyrhincus	Actinopterygii	Surface of rivers and streams (O)	9.40; 5.90		HS-V; AC-T	Collin and Northcutt (1993)
Aspius aspius	Actinopterygii	Pelagic carnivore	9.9	5.1 cones	AC-TV; AC-ND	Zaunreiter *et al.* (1991)
Rutilus rutilus	Actinopterygii	Pelagic omnivore	6.30	5.5 cones	AC-T; AC-N	Zaunreiter *et al.* (1991)
Salmo trutta trutta	Actinopterygii	Benthopelagic		14.8 dc	AC-TV	Ahlbert (1976)
Carassius auratus	Actinopterygii	Pelagic omnivore	5.96	14.0 cones	AC-T	Mednick and Springer (1988); Mednick *et al.* (1988)
Lienardella fasciata	Actinopterygii	Open areas of coral reef (O)	7.4; 6.0		AC-TD; HS-C	Collin and Pettigrew (1988c)
Plectropoma leopardus	Actinopterygii	Coral caves	15.0		AC-TV	Collin (1989)
Balistoides conspicillum	Actinopterygii	Open areas of reef crest (O)	51; 42; 35		AC-T; AC-TC; HS-C	Collin and Pettigrew (1988b)
Choerodon albigena	Actinopterygii	Open areas of coral rubble (O)	83; 80; 50		AC-TD; AC-T; HS-C	Collin and Pettigrew (1988b)
Lethrinus chrysostomus	Actinopterygii	Open areas of reef (O)	50; 40; 35		AC-T; HS-C; AC-TD	Collin and Pettigrew (1988b)
Gymnocranius bitorquatus	Actinopterygii	Open areas of coral rubble (O)	63; 20		AC-TD; HS-N	Collin and Pettigrew (1988b)
Amblyglyphidodon curacao	Actinopterygii	Seeks protection of branches of staghorn coral	30; 25; 23		AC-TV; AC-T; AC-ND	Collin and Pettigrew (1988a)
Cephalopholis miniatus	Actinopterygii	Coral caves	47; 22		AC-T; AC-N	Collin and Pettigrew (1988a)
Semotilus atromaculatus	Actinopterygii	Crepuscular omnivore	16.6; 12.5; 11.4		AC-T; AC-N; AC-ND	Collin and Ali (1994)
Limnichthyes fasciatus	Acitnopterygii	Benthic feeder of plankton	150.0		F-C	Collin and Collin (1988)
Ictalurus punctatus	Actinopterygii	Benthic forager	1.5		HS-C	Dunn-Meynell and Sharma (1987)
Navodon modestus	Actinopterygii	Amongst coral grottos	16.0; 16.0		AC-T; HS-NC	Ito and Murakami (1984)

Continues on next page

Table 1: Continued

Species	Class (Order)	Habitat	Peak retinal ganglion cell density	Peak photoreceptor density	Retinal specialisation and retinal location	Reference
Sebastiscus marmoratus	Actinopterygii	Benthic lie-in-wait predator	16.0		HS-C	Ito and Murakami (1984)
Trichiurus lepturus	Actinopterygii	Pelagic with vertical orientation in water column		60.0 cones	RI-TV	Kawamura and Ohashi (1988)
Howella sherborni	Actinopterygii	Benthic or benthopelagic	24.5; 21.0		F-T; VS-T	This study
Paralabrax nebulifer	Actinopterygii	Cryptic amongst kelp and reef	33.4		F-T	Schwassmann (1968)
Bufo marinus	Amphibia	Canefields and forests (O)	29; 25; 16		RI-P; HS-C	Zhang and Straznicky (1991)
Heleioporous eyrei	Amphibia	Burrows and sandy swamps	6.7(tad); 3.7(met); 4.0(juv); 4.0(juv); 3.1(adult)		RI-P; RI-P; AC-T; AC-N; HS-C	Dunlop and Beazley (1981)
Limnodinastes dorsalis	Amphibia	Permanent swamps and pools (O)	1.78		HS-C	Coleman *et al.* (1984)
Littoria moorei	Amphibia	Permanent pools with bullrushes (O)	2.3		HS-C	Humphrey and Beazley (1985)
Anolis carolinensis	Reptilia	Crowns of tree trunks		290; 16 cones	F-C; F-T	Makaretz and Levine (1980)
Thamnophis sirtalis	Reptilia	Open grassland (O)	10.0	20pdc+lsc;4sc	HS-C	Wong (1989)
Pseudemys scripta elegans	Reptilia	Near water's edge (O)	20.1	26.7 cones	HS-C	Peterson and Ulinski (1979)
Ctenophorus nuchalis	Reptilia	Open arid areas (O)	13.0		HS-C	Wilhelm and Straznicky (1992)
Halcyon sancta	Aves	Perched lie-in-wait predator	180; 25		F-C; F-T	Moroney and Pettigrew (1987)
Dacelo gigas	Aves	Perched lie-in-wait predator	140; 20		F-C; F-T	Moroney and Pettigrew (1987)
Carogyps atratus	Aves	Predatory raptor	25; 16		F-N; AC-T	Inzunza *et al.* (1989)
Couturnix couturnix japonica	Aves	Open areas of grassland (O)	40; 38; 26		AC-T; AC-N; HS-C	Budnik *et al.* (1984)
Gallus domesticus	Aves	Terrestrial ground feeder	25		AC-C	Ehrlich (1981)
Speotyto cunicularia	Aves	Perching in open areas (O)	23.0; 15.5		F-T; HS-C	Bravo and Pettigrew (1981)
Bradypus variegatus	Mammalia	Hanging upside down on horizontal or inclined tree branches (arboreal)	1.0; 0.6		AC-T; VS-T	Costa *et al.* (1989)
Choloepus didactylus	Mammalia	Hanging upside down on horizontal or inclined tree branches (arboreal)	1.7; 1.1		AC-V; VS-T	Costa *et al.* (1987)
Megalia rufa	Mammalia	Open plains (O)	6.0		HS-C	Hughes (1974); Hughes (1977)
Dendrolagus doriana	Mammalia	Arboreal	4.0		AC-T	Hughes (1974)
Loxodonta africana	Mammalia	Open arid plains (O)	0.61; 0.58		AC-T; HS-V	Stone and Halasz (1989)
Phascolarctos cinereus	Mammalia	Upper regions of eucalyptus trees (O)	1.5		HS-V	Schmid *et al.* (1992)
Tursiops truncatus	Mammalia (Cetacea)	Pelagic	0.83; 0.78; 0.30		AC-T; AC-N; HS-V	Mass and Supin (1995)
Callorhinus ursinus	Mammalia (Cetacea)	Reef areas	1.3		AC-T	Mass and Supin (1992)

may be comprised of either single, unequal double, equal double, triple or quadruple cones or rods which are predominantly arranged in single banks with the exception of some deep-sea teleosts and some snakes which possess multiple banks of photoreceptors (see Chapter 5). The ganglion cells are also of various types and predominantly lie within a number of sublaminae in the ganglion cell layer immediately sclerad of the optic fibre layer. In most vertebrates, an ectopic population of ganglion cells and amacrine cells also lies within the inner nuclear and ganglion cell layers, respectively. Retrograde labelling from the optic nerve is necessary for definitive identification of the ganglion cell population but the topography and peak densities remain relatively unchanged upon inclusion of the displaced amacrine cell population. The axons of the optic fibre layer abut the inner limiting membrane, except in the agnatha (see below), and may either form a continuous sheet across the retina or form columns or fascicles radiating from the optic nerve head.

In order to encompass sufficiently the large degree of variation found throughout the vertebrate phyla, the adult pattern of selected species will be discussed except where the retinal topography alters markedly in response to obvious changes in habitat during development.

16.2 Agnathans

The jawless fishes or the agnathans are split into two orders; the Myxiniformes (hagfishes) and the Petromyzontiformes (lampreys). The hagfishes are entirely marine and do not have a larval phase, hatching as small adults to scavenge the deep-sea for dead or dying fish and benthic invertebrates which they immobilise by encapsulating them in copious amounts of mucus. Hagfishes possess only rudimentary eyes with each of the three extant genera showing different stages of eye degeneration. In *Myxine*, the eye lacks a lens and vitreous body and is covered by pigmented skin. In *Eptatretus*, the eye lies beneath an opaque epithelial covering, with a vitreous body but no lens. The retina in both groups is not well differentiated and comprises a photoreceptor layer with two ill-defined neuronal layers interspersed between two fibrous layers. Although both groups possess functional photoreceptors, these cells occur in low numbers with the number of axons in the optic nerve (and therefore an estimate of the number of ganglion cells excluding retinopetal neurons) found to be between 800 and 1400 (Holmberg, 1972). Further work is needed to identify neuronal populations but the topography of both photoreceptors and ganglion cells may not reveal any significant changes in density. This lack of retinal specialisation may be a product of the inability of the eye to focus an image on the retina. In fact, in order to compensate for its poor visual capabilities, this group has evolved light receptors in its tail which elicit a negative phototaxic response (Patzner, 1978).

Lampreys have a protracted development where they spend up to three years as an ammocoete buried in the soft substrata of freshwater streams feeding on algae, detritus and micro-organisms. Metamorphosis into a juvenile lasts between three and six months after which time they migrate downstream to the sea to become parasitic on fish. The eyes of adult lampreys are well developed, possessing a non-spherical lens (Collin and Fritzsch, 1993) and a differentiated retina. In contrast to all other gnathostome vertebrates, the nerve fibre layer in agnathan chordates is situated in the sclerad region of the inner plexiform layer. Therefore, most of the (orthotopic) ganglion cells lie in the inner nuclear layer. Retrograde labelling from the optic nerve with the carbocyanine dye DiI reveals an increase in ganglion cell density (5.09×10^3 cells per mm^2) in the dorsal periphery of the retina in a postmetamorphic juvenile (Fritzsch and Collin, 1990). Therefore, this dorsally located area centralis is sampling the animal's ventral visual field with increased spatial resolving power. However, if one divides the ganglion cells into orthotopic (in the inner nuclear layer) and ectopic (in the ganglion cell layer) populations, the topography of the two populations are non-overlapping, sampling different regions of visual space and probably projecting to different central targets. The dorsal area centralis of ectopic ganglion cells samples latero-ventral visual space but the temporal area centralis of orthotopic ganglion cells subtends its small frontal binocular field. Therefore, even in the most primitive vertebrate retinae, at least on morphological grounds, there is the possibility that parallel processing exists, where different components of the visual scene observed may be sampled by subsets of ganglion cells with different receptive fields.

16.3 Chondrichthyes

The chimaeras, rays and sharks comprise the class Chondrichthyes which all possess a cartilaginous skeleton and are predominantly marine. Over 950 species exist throughout the world in a number of different habitats. Very little is known of the visual systems of these impressive predators. Despite the large number of species in this group there are relatively few studies on retinal topography.

The chimaera or ratfish, *Hydrolagus collei,* is a benthic deep-water predator with large eyes and an intensely reflective retinal tapetum that increases retinal sensitivity at the low light levels in which it survives. No studies have examined the entire population of retinal ganglion cells but injections of rhodamine labelled dextran into the rostral pole of the contralateral optic tectum reveals a cohort of both orthotopic and ectopic ganglion cells distributed in a visual streak across the retinal meridian from the temporal margin to a point just beyond the level of the optic nerve head. At least three ganglion cell classes exist (mean soma diameters between 15 and 30 μm) to mediate panoramic vision in this species (Collin, unpublished results).

In the shovel-nosed ray, *Rhinobatos batillum*, retrograde labelling from the optic nerve with cobaltous lysine has determined that approximately 50% of the cells in the ganglion cell layer are displaced amacrine cells (Collin, 1988). Three subpopulations of ganglion cells were characterised on the criteria of soma size and soma position; small (20-50 μm^2), large (80 – 300 μm^2) and giant (> 300 μm^2) ganglion cell populations each possessing distinct topographic specialisations in the central (3.00×10^3 cells per mm^2), dorso-nasal (1.25×10^3 cells per mm^2) and dorsal (1.57×10^3 cells per mm^2) regions of the retina (Figure 1).

The California horned shark, *Heterodontus francisci* is a sluggish swimmer which feeds on benthic invertebrates at night and rests in caves and crevices during the day. This species possesses a horizontal visual streak of 0.5×10^3 cells per mm^2 located in the ganglion cell layer and a temporal area centralis of 0.2×10^3 cells per mm^2 located in the inner plexiform layer (Peterson and Rowe, 1980). Similarly, the 'blind' shark, *Brachaelurus waddi,* shows distinct distributions of different ganglion cell classes. The total ganglion cell population possesses a relatively uniform density of 0.21×10^3 ganglion cells per mm^2 in contrast to the giant ganglion cell population which shows a temporal area centralis (1.75×10^3 cells per mm^2) and a visual streak (1.5×10^3 cells per mm^2) extending to the nasal margin (Collin, unpublished results; Figure 1). The obvious differences in ganglion cell soma size and relative topography in the chondrichthyes suggest that functionally separate classes of cells may analyse the visual environment.

In the black-tipped reef shark, *Carcharhinus melanopterus,* and the epaulette shark, *Hemiscyllium ocellatum*, a horizontal streak of 0.20×10^3 ganglion cells per mm^2 and 0.19×10^3 ganglion cells per mm^2 respectively, lies across the retinal meridian emphasising the importance of increased sampling of the horizon (Collin, unpublished results; Figure 1). In the smooth dogfish, *Mustelis canis*, Stell and Witkovsky (1973) found a visual streak (density not examined) across the horizontal meridian of the dorsal retina and described three populations of giant ganglion cells at various levels of the inner plexiform layer. The dorsal visual streak of *M. canis* has since been re-examined and found to possess over 1.6×10^3 cells per mm^2 (Figure 1).

The visual streak is a common trait of almost all chondrichthyes studied thus far. Due to the emphasis of the horizon (or the sand-water interface) to most species, it is suggested that the monocular visual field of each laterally placed eye can scan a panoramic field without resorting to a large degree of eye movement. In the lemon shark, *Negaprion brevirostris*, a horizontal visual streak with an increased density of both ganglion cells (1.60×10^3 cells per mm^2) and cone photoreceptors (6.50×10^3 cones per mm^2) representing 26% of the total visual field is represented centrally by 52% of the surface of the optic tectum (the main visual centre in this species, Hueter, 1991). With this degree of central representation, the visual streak is an important retinal specialisation providing maximum resolving power in a part of the visual field into which the majority of new visual targets appear. The compensatory eye movements of some species to the horizon may also be mediated by the visual streak which could provide orientation cues. Moreover, in species that are constantly patrolling over sand or at the edge of a

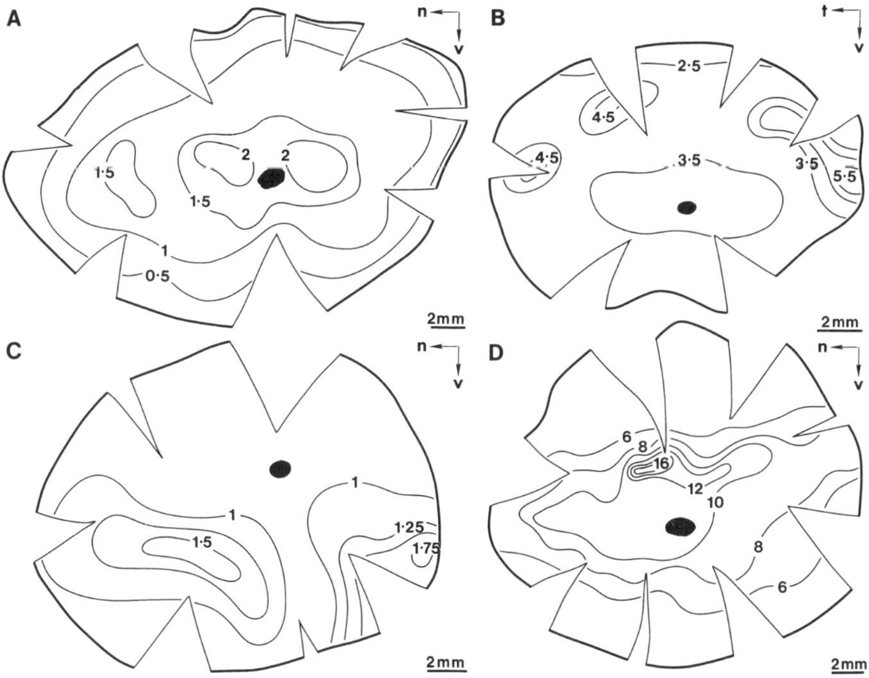

Figure 1 Range of retinal ganglion cell topographies of species in the Chondrichthyes. **A**. The black-tipped reef shark, *Carcharhinus melanopterus*. **B**. The shovel-nosed ray, *Rhinobatos batillum* redrawn from Collin (1988). **C**. The 'blind' shark, *Brachaelurus waddi*. **D**. The smooth dogfish, *Mustelus canis*. All maps except **B**, which shows the topography of the giant ganglion cell population, show the distribution of the total ganglion cell population. All species possess a retina with a horizontal increase in cell density and inhabit an 'open' environment where the sand-water interface predominates the visual field. The black disc denotes the optic nerve head. Densities (**A-C**) × 10^3 cells per mm^2. Densities (**D**) × 10^2 cells per mm^2. n, nasal; t, temporal; v, ventral

reef crest, visual disturbances within the region of the visual field subtended by the streak may be accentuated due to the predominantly horizontal component of its constant motion.

16.4 Actinopterygians

The ray-finned fishes or actinopterygians comprising the Chondrostei (polypteriformes, sturgeons and paddlefishes), the Ginglymodi (the garfishes) and the Halecostomi (bowfins and teleosts) are the largest vertebrate radiation boasting over 20,000 species. The ecological niches occupied by species in this group vary considerably. Some species survive in the deepsea where no sunlight penetrates while others inhabit the well-lit marine waters of a coral reef. In response to the diverse habitats available, the visual system has undergone incredible selection pressure due to the important role vision plays in the survival of this group. Like the other optical components of the visual system, the topography of the retinal cells, and therefore resolving power, has altered in accordance with the ecological constraints placed upon it. Light intensity, visibility, the size and mobility of prey, the size and feeding strategies of potential predators and the symmetry of each species' ecological niche have contributed to the adaptive nature of the arrangement of retinal cells across the retina. Concomitant changes in retinal cell packing, eye size and shape, accommodatory apparatus, the size of the visual field and eye mobility have all been selected for if a species is to be successful.

In all species studied thus far, topographic variations in ganglion and photoreceptor densities provide an important clue to the visual capabilities and the way in which each species samples its environment. In ecosystems where there is a clear gradation of habitats, for example on a coral reef, the symmetry of a species' perceived world is reflected in the topographic distribution of the retinal receptors and these specialisations can take three basic forms; visual streaks, areae centrales and foveae.

16.4.1 Areae centrales

It has long been recognised that many species of actinopterygians, along with most species of vertebrates, possess an area of acute vision (Slonaker, 1897). The term *area* was adopted by the early German investigator Chievitz (1891) to describe the macula lutea of humans and this was later termed an *area nasalis, area temporalis* or *area centralis* in other vertebrates according to the region of retina in which it lay. In these earlier studies, areae have previously been described simply as a part of the retina which is relatively thick (Kahmann, 1935) which, upon quantitative analysis, has translated to a localised increase in the number of retinal neurons. Although Slonaker did not consider *areae centrales* or *areae retinae* to be widespread in teleosts, Ali and Anctil (1976) have described these specialisations in over 47 families predominantly described with respect to loci of increased photoreceptor density. In the pacific sardine, *Sardinops caerulea,* and the northern anchovy, *Engraulis mordax*, O'Connell (1963) found a distinct area temporalis of 66.41×10^3 cones per mm^2 (13.91×10^3 ganglion cells per mm^2) and 26.88×10^3 double cones per mm^2 (16.10×10^3 ganglion cells per mm^2), respectively and a concomitant increase in all other retinal neurons of the inner nuclear layer. However, the density of rods within the area temporalis in these species and others studied by O'Connell (1963) varied greatly which signifies that each species may have a different affinity for dim or crepuscular vision. O'Connell also noted that the degree of eye mobility, and therefore the development of the extraocular eye muscles, may be related to each species' ability to stabilise a rapidly shifting visual environment and to their acuity. He also mentions that the retinal regions not occupied by the area temporalis would be hypermetropic and with the elliptical shape of the eye, the area temporalis with a longer lens-to-retina distance would be myopic. Therefore, small objects in front of and close to the fish subtended by the area temporalis would thus be in focus and would remain in focus by accommodation of the spherical lens in the temporo-nasal axis while large objects subtended by the remainder of the visual field would be in focus at infinite distance.

The relationship between eye mobility and the occurrence and location of retinal acute zones seems to be closely related in actinopterygians. Active predators without a visual streak often have a well developed area centralis and a foveal depression (see below) in conjunction with increased eye mobility. The eyes of the predatory coral trout, *Plectropoma leopardus,* (Collin, 1989), and the coral cod, *Cephalopholis miniatus,* (Collin and Pettigrew, 1988a), both possess temporal areae centrales, subtend a binocular overlap of 36° and 54° respectively and accurately align their areae centrales upon prey objects by conjugated saccadic eye movements before striking. Without the advantage of perceiving a panoramic visual field with high acuity using a horizontal visual streak, species with an 'enclosed' environment must rely on increased eye mobility to capture prey and observe potential predators.

Areae centrales lie predominantly in temporal retina to subtend some part of the frontal binocular field. Ocular adaptations to increase this binocular field include a rostral tapering of the pupil leaving an aphakic gap, a binocular sighting groove etched into the snout, conjugated frontal eye rotation and an invagination of the iris in some species of deep-sea fish with tubular eyes (Collin *et al.*, 1997). The positioning of the eyes in the head also governs the degree of binocular overlap and therefore the location of an areae. In the Australian frogfish, *Halophryne diemensis,* the eyes are frontal, they subtend a large binocular overlap with an acute zone of $5.00 - 6.00 \times 10^3$ cells per mm^2 in the central retina while in the coral cod, *C. miniatus,* with a smaller degree of binocular overlap the acute zone of 47.00×10^3 cells per mm^2 is located in the temporal retina (Collin and Pettigrew, 1988a; Figure 2).

Although the temporal retina is the most common zone for acute vision, some species possess multiple zones of varying spatial resolution. Once thought to be devoid of any retinal specialisation (Johns and Easter, 1977), the goldfish, *Carassius auratus,* has since been found to possess a temporal area centralis with respect to both ganglion cell (Mednick and Springer, 1988) and photoreceptor (Mednick *et al.*, 1988) densities with many other species of cyprinids possessing up to three areae (Zaunreiter *et al.*, 1991; Collin and Ali, 1994). The staghorn damselfish, *Amblyglyphidodon curacao,* also boasts three acute zones with an area in the temporal, ventro-temporal and dorso-nasal regions of the retina. The temporal and ventro-temporal zones receive input from frontal visual space and may aid in the detection of small micro-organisms that slowly drift past while the dorso-nasal zone may enable the fish to negotiate the finger-like projections of the staghorn coral during its escape response (Collin and Pettigrew, 1988a).

16.4.2 Visual streaks

Chievitz (1889; 1891) and Slonaker (1897) were the first to describe horizontal band-shaped retinal areae in actinopterygians. These horizontal specialisations or visual streaks can take two forms; either a band-shaped increase in retinal thickness across the retinal meridian as found in a cyprinodontid, *Fundulus heteroclitus* (Butcher, 1938), and two species of mudskippers, *Boleophthalmus* and *Periophthalmus* (Munk, 1970), or simply a marked increase in ganglion/ photoreceptor cell density as found in the balistids, *Navodon modestus* (Ito and Murakami, 1984), and *Balistoides conspicillum* (Collin and Pettigrew, 1988b), and the yellow-finned trevally, *Caranx ignobilis* (Figure 2; Collin, 1987). In the channel catfish, *Ictalurus punctatus*, a species with 17 optic discs, the retina possesses a visual streak of over 1.50×10^3 cells per mm^2 (Dunn-Meynell and Sharma, 1987). As in the Chondrichthyes, the horizon or sand-water interface predominates the visual fields of these species and the position of the visual streak across the central region of the retina suggests that higher sampling of a visual target is possible without the saccadic eye movements necessary with an area centralis. However, the visual streak may not always fall across the central retina but may be either above or below the retinal meridian sampling ventral and dorsal visual fields, respectively.

The Florida garfish, *Lepisosteus platyrhincus,* possesses a pronounced visual streak (9.40×10^3 ganglion cells per mm^2) across the ventral meridian of the eye which subtends the surface of the water which it uses as a background to prey upon live fish with its long snout armed with needle-like teeth (Collin and Northcutt, 1993). The distribution of the ectopic ganglion cells situated in the inner nuclear layer in the garfish retina also reveals a ventral visual streak (6.25×10^3 cells per mm^2) inflected 20° from the horizontal and a temporal area centralis (6.25×10^3 cells per mm^2). The inflected horizontal streak may provide a finer control for the stabilisation of the retina during compensatory eye movements, while the temporal area centralis could provide an increase in visual acuity for prey observed in frontal visual space.

As in the garfish (Ginglymodi), some species of euteleosts possess more than one retinal specialisation. In addition to a horizontal streak, a temporal area centralis has been discovered in the blue tuskfish, *Choerodon albigena,* and the red-throated emperor, *Lethrinus chrysostomus* (Collin and Pettigrew, 1988b; Figure 2). In the collared sea bream, *Gymnocranius bitorquatus,* a dorsal area centralis also lies above a nasal horizontal streak (Collin and Pettigrew, 1988b). The existence of two specialisations subtending different regions of the visual field suggests each may provide different sampling, especially when each specialisation is comprised of cells of different morphology and receptive fields. The temporal position of the area centralis of *L. chrysostomus* and *C. albigena* may reflect their strategies of feeding on prey approaching from the front while the dorsal position of the area centralis in *G. bitorquatus,* may reflect its

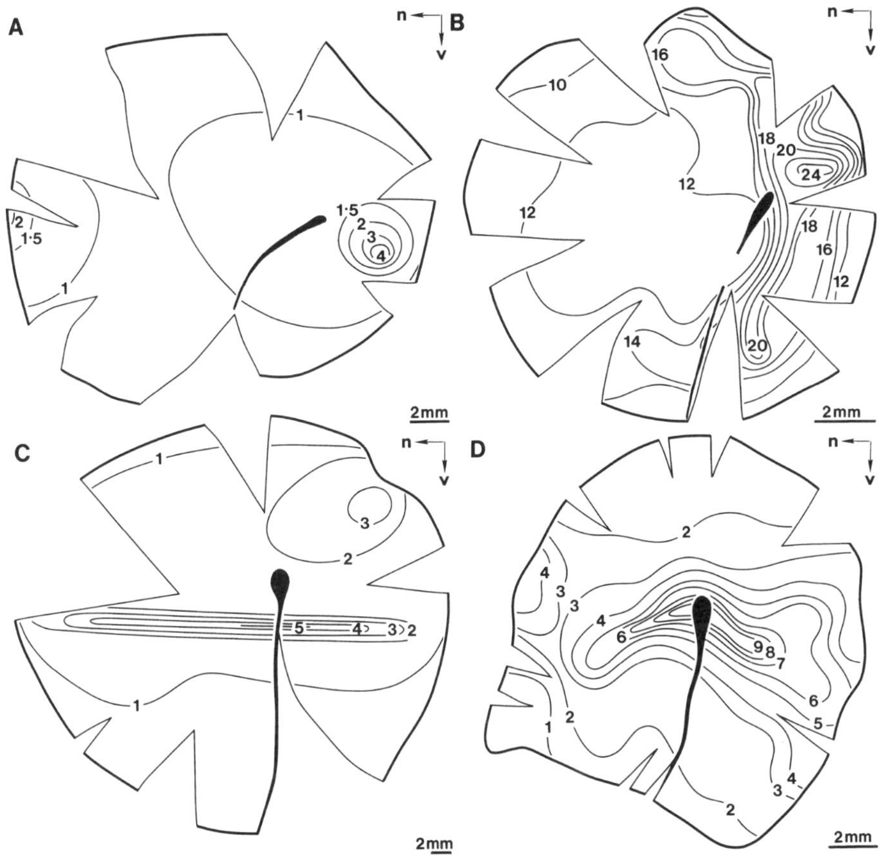

Figure 2 Range of retinal ganglion cell topographies of species in the Actinopterygii. **A**. The coral cod, *Cephalopholis miniatus* redrawn from Collin and Pettigrew (1988a). **B**. The deep-sea bass, *Howella sherborni*. redrawn from Collin and Partridge (1996). **C**. The red-throated emperor, *Lethrinus chrysostomus* redrawn from Collin and Pettigrew (1988b). **D**. The yellow-finned trevally, *Caranx ignobilis*. All maps show the distribution of the total ganglion cell population. A and B inhabit dark coral caves and the deep-sea, respectively and possess concentric or vertical iso-density contours. **C** and **D** inhabit 'open' areas of a coral ecosystem with an uninterrupted view of their horizon or the sand-water interface which is reflected in the topography of their retinae by a pronounced horizontal streak across the retinal meridian. The black elongated disc denotes the optic nerve head and falciform process. Densities × 10^3 cells per mm^2. n, nasal; v, ventral

continual foraging through the coral debris on the substrate at an angle of approximately 45°.

In *Aplocheilus lineatus* (Engström, 1963), and *Epiplatys grahami,* which prey upon insects that lie trapped in the surface film of the water (Munk, 1970), there are two band-shaped thickenings across the horizontal meridian of each retina. It has been suggested that these central band-shaped areae, although both oriented parallel to the surface of the water, are separated by 40° and subtend different regions of the visual field. The upper band is able to view an insect above the water by surface refraction and the lower band is able to visualise the same insect from below the water.

The visual streak may have various functions. Thus far, the horizon (either the sand-water or air-water interface) has predominated the visual field of each species of ray-finned fish found to possess a horizontal streak. If each of these species is able to receive a panoramic view of their visual field, as in the chondrichthyes, disturbances in this region will be emphasised, implying a lower threshold for the perception of movement and the signalling of either prey or predator. A cell density peak within a visual streak may also be utilised behaviourally. For example in the balistid, *N. modestus,* a peak ganglion cell density of over 16.00×10^3 cells per mm^2 located on the nasal margin of the visual streak may enable the fish to manoeuvre backwards successfully which it often performs using its pectoral fins (Ito and Murakami, 1984). A streak may also provide an important cue for the stabilisation of a species in the water column.

Few studies have estimated the *magnification factor* of the visual streak onto the optic tectum. A magnified representation of a 25° wide horizontal band across the retinal meridian has been found in the optic tectum of the four-eyed fish, *Anableps microlepis* (Schwassmann and Kruger, 1965). Qualitative comparisons between the optic tecta of species with comparably-sized eyes but no visual streak also reveals that the optic tectum in species with a visual streak are appreciably larger (Ito and Murakami, 1984; Collin, 1987). Therefore, it has been suggested that the visual streak projects to the optic tectum and the temporal area centralis projects to the thalamus, a comparable scenario to that found in birds (Bravo and Pettigrew, 1981).

16.4.3 Foveae

Kahmann (1934) was the first to compare the presence of foveae in the retinae of a large number of bony fishes although Slonaker (1897) noted the occurrence of five species that possessed shallow foveae. Kahmann's study suggested that a fovea is predominantly found in temporal retina and mediates acute binocular vision but may also be used for monocular vision. As in species with temporal areae centrales, species that possess a fovea have fixating eye movements and foveal location varies with the position of the eyes in the head.

The teleost fovea appears as an indentation of the retina and may be of two types, convexiclivate with steep sloping sides or concaviclivate with a shallow depression. This retinal specialisation is characterised by a displacement of the inner layers of the retina and is often associated with an increased packing of photoreceptors with elongated and smaller diameter outer segments beneath the foveal pit and high densities of ganglion cells in the perifoveal region. Convexiclivate foveae are found in the seahorse, *Hippocampus,* and the pipefish,

Corythoichthys paxtoni (Collin, 1997; Collin and Collin, 1998), while a concaviclivate fovea is found in the banded toado, *Sphaeroides pleurostictus* (Collin, 1987), and the deep-sea teleost, *Bathylagus benedicti* (Vilter, 1954). A single bank of cones packed into a regular array comprise the foveae of shallow water species but in deep-sea species, foveae may comprise only long single rods as in the searsid, *Platytroctes apus* (Locket, 1971), multiple bank rods as in the deepsea bass, *Howella sherborni* (Locket, 1992; Figure 16.2), or form as many as 28 banks as in the alepocephalid, *Bajacalifornia drakei* (Locket, 1985; Chapter 5).

The precise function of the fovea is still not known. Walls (1942) suggested that the refraction of light at the vitreous-retina border in conjunction with the steep walls of the pit would produce a magnification of the image on the photoreceptor array and therefore enhance resolution. This magnification hypothesis was disputed by Pumphrey (1948) who found that the sharp change of curvature at the base of the foveal pit would distort the image thereby reducing resolution. Pumphrey suggested that the optical exaggeration of an image of a point source traversing the fovea, lagging as it approached and leading as it left, would provide an effective means of maintaining fixation and exaggerating small movements of objects in the visual field. This is exemplified in the sandlance, *Limnichthyes fasciatus,* which possesses a convexiclivate fovea with extremely high densities of ganglion cells (150.0×10^3 cells per mm^2) in the perifoveal region which it uses to track small mobile prey with its independent eyes (Collin and Collin, 1988). The species also boasts a unique lens system which comprises a refractive cornea and a non-spherical lens which is used as a lens doublet to provide non-uniform magnification with eccentricity of tiny prey on the visual axis (Pettigrew and Collin, 1995). Other teleosts with foveae include the clingfish, *Gobiesox strumosus* (Wagner *et al.*, 1976), the parrotfish, *Chryptotomus roseus* (Ali *et al.*, 1973), and the barred sandperch, *Parapercis nebulosus* (Easter, 1992).

The foveal depression is circular in the basses, *Paralabrax clathratus* and *P. maculofasciatus,* but appears elliptical in large individuals of *P. nebulifer* (Schwassmann, 1968), and even appears elongated across the retina from the optic nerve head to the temporal margin in *Pseudocheilinus hexataenia* (S. L. Kondrashev pers. comm.). The ganglion cell density (33.44×10^3 cells per mm^2) in the perifoveal region of *P. clathratus* which subtends only 10 –15° of arc in the frontal visual field occupies a disproportionately large area (five times larger than a comparable region of the ventral or dorsal visual field) of the optic tectum (Schwassmann, 1968).

16.5 Amphibians

The amphibia are a diverse group of vertebrates with well developed sensory systems. There are three orders; the Urodela (newts and salamanders), the Anura (frogs and toads) and the Caecilia (caecilians). The caecilians are burrowing species with vestigial eyes but of the other two orders, it is the visual system of the frogs and toads that has received the most attention.

Slonaker (1897) examined the retinae of fourteen species of amphibia and found 2 species with areae centrales (two urodeles; *Salamandra atra* and *Triton cristatus*), 8 species with visual streaks (various species of *Hyla*, *Bufo* and *Rana*) and 2 species with a simple fovea (*Bufo calamitra* and *B. vulgaris*). More quantitative studies by Kalinina (1976) and Graydon and Giorgi (1984) found a peak of 10.7×10^3 ganglion cells per mm^2 in central retina (comprising a mixture of five morphological subtypes of ganglion cells) and a weak visual streak of 4.0×10^3 cells per mm^2 in the adult lake frog, *Rana ridibunda,* and the clawed toad, *Xenopus laevis*, respectively. It is suggested that, although *X. laevis* may rely heavily on its lateral line system, the weak horizontal visual streak may be effective for the avoidance of predators while floating at the surface to breathe.

A visual streak has been described in a number of species of adult frogs, including *Heleioporus eyrei*, *H. albopunctatus*, *Hyla moorei*, *Limnodynastes dorsalis* and *Neobatrachus pelobatoides,* but not in the tadpoles (Dunlop and Beazley, 1981). In the moaning frog, *H. eyrei,* and the pobblebonk, *L. dorsalis*, the retinal topography of the ganglion cell population was found to change from radial iso-density contours in the tadpoles and juveniles to a horizontal visual streak with a peak ganglion cell density in temporal retina in the adults (Dunlop and Beazley, 1981; Coleman *et al*, 1984). Therefore, a transition from an aquatic habitat where tadpoles browse for food in a three-dimensional environment to a ground-dwelling ‘open’ existence where postmetamorphic juveniles and adults actively seek worms and flies is closely reflected by changes in retinal topography.

The orthotopic population of ganglion cells shows a temporal area centralis and a horizontal visual streak in the frog, *Rana pipiens* (Humphrey, 1988), with less than 1% of the ganglion cell population displaced to the inner nuclear layer (Frank and Hollyfield, 1987). These ectopic ganglion cells possess large soma and a wide dendritic field, are useful for movement detection and the initiation of the escape response, and their distribution mirrors that of the orthotopic population with a distinct horizontal visual streak just dorsal of the optic nerve head. The orthotopic ganglion cell populations of both *Hyla raniceps* (Bousfield and Pessoa, 1980), and *Bufo marinus* (Nguyen and Straznicky, 1989), also show a pronounced streak. In *B. marinus*, the location of the visual streak is mirrored by both the ganglion cell (16.0×10^3 cells per mm^2) and cone photoreceptor (25.0×10^3 cells per mm^2) populations but the density of rods is highest in the peripheral margins of the retina. It is suggested that high resolving power is maintained by a low summation ratio from photoreceptors to ganglion cells ($< 3:1$, Zhang and Straznicky, 1991).

16.6 Reptiles

The reptilia are a large group of poikilothermic air-breathing vertebrates that are either terrestrial or aquatic. The eyes are well-developed and possess eyelids which can be either free and moveable or fused and transparent. When of the moveable type, a third inner eyelid or nictitating membrane exists to protect the eye. Underwater, the eyes of some reptiles compensate for the loss of refractive power of the cornea by increasing lens power. This is accomplished by changing the shape of the soft lens by contraction of the iris and ciliary muscles. The retina may be highly specialised and, in most species examined, an area centralis is found in conjunction with a foveal pit.

In fourteen species of Anolis lizards, Fite and Lister (1981) found two foveae in each eye, a convexiclivate fovea in central retina and a concaviclivate fovea in temporal retina. They found that all temporal foveae were specialised for maximal spatial resolution (marked by a concomitant increase in both photoreceptor and ganglion cell density and therefore providing a low convergence ratio). The deep central fovea is thought to be for image magnification due to the refraction of light at the vitreo-retinal boundary (Snyder and Miller, 1978), for the detection and maintenance of accurate fixation of moving objects (Pumphrey, 1948), and/or for sensitive indication of monocular focus (Harkness and Bennett-Clark, 1978). Anolis lizards require a sensitive movement detector and rely on a low tolerance of image distortion and aberration for their successful survival. This arrangement would be particularly useful when the animals scan their surrounds with both head and eye movements before making rapid, accurate strikes at small prey objects such as ants. Other adaptations of foveal vision in lizards include the tighter packing and lengthening of the central foveal photoreceptors which not only increases visual resolution, ensures a more efficient photon capture and greater sensitivity at low light levels but also allows for small variations or errors in accommodation thereby allowing some species to scan for prey over longer distances.

The chameleon is a well known example of a visual predator with a well developed convexiclivate fovea which is able to mediate high visual acuity. This retinal specialisation is useful for rapid strikes at prey by its projectile tongue. The chameleon enjoys independent eye movements (both pursuit and saccadic) and depth is perceived using monocular accommodatory cues (Collett and Harkness, 1982) from its negatively powered lens (Ott and Schaeffel, 1995). The topography of the retinal cells shows a radial arrangement of iso-density contours surrounding the fovea which reflects its three-dimensional environment amongst the branches of highly foliated trees.

In addition to the deep convexiclivate foveae found in most species of lizards, some species also possess a horizontal streak. In the ornate dragon, *Ctenophorus ornatus*, the density of retinal ganglion cells increases across the retinal meridian and the photoreceptors in this region possess yellow oil droplets which may provide a coloured filter to increase sensitivity to specific wavelengths of light along its visual horizon (Beazley *et al.*, 1997). Both the photoreceptors (24.0×10^3 cones per mm^2) and ganglion cells (10.0×10^3 cells per mm^2) of the garter snake, *Thamnophis sirtalis,* peak in a temporal area centralis which is extended across the retinal meridian as a visual streak (Wong, 1989). Presumably, the temporal area centralis is used in feeding on prey within its frontal visual field and the visual streak to perceive its horizon. In this diurnal species, there are three types of cones; large and small single cones and unequal double cones, but there are no rods. Most of the cones in the garter snake contain a collection of microdroplets in their ellipsoid region which may act as a condenser of light rays thereby increasing light capture of rays entering this small eye at oblique angles and reducing the Stiles-Crawford effect (Stiles and Crawford, 1933). This reduction may be an advantage for such a small eye which needs to maximise light capture under scotopic conditions. Alternatively, the snake, *Leptodiera* has a tiered arrangement of rods and cones for optimal light capture by the cones (Miller and Snyder, 1976).

Foveae have been found in only two genera of snakes; the east indian long-nosed tree snake, *Dryophis mycterizans,* and the african bird snake, *Thelotornis kirtlandi* (Walls, 1942), and one genera of turtle: the soft-shelled turtle, *Amyda* (Gillett, 1923). The western painted turtle, *Chrysemys picta marginata,* possesses only an area centralis (Walls, 1942). However, a pronounced linear area centralis or visual streak has been found in the red-eared slider turtle, *Pseudemys scripta elegans* (Peterson and Ulinski, 1979). In *P. scripta elegans*, the ganglion cell (20.1×10^3 cells per mm^2) and photoreceptor (26.7×10^3 cones per mm^2) densities both peak in central retina but extend from temporal to nasal margins subtending a sector of visual field approximately 143° (horizontal) by 23.5° (vertical) and 4.7° (horizontal) by 2.4° (vertical), respectively. According to the 'terrain theory' of Hughes (1977), the visual streak of *P. scripta elegans* is important in maintaining a fixed orientation between an 'overt' terrain, where the horizon is a predominant feature, and the retina. Although compensatory eye movements occur, Brown (1969) reports that the visual streak is held parallel to the horizon by non-visual reflexes (probably vestibular) of the head and neck since compensatory alignment of the visual streak to the horizon was observed in darkness with infrared illumination. However, it is obvious that the turtles' visual streak is important for panoramic high acuity vision into the upper lateral visual field especially while basking or moving about near the water's edge where predator avoidance is crucial.

16.7 Birds

The Aves are a large and diverse group with some species of predatory birds boasting the highest visual acuity of any vertebrate. It has long been known from Wood's monograph in 1917 that the fundus oculi of birds possess a range of retinal specialisations including single and double foveae and a pronounced visual streak. Since these qualitative studies, a plethora of retinal specialisations have been described which do not necessarily correlate with phylogeny but rather with habitat and style of food acquisition. Some species possess a single area (the diurnal chick, Erhlich, 1981 and the nocturnal owl, Bravo and Pettigrew, 1981) while others possess two areas of increased retinal cell density either in the form of an area centralis (i. e. in the pigeon) or a fovea (predatory birds such as hawks). In the quail, *Couturnix couturnix japonica*, two regions of high ganglion cell density were found: an area temporalis (40.0×10^3 cells per mm^2) and an area nasalis (38×10^3 cells per mm^2) both situated in the dorsal retina with a region of high density joining the two areae (Budnik *et al.*, 1984). This elongate region of high density or horizontal visual streak between the two areae centrales is similar to that found in the albatross and suggests that the horizon is also an important visual cue in both species. Therefore, as in fishes, high acuity vision is important in two major regions of the retina in birds, in the frontal and lateral visual fields.

The importance of these regions is confirmed in a study of visual fixation in a variety of birds by Maldonado *et al.*, (1988) where two types of visual behaviour were found: a frontal fixation in relation to feeding and perching behaviours and a lateral fixation in relation to vigilance and flying behaviours. Similarly, Bloch *et al.*, (1984) found pigeons to possess two modes of vision: a frontal binocular mode mediated by the dorsal region and specialised in the fine resolution of static near objects and a monocular lateral mode mediated by the central region and specialised in the detection of distant moving objects.

In the sacred kingfisher, *Halcyon sancta,* and the kookaburra, *Dacelo gigas,* (in addition to bee-eaters, raptors and most passerine birds) there are two foveal pits; a temporal binocular fovea which subtends binocular visual space and a nasal or monocular fovea whose axis projects close to the laterally-pointing optical axis (Moroney and Pettigrew, 1987). The monocular fovea in the sacred kingfisher and the kookaburra was found to be 180.0×10^3 cells per mm^2 and 140.0×10^3 cells per mm^2, respectively while the respective binocular foveal densities were 25.0×10^3 cells per mm^2 and 20.0×10^3 cells per mm^2. The monocular fovea may enhance acuity by magnifying the image (Walls, 1942). The binocular fovea may aid in the detection of movement by distorting the image as it changes eccentricity (Pumphrey, 1948) or in the judgement of plane of focus as suggested for the chameleon (Collett and Harkness, 1982). Despite major differences in evolutionary history, taxonomy and behavioural ecology, the bifoveal visual systems of Anolis lizards (described earlier) and diurnal birds of prey are good examples of convergent evolution.

In birds with two foveae, the nasal fovea always possesses the highest density of photoreceptors and ganglion cells. Associated with this increase in retinal cell density in either fovea, are retinal asters which are radial arrangements of inner nuclear layer cells (bipolar cells) emanating from the foveal pit towards the periphery (Inzunza *et al.*, 1989). The function of these asters is unknown but it is thought that, in conjunction with the photoreceptors and ganglion cells to which they synapse, they may represent functional channels for the mediation of high visual acuity in both the binocular (subtended by the temporal fovea and used for feeding) and monocular fields (subtended by the nasal fovea and used for tracking of prey).

In many bifoveate birds, the density of retinal cells also increases in the region between the two foveae which may often be inflected from the horizontal meridian, the degree being species-specific. Moroney and Pettigrew (1987) suggest that the degree of inflection alters the tilt of the head so that the ground plane is aligned while the image of a prey object is transferred from the monocular fovea to the binocular one. This would facilitate binocular fixation at a variety of eccentricities by reducing the need for complicated rotations of the head and eyes.

The function of a horizontal visual streak in birds has also been suggested to aid in navigation i.e. by aiding in the accurate fixation of the horizon which may serve as a base line to which other objects in the visual field can be referred (Mathews, 1968). However, due to the absence of a visual streak in many migrating birds, Dujim (1958) suggests a streak may be used in the accurate establishment of the normal position of the eyes and therefore be used for compensatory eye movements and spatial orientation.

Peripheral myopia is an adaptation for monitoring the distant horizon and sky for predators while permitting nearby objects on the ground to remain in focus. Such myopia is limited to the lower visual field where a magnified retinal image is obtained without changing the accommodative state of the eye (Hodos and Erichsen, 1990). This lower-field myopia is described in foraging species such as chickens, quails and cranes and the amount of myopia is determined by the distance from the pupil to the ground. Similarly, the retina of the pigeon is divided into a yellow field (coloured due to the presence of yellow oil droplets, Mariani and Leure-Dupree, 1978), with a central fovea near the optic axis that subtends the lateral monocular field providing emmetropic optics, and a red field possessing an area dorsalis of higher acuity which serves lower frontal vision along and below the beak in order to estimate accurately the size and distance of the grain during pecking (Nalbach *et al.*, 1990). In all these birds, the height of the eye above the ground provides sharp imaging of the ground independent of elevation of the line of sight and may assist in monocular distance estimation (Hughes, 1977).

The cornea of manx shearwaters (*Puffinus puffinus*) is relatively flat which reduces the amount of compensation required to increase the refractive power of the lens during immersion (since the refractive power of the cornea is lost in water). Since these birds fly at the sea surface and dive to a depth of at least one meter to capture prey, a flattened cornea would reduce lenticular accommodation under the water (Martin and Brooke, 1991) but not necessarily aid in the detection of prey above the water. However, upon examination of the retina of this species and other procellariform species one finds an area of giant alpha-like ganglion cells (*an area giganto cellularis*) subserving binocular vision located in temporal retina (Hayes *et al.*, 1991). Due to the large soma size, extensive dendritic field and their morphological similarity to the physiological class of Y cells in mammals, they are most likely motion sensitive and function as a rapidly acting system for the detection of prey on or beneath the sea surface. It may also be important for the maintenance of the visual control of the bird's low and fast flight path over the constantly changing contours of the sea. Interestingly, a similar area has been found in the tubular eye of the deepsea pearl-eye, *Scopelarchus michaelsarsi* which possesses a similar zone of giant ganglion cells in temporal retina (between the accessory and main retinae, Collin *et al.*, 1998). These zones could have similar functions in these two phylogenetically different groups and could be used to lower the threshold for the detection of movement within the binocular visual field (which is extensive in the case of the pearl-eye) providing an effectively reduced reaction time for the capture of small moving prey items.

In the Australian brown falcon, *Falco berigora*, a central binocular fovea maximises the posterior nodal distance providing a very fine photoreceptor grain of 186.4×10^3 cones per mm^2 and a maximum resolving power of approximately 73 cycles per degree (measured both behaviourally and anatomically) which is used to discriminate near objects (Reymond, 1987). The shallow temporal fovea (approximately 28° from the central fovea), is specialised for panoramic monocular vision and the discrimination of distant objects. The central and temporal regions of the retina project to the optic tectum and thalamus, respectively (Bravo and Pettigrew, 1981).

16.8 Mammals

One of the first studies of visual fields in mammals is discussed by Kahmann (1934) where he points out the various degrees of binocular overlap in relation to the position of a retinal fovea. Although not all mammals possess a fovea, species with a binocular overlap (and potentially functional binocular vision) demand a degree of visual acuity sufficient for the small differences between the two retinal images to be measured. Moreover, the presence of a retinal specialisation in each eye is a strong indication of the presence of functional binocular vision while the degree of specialisation depends upon species-specific factors such as the distance at which binocular vision is normally used and the size of the targets to be detected (Pettigrew, 1990).

Several studies of retinal topography have examined various species of mice (Drager and Olsen, 1981), rats (Perry *et al.*, 1983), rabbits (Vaney, 1980), cats (Stone, 1965; Wong and Hughes, 1987), marsupials (Dunlop *et al.*, 1994; Dunlop and Beazley, 1985), microchiropteran bats (Pettigrew, *et al.*, 1988) and primates (Fischer and Kirby, 1991). A plethora of retinal specialisations is found in these species including horizontal visual streaks, areae centrales and foveae. A comparison of retinal topography with the behavioural ecology of many of these species by Hughes (1977) proposes the 'terrain theory' which suggests that the distribution of ganglion cells across the retina depends on the symmetry of the perceived world, and that an increase in ganglion cell density across the horizontal meridian is a specialisation common to terrestrial species whose field of view is not completely obscured by nearby vegetation. Further, it is suggested that animals occupying such a habitat maintain a fairly fixed orientation between the retina and the horizon. As noted previously, the horizon may also comprise the sand-water or air-water interfaces but, in all cases, the symmetry of the perceived environment is reflected in the arrangement of retinal cells. Some notable examples include the retina of the plains kangaroo, *Megalia rufa* which possesses a pronounced horizontal streak of high ganglion cell density (6.0×10^3 cells per mm^2) in conjunction with an uninterrupted view of the horizon in contrast to the retina of the arboreal kangaroo, *Dendrolagus doriana* with a concentric arrangement of iso-density contours (4.0×10^3 cells per mm^2) reflecting the three dimensional nature of its foliated environment (Hughes, 1974; Figure 3).

Similarly, the topography of the retinal ganglion cells in the african elephant, *Loxodonta africana* also supports the terrain theory since the retina possesses a region of high ganglion cell density across the horizontal meridian in keeping with its open habitat where it scans its horizon monocularly. In addition to a visual streak, the retina of this species also possesses a second concentration of ganglion cells in the upper temporal retina which it uses to monitor the activity of its trunk (Stone and Halasz, 1989).

It appears that even in similar ecological niches, the retinal topography of two closely related species can differ according to their behaviour. In the retina of the two-toed sloth, *Choloepus didactylus* a temporal area centralis with a peak of 1.0×10^3 ganglion cells per mm^2 shows an extension of iso-density contours to form a vertical visual streak from temporal to inferior retina (Costa *et al.*, 1989; Figure 3). The temporal area centralis allows the claws to be placed carefully around the branch above its head thereby aligning its vertical streak on the branch while hanging upside down. In contrast, the three-toed sloth, *Bradypus variegatus*, although a closely related species with a similar habit of hanging upside down on horizontal or inclined branches, possesses a vertical streak with an area centralis in the inferior retina. It is suggested that the area centralis in *B. variegatus* has moved its position due to its ability to rotate its head 180° about the midline axis (Costa *et al.*, 1987).

All these studies emphasise that retinal topography is closely related to an animal's visual behaviour. Species which possess a horizontal streak allow scanning of a broad area of visual space without the saccadic eye movements necessary with an area centralis. However, the location of the streak within the retina may also be adaptive. This is exemplified in studies of the Australian koala, *Phascolarctos cinereus* (Schmid *et al.*, 1992) and the wombat, *Lasiorhinus latifrons* (Tancred, 1981) which possess a horizontal streak inferior and superior to the optic nerve head, respectively, which correlates to the part of their visual field most used during feeding. Similarly, the displacement of the visual streak temporal to the optic nerve head in most marsupials may be characteristic of species in which vision plays a greater role in feeding than in predatory avoidance.

16.9 Marine mammals

The eyes of marine mammals (i.e. cetaceans and pinnepeds) tend to be larger than terrestrial eyes, of limited mobility with similarly larger retinal ganglion cells (up to a mean diameter of 60 μm in the bottlenosed dolphin and up to 90 μm in the toothed whale). In water, the cornea has little or no refractive power, leaving the non-spherical lens to focus light on the retina and attain emmetropia. However, in air the eyes are 16 to 20 dioptres myopic due to the addition of the refractive power of the cornea (Dral, 1972). Studies of retinal ganglion cell distribution reveals either a single area centralis predominantly in temporal retina (i.e. in the fur seal, *Callorhinus ursinus* with 1.33×10^3 cells per mm^2) or two areae centrales, one in temporal and one in nasal retina (i.e. in the bottlenosed dolphin, *Tursiops truncatus* with 0.67×10^3 cells per mm^2 and 0.68×10^3 cells per mm^2, respectively). In *T. truncatus,* a weak horizontal streak along the retinal meridian joins the two areae centrales (Figure 3). Although the streak may provide some advantage for stabilising vision to the surface of the water, the temporal area centralis is thought to be used for aerial vision while the nasal area centralis is used for aquatic vision (Dral, 1977). Theories of decreased aerial myopia for light rays on the axis of the temporal area centralis, include a ramped temporal retina (Waller, 1984), a decreased corneal curvature in the frontal region of the eye (Dawson, 1987) and a reduced refractive power for rays striking the temporal retina through a pinhole aperture made by a strong pupillary constriction (Rivamonte, 1976). Mass and Supin (1995) consider both areae may be used underwater, the temporal area for frontal vision during rapid swimming and the nasal area to increase the visual field due to the restricted mobility of the head and eyes.

16.10 Conclusions

It is clear that the retina is specialised for various visual tasks and the degree and location of this specialisation depends upon which part of an animal's visual environment they need to sample. Studies of retinal topography provide valuable information about the location of areas of acute vision and therefore which part of the visual field they subtend. Accurate measurements of cell spacing in regions of peak photoreceptor and ganglion cell density also provide estimates of

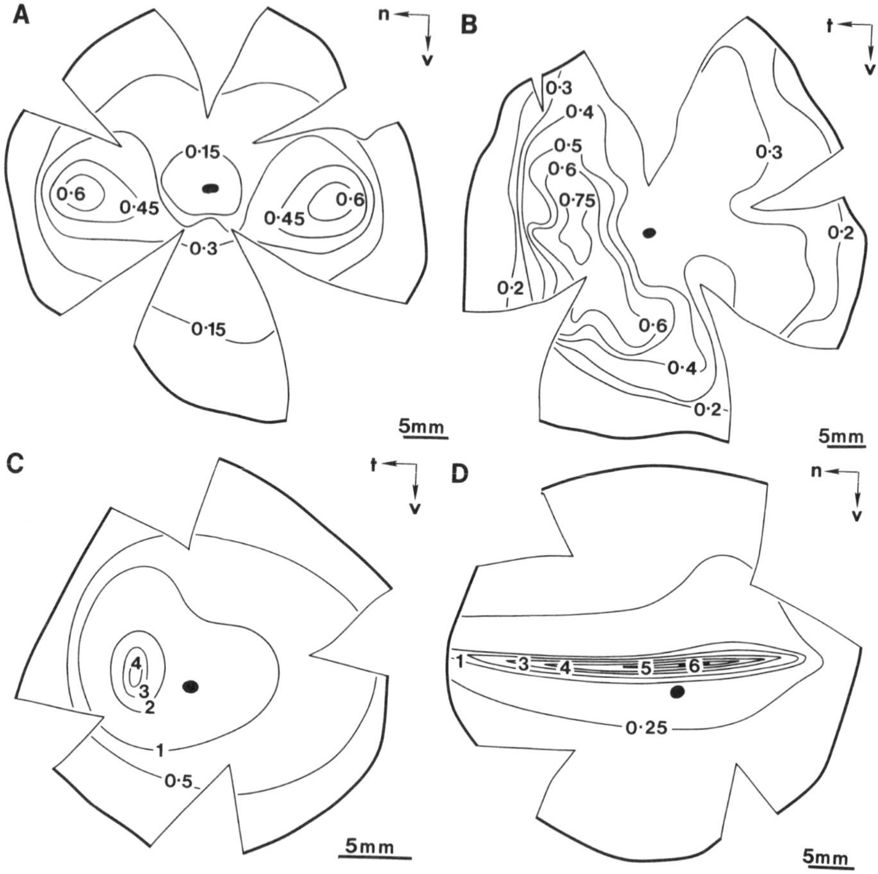

Figure 3 Range of retinal ganglion cell topographies of species in the Mammalia. **A**. The bottlenosed dolphin, *Tursiops truncatus* redrawn from Mass and Supin (1995). **B**. The two-toed sloth, *Choloepus didactylus* redrawn from Costa *et al*., (1989). **C**. Doria's tree kangaroo, *Dendrolagus doriana* redrawn from Hughes (1974). **D**. The red or plains kangaroo, *Megalia rufa* redrawn from Hughes (1974). The temporal and nasal areae in the dolphin are thought to subserve frontal and lateral aquatic vision, respectively while the vertical streak in the retina of the sloth reflects the symmetry of its arboreal habitat where it climbs upside down underneath horizontal branches. **C** and **D** illustrate the 'terrain theory' of Hughes (1977) where the three dimensional habitat of the arboreal kangaroo is reflected by its concentric retinal iso-density contours while the predominance of the horizon in the habitat of the plains kangaroo is reflected by a pronounced horizontal streak. The black disc denotes the optic nerve head. Densities $\times 10^3$ cells per mm^2. n, nasal, t, temporal; v, ventral

the spatial resolving power of each specialisation. The topography of each species examined becomes its unique 'fingerprint' which reflects the symmetry of its perceived world and may explain other visual behaviours such as eye movements and adaptations to increase the visual field along certain axes.

The findings of non-overlapping topographies of different classes of photoreceptor and retinal ganglion cells suggests that, at least on morphological grounds, there is the possibility that parallel processing exists where different components of the visual scene may be sampled by subsets of cells with different receptive fields. Further detailed tract tracing studies and electrophysiological research will undoubtedly pave the way in establishing these parallel pathways, especially in the large vertebrate radiations such as the Actinopterygii and the Aves where one finds a plethora of visual adaptations.

References

Ahlbert, I.-B. (1976) Organization of the cone cells in the retinae of salmon (*Salmo salar*) and trout (*Salmo trutta trutta*) in relation to their feeding habits. *Acta Zoologica (Stockholm)*, 57, 13 – 35.

Ali, M. A. and Anctil, M. (1976) *Retinas of fishes. An atlas.* Springer-Verlag, Berlin.

Ali, M. A., Anctil, M. and Raymond, M. (1973) La rétiné de quelque poissons marins du littoral bresilién. *Rev. Biol. Lisbon*, 9, 101–114.

Beazley, L. D., Sheard, P. W., Tennant, M., Starac, D. and Dunlop, S. A. (1997) Optic nerve regenerates but does not restore topographic projections in the lizard *Ctenophorus ornatus*. *The Journal of Comparative Neurology*, 377, 105 –120.

Bloch, S., Rey, J. and Martinoya, C. (1984) Visual acuity as a function of distance for frontal and lateral viewing in the pigeon. III. Different patterns of eye movements for binocular and monocular fixation. *Brain Behavior and Evolution*, 13, 173 –182.

Bousfield, J. D. and Pessoa, V. F. (1980) Changes in ganglion cell density during the post-metamorphic development in a neotropical frog, *Hyla raniceps*. *Vision Research*, 20, 501– 510.

Bravo, H. and Pettigrew, J. D. (1981) The distribution of neurons projecting from the retina and visual cortex to the thalamus and tectum opticum of barn owl, *Tyto alba*, and the burrowing owl, *Speotyto cunicularia*. *The Journal of Comparative Neurology*, 199, 419 – 441.

Brown, K. T. (1969) A linear area centralis extending across the turtle retina and stabilized to the horizon by non-visual cues. *Vision Research*, 9, 1053 –1062.

Budnik, V., Mpodozis, J., Varela, F. J. and Maturana, H. R. (1984) Regional specialization of the quail retina: ganglion cell density and oil droplet distribution. *Neuroscience Letters*, 51, 145 –150.

Butcher, E. O. (1938) The structure of the retina of *Fundulus heteroclitus* and the regions of the retina associated with the different chromatophoric responses. *Journal of Experimental Zoology*, 79, 275 – 293.

Chievitz, J. H. (1889) Untersuchungen über die Area centralis retinae. *Arch. Anat. Physiol. Anat. Abt. Suppl*, 1889, 139 –194.

Chievitz, J. H. (1891) Ueber das Vorkommen der Area centralis retinae in den vier höheren Wirbelthierklassen. *Arch. Anat. Physiol. Anat. Abt*, 1891, 311– 333.

Coleman, L.-A., Dunlop, S. A. and Beazley, L. D. (1984) Patterns of cell division during visual streak formation in the frog *Limnodynastes dorsalis*. *Journal of Embryology and Experimental Morphology*, 83, 119 –135.

Collett, T. S. and Harkness, L. I. K. (1982) Depth vision in animals, in *Analysis of visual behavior* (eds. Ingle, D. J., Goodale, M. A. and Mansfield, R. J. W.). MIT Press, Cambridge, Massachusetts, pp. 111–176.

Collin, S. P. (1987) *Retinal topography in reef teleosts,* PhD Thesis, University of Queensland, Australia.

Collin, S. P. (1988) The retina of the shovel-nosed ray, *Rhinobatos batillum* (Rhinobatidae): morphology and quantitative analysis of the ganglion, amacrine and bipolar cell populations. *Experimental Biology*, 47, 195 – 207.

Collin, S. P. (1989) Topography and morphology of retinal ganglion cells in the coral trout *Plectropoma leopardus* (Serranidae): A retrograde cobaltous-lysine study. *The Journal of Comparative Neurology*, 281, 143 –158.

Collin, S. P. (1997) Specialisations of the teleost visual system: Adaptive diversity from shallow-water to deep-sea. *Acta Physiologica Scandinavica,* 161 Supplement, 5 – 24.

Collin, S. P. and Ali, M. A. (1994) Multiple areas of acute vision in two freshwater teleosts, the creek chub, *Semotilus atromaculatus* (Mitchell) and the cutlips minnow, *Exoglossum maxillingua* (Lesueur).

Canadian Journal of Zoology, 72, 721–730.
Collin, S. P. and Collin, H. B. (1988) Topographic analysis of the retinal ganglion cell layer and optic nerve in the sandlance *Limnichthyes fasciatus* (Creediidae, Perciformes). *The Journal of Comparative Neurology*, 278, 226 – 241.
Collin, S. P. and Collin, H. B. (1998) The foveal photoreceptor mosaic in the pipefish, Corythoichthys paxtoni (Syngnathidae, Teleostei). *Histology and Histopathology* (in press).
Collin, S. P. and Fritzsch, B. (1993) Observations on the shape of the lens in the eye of the silver lamprey, *Ichthyomyzon unicuspis*. *Canadian Journal of Zoology*, 72, 721–730.
Collin, S. P. Hoskins, R. V. and Partridge, J. C. (1997) Tubular eyes of deep-sea fishes: A comparative study of retinal topography. *Brain Behavior and Evolution*, 50, 335 – 357.
Collin, S. P., Hoskins, R. V. and Partridge, J. C. (1998) Seven retinal specializations in the tubular eye of the deep-sea pearleye, *Scopelarchus michaelsarsi*: A case study in visual optimisation. *Brain Behavior and Evolution*, 51, 291– 314.
Collin, S. P. and Northcutt, R. G. (1993) The visual system of the Florida garfish, *Lepisosteus platyrhincus*. (Ginglymodi): III. Retinal ganglion cells. *Brain Behavior and Evolution*, 42, 295 – 320.
Collin, S. P. and Partridge, J. C. (1996) Retinal specializations in the eyes of deep-sea teleosts. *Journal of Fish Biology*, 49 (Supplement A), 157–174.
Collin, S. P. and Pettigrew, J. D. (1988a) Retinal topography in reef teleosts. I. Some species with well developed areae but poorly developed streaks. *Brain Behavior and Evolution*, 31, 269 – 282.
Collin, S. P. and Pettigrew, J. D. (1988b) Retinal topography in reef teleosts. II. Some species with prominent horizontal streaks and high density areae. *Brain Behavior and Evolution*, 31, 283 – 295.
Collin, S. P. and Pettigrew, J. D. (1988c) Retinal ganglion cell topography in teleosts: a comparison between Nissl-stained material and retrograde labelling from the optic nerve. The *Journal of Comparative Neurology*, 276, 412 – 422.
Costa, B. L. S. A., Pessoa, V. F., Bousfield, J. D. and Clarke, R. J. (1987) Unusual distribution of ganglion cells in the retina of the three-toed sloth (*Bradypus variegatus*). *Brazilian Journal of Medical and Biological Research*, 20, 741–748.
Costa, B. L. S. A., Pessoa, V. F., Bousfield, J. D. and Clarke, R. J. (1989) Ganglion cell size and distribution in the retina of the two-toed sloth (*Choloepus didactylus* L.). *Brazilian Journal of Medical and Biological Research*, 22, 233 – 236.
Dawson, W. W. (1987) Corneal surface properties of two marine mammal species. *Marine Mammal Science*, 3, 186 –197.
Drager, U. C. and Olsen, J. F. (1981) Ganglion cell distribution in the retina of the mouse. *Investigative Ophthalmology*, 20, 285 – 293.
Dral, A. D. G. (1972) Aquatic and aerial vision in the bottlenosed dolphin. *Netherlands Journal of Sea Research*, 5, 510 – 513.
Dral, A. D. G. (1977) On the retinal anatomy of cetacea (mainly *Tursiops truncatus*), in *Functional anatomy of marine mammals* (ed Harrison, R. J.), Academic Press, London, pp. 81–134.
Dujim, M. (1958) On the position of a ribbon-like central area in the eyes of some birds. *Archs néer. Zool. Supplement*, 13, 128 –145.
Dunlop, S. A. and Beazley, L. D. (1981) Changing retinal ganglion cell distribution in the frog *Heleioporus eyrei*. *The Journal of Comparative Neurology*, 202, 221– 236.
Dunlop, S. A. and Beazley, L. D. (1985) Changing distribution of ganglion cells during area centralis and visual streak formation in the marsupial *Setonix brachyurus*. *Developmental Brain Research*, 23, 81– 90.
Dunlop, S. A., Ross, W. M. and Beazley, L. D. (1994) The retinal ganglion cell layer and optic nerve in a marsupial, the honey possum (*Tarsipes rostratus*). *Brain Behavior and Evolution*, 44, 307– 323.
Dunn-Meynell, A. A. and Sharma, S. C. (1987) Visual system of the channel catfish (*Ictalurus punctatus*): II. The morphology associated with the multiple optic papillae and retinal ganglion cell distribution. *The Journal of Comparative Neurology*, 257, 166 –175.
Easter, S. S. Jr. (1992) Retinal growth in foveated teleosts: Nasotemporal asymmetry keeps the fovea in temporal retina. *The Journal of Neuroscience*, 12, 2381– 2392.
Ehrlich, D. (1981) Regional specialization of the chick retina as revealed by the size and density of neurons in the ganglion cell layer. *The Journal of Comparative Neurology*, 195, 643 – 657.
Engström, K. (1963) Cone types and cone arrangements in teleost retinae. *Acta Zoologica (Stockholm)*, 44, 179 – 243.
Fischer, Q. S. and Kirby, M. A. (1991) Number and distribution of retinal ganglion cells in anubis baboons (*Papio anubis*). *Brain Behavior and Evolution*, 37, 189 – 203.

Fite, K, V. and Lister, B. C. (1981) Bifoveal vision in anolis lizards. *Brain Behavior and Evolution*, 19, 144 –154.

Frank, B. D. and Hollyfield, J. G. (1987) Retinal ganglion cell morphology in the frog, *Rana pipiens*. *The Journal of Comparative Neurology*, 266, 413 – 434.

Fritzsch, B. and Collin, S. P. (1990) Dendritic distribution of two populations of ganglion cells and the retinopetal fibres in the retina of the silver lamprey (*Ichthyomyzon unicuspis*). *Visual Neuroscience*, 4, 533 – 545.

Gillett, W. G. (1923) The histologic structure of the eye of the soft-shelled turtle. *American Journal of Ophthalmology*, 6, 955 – 973.

Graydon, M. L. and Giorgi, P. P. (1984) Topography of the retinal ganglion cell layer of *Xenopus*. *Journal of Anatomy,* 139, 145 –157.

Harkness, L. and Bennett-Clark, H. C. (1978) The deep fovea as a focus indicator. *Nature (London)*, 272, 814 – 816.

Hayes, B. P., Martin, G. R. and Brooke, M. de L. (1991) Novel area subserving binocular vision in the retinae of Procellariform seabirds. *Brain Behavior and Evolution*, 37, 79 – 84.

Hodos, W. and Erichsen, J. T. (1990) Lower-field myopia in birds: An adaptation that keeps the ground in focus. *Vision Research*, 30, 653 – 657.

Holmberg, K. (1972) Fine structure of the optic tract in the Atlantic hagfish, *Myxine glutinosa*. *Acta Zoologica (Stockholm)*, 53, 155 –171.

Hueter, R. E. (1991) Adaptations for spatial vision in sharks. *The Journal of Experimental Zoology Supplement,* 5, 130 –141.

Hughes, A. (1974) A comparison of retinal ganglion cell topography in the plains and tree kangaroo. *Journal of Physiology*, 244, 61– 63P.

Hughes, A. (1977) The topography of vision in mammals of contrasting lifestyles: comparative optics and retinal organization in *Handbook of sensory physiology*, Volume VII/5 (eds Autrum, H., Jung, R., Loewenstein, W. R., MacKay, D. M. and Teuber, H. L.), Springer-Verlag, New York, pp. 613 –756.

Humphrey, M. F. (1988) A morphometric study of the retinal ganglion cell response to optic nerve severance in the frog, *Rana pipiens*. *Journal of Neurocytology*, 17, 293 – 204.

Humphrey, M. F. and Beazley, L. D. (1985) Retinal ganglion cell death during optic nerve regeneration in the frog Hyla moorei. *The Journal of Comparative Neurology*, 236, 382 – 402.

Inzunza, O. Bravo, H. and Smith R. L. (1989) Foveal regions of bird retinas correlate with the aster of the inner nuclear layer. *The Anatomical Record*, 223, 342 – 346.

Ito, H. and Murakami, T. (1984) Retinal ganglion cells in two teleost species, *Sebastiscus marmoratus* and *Navodon modestus*. *The Journal of Comparative Neurology*, 229, 80 – 96.

Johns, P. R. and Easter, S. S. Jr. (1977) Growth of the adult goldfish eye. II. Increase in retinal cell number. *The Journal of Comparative Neurology*, 176, 331– 342.

Kahmann, H. (1934) Ueber die Vorkommen einer Fovea centralis im Knochenfischauge. *Zoologischer Anzeiger*, 106, 49 – 55.

Kahmann, H. (1935) Ueber das foveale Sehen der Wirbeltiere. II. Gesichtsfeld und Fovea centralis. *Sitz. Ges. naturf. Freunde*, pp. 361– 376.

Kalinina, A. V. (1976) Quantity and topography of frog's retinal ganglion cells. *Vision Research,* 16, 929 – 934.

Kawamura, G. and Ohashi, S. (1988) The habit of cutlassfish as inferred from the retina. *Nippon Suisan Gakkaishi*, 54, 889.

Locket, N. A. (1971) Retinal structure in *Platytroctes apus*, a deep-sea fish with a pure rod fovea. *Journal of the Marine Biological Association of the United Kingdom*, 51, 79 – 91.

Locket, N. A. (1985) The multiple bank rod fovea of *Bajacalifornia drakei*, an alepocephalid deep-sea teleost. *Proceedings of the Royal Society of London B*, 224, 7– 22.

Locket, N. A. (1992) Problems of deep foveas. *Australian and New Zealand Journal of Ophthalmology*, 20, 281– 295.

Makaretz, M. and Levine, R. L. (1980) A light microscopic study of the bifoveate retina in the lizard *Anolis carolinensis*: General observations and convergence ratios. *Vision Research*, 20, 679 – 686.

Maldonado, P. E., Maturana, H. and Varela, F. J. (1988) Frontal and lateral visual system in birds: Frontal and lateral gaze. *Brain Behavior and Evolution*, 32, 57– 62.

Mariani, A. P. and Leure-Dupree, A. E. (1978) Photoreceptors and oil droplet colors in the red field of the pigeon retina. *The Journal of Comparative Neurology*, 182, 821– 837.

Martin, G. R. and Brooke, M. de L. (1991) The eye of a procellariform seabird, the manx shearwater, *Puffinus puffinus:* Visual fields and optical structure. *Brain Behavior and Evolution,* 37, 65 –78.

Mass, A. M. and Supin, A. Y. (1992) Peak density, size and regional distribution of ganglion cells in the retina of the furseal *Callorhinus ursinus*. *Brain Behavior and Evolution*, 39, 69 –76.

Mass, A. M. and Supin, A. Y. (1995) Ganglion cell topography of the retina in the bottlenosed dolphin, *Tursiops truncatus*. *Brain Behavior and Evolution*, 45, 257– 265.

Mathews, G, V. T. (1968) *Bird navigation*, Cambridge Monographs in Experimental Biology No. 3, Cambridge, pp. 1–197.

Mednick, A. S. and Springer, A. D. (1988) Asymmetric distribution of retinal ganglion cells in goldfish. *The Journal of Comparative Neurology*, 268, 49 – 59.

Mednick, A. S., Berk, M. F. and Springer, A, D. (1988) Asymmetric distribution of cells in the inner nuclear and cone mosaic layers of the goldfish retina. *Neuroscience Letters*, 94, 241– 246.

Miller, W. H. and Snyder, A. W. (1976) The tiered vertebrate retina. *Vision Research*, 17, 239 – 255.

Moroney, M. K. and Pettigrew, J. D. (1987) Some observations on the visual optics of kingfishers (Aves, Caraciformes, Alcedinidae). *Journal of Comparative Physiology A*, 160, 137–149.

Munk, O. (1970) On the occurrence and significance of horizontal band-shaped retinal areae in teleosts. *Videnskabelige Meddelelser fra Dansk Naturhistorik Forening i Kjobenhavn*, 133, 85 –120.

Nalbach, H.-O., Wolf-Oberhollenzer, F. and Kirschfeld, K. (1990) The pigeon's eye viewed through an ophthalmoscope microscope: orientation of retinal landmarks and significance of eye movements. *Vision Research*, 30, 529 – 540.

Nguyen, V.-S. and Straznicky, C. (1989) The development and the topographic organisation of the retinal ganglion cell layer in *Bufo marinus*. *Experimental Brain Research*, 75, 345 – 353.

O'Connell, C. P. (1963) The structure of the eye of *Sardinops caerulea*, *Engraulis mordax*, and four other pelagic marine teleosts. *Journal of Morphology*, 113, 287– 330.

Ott, M. and Schaeffel, F. (1995) A negatively powered lens in the chameleon. *Nature*, 373, 692 – 694.

Patzner, R. A. (1978) Experimental studies on the light sense in the hagfish, *Eptatretus burgeri* and *Paramyxine atami* (Cyclostomata). *Helgoländer Wiss. Meeresunters*, 31, 180 –190.

Perry, V. H., Henderson, Z. and Linden, R. (1983) Postnatal changes in retinal ganglion and optic axon populations in pigmented rat. *The Journal of Comparative Neurology*, 219, 356 – 368.

Peterson, E. H. and Rowe, M. H. (1980) Different regional specializations of neurons in the ganglion cell layer and inner plexiform layer of the California horned shark, *Heterodontus francisci*. *Brain Research*, 201, 195 – 201.

Peterson, E. H. and Ulinski, P. S. (1979) Quantitative studies of retinal ganglion cells in a turtle *Pseudemys scripta elegans*. I. Number and distribution of ganglion cells. *The Journal of Comparative Neurology*, 186, 17– 42.

Pettigrew, J. D. (1990) Evolution of binocular vision, in *Vision and visual dysfunction, Volume III. Evolution of the eye and visual system.* (eds Gregory, R. and Cronley-Dillon, J. R.), CRC Press, Florida, USA, pp. 271– 283.

Pettigrew, J. D. and Collin, S. P. (1995) Terrestrial optics in an aquatic eye: The sandlance, *Limnichthyes fasciatus* (Creediidae, Teleostei). *Journal of Comparative Physiology A*, 177, 398 – 408.

Pettigrew, J. D., Dreher, B., Hopkins, C. S., McCall, M. J. and Brown, M. (1988) Peak density and distribution of ganglion cells in the retinae of microchiropteran bats: implications for visual acuity. *Brain Behavior and Evolution*, 32, 39 – 56.

Pumphrey, R. J. (1948) The theory of the fovea. *Journal of Experimental Biology*, 25, 299 – 312.

Pumphrey, R. J. (1961) Sensory organs. Part 1, in *Biology and comparative physiology of birds* Volume II (ed Marshall, A. J.), Academic Press, New York, pp. 55 – 68.

Reymond, L. (1987) Spatial visual acuity of the falcon, *Falco berigora*: A behavioral, optical and anatomical investigation. *Vision Research*, 27, 1859 –1874.

Rivamonte, A. (1976) Eye model to account for comparable aerial and underwater acuities of the bottlenose dolphin. *Netherlands Journal of Sea Research*, 10, 491– 498.

Schmid, K. L., Schmid, L. M., Wildsoet, C. F. and Pettigrew, J. D. (1992) Retinal topography in the koala (*Phascolarctos cinereus*). *Brain Behavior and Evolution*, 39, 8 –16.

Schwassmann, H. O. (1968) Visual projection upon the optic tectum in foveate marine teleosts. *Vision Research*, 8, 1337–1348.

Schwassmann, H. O. and Kruger, L. (1965) Experimental analysis of the visual system of the four-eyed fish *Anableps microlepis*. *Vision Research*, 5, 269 – 281.

Slonaker, J. R. (1897) A comparative study of the area of acute vision in vertebrates. *Journal of Morphology*, 13, 445 – 492.

Snyder, A. W. and Miller, W. H. (1978) Telephoto system of falconiform eyes. *Nature*, 275, 127–129.

Stell, W. K. and Witkovsky, P. (1973) Retinal structure in the smooth dogfish *Mustelus canis*: general

description and light microscopy of giant ganglion cells. *Journal of Comparative Neurology*, 148, 1–32.
Stiles, W. S. and Crawford, B. H. (1933) The luminous efficiency of rays entering the eye pupil at different points. *Proceedings of the Royal Society of London (Biology)*, 112, 428 – 450.
Stone, J. (1965) A quantitative analysis of the distribution of ganglion cells in the cat retina. The *Journal of Comparative Neurology*, 124, 337– 352.
Stone, J. and Halasz, P. (1989) Topography of the retina in the elephant *Loxodonta africana. Brain Behavior and Evolution*, 34, 84 – 95.
Tancred, E. (1981) The distribution and sizes of ganglion cells in the retinae of five Australian marsupials. *The Journal of Comparative Neurology*, 196, 585 – 603.
Vaney, D. I. (1980) A quantitative comparison between the ganglion cell populations and axonal outflows of the visual streak and periphery of the rabbit retina. *The Journal of Comparative Neurology*, 189, 215 – 233.
Vilter, V. (1954) Différenciation fovéale dans l'appareil visuel d'un poisson abyssal, le *Bathylagus benedicti. Compte Rendu des Séances de la Société Biologie, Paris*, 148, 59 – 63.
Wagner, H.-J., Menezes, N. A. and Ali, M. A. (1976) Retinal adaptations in some Brazilian tide pool fishes. *Zoomorphology*, 83, 209 – 226.
Waller, G. N. H. (1984) The ocular anatomy of cetacea: an historical perspective, in *Investigation of cetacea*, Vol. XVI (ed Pillery, G.). pp. 138 –148.
Walls, G. (1942) *The vertebrate eye and its adaptive radiation*, New York, Hafner.
Wilhelm, M. and Straznicky, C. (1992) The topographic organisation of the retinal ganglion cell layer of the lizard, *Ctenophorus nuchalis. Archives of Histology and Cytology*, 55, 251– 259.
Wong, R. O. L. (1989) Morphology and distribution of neurons in the retina of the american garter snake *Thamnophis sirtalis. The Journal of Comparative Neurology*, 283, 587– 601.
Wong, R. O. L. and Hughes, A. (1987) The morphology, number, and distribution of a large population of confirmed displaced amacrine cells in the adult cat retina. *The Journal of Comparative Neurology*, 255, 159 –177.
Wood, C. A. (1917) *The fundus oculi of birds especially as viewed by the ophthalmoscope*. The Lakeside Press, Chicago, pp. 1–180.
Zaunreiter, M., Junger, H. and Kotrschal, K. (1991) Retinal morphology of cyprinid fishes: a quantitative histological study of ontogenetic changes and interspecific variation. *Vision Research*, 31, 383 – 394.
Zhang, Y. and Straznicky, C. (1991) The morphology and distribution of photoreceptors in the retina of *Bufo marinus. Anatomy and Embryology*, 183, 97–104.

17
Flower advertisement for insects: Bees, a case study

M. VOROBYEV and R. MENZEL
Institut für Neurobiologie, Freie Universität, Berlin, Königin-Luise-Str. 28 – 30, 14195 Berlin, Germany

17.1 Introduction

Visual systems are generally highly adaptive, and are likely to be a reflection of the selective pressure brought to bear on the evolution of modern animals (Wald, 1940; Lythgoe, 1979). At the same time, we believe that signals, such as the colours and patterns of plants and animals, were adapted to these visual systems. Although the hypothesis of the mutual evolution of the visible environment and visual systems is based on many observations (Lythgoe, 1979), it has not yet been tested by quantitative data. To base the speculation on quantitative grounds we must specify: 1) the objects that played the *main* role in the evolution of the visual system of a particular animal, and 2) the visual systems to which the coloured signals were adapted.

The majority of angiosperm plants rely on animals as pollen vectors. In order to attract pollinators they have evolved flowers as brightly-coloured advertisements. Flowers contain nectar and pollen – which is the main food source for many insects (Darwin, 1876). Insects are often attracted by colours, and some insects, like bees and flies, can learn colours (Menzel & Backhaus, 1991; Troje, 1993). To make efficient choices an animal must select those flowers which give the greatest reward, detect these flowers, identify them and discriminate them from others. Because visual cues are important for all these tasks, one may expect colour vision of animals foraging on flowers to be adapted to flower colours. The flowers whose colours are easily learned or preferred, identified, detected and discriminated from others would be effectively pollinated. The colours of flowers are, in turn, very likely to be adapted to the visual systems of the animals on which they depend. This view appears to be supported by the observation that flowers specially adapted for pollination by birds, which have a red receptor, are usually red or orange, whereas flowers pollinated by bees, animals whose visual range does not extend into the red part of the spectrum, are often blue, yellow or white (Lythgoe, 1979, Menzel and Shmida, 1993). However, is the intuitive view supported by quantitative data, and what are the consequences of co-evolution of flower signals and the visual systems of pollinators? Bees are the most abundant group of insect pollinators, foraging predominantly

S.N. Archer et al. (eds.), Adaptive Mechanisms in the Ecology of Vision, 537–553

on flowers. Our analysis focuses on the mutual evolutionary adaptation of bee colour vision and of flower colours. To quantitatively formulate the relation between the properties of colour vision and the spectral properties of flowers we use the classical colour theory, originally developed for the visual system of man (Wyszecki and Stiles, 1982). This theory should be extended to animals other than man only with caution, but the basic ideas can be useful.

17.2 Receptor spectral sensitivity functions and colour vision systems

Where animals have several types of photoreceptor cells whose sensitivities are tuned to light of different wavelengths, comparison of the signals from these cells makes it possible to perceive colours. The spectral sensitivity functions of photoreceptor cells depend on the absorption spectrum of visual pigments, which consist of a protein (opsin) bound to a chromophore, retinal or a related molecule. Small variations in the amino-acid sequence of the protein yield pigments with a peak sensitivity in different parts of the spectrum (Mollon, 1992). Thus, the evolution of colour vision is to a large degree determined by mutations in the opsin coding genes.

The shape of receptor spectral sensitivity function can be predicted on the basis of the wavelength position of the peak sensitivity of visual pigment (Dartnall, 1953; Maximov, 1988). This allows us to estimate parameters of visual systems depending on wavelength position of the peaks of visual pigments. Performance of hypothetical visual systems with variable spectral sensitivity functions has been first analysed for dichromatic vision (Lythgoe and Partridge 1989), and later for trichromatic and tetrachromatic vision (Govardovskii & Vorobyev 1989; Chittka & Menzel, 1992; Chittka *et al.*, 1994; Osorio & Vorobyev 1996). Two visual tasks have been considered in the calculations: 1) discrimination of objects from each other (Govardovskii & Vorobyev 1989; Chittka *et al.*, 1994; Osorio & Vorobyev, 1996), 2) detection of objects against background (Lythgoe and Partridge 1989; Osorio & Vorobyev 1996).

To evaluate the ability of colour vision to detect or discriminate objects we must introduce a mathematical definition of colour. In the case of trichromatic vision, colour can be presented as a point (Q) in the three-dimensional receptor space, where quantum catches of the long-wavelength (Q_L), middle-wavelength (Q_M) and short-wavelength (Q_S) receptors are placed along the co-ordinate axes. For an eye viewing the surface of an object the quantum catches are given by the following expression:

$$Q_i \equiv \int_{\lambda} I(\lambda)\, S(\lambda)\, R_i(\lambda)\, d\lambda \qquad [1]$$

where λ defines the wavelength, $S(\lambda)$ is the reflectance spectrum of an object, $R_i(\lambda)$ is the spectral sensitivity function of a receptor ($i = S,M,L$), $I(\lambda)$ is the illumination spectrum, and integration is performed over the spectral range where visual system is sensitive (Wyszecki and Stiles, 1982).

Instead of the three-dimensional receptor space, two-dimensional 'chromaticity' diagrams are often used (Wyszecki and Stiles, 1982). A common two-dimensional representation is obtained in the unit plane, $Q_S + Q_M + Q_L = 1$, of the receptor space as shown in Figure 1a. A line connecting the origin with the point Q or its extension must intersect the unit plane in a point q. This intersect gives a *steriographic* projection of the point Q onto the unit plane. The location of that point determines the 'chromaticity' of the colour, whereas the length of the vector characterises luminosity or brightness of colour. The chromaticity diagram shown in Figure 1a is in a shape of an equilateral triangle, and is referred to as the Maxwell triangle. Generalisation of the Maxwell triangle for tetrachromatic vision gives a three-dimensional projection of the four-dimensional colour space. This three-dimensional chromaticity diagram has a shape of a tetrahedron (Goldsmith, 1991).

A modified chromaticity diagram is obtained in the space where so-called receptor 'excitations' rather than quantum catches are placed along the co-ordinate axes (Chittka, 1992). 'Excitations' (E_i) are related to quantum catches by the following equations:

$$E_i = \frac{R_i Q_i}{1 + R_i Q_i} \qquad [2]$$

where, i = S,M,L, and the factor R_i is found from the condition that $E_i = 0.5$ when the eye is viewing the background (Backhaus, 1991). From Eq. 2 it follows that E_i may change from zero to unity. Thus colour points are restricted to the unit cube (Figure 1b). A two dimensional presentation is now obtained by an *orthogonal* projection of the colour point E onto the unit plane $E_S + E_M + E_L = 1$. This chromaticity diagram is in the shape of an equilateral hexagon, and is referred to as a colour hexagon. The point corresponding to the colour of the background is situated in the centre of the diagram and is by definition 'achromatic' (Chittka, 1992).

'Chromaticity' diagrams necessarily reduce information contained in receptor quantum catches. Generally, these diagrams are useful if and only if an animal uses 'chromatic' cues independently from 'achromatic' ones. Although man distinguishes chromatic aspects of colour (hue and saturation) from an achromatic one (luminosity or brightness), in general we do not know if animals have independent 'chromatic' and 'achromatic' visual mechanisms.

17.3 Visual systems of bees

Colour vision of the honeybee has been investigated in more detail than any other animal except man. Behavioural (Daumer, 1956) and electrophysiological (Autrum and Zwehl, 1964; Menzel and Blakers, 1976; Peitsch *et al.*, 1992)

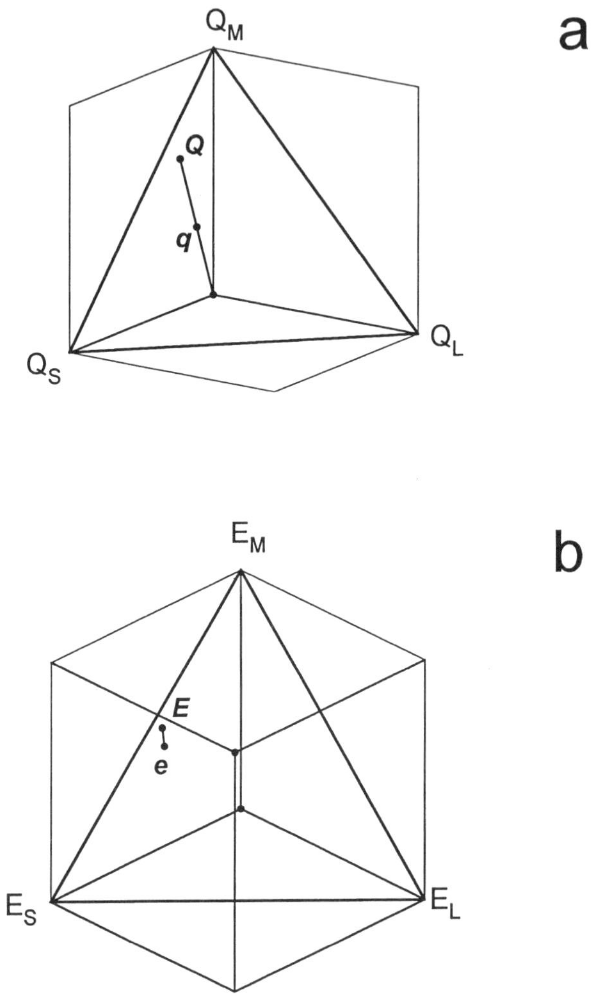

Figure 1 Receptor space and 'chromaticity' diagrams. **a** – Maxwell triangle. Receptor quantum catches, Q_S, Q_M and Q_L are placed along co-ordinate axes. *q* , the triangle co-ordinates of a point *Q* are given by a steriographic projection of *Q* onto the unit plane, $Q_S + Q_M + Q_L = 1$. **b**. Colour hexagon. Receptor 'excitations' (Backhaus, 1991), E_S, E_M and E_L are placed along co-ordinate axes. *e* , the hexagon co-ordinates of a point *E* are given by an orthogonal projection of *E* onto the unit plane, $E_S + E_M + E_L = 1$

studies have revealed that bees have a trichromatic colour vision system; its photoreceptors have sensitivity maxima (λ_{max}) in the UV (S, λ_{max} = 344 nm) , blue (M, λ_{max} = 436 nm) and green (L, λ_{max} = 544 nm; Figure 2). Theoretical analysis of colour training experiments suggests that colour in the honey bee is coded by two opponent neural subsystems forming chromatic visual pathways

(Backhaus, 1991; Menzel and Backhaus, 1991; Brandt and Vorobyev, 1996). An achromatic channel is mediated exclusively by the L receptor (Lehrer, 1994). In man, both chromatic and achromatic pathways contribute to the sensation of colour, i.e. colour is described by hue, saturation and brightness (Wyszecki and Stiles, 1982). In contrast chromatic and achromatic vision in the honeybee are used for different purposes, and strikingly, our recent observations show that they are mutually exclusive (Giurfa *et al.*, 1996). Learning, discrimination and detection of large targets (angle > 15°) is mediated exclusively by chromatic vision (Menzel and Backhaus 1991; Giurfa *et al.*, 1996), whereas motion perception, pattern recognition, and detection of small targets (subtended angle < 15°) is mediated exclusively by the achromatic channel, i.e by the L receptor (Lehrer, 1994; Giurfa *et al.*, 1996).

In addition to studies of the honey bee itself, comparative observations from other species are important if we are to understand the evolution of colour vision. Intracellular recordings from receptors of 40 hymenopteran species (bees and wasps) have revealed that the peak positions of the spectral sensitivity functions in all of these animals are similar to those of the honeybee (Peitsch *et al.*, 1992). Moreover, colour discrimination in several species of bees can be described by the model assuming that colour is coded by two colour opponent chromatic subsystems (Chittka *et al.*, 1992). Thus the two dimensional 'chromatic' diagrams are likely to give an adequate representation of colour for many Hymenoptera.

17.4 The colours of flowers

Flower colours are due to pigments of several chemical classes, the anthocyanins, flavinoids, carotinoids, betalaines and chlorophylls (Harborne, 1988). Pigments can be located in different arrangements giving subtractive and additive colour mixtures, hence it is in principle possible to construct a large number of different

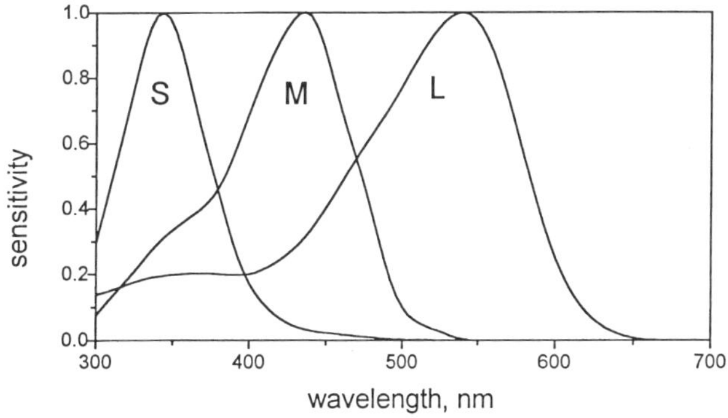

Figure 2 Spectral sensitivities of the three photoreceptor types in honeybee. The functions are scaled to give unity sensitivity at their peaks which are at 344 nm (S), 436 nm (M) and 544 nm (L)

spectral reflectances (SRs) from a few pigment types. However, analysis of 1056 SRs of flowers collected in Israel, the Austrian Alps, Germany, Norway and Brazil (Menzel and Shmida, 1993; Chittka *et al.*, 1994), shows that only limited types of SRs are used by flowers; this is indicative of some constraint on flower SRs. None of the flowers have spectral reflectances which change markedly more than three times in the spectral range 300 –700 nm. Frequently the SRs resemble cut-off filters that absorb short wavelengths and reflect long wavelengths, while none have the inverse shape, absorbing long and reflecting short wavelengths. It has been proposed that flower reflectance spectra (not bee colours) fall into six categories (Chittka *et. al.*, 1994). To check whether flower spectra are indeed clustered, rather than being uniformly spread, in the multidimensional space of the SRs we performed a cluster analysis. SRs were classified into several groups (clusters) where the distance between the SRs within each cluster is small compared to the distance between the centres of clusters. The distance between two spectra, $S_1(\lambda)$ and $S_2(\lambda)$, was calculated as the mean square difference between them in the spectral region 300 –700 nm:

$$\int_{300}^{700} (S_1(\lambda)-S_2(\lambda))^2 d\lambda$$

This method revealed three groups of flower SRs (Figure 3): a) cut-off filters, which are white, yellow, orange or red for the human observer; b) SRs with a broad maximum in the UV or blue part of the spectrum in addition to a cut-off type reflectance in the red part of the spectrum which produce blue or pink flowers; and c) flowers where the reflectance gradually increases in the long wavelength part of the spectrum, and have low overall reflectance. These 3 groups are relatively uniform and they can not be reliably subdivided into smaller clusters.

Another method of classification of flower SRs is based on the spread of flower colours in the colour hexagon (see above). Chittka *et al.*, (1994) named the six sectors of the hexagon – i.e. ‘blue’ for the colours which correspond to the strong excitation of the M (blue) receptor, ‘blue-green’ for the colour which correspond to weak excitation of the S (UV) receptor, compared to that of L (green) and M (blue) receptors (Figure 4a). The central loci correspond to achromatic colours, which have similar excitations for all 3 receptor types. Flower colours are assumed to be viewed against green foliage, thus the loci of leaves that are situated in the centre of the coloured leaves, are by definition achromatic. Flowers tend to be different from leaves, most flowers are bee ‘blue-green’, and UV colours are rare (Chittka *et al.*, 1994).

Flower colours are related to the amount of nectar they produce: 80% of ‘UV-blue’ flowers (mostly blue or purple for humans) offer on average a high reward; ‘blue-green’ (usually human white) and ‘UV-green’ (yellow for humans) flowers offer low rewards (20% and 5% respectively); and others lie between these

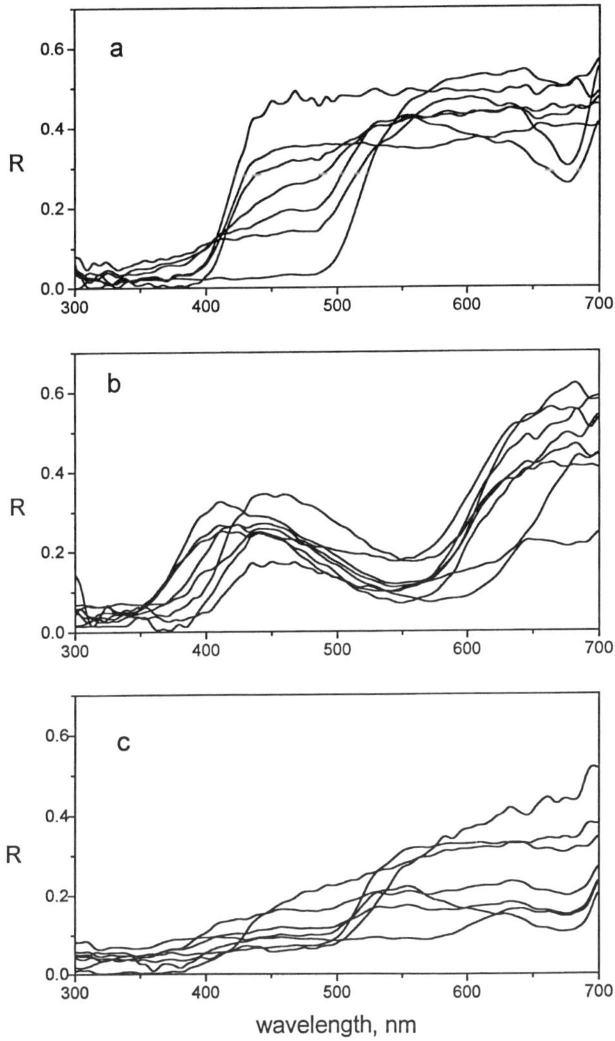

Figure 3 Examples of the three types of the flower spectral reflectance functions. R denotes reflectance relative to the BaSO4 white standard. **a** – cut-off type shapes; **b** – broad maximum in the UV or blue part of the spectrum and cut-off type shape with a sharp transition in the red part of the spectrum; **c** – low reflecting flowers

extremes (Giurfa *et al.*, 1995). This makes it possible for inexperienced pollinators to predict the amount of reward on the basis of flower colour; 'UV-blue colours' are indeed preferred by flower-naive bees (Giurfa *et al.*, 1995), and are also most easily learned (Menzel, 1967). Flowers with colours of different categories tend to advertise themselves to different groups of insect pollinators: 'UV-blue' and 'blue' flowers are more frequently visited by bees, whereas 'green' and 'UV-green' flowers are more frequently visited by flies and beetles (Menzel and Shmida, 1993). However, bees can learn all chromatic colours (Menzel, 1967) and they pollinate flowers belonging to all colour categories (Menzel and Shmida, 1993).

What use has flower colour for plants? In order to be effectively pollinated, the plant has to advertise itself so that it can be easily learned or chosen by inexperienced pollinators, and detected, recognised and discriminated from the advertisements of other plant species. Generally, these tasks may make different demands on flower colours. The flowers which are most effectively learned and preferred by flower-naive bees, the 'UV-blue' flowers (Menzel, 1967, Giurfa *et al.*, 1995), have, by definition, low contrast to the L (green) receptor which mediates the achromatic vision necessary for detection of small targets, and thus they may be less effectively detected from far away (Giurfa *et al.*, 1996). It is advantageous for flowers to be discriminable from each other to minimise the risk of cross-pollination, and this requirement predicts that their colours should be uniformly distributed in the bee's colour space. While it is generally beneficial for flowers to be different from each other and from the background, there are exceptions, for example, flowers which offer little or no reward mimic others which are more profitable (Dafni, 1984).

17.5 Discrimination of flower colours

Generally, the tasks of target vision are detection, recognition and discrimination. Increasing, the discriminability of flower colours is, in most cases, beneficial for plants and pollinators. It should be noted that the demands of different visual tasks may be conflicting. For example, to identify flowers in conditions of patchy illumination it is beneficial to discard the luminosity signal (Osorio & Vorobyev 1996), whereas discrimination of flowers blooming under uniform illumination would be most accurate if chromatic *and* achromatic signals were used.

Are the peak positions of the receptor sensitivities in bees optimal for discrimination of flowers from each other? In order to answer this question we must introduce a quantitative criterion for discriminability of coloured objects. Our criterion is the number of discriminable colours (N), or, conversely, the percentage of objects (P) that are indistinguishable (Govardovskii and Vorobyev, 1989, Vorobyev *et al.*, 1993). Thus, to specify the quality of a visual system with a particular set of photoreceptors one needs to estimate P or N, and to compare these values with those of hypothetical animals with different sets of photoreceptors. While P and N cannot be obtained experimentally, they can be calculated theoretically.

17.5.1 Colours of natural objects

Because receptor spectral sensitivities overlap (Figure 2) some parts of the space are not occupied by physically realisable colours. The points corresponding to the colours of all theoretically possible objects fill a part of the colour space – an object-colour solid (Luther, 1927, Nyberg, 1928). The points corresponding to the colours of real objects (flowers) form a cloud which lies inside this solid (Figure 4b). The boundaries of the solid are formed by optimal colour stimuli

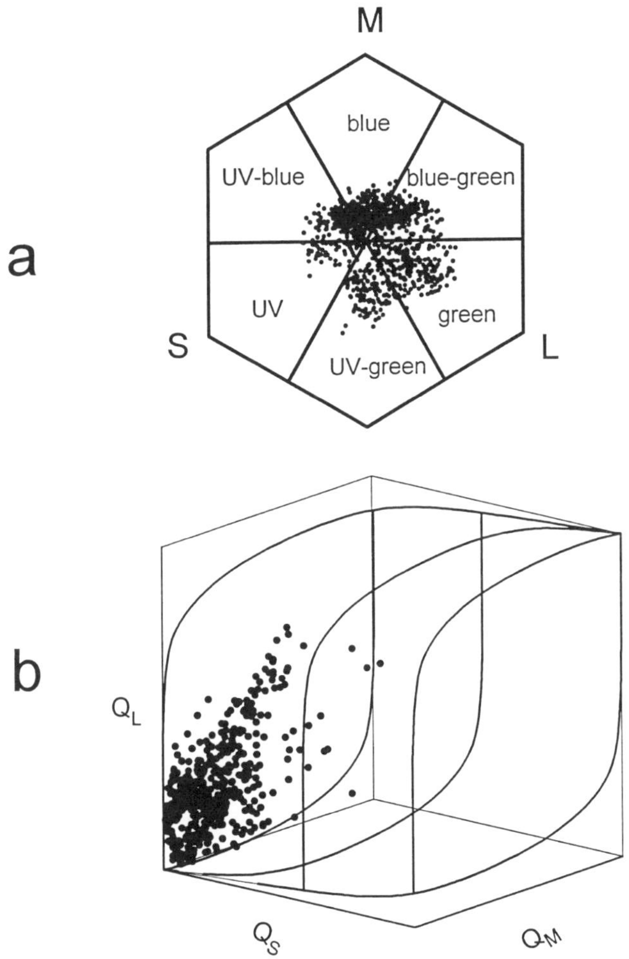

Figure 4 The spread of flower colours in the receptor space of the honeybee. **a** – a two-dimensional chromatic diagram, the colour hexagon. The sectors of the hexagon give the names for chromatic colours used by Chittka *et al.*, (1994). **b**. – the loci of the spectra in the three-dimensional colour space. The boundaries of the object-colour solid are shown by lines. Flower colours form a cloud which lie inside the object-colour solid

(corresponding to colours of high saturation for humans). Optimal colour stimuli have spectral reflectances of a very simple type: the spectral reflectance is either zero or unity and across the spectrum there are generally not more than two transitions between these two values. Optimal colour stimuli are imaginary stimuli in a sense that no real reflectance spectra have abrupt transitions of this kind (Wyszecki and Stiles, 1982). Substitution of the reflectance of optimal colour stimuli into Eq. 1 gives the co-ordinates of the boundaries of colour body (Q^b_i). Let λ_a and λ_b be the wavelength positions of the abrupt changes ($\lambda_a > \lambda_b$), and λ_{min} and λ_{max} be the limits of a visual system's spectral range, then

$$Q^b_i = \int_{\lambda_a}^{\lambda_b} I(\lambda)\, R_i(\lambda)\, d\lambda, \qquad [3a]$$

if λ_a corresponds to a transition from zero to one, and

$$Q^b_i = \int_{\lambda_{min}}^{\lambda_a} I(\lambda)\, R_i(\lambda)\, d\lambda + \int_{\lambda_b}^{\lambda_{max}} I(\lambda)\, R_i(\lambda)\, d\lambda \qquad [3b]$$

if λ_a corresponds to a transition from one to zero (Maximov, 1984). The object-colour solid so defined is an asymmetric torpedo-shaped body, with one apex at the origin which corresponds to the colour 'black' and the other at the point which corresponds to the colour 'white' (Wyszecki and Stiles, 1982, Maximov, 1984).

The shape of the colour solid depends on receptor spectral sensitivities, and it changes substantially as the wavelength positions of the peak of receptor spectral sensitivities shift. When the wavelength position of the M receptor approaches that of the L or S, the colour solid flattens and its volume decreases (Govardovskii & Vorobyev 1989; Govardovskii & Vorobyev 1992).

How can we calculate the number of distinguishable colours? Each point in colour space is surrounded by a small elementary volume of colours that are indistinguishable from it. The total number of distinguishable colours (N) is equal to the number of these elementary volumes filling the colour solid, and similarly, the number of distinguishable flower colours is equal to the number of elementary volumes filling the cloud of flower colours. If the elementary volume of indistinguishable colours is independent of location in the colour space, then the number of distinguishable colours is related to the volume occupied by colours (V) and to the volume of indistinguishable colours (u) by:

$$N = \frac{V}{u} \qquad [4]$$

However the volume of indistinguishable colours is generally dependent on the position in the receptor space, i.e. u= $u(Q_S, Q_M, Q_L)$. Thus Eq. 4 should be modified:

$$N = \int_V \frac{1}{u(Q_S, Q_M, Q_L)} \, dQ_S dQ_M dQ_L, \tag{5}$$

where integration is perfomed over the volume V. Obviously, the number of distinguishable colours increases with increasing V or with decreasing u. To estimate u we use classical metric theory of colour discrimination (Helmholtz, 1896; Schrödinger 1920). According to this theory any two colours have a particular distance between them. If the distance between the colours is smaller than a given threshold, then the colours are not distinguishable. This distance (Δs) is commonly described by line element:

$$\Delta s^2 = \sum_{i,k = S,M,L} g_{ik} \, \Delta Q_i \, \Delta Q_k, \tag{6}$$

where ΔQ_i defines the difference in co-ordinates of these points, g_{ik} are the components of metric tensor (Wyszecki and Stiles, 1982). Generally g_{ik} depends on the position of a point in receptor space. Contours corresponding to a threshold distance form an ellipsoid in receptor space, and the volume of this ellipsoid is equal to:

$$u = C\sqrt{\frac{1}{Det[g_{ik}]}} \tag{7}$$

where $Det[g_{ik}]$ defines the determinant of a tensor with components g_{ik}, C is a constant which depends on the value of the threshold distance. Substitution of Eq. 7 into Eq. 5 gives

$$N = 1/C \int_V \sqrt{Det[g_{ik}]} \, dQ_S dQ_M dQ_L, \tag{8}$$

where integration is performed over the volume of object-colour solid or over the volume of the cloud of flowers (Govardovskii and Vorobyev, 1989).

The mean percentage of objects from which any given object is indistinguishable depends on the mutual positions of the points corresponding to different objects. Let $f(Q_S,Q_M,Q_L)$ be the probability distribution of the object colours in the receptor space. The number of objects indistinguishable from one with co-ordinates $\{Q_S, Q_M, Q_L\}$, $n(Q_S, Q_M, Q_L)$ is given by the following equation:

$$n(Q_S,Q_M,Q_L) = K \, f(Q_S,Q_M,Q_L) \, u(Q_S,Q_M,Q_L), \tag{9}$$

where K is the total number of objects. If the positions of the points in the colour space are independent, the mean number of indistinguishable objects (P) is, by definition, given by

$$P = K \int_V f(Q_S, Q_M, Q_L) n(Q_S, Q_M, Q_L) \, dQ_S dQ_M dQ_L) \tag{10}$$

substitution of Eq. 9 and 7 into Eq. 10 gives:

$$P=K\,C\int_V \frac{f^2(Q_S, Q_M, Q_L)}{\sqrt{Det[g_{ik}]}}\, dQ_S dQ_M dQ_L \qquad [11]$$

To estimate either the number of distinguishable colours or the number of indistinguishable objects the metric of colour space should be introduced. This metric can be derived either from behavioural data, or theoretically. The limit for colour discrimination is set by quantum fluctuations of absorbed light. Thus the metric based on the quantum fluctuations of the light absorbed in receptors allows us to estimate the upper limit of distinguishable colours. Generally the metric tensor is defined as a matrix inverse to the correlation matrix of uncertainties in colour channels (Wyszecki and Stiles, 1982). The amplitude of noise caused by quantum fluctuations is proportional to the square root of the number of effectively absorbed quanta $\sqrt{Q_i}$. Because the noises in different receptors are not correlated with each other the tensor is diagonal with components:

$$g_{ik} = \frac{C_i}{Q_i} \qquad [12]$$

if $i = k$, and $g_{ik} = 0$ if $k \neq i$, where C_i is the proportionality factor. Substitution of Eq. 12 into Eq. 7 gives the expression for the volume of indistinguishable colours

$$u=C\sqrt{Q_L Q_M Q_S} \qquad [13]$$

Then the number of distinguishable colours can be calculated as:

$$N=1/C\int_V \frac{1}{\sqrt{Q_L Q_M Q_S}}\, dQ_S dQ_M dQ_L \qquad [14]$$

(see Eq. 8). The number of absorbed quanta is proportional to light intensity. From Eq. 14 it follows that the number of distinguishable colours is proportional to the light intensity raised by the power of 1.5 (Govardovskii and Vorobyev, 1992; Vorobyev and Govardovskii, 1993).

Because Hymenoptera use chromatic cues exclusively for colour discrimination, their colour space can be described by a two-dimensional diagram (Backhaus, 1991). It has been proposed that distance in the colour hexagon is a measure of discriminability for Hymenoptera (Chittka *et. al.*, 1993). In this case projection of the cloud of flower colours onto the colour hexagon gives the number of distinguishable colours (Vorobyev *et al.*, 1993; Chittka *et al.*, 1993).

17.5.2 Colour discrimination and the photoreceptors' wavelength positions

Figure 5a shows the number of all theoretical object colours which can be distinguished where performance is limited by quantum fluctuations of absorbed light.

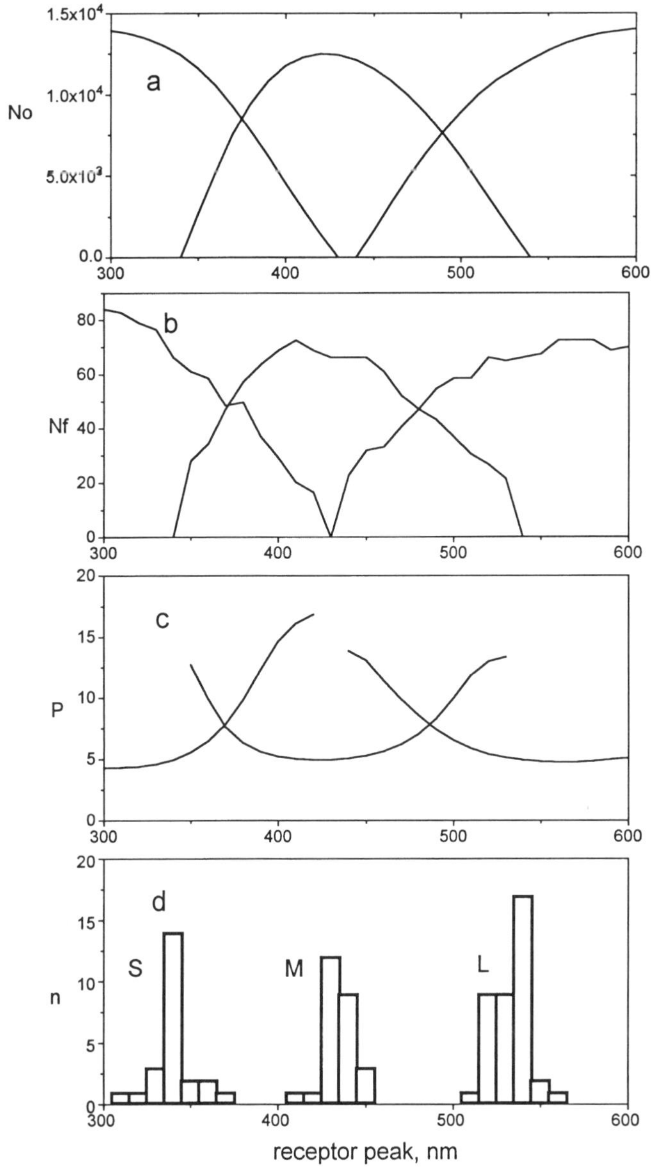

Figure 5 The number of distinguishable colours and percentage of indistinguishable flowers depending on the peak positions of S, M, and L receptors. **a** – the number of all theoretically possible colours (No), i.e all colours which fill the object-colour solid; discrimination was assumed to be limited by the quantum fluctuations of absorbed light. The illumination corresponds to 10000 effectively absorbed photons per receptive field during the integration time. **b** – the number flower colours (Nf) discriminable on the basis of their chromaticity. **c** – percentage of indistinguishable flowers (P), the colours were assumed to be discriminated on the basis of their chromatic properties. **d** – histogram of the distribution of spectral sensitivity peaks for photoreceptors (n) from 40 Hymenopteran species (Peitsch *et al.*, 1992)

Each curve shows the consequences of varying the peak of one receptor, while the other two are kept constant at the positions corresponding to those of bees. The total number of object colours which a trichromatic visual system can discriminate increases when the S receptor is shifted to shorter wavelengths, or the L receptor to longer wavelengths, and the M receptor should be equidistant from these two (Figure 5a). Thus increase of receptor spectral separation is beneficial for the discrimination of theoretical object colours.

Because the shape of SRs of flowers is constrained by properties of the pigments and by flower morphology, their colours do not fill the whole colour solid, and the task of flower discrimination may require a set of receptors different from that optimal for the discrimination of all theoretically possible colours. Figure 5b shows the number of distinguishable flower colours and Figure 5c the percentage of indistinguishable colours as a function of receptor spectral positions. In these calculations the colours were assumed to be discriminated on the basis of their chromatic properties, i. e. distance in the colour hexagon was used as the measure of discriminability. In practice, the number of all theoretically possible distinguishable colours and the number of flower colours show very similar dependencies on receptor wavelength positions (compare Figure 5a and b). It is always beneficial to reduce the overlap between the receptor sensitivities. The same conclusion applies if the mean of objects of which any one is indistinguishable (P) is considered (Figure 5c). Thus the optimal photo receptor peak positions do not depend on the colours used in the analysis nor on the assumptions about the factors limiting the discrimination.

Figure 5d shows the distribution of receptors' peak positions of 40 hymenopteran species (Peitsch *et al.*, 1992). Bees and other hymenopterans have a set of photoreceptors which is very close to optimal. The peak position of the M receptor corresponds to that of the maximum on the curve of distinguishable colours and to the minimum of indistinguishable objects, whereas movement of S and L receptors can improve colour discrimination only slightly.

Although colour discrimination would be maximised if flower colours were uniformly distributed in the colour space, the SRs of flowers are, as we have seen, not uniformly distributed. Our calculations show that there is only a 5% probability that the colours of two arbitrarily picked flowers will not be discriminable (Figure 5c), so in practice there may be little evolutionary pressure to optimise this distribution.

17.6 Conclusions

Do our calculations give any insight into the co-evolution of flowers and bee vision? The analysis indicates that bees' receptors are well-adjusted to flower discrimination. It is tempting to speculate that the receptors in flower foraging animals were evolutionarily adapted to the flower colours (Chittka and Menzel, 1992). However, this speculation meets two substantial difficulties: 1) bee receptors allow for good discrimination of all (theoretical) colours, not only those of flowers; 2) a similar set of receptors have been found in invertebrate species that do not forage on flowers (Menzel and Backhaus, 1991; Peitsch *et al.*, 1992, Chittka 1996). Thus one can infer that the ancestors of bees had a similar set of receptors before they started to forage on flowers, and this set has since been preserved.

Given that hymenopteran colour vision seems to be conservative, were flower colours adapted to the vision of animals with a set of receptors similar to that of modern bees? If the flower colours were adapted to a particular visual system they would occupy a large part of the object-colour solid in the colour space of this visual system, whereas the part of the object-colour solid which they would occupy in the colour space of a different visual system might be smaller. Our analysis shows that the number of all theoretically possible distinguishable colours, i.e. the volume of object-colour solid, and the number of flower colours, i.e. the volume of the cloud of flower colours, have nearly the same dependence on the receptor wavelength position (compare Figure 5a and b). This indicates that flower colours are generally discriminable for any potential pollinator and, therefore, the flower colours could, in principle, have evolved before the ancestors of modern bees started to pollinate them.

Thus, although the visual system of bees are optimal for discrimination of flowers and flower SRs have sufficient variability for effective pollination by bees, it is difficult to say what if any mutual adaptation explains the origin of flower colours and of the colour vision of pollinators.

Acknowledgements

We thank Daniel Osorio for valuable discussions, critical comments, and corrections of the manuscript, Andreas Gumbert for fruitful discussion, Robert Brandt for corrections in the manuscript and Philomena Maher for the linguistic corrections. This work was supported by the Alexander von Humboldt Foundation and by the Deutsche Forschungsgemeinschaft.

References

Autrum, H. and von Zwehl, V. (1964) Die spectrale Empfindlichkeit einzelner Sehzellen des Bienenauges. *Z. Vergl Physiol.* 48, 357–384

Backhaus, W. (1991) Color opponent coding in the visual system of the honeybee, *Vision Res.* 31, 1381–1397.

Brandt, R. and Vorobyev, M. (1996) Metric analysis of threshold spectral sensitivity in the honeybee. *Vision Res.* (in press)

Chittka, L. (1992) The color hexagon: a chromaticity diagram based on photoreceptor excitations as a generalized representation of colour opponency, *J Comp Physiol A*, 170, 533 – 543.

Chittka, L. (1996) Does bee color vision predate the evolution of flower color? *Naturwissenschaften*, 83, 136 –138.

Chittka, L., Beier, W., Hertel, H., Steinmann, E. and Menzel, R. (1992) Opponent colour coding is a universal strategy to evaluate the photoreceptor inputs in Hymentoptera, *J Comp Physiol A*, 170, 545 – 563.

Chittka, L. and Menzel, R. (1992) The evolutionary adaptation of flower colors and the insect pollinators' color vision systems, *J. Comp Physiol A*, 171, 171–181.

Chittka, L., Shmida, A., Troje, N. and Menzel, R. (1993) Ultraviolet as a component of flower reflections, and the colour perception of Hymenoptera. *Vision Res.* 34, 1489 –1508.

Chittka, L., Vorobyev, M., Shmida, A. and Menzel, R. (1994) Bee colour vision – the optimal system for the discrimination of flower colours with three spectral photoreceptor types? In: *Sensory systems of Arthropods* (eds. Wiese, K., Gribakin S. G., Popov A.V. and Renninger G.), Birkhauser, Berlin. Verlag, pp. 211– 218.

Dafni, A. (1984) Mimicry and deception in pollination. *Ann. Rev. Ecol. Syst.* 15, 259 – 278.

Dartnall, H. J. A. (1953) The interpretation of spectral sensitivity curves. *Brit. Med. Bul.* 9, 24.

Darwin, C. (1876) *Cross and self fertilisation in the vegetable kingdom*, Murray, London.

Daumer, K. (1956) Reizmetrische Untersuchung des Farbensehens der Bienen. *Z. Vergl Physiol.* 38, 413 – 478.

Giurfa, M., Nunez, J., Chittka, L. and Menzel, R. (1995) Colour preferences of flower-naive honeybees. *J Comp. Physiol A*, 177, 247– 259.

Giurfa, M., Vorobyev, M., Kevan, P. and Menzel, R. (1996) Detection of coloured stimuli by honeybees: minimum visual angles and receptor specific contrasts. *J Comp. Physiol A*, 178, 699 –709

Goldsmith, T. H. (1991) The evolution of visual pigments and colour vision. In: *Vision and Visual Disfunction. vol. 6: The Perception of Colour,* (ed. Gouras, P.) MacMillan, London, pp. 62 – 89.

Govardovskii, V. I. and Vorobyev, M. V. (1989) The role of colored oil drops of cones in the color vision. *Sensory systems*, 3, 150 –159. (In Russian).

Govardovskii, V. I. and Vorobyev, M. V. (1992). Photoreceptor spectral sensitivity curves and discrimination of natural colours. *Abstr. of 9th International Neuroethological Congress, Montreal*, pp. 24 – 25.

Harborne, J. B. (1988) *Introduction to Ecological Biochemistry*, (3rd. ed.), Academic Press, London.

Helmholtz, H. L. F. (1896) *Hanbuch der Physiologischen Optik*, (2nd ed.), Voss, Humburg.

Lehrer, M. (1994) Spatial vision in the honeybee: the use of different cues in different tasks. *Vision Res.* 34, 2363 – 2385.

Lythgoe, J. N. (1979) *The Ecology of Vision*, Claredon Press, Oxford.

Lythgoe, J. N. Partridge, J. C. (1989) Visual pigments and the acquisition of visual information. *J. Exp. Biol.* 146, 1– 20.

Luther, R. (1927) Aus dem Gebiet der Farbreizmetrik. *Z. Techn. Phys.* 8, 540 – 558.

Maximov, V. (1984) Transformation of colour under the changing illumination. (ed. Bysov, A. L.) Nauka, Moskva (in Russian).

Maximov, V. (1988). An approximation of visual absorption spectra. *Sensornye Systemy,* 2, 3 – 8.

Menzel, R. (1967) Untersuchungen zum Erlernen von Spectralfarben durch die Honigbienen (*Apis mellifera*). *Z. Vergl. Physiol*, 56, 22 – 62.

Menzel, R. and Blakers, M. (1976) Colour receptors in the bee eye – morphology and spectral sensitivity. *J Comp. Physiol.* 108, 11– 33.

Menzel, R. and Backhaus, W. (1991) Color vision in insects, in *Vision and Visual Disfunction, vol. VII. Perception of Color*, (ed. Gouras, P.), MacMillan Press, Haundsmills, pp. 262 – 293.

Menzel, R. and Shmida, A. (1993) The ecology of flower colours and the natural colour vision of insect pollinators: The Israeli flora as a study case. *Biol Rev.* 68, 81–120.

Mollon, J. (1992) Worlds of difference. *Nature*, 356, 410 – 411.

Nyberg, N. (1928) Zum Aufbau des Farbenkörpers im Raume aller Lichtempfindungen, *Z. Phys.* 52, 406 – 419.

Osorio, D. and Vorobyev, M. (1996) Colour vision as an adaptation to frugivory in primates. *Proc. R. Soc. Lond. B.* 263, 593 – 599.

Peitsch, D., Fietz, A., Hertel, H., de Souza, J., Ventura, D. F. and Menzel, R. (1992) The spectral input systems of hymenopteran insects and their receptor-based colour vision. *J Comp. Physiol A*, 170, 123 – 40.

Schrödinger, E. (1920) Grundlinien einer Theorie der Farbenmetrik im Tagessehen. *Ann. Phys.* 63, 397– 520.

Troje, N. (1993) Spectral categories in the learning behaviour of blowflies. *Z. Naturforsch.* 48c, 96 –104.

Vorobyev, M., Chittka, L. and Menzel, R. (1993) How many colors can a bee discriminate? Are the wavelength positions of the bee color receptors optimal for this task? in *Proceeding of the Göttingen Neurobiology conference 1993*, p. 339.

Vorobyev, M. V. and Govardovskii, V. I. (1993) At low illuminance levels monochromats are able to distinguish more shades than dichromats, and dichromats more than trichromats. In: *Proceeding of the Gottingen Neurobiology conference 1993*, p. 397.

Walls, G. L. (1942) *The Vertebrate Eye and Its Adaptive Radiation*. Granbrook Institute of Science, MI: Bloofield Hills.

Wyszecki, G. and Stiles, W. S. (1982) *Color science: Concepts and methods, quantitative data and formulae*, (2nd ed.), Wiley, New York.

18
Bioluminescence

E. A. WIDDER
Senior Scientist, Harbor Branch Oceanographic Institution, 5600 U.S. 1 North, Fort Pierce, FL 34946, USA

18.1 Introduction

Bioluminescence is visible light made by organisms. It is one form of *chemiluminescence,* which is light generated by a chemical reaction. Unlike *incandescence,* where the electronically excited state is a consequence of thermal excitation of electrons, chemiluminescence is essentially heatless. Although bioluminescence is often referred to as *phosphorescence,* as in the quote from Darwin in the next section, this is a misnomer. In phosphorescence the electronically excited state is generated by light absorption, rather than a chemical reaction, and light emission persists for a limited period of time after removal of the excitation source. Many glow-in-the-dark posters, T-shirts and toys, including those depicting bioluminescence, are created with phosphorescent paint. Bioluminescence is also occasionally confused with *fluorescence*. Like phosphorescence, fluorescence is dependent on the absorption of light, but differs from phosphorescence in that fluorescence does not persist when the excitation source is removed. Although some bioluminescent emitters may also be fluorescent, this is by no means a universal attribute. Therefore, the presence or absence of fluorescent sites in an organism can not be used as an indication of the presence or absence of bioluminescent cells or organelles.

18.2 Distribution and abundance

> 'While sailing a little south of the Plata on one very dark night, the sea presented a wonderful and most beautiful spectacle. There was a fresh breeze, and every part of the surface, which during the day is seen as foam, now glowed with a pale light. The vessel drove before her bows two billows of liquid phosphorus, and in her wake she was followed by a milky train. As far as the eye reached, the crest of every wave was bright, and the sky above the horizon, from the reflected glare of these livid flames, was not so utterly obscure as over the vault of the heavens.' (Darwin, 1933)

Although Darwin attributed 'the *phosphorescence* of the sea' to 'organic beings' and noted that 'torn and irregular particles of gelatinous matter' were a common

S.N. Archer et al. (eds.), Adaptive Mechanisms in the Ecology of Vision, 555–581

source of luminescence in his nets, the organisms responsible for the displays he witnessed from the deck of the *Beagle* were never identified. Much progress has been made in describing and cataloging sources of bioluminescence since Darwin's voyage. Herring (1987) provides a listing of the more than 700 genera known to contain luminous species, and based on this list Hastings and Morin (1991) estimated the percentage of luminescent genera in each taxon. Their analysis indicates that in terms of numbers of species, bioluminescence is rare. In the terrestrial environment bioluminescence occurs primarily among insects, with fireflies (order Coleoptera) and New Zealand glow-worms (order Diptera), being two of the best known examples. Within the Coleoptera alone, there are 119 luminescent genera, but these represent less than half of a percent of the total number of genera in this taxon. Other terrestrial light emitters include fungi (Basidomycetes), earthworms (Oligochaeta), millipedes (Diplopoda), centipedes (Chilopoda), a land snail (*Quantula striata*), and a nematode (*Heterorhabditis bacteriophora*). These too make up very small percentages of their particular taxa with earthworms having the highest percentage of luminescent genera, at over 7 percent. Bioluminescence is virtually absent from freshwater environments with only one confirmed case, a limpet, *Latia neritoides*, found exclusively in New Zealand.

It is in the oceans that bioluminescence is at its most bountiful. Although the percentages are generally low in coastal waters, where Morin (1983) estimates that only about 1–2% of the species are luminescent, numbers increase dramatically in the open ocean where bioluminescence seems to be the rule rather than the exception. The faunal composition of mesopelagic micronekton is dominated by fish and crustaceans, both in terms of numbers and biomass (Maynard *et al.,* 1975) and, these in turn are dominated by bioluminescent representatives (Table 1). Additionally, there are many gelatinous organisms in the open ocean which, due to their fragility, are inadequately sampled by nets (Pugh, 1989) but which, nonetheless, contribute significantly to the bioluminescence in the water column (Morin, 1974; Widder *et al.,* 1989; Widder *et al.,* 1992).

Measurements of stimulated bioluminescence in the world's oceans corroborate the view gained from net sampling, that bioluminescence is ubiquitous in the marine environment. The first radiometers designed to measure the penetration of sunlight below the euphotic zone detected flashes of light which were presumed to be from luminescent organisms (Clarke and Wertheim, 1956). These open field detectors, which depended on the movement of the instrument through the water to stimulate bioluminescence, have been largely superseded by detector systems which view an enclosed volume where turbulent flow stimulates bioluminescence (see Case *et al.,* 1993; Swift *et al.,* 1985; and Widder *et al.,* 1993 for discussions of bathyphotometer systems). Some values for the average bioluminescence per cubic meter measured in surface waters with these instruments are given in Table 2. The former USSR devoted considerable scientific effort to measuring distribution patterns of bioluminescence and they found that

Table 1 Percent luminous organisms in pelagic collections

Taxon	Depth range (m)	Percent Luminescent Species	Percent Luminescent Individuals	Source
Fish	0 – 500	71	90	(Badcock and Merrett, 1976)
	500 –1000	73	97	
	>1000	71	91	
	0 – 2560	66	97	(Beebe, 1937)
Shrimp	0 – 500	67–100	32 –100	(Herring, 1976)
	500 –1000	60 –78	23 – 80	
	>1000	39 – 73	55 – 87	

Table 2 Bathyphotometer measurements of average bioluminesence per litre in different oceans

Location	Depth[a] (m)	Bioluminescence/l[b]	Source
Sargasso Sea	0 –100	1×10^{10}	(Widder *et al.*, 1993)
Greenland Sea	0 –150	1.1×10^{10}	(Buskey, 1992)
Pacific Ocean (Alaska to Hawaii)	surface	$1–5 \times 10^{10}$[c]	(Losee *et al.*, 1985)
North Atlantic Ocean (MLML)	12 – 80	$0.7–2.8 \times 10^{10}$	(Neilson *et al.*, 1995)
Caribbean Sea	0 –120	4.5×10^{10}	(Swift *et al.*, 1985)
Beaufort Sea	0 – 60	$5 \times 10^{8} – 1.2 \times 10^{11}$[c]	(Laopta *et al.*, 1992)
Vestfjord, Norway	0 –100	$0.9 – 2.7 \times 10^{11}$[c]	(Lapota *et al.*, 1989)
Arabian Sea	0 –70	$2 – 9 \times 10^{11}$[c]	(Lapota and Rosenberger, 1990)

[a]Depth of bathyphotometer profiles over which bioluminescence in next column was averaged
[b]Units of Photons l^{-1}
[c]Units of Photons s^{-1} l^{-1} for short residence time bathyphotometers

bioluminescence was present 'from the North Pole to the Antarctic coastal waters' (Gitel'zon and Levin, 1989).

Deeper measurements, made from a submersible, down to depths of 550 m in the highly productive waters of Monterey Submarine Canyon, were in a comparable range to many surface measurements ($7.6 \times 10^{11} – 1.1 \times 10^{13}$ photons s^{-1} m^{-3}). However the densities of luminescent sources were much lower (43 to 175 m^{-3}) (Widder *et al.,* 1989) than those reported for surface waters where numbers of bioluminescent sources often range between thousands and millions per cubic meter. This is indicative of a tendency for source intensities to increase below the euphotic zone, primarily due to a shift from dinoflagellate- to zooplankton-dominated luminescence. The deepest bathyphotometric measurements of bioluminescence were made at 7.2 km where bioluminescent flashes were detected approximately every three minutes (Gitel'zon and Levin, 1989). Bioluminescence was visually observed from the bathyscaph *Trieste* during the record-breaking dive by Piccard and Walsh in 1960 to the bottom of the Mariana Trench. During their descent they noted a streak of luminescence outside their porthole at 6 km. But perhaps even more significant is what they saw at the very deepest part of their dive, at 10.9 km:

> 'As we were settling this final fathom, I saw a wonderful thing. Lying on the bottom just beneath us was some type of flatfish, resembling a sole, about one foot long and six inches across. Even as I saw him, his two round eyes on top of his head spied us – a monster of steel – invading his silent realm. Eyes? Why should he have eyes? Merely to see phosphorescence?' (Piccard and Dietz, 1961)

All the evidence seems to indicate that bioluminescence occurs throughout the depth and breadth of the oceans and, judging by the numbers of organisms which retain eyes while living at depths where sunlight never penetrates, luminescence has a profound impact on the photoecology of the marine environment.

The view that bioluminescence is rare is therefore based on the total number of luminescent compared to non-luminescent species, not on the actual number of luminescent individuals in the world. Therefore, while it is true that of the 5,354 recognized crustacean genera, only 1% contain luminous species (Herring, 1985) it should also be kept in mind that included in that 1% are bioluminescent euphausiids which constitute a major fraction of the total plankton biomass in many of the oceans of the world (Mauchline and Fisher, 1969b), as well as similarly abundant bioluminescent decapods (Table 1) like the oplophorids and sergestids, bioluminescent copepods which include all species of the Metridiidae (*Metridia, Pleuromamma*, and *Gaussia*), bioluminescent ostracods which are ubiquitous and may be responsible for brilliant surface displays of bioluminescence (Herring, 1985 and refs) and bioluminescent amphipods like *Cyphocaris* which is one of the most abundant of the oceanic Gammaridea (Herring, 1981). The notion of rarity also stems from our anthropocentric, terrestrial-based view of the world. We all too easily lose sight of the fact that Earth is an ocean planet and that some of the most successful species on this planet are bioluminescent (Figure 1).

There are also undoubtedly many sources of bioluminescence that remain to be identified. Because so much of our information about the extent of bioluminescence in the ocean depends on net captured material, organisms that are destroyed by nets, such as the abundant gelatinous zooplankton, or organisms that avoid net capture or that exhaust their luminescence during net capture, will often go unrecognized as light producers. However, thanks to more direct access to the oceans made possible by SCUBA, submersibles and ROVs (Remotely Operated Vehicles), adjustments to species lists are occurring all the time. For example, a fortuitous observation made from a submersible of what appeared to be a chaetognath releasing a luminescent plume, lead to the addition of a new phylum to the bioluminescent species list (Haddock and Case, 1994). Prior to this identification of the cosmopolitan, bathypelagic chaetognath *Caecosagitta macrocephala* as luminescent, the Chaetognatha was the only major zooplanktonic phylum without a luminous representative.

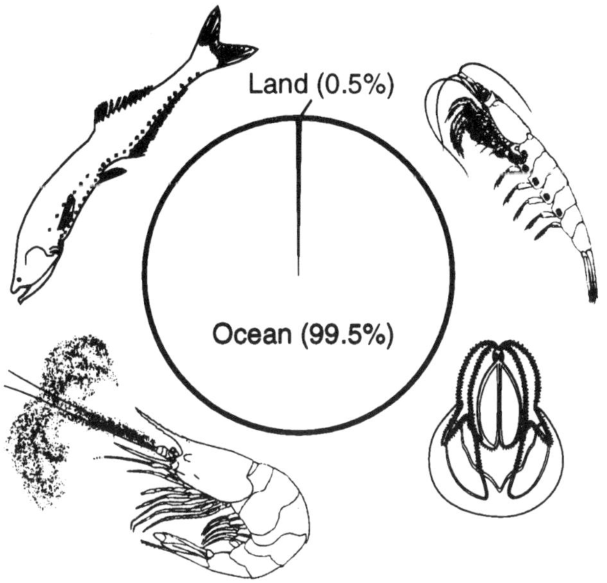

Figure 1 Volume of habitats on earth (Redrawn from Childress, 1982) and a few common oceanic inhabitants which are bioluminescent. Top left: Stomiatoid fish, *Cyclothone*, possibly the most common vertebrate on earth. Top right: Euphausiid (Redrawn from Baker, 1990). Bottom left: Oplophorid shrimp. Bottom right: Ctenophore (Redrawn from Herring, 1978)

In most discussions on the phylogenetic distribution of bioluminescence, there is usually some reference to the analogy made by E. Newton Harvey regarding the apparently random phylogenetic distribution of bioluminescence: 'It is as if the various groups had been written on a black board and a handful of sand cast over the names. Where each grain of sand strikes, a luminous species appears. The Coelenterates have received most sand ... Several phyla contain no luminous forms whatever. It is an extraordinary fact that one species in a genus may be luminous and another closely allied species contain no trace of luminosity,' (Harvey, 1920). The implication has been that there is no clear rationale for this randomness. However, there are many examples among terrestrial organisms of profound differences between closely related species which were just as perplexing until more was learned about their ecosystems and interactions. For example, the reason why some birds sing while closely related species do not has been found to be a function of niche occupancy: singers occupy heavily vegetated regions whereas non-singers do not and can therefore depend on visual displays (Hailman, 1977). An obvious counter to this example is that the pelagic environment, which is the habitat of so many luminescent individuals, is characterized by gradients rather than boundaries (Herring and Roe, 1988). A comparable terrestrial example might therefore be bees which occupy the same habitat but where one species relies on pheromonal courtship signals while their congeners rely on visual signals. Because courtship signals, such as pheromones and sound, radiate indiscriminately, the improved signal-to-noise which results from visual signals that can be directed at specific receivers may favor the evolution of visual communication (Hailman, 1977). Additionally, it should be kept in mind that the pelagic environment is not wholly lacking in boundaries. For example, although it was once thought that the majority of hyperiid amphipods were free-living plankton, it is now recognized, thanks to *in situ* sampling techniques and observations, that these amphipods live in highly specific and stereotyped associations with gelatinous organisms such as jellyfish, ctenophores, siphonophores, colonial radiolarians, and pelagic tunicates (Harbison *et al.,* 1977). Therefore, in looking for a pattern to the distribution of bioluminescence among organisms we should look for selection pressures imposed by intergroup competition as well as by the environment, and in assessing environmental pressures, we need to keep in mind that our information about the pelagic environment has been biased by our remote sampling techniques.

18.3 Chemistries

In all bioluminescent reactions the excited electron state which precedes light emission is generated when an enzyme catalyzes the oxidation of a substrate by molecular oxygen. Although the enzyme and substrate in a bioluminescent reaction are called luciferase and luciferin respectively, these are generic terms. A taxon prefix is required in order to differentiate the different chemicals which have been isolated from different organisms (Figure 2). The chemical differences between these systems are taken as evidence that the ability to make light has evolved independently many times. In fact, based on the phylogenetic diversity of bioluminescence, it has been estimated that bioluminescence may have arisen as many as 30 times (Hastings, 1983). As with photoreceptors, which may have originated 40 or more times during evolutionary history (Salvini-Plawen and Mayr, 1967), these multiple acquisitions of the same trait are taken as clear evidence of the extreme selective advantages afforded by that trait.

18.4 Evolutions

The evolutionary origins of bioluminescence has been the subject of considerable speculation and many authors have addressed the subject (see, for example: Lloyd, 1977; Buck, 1978; Hastings, 1983; McCapra, 1990; Seliger, 1993 and Case *et al.,* 1994). Although there has been some controversy about when, during evolutionary history, bioluminescence may have arisen, there seems to be an emerging consensus that disparate chemical systems and diverse phylogenetic distribution patterns are most easily explained by the late emergence of bioluminescence, after the evolution of photoreceptors. Since bioluminescence

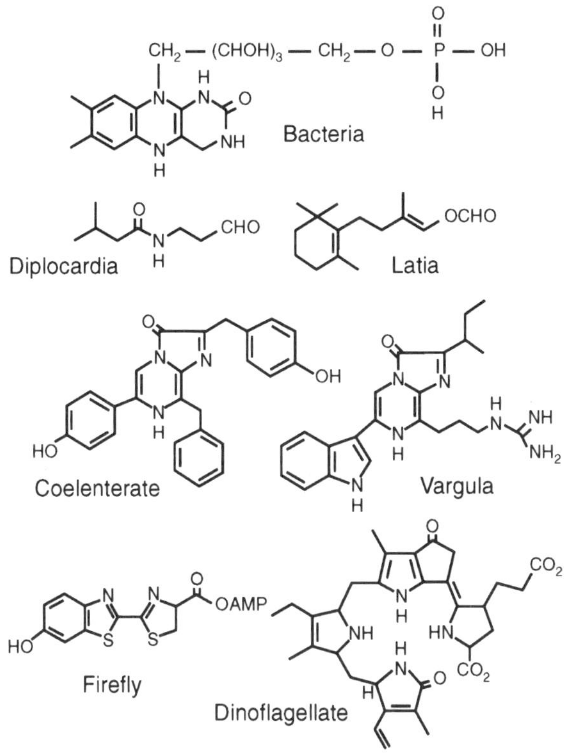

Figure 2 Chemical structures of different luciferins (Redrawn from Hastings and Morin, 1991)

presumably has its greatest utility in dimly-lit environments, it follows that the ability to produce light must have first evolved where organisms had adopted nocturnal or crepuscular lifestyles and therefore had highly sensitive photoreceptors. Lloyd (1977) suggests that in such habitats, bioluminescence might serve to enhance and amplify already established visual signals. For example, fish colonizing the twilight depths of the ocean and dependent on visual displays, such as raising the opercula to look like eyes, might evolve highly reflective patches to maximize the effect. If these patches were then to develop a chemiluminescent capacity, the effect would be amplified, resulting in an immediate selective advantage for the light emitter. Seliger (1993) has proposed that such a fortuitous mutation might occur in individuals possessing enzymes involved in oxidative degradation of pigment molecules. He suggests that oxidative breakdown of pigments, such as melanin and porphyrin, could produce electronically excited states capable of emitting visible light. Since such pigment molecules are responsible for the coloration patterns of many animals, the emission would therefore occur at anatomical sites already adapted for visual signaling.

Bioluminescence was well established in the oceans when life first invaded terrestrial and freshwater environments, however, evolutionary history suggests that early colonists of these realms were not luminescent. Bioluminescence, therefore, had to be reinvented and because of the ease with which prey could find cover in these environments the evolutionary pressures for optical concealment were not as strong and selection for light producing capabilities were diminished compared to in the open ocean (see next section). Case *et al.,* (1994) suggest that these factors may account for the relative rarity of bioluminescence on land and in freshwater.

18.5 Functions and the need for optical concealment

Bioluminescence is an adaptation to life in dim-light environments. Since the colonization of such environments is primarily a consequence of selection pressures on prey to avoid detection by predators, this may be a useful context for exploring some of the presumed selective advantages afforded by light production.

Many functions have been attributed to bioluminescence (Buck, 1978; Tett and Kelly, 1973; Young, 1983). These fall into the three major categories of: 1) attracting mates, 2) finding food and 3) avoiding predators. The body of literature associated with each of these categories is far too vast to be addressed in detail here. I will therefore touch on each of these topics only briefly, within the context of the need for optical concealment, and place the greatest emphasis on the most common province of bioluminescence, which is counterillumination.

In the open ocean, where bioluminescence is most prevalent, there are no trees or bushes to hide behind and prey strategies for avoiding detection by predators involve both behavioral and morphological adaptations. The massive vertical migration of animals, which occurs daily in the world's oceans, from surface waters during the night to dark depths during the day, is believed to be a means of allowing organisms to feed in food-rich surface waters while reducing light-dependent mortality risks (Lampert, 1989). Camouflage is another common predator avoidance strategy, either by means of transparency, reflectivity or luminescence in organisms inhabiting the twilight regions or by dark pigmentation in species living below the depths where sunlight is detectable (McFall-Ngai, 1990). In this visually featureless environment where photoreceptors are adapted for maximum sensitivity (Denton, 1990), a single flash of bioluminescence is an attention-riveting event. In the open ocean there is apparently an enormous disparity between the levels of natural background bioluminescence (i.e. bioluminescence which is unstimulated by the observer) and the actual number of potentially bioluminescent sources in the water column. For example, while a submersible moving through the water may be surrounded by hundreds of flashes per cubic meter, once it is trimmed to neutral buoyancy and mechanical stimulation ceases, the flashing stops (Widder *et al.,* 1992; Widder *et al.,* 1989) and, to quote William Beebe, 'it seems to show as blacker than black' (Beebe, 1951). These observations suggest that while the ability to make light is ubiquitous in the open ocean, it is employed very conservatively.

Observations of spontaneous bioluminescence are primarily confined to coastal and terrestrial ecosystems. This may, in part, reflect the greater opportunity for long-term observations in these environs but may also reflect a lower risk factor for the use of predator-attracting bioluminescent displays under circumstances where the habitat affords possible points of refuge.

18.5.1 Attracting mates

Bioluminescence is one possible solution to the problem faced by animals living in dim light environments of how to attract mates. Some of the best evidence for this use of bioluminescence lies with fireflies (Case, 1984; Lloyd, 1977), ostracods (*Vargula*) (Morin, 1986), and Bermuda fireworms (*Odontosyllis*) (Huntsman, 1948; Tsuji and Hill, 1983). The simplest type of courtship display is a continuous glow emitted by one sex to attract the other. The female firefly, *Lampyris noctiluca*, employs such a display to attract the non-luminous male (Case, 1984). Since such displays also attract predators, such as wolf spiders which prey upon fireflies, Lloyd (1977) suggests that selection pressure is strong to shorten the glow into species specific flash patterns such as those employed by most fireflies. In the marine polychaete, *Odontosyllis* and Caribbean ostracods of the genus *Vargula*, extrinsic luminescent displays are used to attract mates. They release long-lasting clouds of luminescence and remain separate from the display, which might also serve as a means of reducing predation. However, in the case of *Odontosyllis*, fish which were apparently attracted by the bioluminescence have been observed to gorge on the worms (Huntsman, 1948), while in *Vargula*, Morin (1986) indicates he has never observed attacks on the ostracods during mating displays and attributes this to the burglar alarm defense (see below).

The light organs of fireflies, fireworms and ostracods are relatively simple. Therefore, the information content of the bioluminescent signals is primarily temporal, although a spatial component may be introduced by flight or swimming patterns. For example, the flight pattern of *Photinus pyralis* during luminescent displays results in a J pattern (Case, 1984) while the luminescent secretions of *Odontosyllis* appear as glowing rings (Huntsman, 1948) and those of *Vargula* appear as complex trains of dots spaced in species-specific patterns (Morin, 1986).

Bioluminescent courtship strategies among fireflies are generally of two distinct types. In the first of these, solitary males seek out females by patrolling an area while emitting species-specific flash patterns and looking for an appropriate response from a stationary female (eg. *Photinus, Photuris* and *Luciola*). In the second strategy, males remain sedentary but display synchronously in a species-specific pattern which attracts motile females (eg. *Pteroptyx*). The former strategy is associated with low population densities while the latter occurs where local densities are high and synchrony helps ensure that species-specific flash patterns don't get lost in the noise (Case, 1984; Lloyd, 1983).

In the marine environment most species of *Vargula* are found in high local densities in reefs or seagrass beds and males synchronize their displays, although, unlike fireflies, both males and females remain motile. There is no luminescent response from females during courtship and Morin (1986) suggests that this may be a predator avoidance strategy.

18.5.2 Finding food

Bioluminescence is used both to seek out and to attract prey. To find prey in the dark, some predators employ ocular light organs. For example, the aptly-named flashlight fishes of the family Anomalopidae, have very large, high intensity light organs beneath each eye. These light organs contain symbiotic bioluminescent bacteria. Bacteria are unusual among luminous marine organisms in that light emission is continuous whereas in most luminous systems light emission is discontinuous and regulated physiologically. Flashlight fishes regulate light organ emissions by means of elaborate mechanical shuttering strategies which permit them to shut the light off, presumably to avoid detection by predators and possibly premature detection by prey (Herring, 1982 and refs).

A prime example of an optical concealment strategy involving bioluminescence may be found in certain malacosteid and stomiatid fishes that emit red light from subocular light organs (Denton *et al.,* 1985, Latz *et al.,* 1988; Widder *et al.,* 1984 and refs) and which, unlike the vast majority of oceanic inhabitants, possess long-wave visual pigments. This sniper-scope system allows covert detection of prey over ranges 10 times that of lateral line systems (Partridge and Douglas, 1995).

Bioluminescence may also be used as a lure to attract prey. Deep-sea angler fishes are proficient at this strategy, using bacterial light production in light organs, called escae, to lure prey within striking distance of their capacious jaws. As with the bacterial light systems used by flashlight fish, angler fish can shut off the light, although the methods used to accomplish this are still unknown (Herring and Morin, 1978 and refs). Interestingly, although escal light organs come in a remarkable variety of forms, this diversity does not appear to be directed at attracting particular prey, as judged by stomach content analyses, but may, instead, be a sexual cue in certain species to aid the dwarf males, which have well-developed eyes, in locating an appropriate female (Marshall, 1979 and refs).

18.5.3 Avoiding predators

Of the multitude of functions attributed to bioluminescence, probably the most common and all-encompassing is predator avoidance.

a) Blinding and distraction

In dim light environments where eyes have evolved to maximize sensitivity and therefore often lack adaptations such as eyelids or screening pigments, dazzling flashes of light may be a very effective predator deterrent. Prey which employ such strategies are generally active swimmers. If the display is intrinsic then the flash is usually brief and bright and accompanied by a burst of swimming. Some myctophids probably employ this strategy since the caudal light organs produce a brief (400 ms) bright (3×10^{11} photons s^{-1}) flash which is generally accompanied by a tail flick (Mensinger and Case, 1990). If the display is extrinsic then the emissions are generally longer lasting. Many copepods release long-lasting (up to 30 s) clouds of luminescence from glands located on appendages, such as the tail or swimming legs, which generate thrust during escape (Figure 3). This coupling of emission sites with the appendages which generate the power stroke helps ensure that the luminescent material is ejected beyond the boundary layer of the escaping copepod (Herring, 1988; Widder, 1992).

There are many animals that are capable of the extrinsic release of bioluminescence including ctenophores (Harbison *et al.,* 1978), ostracods (Morin, 1986), amphipods (Herring, 1981), copepods (Herring, 1988), mysid and decapod shrimps (Herring, 1976), squid (Herring, 1978), searsiid fishes and some ceratioid fishes (Herring and Morin, 1978). In some cases luminescent emissions are cohesive and may act as a 'false clue' to distract the predator, while in other cases the luminescence is bright but diffuse and may create 'false noise' which would decrease the signal-to-noise ratio of any light clues which the predator might normally employ during an attack (Lloyd, 1977) or the emission might serve to illuminate the predator making it more vulnerable to attack (see next section).

b) Burglar alarms

Once the prey is in the clutches of a predator its only chance for survival may be to attract the attention of a second order predator, which will attack the primary predator and afford the prey an opportunity for escape. Known as bioluminescent burglar alarms (Burkenroad, 1943) these displays are particularly effective in an environment where the natural levels of bioluminescence are low and every eye is tuned to detect a luminescent event. Some of the best evidence for this function comes from laboratory experiments that have demonstrated reduced grazing by copepods on luminescent as compared to non-luminescent dinoflagellate cultures (Esaias and Curl, 1972; White, 1979). Reduced grazing occurs because flashes stimulate swimming bursts in marine copepods, thereby limiting the time available for grazing (Buskey and Swift, 1983; Buskey *et al.,* 1983; Buskey *et al.,* 1985). Since flash-induced burst swimming does not occur in copepods from freshwater environments, where bioluminescence is absent, this behavior is not a universal startle response (Buskey *et al.,* 1987). Rather, flash-induced burst swimming by marine copepods is apparently a means of evading visual predators alerted to the presence of the grazers by dinoflagellate flashing. The multi-

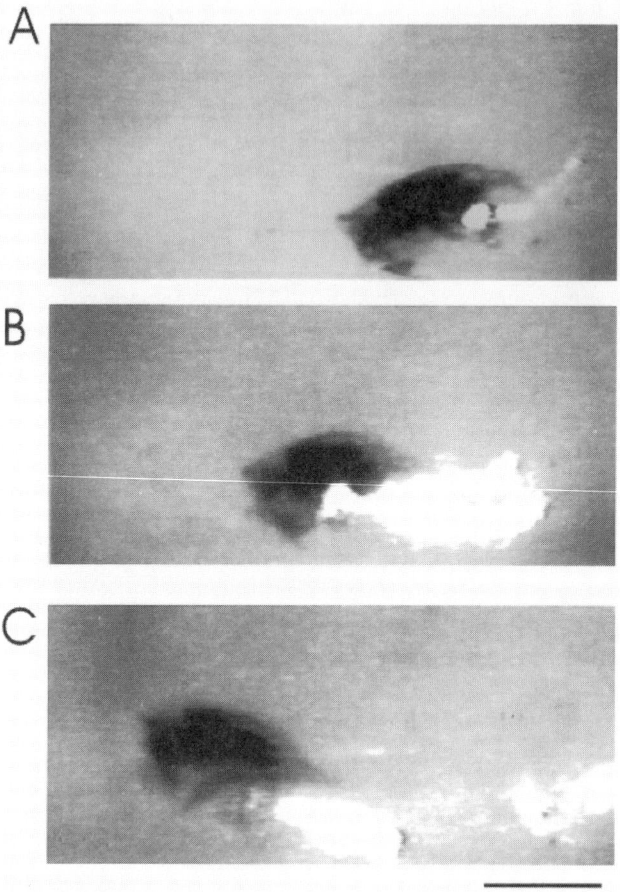

Figure 3 A sequence of three video frames showing the calanoid copepod *Euaugaptilus magnus* releasing clouds of luminescence from glands located on the swimming legs. As the copepod swam from right to left at an average speed of 1.3 cm s^{-1}, the power stroke of the swimming legs served the dual function of providing forward momentum and flicking the luminescent material beyond the boundary layer of the escaping calanoid. The luminescent exudate was cohesive and formed a glowing bolus that was visible for 17 s following release. Scale bar, 5 mm. (Widder, 1992)

trophic basis for this theory has been borne out by recent laboratory studies showing increased susceptibility of zooplankton to visual predators in the presence of bioluminescent dinoflagellates (Abrahams and Townsend, 1993; Fleisher and Case, 1995; Mensinger and Case, 1992). It has been further substantiated by *in situ* observations of attacks on luminescent ostracods (not during mating displays); the luminescence released by the ostracod in response to the attack by a primary (small fish) predator has attracted larger, second order predators, such as snappers, jacks and squirrelfish which then execute successful attacks on the primary predator (Morin, 1986).

Some of the most elaborate and eye-catching bioluminescent displays in the ocean are produced by animals without eyes, the gelatinous zooplankton. These displays, which are presumably burglar alarms, involve at least two distinct methods of drawing attention to the predator. Sticky glowing secretions may be used to mark a predator or brilliant intracellular emissions, often involving propagated waves of light, may be used like the lights on a theater marquee to attract attention (Widder *et al.,* 1989; Herring, 1990; Mackie, 1991) (Figure 4). While the release of luminescent secretions occurs on initial contact, neural facilitation is required to initiate the propagated emissions which can be self sustaining. Therefore jellies like *Atolla*, which are capable of both types of emissions, may use them in multiple lines of defense with the harder-to-stimulate, intracellular response held in reserve for predators which are not dissuaded when marked by initial contact.

c) Counterillumination

Nowhere is the need for optical concealment in the open ocean better documented than in the myriad examples of evolutionary convergence associated with the use of bioluminescence for camouflage. In the twilight regions of the ocean the silhouette of an opaque animal against the downwelling lightfield is an easy target for a visual predator. The unpatterned, yet highly directional and frequently fluctuating nature of this lightfield makes the common predator avoidance strategy of blending-in-with-the-background a consummate challenge. Perfect

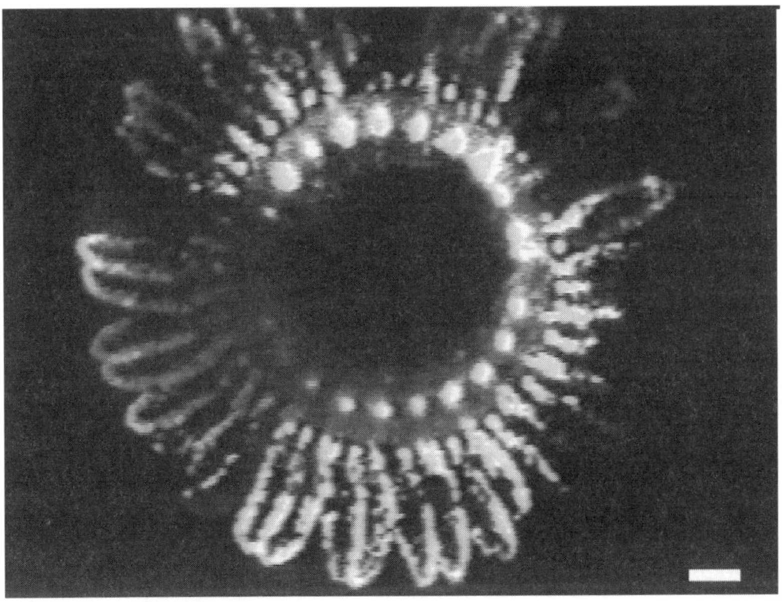

Figure 4 Self-sustaining luminescent display in the scyphozoan jellyfish *Atolla wyvillei* (Herring and Widder, unpublished). Scale bar, 5 mm

camouflage requires that the animal replace any light absorbed or reflected by its tissues by light of comparable color, intensity and angular distribution. A great number of mesopelagic animals have evolved an exceptional variety of strategies using bioluminescence to meet this challenge.

Color

Spectral measurements of the ventrally directed bioluminescence characteristic of so many mesopelagic animals have revealed that their luminescence is frequently an excellent match for the predominantly blue and spectrally narrow ambient light in the deep-sea (Herring, 1983; Widder *et al.,* 1983). A variety of mechanisms have evolved to accomplish this. In euphausiids the narrow bandwidth of the bioluminescence is apparently a direct consequence of the chemiluminescent reaction itself, since none of the complex optical components which reflect and refract the light out of the ventral photophores appear to alter the spectral character of the emission (Herring and Locket, 1978). In the deep-sea squid *Pterygioteuthis microlampas*, iridophores that line the posterior wall of the light organs are quarter wavelength reflectors that both redirect and narrow the spectral bandwidth of light produced by the photogenic tissue (Arnold *et al.,* 1974). Similar colored reflective surfaces line the photophores of the myctophid fish *Diaphus rafinesquii* and the decapod shrimp *Oplophorus spinosus* (Denton *et al.,* 1985). In many other animals light-absorbing colored filters are located between the point of origin of the luminescence and its point of exit from the body (Figure 5). Spectral absorbance measurements of these structures have revealed that although filters from different animals contain very different pigments or pigment mixtures they all possess transmission bands in the blue centered around 475 nm. Since the absorption of light by these filters results in a considerable loss of light energy (as much as 90% in *Opisthoproctus* and 75% in *Argyropelecus*, Denton *et al.,* 1985) it is thought that any disadvantages associated with these losses must be compensated by the selective advantages associated with an improved spectral match with the downwelling light field. However, selective adaptations in prey, which lead to improved crypticity also result in increased selection pressure on predators to 'break' that camouflage (Herring, 1982). For example, Muntz (1976) suggested that the yellow lenses found in many deep-sea predators, both fish and squid, might aid in the detection of counter-illuminators if their spectral match to the background was less than perfect. Since yellow pigments reduce the overall sensitivity of predator vision, this is a corollary of the compromises associated with bioluminescence filtering and, in this case, the disadvantages are presumably offset by improved prey detection.

Achieving a perfect spectral match to the downward irradiance is a great deal more difficult in shallow water than in deep water. Below 100 m, logarithmic plots of wavelength specific attenuation approach straight parallel lines, indicating that there is little change in spectral distribution with depth. However,

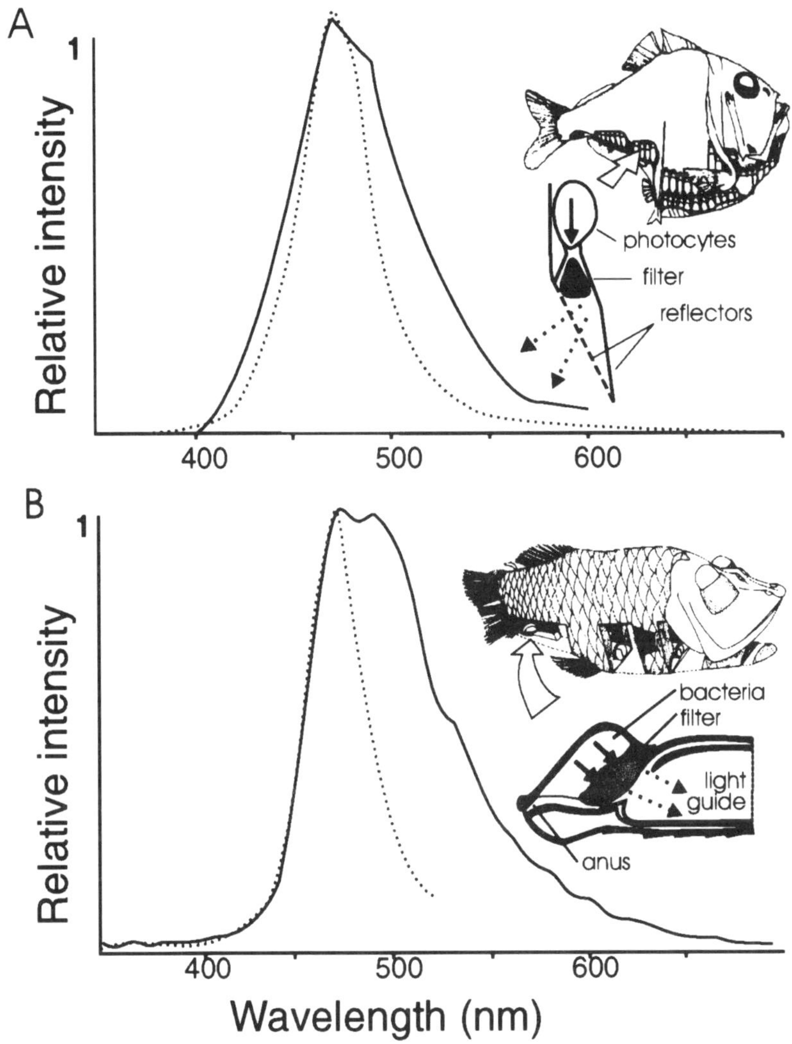

Figure 5 **A**. *Argyropelecus*. Spectral distribution of light before (solid line) and after (dotted line) passing through filter. Diagrams showing photophore distribution pattern, and vertical section through single photophore. (Redrawn from Denton *et al.,* 1985). The filter is located in the neck of the photocyte chamber. Light passing through the filter reflects off the silvered inner surfaces of the tubular photophore and is emitted through the half-silvered outer surface of the tubule. **B**. *Opisthoproctus*. Spectral distribution of light from bacterial light organ (solid line) and from ventral sole (dotted line), Widder, Latz, Case and Herring unpublished data. Diagrams of whole animal showing light guide system and light organ (Redrawn from Denton *et al.,* 1985). Light generated by the bacterial light organ near the anus passes through the filter and into the light guide system which distributes the light over the flattened ventral sole of the fish

in the upper layers of the ocean the spectral distribution changes radically, being very broad near the surface and narrowing rapidly with increasing depth, due to wavelength-selective attenuation mechanisms (Jerlov, 1976). Therefore, in order to be able to counterilluminate effectively in surface waters against moonlight or starlight an animal must be able to broaden the spectral distribution of its ventrally directed bioluminescence. Two species of squid and possibly one myctophid fish have this ability (Young, 1983). In the squid, *Abralia trigonura*, spectral broadening is accomplished in gradual stages, first by altering the color of light emitted by one set of photophores through adjustments of variable interference filters and reflectors (Young and Arnold, 1982) and then by turning on a second class of photophores (Figure 6). This remarkable spectral control is not accomplished with a feedback control loop, such as that used for intensity regulation (see next section), but rather, spectral shifts are triggered by changes in water temperature, corresponding to changes that the animal would normally encounter during vertical migration (Young and Mencher, 1980). As these authors point out, this very high degree of spectral control can not be taken as evidence that natural predators on these squid have color vision since the improved color

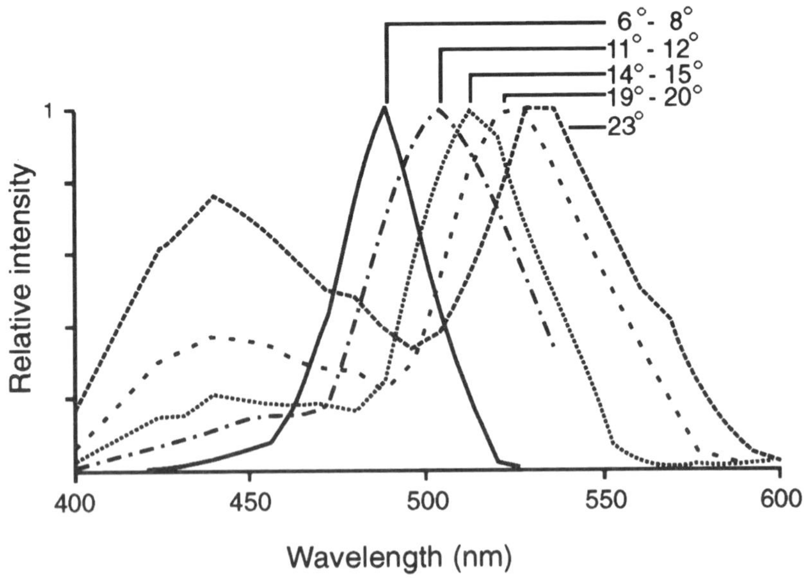

Figure 6 Bioluminescence emission spectra of ventral photophores from *Abralia trigonura*. Surface light was held constant as water temperature was adjusted. The shift in the long-wave peak was due to a spectral shift within one set of photophores while the short-wave peak originated from a separate set of photophores (Redrawn from Young and Arnold, 1982)

match may simply ensure an optimum intensity match for predators with different spectral sensitivities.

Intensity

For counterillumination to work, it is critical that the intensity of the bioluminescence match that of the constantly fluctuating downwelling lightfield. To accomplish this, animals must be able to monitor the environmental irradiance and either adjust their bioluminescence or adjust their depth to effect a match. In theory such a control system could be hardwired so that a given excitation of the photoreceptors produces a comparable activation of the photophores and/or depth adjustment mechanisms. This is apparently the case in the midwater shrimp *Sergestes similis* where downward-directed illumination of the eyes or adjacent tissues is essential to effect the light-matching response (Warner *et al.*, 1979). Since there is no apparent means by which the shrimp can see its own ventrally directed bioluminescence, there would appear to be no means for comparison between the environmental light and bioluminescence. Interestingly, these shrimp do not appear to counterilluminate when their eyes are fully dark-adapted, as evidenced by conspicuous eye glow, but require a period of about 25 minutes of dim light adaptation to induce counterillumination (Latz and Case, 1992). Perhaps this induction process helps circumvent the difficulties inherent with any hard-wired control system resulting from variable sensitivity of photoreceptors as a function of light adaptation. Effective light matching in *Sergestes similis* may also require a combination of bioluminescence control and depth adjustments, since their ability to counterilluminate has been demonstrated over an intensity range of only half an order of magnitude (Warner *et al.*, 1979) as compared to four and a half orders of magnitude in some squid (Young *et al.*, 1980).

A much finer degree of control is possible if there is some sort of feedback loop that allows the animal to monitor the intensity of its bioluminescence relative to that of the downwelling irradiance. One means of accomplishing this is to have photophores located in the vicinity of the eye so that a direct comparison is possible between the bioluminescence and the downwelling irradiance. This is the control system which has been proposed for the myctophid fish *Tarletonbeania crenularis* which has supraorbital photophores that shine into the eye (Figure 7) (Lawry, 1974). Some other examples of light organs which seem to be positioned to shine light into the eye and might therefore function in a feedback control system are found in *Chauliodus sloani* (Tchernavin, 1953), in hatchet fish (Herring and Morin, 1978), in the deep-sea squaloid shark *Centroscyllium ritteri* (Iwai, 1960) and in euphausiids (Mauchline and Fisher, 1969a). In the midwater squid *Abralia* and *Abraliopsis* downwelling irradiance is monitored by photosensitive vesicles located on the dorsal surface of the head, while bioluminescence is monitored by another set of ventrally directed photosensitive vesicles which monitor the emissions of photophores situated on the dorsal surface of the siphon (Figure 8) (Young *et al.*, 1979).

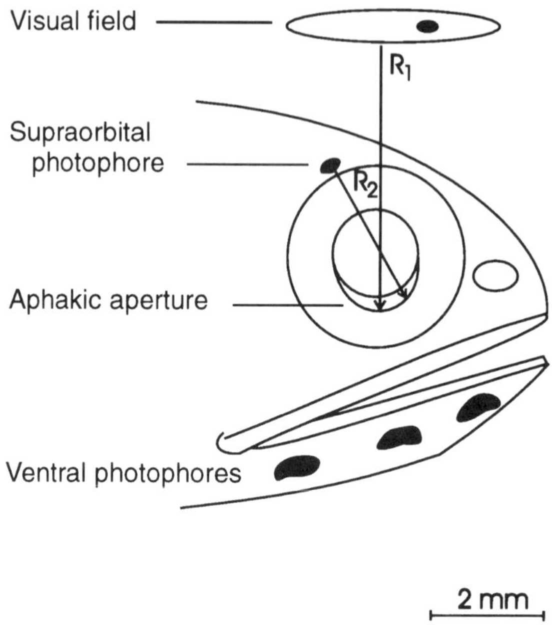

Figure 7 Feedback control system between the bioluminescent and visual systems of *Tarletonbeania crenularis*. Light rays from the dorsal visual field (R_1) and the supraorbital photophore (R_2) enter the eye through the aphakic aperture. By comparing the radiance of the supraorbital photophore with that of the downwelling light field, photophore emission, both supraorbital and ventral, can be adjusted to match the environmental light. (Redrawn from Lawry, 1974)

Direction

Light distribution in the ocean is highly directional (Jerlov, 1976) and effective counterillumination dictates that the ventrally directed bioluminescence exhibit the same directionality. This is apparently the function of many of the very elaborate optical structures associated with the ventral photophores of shrimp, squid and fish. Pigmented and/or reflective layers at the back of a photophore ensure that no light escapes upward, while refractile and reflective elements shape the intensity profile of the emitted light so that it has the same angular distribution as the downwelling lightfield (Herring, 1986). The apparent effectiveness of these various radiance matching strategies has been examined by comparing measurements of the radiance distribution of ventrally directed bioluminescence with the asymptotic radiance distribution of light in the deep-sea. These measurements, which have been performed on the euphausiids *Meganyctiphanes norvegica*, *Nematobrachion sexspinosum*, *Nematoscelis tenella* and *Thysanopoda cristata* (Herring and Locket, 1978), the decapod shrimp *Oplophorus spinosus*, *Systellaspis debilis* (Herring, 1976) and *Sergestes similis*

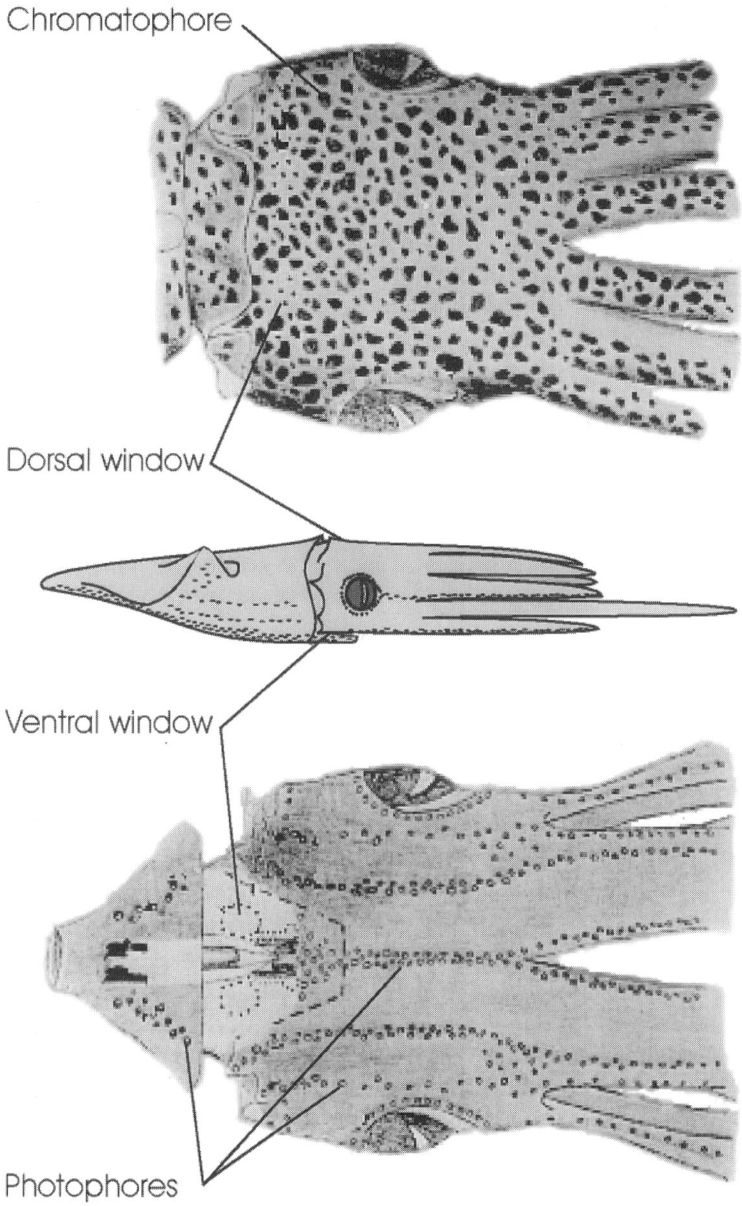

Figure 8 From top to bottom, dorsal, lateral and ventral views of *Abralia trigonura* showing the dorsal and ventral windows through which light penetrates to the extraocular photoreceptors. Bioluminescent feedback mechanism for controlling counterillumination is based on comparison of downwelling irradiance detected by the dorsal photosensitive vesicles and photophore bioluminescence detected by the ventral photosensitive vesicles which are in contact with photophores on the dorsal surface of the funnel. Funnel is folded back in ventral view and normal funnel position is indicated by dotted line. (Redrawn from Young *et al.*, 1979)

(Latz and Case, 1982), the hatchet fish *Argyropelecus affinis*, and the viper fish *Chauliodus sloani* (Denton *et al.,* 1972), all indicate a high degree of matching. The magnitude of the selective advantage afforded by such matching is manifest in the extreme degree of evolutionary convergence that these various organisms exhibit. As with the situation of spectral matching mentioned earlier, it follows that there must be comparable evolutionary pressure on predators to develop some means of breaking this camouflage. Areas of retina with enhanced resolving power, such as the fovea of notosudid fishes, may have evolved for this purpose (Herring, 1982 and refs, see also Collin's Chapter).

The extreme directionality of the light emitted by ventral photophores suggests that the orientation of a counterilluminating animal relative to the incident light distribution must be rigidly controlled since any misalignment of the emitter with the background would undermine the whole point of this camouflage strategy. As animals move about they must therefore maintain their bodies, or at least their photophores, such that the incident light distribution remains symmetrical around them. Euphausiids and sergestid shrimps, which tilt their bodies to swim up or down in the water column, maintain the vertical orientation of their light organs by rotating them, as much as 140° in *Sergestes similis* (Latz and Case, 1982) and 180° in *Meganyctiphanes norvegica* (Hardy, 1962). On the other hand, the hatchetfish *Argyropelecus* maintains the vertical orientation of its photophores by maintaining a perfectly horizontal orientation of its body even while changing depths at considerable speed (50 cm/s) (Janssen *et al.,* 1986).

Pattern

Very different patterns of ventral light organs are found in different animals (Figure 9). For example, shallow water counterilluminators often exhibit diffuse patterns of ventrally directed bioluminescence, as compared to midwater organisms which generally have discrete ventral photophores (Morin, 1981). In addition, trends have been noted among animals with photophores, toward a reduction in photophore complexity and size with increasing depth of habitat (Young and Roper, 1977 and refs). It is thought that these differences are a consequence of the improved visual acuity of predators under the higher ambient light conditions associated with shallower habitats. This improved resolving power of the eyes of predators places greater demands on the counterilluminator to blend in with the background (Morin, 1981; Young and Roper, 1977). It is noteworthy, in this regard, that many shallow water counterilluminators possess gut-associated bacterial light organs (Morin, 1981) and luminescence is expressed externally in a diffuse mottled pattern which may be a more effective form of camouflage in shallow water environments, which are spatially and temporally heterogeneous (McFall-Ngai and Morin, 1991), and may also permit counter-illumination at much higher intensities than in animals that depend on autogenic chemical systems.

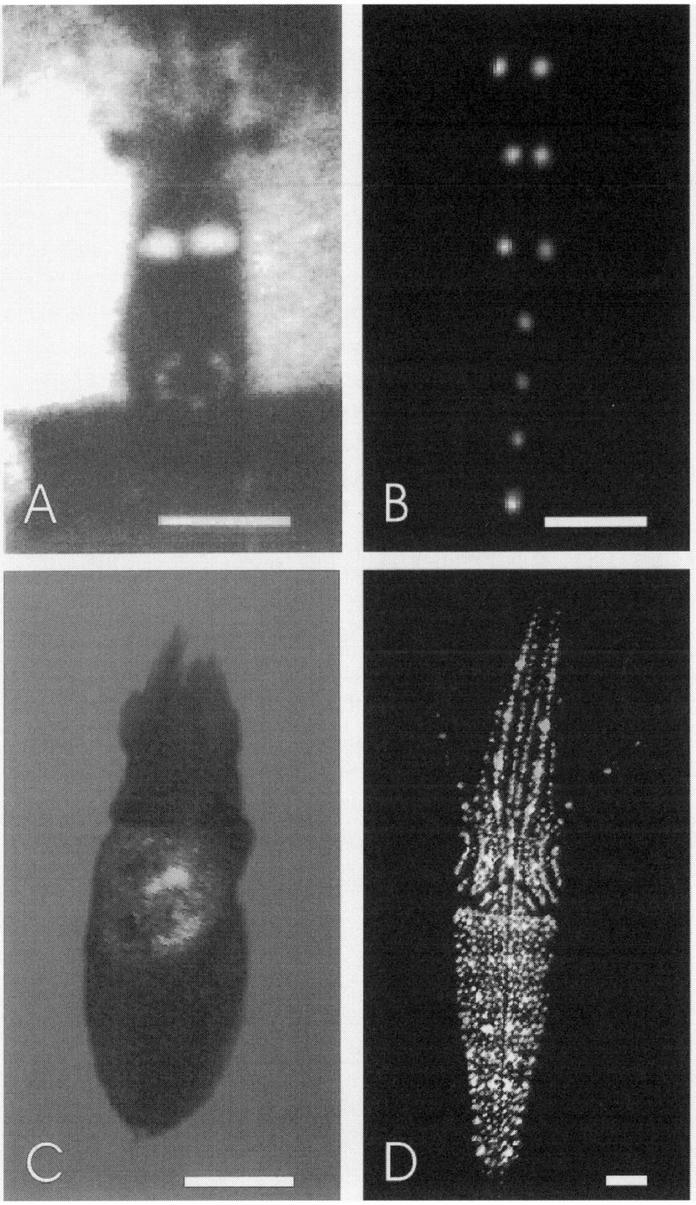

Figure 9 Patterns of counterillumination. **A**. In the decapod shrimp *Sergestes similis* bioluminescence originates from the organs of Pesta which are associated with the digestive glands in this largely transparent shrimp (Adapted from Latz and Case, 1992). **B**. In the euphausiid *Meganyctiphanes norvegica* bioluminescence (Widder, unpublished) originates from 10 ventral photophores, one on each eye, two pairs of thoracic photophores and 4 serial abdominal photophores. **C**. Ventral bioluminescence in the small, shallow-living squid *Heteroteuthis* sp. (Widder and Herring, unpublished) originates from a pair of glands associated with the ink sac. **D**. Photophore pattern in *Abralia veranyi* (Herring *et al.,* 1992. Scale bars, 5 mm

Many upward looking predators in the pelagic zone are themselves counter-illuminators, presumably to evade detection by larger predators beneath them. Although in many environments predators employ crypsis to evade detection by their prey, this use of counterillumination in the pelagic zone seems unlikely since the prey at greatest risk are those above the predator, not below where the bioluminescence is directed. One possible example of a predatory use of counter-illumination may be found in the bathypelagic shark, *Isistius brasiliensis*, which has thousands of tiny (0.03 – 0.05 mm) photophores covering its entire ventral surface except for a dark collar at the throat (Parin, 1966) (Figure 10). This 'dog collar' is not only wholly lacking in photophores but is also darkly pigmented and would therefore produce the effect of a dark and narrow silhouette against the downwelling lightfield. This apparent mimicry of the search image employed by many upward looking predators and the location of the collar right behind the mouth suggest that this unusual counterillumination pattern may function as a lure (Widder, 1998). This hypothesis is supported by gut content analyses which indicate that large squid (15 – 30 cm) are a common prey item for these small (29 – 46 cm) sharks. Since these sharks are apparently slow-swimmers that are caught in slow-moving nets (2.5 –7 kts) it is difficult to imagine how they could capture such large and fast-moving prey except by luring them within easy reach of their large jaws (Strasburg, 1963). It has also been difficult to imagine how these sharks could inflict the crater-like wounds found on large, fast-swimming marlins, tunas and porpoises. The match between the odd shape of these wounds and the unusual jaw structure of *Isistius brasiliensis*, is the basis for its common name of cookie-cutter shark. Although it has been suggested previously that these peculiar sharks might somehow be using their bioluminescence as a lure, the suggestion was that they were mimicking the luminescence of other organisms, such as squid (Jones, 1971). By contrast, I suggest that the bioluminescence of *Isistius brasiliensis* serves as camouflage against the downwelling light field and the lure is the strategically placed non-luminescent band behind

Figure 10 Bioluminescent shark, *Isistius brasiliensis* with a patch of skin showing scales and the many tiny photophores which cover the entire ventral surface except for the darkly pigmented 'dog collar' behind the jaw. (Reproduced with permission from Parin, 1966)

the jaws. In this context it is noteworthy that *Isistius brasiliensis* both schools and vertically migrates (Strasburg, 1963). By vertically migrating these sharks can occupy a lightfield which optimizes the silhouette illusion and by schooling they will enhance the illusion by appearing as a school of small fish.

18.6 Conclusions

The degree of selective advantage associated with bioluminescence is chronicled in the large number of independent acquisitions of the trait during evolutionary history, in the remarkable degree of convergence exhibited between very different organisms using bioluminescence for similar purposes, in the extraordinary success of bioluminescent organisms in the open ocean, and in the visual adaptations in predators that have apparently been driven by the improved adaptedness of bioluminescent prey. All of these factors indicate that bioluminescence is of critical importance to the ecology of the largest ecosystem on the planet. However data acquisition, particularly that related to the natural history, has been severely impeded by limited access to the organisms and to their environments. As technical advancements permit greater access and greater observational capabilities in dim light environments, it is hoped that a much greater appreciation and understanding may be gained of the multitude of bioluminescent creatures that populate our planet.

Acknowledgments

I thank Jim Case, Tammy Frank and Peter Herring for their helpful comments on the manuscript and I am very grateful to David Smith for his valuable assistance in the preparation of the figures.

References

Abrahams, M. V. and Townsend, L. D. (1993) Bioluminescence in dinoflagellates: a test of the burglar alarm hypothesis. *Ecol.,* 74, 258 – 260.

Arnold, J. M., Young, R. E. and King, M. V. (1974) Ultrastructure of a cephalopod photophore. II. Iridophores as reflectors and transmitters. *Biol. Bull.,* 147, 522 – 534.

Badcock, J. R. and Merrett, N. R. (1976) Vertical distribution of midwater fishes. *Prog. Oceanogr.,* 7, 3 – 58.

Baker, A. de C., Boden, B. P. and Brinton, E. (1990) *A Practical Guide to the Euphausiids of the World,* London: Natural History Museum Publications.

Beebe, W. (1951) *Half-Mile Down*, 2nd edn. New York: Duell, Sloan and Pearce (affiliate of Meredith Press).

Beebe, W. M. (1937) Preliminary list of Bermuda deep-sea fish. *Zoologica, N. Y.,* 22, 197– 208.

Buck, J. B. (1978) Chapter 13: Functions and evolutions of bioluminescence. In: Herring, P. J. (Ed.) *Bioluminescence in Action*. Academic Press

Burkenroad, M. D. (1943) A possible function of bioluminescence. *J. Mar. Res.,* 5, 161–164.

Buskey, E. J. (1992) Epipelagic planktonic bioluminescence in the marginal ice zone of the Greenland Sea. *Mar. Biol.,* 113, 689 – 698.

Buskey, E. J., Mann, C. G. and Swift, E. (1987) Photophobic responses of calanoid copepods: possible adaptive value. *J. Plank. Res.*, 9, 857– 870.

Buskey, E. J., Mills, L. and Swift, E. (1983) The effects of dinoflagellate bioluminescence on the swimming behavior of a marine copepod. *Limnol. Oceanogr.,* 28(3), 575 – 579.

Buskey, E. J., Reynolds, G. T., Swift, E. and Walton, A. J. (1985) Interactions between copepods and bioluminescent dinoflagellates: direct observations using image intensification. *Biol. Bull.*, 169, 530

Buskey, E. J. and Swift, E. (1983) Behavioral responses of the coastal copepod *Acartia hudsonica* (Pinhey) to simulated dinoflagellate bioluminescence. *J. Exp. Mar. Biol. Ecol.,* 72, 43 – 58.

Case, J. F. (1984) Vision in mating behavior of fireflies. In: Lewis, T. (Ed.) *Insect Communication,* pp. 195 – 222. New York: Academic Press.

Case, J. F., Haddock, S. H. D. and Harper, R. D. (1994) The ecology of bioluminescence. In: Campbell, A. K., Kricka, L. J. and Stanley, P. E. (Eds.) *Bioluminescence and Chemiluminescence. Fundamentals and Applied Aspects.* pp. 115 –122. New York: John Wiley & Sons.

Case, J. F., Widder, E. A., Bernstein, S., Ferrer, K., Young, D., Latz, M. I., Geiger, M. and Lapota, D. (1993) Assessment of marine bioluminescence. *Naval Research Reviews XLV,* 31– 41.

Childress, J. J. (1982) Oceanic biology: Lost in space? In: Brewer, P. G. (Ed.) *Oceanography. The Present and Future*, pp. 127–135. New York: Springer-Verlag.

Clarke, G. L. and Wertheim, G. K. (1956) Measurements of illumination at great depths and at night in the Atlantic Ocean by means of a new bathyphotometer. *Deep-sea Res.* 3, 189 – 205.

Darwin, C. R. (1933) *The Voyage of the Beagle*, December 1831 – October 1836, Anchor edn. New York: Doubleday and Company.

Denton, E. J. (1990) Light and vision at depths greater than 200 meters. In: Herring, P. J., Campbell, A. K., Whitfield, M. and Maddock, L. (Eds.) *Light and Life in the Sea.* pp. 127–148. Cambridge: Cambridge University Press.

Denton, E. J., Gilpin-Brown, J. B. and Wright, P. G. (1972) The angular distribution of the light produced by some mesopelagic fish in relation to their camouflage. *J. Physiol.*, 182, 145 –158.

Denton, E. J., Herring, P. J., Widder, E. A., Latz, M. F. and Case, J. F. (1985) The roles of filters in the photophores of oceanic animals and their relation to vision in the oceanic environment. *Proc. R. Soc. Lond. B,* 225, 63 – 97.

Esaias, W. E. and Curl, H. C. (1972) Effect of dinoflagellate bioluminescence on copepod ingestion rates. *Limnol. Oceanogr.,* 17, 901– 906.

Fleisher, K. and Case, J. F. (1995) Enhancement of cephalopod predation by dinoflagellate luminescence. *Biol. Bull.,* 189, 263 – 271.

Gitel'zon, I. I. and Levin, L. A. (1989) Bioluminescence in oceanology. *J. Biolum. Chemilum.* 4, 555 – 562.

Haddock, S. H. D. and Case, J. F. (1994) A bioluminescent chaetognath. *Nature,* 367, 225 – 226.

Hailman, J. P. (1977) Communication by reflected light. In: Sebeok, T.A. (Ed.) *How animals communicate*, pp. 184 – 210. Bloomington: Indiana Univ. Press.

Harbison, G. R., Biggs, D. C. and Madin, L. P. (1977) The associations of Amphipoda Hyperiidea with gelatinous zooplankton. II. Associations with Cnidaria, Ctenophora and Radiolaria. *Deep-sea Res.*, 24, 465 – 488.

Harbison, G. R., Madin, L. P. and Swanberg, N. (1978) On the natural history and distribution of oceanic ctenophores. *Deep-sea Res.*, 25, 233 – 256.

Hardy, M. G. (1962) Photophore and eye movement in the euphausiid *Meganyctiphanes norvegica* (Sars, G. O.). *Nature,* 196, 790 – 791.

Harvey, E. N. (1920) *The Nature of Animal Light*, Philadelphia.

Hastings, J.W. (1983) Biological diversity, chemical mechanisms, and the evolutionary origins of bioluminescent systems. *J. Mol. Evol.*, 19, 309 – 321.

Hastings, J. W. and Morin, J. G. (1991) Bioluminescence. In: Prosser, C. L. (Ed.) *Comparative Animal Physiology. Neural and Integrative Animal Physiology*. 4th edn. pp. 131–170. New York: Wiley-Liss.

Herring, P. J. (1976) Bioluminescence in decapod crustacea. *J. Mar. Biol. Ass. U. K.*, 56, 1029 –1047.

Herring, P. J. (1978) Bioluminescence in invertebrates other than insects. In: Herring, P. J. (Ed.) *Bioluminescence in Action*, pp. 199 – 240. New York: Academic Press.

Herring, P. J. (1981) Studies on bioluminescent marine amphipods. *J. Mar. Biol. Ass. U.K.,* 61, 161–176.

Herring, P. J. (1982) Aspects of the bioluminescence of fishes. *Oceanogr. Mar. Biol. Ann. Rev.*, 20, 415 – 470.

Herring, P. J. (1983) The spectral characteristics of luminous marine organisms. *Proc. R. Soc. Lond. B,* 220, 183 – 217.

Herring, P. J. (1985) Bioluminescence in the crustacea. *J. Crus. Biol.,* 5(4), 557– 573.

Herring, P. J. (1986) How to survive in the dark: bioluminescence in the deep-sea. *Symposia of the Society for Experimental Biology*, 39, 323 – 350.

Herring, P. J. (1987) Systematic distribution of bioluminescence in living organisms. *J. Biolum. Chemilum.*, 1, 147–163.

Herring, P. J. (1988) Copepod luminescence. *Hydrobiol.,* 167/168, 183 –195.

Herring, P. J. (1990) Bioluminescence responses of the deep-sea scyphozoan *Atolla wyvillei. Mar. Biol.,* 106, 413 – 417.

Herring, P. J. and Locket, N. A. (1978) The luminescence and photophores of euphausiid crustaceans. *J. Zool., Lond.,* 186, 431– 462.

Herring, P. J. and Morin, J. G. (1978) Bioluminescence in fishes. In: Herring, P. J. (Ed.) *Bioluminescence in Action*, pp. 273 – 329. New York: Academic Press.

Herring, P. J. and Roe, H. S. J. (1988) The photoecology of pelagic oceanic decapods. *Symp. Zool. Soc. Lond.,* 59, 263 – 290.

Herring, P. J., Widder, E. A. and Haddock, S. (1992) Correlation of bioluminescence emissions with ventral photophores in the mesopelagic squid *Abralia veranyi. Mar. Biol.,* 112, 293 – 298.

Huntsman, A. G. (1948) *Odontosyllis* at Bermuda and lunar periodicity. *J. Fish. Res. Bd Can.* 7, 363 – 369.

Iwai, T. (1960) Luminous organs of the deep-sea squaloid shark *Centroscyllium ritteri* Jordan and Fowler. *Pacif. Sci.,* 14, 51– 54.

Janssen, J., Harbison, G. R. and Craddock, J. E. (1986) Hatchetfishes hold horizontal attitude during diagonal descents. *J. Mar. Biol. Ass. U. K.*, 66, 825– 833.

Jerlov, N. G. (1976) *Marine Optics*, New York: Elsevier Scientific Publishing Company.

Jones, E. C. (1971) *Isistius brasiliensis*, a squaloid shark, the probable cause of crater wounds on fishes and cetaceans. *Fish. Bull.*, 69, 791–798.

Lampert, W. (1989) The adaptive significance of diel vertical migration of zooplankton. *Func. Ecol.,* 3, 21– 27.

Lapota, D., Geiger, M. L., Stiffey, A. V., Rosenberger, D. E. and Young, D. K. (1989) Correlations of planktonic bioluminescence with other oceanographic parameters from a Norwegian fjord. *Mar. Ecol. Progr. Ser.,* 55, 217– 227.

Lapota, D. and Rosenberger, D. E. (1990) Bioluminescence measurements and light budget analysis in the Western Arabian Sea. *EOS,* 71, 97– 98.

Lapota, D., Rosenberger, D. E. and Lieberman, S. H. (1992) Planktonic bioluminescence in the pack ice and the marginal ice zone of the Beaufort Sea. *Mar. Biol.,* 112, 665 – 675.

Latz, M. I. and Case, J. F. (1982) Light organ and eyestalk compensation to body tilt in the luminescent midwater shrimp, *Sergestes similis. J. Exp. Biol.,* 98, 83-104.

Latz, M. I. and Case, J. F. (1992) Slow photic and chemical induction of bioluminescence in the midwater shrimp, *Sergestes similis* Hansen. *Biol. Bull.,* 182, 391– 400.

Latz, M. I., Frank, T. M. and Case, J. F. (1988) Spectral composition of bioluminescence of epipelagic organisms from the Sargasso Sea. *Mar. Biol.,* 98, 441– 446.

Lawry, J. V. (1974) Lantern fish compare downwelling light and bioluminescence. *Nature,* 247, 155 –157.

Lloyd, J. E. (1977) Bioluminescence and communication. In: Sebeok, T. A. (Ed.) *How Animals Communicate*, pp. 164 –183. Bloomington: Indiana Univ. Press.

Lloyd, J. E. (1983) Bioluminescence and communication in insects. *Ann. Rev. Entomol.,* 28, 131–160.

Losee, J., Lapota, D. and Lieberman, S. H. (1985) Bioluminescence: A new tool for oceanography. In: Zirino, A. (Ed.) *Mapping strategies in chemical oceanography*. 209th edn. pp. 211– 234. Advances in Chemistry Series: American Chemical Society.

Mackie, G. O. (1991) Propagation of bioluminescence in *Euphysa japonica* hydromedusae, (*Tubulariidae*). *Hydrobiol.,* 216/217, 581– 588.

Marshall, N. B. (1979) *Deep-Sea Biology. Developments and Perspectives.* New York and London: Garland STPM Press.

Mauchline, J. and Fisher, L. R. (1969a) The biology of euphausiids. Chapter 8: Vision and Bioluminescence. *Adv. Mar. Biol.,* 7, 227– 257.

Mauchline, J. and Fisher, L. R. (1969b) The biology of euphausiids. Chapter 13: Euphausiids in the marine economy. *Adv. Mar. Biol.,* 7, 374 – 391.

Maynard, S. D., Riggs, R. V. and Walters, J. F. (1975) Mesopelagic micronekton in Hawaiian waters: faunal composition, standing stock and diel vertical migration. *Fish. Bull.* 73, 726 –736.

McCapra, F. (1990) The chemistry of bioluminescence: Origins and mechanism. In: Herring, P. J., Campbell, A. K., Whitfield, M. and Maddock, L. (Eds.) *Light and Life in the Sea,* pp. 265 – 278. Cambridge: Cambridge University Press

McFall-Ngai, M. J. (1990) Crypsis in the pelagic environment. *Amer. Zool.*, 30, 175 –188.

McFall-Ngai, M. J. and Morin, J. G. (1991) Camouflage by disruptive illumination in Leiognathids, a family of shallow-water, bioluminescent fishes. *J. Exp. Biol.*, 156, 119 –137.

Mensinger, A. F. and Case, J. F. (1990) Luminescent properties of deep sea fish. *J. Exp. Mar. Biol. Ecol.*, 144, 1–15.

Mensinger, A. F. and Case, J. F. (1992) Dinoflagellate luminescence increases susceptibility of zooplankton to teleost predation. *Mar. Biol.,* 112, 207– 210.

Morin, J. G. (1974) Coelenterate bioluminescence. In: Muscatine, L. and Lenhoff, H. (Eds.) *Coelenterate Biology: Reviews and Perspectives*, pp. 397– 438. New York: Academic Press.

Morin, J. G. (1981) Bioluminescent patterns in shallow tropical marine fishes. *Proc. 4th Int. Coral Reef Symp.,* 2, 569 – 574.

Morin, J. G. (1983) Coastal bioluminescence: patterns and functions. *Bull. Mar. Sci.,* 33(4), 787– 817.

Morin, J. G. (1986) 'Firefleas' of the sea: Luminescent signaling in marine ostracode crustaceans. *Florida Entomologist,* 69, 105 –121.

Muntz, W. R. A. (1976) On yellow lenses in mesopelagic animals. *J. Mar. Biol. Ass. U. K,.* 56, 963 – 976.

Neilson, D. J., Latz, M. I. and Case, J. F. (1995) Temporal variability in the vertical structure of bioluminescence in the North Atlantic Ocean. *J. Geophys. Res.* 100, 6591– 6603.

Parin, N. V. (1966) Data on the biology and distribution of the pelagic sharks, *Euprotomicrus bispinatus* and *Isistius brasiliensis* (Squalidae, Pisces). In: Rass, T. S. (Ed.) *Fishes of the Pacific and Indian Oceans. Biology and Distribution*, pp. 173 –195. Jerusalem: Israel Program for Scientific Translations.

Partridge, J. C. and Douglas, R. H. (1995) Far-red sensitivity of dragon fish. *Nature,* 375, 21– 22.

Piccard, J. and Dietz, R. S. (1961) *Seven Miles Down*, New York: G. P. Putnam's Sons.

Pugh, P .R. (1989) Gelatinous zooplankton – the forgotten fauna. *Prog. Underwat. Sci.*, 14, 67–78.

Salvini-Plawen, L. V. and Mayr, E. (1967) On the Evolution of Photoreceptors and Eyes. In: Hecht, M. K., Steere, W. C. and Wallace, B. (Eds.) *Evolutionary Biology*, pp. 207– 263. New York and London: Plenum Press.

Seliger, H. H. (1993) Bioluminescence: Excited states under cover of darkness. *Naval Research Reviews XLV*, 5 –11.

Strasburg, D. W. (1963) The diet and dentition of *Isistius brasiliensis*, with remarks on tooth replacement in other sharks. *Copeia,* 1, 33 – 40.

Swift, E., Biggley, W. H. and Lessard, L. J. (1985) Distributions of epipelagic bioluminescence in the Sargasso and Caribbean Seas. *Advances in Chemistry Series #209: Mapping Strategies in Chemical Oceanography.*

Swift, E., Sullivan, J. M., Batchelder, H. P., Van Keuren, J., Vaillancourt, R. D. and Bidigare, R. R. (1995) Bioluminescent organisms and bioluminescence measurements in the North Atlantic Ocean near latitude 59.5° N, longitude 21° W. *J. Geophys. Res.,* 100, 6527– 6547.

Tchernavin, V. V. (1953) *The feeding mechanisms of a deep sea fish*, London: The Trustees of the British Museum.

Tett, P. B. and Kelly, M. G. (1973) Marine bioluminescence. *Oceanogr. Mar. Biol.,* 11, 89 –173.

Tsuji, F. I. and Hill, E. (1983) Repetitive cycles of bioluminescence and spawning in the polychaete, *Odontosyllis phosphorea. Biol. Bull.,* 165, 444 – 449.

Warner, J. A., Latz, M. I. and Case, J. F. (1979) Cryptic bioluminescence in a midwater shrimp. *Science,* 203, 1109 –1110.

White, H. H. (1979) Effects of dinoflagellate bioluminescence in the ingestion rates of herbivorous zooplankton. *J. Exp. Mar. Biol. Ecol.*, 36, 217– 225.

Widder, E. A. (1992) Mixed light imaging system for recording bioluminescence behaviours. *J. Mar. biol. Ass. U. K.*, 72, 131–138.

Widder, E. A. (1998) A predatory use of counterillumination by the squaloid shark, *Isistius brasiliensis. Env. Biol. Fish.* (in press).

Widder, E. A., Bernstein, S. A., Bracher, D. F., Case, J. F., Reisenbichler, K. R., Torres, J. J. and Robison, B. H. (1989) Bioluminescence in Monterey Submarine Canyon: image analysis of video recordings from a midwater submersible. *Mar. Biol.,* 100, 541– 551.

Widder, E. A., Case, J. F., Bernstein, S. A., MacIntyre, S., Lowenstine, M. R., Bowlby, M. R. and Cook, D. P. (1993) A new large volume bioluminescence bathyphotometer with defined turbulence excitation. *Deep-Sea Res.*, 40, 607– 627.

Widder, E. A., Greene, C. H. and Youngbluth, M. J. (1992) Bioluminescence of sound-scattering layers in the Gulf of Maine. *J. Plank. Res.*, 14, 1607–1624.

Widder, E. A., Latz, M. I. and Case, J. F. (1983) Marine bioluminescence spectra measured with an optical multichannel detection system. *Biol. Bull.,* 165, 791– 810.

Widder, E. A., Latz, M. I., Herring, P. J. and Case, J. F. (1984) Far red bioluminescence from two deep-sea fishes. *Science,* 225, 512 – 514.

Young, R. E. (1983) Oceanic bioluminescence: an overview of general functions. *Bull. Mar. Sci.,* 33(4), 829 – 845.

Young, R. E. and Arnold, J. M. (1982) The functional morphology of a ventral photophore from the mesopelagic squid, *Abralia trigonura. Malacol.*, 23(1), 135 –163.

Young, R. E., Kampa, E. M., Maynard, S. D., Mencher, F. M. and Roper, C. F. E. (1980) Counter-illumination and the upper depth limits of midwater animals. *Deep-sea Res.*, 27A, 671– 691.

Young, R. E. and Mencher, F. M. (1980) Bioluminescence in mesopelagic squid: Diel color change during counterillumination. *Science,* 208, 1286 –1288.

Young, R. E. and Roper, C. F. E. (1977) Intensity regulation of bioluminescence during countershading in living midwater animals. *Fish. Bull.,* 75, 239 – 252.

Young, R. E., Roper, C. F. E. and Walters, J. F. (1979) Eyes and extraocular photoreceptors in midwater cephalopods and fishes: Their roles in detecting downwelling light for counterillumination. *Mar. Biol.,* 51, 371– 380.

19
The behavior of animals around twilight with emphasis on coral reef communities

W. McFARLAND[1], C. WAHL[2], T. SUCHANEK[3] and F. McALARY[1]
[1]Friday Harbor Marine Laboratory, University of Washington, 620 University Drive, Friday Harbor, WA 98250.
[2]Department of Anatomy, Veterinary College, Cornell University, Ithaca, NY 14853.
[3]Division of Environmental Studies, University of California, Davis, CA 95616.

19.1 Introduction

Light, which represents a limited range of the electromagnetic spectrum, is characterized by three primary qualities – spectral frequency, intensity and polarization. Natural ambient light that organisms respond to emanates primarily from astronomical light sources and to a lesser extent from bioluminescence. For both sources, be it due to the earth's daily rotation and the moon's monthly procession about earth, or the flashing of fireflies on land or fishes in the deep sea, light is presented to animals with a distinct temporal component. From midday through night-time light intensities on earth's surface can vary 8 to 9 orders of magnitude, with the most rapid changes of 5 to 6 decades of intensity occurring during twilight (Figure 1). Animals exhibiting visual behaviors over this dynamic range are few, and those that do usually possess duplex visual mechanisms, one sensitive to the 'bright' light of day (photopic), and the other sensitive to the 'dim' light of night (scotopic).

Temporal separation of daily activities such as feeding is widespread in animal communities, and recognized in the designation of animal species as diurnal, as nocturnal, and more rarely as crepuscular. There are also arrhythmic species that show no demonstrable change in activity patterns from day to night (Yamanouchi 1956, Southward and Crisp 1965, Curio 1976). Most animals show structural adaptations suited to the light characteristics of their activity cycle. These include, for example, cryptic body coloration in diurnal species (Cott 1940), or optically specific visual structures (invertebrates-Land 1981, vertebrates-Walls 1942, Crescitelli 1972, Lythgoe 1979). Often such visual adaptations also reflect phylogenetic lineages, as primarily diurnal birds and nocturnal mammals (Charles-Dominique 1975) or Lepidopterans, divided into primarily diurnal butterflies and nocturnal moths (Kennedy 1928, Cloudley-Thompson 1960).

Shifts in the daily activities of a variety of animals are often dramatic, usually coincide with the twilight periods, and usually involve movement from active feeding to a quiescent mode of activity. Extensive literature describes the

S.N. Archer et al. (eds.), Adaptive Mechanisms in the Ecology of Vision, 583–628

temporal patterns for different animals from varied ecosystems (e.g., Calhoun 1944 and 1945, Harker 1958 and 1964, DeCoursey 1960, Aschoff and Wever 1962, Beck 1968, Aschoff 1969, Aschoff *et al.,* 1970, Blaxter 1970, Segal 1970, Charles-Dominique 1975, Thorpe 1978, Naylor and Hartnall 1979, Mills 1983). Most investigators have considered only a single or related species, and provide little information about other community members or specific light levels. As a consequence, relationships of precisely when and how the constituent species of animal communities slip into and out of their daily behavior modes is poorly documented. Ebeling *et al.,* (1966) investigated the behaviors of several benthic temperate marine invertebrates and concluded that their stereotyped movements around twilight functioned to separate prey from predators, a conclusion also applied forcefully to the twilight transitions of coral reef fishes (see Hobson 1979, and following citations). Several studies document the timing of diel transitions in specific taxonomic categories and, importantly, many provide measures of changing light levels with which the behavior changes can be correlated (vertical plankton movements – Clarke and Backus 1964, Boden and Kampa 1967, Kampa 1976; tropical marine fishes – Hobson 1965, 1968 and 1972, Collette and Talbot 1972, Domm and Domm 1973, Sbikin 1976; temperate marine fishes – Ebeling and Bray 1976, Hobson *et al.,* 1981; temperate freshwater fishes – Emery 1973, Helfman 1979, 1981 and 1993; non-primate mammals – Kavanau 1971, Voute *et al.,* 1974, Zawhlen 1975, Kenagy 1976; primates – Charles-Dominique 1975, Pariente 1974, 1979 and 1980, Pages and Petter-Rousseaux 1980). The temporal precision in these daily behavior transitions introduces ordered chronology into the changes in communal structure that occur at dusk and dawn, and presumably benefits each individual – e.g. increasing foraging time vs. reducing predation.

Because these tightly orchestrated shifts in behavior usually occur at twilight, it is generally inferred that the changes in light, in some manner, act as a signal to elicit the transitions (McFarland *et al.,* 1979). As a result it is the transitional behavior of animals at twilight and their relationship to changes in ambient light that we consider in this chapter. We focus on the transition behaviors in coral reefs because of our familiarity with these ecosystems and, especially, because of their high faunal diversity, the leading role that predation plays in molding community interactions (Grigg *et al.,* 1984, Huston 1985, Sorokin 1993), and the relative constancy of photoperiod and the duration of twilight in the tropics. What then is the temporal precision in the daily behavior transitions of the highly diverse invertebrates and fishes that inhabit coral reef communities?

19.2 Diel changes in coral reefs

19.2.1 Light

Twilight represents the period during dawn and dusk when the sun continues to illuminate the sky even though it is below the horizon. The optical causes of twilight, which result in the vivid color changes that inspire many of us in various ways, are described in several works (Minnaert 1954, Rosenberg 1966, McCartney 1976, Lynch and Livingston 1995). Under ideal clear skies the change in light intensity and spectral phenomena are mirror images during dawn and dusk. Twilight is classified by the altitude of the sun below the horizon as follows: the sun at –6 degrees is civil twilight (CT), the sun at –12 degrees is nautical twilight (NT), and the sun at –18 degrees is astronomical twilight (AT) (Figure 1). Because of the earth's tilted axis, relative to the sun, the duration of twilight varies with latitude and season. Seasonal shifts are minimal in the tropics and increase poleward, especially, at latitudes above 30 – 40 degrees (Figure 2).

The light intensities cast on the earth's surface during twilight under clear sky conditions is a function of the sun's altitude below the horizon and is independent of latitude and season (Figure 1). Thus, at sunset (SS) and at sunrise (SR) the

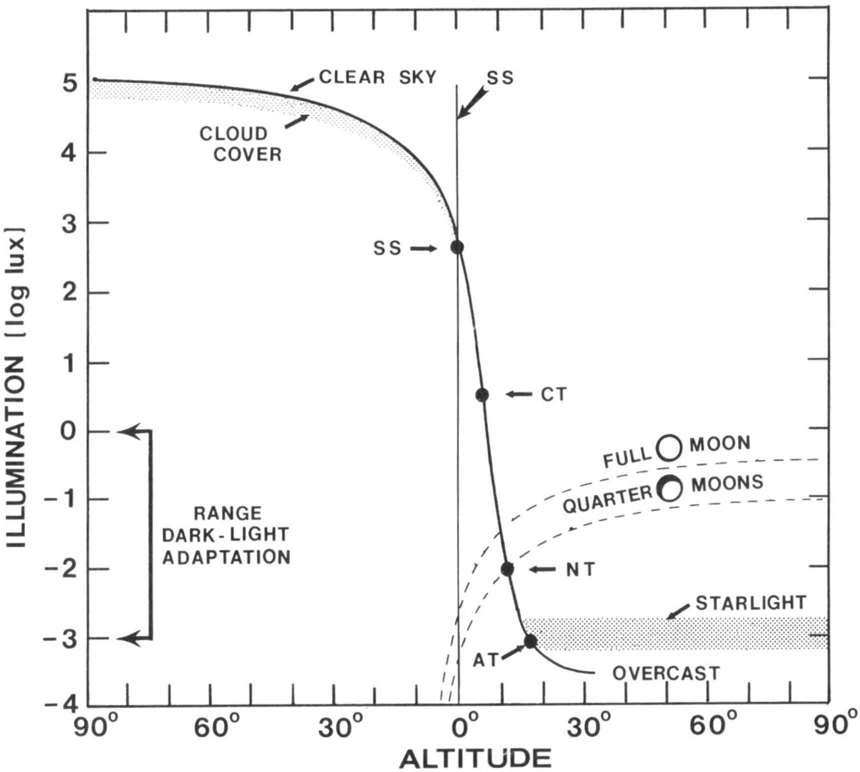

Figure 1 Relationships of illumination at earth's surface with respect to solar and lunar altitude above and below the horizon. SS = sunset, CT = civil twilight with sun at 6° below horizon, NT = nautical twilight with sun at 12° below horizon, AT = astronomical twilight with sun at 18° below horizon. Dark-light adaptation represents the intensity range over which retinal motor movements occur in most fishes. Modified from Brown (1952) and Blaxter (1970)

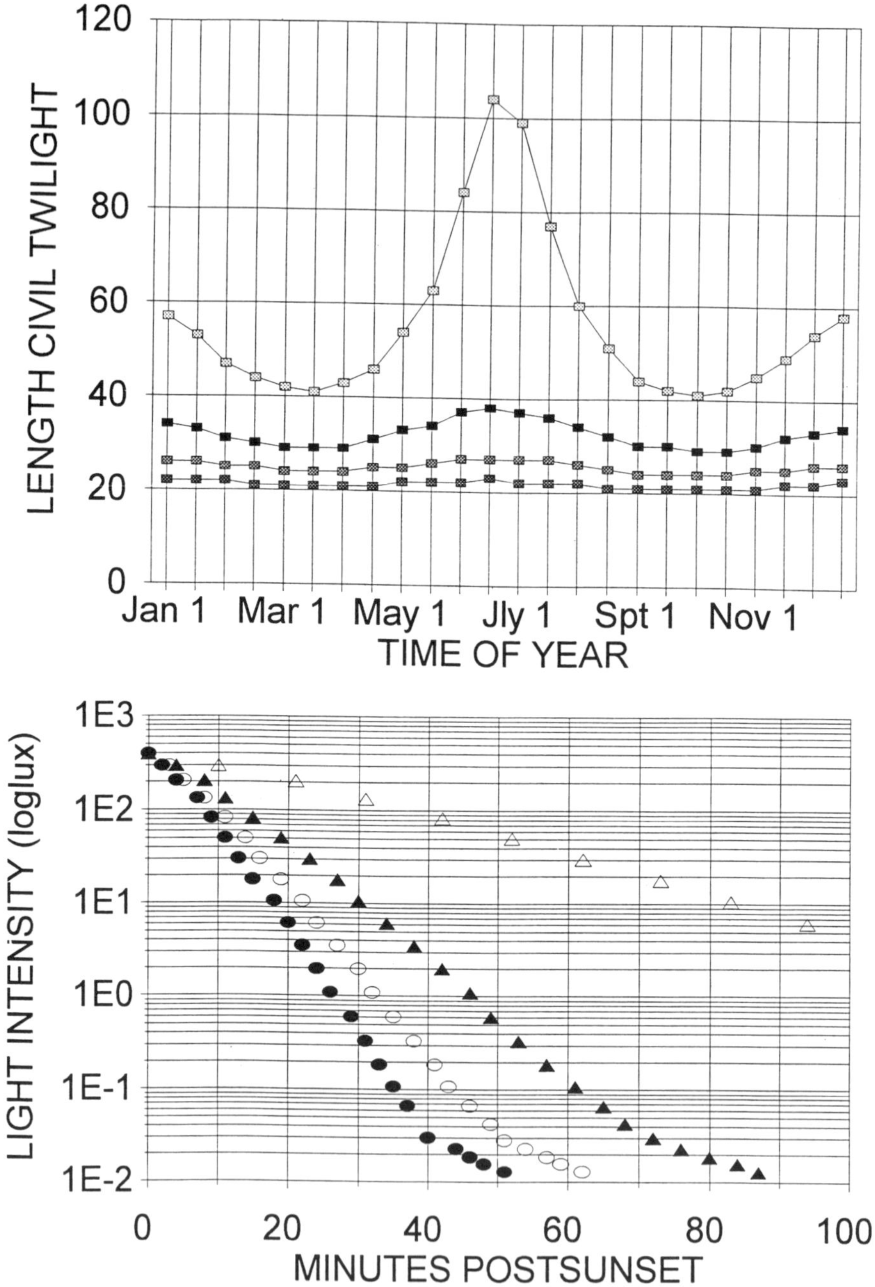

Figure 2 Relation of the duration in minutes of civil twilight to time of year and latitude (upper graph) and decline in light intensity following sunset at different northern latitudes during June (lower graph). Data in both graphs are sequential from bottom to top and represent 0°, 30°, 45° and 60° N L

light intensity on a flat surface approximates to typical fluorescent room light (400 lux, although values vary a few lux between reports), and is approximately 3.55 lux at the end of CT (Nielsen 1961 and 1963). At NT intensity is close to 0.01 lux, and as night begins or ends at AT the intensity is less than 0.00007 lux (Figure 1). Meteorological conditions modify these values, but by only a few minutes, at least in tropical and subtropical latitudes (see Figure 7, and Nielsen 1963, Munz and McFarland 1973, McFarland 1985). The rate of change of intensity reaches a maximum at CT (Figure 3). Illumination from moonlight provides intensities at the earth's surface from approximately 0.3 lux during full moon to 0.1 lux for the first and third quarter moons, and about 0.01 lux for a one-quarter disc moon (Figure 1). It is only when the altitude of the moon is above 15 to 20 degrees during twilight that moonlight contributes a significant amount of illumination (Rosenberg 1966). Under these conditions the intensity of moonlight at the earths' surface falls between the intensity of solar illumination at CT to NT. Because twilight is accompanied by a spectral reduction in green to orange-red light (Figure 4), the use of photometric, rather than spectroradiometric, units to express absolute changes in intensities can introduce error. Since the spectral shifts occur slightly before SS and after SR and continue into NT photometric values obtained with an inexpensive photometer provide a

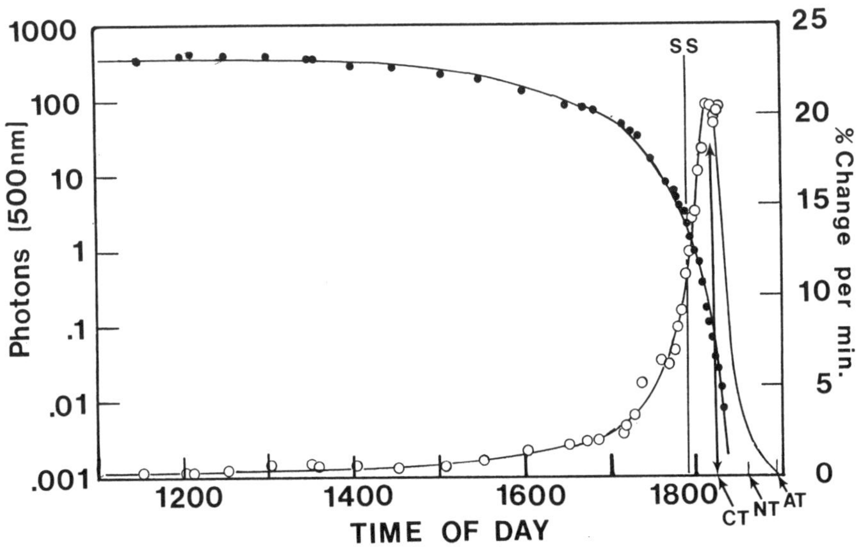

Figure 3 Decline in photons and their rate of change from midday to astronomical twilight at St. Croix. Values of photons at 500 nm were obtained from a series of spectral irradiance scans recorded on October 16, 1976. Filled circles are photons x 10^{12} cm^{-2} nm^{-1} sec^{-1}; open circles are the rate of change of photons min^{-1}. SS = time sunset, CT, NT and AT are the time of civil, nautical and astronomical twilight. The right limb of the percent change curve was generated from the equation analysis program TABLE CURVE (Jandell Corp.). This range was below the spectroradiometers sensitivity. Evaluation of the equation for zero percent change provides fair agreement with the time of AT, when the rate of change should approach zero

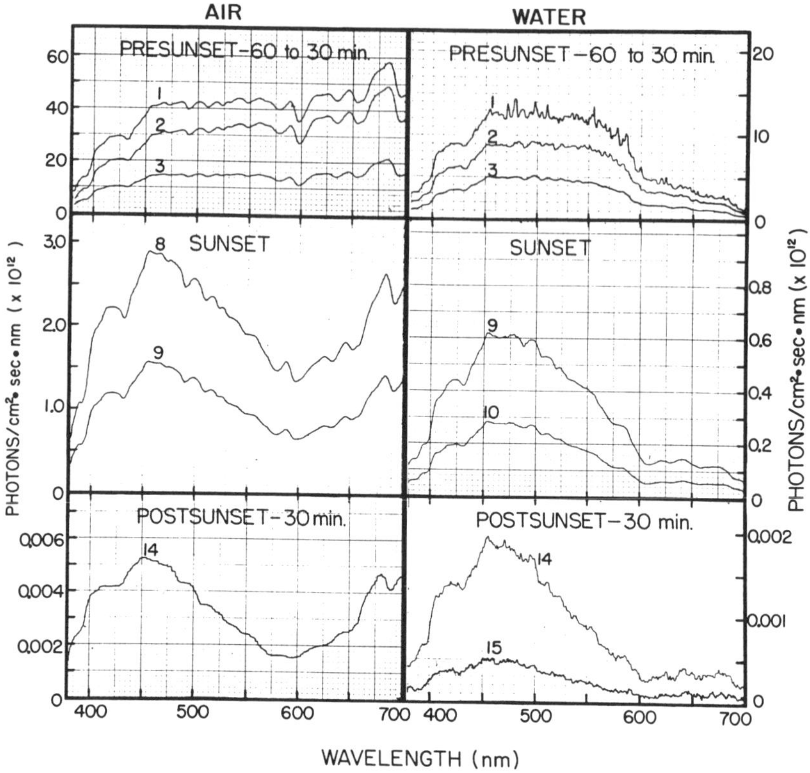

Figure 4 Shifting of the spectral distribution of surface and downwelling light intensity at 3 m depth before and during dusk at Enewetak Atoll, Marshall Islands. Times for each curve are: pre-sunset, 1 = 57 min, 2 = 42 min, 3 = 26 min, 8 = 7 min, 9 = 2 min; post sunset, 10 = 3 min, 14 = 34 min, 15 = 39 min. Reproduced with permission from Munz and McFarland (1973)

useful, if not absolute, measure of the rate of change of intensity around dusk and dawn. The percentage sky polarization under clear conditions reaches 75 percent with the sun just above the horizon, and increases to as much as 85 percent when the sun is 4 to 5 degrees below the horizon (Waterman 1981). Polarization values of 60 percent are typical during the day in clear seas near the surface. The percentage polarization during early twilight, as in air, should reach a maximum (Waterman 1954 and 1981). But as far as we are aware submarine measures of the percentage polarization around twilight have yet to be reported.

Based on this information submarine twilight in coral reefs would have the following characteristics due to their tropical latitudinal range:

1. Photoperiod at the equator varies annually by only 2 minutes. At the northern and southern limits generally accepted for coral reefs (i.e. 20° N and S) photoperiod varies annually by 2 hours and 25 min (from Beck 1968).

2. The duration of CT varies only minimally throughout the range of coral reefs from 21– 23 min at the equator and from 22 – 24 min at 20°, similarly, NT at the equator varies from 45 – 49 min and from 47– 54 min at 20°. Thus annual changes in the rate of change of light intensity are negligible over the range of coral reef ecosystems (see Figures 8.11 and 8.13 in Blaxter 1976).
3. Downwelling underwater light around twilight will be shifted to shorter wavelengths, except, perhaps, under severe overcast skies.
4. Submarine light most probably will be more highly polarized around twilight than during the day.

19.2.2 Fish

The daily migrations and changeover between nocturnal and diurnal species of fish on the coral reef occur during twilight in a predictable fashion according to family, diel habits, and relative size of species (Hobson 1972, Collette and Talbot 1972, Domm and Domm 1973, Galzin 1987, and Figure 5), and over roughly the same range of light intensities at dawn and dusk. The transition between diurnal and nocturnal assemblages is separated by a quiet period at just before SR and just after SS (Hobson 1972), in which there is very little fish activity (Figure 5). Although transition behaviors near dusk and dawn usually are summarized and reported as a single event (the last or first individual observed, as in Figure 5), most species undergo a progressive suite of behaviors the sequence of which can cover a considerable period of time. This is especially so for fishes, such as herrings, silversides, or grunts that migrate from daytime resting sites in, on, or near the reef to nocturnal foraging sites (Hobson 1973), and some parrotfish that after foraging during the day migrate to night resting sites (Dubin and Baker 1982, personal observations at St. Croix). At St. Croix, the sequence of diel transition behaviors associated with twilight is vividly portrayed by copper sweepers *Pempheris schomburghi*, that initiate evening activities before SS and end their precarious migrations across a barrier reef to deeper water after CT (Gladfelter 1979) and by large juvenile French grunts (*Haemulon favolineatum*, McFarland *et al.*, 1979) that begin transitional behaviors around SS and migrate from reefs across seagrass beds to feeding areas at CT (Figure 6). In general the sequence of behaviors at dawn, especially in French grunts, is somewhat truncated. Similar complex sequences of diel transition occur in schooling reef invertebrates, as described later. At Barbados, princess parrotfish *Scarus taeniopterus* exhibit two types of transition behavior at twilight in which individuals either remain on their territory or migrate along fixed routes to a sleeping spot elsewhere on the reef (Dubin and Baker 1982). At least one hour before SS, feeding activity declines and swimming increases. About 20 to 25 min before SS, both territorial and migratory terminal phase males commence cover seeking or migrate, and exhibit bouts of a peculiar 'tail standing' behavior. Interspecific aggression, noticeably elevated when cover-seeking begins, ceases within 5 min

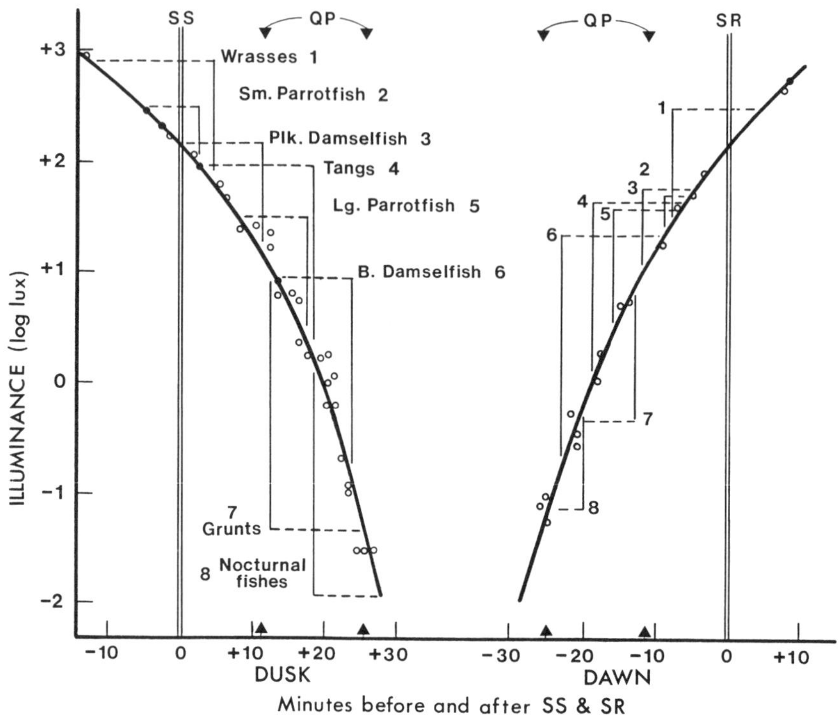

Figure 5 Changes at dusk and dawn in the activities of coral reef fishes at St. Croix, U.S.V.I. Data are unpublished results of G. Helfman and W. McFarland obtained using scuba over five consecutive evenings and three dawns. QP indicates the quiet period during which fish activity in the water above the reef was virtually non-existent, except for the passage of occasional predatory fish (small carangids). The range for each group of fishes represents the intensities (vertical lines) and the times (horizontal dashed lines) over which the last sighting of diurnal fishes and the first emergence of nocturnal fishes at dusk occurred over the five dusk dives; and the descent into the reef of the last nocturnal fishes and the first emergence of diurnal fishes during the three dawn dives. Species observed were: 1. Wrasses – Spanish Hogfish, *Bodianus rufus,* Slippery Dick, *Halichoeres bivittatus,* Bluehead Wrasse, *Thalassoma bifasciatum* 2. Small Parrotfish – Stripped Parrotfish, *Scarus croicensis,* Redtail parrotfish, *Sparisoma chrysopterum* 3. Planktivorous Damselfish – Blue Chromis, *Chromis cyaneus,* Brown Chromis, *Chromis multilineatus*, Bicolored Damselfish, *Stegastes partitus* 4. Tangs – Ocean Surgeon, *Acanthurus bahianus,* Blue Tang, *A. coeruleus*, Doctorfish, *A. chirugus* 5. Large Parrotfish – Bucktooth Parrotfish, *Sparisoma radians,* Stoplight Parrotfish, *S. viride* 6. Bottom Damselfish – Sergeant Major, *Abudefduf saxatilis*, Dusky Damselfish, *Stegastes dorsopunctatus*, Beaugregory, *S. leucostictus,* Three-spot Damselfish, *S. planifrons* 7. Grunts – medium and large juvenile French Grunts, *Haemulon flavolineatum* 8. Nocturnal Fishes – Squirrelfish, *Holocentrus rufus*, Blackbarred Soldierfish, *Myripristes jacobus,* Glasseye, *Priacanthus cruentatus,* Cardinalfishes, *Apogon* spp., mostly the Two-spot Cardinalfish *A. pseudomaculatus*. The startling migrations of the nocturnal coppersweepers, *Pempheris schomburgk*i from and return to discrete sites in the reef, as described by Gladfelter (1979), occur in conjunction with the other nocturnal fishes (see Figure 6). Note the staircase order of the transitional changes at dawn and dusk, and the tendency for the shifts at dawn to occur over a slightly truncated range of intensities

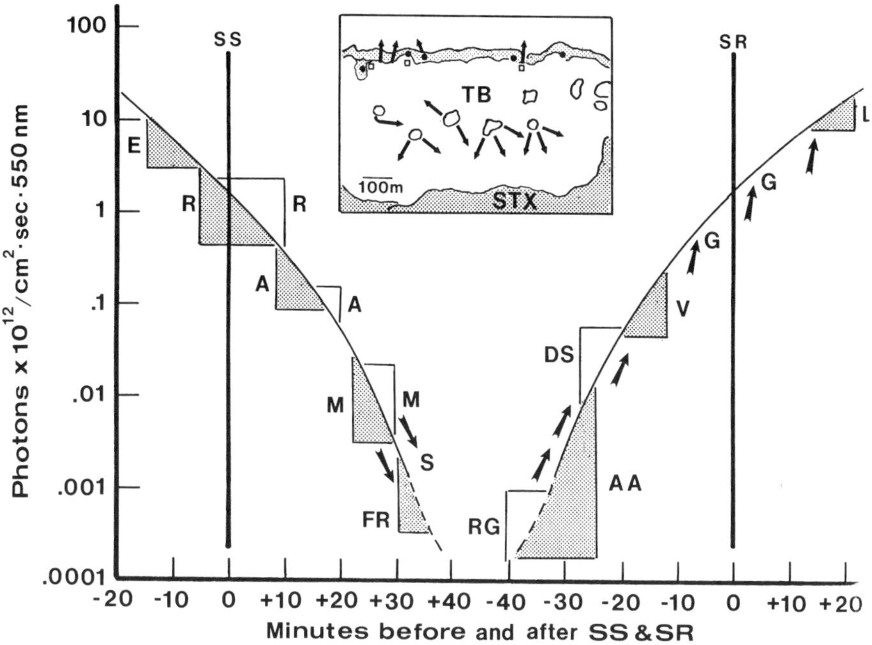

Figure 6 Comparison of the evening and dawn sequential changes in transition behaviors of two migratory teleosts, the copper sweeper *Pempheris schomburgki* and juveniles of the French grunt *Haemulon flavolineatum*. The map inset illustrates a portion of Tague Bay located toward the northeastern end of St. Croix, U.S.V.I. The bay is enclosed by an extensive barrier reef, in which copper sweepers find refuge during the day along the backreef in deep grottos (closed circles) formed primarily by the elkhorn coral, *Acropora palmata*. Within the bay several coral patch reefs provide protective daytime schooling sites for juvenile grunts, of which the French grunt is particularly abundant. Small arrows indicate several of the traditional evening migration routes used by the copper sweepers to gain access to the forereef, larger arrows indicate the traditional routes from the patch reefs traversed by juvenile French grunts to reach the adjacent seagrass beds, where each species forages during the night. The same routes are used at dawn to return to the daytime refuge sites. The triangular areas, located along the light intensity vs time curves, represent the suite of transitional behaviors that the grunts (open areas) and the copper sweepers (stippled areas) display each day. Letter symbols are: SS and SR = time of sunset and sunrise, E = emergence of sweepers from grottos into fronds of elkhorn coral, R = noted agitation of sweepers in and near the fronds and grunts on schooling sites, characterized by a rolling of the body that produces bright flashes of light, A = assembly – a distinctive movement by sweepers over coral sand where they assemble into a definitive school (open rectangles), and the descent of grunts from their day schooling sites on the patch reef to its base. The assembly movements by each species are usually followed by ambivalent attempts to migrate, which are often thwarted by the presence of fish predators. M = actual migrations of each species in the direction indicated by the map arrows and general time and light levels by tailed arrows, FR = last observations of sweepers as they leave the forereef to forage on zooplankton in open water. During this period the grunts seek solitary sites on the bay bottom where they forage on benthic invertebrates. RG = solitary grunts begin to regroup over the grass beds following first morning light, and begin their return migration to the daytime schooling site (DS), AA = first to last morning arrival of sweepers in the vestibules provided by the elkhorn coral leading to the grottos, V = first disappearance of sweepers into day time refuges, G = disappearance of small groups of sweepers, L = last disappearance of sweepers. Graph is a summary of information derived from Gladfelter 1979, McFarland *et al.,* 1979, McFarland and Hillis 1982). The chronology for each species were established from the times of occurrence of each behavior relative to SS and SR. Light intensities represent the values obtained for the French grunt migrations obtained at 3 metres depth. Light intensities for the sweepers were measured at 1-meter depth and in water of higher clarity than in the bay proper, and were about one-half a log unit higher than indicated

before SS as fish select a sleeping site. Several sites may be investigated and repeatedly entered before a final choice is made. By 8 –11 min after SS, all individuals have retired. Immatures and females leave daytime areas earlier (40 to 50 min before SS) than terminal phase males and small princess parrotfish (< 100 to 120 mm SL) do not migrate. Thus the actual array of transition behaviors, as in these examples, occurs over as many as five decades of diminishing or increasing intensities of light and follows a more or less predictable temporal pattern. In these instances, and those described by Hobson and Chess (1973) for the silverside *Pranesus penguis* in the Marshall Islands, predatory fishes can and do interrupt the sequences with attacks. Nevertheless, when a given low intensity is achieved, as in the French grunts (Figure 7), the evening migration proceeds even in the presence of threatening predators. Are specific light levels or changes in light intensity the proximate cause of each event in the behavioral sequence? Surely they must be related, but whether the cause is due to the rate of change, brightness discrimination, or even a reduction in spectral discrimination is uncertain.

19.2.3 Benthic invertebrates

Compared to fishes, the daily transitions in activity and twilight behavior among reef invertebrates has received less attention, particularly with regard to community organization. We assessed the precision of twilight behavior for a few, easily observed species of invertebrates and reef plankton at Tague Bay, St. Croix, U.S. Virgin Islands, and integrated these observations with selected examples for other species from the literature. From this collection of information, we found, with few exceptions, that animals within a taxon did not generally exhibit both strictly diurnal and nocturnal individuals and that, over a wide range of taxa, many reef invertebrates exhibit predictable behaviors at twilight and execute them with a high degree of precision in timing relative to SS and SR. For many species, transitional behaviors coincide with the rapid changes in light intensity at SS and SR, suggesting light as the controlling factor. For some, transitions seem to follow the diel cycle of prey availability while for others, the timing of twilight transitions is best explained as predator avoidance. It should be noted as caveat that rigorously quantified investigations of the daily activities of tropical marine invertebrates is far from complete at this writing.

Sponges. In sponges, water-pumping activity may be constant (*Mycale* sp.) or periodic and asynchronous (*Verongia gigantea*) occurring without regard to recognizable environmental events including twilight (Reiswig 1971). However, in shallow-water populations of the sponge *Tethya crypta*, water-pumping begins at SR and exhalent volumes increase rapidly reaching a maximum within 4 to 6 hours. In the late afternoon pumping declines and, at SS, exhalent velocity drops suddenly (decreased choanocyte flagellar activity). A rapid increase in diameter and slow contraction of exhalent pores compensates resulting in a gradual

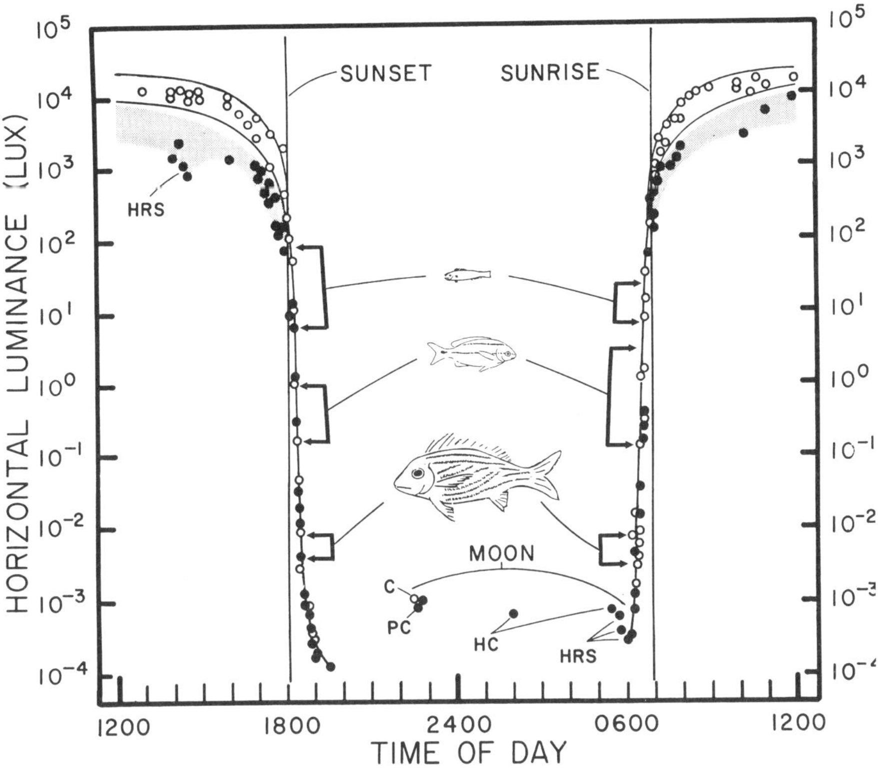

Figure 7 Underwater horizontal luminance at 1 meter depth during a clear 24 hour period, 19 – 20 January and during an overcast day 21– 22 January, C = clear, HC = heavy overcast, HRS = heavy rain showers, PC = partly cloudy, lines and open circles are for clear skies, stippled area and filled circles are for overcast skies. Values bracketed by MOON are for period of moonrise during a full moon on the night of 19 – 20 January. The fishes from the smallest to largest represent the three size classes of juvenile French grunt *Haemulon flavolineatum*. The brackets associated with each size class are the light levels over which each size of grunt migrate from and back to its daytime schooling sites, even when threatened by predators (see text for details)

decline in exhalent volume throughout twilight and into the night. The synchronized diurnal cycle of water-pumping activity in *T. crypta* coincides with periods when water circulation in the habitat are highest. In nature, illumination at night (flashlight) inhibits predawn contraction suggesting that the flagellar activity cycle is controlled by changes in ambient light.

Cnidarians. Among anemones and hermatypic corals, polyp activity is typically nocturnal, but strongly influenced by the capacity for photosynthesis (Johannes and Tepley 1974, Pearse 1974, Sebens and DeReimer 1977, Lasker 1979). Species that lack symbiotic zooxanthellae such as the anemones *Bunodactis sp., Telmatactis cricoides*, and *T. vernonia*, extend their tentacles and feed only at night, whereas some species with zooxanthellae extend polyps during the day as well as night (*Condylactis gigantea* and *Stoichactis helianthoidies*). Among corals in the Caribbean, Porter (1974) found that 10 of 30 species scored for polyp expansion remained fully extended throughout the day and night. On shallow reefs around the Banda Islands in Indonesia, about 20% of corals had open polyps during the daytime (Sweeney 1976). At Tague Bay, *Porites porites* and *P. astreoides*, species that remained extended at all times, showed no discernible changes during twilight. In contrast, polyps of corals that extended only at night including *Dichocoenia stokesii, Diploria strigosa, Eusmilia fasciata, Meandrina meandrites* and *Montastrea cavernosa*, showed strong changes in activity near SR and SS (Figure 8).

Along the Caribbean coast of Panama, Lasker (1979) described a shallow water (< 20 m) diurnal morph of *Montastrea cavernosa* in which large numbers of polyps remained partially expanded during the day. Strictly nocturnal morphs of *M. cavernosa* could be induced to expand during the day by temporary shading. Diurnal morphs, however, remained expanded unless their zooxanthellae were bleached or lost. Expansion reappeared when bleached polyps regained pigmentation. Experimental application of DCMU, an inhibitor of photosynthesis, did not reduce daytime expansion suggesting that zooxanthellae do not control polyp responses to light on a day to day basis (Lasker 1979). Instead, polyps may express a circadian day-time expansion in response to resident zooxanthellae (Pearse 1974). The full expansion of polpys during twilight coincides with the increased availability of zooplankton and higher feeding rates at twilight (Glynn 1973, Porter 1974, Johannes and Tepley 1974). The contraction of polyps at dusk in a few strictly diurnal corals such as *Tubipora musica* presents an interesting exception (Sweeney 1976). Using a laser based measurement system *in situ* on the Great Barrier Reef, Vago *et al.,* (1997) demonstrated that skeletal extension in the corals *Acropora grandis* and *Porites lutea* follows a strong diel pattern centered on twilight. Linear extension of the skeleton increased in the late afternoon from levels of < 5 μm hr^{-1} (limits of laser detection) to peak values as high as 35 μm hr^{-1} about 30 min before SS and remained high for two to three hours before declining to 15 μm hr^{-1} and ceased soon after SR. The interactions between symbiotic zooxanthellae, skeletal calcification in corals and environmental factors are not well understood (Gladfelter 1985), but this example points to the influence of photic cycles in the environment on basic physiological processes, such as growth, as well as on behavioral activities around the times of dusk and dawn.

Among gorgonians in which polyps are extended principally during the daytime transition activity occurs near twilight (Wainwright 1967). For many species, responses are most dramatic at dawn or shortly thereafter when the proportion of extended polyps, near 0% at 0400 hrs, rapidly increases; a decline in activity shortly before dusk is followed by a second, lower activity peak between 2000 and midnight (see Figures 2 – 5 in Preston and Preston 1975). As with corals, changes in polyp activity among gorgonians, including hermatypic

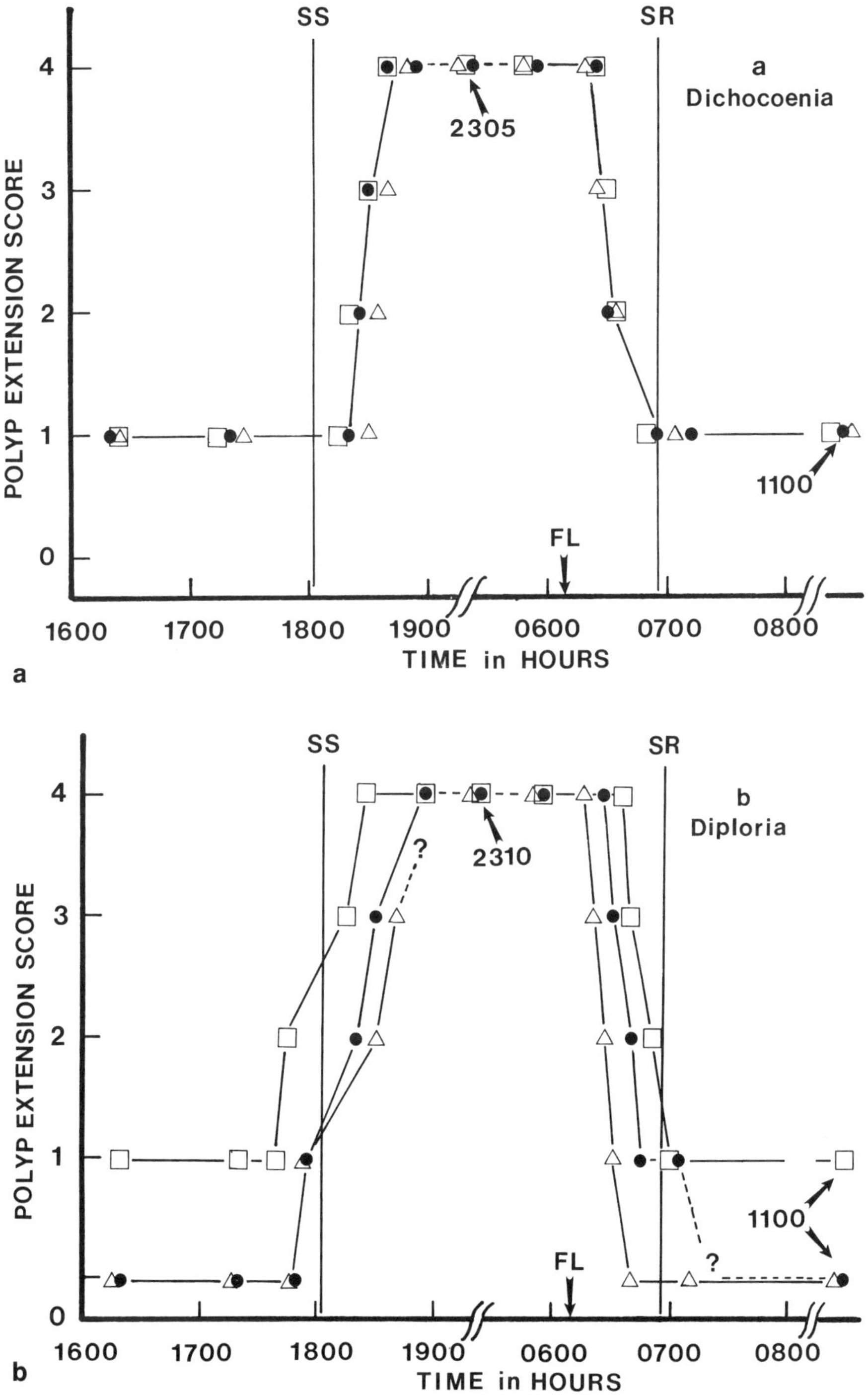
SS
SR
a
Dichocoenia
2305
1100
FL
POLYP EXTENSION SCORE
0
1
2
3
4
1600
1700
1800
1900
0600
0700
0800
TIME in HOURS
a
SS
SR
b
Diploria
?
2310
1100
?
FL
POLYP EXTENSION SCORE
0
1
2
3
4
1600
1700
1800
1900
0600
0700
0800
TIME in HOURS
b

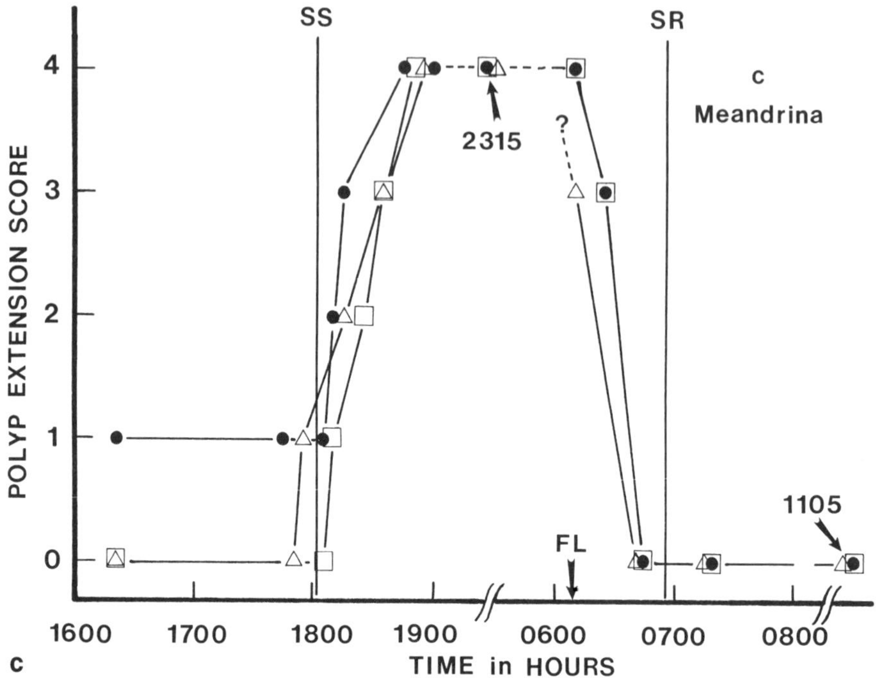

Figure 8 Polyp extension and retraction of different colonies for three species of scleractinian corals over a 24-hour period. Polyp extension scores are: zero = polyps retracted with tentacle tips just barely visible if at all, 1 = tentacle tips visible, 2 = tentacles flush with skeletal ridges, 3 = tentacle tips above ridges, 4 = polyp tips fully extended. Filled circles, open triangles and open squares are scores of three separate colonies of each coral species. 6a *Dichocoenia stokesii*, 6b *Diploria strigosa*, 6c *Meandrina meandrina*. FL = first light

species, seem to coincide with temporal patterns of plankton abundance (Figure 9). Variations in diurnal activity patterns between species suggest niche separation and varying dependence on zooxanthellae (Preston and Preston 1975). We attempted to document the precision of such changes in colonies of *Muricia* sp. at Tague Bay, but found that polyp expansion varied day and night without a discernable temporal pattern.

On the reef flat at Galeta, Panama and Sao Sebastiao, Brazil, Sebens (1977) observed that polyps of mat forming zooanthids *Zoanthus sociatus* and *Z. solandri* were continuously expanded day and night. Periodic closures occurred in response to tides and disturbance, and did not appear to follow a diel cycle. In contrast, *Palythoa caribaeorum* and *P. variabilis* showed clear diel cycles of nocturnal activity with changes in polyp extension occurring at dusk and dawn (see Figures 2 and 3, Sebens 1977). In Tague Bay, St Croix, the percentage of extended polyps in colonies of *Z. sociatus* and *Z. solandri*, as well as *Palythoa caribaeorum*, was always high in the daytime yet increased significantly at night

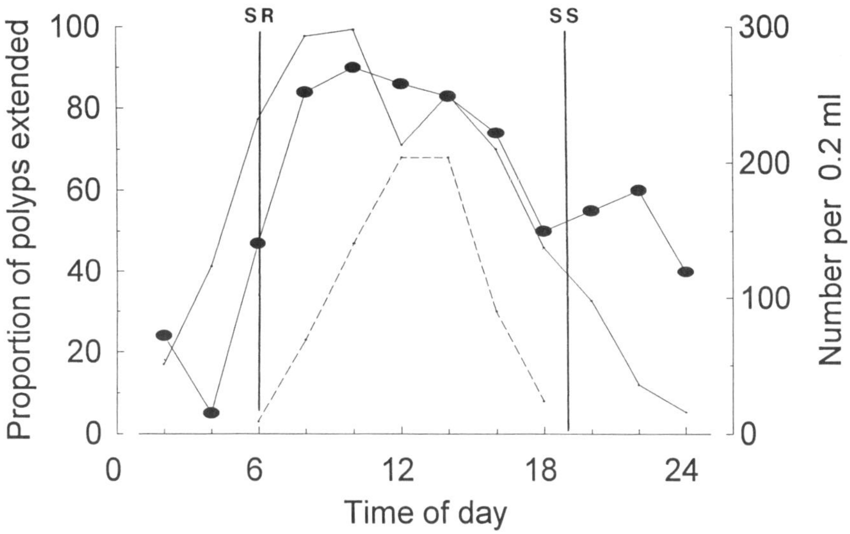

Figure 9 Composite diel activity pattern for 15 species of gorgonians at Puerto Rico. Solid circles represent the proportion of polyps extended, summed for all species at each time interval. Solid line gives the mean abundance of copepods at different times of the day. Dashed line represents the diurnal light cycle. Redrawn from Figure 6, Preston and Preston (1975)

(Figure 10). Transitions in polyp activity centered around twilight: extension of individual polyps commenced at least 15 min before SS and retraction at 30 min before SR.

In the anemones *Lebrunia coralligens* and *L. danae* the timing of diel transition at twilight is discrete for diurnal symbiont-rich pseudo-tentacles and regular tentacles that extend only at night (Figure 11). In the laboratory, differential responses to light and darkness can be induced independent of the time of day (Gladfelter 1975); however, the graded extension and retraction of the two polyp types scored during prolonged exposure to specific light intensities in the laboratory broadly overlapped, unlike behavior in the field. A similar differentiation between tentacles, which extend at dusk and retract at dawn, and the modified tentacles termed 'bubbles', that enlarge at dawn and deflate at dusk, is described for the coral *Plerogyra sinuosa* (Fricke and Vareschi, 1982). The bubbles contain high concentrations of zooxanthellae that generate oxygen photosynthetically in response to changes in light intensity throughout the day. When placed in darkness during midday the tentacles expand and the bubbles deflate; illuminated at night the tentacles retract and the bubbles expand, indicative of a primary control of the movements by light. Bubbles exposed to full sunlight on the upper surface of *P. sinuosa* colonies are less expanded than shaded bubbles, and the expansion size of shaded and non-shaded bubbles increases with depth;

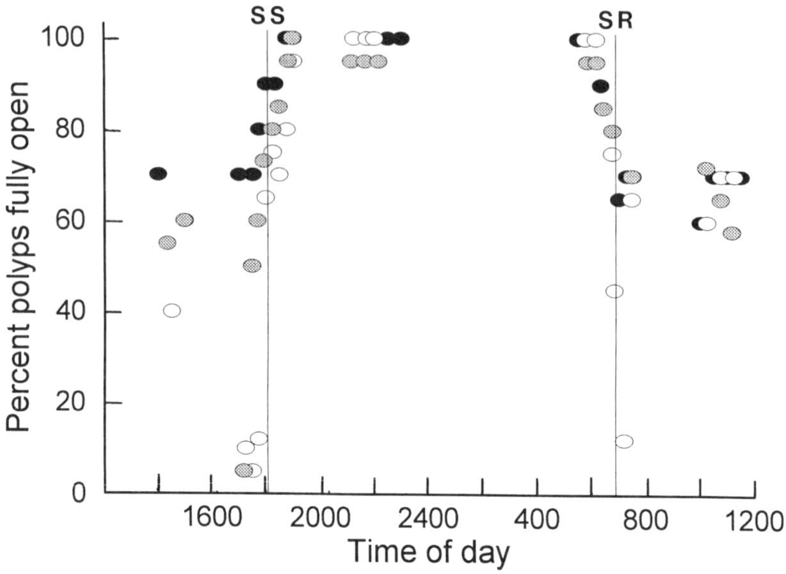

Figure 10 Percent of polyp expansion in three colonies of *Palythoa caribaeorum* (symbols are indivd-ual colonies, n = ca. 1,000 polyps per colony). The closure of most polyps in two of the colonies near SS and SR likely resulted from sand disturbance

behaviors that correlate with an increased capacity for photosynthesis as light intensity decreases with depth and shading.

Annelids. Although many sessile tubicolous and free-living marine polychaetes display short-term cyclic events such as changes in the rates of ventilation (Wells 1950) and, in some species, diel rhythms of metabolism (Magnum and Miyamoto 1970), the sessile polychaetes that we observed lacked any discernable diel pattern of tentacle activity. *Sabella pavonnina* and *Sabellastarte magnifica* were fully expanded on 39 out of 42 observations and *Spirobranchus giganteus* on all occasions from predawn to 2200 hours. To assess possible changes in activity of the terebellid, *Eupolymnia nebulosa*, the length and orientation of the feeding filaments and the rate of food transport along the filaments were determined at different times of the day. No diel trend existed. Food particle transport rates along the tentacle grooves varied from 0.03 and 0.33 cm sec^{-1}. But all *E. nebulosa* varied this much at all times of the day. In the laboratory, Gremare (1988) observed that tentacular activity increased ca. 25% at SS (1800 hrs) and reached a maximum prior to nocturnal emergence between 2100 and 2200 hrs (Gremare 1988). Manipulating the light dark cycle showed that the onset of light immediately induced a reduction in activity suggesting that light is probably the exogenous cue which controls activity rhythms in this species (Gremare 1988). The coral-eating fireworm *Hermodice carunculata* (anphinomid polychaete) is

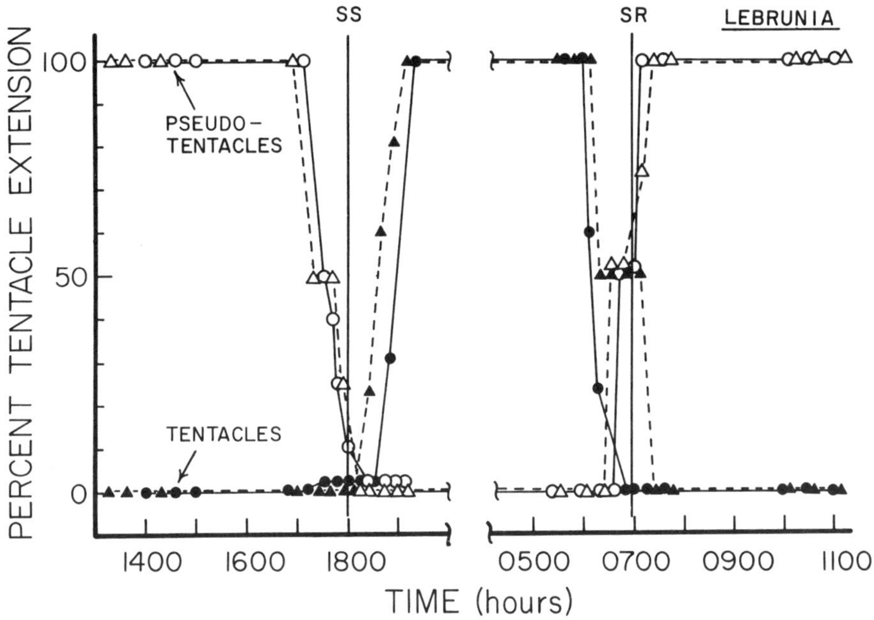

Figure 11 Percent tentacle extension in two species of sea anemones, *Lebrunia coalligens* (circles) and *L. danae* (triangles) over 24 hours. Open circles and triangles are the pseudotentacles that contain high numbers of photosynthetic zooxanthellae, filled circles and triangles the feeding tentacles that contain nematocysts

inactive most of the day and night (Marsden 1962, Ott and Lewis 1972). It leaves its burrow only in the late afternoon ca. 1500 hrs and returns after feeding at or before SS. Although *Hermodice* may be out in the early morning before 0900 hrs, it was never seen at night.

Crustaceans. Reef crustaceans exhibit a wide variety of responses to twilight conditions. The stomatopods (mantis shrimp) have well developed photopic vision and are generally active only during the daytime. At dusk, they seal themselves in burrows for the night. In two species, *Gonodactylus oerstedii* and *G. spinulosus*, individuals emerged for longer periods of time and moved further from their burrows at dawn and dusk than during midday (Dominguez and Reaka 1988). Food availability and foraging success were not related to time of day, but the increased activity observed at twilight coincided with the quiet period when the number and size of diurnal invertivorus fish was lowest.

Focal observations of the Caribbean arrow crab *Stenorhynchus seticornis* throughout the diel cycle revealed that most crabs remained in contact with stony corals during the daytime in protected crevices, holes and overhangs. Near dusk between 1700 and 1900 hrs crabs emerged onto the exposed surface of stony corals. Over a short period between 1900 and 1930 hrs, crabs ascended a meter or more to nocturnal perches on the branches of soft corals and gorgonians (Barr 1975). At 0500 to 0600 hrs, crabs descended and returned to concealment. Arrow crabs are opportunistic scavengers that feed mostly by picking detritus off their body surface. Self-cleaning is most intense in the later afternoon (1700 –1800 hrs) and ceases prior to migration. Barr (1975) speculates that diel migration and nocturnal exposure of non-feeding individuals facilitates the collection of detritus on the body. In daytime feeding experiments, fish rejected intact arrow crabs after contacting the legs, but readily consumed carapaces after legs were removed. Thus, the spindly legs, sequestering behavior, and the timing of diel migration may deter attacks by diurnal fish predators.

The spiny lobster *Panulirus argus* does not emerge from daytime dens to forage until one or two hours after SS, although locomotor activity increases significantly just before SS (Kanciruk and Herrnkind 1973) and they often appear at the entrance to their dens in the late afternoon. During the autumn this nocturnal activity pattern is interrupted when lobsters undertake extensive diurnal migration (Herrnkind *et al.,* 1975).

Echinoderms. Responses of echinoderms to light have been intensely investigated, particularly for the largely tropical genus of the sea urchin *Diadema* (Millot, 1955, see Reese 1966 and Yoshida 1979 for reviews). In the Caribbean, the activity of *Diadema antillarum*, like many shallow-water echinoderms, is largely nocturnal (Crump 1965, Reese 1966, Magnus 1967, Clark 1976, Hammond 1982, Hendler 1982, 1984a and b, Meyer 1982). As night falls and diurnal predators retreat (principally trigger fish), *Diadema* emerge from hiding and migrate to feed on algae and seagrasses (Thornton 1956, Sinclair 1959, Ogden *et al.,* 1973). Increased activity in late afternoon and early morning suggest that diel transition may begin long before SS or SR, although the urchins moved most rapidly during twilight (Figure 12). The detritus feeding urchins *Meoma ventricosa* and *Clypeaster rosaceus* are largely nocturnal and show changes in locomotory and feeding activity at dusk and between first light and SR (Hammond 1982). In *M. ventricosa*, from poorly aerated sediments, emergence at dusk and burrowing at dawn were coincident with the diel cycle of sediment oxygen tensions. Dusk emergence did not occur in coarse, well aerated sediments (Chesher 1969a, Hammond 1982). In the Caribbean, *Cassis tuberosa*, a common large gastropod predator on sea urchins, follows the general activity pattern of its prey; hunting begins at dusk (1800 hrs) and ceases before dawn (0600 hrs) (Hughes and Hughes 1971).

Like echinoids, tropical asteroids and ophiuroids are mostly nocturnal. Morin (1988) in observing the twilight activity patterns of fish and echinoderms around a coral reef in the central Philippines noted that *Luidia magnifica*, a large asteroid that preys on sea urchins, appears on the reef at SS but does not become active until relatively late in the quiet period ca. 30 min after SS and returns to cover within 40 to 15 min before SR. The sandstar *Astropectin* is another asteroid that buries in sandy areas adjacent to coral reefs. *Astropectin polyacanthus* from

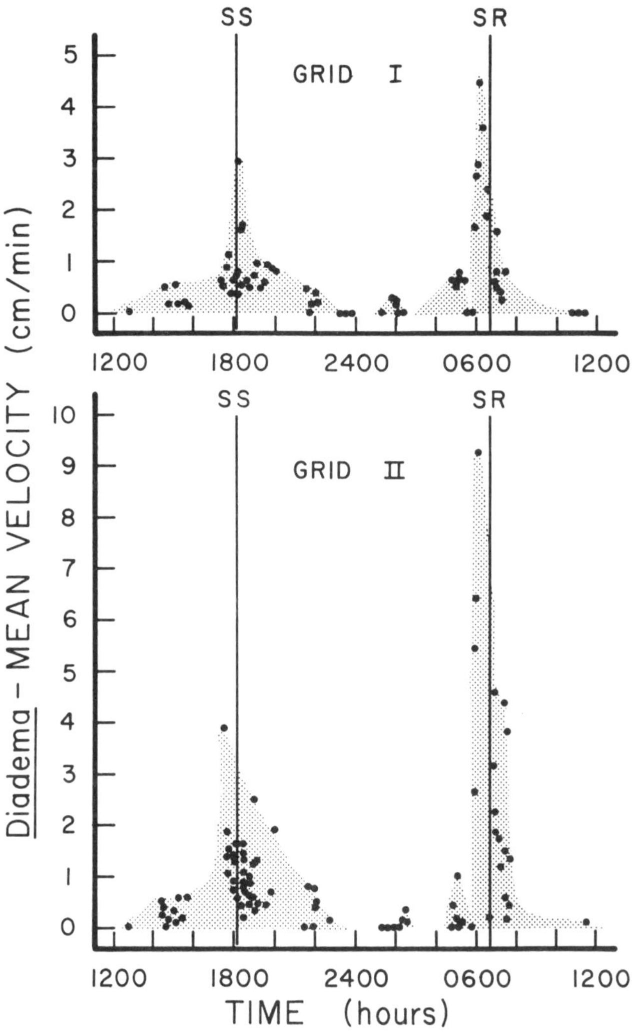

Figure 12 Mean velocity of movement for *Diadema antillarum* at different times over three 24 hour periods. Grids I and II are for separate 6 m × 6 m grids. Each filled circle is the average rate of movement for between 4 and 24 urchins measured over three days

southern Japan exhibits two peaks of daily locomotory activity at dawn and at dusk that broaden during periods of experimental starvation (Mori and Matutani 1952). Exposing individuals to sudden decreases in illumination at different times of the day revealed that *A. polyacanthus* was most responsive to changes in illumination near the time of SR and SS. At normal low densities, the coral-eating asteroid *Acanthaster planci* behaves as a nocturnal predator; however, at high population densities. *A. planci* does not return to shelter at dawn, but continues feeding throughout the day (Chesher 1969b). Thus, twilight behaviors may be shaped not only by prey availability and changing light conditions but also by conspecific effects of population structure.

In the Philippines, *Ophiarachna incrassata*, a large piscivorous ophiuroid, emerges from reef crevices 10 –15 min after SS and remains active until 15 – 20 min before SR (Morin 1988). During the twilight changeover, *O. incrassata* position themselves near shelter sites used by fishes and take up a distinctive posture with the oral disc elevated a few cm above the substrate. A rapid body-spiraling enables *O. incrassata* to capture small fish (damsels, cardinals) that mistakenly seek shelter in the gap beneath its disc. The nocturnal activities of smaller deposit or suspension feeding ophiuroids begin later, 30 to 40 min after SS, and end before SR (Morin 1988, Sides and Woodley 1985, Hendler 1984a and b). In the Caribbean, *Ophiocoma echinata* and *O. pumula* extend their arms to feed only at night when diurnal ophiuroid predators, particularly wrasses, are absent (Hendler 1982). Protraction of the arms from crevices occurs rapidly over a period of 30 to 40 min beginning ca. 10 min before SS. Retraction starts 50 to 60 min before SR and is complete by SR (Figure 13). In *Ophiocoma wendtii*, another Caribbean ophiuroid, individuals may be poised at crevice openings before SS, but do not complete emergence until 30 to 40 min after SS (Sides and Woodley 1985). At emergence, *O. wendtii* is distinctly paler in coloration than at midday. In captive individuals, diel changes in coloration occurred around the time of dusk (1700 to 1800 hrs) and dawn (0630 to 0700 hrs) (Hendler 1984b), similar to changes that occur in French grunts at dusk and dawn (McFarland *et al.*, 1979). Paling is accompanied by an increased sensitivity to light which may help exposed individuals avoid predators especially in moonlight (Hendler 1984b). The basketstar *Astrophyton muricatum* makes a twilight ascent to feeding posts atop the reef just after SS. Individuals, observed directly and with remote cameras, appeared within the same 5 –10 min period and followed the same route on successive evenings taking ca. 15 to 20 min to reach the feeding post at about 40 min post-SS (Macurda 1976, Hendler 1982). Thus, the diel transition behaviors of many nocturnal tropical echinoderms are temporally predictable and center on twilight. Their punctuality corresponds not only to specific levels of illumination and rates of change in light intensity but, in some instances, to the relative abundances of potential prey, or predators, as in the case of small ophiuroids to the absence of wrasses after CT and before SR (Hendler 1984a and b, Morin 1988).

Among the holothuroidea or sea cucumbers, behavior at twilight is varied depending on the species pattern of locomotory and feeding behavior (Yamanouchi 1939 and 1956, Berrill 1966, Hammond 1982). As with previous examples, diel transition behaviors are most distinct among cryptic species that feed only at night. In Discovery Bay Jamaica, *Euapta lappa* remain hidden during the daytime and leave concealment rapidly after sunset over a period of 15 – 20 min

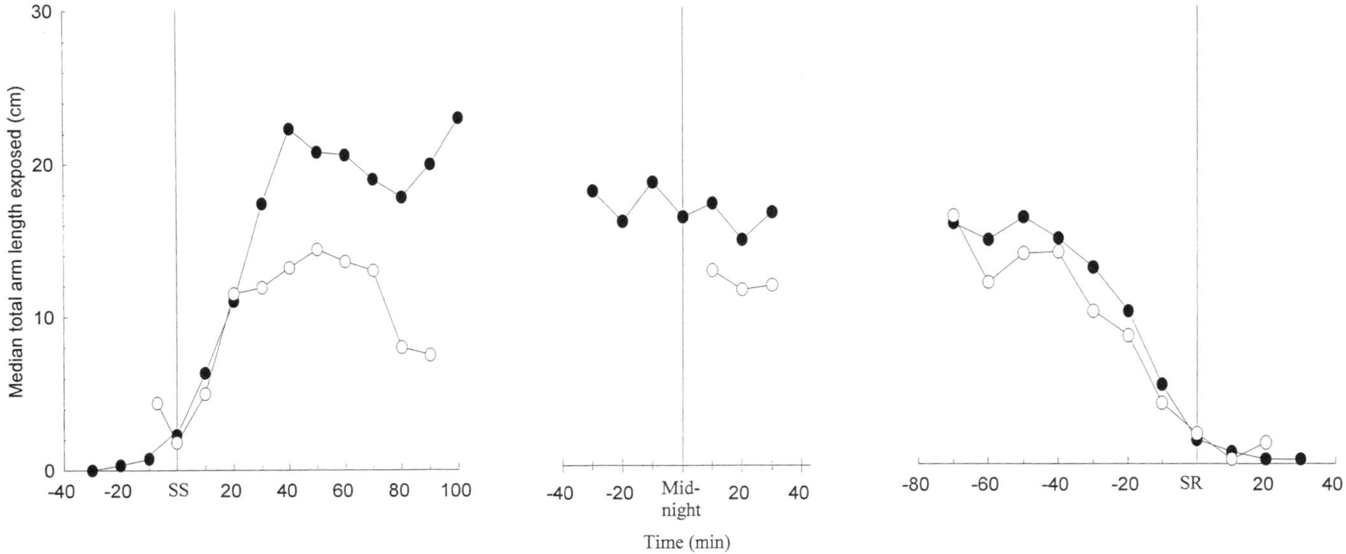

Figure 13 The median total arm length exposed by individuals of *Ophiocoma echinata* (solid circles) and *O. pumila* (open circles) before and after sunset (SS), midnight, and sunrise (SR). Redrawn from Figure 5, Sides and Woodley (1985)

between the hours of 1915 and 1945 and return before first light at about 0345 – 0415; *Holothuria thomasi* and *Actinopyga agassizi*, species that are not strictly nocturnal, emerge throughout the afternoon with a peak immediately following the fall of darkness and return to crevices or become inactive at SR, although a few *H. thomasi* remained out in late morning. Among holothuroids that remain more or less exposed throughout the 24-hour cycle, *H. mexicana* and *Isostichopus badionotus* exhibit diel rhythms of activity in which feeding and movement increase throughout the afternoon, reach a peak at or before midnight and then gradually diminish toward mid-day as individuals become inactive without significant changes at dawn or dusk. *H. arenicola*, a strictly infaunal species, lacks any apparent diel rhythm of activity. In Hawaii, the giant synaptid *Opheodesoma spectabilis* spends the daytime curled within the beds of the alga *Sargassum* or among coral rubble (Berrill 1966). Within 1– 2 hours of SS, 80% of *O. spectabilis* aggregated in a *Sargassum* bed had moved to the surface of the alga to feed and, by SS, 95% were at the surface engaged in tentacular feeding, with the greatest activity from shortly before SS to about 2100 hrs. Gradually, following SR and throughout the morning, animals plunged back into the *Sargassum*; 10% of the aggregation disappeared just after SR, 50% by mid-morning, and by noon, only 2 – 20% were still exposed. Thus, as a group, holothurians exhibit varied patterns of response to changing illumination at twilight and, unlike other reef inhabitants, populations of some holothurians exhibit diel transition behavior over considerable periods of time. A causal relationship between susceptibility to predation, toxicity, and rates of diel transition is not easy to establish, although holothurian toxicity is purported to be a major predator deterrent (Bakus 1968 and 1974, Hammond 1982). Diel rhythms and varied twilight behavior observed among some species in modern reef communities, in this case among holothurians, may represent ecological vestiges of past selection pressures (Bakus 1968, Hobson 1974).

19.2.4 Reef plankton

Many plankton that inhabit the reef during the daytime swim freely at night, including species of hydromedusae, polychaetes, cumaceans, zoea, ostracods, amphipods, copepods,and mysids (Emery 1968, Larson 1986). The appearance of zooplankton in the water column at night including species rarely seen in the daytime is well documented (Thomas and Jelly 1972, Hobson and Chess 1976 and 1978, Porter *et al.,* 1977, Aldridge and King 1980, Robichaux *et al.,* 1981, Ohlhorst 1982). At Tague Bay the number of zooplankters rose steeply within 30 min post SS, and peaked one or more hours later, when maximum darkness occurred (Figure 14; see also Figure 3 in Hamner and Carleton 1979, Figures 3.16 and 3.17 in Sorokin 1993). On evenings when the moon was at 1st quarter or full during twilight, the numbers of zooplankton remained low (Figure 15). The increase in reef plankton following moonset and their suppression following

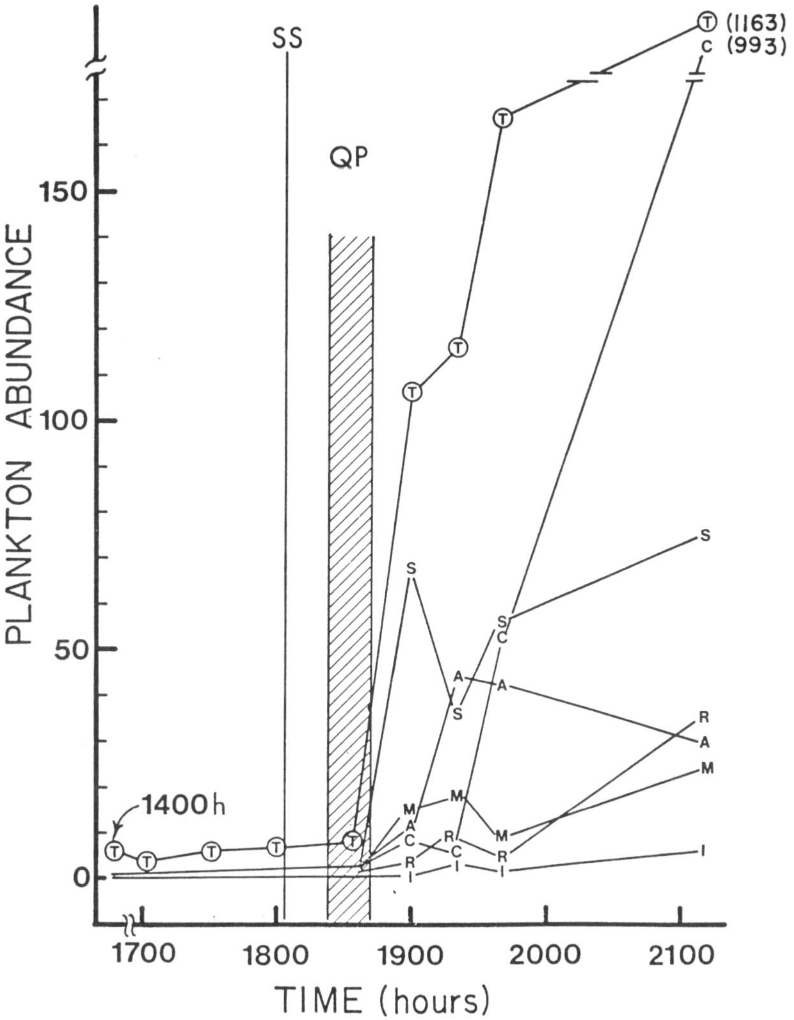

Figure 14 Changes in the abundance of total reef plankton and six major taxa for the night of 22 – 23 January. T = totals for all categories, C = copepods, S = shrimp (mostly mysids), R = crab larvae, A = amphipods, I = isopods, M = miscellaneous, QP = quiet period (see Figure 5). Values for individual categories are not indicated before 1900 hours

moonrise suggest that emergence is directly light-related, and coupled to intensities less than that of moonlight (Aldridge and King 1980). Presumably descent occurs before dawn and may begin as early as 0200 – 0400 hrs (Aldridge and King 1980, Ohlorst 1982, Sorokin 1993).

Although the abundance of reef plankton increases significantly after SS, not all taxa exhibit the same diel pattern. For example, Ohlhorst (1982) observed that

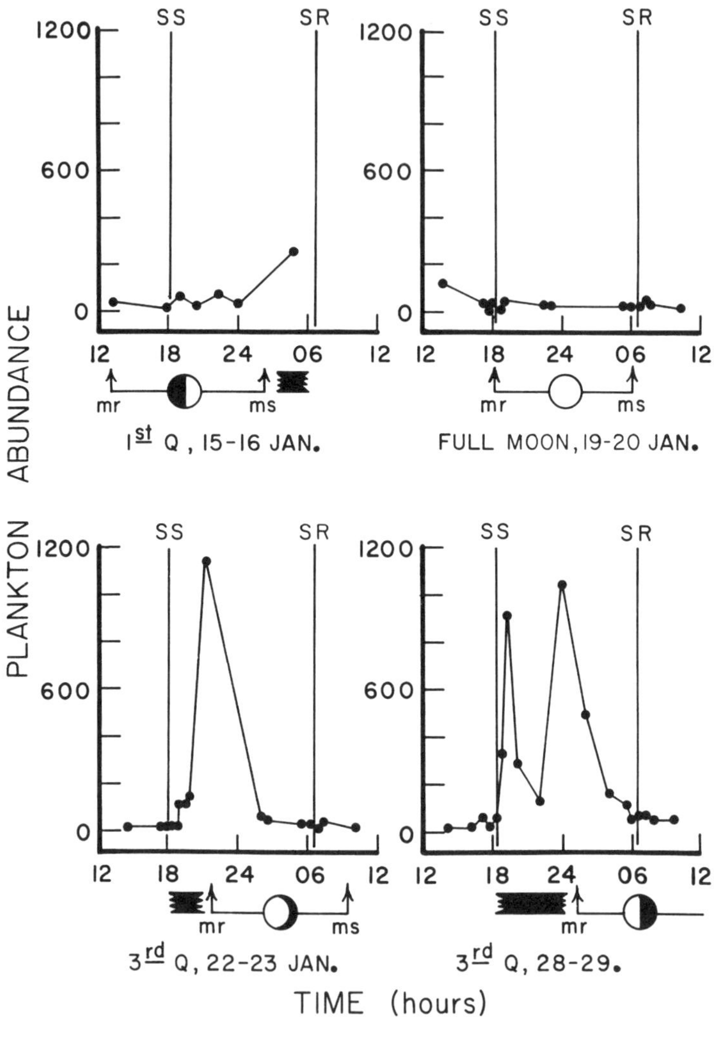

Figure 15 Total net-collected zooplankton abundance for all taxonomic categories (see text and Figure 14) over four 24-hour sampling days. Volume sampled for each datum point = 3.7 m^3 along a 75 m transect at 1 m below the surface. Symbols represent phases of the moon, times of moonrise = mr and moonset = ms, and darkness (black bar) during each sampling day

polychaetes were most abundant in the first hour after SS and isopods in the second. Similarly, the bioluminescent displays of syllid polychaetes begin about 20 min after SS and continue for about an hour whereas bioluminescent displays of emergent ostracods (*Kornickeria hastingsi* and four other undescribed Caribbean species) begin slightly later, 35 – 40 min after SS, and continue for about 45 min (J. Morin, personal communication). Thus, subtle differences exist in the timing of diel transition and twilight behavior among different taxa (or species) of reef plankton. Methods of collection and geographic location may also affect diel patterns of abundance (Aldridge and King 1980, Hobson and Chess 1979, Olhorst 1982).

The diel transition of many plankters is complex. During the day the hydromedusae *Olindias tenuis* attach to seagrass with their tentacles contracted. At dusk (0 to 15 min after SS) individuals swim upward briefly with their tentacles still contracted and then sink to the bottom. Within 45 min to 1 hr after SS the medusae extend their tentacles and become planktonic (Larson 1986). Captive medusae placed in darkness during daytime swam freely, but swimming (and buoyancy) could be inhibited repeatedly by switching a 100 watt lamp on or off. Thus, swimming responses seem to be under the direct control of ambient light.

A number of zooplankters form dense swarms near the reef during the daytime and disperse into the plankton at dusk, including several species of reef mysids and the pelagic copepods, *Acartia australis, A. bispinosa, A. tonsa, Oithona nana,* and *O. oculata* (Emery 1968, Hamner and Carleton 1979). On Tague Bay backreef we observed three species of mysids through twilight. In *Mysidium gracile,* which forms discrete diurnal schools in coral grottos (Emery 1968), schools broke up into small groups of 50 to 100 individuals and left daytime schooling stations 1 to 6 min before SS (n = 4 schools). These highly polarized groups remained 5 to 15 cm above the reef as they travelled 1 to 10 m toward the reef crest. At 6 to 8 min after SS, they halted and withdrew into shelter, where individuals dispersed and continued to swim 3 to 5 cm above the substrate for an hour or more. Other groups that remained on the reef did not migrate, but expanded and then dispersed in the immediate vicinity. Individual mysids remained within 30 cm of reef structure. Expansive shoals of *M. columbiae* that occur during daytime over sand had already fragmented into smaller shoals (1000 – >2000 individuals) and condensed at the base of the reef 30 min before SS (n – 5 shoals). At discrete intervals each shoal rose in the water column, and around 40 – 50 min after SS, expanded and dispersed (Figure 16). A third unidentified species of mysid was observed on five consecutive evenings. Small diffuse schools swam just below the sea surface during the daytime, coalescing into large polarized groups (~200 – 300 mysids) near the tips of *Acropora palmata* 40 to 50 min before SS. Within 16 to 38 min before SS the school departed the *Acropora* and swam toward the reef crest. On each evening, the mysids followed the same general route for a distance of 7 to 10 m where, just after SS, they descended to a depth of ca. 30 cm (midwater) and split into two groups. Each group moved a short distance to a coral head and remained in place until approximately CT, then gradually dispersed.

The diurnal mode of swimming among these mysids was characterized by pale coloration and a high degree of orientation to other mysids, water motion, and reef structure. In the final stage of nocturnal emergence into the plankton, all

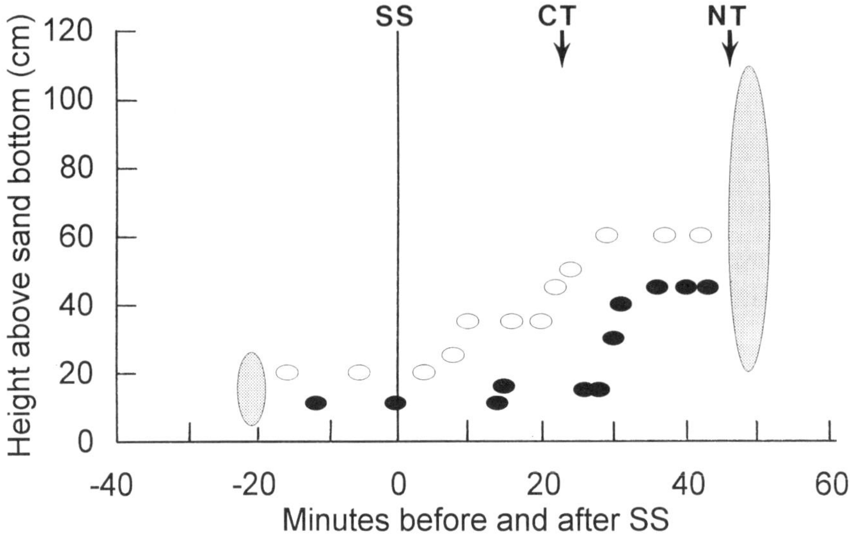

Figure 16 Dusk migratory movements of the mysid *Mysidium columbiae*. The transition behavior is characterized by a sequence of horizontal movements that ended in ascent and general dispersion of the mysids above the reef. The small and large stippled ellipses indicate the depth range of the mysid shoals above the substrate at the beginning and the end of the transitions. The small open and closed ellipses represent the average depth at which two shoals occurred during the transitions (see text for details)

three species of mysids darkened, and their eyes dark adapted as in *Neomysis integer* (Hallberg *et al.,* 1980). They shifted to a very slow gliding mode in which individuals simply floated away from one another. Similar transition behaviors and species specific nocturnal migratory behavior have been described for mysids at higher latitudes during twilight (Clutter 1969, Macquart-Moulin and Maycas 1995). Graded changes in schooling, locomotory mode, and pigmentation can be induced during the day by placing mysids in darkness.

We observed four diurnal schooling sites at dawn. At two sites, schools of *M. gracile* reformed as individuals descended from the water column over a period of 15 min beginning 9 to 11 min before SR. In the other two locations, *M. columbiae* appeared 22 min before SR and returned to daytime locations within 7 to 8 min, except for a few tardy individuals that delayed arrival until 8 min after SR.

The effect of moonlight during twilight on the timing of mysid migrations was not apparent. Although moon phase and rise time varied on different nights, the timing of migrations was consistent relative to SS for new and full moon and for different times of moon rise during twilight. The effect of moonlight on the final dispersion of mysids into the water column was less clear. *M. gracile* that dispersed at SS in an outdoor tank regrouped when moonlight illuminated the tank at 2000 hrs. In nature, observations suggest that loose aggregations (drifting individuals) persisted longer after SS and remained deep when there is moonlight.

19.2.5 Light as a trigger for diel transitions

What characteristic(s) of light, if any, might trigger diel changes? Two hypotheses have prevailed: 1) that the rate of change of light intensity triggers transition behavior(s), and/or 2) that a threshold intensity induces these changes (see Segal 1970). Each attribute can affect behavior (as in urchins, Yoshida 1979) and they need not always be mutually exclusive. In some species, especially fishes such as grunts, both changing brightness and hue discrimination abilities may be involved. Changes in the underwater polarization field that occur around twilight (Waterman 1981), might also act as a signal for polarization-sensitive animals, particularly where water clarity is high (as in coral reefs). How then do the transition behaviors of coral animals fit these hypotheses?

Resting schools of silversides in the Marshall Islands form tight groups in the late afternoon in preparation for evening migration to and dispersal in the lagoon of the atoll (Hobson and Chess 1973). We observed similar changes in evening behavior in several reef organisms, beginning at least one hour before SS (silversides, parrotfishes, mysids, and some urchins). Similar transition behaviors occur in the highly social spinner porpoises of Hawaii which, like the silversides, begin to move offshore from their daytime resting sites near shore in the late afternoon, and around SS head for open water to feed (Norris *et al.,* 1994). At this time the rate of change in light intensity is approximately 1% min^{-1}, or about 5% of the maximum rate of change of 20% at CT (Figure 3). Many fishes in the range of light intensities from 1000 to 100 lux require a 10% or greater change in intensity to discriminate a change in 'brightness', i.e. to distinguish or compare one intensity with another intensity (Kobayashi 1962, Blaxter 1976). Brightness discrimination is usually measured experimentally under conditions where a reference intensity is present, constant, and not changing, as the general irradiance does during twilight. In order to evaluate a rate change of 1% min^{-1}, a fish must be able to gauge the intensity and store it for 10 min as a reference. Although this may be possible, it seems unlikely, and in our own experience as humans is, at best, an uncertain ability. At SS the rate change is 10% min^{-1} and a fish need hold a reference intensity for only 1 min, and at CT for only 30 seconds. For fishes, the percent change in light intensity required for brightness discrimination is inversely proportional to the reference around twilight, when intensities are lower. Thus, assessing rate of change in brightness should reduce required integration time (Figure 3). At this time we cannot with any certainty evaluate whether noted changes in behavior that occur when light is changing slowly (albeit faster than during midday) are directly due to the rate of change of light, or are driven by other factors such as hunger or, perhaps, entrained endogenous circadian rhythms. We were unable in our search of the literature to uncover studies that specifically measured the thresholds of detection for humans and other animals as a function of constantly changing levels of illumination.

As cumulus clouds drift overhead surface intensities may change dramatically over seconds to minutes, and produce significant oscillations in the rate of change of light intensity. Although the passage of clouds overhead at St. Croix was variable, it nevertheless was common. One must wonder, therefore, if the slow increase in the rate of change of light intensity prior to SS under such changeable ambient light conditions can provide an adequate signal for the initiation of transition behaviors? At dusk dramatic changes in behaviors did not occur until the light intensity exceeded 4 – 5 percent min^{-1}, a rate reached at about 30 min before SS (Figure 3). Fluctuations in light intensity associated with meteorological conditions can produce changes in intensity that range from seconds through minutes to hours. The passage of clouds over the solar beam at St. Croix, over a time span of approximately 30 seconds (which was most typical), yielded a rate of reduction in intensity slightly greater than 60 percent min^{-1}. But fast drifting clouds can produce even higher rates (in one instance a decreasing rate of 3.8% sec^{-1}, and with passage of the cloud an intensity increase of 5.5%). Waves also induce extremely rapid changes in underwater light and can produce changes in intensity that exceed several hundred percent min^{-1} (Schenck 1957). Fish and mysid schools around the reefs show slight shifts in school positions to passing clouds, like shadow responses (personal observation, McAlary and McFarland). But these behavior shifts do not initiate characteristic pre-migratory behaviors associated with dusk. Furthermore, reef fishes show no apparent response to the very rapid flickering intensities produced by waves (McFarland and Loew 1983). The maximum rate of change in light intensity during twilight of 18 – 20 percent min^{-1} occurs at CT (Figure 3), and is considerably less than the rates associated with passing clouds or wave produced underwater flicker. Overcast skies at dusk did initiate transition behaviors in fishes a few minutes earlier than under clear skies (Hobson 1972, McFarland *et al.,* 1979), but at the same light intensities as on clear days. Eleven of the 21 invertebrates we identified to species in Tague Bay were invariably active around twilight. In only two instances were behavior changes initiated near the time of the highest rate of light intensity change: i.e., tentacle extension in the anemone species of *Lebrunia* and the coral *Dichocoenia*. Cniderians exemplified the more common transitions, in which, polyp extension began near SS and was only 50 percent complete during the maximum rate of change of light intensity at dusk. Polyp retraction near dawn did not commence until first morning light, reached 50 percent before the maximum rate of change in intensity, and were completed by SR.

Whether a specific light intensity, the rate of change in intensity, or both trigger transition behaviors in coral reef animals requires specific experiments designed to gauge the importance of each factor. These experiments have not been executed, but inferences about the role of each factor can be drawn from available information on plankton, invertebrates, and fishes.

Changes in diel activity with depth emphasize the role of light intensity in determining activity rhythms. Among some reef invertebrates, including the crinoids *C. multiradiatus* and *Herterometra savignii* and the carnivorous gastropod *Charonia variegata*, shallow-water populations are nocturnal and hide during the daytime in various crevices. Their activity begins at SS. In deeper-water (8 to > 30 m), however, individuals are less influenced by the diel rhythm and remain active during the daytime, as well as night (Rutman and Fishelson 1969, Percharde 1972). Shifts to diurnal and continuous activity also occur among normally nocturnal fish in the Arctic when daylight intensities are low but continuous throughout each day (Eriksson 1978, Muller 1978).

That an intensity threshold can affect transition behaviors is suggested by suppression of the emergence of reef plankton by moonlight (Aldridge and King 1980, and Figure 15). At St. Croix reef, plankton began to rise into the water column above the reef about 20 min following SS, but only when the moon was set. Descent from the water above the reef occurred whenever the moon rose. This effect is associated with surface moonlight illumination levels between CT and NT (i.e., ca. 0.1 lux). Presumably this response serves to remove individual plankton from the presence of nocturnal fish predators (Jerling and Wooldridge 1992). However, Gliwicz (1986) suggested that the sudden change from increasing darkness at dusk by the rise of a full moon was used by planktivorous fishes as a ‘moon trap’ to feed on zooplankton. A similar suppressive effect of moonlight on nocturnal behavior has been shown among terrestrial animals (e.g. some mammals – Lockard and Owings 1974, but not on others – Kenagy 1976). In the opposite sense, some diurnal coral reef fishes become active in the presence of moonlight (Hobson 1972). For those invertebrates that display distinct twilight transitions, the time of the changes in behavior often correlate with the avoidance of potential predators and feeding requirements, as in the evening extension of polyps of most corals (Lasker 1979). In nocturnal corals, the dusk and dawn rhythm may relate to improved feeding success with the increased availability of plankton at twilight and night. But why many of the polyps within colonies of some species remain partly open or extended and not fully closed or withdrawn during the daytime is unclear (see Newell and Northcroft 1967, Johannes and Tepley 1974, Preston and Preston 1975, Sebens 1977, Lasker 1979).

An example of critical predator-prey interactions during twilight transitions may be found at St. Croix, where the bluehead wrasse *Thalassoma bifasciatum*, an abundant diurnal planktivore, retreats into the reef around SS (Figure 5). The last bluehead seen near mysid schooling sites was from −9 to +9 min from SS (n = 20 sites), over the time when small groups of *M. gracile* and the unidentified mysids were migrating up-reef. The first cardinal fish, an abundant nocturnal planktivore that sequesters in the reef during the day, emerged in the vicinity of migrating mysids 12 – 36 min after SS (n = 12 sites), the time when small groups of mysids 'dissolved' into the plankton. We observed encounters and attacks by both of these predators. Early in the dusk migration, small groups of mysids escaped wrasses by actively avoiding them. When confronted, schools retraced their migration path, changed directions or, with direct attack, exploded in all directions and reassembled further up the reef to proceed with migration. Mysids seemed less responsive to the presence of cardinal fish, which generally attacked from below at a time when mysids were more or less drifting apart. Descent from the plankton occurred in the narrow window when most cardinal fish had retreated into the reef (last cardinal fish seen –20 to 23 min before SR) and while bluehead wrasse and other small invertivorous fish were not yet in the water column (first wrasse or other small invertivore seen –9 to 18 min after SR). The tight coupling of mysid migrations to a window of relative safety during twilight emphasizes the potential value of appropriate timing of diel transition behaviors. Still, for most zooplankters, especially those grazed by corals, the threat of predation is probably highest at twilight (Porter 1974). The twilight behavior of microzooplankters (zooflagellates and ciliates) that undergo diel changes in abundance (Sorokin 1993) are unknown. These forms compose a significant proportion of the total animal plankton biomass and serve as an important resource for the plankton community, especially larval fish. An understanding of their behavior at twilight is important to a full interpretation of community dynamics.

In many of the coral reef invertebrates that are arrhythmic explanations other than the direct action of light, endogenous rhythms, or vulnerability to predators could account for the absence of transitional behaviors. Many corals and sessile polychaete worms are capable of rapidly retracting their polyps or tentacles when threatened. The presence of noxious compounds in corals and a diversity of other reef inhabitants may well provide protection from potential predators (Moore and Scheuer 1971, Endler 1986, Harvell and Fenical 1989, Duffy and Paul 1992, Paul 1992, Pawlik and Fenical 1992, VanAlstyne and Paul 1992, Pawlik 1993, Faulkner 1994). In some coral communities sponges out compete corals for Lebens-raum (Suchanek *et al.,* 1983), a competitive advantage attributable to structural defenses (e.g. spicules) and the largest number of secondary metabolites amongst marine invertebrates, many of which are known anti-predator compounds (Pawlik *et al.,* 1995).

The often extensive diel vertical migrations (DVM) of marine and freshwater plankton and fishes have focused investigations on the proximal and ultimate cause(s) of the migrations (e.g. light, antipredator response, foraging, metabolic costs, endogenous rhythms; see Enright and Hamner 1967, Segal 1970 and Longhurst 1976 for engaging reviews). It was suggested by Russell (1926) that vertically migrating organisms were attempting to remain within a 'preferred' range of light intensities. Several investigations of the vertical movements of herring shoals at dusk and dawn, revealing that fish tended to follow isolumes between 1 and 10 lux, bolstered this hypothesis (Richardson 1952, Postuma 1958). When Boden and Kampa (1967) established that the 400 meter dusk ascent and dawn descent of a deep scattering layer (i.e., acoustical sonar reflections of dense concentrations of invertebrates and fish) was restricted to an isolume (5×10^{-4} μW cm^{-2} at 474 nm) the concept that light intensity controls vertical migrations became generally accepted. Subtle changes in light intensity can modify the depth at which organisms reside at different times of the day (Worthington 1931, Dietz 1962, Backus *et al.,* 1965, Blaxter and Currie 1967). Lights lowered to the depth of a scattering layer off Scotland induced the layer to descend at night and delayed its rise at dusk, as did passing clouds (Blaxter and Currie 1967). This intriguing finding, and his elegant studies on the visual behavior of marine fishes, led John Blaxter (1973 and 1976) to champion the concept that many fishes can respond to a specific light intensity. In his 1976 review of vertical migrations, Blaxter evaluated the echograms for several ascending scattering layers and indicated that the ascent at dusk did not follow specific isolumes precisely, but rather crossed over and under isolumes, as if the organisms were 'hunting' for a preferred intensity. Calculating the intensity changes that the organisms in each scattering layer must have encountered, Blaxter established that the percentage change in intensity was similar to that required by several fishes to discriminate a change in 'brightness' (i.e., the Weber fraction varied from 10 to 100 percent; see Figures 8.14, 8.15 and 8.16 in Blaxter 1976). Furthermore, Blaxter emphasizes that changes in 'brightness' discrimination will be enhanced at higher rates of changing intensity (as in the tropics during twilight, Figure 2), and that some specific status of the detector would be required to serve as an intensity reference. In fishes this could be served by the rod threshold and the intensities at which dark/light adaptation occurs. Unfortunately for our coral reef discussion, only deep sea fishes can make use of the rod threshold as a reference, since it is seldom achieved even at night in the upper reaches of the photic zone (10^{-6} to 10^{-8} lux). Actual dark/light adaptation in other fishes occurs during the intensity range within which many vertical migrations seem to take place, from 10° to 10^{-3} lux (Figure 1). Not surprisingly, this is also the intensity range over which coral reef fishes undertake diel transition behaviors.

Field and laboratory results by Forward *et al.,* (1984) reveal that larval estuarine crabs, that consistently migrate over a few meters and ascend at dusk and descend at dawn, respond to specific intensity thresholds. Zoea of *Rhithropanopeus harrisii* respond to a particular threshold intensity by initiating a sinking response. At lower intensities and in darkness they develop a strong negative geotaxis and swim upward, a behavior also described for a chaetognath (Pearre 1973). Early stage I zoea respond at higher intensity thresholds than each later stage, and fine tuning of ascent and descent along an isolume occurs primarily in older stage IV larvae. Each larval stage associates sinking behavior with a specific isolume during the day, and therefore tends to rise in the water and sink during the twilight periods at dusk and dawn. In other zooplankton, results suggest that the direction and rate of change of light level from an adaptation intensity cue DVMs (Ringelberg 1964, Daan and Ringelberg 1969, Haney and Hall 1975, Buchanan and Haney 1980, Kerfoot 1980).

In the oceans, vertical migrations typically involve a rise toward the surface at dusk and descent at dawn. Reverse vertical migrations and the absence of vertical movements have been documented in plankton and fishes (see Longhurst 1976, Neilsen and Perry 1990). Dabob Bay, a semi-isolated fjord in Puget Sound, Washington, characterized by a shallow headwater (50 m) and a deeper basin toward the mouth (185 m), provides clear examples of all three types of migration (Ohman 1990). In the shallow headwater during summer the copepods *Pseudocalanus newmani* are near the surface at night and deeper during the day, whereas a reverse migration occurs in the deeper basin. In winter DVM ceases in both basins, and the copepods, now at lower densities, are found in the upper 25 metres. Ohman provides evidence that the reverse migration in the deeper basin is an antipredator response to the nocturnal ascent and increased abundance of several non-visual (but light-sensitive) carnivorous zooplankton during the summer (verified by Frost and Bollens, 1992). A typical DVM (downward migration at dawn) occurs in the shallow basin where, near the surface, the summer abundance of nocturnal carnivorous zooplankton is low, but daytime abundance of planktivorous fishes is high. Frost and Bollens (1992) noted that a typical DVM of *P. newmani* did occur on one occasion in the deep basin of Dabob Bay, which correlated with a peak abundance of the predatory sand lance *Ammodytes hexapterus*.

The plasticity of DVM of *P. newmani* fits the anti-predator hypothesis. But is the mechanism underlying migration plasticity a phenotypic response of individual copepods, or a polymorphic trait in the populations? Ohman indicates that the generation time of the copepods ($8 - 9\ yr^{-1}$) is sufficient to select for genotypes favoring the appropriate type of migration to avoid predators that are either non-visual (descent during night) or visual (descent during the day). This polymorphism has been shown in *Daphnia pulex* (Weider 1984 and 1985), and for coexisting clones of *Daphnia magna* that show either positive or negative phototaxis and have different vertical distributions (DeMeester and Dumont 1988 and 1989). The very rapid switch over, about 4 weeks from reverse to normal DVMs, may also be the result of phenotypic flexibility (Frost and Bollens 1992). Identification of the responses underlying DVM switches in *P. newmani* awaits further studies. The *Daphnia* results reinforce the concept that control of a diel transition behavior requires an environmental trigger such as light, but the behavior is selected and modified in the context of long-term survival in the organism's community – in most instances, predator-prey interactions. Responses to light may thus be considered a facultative rather than obligatory behavior, given the suggestive correspondence between the direction of migrations and other factors including the abundance of prey and predators (Neilson and Perry 1990).

In fishes, transition behaviors are very likely constrained by the level of light/dark adaptation in the retina (retinal motormovements). The relationships of light/dark adaptation in the retina to transition behaviors have been most thoroughly investigated in the French grunt *Haemulon flavolineatum* (McFarland *et al.,* 1979, McFarland and Wahl 1996). In large juveniles and adults, the evening migration commences after the melanin screening pigment has withdrawn to the base of the pigment epithelium. This process exposes the thick rod layer to light at a time when the cones are still in their light adapted position in the focal region adjacent to the outer limiting membrane. Migration in these large juveniles and the adults is limited to one-log unit of intensity and initiated near CT when the peak rate of change in light intensity is maximal. A comparison of the times of migration for different sized french grunts reveals that smaller grunts imitate migration at higher light intensities than larger grunts (see Figure 7, and Helfman *et al.,* 1982, McFarland and Kotchian 1982). Thus, smaller fish leave and return to the reef at light intensities that are 'brighter' than the intensities at which large juveniles migrate.

Coral communities are noted for high levels of competition and predation (Grigg *et al.,* 1984, Huston 1985, Hixon 1991). Attacks by piscivorous fishes reach a peak during the crepuscular periods of each day and are especially intense in tropical reef systems (Hobson 1972 and 1979, Major 1977 and 1978, McFarland *et al.,* 1979, Potts 1980 and 1981, Endler 1986, and see Helfman 1993 for review). This is the case for juvenile grunts at St. Croix that must traverse a gauntlet of piscivores during their migrations away from and back to their daytime schooling sites (e.g., groupers, trumpetfish and lizard fish). The ability to 'see' these lurking predators is critical to each size class of grunt to assure safe passage to and from the daytime schooling sites. In French grunts, as in many teleosts, visual acuity improves with size (Wahl *et al.,* 1993, Wanzenbock *et al.,* 1996). Furthermore, the image cast on the retina by a specific target is enlarged with fish growth and this increases the number of cones illuminated. As a result, large juvenile French grunts can detect similar targets at lower light intensities than smaller grunts (McFarland and Wahl 1996). We infer from this particular example that the capacity to visualize critical objects, such as potential predators, is influenced by eye size and the light intensities over which cones enter or leave the visual plane (retinal motormovements). Each acts as a visual constraint on the absolute light intensities at which targets are adequately perceived, and hence the time at which different sizes of grunts can safely migrate. The tendency for smaller fishes within a species to initiate transition behaviors earlier than larger individuals during dusk and vice versa during dawn appears to be general within fish communities (Hobson 1972, Hobson *et al.,* 1981).

In one instance we examined the effect of differing rates of change of illumination versus absolute light intensity on the substrate burrowing behavior of the California seniorita, *Oxyjulis californica* (McAlary and McFarland, unpubl. ms). Under field conditions this species of wrasse buries beneath the sand substrate during dusk and emerges again near SR. The burying behavior can be experimentally controlled by diminishing the illumination. Doubling or halving the rate of change at which the illumination was decreased near SS, for example, had no significant effect on modifying the absolute intensity range over which the wrasses buried in the sand. Thus, although in most instances the primary mechanisms governing twilight behaviors remain unclear, in this single instance we can conclude that the seniorita wrasse responds behaviorally to a specific intensity at twilight and not the rate of change in illumination. In the case of French grunts, confirmation of this inference, however, requires specific experiments in which each juvenile size class of grunt is subjected to varied changing rates of illumination to determine whether each size class responds at the same specific light intensity or not.

19.2.6 Transition behaviors and the duration of twilight

Studies of tropical and subtropical marine communities reveal that fish assemblages shift from diurnal to nocturnal species over a period of 15 – 20 min from SS (Hobson 1968 and 1972, Starck and Davis 1966, Collette and Talbot 1972, Domm and Domm 1973). At increasing latitudes twilight lengthens and lower light levels are reached earlier before SS and later after SR (Figure 2). How might the duration of twilight affect diel transition behaviors?

A comparison of diel change-over among fish from freshwater tropical and temperate lakes suggests that the period of diel transition is prolonged at higher latitudes (Helfman 1993). But the data for diel transitions for marine communities is not as easily interpreted. However, at 56° N in Steamer Bay, Etolin Island, Alaska CT lasted 57 min (Hobson 1986) compared to 23 min at St. Croix. At Steamer Bay, the sand lance (*Ammodytes hexapterus*) shifted from daytime schooling to burying in the sand over a period of nearly two hours. This prolonged retreat becomes the focus of crepuscular predators for a period that is twice as long as that observed for crepuscular piscivorous fish in the tropics. Although most of the sand lances burrowed before SS, retreat wasn't completed until 30 min or so after SS (Hobson 1986). Comparison of the evening fish transitions between temperate latitudes (Catalina Island and Santa Barbara, Southern California) and coral reefs (Kona Coast, Hawaii and St. Croix) reveals general differences in the timing of the transitions. In Southern California (33 – 34° N L.) CT is reached from 5 to 6 min later than it is in Hawaii (19°30′ N L.) or St. Croix (16°45′ N. L). At Catalina and Santa Barbara, the diel transition among the diurnal kelp forest fish overlaps the emergence of nocturnal species such that a quiet period, as observed in the tropics, is poorly defined (Ebeling and Bray 1976, Hobson *et al.,* 1981). At Hawaii, St. Croix and Southern California wrasses are the very first group of diurnal fishes to seek cover and provide, therefore, a cohesively related taxon for which we can evaluate the effect of a slightly longer duration of twilight. The slightly later retreat of similarly sized wrasses at Catalina (SS + ca. 20 min, Hobson *et al.,* 1981 and our unpublished observations) compared to Hawaii and St. Croix (SS + 7–10 min) can be attributed to the achievement of similar light intensities at the different times after SS at the different latitudes (Hawaii 50 –105 lux, St. Croix 50 – 80 lux, Catalina 40 –125 lux).

19.2.7 Endogenous rhythms and twilight behavior

The ubiquity of cellular endogenous circadian rhythms in eukaryotes from bacteria to invertebrates, and even in some prokaryotes, suggests that the absence of a daily cellular rhythm would be unique (see Pittendrigh 1993 for an enlightening review). The presence of genetic molecular clocks is now well established in several organisms, including *Neurospara, Drosophila* and mice, and the list grows (see Takahashi and Hoffman 1995). There is little reason that similar clocks are not present in most fishes and marine invertebrates. The rush of studies over the last few years on cloning and defining molecular clocks have demonstrated that multiple autoregulated clock genes (e.g. timeless (*tim*) and period (*per*) in *Drosophila*) exist (Crosthwaite *et al.,* 1995, Kay and Millar 1995). These clock genes in turn regulate the transcription of other genes that ultimately must produce the multitude of biochemical and physiological cycles found in diverse organisms (Pittendrigh 1993, Hall 1995, Millar *et al.,* 1995, Bell-Pedersen 1996, for fishes see Schwassman 1971). Understanding the mechanisms of molecular clocks is central in demystifying endogenous rhythms, but tying function from clock genes to the twilight behavioral ecology of diverse organisms of different ecosystems, will not be an easy task. What function might endogenous circadian rhythms serve in daily transition behaviors?

The difficulty in unraveling the relative role of light versus endogenous circadian rhythms in initiating diel transition behaviors, and especially in tropical animals, is revealed by John Blaxter's skepticism as expressed in the *Rhythmic Activity of Fishes* (Blaxter 1978):

'I am far from convinced of the value to the fish of an endogenous rhythm unless the fish requires to be *pre*-adapted to some change in the environment. If the *zeitgeber* is always present why have endogenous rhythms developed? It may be desirable for a fish to respond to the onset of a stimulus like light if it is removed from it, say in a cave or a burrow, but do fish really have to be pre-adapted to awake *before* dawn and so escape their predators? One might have thought the predators would have evolved the same mechanism themselves.'

Blaxter's point is not trivial. However, Ali's comment (1992) emphasizes the importance of rhythmic cycles from an ecological perspective – 'In as much as animals are dependent upon other organisms for food, they must be able to live with not only the rhythms of their inanimate surroundings, they must also adjust to the rhythms of the organisms upon which they feed. Maybe this answers, at least in part the question 'What good are rhythms?'.'

Certainly free-running endogenous cycles that produce transitional behaviors exist in coral reef fish (Livingston 1971) and invertebrates, but their presence is not ubiquitous (as in some corals, Sweeney 1976). Thus in *Lebrunia* neither the pseudo-tentacles nor the tentacles display endogenous rhythms when placed in constant photic conditions (Gladfelter 1975). Indeed, Gladfelter obtained graded extension and retraction of the pseudo-tentacles and tentacles to specific light levels in the laboratory, suggesting the behavior is a direct response to light intensities. Similarly, the diel migrations of hydromedusae at twilight are light dependent, since they show no intrinsic rhythm when maintained in CL or CD (Passano 1973, Mackie *et al.,* 1981, Mills 1983).

The diel activity of other reef invertebrates including the asteroids, holothuroidians, and crustaceans do exhibit light-dependent endogenous circadian rhythmicity (Mori 1943 and 1960, Mori and Matutani 1952, Yamanouchi 1956, Lipcius *et al.,* 1979), as does the tropical lobster *Panuliris argus* (Lipcius *et al.,* 1979). In response to a shift in twilight, the Norwegian lobster *Nephrops norvegicus* adjusts its nocturnal activity peak, an effect that seems to be related to relative rather than absolute intensities (Hammond and Naylor 1977). Examination of the daily movements of *Diadema* revealed that the predusk movements were more prolonged than at dawn, even though maximum movements occurred close to SR (Figure 12). The asymmetrical prolonged predusk activity could result from hunger, for the urchins have presumably not fed since the previous night. The existence of an endogenous rhythm also seems likely, because the initial movement peak at 0500 hours occurred 1 hour before we could detect first light. In general, the existence and potential contribution of endogenous circadian rhythms to diel transitions amongst coral reef invertebrates have not been thoroughly tested. As a result, scattered information on endogenous activity patterns, responses to light, and diel activity patterns of various species lack a definitive synthesis.

Studies that have defined free-running rhythms have utilized behavioral attributes such as locomotion or feeding periods to define periods of activity (Calhoun 1944 and 1945, Harker 1958, Cloudley-Thompson 1961, Bunning 1967, Enright 1975). Earlier we emphasized the importance of the status of the retinomotor movements (RM) of fishes in relation to transition behaviors. There are many experimental data that clearly demonstrate that the RMs are directly influenced by light and darkness (Ali and Wagner 1975). In most teleosts a specific light threshold exists above which the fish retina light adapts. Below the threshold the retinal elements dark adapt. The powerful influence of light in inducing light adaptation of the retina was demonstrated by flashing a 3,000 lux stimulus for one-minute at four different species of teleosts that were maintained in darkness (Wagner and Douglas 1983). Within 10 min all four species began retinal light adaption, which was virtually complete for the cones by 20 min, but not by the screening pigment granules. Within 20 to 40 min dark adaptation recovery began. Even light pulses of very short duration (< 0.003 seconds) can initiate the light adaptation sequence in dark-adapted trout (Muntz and Richard 1975). That the effect is intra-retinal comes not only from the rapidity of the onset of light adaptation, but also from the clever demonstration that the light adaptative movements can be localized by focusing a beam of light on a specific small area of the retina (Easter and Macy 1978). So, if the retinal position of the cones and/or rods is critical in the diel transitions of diurnal or nocturnal fishes, as we and others maintain, how much is a fish's retinal state influenced by its endogenous rhythm?

The evidence on this is not clear, because different species of teleosts have yielded varied results. Douglas and Wagner (1982) indicated that definable RM rhythms were present in the nine species they investigated when held under constant darkness. In general the melanin pigment granules did not achieve complete movement to the light adapted position, nor did the cones, except in three of the nine species. Under constant illumination only two species showed evidence of a circadian rhythm. Douglas and Wagner classified three of the species as strongly rhythmic (*Astyanax mexicanus* Welsh and Osborn 1937, Arey and Mundt 1941; *Nannacara anomola* Douglas and Wagner 1982; *Cichlasoma citrinellum* Levinson and Burnside 1981), all of which have in common a distribution in tropical and subtropical freshwaters. They suggest that the constancy of the 'light regime' at these latitudes may cause a fish to be more rhythmic, than those exposed to the constantly changing light regimes of temperate and arctic latitudes. Although this may in general be so, it requires comparison of more fishes from the tropics and higher latitudes. Juvenile French grunts do show evidence of an endogenous circadian rhythm in the RMs under constant darkness, but the cone movements are sporadic between individuals and certainly do not achieve complete contraction to the light adapted position at the outer limiting membrane (McFarland and Wahl 1996). What seems to ring clear from these studies is the fact that RMs to varying degrees are under endogenous clock control, but require the presence of light to achieve complete light adaptation, and some degree of darkness to achieve dark adaptation (Levinson and Burnside 1981). The underlying circadian rhythm in RM movements may well poise each fish's visual system in anticipation of a behaviorally critical time. It remains unclear as to whether a fish could behave appropriately without endogenous RM rhythmic movements. The entrainment of the diel rhythm of an organism's internal milieu by changes in light has not only been revealed through experiment, but seems intuitively logical. The *zeitgeber* that entrains a rhythm must be important to tropical organisms even though variation in photoperiod and the duration of twilight is minimal over a year, if for no reason other than that molecular clocks seldom precisely track a 24-hour day. At latitudes above 20 – 30 degrees and, especially, above the Arctic Circle where the photoperiod and duration of twilight are highly variable at different seasons (Figure 2), the *zeitgeber* must have a more critical role in maintaining synchronization of any endogenous oscillations to the daily cycle.

19.2.8 Conclusions

In this chapter we have tried to impart an appreciation of the complexity of animal behaviors and community interactions that occur in response to changes at twilight. Our emphasis on tropical coral communities results from our primary experience with these systems. The substantial literature amassed for coral reef communities, from which we have drawn examples, clearly indicates that twilight is a critical period of the day during which many, but not all, community members undergo significant behavioral changes and these changes are tightly coupled to light. For a few species, experiments have shown that light directly effects behavioral changes, but for the majority of species the role of light, direct or indirect, is inferred from field data and remains to be quantified in the laboratory. Perhaps the greater challenge lies in elucidating the mechanisms underlying the response of the animal to changes in its environment at twilight. Thus, although many reef organisms clearly respond to light, what aspects of change at twilight are they actually measuring – absolute intensities, the rate of intensity, shifts in the ambient spectrum or polarization. In reef organisms dependent on vision, changes in the perception of targets may be important and involve varied combinations of all these photic attributes. How critical are endogenous rhythms to the precision of twilight transition behaviors? At this writing we find no compelling answers. Coral reef communities provide the opportunity to bridge the gap that exists between the role of light and endogenous circadian rhythms without the additional complications of dramatic seasonal changes in photic conditions.

References

Aldridge, A. L. and King, J. M. (1980) Effects of moonlight on the vertical migration patterns of demersal zooplankton. *J. Exp. Mar. Biol. Ecol.*, 44, 133 –156.

Ali, M. A. (1992) Introduction, pp. 1– 5. In Ali, M. A. (ed.) *Rhythms in Fishes*. NATO ASI Series. Plenum Press, New York. 348 pp.

Ali, M. A. and Wagner, H. J. (1975) Distribution and development of retinomotor responses. pp. 369 – 396. In Ali, M. A. (ed.) *Vision in Fishes*. NATO ASI Series. Plenum Press, NY. 348 pp.

Arey, L. B. and Mundt, G. H. (1941) A persistent diurnal rhythm in visual cones. *Anat. Rec.*, 79 (suppl) 41.

Aschoff, J. (1969) Phasenlage der Tagesperiodik in abhangigkeit von Jahreszeit and Breitengrad. *Oecologia,* 3, 125 – 265.

Aschoff, J., Gwinner, E., Kureck, A. and Miller, R. (1970) Diel rhythms of chaffinches *Fringilla coelebs* L, tree shrews *Tupata glis* L. and hamsters *Mesocricetus auratus* L. as a function of season at the Arctic circle. Oikos suppl. 13, 91–100.

Aschoff, J. and Wever, R. (1962) Beginn und ende der taglichen aktivitat freilebender Vogel. *J. Ornithol.*, 103, 2 – 27.

Bakus, G. J. (1968) Defensive mechanisms and ecology of some tropical holothurians. *Mar. Biol.*, 2, 23 – 32.

Bakus, G. J. (1974) Toxicity in holothurians: a geographical pattern. *Biotropica* 6, 229 – 236.

Backus, R. H, Clark, R. C. and Wing, A. S. (1965) Behaviour of certain marine organisms during the solar eclipse of July 20th 1963. *Nature* 205, 989 – 981.

Barr, L. (1975) Biology and behavior of the arrow crab *Stenorhyncus seticornis* (Herbst) in Lameshur Bay, St. John, Virgin Islands. *Bull. Nat. Hist. Mus.*, Los Angeles County 20, 47– 56.

Beck, S. D. (1968) Insect Photoperiodism. *Academic Press*, New York. 288 pp.

Bell-Pedersen, D., Dunlap, J. C. and Loros, J. J. (1996) Distinct cis-acting elements mediate clock, light, and developmental regulation of the *Neurospora crassaeas* (ccg-2) gene. *Molecular and Cellular Biology,* 16, 513 – 521.

Berrill, M. (1996) The ethology of the synaptid holothurian, *Opheodesoma spectabilis. Can. J. Zool.*, 44, 457– 482.

Blaxter, J. H. S. (1970) Light: Animals-Fishes. pp. 213 – 320. In Kinne, O. (ed). *Marine Ecology, Vol. 1*, Environmental Factors, Part 1. John Wiley & Sons, London. 681 pp.

Blaxter, J. H. S. (1973) Do fishes have an absolute sense of light intensity? *Vision Res.*, 13, 719 –729.

Blaxter, J. H. S. (1976) The role of light in the vertical migration of fish-a review. Ch. 8, pp. 189 – 210. In Evans, G. C., Bainbridge, R. and Rackham, O. (eds). *Light as an Ecological Factor: II.* Blackwell Scientific Publ., Oxford. 616 pp.

Blaxter, J. H. S. (1978) Summary on 'Rhythmic Activity in Fish.' pp. 285 – 290. In Thorpe, J. E. (ed), *Rhythmic Activity of Fishes*. Academic Press, NY. 312 pp.

Blaxter, J. H. S and Currie, R. I. (1967) *T*he effect of artificial lights on acoustic scattering layers in the ocean. *Symp. Zool. Soc. Lond.,* 19, 1–14.

Boden, B. P. and Kampa, E. M. (1967) The influence of natural light on the vertical migrations of an animal community in the sea. *Symp. Zool. Soc. Lond.*, 19, 15 – 26.

Brown, D. R. E. (1952) *Natural illumination charts.* U. S. Dept. Navy, Bureau of Ships, project NS 714 –100. Rept. No. 374 –1.

Buchanan, C. and Haney, J. F. (1980) Vertical migration of zooplankton in the arctic: A test of the environmental controls. pp. 69 –79. In Kerfoot, W. C. (ed), Evolution and Ecology of Zooplankton Communities. *Am. Soc. Limnol. Oceanogr. Spec.*, Symp. 3 793 pp.

Bunning, E. (1967) *The Physiological Clock.* 2nd Edition, Springer-Verlag, New York. 167 pp.

Calhoun, J. B. (1944) Twenty-four hour periodicities in the animal kingdom. Part I, The invertebrates. *J. Tenn. Acad. Sci.,* 19, 179 – 200, 252 – 262.

Calhoun, J. B. (1945) Twenty-four hour periodicities in the animal kingdom. Part II, The vertebrates. Tenn, *J. Acad. Sci.,* 20, 228 – 232, 291– 308, 373 – 378, and 21, 208 – 216, 281– 282.

Charles-Dominique, P. (1975) Nocturnally and diurnality - An ecological interpretation of these two modes of life by an analysis of the higher vertebrate fauna in tropical forest ecosystems. pp. 69 – 88. In Luckett, W. P. and Szalay, F. S. (eds), *Phylogeny of the Primates: A Multidisciplinary Approach.* Plenum Press, New York. 483 pp.

Chesher, R. H. (1969a) Contributions to the biology of *Meoma ventricosa* (Echinoidea: Spatangoida). *Bull. Mar. Sci.*, 19, 72 –110.

Chesher, R. H. (1969b) Destruction of Pacific coral by the sea star *Acanthaster planci. Science,* 165, 280 – 283.

Clark, A. M. (1976) Echinoderms of coral reefs. pp. 95 –123. In, Jones, O. A. and Endean, R. (eds). *Biology and Geology of Coral Reefs*. Vol 3 Biology 1. Academic Press, New York. 435 pp.

Clarke, G. L. and Backus, R. H. (1964) Interrelations between vertical migration of deep scattering layers, bioluminescence and changes in daylight in the sea. *Bull. Inst. Oceanogr.,* Monaco 64, 1– 36.

Cloudley-Thompson, J. L. (1960) Adaptive functions of circadian rhythms. Cold Spring Harbor Symp. *Quant. Biol.,* 25, 345 – 355.

Cloudley-Thompson, J. L. (1961) *Rhythmic Activity in Animal Physiology and Behavior*. Academic Press, New York. 236 pp.

Clutter, R. I. (1969) The microdistribution and social behavior of some pelagic mysid shrimp. *J. Exp. Mar. Biol. Ecol.,* 3, 125 –155.

Collette, B. B. and Talbot, F. H. (1972) Activity patterns of coral reef fishes with emphasis on nocturnal-diurnal changover. *Bull. Nat. Hist. Mus.*, Los Angeles County 14, 98 –124.

Cott, H. B. (1940) Adaptive Coloration in Animals. Metheun, London. 508 pp.

Crescitelli, F. (1972) The visual cells and visual pigments of the vertebrate eye. pp. 245 – 363. In Dartnall, H. J. A. (ed) Photochemistry of Vision 8(VII/I), *Handbook of Sensory Physiology*, Springer-Verlag, New York. 810 pp.

Crosthwaite, S. K., Loros, J. J. and Dunlap, J. C. (1995) Light-induced resetting of a circadian clock is mediated by a rapid increase in frequency transcript. *Cell,* 81, 1003 –1012.

Crump, R. G. (1965) The diurnal activity of holothurians. *Symp. Underwater Assn. Malta*, 1965, 43 – 45.

Curio, E. (1976) The Ethology of Predation. *Zoophysiol. and Ecol.,* 7. Springer-Verlag, Berlin. 250 pp.

Daan, N. and Ringelberg, J. (1969) Further studies of the positive and negative phototactic reactions of *Daphnia magna* Straus. *Neth. J. Zool.,* 19, 525 – 540.

DeCoursey, P. J. (1960) Phase control of activity in a rodent. *Cold Spring Harbor Symp. Quant. Biol.*, 25,

49 – 55.
DeMeester, L. and Dumont, H. J. (1988) The genetics of phototaxis in *Daphnia magna*, existence of three phenotypes for vertical migration among parthenogenetic females. *Hydrobiologia,* 162, 47– 55.
DeMeester, L. and Dumont, H. J. (1989) Phototaxis in *Daphnia*: on the interaction of hunger and genotype. *Limnol. Oceanogr.,* 34, 1322 –1325.
Dietz, R. S. (1962) The sea's deep scattering layers. *Sci. Amer., Freeman Sci.*, San Francisco.
Dominguez, J. H. and Reaka, M. L. (1988) Temporal activity patterns in reef-dwelling stomatopods: a test of alternative hypotheses. *J. Exp. Mar. Biol. Ecol.*, 117, 47– 69.
Domm, S. B. and Domm, A. J. (1973) The sequence of appearance at dawn and disappearance at dusk of some coral reef fishes. *Pacific Sci.,* 27, 128 –135.
Douglas, R. H. and Wagner, H. J. (1982) Endogenous patterns of photomechanical movements in teleosts and their relation to activity rhythms. *Cell Tissue Res.*, 226, 133 –144.
Dubin, R. E. and Baker, J. D. (1982) Two types of cover-seeking behavior at sunset by the princess parrotfish, *Scarus taeniopterus*, at Barbados, West Indies. *Bull. Mar. Sci.*, 32, 572 – 583.
Duffy, J. E. and Paul, V. J. (1992) Prey nutritional quality and the effectiveness of chemical defenses against tropical reef fishes. *Oecologia,* 90, 333 – 339.
Easter, S.S. and Macy, A. (1978) Local control of retinomotor activity in the fish retina. *Vision Research,* 18, 937– 942.
Ebeling, A. W. and Bray, R. N. (1976) Day versus night activity of reef fishes in a kelp forest off Santa Barbara, California. *U. S. Fish. Bull.,* 74, 703 –717.
Ebeling, F. J., Hawkins, A. D., Kitching, L. Muntz and Pratt, V. M. (1966) The ecology of Lough Ine. 16. Predation and diurnal migration in the *Paracentrotus* community. *J. Anim. Ecol.,* 35, 559 – 666.
Emery, A. R. (1968) Preliminary observations on coral reef plankton. *Limnol. Oceanogr.,* 13, 293 – 303.
Emery, A. R. (1973) Preliminary comparisons of day and night habits of freshwater fish in Ontario lakes. *J. Fish. Res. Board. Can.*, 30, 761–764.
Endler, J. A. (1986) Defense against predators. pp. 109 –134. In Feder, M. E. and Lauder, G. V. (eds), *Predator-prey Relationships*. Univ. Chicago press, Chicago. 198 pp.
Enright, J. T. (1975) Orientation in time: Endogenous clocks. Ch. 9, pp. 917– 944. In O. Kinne (ed) Marine Ecology, Vol. II, Physiological Mechanisms, Part 2. John Wiley & Sons, London. 992 pp.
Enright, J. T. and Hamner, W. N. (1967) Vertical diurnal migration and endogenous rhythmicity. *Science,* 157, 937– 941.
Eriksson, L. O. (1978) Nocturnalism versus diurnalism - dualism within fish individuals. pp. 69 – 89. In Thorpe, J. E. (ed) *Rhythmic Activity of Fishes*, Academic Press, NY 312 pp.
Faulkner, D. J. (1994) Marine natural products. *Nat. Prod. Rep.*, 11, 355 –394.
Forward, R. B., Cronin, T. W. and Stearns, D. E. (1984) Control of diel vertical migrations: Photoresponses of a larval crustacean. *Limnol. Oceanogr*. 29, 146 –154.
Fricke, H. and Vareschi, E. (1982) A scleractinian coral (*Plerogyra sinuosa*) with photosynthetic organs. *Mar. Ecol. Prog. Ser*. 7, 273 –278.
Frost, B. W. and Bollens, S. M. (1992) Variability of diel migration in the marine planktonic copepod *Pseudocalanus newmami* in relation to its predators. *Can. J. Fish. Aq. Sci.*, 49, 1137–1141.
Galzin, R. (1987) Structure of fish communities of French Polynesia coral reefs II. Temporal scales. *Mar. Ecol. Prog. Ser.,* 41, 137–145.
Gladfelter, E. H. (1985) Metabolism, calcification and carbon production II. Organism-level studies. *Proceedings 5th Int. Coral Reef Congr.,* 527– 539.
Gladfelter, W. B. (1975) Sea anemone with zooxanthellae: Simultaneous contraction and expansion in response to changing light intensity. *Science,* 189, 570 – 571.
Gladfelter, W. B. (1979) Twilight migrations and foraging activities of the copper sweeper *Pempheris schomburgki* (Teleostei: Pempheridae). *Mar. Biol.*, 50, 109 –119.
Gliwicz, Z. M. (1986) A lunar cycle in zooplankton. *Ecology,* 67, 883 – 897.
Glynn, P. W. (1973) Ecology of a Caribbean coral reef, the *Porites* reef flat biotope. Part II Plankton community with evidence for depletion. *Mar. Biol.*, 20, 297– 318.
Gremare, A. (1988) Feeding, tube building and particle-size selection in the terebellid polychaete *Eupolymnia nebulosa. Mar. Biol.,* 97, 243 – 252.
Grigg, R. W., Polovina, J. J. and Atkinson, M. J. (1984) Model of a coral reef ecosystem. III. Resource limitation, community regulation, fisheries yield, and resource management. *Coral Reefs,* 3, 23 – 27.
Hall, J. C. (1995) Tripping along the trail to the molecular mechanisms of biological clocks. *Trends Neurosci.*, 18, 230 – 240.
Hallberg, E. Andersson, M. and Nilsson, D. E. (1980) Responses of the screening pigments in the

compound eye of *Neomysis integer* (Crustacea: Mysidacea). *J. Exp. Zoo.*, 212, 397–402.
Hammond, L. S. (1982) Patterns of feeding and activity in deposit-feeding holothurians and echinoids (Echinodemata) from shallow back-reef lagoon, Discovery Bay, Jamaica. *Bull. Mar. Sci.*, 32, 549 – 571.
Hammond, R. D. and Naylor, E. (1977) Effects of dusk and dawn on locomotor activity rhythms in the Norway lobster *Nephrops norvegicus*. *Mar. Biol.*, 29, 253 – 260.
Hamner, W. M. and Carleton, J. H. (1979) Copepod swarms: attributes and role in coral reef ecosystems. *Limnol. Oceanogr.*, 24, 1– 24.
Haney, J. F. and Hall, D. J. (1975) Diel vertical migration and filter-feeding activities of *Daphnia*. *Arch. Hydrobiol.*, 75, 413 – 441.
Harker, J. E. (1958) Diurnal rhythms in the animal kingdom. *Biol. Rev.*, 33, 1– 52.
Harker, J. E. (1964) The Physiology of Diurnal Rhythms. Cambridge Univ. Press, Cambridge. 114 pp.
Harvell, C. D. and Fenical, W. (1989) Chemical and structural defenses of Caribbean gorgonians (*Pseudopterogorgia* spp.): intracolony localization of defense. *Limnol. Oceanogr.*, 34, 382 – 389.
Helfman, G. S. (1979) Twilight activities of yellow perch, *Perca flavescens*. *J. Fish. Res. Board Can.*, 36, 173 –179.
Helfman, G. S. (1981) Twilight activities and temporal structure in a freshwater fish community. *Can. J. Fish. Aquat. Sci.*, 43, 11595 –1605.
Helfman, G. S. (1993) Fish behaviour by day, night and twilight. In Pitcher, T. J. (ed) *Behaviour of Teleost Fishes*. 2nd Edition. Chapman & Hall, London. 715 pp.
Helfman, G. S., Meyer, J. L. and McFarland, W. N. (1982) The ontogeny of twilight migration patterns in grunts (Pisces, Haemulidae). *Anim. Behav.* 30, 317– 326.
Hendler, G. (1982) Slow flicks show star tricks, elapsed-time analysis of basketstar (*Astrophyton muricatum*) feeding behavior. *Bull. Mar. Sci.*, 34, 909 – 918.
Hendler, G. (1984a). Association of *Ophiothrix lineata* and *Callyspongia vaginalis*: a brittlestar sponge cleaning symbiosis? P. S. Z. N. I.: *Mar. Biol.*, 5, 9 – 27.
Hendler, G. (1984b) Brittlestar color-change and phototaxis (Echinodermata, Ophiuroidea: Ophiocomidae). P. S. Z. N. I.: *Mar. Biol.*, 5, 379 –401.
Herrnkind, W. F., VanDerWalker, J. A. and Barr, L. (1975) Population dynamics, ecology, and behavior of spiny lobsters, *Panulirus argus*, of St. John. U. S. V. I. *Bull. Nat. Hist. Mus.*, Los Angeles County 20, 31– 45.
Hixon, M. A. (1991) Predation as a process structuring coral reef fish communities, Ch. 17, pp. 475 – 508. In P. F. Sale (ed). The Ecology of Fishes of Coral Reefs. Academic Press, San Diego. 754 pp.
Hobson, E. S. (1965) Diurnal-nocturnal activity of some inshore fishes in the Gulf of California. *Copeia,* 1965, 291– 302.
Hobson, E. S. (1968) Predatory behavior of some shore fishes in the Gulf of California. *U. S. bur. Sport Fish. Wildl. Res. Rep.*, 73, 1– 92.
Hobson, E. S. (1972) Activity of Hawaiian reef fishes during evening and morning transitions between daylight and darkness. *Fish. Bull.*, 70, 715 – 740.
Hobson, E. S. (1973) Diel feeding migrations in tropical reef fishes. *Helgol. Wiss. Meersesunters,* 24, 361– 370.
Hobson, E. S. (1974) Feeding relationships of teleostean fishes on coral reefs in Kona, Hawaii. *Fish. Bull.*, 4, 915 –1031.
Hobson, E. S. (1979) Interactions between piscivorous fishes and their prey. pp. 231– 242. In H. Clepper (ed) Predator-Prey Systems in Fisheries Management. Sport Fishing Institute, Washington, D. C. 504 pp.
Hobson, E. S. (1986) Predation on the Pacific sand lance *Ammodytes hexapterus* (Pisces: Ammodytidae), during the transition between day and night in Southeastern Alaska. *Copeia* 1986, 223 – 226.
Hobson, E. S. and Chess, J. R. (1973) Feeding oriented movements of the atherinid fish *Pranesus pinquis* at Majuro Atoll, Marshall Islands. *Fish. Bull.,* 71, 777–786.
Hobson, E. S. and Chess, J. R. (1976) Trophic interactions among fishes and zooplankters near shore at Santa Catalina Island, California. *Fish. Bull.*, 74, 567– 598.
Hobson, E. S, and Chess, J. R. (1978) Trophic relationships among fishes and plankton in the lagoon at Enewetak Atoll, Marshall Islands. *Fish. Bull.,* 76, 133 –153.
Hobson, E. S. and Chess, J. R. (1979) Zooplankters that emerge from the lagoon at night at Kure and Midway atolls, Hawaii. *Fish. Bull.,* 77, 275 – 280.
Hobson, E. S., McFarland, W. N. and Chess, J. R. (1981) Crepuscular and nocturnal activities of California nearshore fishes, with consideration of their scotopic visual pigments and the photic environment. *Fish. Bull.*, 79, 1– 30.

Hughes, R. N. and Hughes, H. P. I. (1971) A study of the gastropod *Cassis tuberosa* (L.) preying upon sea urchins. *J. Exp. Mar. Biol. Ecol.*, 7, 305 – 314.

Huston, M. A. (1985) Patterns of species diversity on coral reefs. *Ann. Rev. Ecol. Syst.*, 16, 149 –177.

Jerling, H. L. and Woolridge, T. H. (1992) Lunar influence on distribution of a calanoid copepod in the water column of a shallow, temperate estuary. *Mar. Biol.*, 112, 309 – 312.

Johannes, R. E. and Tepley, L. (1974) Examination of feeding of the reef *Porites lobata* in situ using time lapse photography. *Proceedings Second International Coral reef Symp.*, 1, 127–131.

Kampa, E. M. (1976) Photoenvironment and vertical migrations of mesopelagic marine animal communities. pp 257– 272. In P. DeCoursey (ed), Biological Rhythms in the Marine Environment. Univ. South Carolina Press. 283 pp.

Kanciruk, P. and Herrnkind, W. F (1973) Preliminary investigations of the daily and seasonal locomotor activity rhythms of the spiny lobster, *Panulirus argus*. *Mar. Behav. Physiol.*, 1, 351– 359.

Kay, S. A. and Millar, A. J. (1995) New models in vogue for circadian clocks. *Cell,* 83, 361– 364.

Kavanau, J. L. (1971) Locomotion and activity phasing of some medium-sized mammals. *J. Mammal.*, 52, 386 – 403.

Kenagy, G. J. (1976) The periodicity of daily activity and its seasonal changes in free-ranging and captive Kangaroo rats. *Oecologia,* 24, 105 –140.

Kennedy, G. H. (1928) Evolutionary level in relation to geographic, seasonal and diurnal distribution in insects. *Ecology,* 9, 367– 379.

Kerfoot, W. C. (ed). (1980) Evolution and Ecology of Zooplankton Communities. *Am. Soc. Limnol. Oceanogr.,* Spec. Symp. 3. 793 pp.

Kobayashi, H. (1962) A comparative study on electroretinogram in fish, with special reference to ecological aspects. *Shimonoseki College Fisheries* , 353, 408 – 538.

Land, M. E. (1981) Optics and vision in invertebrates. pp. 471– 592. In H. Autrum (ed), *The Handbook of Sensory Physiology*, VII/6B. Springer-Verlag, Heidelberg. 629 pp.

Larson, R. J. (1986) Observations on the light-inhibited activity cycle and feeding behavior of the hydromedusae *Olindas tenuis.* Studies on the Fauna of Curacao and other Caribbean Islands. 68, 191–199.

Lasker, H. R. (1979) Light dependent activity patterns among reef corals: *Montastrea cavernosa. Biol. Bull.*, 156, 196 – 211.

Levinson, G. and Burnside, B. (1981) Circadian rhythms in teleost retinomotor movements: A comparison of the effects of circadian rhythm and light conditions on cone length. *Invest. Opthalmol.,* 20, 294 – 303.

Lipcius, Winter, R. M., Andree, S. and Herrnkind, W. (1979) Endogenous control of behavioral rhythms in solitary and grouped western Atlantic spiny lobster, *Panulirus argus. Amer. Zool.,* 19, 889.

Livingston, R. J. (1971) Circadian rhythms in the respiration of eight species of cardinal fishes (Pisces: Apogonidae); comparative analysis and adaptive significance. *Mar. Biol.*, 9, 253 – 266.

Lockard, R. B. and Owings, D. H. (1974) Seasonal variation in moonlight avoidance by bannertail kangaroo rats. *J. Mammal.*, 55, 189 –193.

Longhurst, A. R. (1976) Vertical migration. pp. 116 –137. In D. H. Cushing and J. J. Walsh (eds). *The Ecology of the Seas.* Saunders, W. B. Philadelphia. 467 pp.

Lynch, P. K. and Livingston, W. (1995) *Color and Light in Nature.* Cambridge University Press, Cambridge. 254 pp.

Lythgoe, J. N. (1979) *The Ecology of Vision.* Oxford, Clarendon. 244 pp.

Mackie, G. O., Larson, R. J. and Passano, L. M. (1981) Swimming and vertical migration of *Aurelia aurita* (L) in a deep tank. *Mar. Behav. Physiol.,* 7, 321– 329.

Macquart-Moulin, C. and Maycas, E. R. (1995) Inshore and offshore diel migrations in European benthopelagic mysids, generas *Gastrosaccus*, *Anchialina*, and *Haplostylus* (Crustacea, Mysidacea). *J. Plankton. Res.,* 17, 531– 555.

Macurda, D. B. (1976) Skeletal modifications related to food capture and feeding behavior of the basketstar *Astrophyton. Paleobiol.,* 2, 1–7.

Magnus, D. B. E. (1967) Ecological and ethological studies and experiments on the echinoderms of the Red Sea. *Stud. Trop. Ocenogr.,* 5, 635 – 664.

Major, P. F. (1977) Predator-prey interactions in schooling fishes during periods of twilight: A study of the silverside *Pranesus insularum* in Hawaii. *Fish. Bull.,* 75, 415 – 426.

Major, P. F. (1978) Predator-prey interactions in two schooling fishes, *Caranx ignobilis* and *Stolephorus purpureus*. *Anim. Behav.*, 26, 760 –777.

Mangum, C. P. and Miyamoto, D. M. (1970) The relation between spontaneous activity cycles and diurnal

rhythms of metabolism in the polychaetous annelid *Glycera dibranchiata*. *Mar. Biol.*, 7, 7–10.
Marsden, J. R. (1962) A coral-eating polychaete. *Nature,*193, 598.
McCartney, E. J. (1976) *Optics of the Atmosphere*. John Wiley and Sons, New York 408 pp.
McFarland, W. N. (1985) *Light in the sea-correlations of fishes and invertebrates. Amer. Zool.*, 26, 389 – 401.
McFarland, W. N. and Hillis, Z. M. (1982) Observations on agonistic behavior between members of juvenile French and white grunt schools, Family Haemulidae. *Bull. Mar. Sci.*, 32, 255 – 268.
McFarland, W. N. and Kotchian, N. M. (1982) Interaction between schools of fish and mysids. Behav. *Ecol. and Sociobiol.,* 11, 71–76.
McFarland, W. N. and Loew, E. R. (1983) Wave produced changes in underwater light and their relations to vision. *Envir. Bio. Fish.,* 8, 173 –184.
McFarland, W. N., Ogden, J. C. and Lythgoe, J. N. (1979) The influence of light on the twilight migrations of grunts. *Env. Biol. Fish.,* 4, 9 – 22.
McFarland, W. N. and Wahl, C. M. (1996) Visual constraints on migration behavior of juvenile French grunts. *Envir. Biol. Fish.*, 46, 109 –122.
Meyer, D. L. (1982) Food and feeding mechanisms: Crinozoa. pp. 24 – 42. In M. Jangoux and J. M. Lawrence (eds) *Echinoderm Nutrition.* A. A. Balkema, Rotterdam. 654 pp.
Millar, A. J., Carre, I. A., Strayer, C. A., Chua, N. H. and Kay, S. A. (1995). Circadian clock mutants in *Arabidapsis* identified by luciferase imaging. *Science* 267, 1161–1163.
Millot, N. (1955). Sensitivity to light and the reactions to changes in light intensity of the echinoid *Diadema antillarum Philippi. Philos. Trans. R. Soc. B.*, 238, 187– 220.
Mills, C. E. (1983) Vertical migration and diel activity patterns of hydromeduase, studies in a large tank. *J. Plankton Res.*, 5, 619 – 635.
Minnaert, M. J. G. (1954) The Nature of Light and Colour in the Open Air. Dover Publications, New York. 362 pp.
Moore, R. E. and Scheuer, P. J. (1971) Palytoxin: a new marine toxin from a coelenterate. *Science,* 172, 495 – 497.
Mori, S. (1943) Daily rhythmic activity of the sea pen, *Caverulana obesa* Valenciennes. II. Activities under constant darkness and illumination. *Zool. Mag. Takyo,* 55, 247– 253.
Mori, S. (1960) Influence of environmental and physiological functions on the daily rhythmic activity of a sea-pen. *Cold Spring Harbor Symp. Quant. Biol.,* 25, 333 – 344.
Mori, S. and Matutani, K. (1952) Studies on the daily rhythmic activity of the starfish, *Astropecten polyacanthus* Muller et Troschel, and the accompanied physiological rhythms. *Publ. Seto Mar. Biol. Lab. II*, 213 – 225.
Morin, J. G. (1988) Piscivorous behavior and activity patterns in the tropical ophiuroid Ophiarachna *incrassata* (Ophiuroidea: Ophiodermatidae). pp. 401– 407. In Burke, R. D. Mladenov, P. V., Lambert, P. and Parsley, R. L. (eds). *Echinoderm Biology*., Balkema, A. A. Brookfield, VT, USA.
Muller, K. (1978) The flexibility of the circadian system of fish at different latitudes. pp. 91–104. In Thorpe, J. E. (ed), *Rhythmic Activity of Fishes*, Academic Press, NY. 312 pp.
Muntz, W. R. A and Richard, D. S. (1975) Photomechanical movements in the trout retina following brief flashes of light. *Vision Res.*, 22, 529 – 530.
Munz, F. W. and McFarland, W. N. (1973) The significance of spectral position in the rhodopsins of tropical marine fishes. *Vision Research,* 13, 1829 –1874.
Naylor, E. and Hartnall, R. G. (eds). (1979) Cyclic Phenomena in Marine Plants and Animals. *Proc. 13th European Marine Biol. Symp.* Pergamon Press, New York. 477 pp.
Neilson, J. D. and Perry, R. I. (1990) Diel vertical migrations of marine fishes: an obligate or facultative process? pp. 115 –168. In Blaxter, J. H. S. and Southward, A. J. (eds), *Advances in Marine Biology*, Vol. 26. Academic Press, London. 314 pp.
Nielsen, E. T. (1961) Twilight and the ‘Crep’ unit. *Nature* 190, 878 – 879.
Nielsen, E. T. (1963) Illumination at twilight. *Oikos,* 14, 9 –21.
Newell, R. C. and Northcraft, H. R. (1967) A re-interpretation of the effect of temperature on the metabolism of certain marine invertebrates. *J. Zool. (Lond.),* 151, 277– 298.
Norris, K. S., Wursig, B., Wells, R. S. and Wursig, M. (1994) *The Hawaiian Spinner Dolphin.* Univ. California Press, Berkeley. 408 pp.
Ogden, J. C., Brown, F. A. and Salesky, N. (1973) Grazing by the echinoid *Diadema antillarum* Philippi: formation of halos around West Indian patch reefs. *Science,* 182, 715 –717.
Ohlhorst, S. L. (1982) Diel migration patterns of demersal reef zooplankton. *J. Exp. Mar. Biol. Ecol.*, 60, 1–15.

Ohman, M. D. (1990) The demographic benefits of diel vertical migration by zooplankton. *Ecol. Monogr.,* 60, 257– 281.
Ott, B. and Lewis, J. B. (1972) Importance of the gastropod *Coralliophila* abbreviata (Lamarck) and the polychaete *Hermodice carunculata* (Pallas) as coral reef predators. *Can. J. Zool.,* 50, 1651–1656.
Pages, E. and Petter-Rousseau, A. (1980) Annual variations in the circadian activity rhythms of five sympatric species of nocturnal prosimians in captivity. pp. 153 –167. In M. Rumbaugh (ed). *Nocturnal Malagasy Primates*. Academic Press, New York. 215 pp.
Pariente, G. F. (1974) Influence of light on the activity rhythms of two Malagasy lemurs: *Phaner furcifer* and *Lepilemur mustelinus leucopus*. pp. 183 –198. In Martin, R. D., Dyle G. A. and Walker, A. C. (eds). *Prosimian Biology.* Duckworth, London. 983 pp.
Pariente, G. F. (1979) The role of vision in prosimian behavior. pp. 411– 459. In Doyle, G. A. and Martin, R. D. (eds). *The Study of Prosimian Behavior.* Academic Press, New York. 696 pp.
Pariente, G. F. (1980) Quantitative and qualitative study of the light available in the natural biotope of malagasy prosimians. pp. 117–136. In Rumbaugh, D. M. (ed). *Nocturnal Malagasy Primates*. Academic Press, New York. 215 pp.
Passano, L. M. (1973) Behavioral control systems in medusae; a comparison between hydro- and scyphomedusae. *Publ. Seto Mar. Biol. Lab.,* 20, 182 – 200.
Paul, V. J. (1992) Chemical defenses of benthic marine invertebrates. pp. 164 –188. In Paul, V. J. (ed). *Ecological Roles of Marine Natural Products*. Comstock Publ., Ithaca, NY. 245 pp.
Pawlik, J. R. (1993) Marine invertebrate chemical defenses. *Chem. Rev*., 93, 1911–1922.
Pawlik, J. R. and Fenical, W. (1992) Chemical defense of *Pterogorgia anceps,* a Caribbean gorgonian coral. *Mar. Ecol. Prog. Ser.,* 87, 183 –188.
Pawlik, J. R., Chanas, B. Toomen, R. J. and Fenical, W. (1995) Defenses of Caribbean sponges against predatory reef fish. I. Chemical deterrency. *Mar. Ecol. Prog. Ser*., 127, 183 –194.
Pearre, S. (1973) Vertical migration and feeding in *Sagitta elegans* (Verrill). *Ecology,* 54, 300 – 314.
Pearse, V. B. (1974) Modifications of sea anemone behavior by symbiotic zooxanthellae expansion and contraction. *Biol. Bull.,* 147, 641– 651.
Percharde, P. L. (1972) Observations on the gastropod *Charonia variagata* in Trinidad and Tobago. *Nautilus,* 85, 84 – 92.
Pittendrigh, C. S. (1993) Temporal organization: Reflections of a Darwinian clock-watcher. *Ann. Rev. Physiol.,* 55, 17– 54.
Porter, J. W. (1974) Zooplankton feeding by the Caribbean reef building coral, *Montastrea cavernosa. Proceedings Second International Coral Reef Symposium,* 1, 111–125.
Porter, J. W., Porter, K. G and Batac-Catalan, Z. (1977) Quantitative sampling of Indo-Pacific demersal reef plankton. Proceedings Third International Coral Reef Symposium 1, 105 –112.
Postuma, K. H. (1958) The vertical migration of feeding herring in relation to light and the vertical temperature gradient. *Int. Coun. Explor. Mer*. Herring Committee., Mimeo Rept. 8 pp. (not seen, from Blaxter 1975).
Potts, G. W. (1980) The predatory behaviour of *Caranx melampygus* (Pisces) in the channel environment of Aldabra Atoll (Indian Ocean). *Zool. J. Lond*., 192, 323 –350.
Potts, G. W. (1981) Behavioral interactions between carangidae (Pisces) and their prey on the fore-reef slope of Aldabra, with notes on other predators. *J. Zool.*, 195, 385 –404.
Preston, E. M. and Preston, J. L. (1975) Ecological structure in a West Indian gorgonian fauna. *Bull. Mar. Sci.*, 25, 248 – 258.
Reese, E. (1966) The complex behavior of echinoderms. pp. 157– 218. In Boolootian, R. A. (ed). *Physiology of Echinoderms*. Wiley, New York. 822 pp.
Reiswig, H. M. (1971) *In situ* pumping activities of tropical Demospongiae. *Mar. Biol.,* 9, 39 –50.
Richardson, L. D. (1952) Some reactions of pelagic fish to light as recorded by echosounding. *Fishery Invest. Lond*., (Ser. 2) 18, 20 pp.
Ringelberg, J. (1964) The postively phototactic reaction of *Daphnia magna* (Straus): A contribution to the understanding of diurnal vertical migration. *Neth. J. Sea Res*., 2, 319 – 406.
Robichaux, D. M., Cohen, A. C., Reaka, M. L and Allen, D. (1981) Experiments with zooplankton on coral reefs, or, will the real demersal plankton please come up? *Mar. Ecol.,* 2, 77– 94.
Rosenberg, G. V. (1966) *Twilight, A Study in Atmospheric Optics.* Plenum Press, New York. 358 pp.
Russell, F. S. (1926) The vertical distribution of marine macroplankton. III. Diurnal observations on the pelagic young of teleostean fishes in the Plymouth area. *J. Mar. Biol. Assoc. UK*, 14, 1178 –1188.
Rutman, J. and Fishelson, L. (1969) Food composition and feeding behavior of shallow-water crinoids at Eilat (Red Sea). *Mar. Biol.,* 3, 46 – 57.

Sbikin, Y. N. (1976) Changes in the behavior of some fish from the Cuban Shelf during morning and evening twilight. *J. Ichthyol.*, 17, 785 –790.

Schenck, H. (1957) On the focusing of sunlight by ocean waves. *J. Opt. Soc. Am (A)*, 47, 653 – 657.

Schwassman, H. O. (1971) Biological rhythms. pp. 371– 428. In Hoar, W. S. and Randall, D. J (eds), *Fish Physiology,* V. 6. Academic Press, NY. 559 pp.

Sebens, K. P. (1977) Autotrophic and heterotrophic nutrition of coral reef zooanthids. *Proceedings Third International Coral Reef Symposium*, May 1977, Miami Florida. Part I, 398 – 404.

Sebens, K. P. and DeReimer, K. (1977) Diel cycles of expansion and contraction in coral reef anthozoans. *Mar. Biol.*, 43, 247– 256.

Segal, E. (1970) Light-invertebrates. pp. 159 – 211. In Kinne, O. (ed). *Marine Ecology*, Part I, Vol 1, Environmental Factors. Wiley-Interscience, New York. 681 pp.

Sides, E. M. and Woodley, J. D. (1985) Niche separation in three species of *Ophiocoma* (Echinodermata: Ophiuroidea) in Jamaica, West Indies. *Bull. Mar. Sci.*, 36, 701–715.

Sinclair, A. N. (1959) Observations on the behavior of sea urchins. *Aust. Mus. Mag.*, 13, 3 – 8.

Sorokin, Y. I. (1993) *Coral Reef Ecology.* Springer-Verlag, Berlin. 465 pp.

Southward, A. J. and Crisp, D. J. (1965) Activity rhythms of barnacles in relation to respiration and feeding. *J. Mar. Biol. Assoc. U.K.*, 45, 161–185.

Starck, W. A. and Davis, W. P. (1996) Night habits of fishes of Alligator Reef, Florida. *Icthyologia,* 38, 313 – 356.

Suchanek, T. H., Carpenter, R. C., Witman, J. D., and Harvell, C. D. (1983) Sponges as important space competitiors in deep Caribbean coral reef communities. pp. 55 – 59. In Reake, M. L. (ed), *The Ecology of Deep and Shallow Coral Reefs*. Symposia Series for Undersea Research 3(1)NOAA/NURP, Rockville, Maryland.

Sweeney, B. M. (1976) Circadian rhythms in corals, particularly Fungiidae. *Biol. Bull.*, 151, 236 – 246.

Takahashi, J. G., and Hoffman, M. (1995) *Molecular biological clocks. Am. Sci.*, 83, 58 – 65.

Thomas, M. L. H. and Jelly, E. (1972) Benthos trapped leaving the bottom in Bildeford River, Prince Island. *J. Fish. Res. Board Can.*, 29, 1234 –1237.

Thornton, I. W. E. (1956) Diurnal migrations of the echinoid *Diadema setosum* (Leake). *Br. J. Anim. Behav.*, 4, 143 –146.

Thorpe, J. E. (ed) (1978) *Rhythmic Activity of Fishes. Academic Press*, New York. 312 pp.

Vago, R., Gill, E. and Collingwood, J. C. (1997) Laser measurements of coral growth. *Nature,* 386, 30 – 31.

VanAlstyne, K. L. and Paul, V. J. (1992) Chemical and structural defenses in the sea for *Gorgonia ventalina:* effects against generalist and specialist predators. *Coral Reefs,* 11, 155 –159.

Voute, A. M., Sluiter, J. W. and Grimm, M. P. (1974) The influence of the natural light-dark cycle on the activity of pond bats (*Myotis clasycneme* Boie 1825) during summer. *Oecologia,* 17, 221– 243.

Wagner, H. J. and Douglas, R. H. (1983) Morphologic changes in teleost primary and secondary cells following brief exposure to light. Invest. *Opthalmol.* 24, 24 – 29.

Wahl, C. M., Mills, E. L., McFarland, W. N. and DeGisi, J. S. (1993) Ontogenetic changes in prey selection and visual acuity of the yellow perch, *Perca flavescens. Can. J. Fish. Aq. Sci.*, 50, 743 –749.

Wainwright, S. A. (1967) Diurnal activity of hermatypic gorgonians. *Nature,* 216, 1041.

Walls, G. L. (1942) *The Vertebrate Eye and It's Adaptive Radiation.* (1963) Edition, Hafner Publ. Co., New York. 785 pp.

Wanzenbock, J., Zaunreiter, M., Wahl, C. M. and Noakes, D. L. G. (1996) Comparison of behavioral and morphological measures of visual acuity during ontogeny of roach (*Rutilus rutilus*) and yellow perch (*Perca flavescens*). *Can. J. Fish. Aqua. Sci.,* 53, 1506 –1512.

Waterman, T. H. (1954) Polarization patterns in submarine illumination. *Science,* 120, 927– 932.

Waterman, T. H. (1981) Polarization sensitivity, Ch. 3, pp. 281– 469. In H. Austrum (ed) *The Handbook of Sensory Physiology*, V. VII/6B, Vision in Invertebrates B, Invertebrate Visual Centers and Behavior I. Springer-Verglag, Berlin. 629 pp.

Weider, L. J. (1984) Spatial heterogeneity of *Daphnia* genotypes: vertical migration and habitat partitioning. *Limnol. Oceanogr.,* 29, 225 – 235.

Weider, L. J. (1985) Spatial and temporal genetic heterogeneity in a natural *Daphnia* population. *J. Plankton Res.*, 7, 101–123.

Welsh, J. J. and Osborn, C. M. (1937) Diurnal changes in the retina of the catfish, *Ameiurus nebulosus. J. Comp Neurol.*, 66, 349 –359.

Wells, G. P. (1950) Spontaneous activity cycles in polychaete worms. *Symp. Soc. Exp. Biol.*, 4, 128 –142.

Worthington, E. B. (1931) Vertical movements of fresh water macroplankton. *Int. Rev. Hydrobiol.*, 25,

394 – 436.
Yamanouchi, T. (1939) Ecological and physiological studies on the holothurians in the coral reef of Palao Islands. *Palao Trop. Biol. Stat. Studies,* 4, 603 – 635.
Yamanouchi, T. (1956) The daily activity rhythms of the holothurians in the coral reef of Palao Islands. *Publ. Seto Mar. Biol. Lab.*, 3, 347– 362.
Yoshida, M. (1979) Extraocular perception. pp. 581– 640. In H. Autrum (ed). *The Handbook of Sensory Physiology,* VII/6A. Springer-Verlag, Heidelberg. 729 pp.
Zawhlen, R. (1975) Die lokomotorishe aktivitat des Eichhornchens (*Sciurus vulgaris*). *Oecologia,* 22, 79 – 98.

20
Vision and behavior in primates

G.H. JACOBS
Neuroscience Research Institute and Department of Psychology, University of California, Santa Barbara, CA 93106, USA

20.1 Introduction

Descriptions of living primates inevitably include an acknowledgement of the overwhelming importance of sight for primate behavior, and they usually identify as unique primate characteristics a number of features of the visual system. These features include frontally-directed eyes and large binocular fields, a high concentration of ganglion cells in the central retina and an expanded representation of the central visual field in the visual system, substantial numbers of uncrossed retinofugal fibers, and a visual cortex that contains multiple maps of the visual field in which inputs from the two eyes may be brought into precise register (Allman, 1982). It has been argued that the relative enlargement of brain size in primates directly reflects an increased emphasis on vision (Martin, 1990), and recent mappings of the projection pathways and functional properties of cortex serve to underline just how extensively visual information must be utilized by the primate brain. In the macaque monkey, for instance, there are a minimum of 32 or more separate visual areas lying beyond the primary visual cortex. These areas expand over a region making up at least 75% of the cortical surface area outside of striate cortex (Sereno and Allman, 1991). Less detailed maps are available for human cortex, but they too show a very substantial visual representation (Sereno *et al.,* 1995; Tootell *et al*., 1996). Although it is an obvious exaggeration, it is almost justified to conclude that in many ways primate neocortex *is* visual cortex.

If primates are such visual animals, then one would expect to find an extensive catalogue of the precise ways in which primates use visual information in the conduct of their normal behavior. But none exists. For example, commentary on primates is essentially absent from John Lythgoe's seminal treatment of the ecology of vision (Lythgoe, 1979). And in the voluminous literature reporting field observations on nonhuman primates, it is virtually impossible to find observations that lead one to anything but the most general inferences about the relationships between the photic world and primate visual capacities. Perhaps this state of affairs is not surprising. After all, as Lythgoe's book and a more recent summary of the literature on visual ecology (Dusenberry, 1992) so ably document, analysis of relationships between photic signals and visual behavior

S.N. Archer et al. (eds.), Adaptive Mechanisms in the Ecology of Vision, 629–650

has so far been most productive in just those cases where behavior is the most stereotyped – movements toward or away from light sources, behaviors released in the presence of particular faunal or floral colors, sensitivity to moving but not stationary objects, and so on. Primate behavior is usually anything but stereotypic; in fact, primate survival results from the versatility and adaptability of its member species. Although these traits are a powerful asset in assuring primate success, they make the task of understanding relationships between vision and behavior in primates particularly challenging. In what follows I have elected to focus on two of the most basic features of vision in nonhuman primates – visual sensitivity and color vision. The motivation for such a restriction is that over the past 15 years there has been substantial advance in our understanding of the low-level biological mechanisms that underlie these visual attributes. That understanding has come, at least in considerable part, from studies of primates.

20.2 Primate evolution

Primates that at least somewhat resemble living forms emerged from a line of placental mammals during the early Eocene 55 to 60 million years ago (Kay *et al.*, 1997). Early primates are assumed to have been small (500 gms or less), arboreal animals that lived in tropical forest ecosystems. Whether these animals were nocturnal, like the common ancestors of placental mammals in general, or diurnal is a matter of debate (see Martin, 1990; Kay *et al.,* 1997). The 200-or-so species that make up the modern primates are divided into two suborders, *Prosimii and Anthropoidea.* The prosimians are the more primitive. Many contemporary prosimians are nocturnal and it has been suggested that these species may be regarded as having preserved many of the features of ancestral primates (Allman, 1982; Fleagle, 1988; Martin, 1990). From the viewpoint of vision and behavior, study of prosimians thus provides the possibility of looking backward in time toward the beginnings of those visual adaptations that might be thought to be unique to primates.

With the exception of the owl monkey (*Aotus*), all of the anthropoid species (typically called 'higher primates') are diurnal. There have been three major anthropoid radiations, the platyrrhine monkeys of the New World and two groups of catarrhines – Old World monkeys and hominoids respectively. The platyrrhine monkeys have an evolutionary history in the New World of 30 to 40 million years. Over the years there has been much debate about the origins of the platyrrhines (e.g. Ciochon and Chiarelli, 1980), but recent fossil evidence supports the claim of an African origin for this line (Flynn *et al.*, 1995). Old World monkeys and hominoids diverged some 30 million years ago with a subsequent, much more recent, divergence (perhaps 5 million years ago) between the great apes and humans.

20.3 Primate photopigments

Photopigments figure centrally in the treatment offered here and thus we require a few summary comments about these remarkable molecules. During the past decade there has been a significant increase in knowledge about primate photopigments. This comes both from an expanded set of photopigment measurements and from vigorous study of the genetic basis and structural organization of photopigments. These topics have been the subjects of a number of reviews (Nathans, 1987; Bowmaker, 1990; Jacobs, 1996; Mollon, 1997; Neitz and Neitz, 1998) and they are treated in detail elsewhere in this volume (see Bowmaker and Hunt, Chapter 13). Here we need only remind the reader that photopigment molecules are composed of a protein, an opsin, that is covalently bound to a chromophore. A single chromophore is common to all primates photopigments, but opsin structures vary and that variation determines the spectral absorption properties of the photopigment. Opsins are specified by the action of single genes. The genes for mammalian rod photopigments are located on chromosome 5; cone pigment genes of mammals are on chromosome 7 and the X-chromosome.

Photopigments are conservative in the sense that structural variation in the opsin appears to allow for the production of only a limited number of different photopigment types. These variations yield discrete differences in the spectral locations of the absorption peaks (λ_{max}) of the photopigments. Primate rod photopigments absorb maximally in the vicinity of 500 nm, but whether or not there are some small systematic variations around this figure for different species remains to be established. Primate cone photopigments show maximum absorption in two different regions of the spectrum. Those absorbing maximally in the middle to long wavelengths (here referred to as M/L pigments) span a λ_{max} range from about 530 nm to 562 nm. Measurements of primate pigments indicate that there are probably six principal varieties of M/L photopigments. The spectra for these pigments are illustrated in Figure 1. Note that they are stepped along the spectrum such that successive λ_{max} values are at spaced intervals of about 5 –7 nm. *In vitro* manipulations of opsin structures yields spectral shifts in photopigments that are smaller than (and thus, potentially, intermediate to) these step sizes (Asenjo *et al.*, 1994; Merbs and Nathans, 1992), but whether these pigment variations exist naturally is not yet conclusively established. The opsins for these M/L pigments are the expression products of genes located on the X-chromosome. The other available primate cone pigments absorb maximally in the short wavelengths (S pigments). The measurements of these pigments are less secure, but it appears that there are only a few types of S-pigment. To date it looks as if there are at least three primate S-pigment types positioned at stepped spectral locations over the range from perhaps about 415 nm to 435 nm (Dartnell *et al.*, 1983; Bowmaker *et al.*, 1991; Tovée *et al.*, 1992). The genes for S-cone pigment opsins are on chromosome 7.

Absorption of light by photopigments results in photoreceptor signals whose characteristics reflect both the photon flux across the receptor array and the

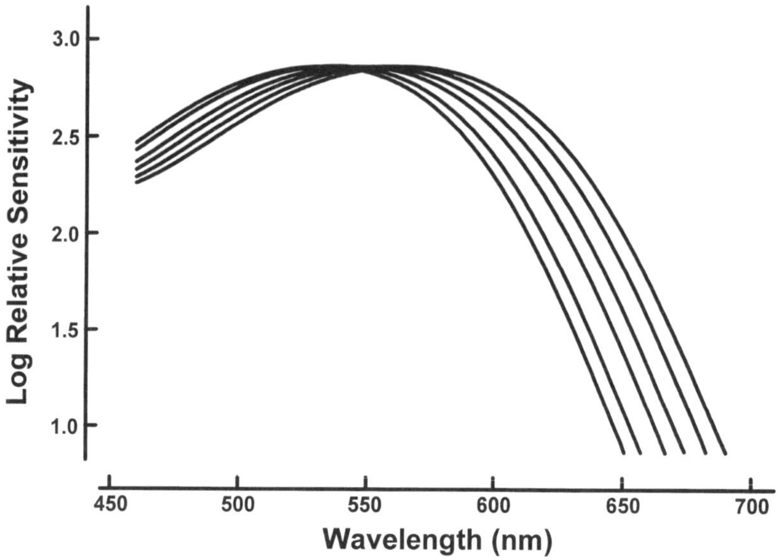

Figure 1 Absorption spectra for the primate M/L cone photopigments. Each of these six different types of photopigment has been found in more than one primate species. The retinas of individual nonhuman primates have any one of these photopigments or a pair of them. The determined λ_{max} values for these pigments vary somewhat depending on the measurement technique; those shown here have respective peak values of 530, 535, 543, 549, 556 and 562 nm

immediate history of light exposure. It has long been appreciated that the neural circuits of the retina deal in two fundamentally different fashions with signals generated by receptors that contain different types of photopigment: inputs from different pigment classes are added to yield nonopponent signals or they are subtracted to yield spectrally-opponent signals (see Chapters 10 and 11). Unlike the retinas of many lower animals, in which spectral opponency is inaugurated in the outer retina, the neural interactions that lead to spectral opponency in the primate occur as a result of signal convergence among bipolar, amacrine and ganglion cells in the inner retina (e.g. Dacey and Lee, 1994). There is a vast literature to support the conclusion that spectral-opponency is an essential feature of the neural fabric underlying color vision (Lennie and D'Zmura, 1988). Nonopponent signals are used for other ends, including strong, sometimes exclusive, contributions to various measurements of spectral sensitivity and to the detection of moving targets.

20.4 Photopigment variations among primates

Although it may well prove an oversimplification, it is convenient to assume that the spectral sensitivity and color vision of all primates fall into one of a small number of patterns. These patterns can be illustrated by reference to the photopigments found in three groups of primates and the potential vision these pigments may permit.

20.4.1 Prosimians

As far as is now known, the retinas of all mammals contain a population of cones (Jacobs, 1993). There are scattered exceptions (e.g., various squirrels and tree shrews), but mammalian cones usually comprise a small proportion (10% or less) of the total number of receptors and contain two types of cone photopigment with absorption peaks displaced 80 or more nm apart. The spectral placement of these two pigment types varies a good bit: one usually has peak sensitivity of 450 nm or shorter; the other pigment peaks to the long side of 500 nm. When animals having such pigment arrangements are properly tested they are typically found to have dichromatic color vision (Jacobs, 1993).

If, as has been argued, present-day prosimians are the species least changed from the earliest primates it would be useful to know how their spectral sensitivity and color vision relates to that of other mammals as well as to other primates. Although we do not possess detailed information about vision and its mechanisms in many of these animals, there are a few studies that illuminate the issue. The half-dozen families of prosimians includes species that are thoroughly nocturnal as well as species that are diurnal or crepuscular in habit. Measurements of photopigments have been made with an electrophysiological technique for two species of lemur that fall into the latter category (the ringtail lemur, *Lemur catta* and the brown lemur, *Eulemur fulvus*) (Jacobs and Deegan, 1993). Much like the pattern described above as typical for many other mammals, the retinas of both species were found to contain two classes of cone photopigment with respective peak values of about 437 and 545 nm. This arrangement predicts dichromatic color vision. Earlier behavioral studies had in fact demonstrated that ringtail lemurs were capable of making some color discrimination, although the nature of these color discriminations made a diagnosis of the nature of their color vision somewhat ambiguous (Blakeslee and Jacobs, 1985).

Similar pigment measurements have recently been completed in a prosimian that is strongly nocturnal, the bushbaby (*Otolemur crassicaudatus*) (Deegan and Jacobs, 1996). Like the two species of lemur, the bushbaby has an M/L cone pigment with a peak of about 545 nm. Surprisingly, that is apparently the only cone pigment in the retina of this primate. Although there are no correlative behavioral measurements, the implication from these pigment measurements is clear – the bushbaby cannot have color vision in the way that capacity is usually defined.

The bushbaby and the lemurs share in common an M/L pigment. Why do the diurnal/crepuscular species have an S-cone pigment while at least one nocturnal species fails to have such a pigment? A recent examination revealed the genetic basis for this difference (Jacobs *et al.*, 1996a). It was discovered that the bushbaby has a gene that is highly homologous to the human gene that produces the opsin for S-cone pigment, but there are deleterious mutational changes in the sequence sufficient to obviate gene expression. One interpretation is that all prosimians have cone pigment genes similar to what must characterize many nonprimate mammals, i.e., two different genes that normally produce an S-pigment and a single type of M/L cone pigment. However, in some of these species (perhaps only those that are nocturnal?) the S-cone pigment gene is nonfunctional and this leaves the animal with no prospect of color vision.

20.4.2 New World monkeys

Studies of New World monkeys have led to substantial insights into the basic biology of visual sensitivity and color vision. From a series of investigations involving both direct pigment measurements and visual discriminations it was established that many species from among this group have polymorphic variations in color vision. These variations can be directly traced to individual variations in cone photopigments and, in turn, the genetic basis of these pigment variations is known. The research backing these conclusions has been the subject of recent reviews (Jacobs, 1995; 1996; 1998). Accordingly, only a brief treatment is necessary.

Direct tests of color vision have been completed on members of three species of New World monkey (the squirrel monkey – *Saimiri sciureus*, the saddle-backed tamarin – *Saguinus fuscicollis*, and the common marmoset – *Callithrix jacchus*). In each case individual monkeys have either dichromatic or trichromatic color vision with several distinct subvarieties existing in each of these categories (Jacobs,1984; Jacobs *et al.*, 1987; Tovée *et al.*, 1992). These color vision variations arise from the fact that a number of different photopigment combinations are available in each species. There are at least two different sets of photopigments found in various species of New World monkeys. This group includes the three species mentioned above along with several additional species for which pigment measurements have now been completed (see Figure 2). Both of these pigment sets contain an S-cone pigment that is common to all members of the species. In addition, each grouping has three different M/L cone pigments. Individual monkeys have any one of the three M/L pigments or they have any pair drawn from the three. The former animals have one of three kinds of dichromatic color vision, the latter have one of three types of trichromatic color vision. A significant feature of the polymorphisms of pigments and color vision in these New World monkeys is its commonality – no single phenotype characterizes more than about 22% of the total population (Jacobs *et al.*, 1993b). Color vision variation is thus a very different matter in these monkeys than it is among human populations. The latter are of course known to show significant

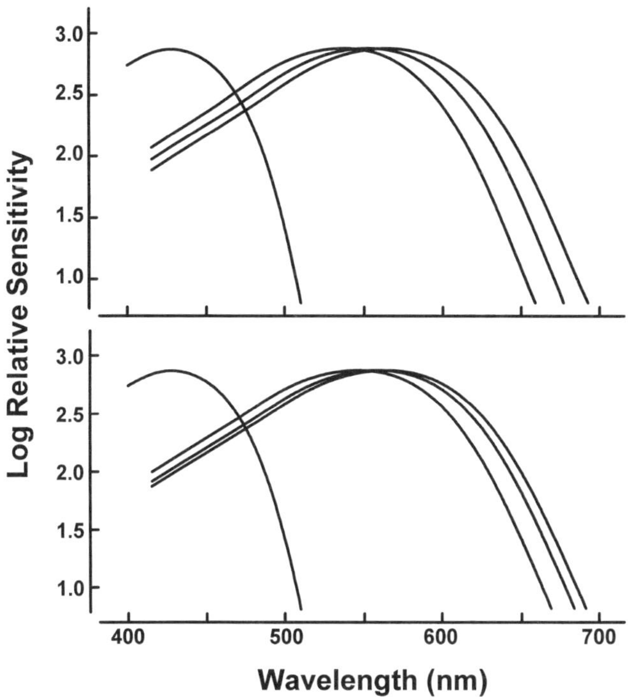

Figure 2 Spectra for the photopigments found in two groups of New World monkeys. Members of each group have the same S-cone pigment, but there is variation in the M/L pigment complement of individuals. The retinas of individuals in each group contain any one of the three M/L pigments or any pair; the former have dichromatic color vision, the latter are trichromats. The pigment set shown at the top has been found in species from the following genera: *Cebus, Saimiri.* Two of these pigments have also been found in *Ateles*. The pigment set at the bottom is known to be present in species from *Callithrix* and *Saguinus*; two of these pigment types have also been seen in *Leontopithecus*

color vision variations; however, unlike the New World monkeys, some 95% of all people have color vision capacities that are essentially the same.

Two other points need to be made about color vision and its photopigment underpinnings among New World monkeys. First is the case of the howler monkey, *Alouatta*. Recent measurements of photopigments and examination of photopigment genes from members of two species of howler monkey (*A. seniculus* and *A. caraya*) reveals that these monkeys are not polymorphic in the manner described for other New World monkeys. Rather, it appears they are routinely trichromatic and their trichromacy appears to be based on the same M and L cone pigments as are found in many catarrhine primates (Jacobs *et al.*, 1996b). Among other things, this discovery underlines the fact that there is still more work to be done with the New World monkeys – there are some 65 species in this group and many remain to be studied. Second stands the intriguing case of the owl monkey, *Aotus*. Unlike all the other simians, the owl monkey is nocturnal. Recent tests of color vision and measurement of photopigments in the owl monkey reveals that it is very similar in these regards to the bushbaby. Like the bushbaby, the owl monkey has only a single type of cone photopigment (an M/L pigment with a 543 nm peak) and thus it lacks color vision (Jacobs *et al*., 1993a). Subsequent molecular genetic examination showed that the owl monkey is also similar to the bushbaby in having an S-cone pigment gene containing deleterious mutational changes. Although the nature of the S-pigment gene defects are different for owl monkey and bushbaby, the outcome is the same: neither produce an S-cone pigment and neither has color vision (an exception might be under viewing conditions where joint signals from rods and cones could be utilized – see Jacobs *et al*., 1993). The loss of a functional S-cone in these nocturnal primates is similar in nature to that detected in one of the human congenital color defects, tritanopia. Like owl monkeys and bushbabies, the S-pigment genes of tritanopes have acquired mutations that obviate the production of functional S-cone pigment (Weitz *et al*., 1992a; 1992b). An intriguing difference between the two cases is their frequency: only about one in 3000 people have suffered a loss of S-cones in this fashion while it appears probable that all owl monkeys and bushbabies lack S-cones.

Why is there so much individual variation in the photopigments (and consequently the visual capabilities) of New World monkeys? The biological mechanisms underlying the variation are known. The key to their understanding was the observation that although individual female New World monkeys may have either two or three classes of cone pigment (and thus various forms of dichromatic and trichromatic color vision), all male New World monkeys are limited to two classes of cone pigment. A model to explain this proposed that there is only a single pigment gene locus on the X-chromosome in these monkeys (Mollon *et al*., 1984; Jacobs and Neitz, 1985). The site was assumed to have three alleles corresponding to the three possible classes of M/L cone pigment (Figure 2) and this means that all males and homozygous females (i.e. those having the same pigment gene allele on each X-chromosome) have only a single M/L cone pigment and dichromatic color vision. However, heterozygous females have two different M/L pigment genes and thus three classes of cone pigment and trichromatic color vision. There is now substantial evidence to suggest that this mechanism correctly accounts for the variations in cone photopigments among New World monkeys (Jacobs *et al*., 1993b). The deeper question as to what functional utility might be served by this variation proves more challenging. Some thoughts on this issue are offered below.

I noted that howler monkeys are different from many other New World monkeys. They seem to be routinely trichromatic and the genetic mechanism underlying their trichromacy is apparently similar to that of catarrhine primates; that is, they have more than one pigment gene locus on their X-chromosomes (Jacobs *et al*., 1996b).

20.4.3 Old World monkeys, apes and people

Compared to the New World monkeys, the photopigment story for the various catarrhines seems straightforward indeed. As far as we know, all of the species in this group routinely have three different classes of cone photopigments and trichromatic color vision (Jacobs, 1995). On average these three pigments have absorption peaks at about 420, 530 and 560 nm. It remains a bit uncertain whether or not the three photopigments are exactly the same for all of these species. We know that in the case of the S-cone pigment there are clear differences between people and various Old World monkeys (Bowmaker, 1990), and there may well be other differences in the spectral positioning of the M and L cone pigments among the catarrhines. However, any such differences seem likely to be small and thus unlikely to have a significant impact on vision.

The basic difference in cone photopigment complements between the catarrhine primates and New World monkeys (with the notable exception of the howler monkey – see above) reflects the fact that the former have multiple pigment gene loci on their X-chromosomes (Nathans *et al.*, 1986). These loci have the genes for both M and L pigments and, thus, the presence in the retina of both pigment types is routine. This makes trichromatic color vision the norm. There are well-known polymorphic variations in cone photopigments in human populations (Neitz and Neitz, 1994). The more significant of these result in clear color vision defects (dichromacies and anomalous trichromacies), some of which resemble the variants of color vision seen among New World monkeys. Although such polymorphic variations are well established among humans, there is so far no evidence that a similar variation is found in any of the other catarrhine species; included in this group is our closest relative, the chimpanzee (Jacobs *et al.*, 1996c). Whether this difference between human and nonhuman primates is merely an artifact of insufficient sampling or whether it may represent something more profound remains uncertain.

20.5 Primate photopigments: prospects for spectral sensitivity and color vision

To understand the substantial thematic variations in the photopigments found in the retinas of various primates, it may be useful to examine in a little more detail how these pigments relate to fundamental features of seeing. It was noted that (a) nervous systems can add or subtract signals delivered from different receptor types, and (b) the potential number of different pigment types that can be manufactured from structural variations in cone opsins appears limited. A reasonable assumption is that the starting point for primate vision involved a photopigment complement similar to that seen in some contemporary prosimians and in many other non-primate mammals, i.e. an S-cone pigment and a single type of M/L cone pigment. Which among the various possible primate M/L pigments (Figure

1) provides the greatest potential for vision? Although the lighting available in visual environments can vary greatly (depending, for instance, on climatic conditions, time of day, and the local vegetation), measurements of a wide variety of terrestrial habitats indicate that most contain abundant amounts of long-wavelength light (Endler, 1993). If the goal is to maximise photon capture it makes sense to position the M/L pigment as far toward the long-wavelengths as possible. Figure 3 illustrates this contention with absorption spectra for the shortest and the longest of the primate M/L pigments (solid lines). Note that the long-wavelength slopes of these pigments are steep and, at the limit, equal (dropping in sensitivity roughly 0.3 log unit every 10 nm). A consequence is that an animal having the shorter pigment is some sixteen times less sensitive to a 650 nm light than one whose retina contains the longer pigment. By comparison, the falloff in absorption efficiency toward the short wavelengths is relatively modest (Figure 2). Other things being equal, one would thus expect that selective advantage would favor animals with longer rather than shorter M/L cone pigments and, indeed, every higher primate species has a representation of the longest of these M/L pigments (λ_{max} = ~562 nm), either routinely (in the Old World anthropoids) or as a variant (in the polymorphic New World monkeys). The exceptions to this rule among the primates are prosimians and the nocturnal owl monkey, all of whom probably have an M/L pigment with a peak of about 543 – 545 nm. As noted above, these are the contemporary species often argued to have preserved many features of the earliest primates. Perhaps one such feature might be the spectral positioning of the ancestral M/L pigment.

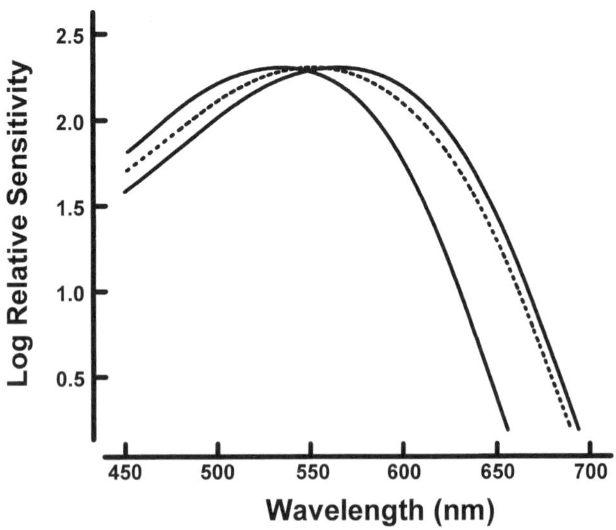

Figure 3 The two solid curves show the absorption spectra for the shortest and the longest primate M/L pigments (λ_{max} of 530 and 560 nm respectively). The dashed line shows the summed sensitivity of these two pigments with a weighing of 1:2. See text for further discussion

No analogous arguments can be offered with respect to the spectral positioning of primate S-cone pigments. Although there does appear to be some clear variation in S-cone spectra among primates, the spectral range is relatively modest. One might suspect that over this range the exact location of the pigment curve is relatively unimportant. There are three facts supporting this conclusion: (1) the M/L cone pigments still have reasonable sensitivity in the short-wavelength portion of the spectrum, (2) in primates, cones containing S-pigment are sparsely represented (<10% of all cones) and thus their contribution to total quantal capture *per se* is likely to be modest, and (3) higher primates have heavily pigmented lenses and macular pigments that severely attenuate short-wavelength lights and thus serve to effectively truncate the spectral range over which primate vision is possible. In the light of all these facts, there is probably only slight advantage to be gained from utilizing one S-cone pigment position versus another.

Of course, a big advantage of having both S and M/L cones in a retina is that their outputs can be combined in subtractive combinations to yield spectral opponency. This provides the possibility of making pure-color discrimination as well as using color as a cue for discrimination when multiple cues may be available. An illustration of the difference in behavior between situations that do and do not allow for the use of these subtractive signals is shown in Figure 4. Plotted in the left panel are discrimination data obtained from a saddle-backed tamarin, a platyrrhine monkey whose retina contained an S-cone pigment (λ_{max} of about 433 nm) and a single M/L cone pigment (λ_{max} = 543 nm) (Jacobs, 1990). The top of Figure 4 shows spectral sensitivity measured in a test involving the discrimination of a rapidly flickering light; at the bottom of the figure is the spectral sensitivity of this same animal when he was confronted with an increment-threshold discrimination. It is known that these two tests maximize, respectively, the contributions made by additive and subtractive signals in the visual system (Boynton, 1982). Note that detection of flickering lights involves little or no contribution from the S cones whereas the increment-detection task has a major contribution from subtractive combinations of the two cone types. So, one consequence of the addition of a second cone type and the utilization of subtractive signals between the cone types is that they can provide substantially heightened sensitivity to short-wavelength lights.

The behavioral tests that yielded the spectral sensitivity functions of Figure 4 each provided two stimulus differences that monkeys might exploit to successfully detect the target: a spectral difference and an intensity difference. The former provides a potential color cue, the latter a lightness or brightness cue. By their design, these two tests emphasized the use of one or the other of these cues and in so doing show the detection capabilities of two different organizational features of primate visual systems. These laboratory tests are very contrived, but they can serve as shorthand approaches to the kinds of discriminations animals make in a world offering a welter of spectral and intensive variation. Thus, the increment-detection test might approximate the challenge offered by focal

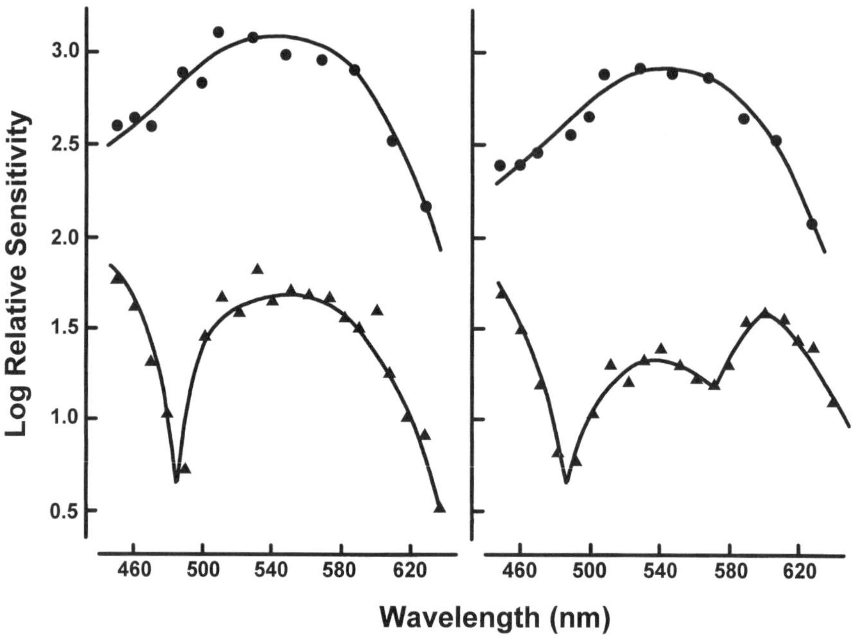

Figure 4 Spectral sensitivity curves for dichromatic (left panel) and trichromatic (right panel) saddle-backed tamarins as measured in behavioral discrimination tasks. The top function (circles) in each case was obtained with a flicker discrimination paradigm in which the animals were required to discriminate a rapidly flickering (30 Hz) monochromatic light from a steady light of the same time-averaged luminance. The bottom functions (triangles) in both panels show spectral sensitivity curves determined in an increment-threshold discrimination task. (These experiments are fully described in Jacobs *et al.*, 1987)

viewing of a preferred food embedded in a sea of vegetation, while the flicker discrimination taps capacities that are required in situations involving rapid spatiotemporal change; for instance, that occasioned by the flight of a tasty insect through the field of view.

Laboratory tests of vision can be designed to isolate cue usage even further. A prime example are tests of color vision in which one assesses the ability to discriminate spectral differences *per se*. Under such constraint, the possibilities offered by a visual system with two active photopigments becomes obvious. Figure 5 (solid line) shows a wavelength discrimination function obtained from a dichromatic monkey. In essence, in the dichromat, the spectrum is split such that animals can successfully discriminate among stimuli whose spectral energy distributions show a shape difference for wavelengths shorter and longer than about 500 nm. It has been argued that such an arrangement will allow color-based discriminations between the spectral reflectance's of two major categories of natural objects: those containing chlorophyll on the one hand and those whose

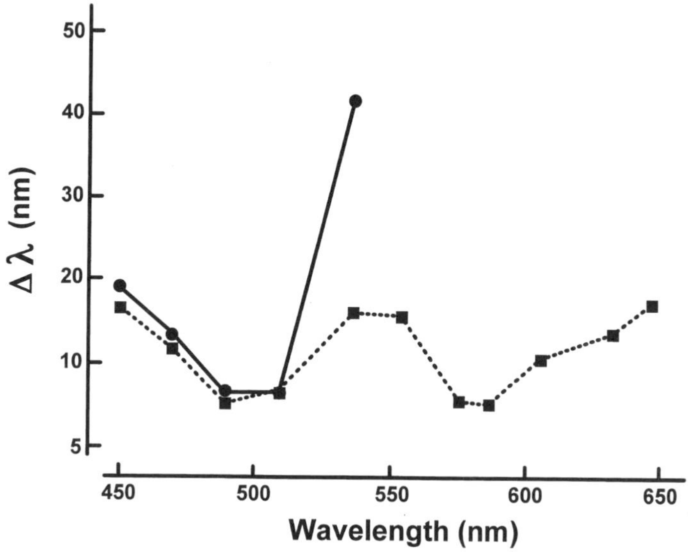

Figure 5 Wavelength discrimination functions measured on a dichromatic (circles) and on a trichromatic (squares) squirrel monkey. (Data from Jacobs, 1984)

spectrum increases monotonically from short to long wavelengths on the other (items that include soil, bark, and melanin-based pigments) (Lythgoe and Partridge, 1989). Along with many other mammals, it seems likely that the diurnal prosimians and perhaps about two-thirds of all individual New World monkeys will have color vision capacities formally similar to this.

Addition of a second class of M/L pigment to the retina yields substantial advantage. First, quantal capture is enhanced over that provided by any single type of M/L pigment. A theoretical illustration of this is given in Figure 3 which shows the summed sensitivities of the two pigments whose individual curves are also displayed. In this summation (dashed curve), the cone sensitivities have been weighted in a ratio of 2L:1M to mimic the relative proportions of the two cone types as they are found in the human retina (see below). Even with this simple assumption about the combination of signals from the two cone types one can see that the pair of pigments effectively combine the advantages for photon capture of either one of the pigments alone. But the real advantage of adding a second M/L cone type is that it allows the nervous system to exploit subtractive signals. Figure 4 (right panel) shows flicker and increment-threshold spectral sensitivity curves from a saddle-backed tamarin whose retina contained two types of M/L pigment (λ_{max} = ~543 nm and 557 nm). As in the case of the dichromatic monkey described previously, subtractive combination of signals allows for an enhanced sensitivity to the short wavelengths over that achieved with additive combination alone. In addition, the figure also shows that subtractive signals from the two M/L cone types yield greatly elevated sensitivity to the long wavelengths.

Reducing the potential cues for discrimination to spectral differences alone illustrates even better the advantage of adding a second M/L pigment. Figure 5 (dashed line) also shows the wavelength discrimination function obtained from a trichromatic monkey. Unlike the animal with only a single M/L pigment, this animal can use subtle spectral differences between long-wavelength stimuli to achieve successful discrimination, e.g. pure-wavelength differences of only a few nm can generate discriminable sensations. By analogy to human color vision, this indicates that a much richer gamut of hues is available to the trichromatic monkey. Some speculations about the utility of this additional capacity are offered below.

This discussion has focused on the number of cone types and their relative spectral positioning because the implications these features hold for seeing have been much studied. Beyond that, there are distributional aspects of the cone types that can impact the nature of spectral sensitivity and color vision; for instance, the overall rod/cone ratio and, particularly, the local cone density. It was noted above that the retinas of some prosimians and some individual New World monkeys contain very similar cone types – each has an S cone and one type of M/L cone (Jacobs, 1993). Despite the similarity of cone types, the acuteness of color vision of these animals differs greatly. The monkeys are able to discriminate spectral variations which are quite invisible to the prosimians. That difference presumably arises from the fact that New World monkeys have a central retinal region of high cone density, while the cones of the prosimian are more scattered across the extent of the retina. The latter arrangement will inevitably decrease the possibilities for local neural interactions of the kind necessary to generate strong color-vision signals.

Another feature of retinal construction beyond the number of cone types and their spectra that appears to vary among the trichromatic Old World primates is the relative numbers of L and M cones. There is a good deal of evidence to indicate that in the human retina L cones outnumber M cones by about 2:1 (Cicerone and Nerger, 1989; Vimal *et al.*, 1989; Jacobs and Neitz, 1993; Hagstrom *et al.*, 1994). In contrast, it appears that these two cone classes are about equally represented in the retinas of various Old World monkeys (Bowmaker, 1991; Mollon and Bowmaker, 1992; Calkins *et al.*, 1994; Jacobs and Deegan, 1997), and there is indication that our closest primate relative, the chimpanzee, may also have an L/M cone ratio closer to 1 than 2 (Jacobs *et al.*, 1996c). The perceptual consequences of these variations in the relative representation of M and L cones are not well understood, but there are two differences in the visual sensitivity and color vision of Old World monkeys and human observers that may be reflective of the differences in M and L cone representation of these species. First, side-by-side spectral sensitivity measurements made on humans and macaque monkeys indicate the latter to be somewhat less sensitive to long-wavelength lights (Harwerth and Smith, 1985; Jacobs and Deegan, 1997); this could quite likely reflect the apparently lower representation of L cones in the monkey retina. Second, comparable measurements of wavelength discrimination of the two species show that discrimination is consistently superior in humans (DeValois *et al.*, 1974; Sperling *et al.*, 1987). There could be any number of explanations for this difference. The relative representation of M and L cones is one possible factor.

20.6 Primate trichromacy: Why? How?

Based on studies of the structure of photopigment genes and comparative studies of color vision and photopigments, a common view is that early mammals had two classes of cone photopigment – an S pigment and a single type of M/L pigment. This provided the basis for dichromatic color vision, a general theme that has been retained in many present-day mammals (Jacobs, 1993). As I noted above, this arrangement is also characteristic of some contemporary diurnal and crepuscular prosimians. A second type of M/L pigment appeared some 30 million years ago leading to routine, trichromatic color vision in Old World monkeys and hominoids. Trichromatic color vision appears in the form of a polymorphic variant in many New World monkeys and at least one genus from among this group seems to have achieved the Old World standard of routine trichromacy. This section examines the utility of primate trichromacy and considers some features of its evolution.

There are several advantages to having color vision that have been enumerated over the years. Principally, color provides a strong cue that can be employed in the task of extracting visual meaning from visual scenes, a process that has among its components object detection, object localization, and object identification. It has been noted that color vision may be particularly useful in the complex visual environments characteristic of real world viewing where cues other than color (for example, lightness) may provide ambiguous information. An example is in the shadowed environments produced by the presence of screening foliage, or where there are rapidly changing light levels. These situations may yield striking local variations in lightness but have relatively little effect on color variation. A consequence of this is that color can provide a more reliable source of information than lightness (De Valois and De Valois, 1988; Mollon, 1991). Another way in which color may be usefully employed is as an aid to perceptual organization. The idea is that color can be used to perceptually link together separable elements in a scene, a process that is prerequisite to object identification. An oft-cited example of this is our ability to detect a digit or a symbol in the pseudoisochromatic plates that are commonly used to test for color vision. These plates are so designed that color differences allow the viewer with color vision to link the separable elements (typically, small dots) and detect the symbol (Mollon, 1991). For purposes of visual evaluation, color also can be used to transmit information about object quality (e.g. the ripeness of a fruit or the identity of a conspecific). Such uses of color have been referred to as the 'signal significance of color' (Jacobs, 1981).

Granted that the presence of color vision is potentially a very useful component of seeing, what is the particular virtue of primate trichromacy? From an historical point of view the answer seems to be that trichromatic color vision permits one to make subtle discrimination among green, yellow and red fruits. In an early statement of the idea, Stephen Polyak put it this way: 'In most instances fruit would not be easily noticed by animals if it were not for its color: The green unripe fruits, for example, blend with the green foliage and remain unnoticed until they become ripe and change color.' (Polyak, 1957). In this view, harvesting of colored fruits by the trichromat fulfills one half of a contract; in exchange the monkey disperses seeds to new locations and thus coloration of some fruits and primate trichromacy could have co-evolved just as is argued in the case of floral colors and the color vision of insect pollinators (Chittka and Menzel, 1992). It is interesting (and probably not coincidental) that one of the common visual difficulties reported by human color defectives is the detection of berries and fruits when they are embedded in foliage and vegetation (Mollon, 1991). This scenario for the evolution of primate trichromacy seems quite plausible. Whether or not it correctly identifies the initiating conditions, once trichromacy has evolved it will prove useful for a variety of purposes far beyond the selection of fruits. For instance, the elaborate variations in facial color patterns of many African monkeys indicates that color vision must be important for the transmission of information about other animals.

Two recent studies are noteworthy for having attempted a more systematic look at the relationship between primate color vision and the discrimination of fruits (Osorio and Vorobyev, 1996; Regan *et al.*, 1996). In each case, large collections of spectrophotometric measurements of the reflectance spectra of fruits and foliage were used to examine how well visual systems utilizing retinas containing various cone photopigment combinations might succeed at discriminating fruits from a foliage background. These attempts necessarily make arbitrary assumptions about the way in which visual systems might process spectral inputs to yield discriminations. Nevertheless, each study concludes that for discriminating fruits from foliage a trichromatic color vision system provides a very significant advantage over any dichromatic color vision system and thus provide support for the idea of the co-evolution of fruit color and primate color vision.

A notable feature of the trichromacy characteristic of Old World monkeys, apes and humans that was mentioned above is the essential invariance of the spectral positioning of the M and L pigments. This seems remarkable given the wide divergence in the photic environments and visual lifestyles of these various species. Why such consistency? One possible answer is suggested by Nagle and Osorio (1993) who argue that the particular positioning of these M and L pigments is such as to *minimize* the size of the spectrally-opponent signal from

M/L cone interactions. This paradoxical fact suggests to them that M/L color vision did not evolve to support a general purpose red/green color vision capacity, but rather for a specific purpose – perhaps the detection and identification of fruit as was explained above. They suggest this to be the case for the explicit reason that the colors of fruits are so qualitatively different from their backgrounds that they can potentially provide large color signals.

Another way to look at primate trichromacy is to ask if there are any disadvantages to being trichromatic. This question classically arose in the context of trying to understand why the incidence of color defective vision is so high in our species (about 2% of all human males are dichromatic), but its answer might also help us to understand why all primates have not become trichromatic. One way to approach this question is to examine the capacities of dichromatic and trichromatic individuals when they are confronted with identical discrimination tasks. Such was the case in an experiment in which dichromatic and trichromatic humans were required to make texture discriminations in situations where the texture elements in a target (defined, for example, by differences in orientation) either did or did not also vary systematically in color (Morgan *et al.*, 1992). In circumstances where color was irrelevant to the detection of the pattern, individuals with trichromatic color vision did much more poorly than did dichromats. We noted above that color may be used to link separable elements in a scene and thus aid in detection of targets. Usually this is to great benefit, but when color varies randomly as it did in this experimental demonstration it serves to mask the intended target and may actually disadvantage the trichromatic viewer. It has sometimes been proposed that the presence of color vision might either aid or impede non-color discriminations. However, a study of New World monkeys revealed that there were *no* differences in the abilities of dichromatic and trichromatic individuals to make discriminations based on luminance differences alone (for example, the detection of flickering lights). On the other hand, in increment-detection tasks where both luminance and chromatic differences could serve as possible cues for discrimination, some combinations were more easily detected by trichromatic animals while others were better detected by dichromats (Jacobs, 1990). Studies of these kinds show that trichromatic color vision cannot be automatically assumed to provide an advantage in all viewing situations. On balance, however, it seems hard to doubt that trichromacy substantially enhances primate vision.

Trichromacy requires three types of cone photopigment. Granting that the usual mammalian plan involves only two types of photopigment, how did the primates achieve three? One scenario is based on the idea that gene/photopigment arrangement seen in many contemporary New World monkeys represents an intermediate stage between the uniform dichromacy of many mammals (including some prosimians) and the uniform trichromacy of Old World monkeys, apes and humans. We know that a mutation producing a single nucleotide substitution in the opsin gene can yield an M/L photopigment spectrally discrete from its progenitor (Neitz *et al.*, 1991). If that novel gene is carried by a heterozygous female she will produce two different M/L photopigments and, through the agency of X-chromosome inactivation, will then have receptors that each contain one of three classes of pigment. Selective advantage will cause the frequency of the new gene to rise in the population and thus trichromacy will appear with increasing frequency among females. The limitations of this arrangement are that not all females will be trichromatic and no males can ever be. To make trichromacy routine in a species requires the addition of a second photopigment gene locus. This could come about as a result of an unequal crossover between chromosomes having different pigment genes; alternatively, two gene loci could be established by gene duplication and subsequent conversion of the duplicated gene. Whether the polymorphic stage with its partial-species trichromacy is essential to the evolution of routine trichromacy is uncertain. Examination of the photopigment gene sequences of the only New World monkey that seems routinely trichromatic – the howler monkey – suggests that, at least in New World primates, a polymorphic stage does precede routine trichromacy (Jacobs *et al.*, 1996b).

In order to have trichromacy, the retina must contain three types of cone pigment and have neural connections to support opponent interactions between the signals generated by various classes of cone. The fact that some female New World monkeys gain trichromacy simply as a result of adding a second type of M/L pigment early suggested that the retinas of these species already have in place the neural organization necessary for trichromatic color vision. Recent anatomical examination of the retinas support the view that, indeed, the analysis of red/green color information utilizes a neural organization required for other aspects of vision. It turns out that the topography and densities of both cones and ganglion cells in common marmosets are very similar to that found in Old World monkeys and, even more to the point, there are no obvious differences in these anatomical arrangements between dichromatic and trichromatic marmosets (Wilder *et al.*, 1996). These observations suggest that the structure of the retina has been shaped by the need to support high spatial resolution, principally through the evolution of the midget-cell system characteristic of the central portion of the primate retina. Once this arrangement is in place it can be immediately utilized for the analysis of trichromatic signals. If so, this also provides one reason why most mammals, including the prosimian primates, have not achieved trichromacy – it is because their retinas do not have the topographic arrangements necessary to support high spatial resolution. The absence of mechanisms for high spatial resolution means that the neural interconnections necessary to assure red/green spectral opponency are similarly absent.

Having specified the mechanisms for achieving routine trichromatic color vision, and discussed the advantages that capacity may offer, is it possible to understand why a significant number of primate species have alternative arrangements? We have identified three photopigment/color vision patterns beyond the routine trichromacy of many Old World primates: the polymorphic color vision of many New World monkeys, the routine dichromacy of some prosimians, and the monochromacy of some nocturnal primates. Prosimians have relatively poor spatial resolution (Langston *et al.*, 1986). Since the neural basis for trichromacy seems to require a retinal construction sufficient to support high spatial resolution, it seems probable that these primates would not be in a position to benefit much from the addition of a second type of L/M cone and the trichromacy it would allow. A similar argument may also explain the absence of trichromacy in vast numbers of other nonprimate mammals. With only one exception, all New World monkeys do have retinal topographies appropriate for high spatial resolution and we have seen that some individuals in these species are trichromats. Why hasn't trichromacy become the norm for these species? Following the discovery that color vision variation in these New World monkeys represents a balanced polymorphism, it was suggested that there were several alternative mechanisms that might be argued to support the maintenance of a variety of kinds of color vision in a species (Mollon *et al.*, 1984). For instance, one possibility is that a group of animals with varying color vision phenotypes provides a strategy for exploiting more kinds of food resources than would be possible if all individuals had the same color capacities. Or, the variation might represent an example of a multiple-niche polymorphism. The argument here would be that the forested environment consists of a number of visually-distinct environments and animals with different phenotypes might be better suited to one or another of these. There are other potential explanations of this same sort for the color vision variation (Mollon, 1991; Jacobs, 1997). A qualitatively different possibility is that, for some unknown reason, the gene-duplication step required to move from dichromacy to routine trichromacy is a low probability event in these primates and has simply not yet occurred in any of these New World species except for the howler monkey. A much better understanding of how New World monkeys use their color vision in natural habitats seems a pre-requisite to providing a better understanding of the remarkable color vision variations in this group of primates.

Finally, what is the explanation for those nocturnal primates that have only a single cone pigment and thus lack conventional color vision? Of the species that we know fall in this category, each has acquired deleterious mutations in the gene that would normally produce the S-cone pigment (Jacobs *et al.*, 1995b). One obvious explanation for the loss of the S-cone might be that its major functional product, color vision, has little or no utility in the low-light environments which these animals characteristically inhabit and thus offers little advantage. I noted above that it seems likely that the loss of S-cone function is species-wide in these nocturnal primates. That fact suggests an even more heretical possibility, that S-cones are in some way detrimental in eyes that become specialized for nocturnal life. If that were so, the loss of S-cones (and color vision) in these nocturnal primates could provide a positive advantage.

20.7 Conclusions

There are large gaps in our understanding of primate vision and, consequently, this chapter has been able to explore only a small portion of the puzzle of primate sight and behavior. From the evidence reviewed it is fair to conclude that good progress has been made toward understanding the basic biological mechanisms that underlie spectral sensitivity and color vision in primates, and that there is reasonable understanding of how these features vary among the various primates. Furthermore, it is also possible to provide some reasonable speculations as to how these variations came about, what visual possibilities they provide, and what limitations they set. Despite these accomplishments, the exact ways in which primates use these capacities in natural settings remain for the most part unknown. To bridge that gap will require much better descriptions of the photic environments in which primates pursue the business of seeing along with accurate catalogues of the sort of visual discrimination that assure primate success. There are signs that such work is underway. If so, we may soon achieve at least a necessary starting point for a real primate visual ecology.

Acknowledgements

The author's research and the preparation of this chapter were supported by grants from the National Science Foundation (IBN93-18770) and the National Eye Institute (EY02052).

References

Allman, J. (1982) Reconstructing the evolution of the brain in primates through the use of comparative neurophysiology and neuroanatomical data. In *Primate Brain Evolution: Methods and Concepts*, (eds E. Armstrong and D. Falk), Plenum Press, New York, pp. 13 – 28.

Asenjo, A. B., Rim, J. and Oprian, D. D. (1994) Molecular determinants of human red/green color discrimination. *Neuron,* 12, 1131–1138.

Blakeslee, B. and Jacobs, G. H. (1985) Color vision in the ring-tailed lemur *(Lemur catta). Brain, Behavior and Evolution,* 26, 154 –166.

Bowmaker, J. K. (1990) Cone visual pigments in monkeys and humans. In *Advances in Photoreception* (eds Committee on Vision), National Academy Press, Washington, D. C., pp. 19 – 30.

Bowmaker, J. K. (1991) Visual pigments and colour vision in primates. In *From Pigments to Perception* (eds A. Valberg and B. B. Lee), Plenum Press, New York, pp. 1– 9.

Bowmaker, J. K., Astell, S., Hurst, D. M. and Mollon, J. D. (1991) Photosensitive and photostable pigments in the retinae of Old World monkeys. *Journal of Experimental Biology*, 156, 1–19.

Boynton, R. M. (1982) Spatial and temporal approaches for studying color vision. In *Colour Vision Deficiencies VI* (ed. G. Verriest). The Hague: W. Junk, pp. 1–14.

Calkins, D. K., Schein, S. J., Tsukamoto, Y. and Sterling, P. (1994) M and L cones in macaque fovea connect to midget ganglion cells by different numbers of excitatory synapses. *Nature,* 371, 70 –72.

Chittka, L. and Menzel, R. (1992) The evolutionary adaptation of flower colours and insect pollinators' colour vision. *Journal of Comparative Physiology A,* 171, 171–181.

Cicerone, C. M. and Nerger, J. L. (1989) The relative numbers of long-wavelength-sensitive to middle-wavelength-sensitive cones in the human fovea centralis. *Vision Research,* 29, 115 –128.

Ciochon, R. L. and Chiarelli, A. B. (1980) *Evolutionary Biology of the New World Monkeys and Continental Drift,* Plenum Press, New York.

Dacey, D. M. and Lee, B. B. (1994) The 'blue-on' opponent pathway in the primate retina originates from a distinct bistratified ganglion cell type. *Nature,* 367, 731–735.

Dartnall, H. J. A., Bowmaker, J. K. and Mollon, J. D. (1983) Human visual pigments: Microspectrophotometric results from the eyes of seven persons. *Proceedings of the Royal Society of London B,* 230, 115 –130.

De Valois, R. L. and De Valois, K. K. (1988) *Spatial Vision*, Oxford University Press, New York.

Deegan II, J. F. and Jacobs, G. H. (1996) Spectral sensitivity and photopigments of a nocturnal prosimian, the bushbaby (*Otolemur crassicaudatus*). *American Journal of Primatology,* 40, 55 – 66.

DeValois, R. L., Morgan, H. C., Polson, M. C., Mead, W. R. and Hull, E. M. (1974) Psychophysical studies of monkey vision. I. Macaque luminosity and color vision tests. *Vision Research,* 14, 53 – 67.

Dusenberry, D. B. (1992) *Sensory Ecology*, W. H. Freeman and Company, New York.

Endler, J. A. (1993) The color of light in forests and its implication. *Ecological Monographs,* 63, 1– 27.

Fleagle, J. G. (1988) *Primate Adaptation and Evolution*, Academic Press, San Diego.

Flynn, J. J., Wyss, A. R., Charrier, R. and Swisher, C. C. (1995) An early miocene anthropoid skull from the Chilean Andes. *Nature,* 373, 603 – 607.

Hagstrom, S. A., Neitz, J., Neitz, M. and Trusk, T. C. (1994) Characterizing the photopigment genes expressed in individual human cone photoreceptors. *Investigative Ophthalmology and Visual Science,* 35, 1710.

Harwerth, R. S. and Smith III, E. L. (1985) The rhesus monkey as a model for normal vision of humans. *American Journal of Optometry and Physiological Optics,* 62, 633 – 641.

Jacobs, G. H. (1981) *Comparative Color Vision*, Academic Press, New York.

Jacobs, G. H. (1984) Within-species variations in visual capacity among squirrel monkeys (*Saimiri sciureus*): Color vision. *Vision Research,* 24, 1267–1277.

Jacobs, G. H. (1990) Discrimination of luminance and chromaticity differences by dichromatic and trichromatic monkeys. *Vision Research,* 30, 387– 397.

Jacobs, G. H. (1993) The distribution and nature of colour vision among the mammals. *Biological Reviews,* 68, 413 – 471.

Jacobs, G. H. (1995) Variations in primate color vision: Mechanisms and utility. *Evolutionary Anthropology,* 3, 196 –205.

Jacobs, G. H. (1996) Primate photopigments and primate color vision. *Proceedings of the National Academy of Sciences USA,* 93, 577– 581.

Jacobs, G. H. (1997) Color-vision polymorphisms in New World monkeys: Implications for the evolution of primate trichromacy. In *New World Primates: Ecology, Evolution and Behavior* (ed. W. G. Kinzey). Aldine de Gruyter, Hawthorne, New York, pp. 45 –74.

Jacobs, G. H. (1998) A perspective on color vision in platyrrhine monkeys. *Vision Research,* 38, 3307– 3313.

Jacobs, G. H. and Deegan II, J. F. (1997) Spectral sensitivity of macaque monkeys measured with ERG flicker photometry. *Visual Neuroscience*, 14, 921– 928.

Jacobs, G. H., Neitz, M. and Neitz, J. (1996a) Mutations in S-cone pigment genes and the absence of colour vision in two species of nocturnal primate. *Proceedings of the Royal Society of London B,* 263, 705 –710.

Jacobs, G. H., Neitz, M., Deegan II, J. F. and Neitz, J. (1996b) Trichromatic colour vision in New World monkeys. *Nature*, 382, 156 –158.

Jacobs, G. H., Deegan II, J. F. and Moran, J. L. (1996c) ERG measurements of spectral sensitivity of common chimpanzee (*Pan troglodytes*). *Vision Research,* 36, 2587– 2594.

Jacobs, G. H. and Deegan II, J. F. (1993) Photopigments underlying color vision in ringtail lemurs (*Lemur catta*) and brown lemurs (*Eulemur fulvus*). *American Journal of Primatology,* 30, 243 – 256.

Jacobs, G. H., Deegan II, J. F., Neitz, J. A., Crognale, M. A. and Neitz, M. (1993a) Photopigments and color vision in the nocturnal monkey, *Aotus. Vision Research,* 33, 1773 –1783.

Jacobs, G. H. and Neitz, J. (1993) Electrophysiological estimates of individual variation in the L/M cone ration. In *Colour Vision Deficiencies XI* (ed. B. Drum), Dordrecht: Kluwer, pp. 107–112.

Jacobs, G. H. and Neitz, J. (1985) Color vision in squirrel monkeys: Sex-related differences suggest the mode of inheritance. *Vision Research,* 25, 141–144.

Jacobs, G. H., Neitz, J.and Crognale, M. (1987) Color vision polymorphism and its photopigment basis in a callitrichid monkey (*Saguinus fuscicollis*). *Vision Research,* 27, 2089 – 2100.

Jacobs, G. H., Neitz, J. and Neitz, M. (1993b) Genetic basis of polymorphism in the color vision of platyrrhine monkeys. *Vision Research,* 33, 269 – 274.

Kay, R. F., Ross, C. and Williams, B. A. (1997) Anthropoid origins. *Science* 275, 797– 804.

Langston, A., Casagrande, V. A. and Fox, R. (1986) Spatial resolution of the galago. *Vision Research,* 26, 791–796.

Lennie, P. and D'Zmura, M. (1988) Mechanisms of color vision. *CRC Critical Reviews in Neurobiology,* 3, 333 – 400.
Lythgoe, J. N. (1979) *The Ecology of Vision*, Oxford University Press, New York.
Lythgoe, J. N. and Partridge, J. C. (1989) Visual pigments and the acquisition of visual information. *Journal of Experimental Biology,* 146, 1– 20.
Martin, R. D. (1990) *Primate Origins and Evolution.* Princeton University Press, Princeton.
Merbs, S. L. and Nathans, J. (1992) Absorption spectra of the hybrid pigments responsible for anomalous color vision. *Science,* 258, 464 – 466.
Mollon, J. D. (1991) Uses and evolutionary origins of primate color vision. In *Evolution of the Eye and Visual System* (eds J. R. Cronly-Dillon and R. L. Gregory), CRC Press, Boca Raton, pp. 306 – 319.
Mollon, J. D. (1997) ...aus dreyerley Arten von Membranen oder Molkulen: George Palmer's legacy. In *Colour Vision Deficiencies XIII,* (ed. C. R. Cavonius), Dordrecht: Kluwer, pp. 3 – 20.
Mollon, J. D. and Bowmaker, J. K. (1992) The spatial arrangement of cones in the primate fovea. *Nature,* 360, 677– 679.
Mollon, J. D., Bowmaker, J. K. and Jacobs, G. H. (1984) Variations of colour vision in a New World primate can be explained by polymorphism of retinal photopigments. *Proceedings of the Royal Society of London B,* 222, 222, 373 – 399.
Morgan, M. J., Adam, A. and Mollon, J. D. (1992) Dichromats detect colour-camouflaged objects that are not detected by trichromats. *Proceedings of the Royal Society of London B,* 248, 291– 295.
Nagle, M. G. and Osorio, D. (1993) The tuning of human photopigments may minimize red-green chromatic signals in natural conditions. *Proceedings of the Royal Society of London B,* 252, 209 – 213.
Nathans, J. (1987) Molecular biology of visual pigments. *Annual Review of Neuroscience,* 10, 163 –194.
Nathans, J., Thomas, D. and Hogness, D. S. (1986) Molecular genetics of human color vision: the genes encoding blue, green and red pigments. *Science,* 232, 193 – 202.
Neitz, J. and Neitz, M. (1994) Color vision defects. In *Molecular Genetics of Inherited Eye Disorders* (eds A. F. Wright and B. Jay), Harwood Academic Publishers, Reading, p. 217– 257.
Neitz, M. and Neitz, J. (1998) Molecular genetics and the biological basis of color vision. In *Color Vision – Perspectives from Different Disciplines* (eds. W. Backhaus, G. K. Werner, R. Kliegl and J. K. Werner). Walter de Gruyter, Berlin, pp. 101–119.
Neitz, M., Neitz, J. and Jacobs, G. H. (1991) Spectral tuning of pigments underlying red-green color vision. *Science,* 252, 971– 974.
Osorio, D. and Vorobyev, M. (1996) Colour vision as an adaptation to frugivory in primates. *Proceedings of the Royal Society of London B,* 263, 593 – 599.
Polyak, S. (1957) *The Vertebrate Visual System,* University of Chicago Press, Chicago.
Regan, B. C., Vienot, F., Charles-Dominique, P. C. Pefferkorn, S. Simmen, B., Juillot, C. and Mollon, J. D. (1996) The colour signals that fruits present to primates. *Investigative Ophthalmology and Visual Science,* 37, S648.
Sereno, M. I. and Allman, J. M. (1991) Cortical visual areas in mammals. In *The Neural Basis of Visual Function,* (ed. A. G. Leventhal), CRC Press, Boca Raton, pp 160 –172.
Sereno, M. L., Dale, A. M., Reppas, J. B., Kwong, K. K., Belliveau, J. W., Brady, T. J., Rosen, B. R. and Tootell, R. B. H. (1995) Borders of multiple visual areas in humans revealed by functional magnetic resonance imaging. *Science,* 268, 889 – 893.
Sperling, H. G., Wright, A. A. and Mills, S. L. (1987) Intense spectral light induced color blindness in rhesus monkeys. *Documenta Ophthalmologica Proceedings Series,* 46, 5 – 20.
Tootell, R. B. H., Dale, A. M., Sereno, M. I. and Malach, R. (1996) New images from human visual cortex. *Trends in Neuroscience,* 19, 481– 489.
Tovée, M. J., Bowmaker, J. K. and Mollon, J. D. (1992) The relationship between cone pigments and behavioural sensitivity in a New World monkey (*Callithrix jacchus jacchus*). *Vision Research,* 32, 867– 878.
Vimal, R. L. P., Pokorny, J., Smith, V. C. and Shevell, S. K. (1989) Foveal cone thresholds. *Vision Research,* 29, 61– 88.
Weitz, C. J., Miyake, Y., Shinzato, K., Montag, E., Zrenner, E. Went, L. N. and Nathans, J. (1992a) Human tritanopia associated with two amino acid subsitutions in the blue sensitive opsin. *American Journal of Human Genetics,* 50, 498 – 507.
Weitz, C. J., Went, L. N.and Nathans, J. (1992b) Human tritanopia associated with a third amino acid substitution in the blue sensitive opsin. *American Journal of Human Genetics,* 51, 444 – 446.
Wilder, H. D., Grunert, U., Lee, B. B. and Martin, P. R. (1996) Topography of ganglion cells and photoreceptors in the retina of the New World marmoset monkey *Callithrix jacchus. Visual Neuroscience,* 13, 335 – 352.

Index

S.N. Archer et al. (eds.), Adaptive Mechanisms in the Ecology of Vision, 651–668